Fabrication Engineering at the Micro- and Nanoscale

The Oxford Series in Electrical and Computer Engineering

Adel S. Sedra, Series Editor

Fabrication Engineering at the Micro- and Nanoscale

Third Edition

Stephen A. Campbell

University of Minnesota

New York   Oxford

OXFORD UNIVERSITY PRESS

2008

Oxford University Press, Inc., publishes works that further Oxford University's objective of excellence in research, scholarship, and education.

Oxford New York
Auckland Cape Town Dar es Salaam Hong Kong Karachi
Kuala Lumpur Madrid Melbourne Mexico City Nairobi
New Delhi Shanghai Taipei Toronto

With offices in
Argentina Austria Brazil Chile Czech Republic France Greece
Guatemala Hungary Italy Japan Poland Portugal Singapore
South Korea Switzerland Thailand Turkey Ukraine Vietnam

Published by Oxford University Press, Inc.
198 Madison Avenue, New York, New York 10016
http://www.oup.com

Oxford is a registered trademark of Oxford University Press

Library of Congress Cataloging-in-Publication Data
Campbell, Stephen A., 1954–
   Fabrication engineering at the micro- and nanoscale / Stephen A. Campbell.—
3rd ed.
      p. cm.—(The Oxford series in electrical and computer engineering)
   Rev. ed. of: The science and engineering of microelectronic fabrication. 2nd. ed.
   Includes bibliographical references and index.
   ISBN 978-0-19-532017-6 (pbk.: alk. paper)   1. Semiconductors—Design and
construction.   I. Campbell, Stephen A., 1954 – Science and engineering of
microelectronic fabrication.   II. Title.
   TK7871.85.C25 2007
   621.3815′2—dc22   2007013450

Printing number: 9 8 7 6 5 4 3 2

Printed in the United States of America
on acid-free paper

Contents

°This section provides background material.

[+]This section contains advanced material and can be omitted without loss of the basic content of the course.

Part III Unit Processes 2: Pattern Transfer 163

Preface

The intent of this book is to introduce microelectronic and some aspects of nano-processing to a wide audience. I wrote it as a textbook for seniors and/or first-year graduate students, but it may also be used as a reference for practicing professionals. The goal has been to provide a book that is easy to read and understand. Both silicon and GaAs processes and technologies are covered, although the emphasis is on silicon-based technologies. The later sections also deal briefly with organic and thin film devices. The book assumes one year of physics, one year of mathematics (through simple differential equations), and one course in chemistry. Most students with electrical engineering backgrounds will also have had at least one course in semiconductor physics and devices including pn junctions and MOS transistors. This material is extremely useful for the last five chapters and is reviewed in the first sections of Chapters 16, 17, and 18 for students who haven't seen it before or find that they are a bit rusty. One course in basic statistics is also encouraged but is not required for this course.

Microelectronics textbooks typically divide the fabrication sequence into a number of unit processes that are repeated to form the integrated circuit. The effect is to give the book a survey flavor: a number of loosely related topics each with its own background material. Most students have difficulty recalling all of the background material. They have seen it once, two or three years and many final exams ago. It is important that this fundamental material be reestablished before students take up new material. Distributed through each chapter of this book are reviews of the science that underlies the engineering. These sections, marked with an "°", also help make the distinction between the immutable scientific laws and the applications of those laws, with all the attendant approximations and caveats, to the technology at hand. Optical lithography, for instance, may have a limited life, but diffraction will always be with us.

A second problem that arises in teaching this type of course is that the solution of the equations describing the process often cannot be done analytically. Consider diffusion as an example. Fick's laws have analytic solutions, but they are valid only in a very restricted parameter space. Predeposition diffusions are done at high concentrations at which the simplifying assumption used in the solution derivation are simply not valid. In the area of lithography even the simplest solutions of the Fresnel equations are beyond the scope of the book. This text uses a widely used suite of simulation tools supplied commercially by Silvaco™. These are "industry standards" and are provided at low cost to educational institutions. For institutions that do not support this software, I plan to replicate the examples using web-based software that is available at no cost on the nanohub website (www.nanohub.org). Information on these examples will be available at www.nano.umn.edu/simexamples. The software is intended to augment, not replace, learning the fundamental equations that describe microelectronic processing. Typical installations include UNIX and Windows-based computers. The book also enriches the basic material with additional sections and chapters on process integration for various technologies and on more advanced processes. This additional material is in sections marked with a "+". If time does not permit covering these sections, they may be omitted without loss of the basic content of the course.

The third edition has added a variety of topics to keep it current. This includes atomic layer deposition, electroplating, immersion lithography, nanoimprint and soft lithography, thin film devices, organic light-emitting diodes, and the use of strain in CMOS. Other topics that are of less current interest were de-emphasized or removed.

Finally, one has to acknowledge that, no matter how many times material is reviewed, it cannot be guaranteed to be free of all (hopefully) minor errors. In the past, publishers have provided errata when errors were sufficiently numerous or egregious. Even when errata are published, they are very difficult to get to people who have already bought the book. This means that the average reader is often unaware of most of the corrections until a new or revised edition of the book is released. This book will have an errata file that anyone can access at any time. We will also provide minor additions to the book that were not available at press time. You can access the file by going to the Oxford University Press website for the book, www.oup.com/us/he/CampbellFabrication. As time goes on I will be adding other minor updates and new topics on this site as well. If you find something that you feel needs correction or clarification in the book, I invite you to notify me at my e-mail address, scampbell@umn.edu. Please be sure to include your justification, citing published references.

Minneapolis   S.A.C.

Fabrication
Engineering
at the Micro-
and Nanoscale

Part I

Overview and Materials

Prediction is very difficult, especially about the future.[1]

This course is unlike many that you may have taken in that the material that will be covered is primarily a number of unit processes that are quite distinct from each other. The book then has the flavor of a survey of topics that will be covered rather than a linear progression. This part of the book will lay the foundations that will be needed to later understand the various fabrication processes.

The first chapter will provide a roadmap of the course and an introduction to integrated circuit fabrication. The processes that are covered are briefly described in a qualitative manner, and the relationships between the various topics are discussed. A simple example of a semiconductor technology, the fabrication of integrated resistors, is used to demonstrate a flow of these processes, which we will call a technology. Extensions of the technology to include capacitors and MOSFETs are also discussed.

The second chapter will introduce the topic of crystal growth and wafer production. The chapter contains basic materials information that will be used throughout the rest of the book. This includes crystal structure and crystal defects, phase diagrams, and the concept of solid solubility. Unlike the other unit processes that will be covered in the later chapters, very few integrated circuit fabrication facilities actually grow their own wafers. The topic of wafer production, however, demonstrates some of the important properties of semiconductor materials that will be important both during the fabrication process and to the eventual yield and performance of the integrated circuit. The differences in the production of silicon and GaAs wafers are discussed.

---

[1]Niels Bohr.

Chapter 1

An Introduction to Microelectronic Fabrication

The electronics industry has grown rapidly in the past four decades. This growth has been driven by a revolution in microelectronics. In the early 1960s, putting more than one transistor on a piece of semiconductor was considered cutting edge. Integrated circuits (ICs) containing tens of devices were unheard of. Digital computers were large, slow, and extremely costly. Bell Labs, which had invented the transistor a decade earlier, rejected the concept of ICs. They reasoned that to achieve a working circuit all of the devices must work. Therefore, to have a 50% probability of functionality for a 20-transistor circuit, the probability of device functionality must be $(0.5)^{1/20} = 0.966$, or 96.6%. This was considered to be ridiculously optimistic at the time, yet today integrated circuits are built with billions of transistors.

Early transistors were made from germanium, but most circuits are now made on silicon substrates. We will therefore emphasize silicon in this book. The second most popular material for building ICs is gallium arsenide (GaAs). Where appropriate, the book will discuss the processes required for GaAs ICs. Although GaAs has a higher electron mobility than silicon, it also has several severe limitations, including low hole mobility, less stability during thermal processing, a poor thermal oxide, high cost, and perhaps most importantly, much higher defect densities. Silicon has therefore become the material of choice for highly integrated circuits. More recently microelectronic fabrication techniques have been used to build a variety of structures including thin film devices and circuits, micromagnetics, optical devices, and micromechanical structures. In some cases these structures have also been integrated into chips containing electronic circuitry. A popular nonelectronic application, microelectromechanical systems (MEMS), will be introduced later in this book.

To chart the progress of silicon microelectronics, it is easiest to follow one type of chip. Memory chips have had essentially the same function for many years, making this type of analysis meaningful. Furthermore, they are extremely regular and can be sold in large volumes, making technology customization for the chip design economical. As a result, memory chips have the highest density of all ICs. Figure 1.1 shows the density of dynamic random access memories (DRAMs) as a function of time. The vertical axis is logarithmic. The density of these circuits increases by increments of $4\times$. Each of these increments takes approximately three years. One of the most fundamental

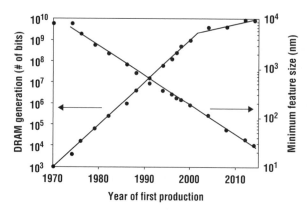

Figure 1.1   Memory and minimum feature sizes for dynamic random access memories as a function of time (data from IC knowledge and ITRS Roadmap, 2005 edition).

changes in the fabrication process that allows this technology evolution is the minimum feature size that can be printed on the chip. Not only does this increase IC density, the shorter distances that electrons and holes have to travel improve the transistor speed. Part of the IC performance improvement comes from this increased transistor performance, and part of it comes from being able to pack the transistors closer together, decreasing the parasitic capacitance. The right-hand side of Figure 1.1 shows that ICs have progressed from 10 microns (μm) (1 μm = 10^{-4} cm) to 0.1 μm, or 100 nm. For sake of reference, Figure 1.2 shows an electron micrograph of a silicon-based IC, along with a human hair. The vertical and horizontal lines are metal wires used to interconnect the transistors. The transistors themselves

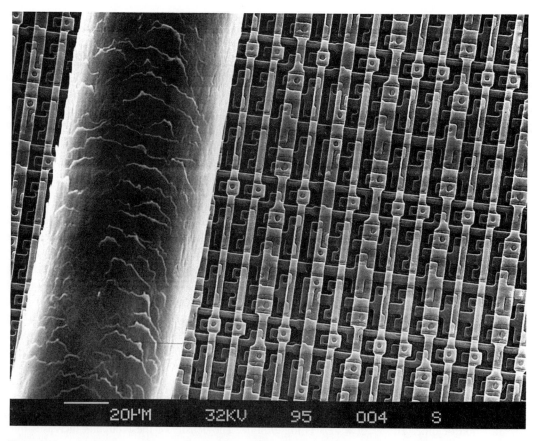

Figure 1.2   Scanning electron micrograph (SEM) of an IC circa mid-1980s. The visible lines correspond to metal wires connecting the transistors.

are below the metal and are not visible in the micrograph. At this rate of progress, chips with 65-nm features will be common by the time you read this book.

At first glance, these incredible densities and the associated design complexity would seem extremely daunting. This book, however, will focus on how these circuits are built rather than how they are designed or how the transistors operate. The fabrication process is similar no matter how many transistors are on the chip. The first half of the book will cover the basic operations required to build an IC. Using mechanical construction as an analogy, these would include steps such as forging, cutting, bending, drilling, and welding. These steps will be called unit processes in this text. If one knows how to do each of these steps for a certain material (e.g., steel), and if the machines and material required are available, they could be used to make a ladder or a high pressure cylinder or a small ship. The required number and order of the steps will clearly depend on what is being built, but the basic unit processes remain the same. Furthermore, once a sequence that produces a good ship has been worked out, other ships of similar design could probably be built with the same process. The design of the ship, that is, what goes where, is a separate task. The shipbuilder is handed a set of blueprints to which he or she must build.

The collection and ordering of these unit processes for making a useful product will be called a *technology*. Part V of the book will cover some of the basic fabrication technologies. Whether the technology is used to make microprocessors, I/O controllers, or any other digital function is largely immaterial to the fabrication process. Even many analog designs can be built using a technology very similar to that used to build most digital circuits. An IC, then, starts with a need for some sort of electronic device. A designer or group of designers translates the requirements into a circuit design: that is, how many transistors, resistors, and capacitors must be used, what values they must have, and how they must be interconnected. The designer must have some input from the fabricator. In the shipbuilding example, the blueprints must somehow reflect the limitation that the shipbuilder cannot put rivets over weld joints or use small rivets and expect them to hold under very high pressures. The builder must therefore give the designer a document that says what can and cannot be done. In microelectronics this document is called the *design rules* or *layout rules*. They specify how small or large some features can be or how close two different features can be. If the design conforms to these rules, the chip can be built with the given technology.

1.1 Microelectronic Technologies: A Simple Example

Instead of blueprints, the circuit designer hands the IC fabricator a set of photomasks. The photomasks are a physical representation of the design that has been produced in accordance with the layout rules. As an example of this interface, assume that a need exists for an IC consisting of a simple voltage divider as shown in Figure 1.3. The technology to build this design is shown in Figure 1.4. Silicon wafers will be used as the substrate since they are flat, reasonably inexpensive, and most IC processing equipment is set up to handle them. The production of these substrates will be discussed in Chapter 2. Since the wafer is at least somewhat conductive, an insulating layer must first be deposited to prevent leakage between adjacent resistors. Alternatively, a thermal oxide of silicon could be grown, since it is an excellent insulator. The thermal oxidation of silicon is covered in Chapter 4. Next a conducting layer is deposited that will be used for the resistors. Several techniques for depositing both insulating and conducting layers will be discussed in Chapters 12 through 14.

This conducting layer must be divided up into individual resistors. This can be done by removing portions of the conducting layer, leaving rectangles of the film that are isolated from each other. The resistor value is given by

$$R = \rho \frac{L}{W \cdot t}$$

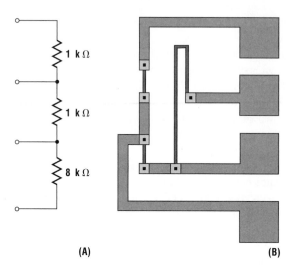

(A)   **(B)**

Figure 1.3   A simple resistor voltage divider. At left is a circuit representation; at right is a physical layout. The layers shown at right are resistor, contact, and low-resistance metal.

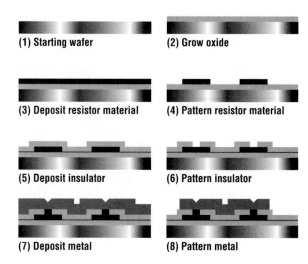

Figure 1.4   The technology flow for fabricating the resistor IC shown in Figure 1.3.

where ρ is the material resistivity, L is the resistor length, W is the resistor width, and t is the thickness of the layer. The designer can therefore select different values of resistors by choosing the width-to-length ratio, subject to the limits specified by the layout rules. The technologist chooses the film thickness and the material (and therefore ρ) to give the designer an appropriate range of resistivities without forcing him to resort to extreme geometries. Since ρ and t are determined during the fabrication and are approximately constant across a wafer, the ratio ρ/t is more often specified than ρ or t individually. This ratio is called the *sheet resistance*, ρ_s. It has units of $\Omega/\square$, where the number of squares is the ratio of the length to width of the resistor line.

The resistor information from the design, namely L and W for each resistor, must be transferred from the photomask to the wafer. This is done using a process called *photolithography*. The most commonly used type of photolithography is *optical lithography*. In this process, a photosensitive layer called *photoresist* is first spread on the wafer (Figure 1.5). Light shining through the mask exposes the resist in the regions of the wafer where some of the metal resistor layer must be removed. In these exposed regions, a photochemical reaction occurs in the resist that causes it to be easily dissolved in a developer solution. After the develop step, the photoresist remains only in the areas where a resistor is desired. The wafer is then immersed in an acid that dissolves the exposed metal layer but does not significantly attack the resist. When the etch is complete, the wafers are removed from the acid bath, rinsed, and the photoresist is removed. The photolithographic process will be covered in Chapters 7 through 9. Chapter 11 will cover etching.

Although the resistors have now been formed, they still need to be interconnected, and metal lines must be brought to the edge of the chip, where they can later be attached to metal wires for contact to the external world. This latter operation, called *packaging*, will not be covered in this text. If the metal lines have to cross over the resistors, another insulating layer must be deposited. To make electrical contact to the resistors, one can open up holes in the insulating layer using the same photolithographic and etch processes we had used for patterning the resistors, although the composition of the acid bath may be different. Finally, the fabrication sequence can be completed by depositing a highly conductive metal layer, applying a third mask, and etching this metal interconnect layer.

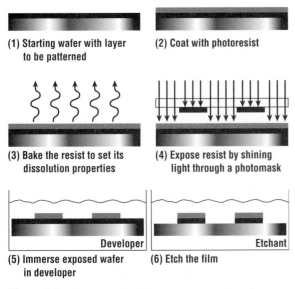

(1) Starting wafer with layer to be patterned

(2) Coat with photoresist

(3) Bake the resist to set its dissolution properties

(4) Expose resist by shining light through a photomask

Developer

Etchant

(5) Immerse exposed wafer in developer

(6) Etch the film

Figure 1.5 Steps required for a pattern transfer using optical lithography.

The technology consists of four layers: the lower insulator, the resistor film, the upper insulator, and the interconnect metal. Photolithography is used to selectively remove some of these layers in certain regions, but any point on the wafer must be made of some subset of these layers in the same order that they were built up. Except for the edges of the patterns, the thickness of these films is constant. The technology uses three photolithography steps, three etch steps, and four thin film deposition (including oxide growth) steps. A very similar set of steps could be used to fabricate a capacitor. With only a few more steps, simple transistors can be built. Notice that the comparison with shipbuilding breaks down in one critical respect. The effort required to build the IC is independent of the number of resistors. The photomask might define one resistor or 1,000,000 resistors (assuming a 1,000,000-resistor circuit could be useful). In fact, two different sets of photomasks, one that defines only a few resistors and one that defines thousands of resistors, could be used interchangeably with exactly the same technology and would require the same amount of work. This is because most of the unit processes described in this book operate in the whole wafer at the same time instead of one rivet at a time.

1.2 Unit Processes and Technologies

Some of the basic steps used in building an IC have already been discussed: photolithography, thin film deposition, and etching. Unit processes for thin film deposition include the processes of sputtering and evaporation. These are physical processes in that they do not generally depend on a chemical reaction. Sputtering is done by using charged atoms of argon called ions (Ar^+) to bombard a target containing the deposition material. The target erodes under this bombardment, and some of the material falls onto the wafers, coating them with a thin film of material. Evaporation involves heating the material to be deposited to a high temperature so that a vapor stream is created. The wafers are placed in this stream for coating. The third thin film deposition process that will be discussed is chemical vapor deposition. In this technique one or more gases are made to flow into a chamber that contain the wafers to be coated. In many cases the wafers are also heated. A chemical reaction occurs that leaves the desired solid product on the surface of the wafer.

A resistor was chosen to simplify the first example. Most semiconductor devices require the formation of doped regions. For example, consider the n-channel MOSFET shown in Figure 1.6. Some familiar layers from the resistor example are recognizable: a blanket insulator and a patterned metal layer. One can also selectively dope the source and drain regions, putting together a technology to make the transistor. Dopants are either donors (n-type) or acceptors (p-type). For silicon, the most common n-type dopants are arsenic, phosphorus, and antimony, and the most common p-type dopant is boron. For gallium arsenide the most common n-type dopants are silicon, sulfur, and selenium, and the most common p-type dopants are carbon, beryllium, and zinc. In early semiconductor technologies, impurities were introduced by exposing heated wafers to a dopant-containing gas. For example, a hydrogen/phosphine (H_2/PH_3) mixture can be used to introduce phosphorus into silicon. The introduction of dopant using this technique and the subsequent movement of the impurities when the wafer is heated is

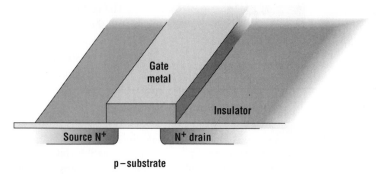

Figure 1.6   Cross section of an MOS transistor showing gate, source, drain, and substrate electrodes. The "+" and "−" indicate very heavy and very light dopings, respectively.

called *diffusion*. This type of dopant introduction method allows the impurities to diffuse deep into the wafer. As a result, it is not desirable for the small devices required in modern fabrication technologies. Ion implantation, which has replaced it, uses a beam of ionized atoms or molecules electrostatically accelerated toward the wafer. This method allows the process technologist to control the amount of impurity introduced (dose) and the depth of the impurity in the wafer (range). To limit the diffusion of impurities, a new class of processes have been developed that allow the rapid heating (to high temperatures) and cooling of the wafer. This type of process is called rapid thermal processing (RTP).

A number of processes that allow the growth of thin layers of semiconductor on top of the wafer will be discussed. These processes are called *epitaxial growth*. They allow the production of patterned dopant regions below the surface of the wafer. The book will first discuss the more traditional techniques of growing silicon on silicon and gallium arsenide on gallium arsenide (homoepitaxy). It will then cover more advanced techniques that allow the growth of extremely thin layers for the fabrication of advanced device structures.

The unit processes can be assembled into functional process modules. These modules are designed to carry out specific tasks such as the electrical isolation of adjacent transistors, low resistance contacts to transistors, and multiple layers of high density interconnect. All of these areas have had dramatic advances over the past few years. Clear trade-offs exist among the various modules in terms of process complexity, circuit density, planarity, and performance. These modules and the basic transistor fabrication are assembled into technologies. Several of the most popular technologies that represent a reasonable cross section of the microelectronics industry will be reviewed. Finally, techniques required for high volume manufacturing of ICs will be discussed.

1.3 A Roadmap for the Course

The various unit processes for fabrication are fairly independent. Each of the next 13 chapters will cover a different unit process. To keep the book to a manageable size, each process can be only briefly introduced. In many cases, the chapters themselves can be expanded into books. The material in the chapter will provide references that will allow you to further investigate each topic if you have an interest. The result of this approach is sometimes called a *survey course*.

Figure 1.7 shows a map of the course chapters. The order that your instructor chooses to follow is completely arbitrary as long as the necessary introductory material is covered. These chapters and sections are marked with a °. The chapters and sections marked with a + are additional material somewhat beyond the basic processes needed to describe simple semiconductor technologies.

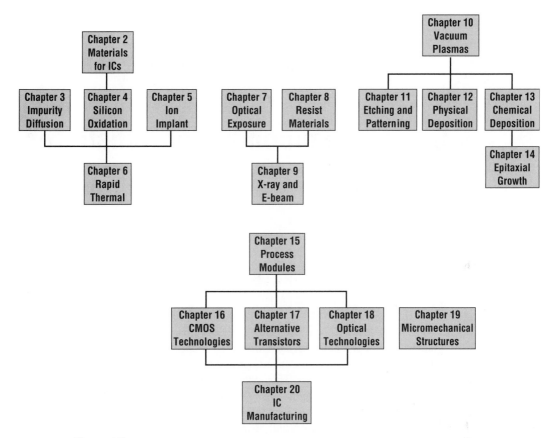

Figure 1.7 A roadmap for the course indicating the relationships between the chapters.

The last six chapters of the book are dedicated to semiconductor technologies. The basic unit processes discussed earlier are brought together to form ICs made from silicon CMOS, bipolar transistors, GaAs field effect transistors, thin film transistors, light-emitting diodes, lasers and micromechanical devices. These technology examples have been chosen because they are popular and because they are representative of many other common technologies. As you might infer from the previous discussion, the same unit processes could be used to fabricate flash memories, charge coupled devices (CCDs), sensors, solar cells, and other microdevices. The only differences are the number, type, and sequence of the processes used to form the technology. You are encouraged to look into one of these other fabrication technologies after you complete the course to see some of the other ways that these unit processes are applied.

1.4 Summary

Integrated circuits have developed with incredible levels of complexity, exceeding 8 billion transistors per chip. Transistor density, as measured by DRAMs, quadruples about every three years, as it has since 1968. This book will introduce the technologies used to fabricate the ICs. The building blocks of these technologies are the unit processes of photolithography, oxidation, diffusion, ion implantation, etching, thin film deposition, and epitaxial growth. The unit processes can be assembled in different order and number, depending on the circuit to be built.

Chapter 2

Semiconductor Substrates

P art II of the book will deal with unit processes that depend strongly on the properties of the semiconductor wafers themselves. Diffusion, for example, depends on the crystalline perfection in the wafers, which in turn depends on the process temperature. We will begin this chapter with a description of phase diagrams. This material is particularly useful for understanding the formation of alloys that will be used later in the book. This topic also leads naturally into a discussion of solid solubility and the doping of semiconductor crystals. Following that, the chapter will concentrate on crystal structures and defects in crystalline materials. The second half of the chapter will discuss the techniques used to fabricate semiconductor wafers. These wafers, which vary in diameter from 1 in for some compound semiconductors up to 300 mm for some silicon wafers, are the basic starting point for device fabrication. Although few fabrication facilities still make their own wafers, the study of the semiconductor substrate makes a good place to begin to develop an understanding of semiconductor processing. For a more complete review of this topic, see Mahajan and Harsha [1].

The materials used for microelectronics can be divided into three classifications, depending on the amount of atomic order they possess. In single crystal materials, almost all of the atoms in the crystal occupy well-defined and regular positions known as lattice sites. In semiconductor products, at most one layer is single crystal. That layer is provided by the wafer or substrate. If there is an additional single-crystal layer, it is grown epitaxially from this substrate (see Chapter 14). Amorphous materials, such as SiO_2, are at the opposite extreme. The atoms in an amorphous material have no long-range order. Instead, the chemical bonds have a range of lengths and orientations. The third class of materials is polycrystalline. These materials are a collection of small single crystals randomly oriented with respect to each other. The size and orientation of these crystals often change during processing and sometimes even during circuit operation.

2.1 Phase Diagrams and Solid Solubility°

Most of the materials of interest to us in this text are not elemental; rather, they are mixtures of materials. Even silicon is not very useful in a pure state. Instead, it is mixed with impurities that affect its electrical properties. A very convenient way to present the properties of mixtures of materials is a *phase diagram*. Binary phase diagrams can be thought of as maps that show the regions of stability for mixtures of two materials as a function of percent composition and temperature. Phase diagrams may also have a pressure dependence, but all of the diagrams of interest in semiconductor device fabrication will be at 1 atm.

Figure 2.1 shows the phase diagram for Ge–Si, an example of the simplest type of system [2]. There are two solid lines on the graph. The upper, or liquidus, curve describes the temperature at which a given mixture will be in a completely liquid state. The lower, or solidus, curve describes the temperature at which the mixture will completely freeze. Between these two curves is a region containing both liquid and solid mixtures. The composition of the melt can be readily determined from the diagram. If a solid mixture with equal atomic concentrations of Si and Ge is heated from room temperature, the material will begin to melt at 1108°C. Assume that the heating proceeds slowly enough so that the system remains in thermodynamic equilibrium. In the region between the two solid lines, the concentration of the melt will be determined by the composition at which the liquidus line crosses the temperature. For example, at 1150°C, the composition of the melt will be 22 atomic percent silicon. The concentration of the solid can be read from the graph as the concentration at which the solidus line crosses the temperature. In this example, the remaining solid will be 58 atomic

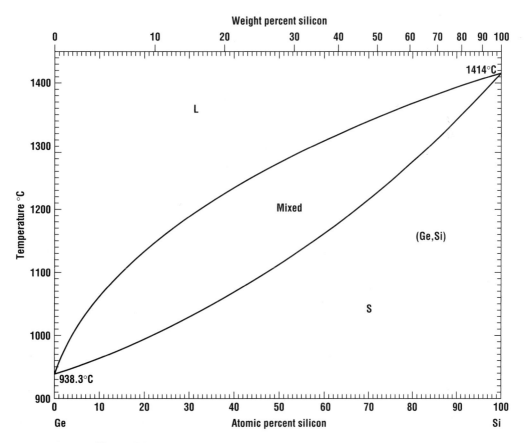

Figure 2.1   Phase diagram of Ge–Si (*courtesy of ASM International*).

percent silicon. Notice that the amount of the solid that will melt can be calculated using these values (see Example 2.1). As the temperature is increased, the composition of the melt moves toward the original value, while the concentration of the solid moves closer to pure silicon. When the temperature reaches the liquidus line the entire charge melts. For the 50% starting mixture this occurs at about 1272°C. Upon cooling, the same processes occur. Although in principle the phase diagram is independent of thermal history, in practice, maintaining thermodynamic equilibrium in the solid is much more difficult than it is in the melt, and so heating through a phase change tends to be closer to equilibrium than cooling at the same rate.

Example 2.1

For the Ge–Si example just discussed, calculate the fraction of the 50% charge that is molten at 1150°C.

Solution

Let x be the fraction of the charge that is molten. Then $1 - x$ is the fraction of the charge that is solid. The fraction of silicon in the melt plus the fraction of silicon in the solid must add to 0.5, the total fraction of Si in the charge

$$0.5 = 0.22x + 0.58(1 - x)$$

Solving for x,

$$0.36x = 0.08$$
$$x = 0.22$$

22% of the charge is molten and 78% is solid.

Figure 2.2 shows the phase diagram for GaAs [3]. Material systems like GaAs that have two solid phases that melt to form a single liquid phase are called *intermetallics*. To examine the phase diagram, start in the lower right-hand corner. This is a solid phase, since it is below the solidus line. The vertical line at the center of the diagram indicates that the compound GaAs will form in this material system. The lower left region is a solid mixture of GaAs and Ga, while the region in the lower right is a solid mixture of GaAs and As. If this As-rich solid is heated to 810°C, the solution will begin to melt. Between this temperature and the liquidus line, the concentration of the melt can be determined as before. For Ga-rich charges, the mixed state begins at about 30°C, only slightly above room temperature. This gives rise to problems associated with the growth of GaAs layers that will be described both later in this chapter as well as in a later chapter on epitaxial growth.

As a final example, consider Figure 2.3, which shows the phase diagram for the As–Si system [4]. While the structure looks quite complicated with several different solid phases mapped out, microelectronic applications are primarily interested in the low arsenic concentration limit. Even very heavily doped silicon is normally less than 5% arsenic. Notice that there is only a very small region in which As will dissolve in silicon as a dopant without forming a compound. The maximum concentration of an impurity that can be dissolved in another material under equilibrium conditions is called the solid solubility. Notice that the solid solubility increases as the temperature approaches 1097°C, where it is about 4 atomic percent, and that in this region a vertical line will actually intersect three curves. The portion of the line that goes from 0 atomic percent As at 500°C to 4 atomic percent As at

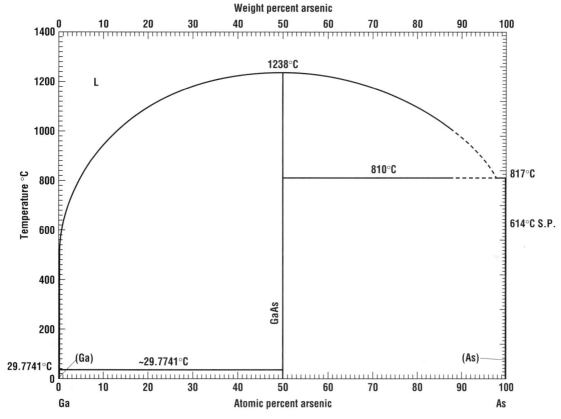

Figure 2.2 Phase diagram for GaAs *(courtesy of ASM International)*.

1097°C is called the *solvus curve*, as it represents a solubility. The remaining two lines are the solidus and liquidus, respectively. As we shall discuss in Chapter 4, dopant atoms must be dissolved in the semiconductor to have the potential to act as electron donors or acceptors. The solid solubility of As in Si is a comparatively large number, and it means that As can be used to form very heavily doped and therefore low resistance regions such as source and drain contacts for MOS transistors (Chapter 16) and emitter and collector contacts to bipolar transistors (Chapter 18). Since it is the solid solubility from the phase diagram that is of primary interest for dopant impurities, and the solid solubility varies by orders of magnitudes for different impurities in silicon, the phase diagram data for a number of common dopants in silicon have been combined [5] on a semilog plot in Figure 2.4.

Consider what happens in the following situation. A silicon wafer is heated to 1097°C and doped to 3.5 atomic percent with As. Doping methods will begin to be discussed in the next chapter. Impurity concentrations are normally expressed in number per unit volume. A 3.5 atomic percent concentration in silicon corresponds to 1.75×10^{21} cm^{-3}. The phase diagram indicates that as the wafer is cooled, it will eventually exceed the maximum concentration that can be in solution. If it is to maintain thermodynamic equilibrium, the excess As must condense out, either by coming out of the surface or, more likely, by forming solid precipitates in the silicon crystal. For this to happen, the As atoms must be mobile in the crystal. If the wafer is cooled rapidly enough, the precipitates cannot form, and a higher concentration of impurities than is thermodynamically allowed can be frozen in. Metallurgists refer to this process as *quenching*. This is an important consideration to keep in mind. Doping concentrations

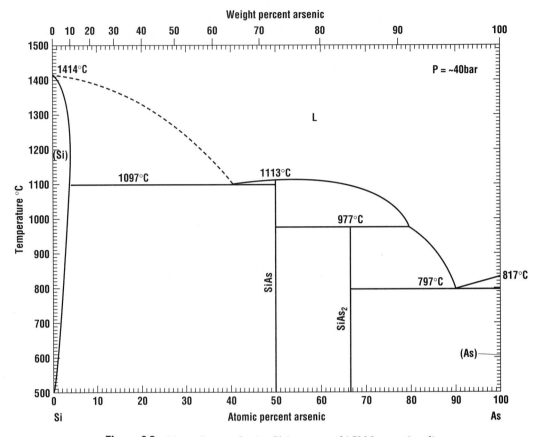

Figure 2.3   Phase diagram for As–Si *(courtesy of ASM International).*

can, and often do, exceed the solid solubility. This is done by heating a wafer with excess dopant atoms and then cooling rapidly. Peak concentrations can exceed the solid solubility by a factor of 10 or more.

2.2 Crystallography and Crystal Structure°

Crystals are described by their most basic structural element: the unit cell. A crystal is simply an array of these cells, repeated in a very regular manner over three dimensions. The unit cells of interest have cubic symmetry with each edge of the unit cell being the same length. Figure 2.5A shows three common types of cubic crystals. The directions in a crystal are identified using a Cartesian coordinate system as [x,y,z]. For a cubic crystal, the faces of the cell form planes that are perpendicular to the axes of the coordinate system. The symbol (x,y,z) is used to denote a particular plane that is perpendicular to the vector that points from the origin along the [x,y,z] direction. Figure 2.5B shows several common crystal directions. The set of numbers x, y, and z that are used to describe planes in this manner are called the *Miller indices* of a plane. They are found for a given plane by taking the inverse of the points at which the plane in question crosses the three coordinate axes, then multiplying by the smallest possible factor to make x, y, and z integers. The notation {x,y,z} is also used to represent crystal planes. This representation is meant to include not only the given plane, but also all equivalent planes. For example, in a crystal with cubic symmetry, the (100) plane will have exactly the same

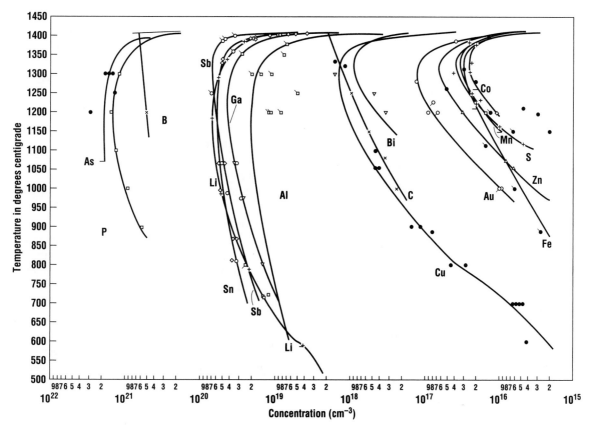

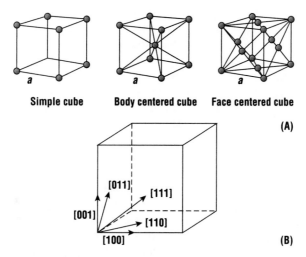

Figure 2.5   (A) Common cubic crystals: simple cubic, body-centered cubic, and face-centered cubic. Lattice sites are indicated by the small circles. (B) Crystal orientations in the cubic system.

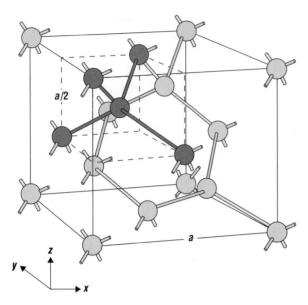

Figure 2.6 The diamond structure. The dark part of the diagram indicates the atom at (a/4, a/4, 3a/4), and its four nearest neighbors (0, 0, a), (a/2, a/2, a), (0, a/2, a/2), and (a/2, 0, a/2).

properties as the (010) and (001) planes. The only difference is an arbitrary choice of coordinate system. The notation {100} refers to all three.

Silicon and germanium are both group IV elements. They have four valence electrons and need four more to complete their valence shell. In crystals, this is done by forming covalent bonds with four nearest neighbor atoms. None of the basic cubic structures in Figure 2.5 would therefore be appropriate. The simple cubic crystal has six nearest neighbors, the body-centered cubic (BCC) has eight, and the face-centered cubic (FCC) has 12. Instead, group IV semiconductors form in the diamond structure shown in Figure 2.6. The unit cell can be constructed by starting with an FCC cell and adding four additional atoms. If the length of each side is a, the four additional atoms are located at (a/4, a/4, a/4), (3a/4, 3a/4, a/4), (3a/4, a/4, 3a/4), and (a/4, 3a/4, 3a/4). This crystal structure can also be thought of as two interlocking FCC lattices. Gallium arsenide also forms in this same arrangement; however, when two elements are present, the crystal has a reduced level of symmetry. The structure is then called zincblende.

2.3 Crystal Defects

Semiconductor wafers are highly perfect single crystals. Nevertheless, crystal defects play an important role in semiconductor fabrication. Semiconductor defects or imperfections, can be divided into four types, depending on their dimensionality. Point defects do not extend in any direction. Line

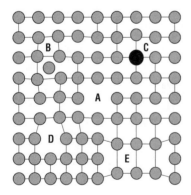

Figure 2.7 Simple 0- and 1-D semiconductor defects include: (A) vacancies, (B) self-interstitials, (C) substitutional impurities, (D) edge dislocations, and (E) dislocation loops.

defects extend in one direction through the crystal. Area and volume defects are 2- and 3-D defects, respectively. Each type of defect impacts different areas of the fabrication process. Point defects are extremely important to the understanding of doping and diffusion. The prevention of line defects is important to any thermal processing, particularly in rapid thermal processing. Volume defects can play a useful role in yield engineering. Figure 2.7 shows a few of the most important semiconductor defects.

One of the most common types of point defect is a lattice site without an atom. This defect is a vacancy. A closely related point defect is an atom that resides not on a lattice site, but in the spaces between the lattice positions. It is referred to as an interstitial. If the interstitial atom is of the same material as the atoms in the lattice, it is a self-interstitial. In some cases, the interstitial comes from a nearby vacancy. Such a vacancy interstitial combination is called a Frenkel defect. The interstitial or vacancy may not remain at the site at which it was created. Both types of defects can move through the crystal, particularly under the high temperatures typically found during processing conditions. Either defect might also migrate to the surface of the wafer where it is annihilated.

Vacancies and self-interstitials are intrinsic defects. Just as one talks about intrinsic carriers in semiconductors, at nonzero temperatures intrinsic defects will tend to occur in an otherwise perfect crystal. Thermal excitation creates a very small percentage of electrons and holes in a semiconductor. It will also remove a small number of atoms from their lattice sites, leaving behind vacancies. Generally, the vacancy concentration is given by an Arrhenius function, which is an equation of the form

$$N_v^o = N_o \, e^{-E_a/kT} \tag{2.1}$$

where N_o is the number density of atoms in the crystal lattice (5.02×10^{22} cm^{-3} for silicon), E_a is the activation energy associated with the formation of a vacancy, and k is Boltzmann's constant. For silicon, E_a is of order 2.6 eV, so at room temperature only one in 10^{44} lattice sites would be vacant in an otherwise perfect crystal. At 1000°C, however, the number of defects rises to 1 in 10^{10}. An identical equation can be used to describe the equilibrium concentration of interstitials in a material. In silicon, the activation energy for interstitials is higher than that for vacancies (about 4.5 eV). Notice from these two equations that, unlike intrinsic electrons and holes, the equilibrium vacancy concentration is not generally the same as the equilibrium interstitial concentration. Several important point defect sources and sinks exist, including extended defects and the surface of the wafer that allow this discrepancy.

GaAs has two types of lattice sites. Gallium sites have four arsenic nearest neighbors. Arsenic sites have four gallium nearest neighbors. The concentration of vacancies in GaAs is given by

$$N_{v,\mathrm{Ga}}^o = 3.3 \times 10^{18} \ \mathrm{cm}^{-3} \, e^{-0.4/kT} \tag{2.2}$$

for gallium and

$$N_{v,\mathrm{As}}^o = 2.2 \times 10^{20} \ \mathrm{cm}^{-3} \, e^{-0.7/kT} \tag{2.3}$$

for arsenic [6].

Example 2.2

At what temperature is

$$N_{v,\,Ga}^o = N_{v,\,As}^o$$

Solution

Setting Equation 2.2 equal to Equation 2.3, we have

$$3.3 \times 10^{18} \, \mathrm{cm}^{-3} \, e^{-0.4/kT} = 2.2 \times 10^{20} \, \mathrm{cm}^{-3} \, e^{-0.7/kT}$$

simplifying,

$$e^{0.3/kT} = 66.7$$

$$\frac{0.3}{kT} = 4.2$$

$$T = \frac{0.3}{4.2k} = 829 \ \mathrm{K} = 556°\mathrm{C}$$

Of course, this picture has considerably simplified the subject of intrinsic point defects. This chapter began by pointing out that in single crystal silicon, each atom has four nearest neighbors to which it covalently bonds. The simplest thing that could happen when a vacancy is created is that all four of the bonds break. This would leave all of the atoms electrically neutral, but it also leaves four unsatisfied valence shells. Alternatively, an electron may remain behind when a vacancy is created, thus completing the valence shell of one of the neighboring atoms, leaving it with a net one negative charge. This situation is described by saying that a -1 vacancy has been created (along with a $+1$ interstitial). The activation energy of this vacancy will be considerably different from that of a neutral vacancy. Furthermore, $-2, -3, -4, +1, +2, +3,$ and $+4$ vacancies are also possible, although triply and quadrupally ionized species are rarely important in practice.

The concentration of charged vacancies at equilibrium is given by

$$N_{v^-}^o = N_v^o \frac{n}{n_i} e^{(E_i - E_v^-)/kT} \tag{2.4}$$

where n is the carrier concentration of free electrons in the semiconductor, n_i is the intrinsic carrier concentration (see Figure 3.4), E_i is the intrinsic energy level (usually close to the center of the bandgap), and E_v^- is the energy level associated with the negatively charged vacancy. The difference in the energy levels $(E_v^- - E_i)$ between the vacancy and the intrinsic level plays the role of a new activation energy, E_a^{v-}. Similarly, the concentration of positively charged vacancies is given by

$$N_{v^+}^o = N_v^o \frac{p}{n_i} e^{(E_v^+ - E_i)/kT} = N_v^o \frac{p}{n_i} e^{-E_a^{v+}/kT} \tag{2.5}$$

The concentration of multiply charged vacancies is similarly proportional to the ratio of the charge density to the intrinsic carrier concentration, raised to the appropriate power. For example, the concentration of $v^=$ is given by

$$N_{v^=}^o = N_v^o \left[\frac{n}{n_i}\right]^2 e^{E_a^{v=}/kT} \tag{2.6}$$

Note that for each of these examples, if the doping concentration is much less than n_i, the semiconductor is instrinsic and $n = p = n_i$. In practice, charged vacancy concentrations are very low in intrinsic silicon.

Example 2.3

A silicon wafer is doped p-type at a concentration of 10^{20} cm^{-3}. At 1000°C, $N_{v^+}^o = 5 \times 10^{11}$ cm^{-3}. Find the value of E_v^+ relative to E_i.

Solution

At 1000°C

$$N_v^o = 5 \times 10^{22} \text{ cm}^{-3} e^{\frac{-2.6\,eV}{1273\,K \cdot k}}$$
$$= 2.6 \times 10^{12} \text{ cm}^{-3}$$

Then

$$N_{v^+}^o = 5 \times 10^{11} \text{ cm}^{-3} = 2.6 \times 10^{12} \text{ cm}^{-3} \left(\frac{10^{20}}{n_i}\right) e^{\frac{-(E_i - E_v^+)}{kT}}$$

From Figure 3.4, $n_i = 10^{19}$ cm^{-3}.

Solving

$$E_i - E_v^+ = 0.43$$

Energy level diagram showing E_c at top, E_i in middle (dashed line), and E_v^+ with $\downarrow 0.43$ down to E_v at bottom.

The second type of point defect that may exist in a semiconductor is an extrinsic defect. This is caused either by an impurity atom at an interstitial site or at a lattice site. In the latter case, it is referred to as a *substitution impurity*. The word "defect" should not necessarily carry with it a negative connotation. For example, dopant atoms that are required to modulate semiconductor conductivity are substitutional defects. While this text will concentrate on substitutional dopant impurities, interstitial impurities can also have a significant impact on device performance. Some impurities that tend to occupy interstitial sites have electronic states near the center of the bandgap. As a result, they are efficient sites for the recombination of electrons and holes. These recombination centers reduce the gain of bipolar transistors and can cause p–n diodes to leak.

Line defects extend in one dimension. The most common example is a dislocation. In this defect, an extra line of atoms is inserted between two other lines of atoms. The simplest type of dislocation is an edge dislocation. Here, an extra plane is terminated on one end by the edge of the crystal. If the extra plane is completely contained in the crystal, the defect is referred to as a dislocation loop. The presence of a dislocation in the crystal is a sign of stress. The bonds just before the insertion of the extra plane are stretched, and bonds just after the plane are compressed. Dislocations are often formed by the agglomeration of point defects. Any defect in a crystal has associated with it a surface energy. The higher the surface area of the defect, the higher the energy stored in the defect. A high concentration of point defects therefore costs much more energy than a dislocation, since the point defects have a larger total surface area. If the concentration of point defects exceeds the equilibrium value, the defects will tend to stick together until a dislocation or other higher dimensionality defect results. The process is called *agglomeration*.

There are various ways that defect inducing stress can occur during processing. If there is a substantial temperature difference across the wafer, the wafer will attempt to expand nonuniformly, and thermoplastic stress will occur in the wafer. Similar stresses can result if the wafer is rigidly clamped and heated or if layers with different thermal expansion coefficients are present when the wafer is heated. The equations governing thermoplastic stress will be covered in Section 6.4, since this problem is particularly important in rapid thermal processing. A second method of inducing dislocations is the presence of high concentrations of substitutional impurities in the crystal. Since these atoms are not the same size as the host atom, stresses will result. The effect of the stress is to lower the energy required to break bonds and form vacancies. The third type of dislocation formation mechanism, which will be reviewed in detail later, is physical damage of the crystal. During some processes, the surface of the wafer is bombarded with other atoms. These atoms transfer enough energy to the lattice to break bonds. Once again the process creates a high concentration of vacancies and interstitials that will tend to agglomerate into dislocations and other higher order defects.

Dislocations move by two primary mechanisms, as illustrated in Figure 2.8: *climb* and *glide*. Glide is the movement of a dislocation line in a direction other than along the line. It is the result of

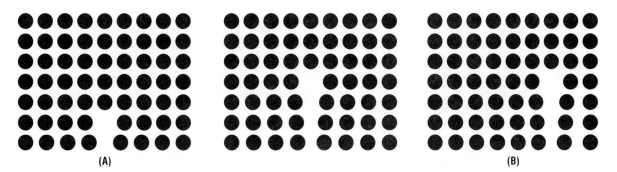

(A)   **(B)**

Figure 2.8   Movement of an edge dislocation (center figure) by (A) climb and (B) glide.

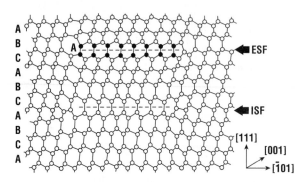

Figure 2.9   An intrinsic stacking fault is the removal of part of a plane of atoms in the {111} directions. An extrinsic stacking fault is the addition of a partial plane of atoms in the {111} directions. The labels A, B, and C correspond to the three different (111) planes in the diamond lattice *(after Shimura)*.

shear stress. In this situation, one of the planes of lattice atoms breaks to form a continuous plane with the previous extra plane, leaving behind a new defect. This process, called *slip*, can continue until the entire width of the wafer has moved by one lattice site. Climb, on the other hand, is simply the growth (or shrinkage) of a dislocation line. Vacancies and/or interstitials are created in response to stress. These point defects then become part of the line defect.

The third type of defect is a *2-D* or *area defect*. The most obvious type of area defect is a polycrystalline grain boundary. An important defect from a device standpoint is a *stacking fault*. Similar to a dislocation line, a stacking fault is an extra plane of atoms. In this case, the pattern is disrupted in two dimensions and is regular only in the third. If, for example, an extra plane of atoms is inserted in the diamond lattice, as shown in Figure 2.9, a stacking fault results [7]. Stacking faults are terminated either by the edge of the crystal or by dislocation lines. Bulk defects are irregular in all three dimensions. A common example of a bulk defect is a *precipitate*. An important class of such precipitates are impurity precipitates, discussed in Section 2.1.

The change in the Gibbs free energy associated with the impurity condensation can be expressed as

$$\Delta G = -V \cdot \Delta G_v + A\gamma + V\varepsilon + R_e \tag{2.7}$$

where V is the volume of the precipitate, ΔG_v is the change in free energy per unit volume associated with the transformation, A is the surface area of the precipitate, γ is the free energy per unit area of the precipitate, ε is the strain energy per unit volume of the precipitate, and R_e is a strain relation term. If the concentration of point defects exceeds the equilibrium values given by Equations 2.2–2.4, the crystal is said to be supersaturated, and there will be a strong driving force toward the formation of extended defects. In the case at hand, as the temperature is lowered, the degree of supersaturation increases, and the first term of Equation 2.7 becomes large enough to make ΔG negative, providing the thermodynamic force necessary to drive the precipitation. When the volume change upon precipitation is large, the precipitate causes strain that may be relaxed through the formation of self-interstitials and/or dislocations [8, 9].

It would seem that defects are undesirable. More precisely, 2-D and 3-D defects are undesirable in active regions (i.e., where the transistors are located). Defects in inactive regions have a beneficial effect as gettering sites. Gettering is a process by which impurities and defects diffuse through the crystal, becoming trapped at the gettering site. Large defects have very low diffusivities and so are unlikely to move once created. Beneficial gettering can be accomplished by providing highly strained or damaged regions away from the active devices, such as the backside of the wafer. This type of process is called *extrinsic gettering*.

Another common application of gettering in silicon technology is the use of oxygen precipitates in the bulk of the wafer. Point defects and residual impurities such as heavy metals become trapped at the precipitate sites, reducing their concentration near the active device regions. Since this process uses oxygen that is intrinsic to the wafer, it is called *intrinsic gettering*. Figure 2.10 shows the

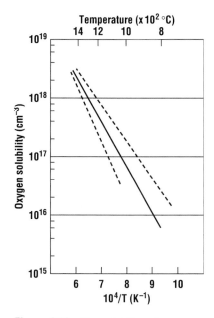

Figure 2.10 The solubility of oxygen in silicon. The dashed lines correspond to the highest and lowest variations *(after Shimura)*.

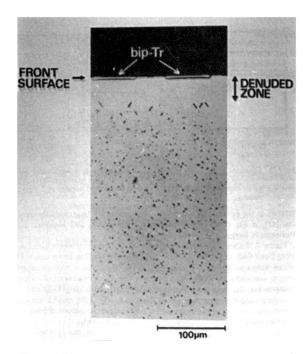

Figure 2.11 Cross-sectional view of denuded wafer with bipolar transistor *(reprinted by permission, Academic Press, after Shimura)*.

solubility of oxygen in silicon [10, 11]. Unlike the dopant curves in Figure 2.4, the solid solubility of oxygen shows a simple Arrhenius behavior, obeying the equation

$$C_{ox} = 2 \times 10^{21} \frac{atoms}{cm^3} e^{\frac{-1.032\,eV}{kT}} \qquad (2.8)$$

As will be discussed later in this chapter, dissolved oxygen results from the process by which the single crystal silicon is grown. Typical oxygen concentrations in silicon wafers are 10 to 40 ppm or about 10^{18} cm^{-3}. If the wafer temperature is less than about 1150°C, the oxygen concentration exceeds the solubility, and so oxygen will tend to precipitate from the crystal, leading to 3-D defects, sometimes called precipitation defect complexes [12]. The shape of the precipitates depends on the temperature of the wafer. If the wafer is cooled rapidly, the excess oxygen is immobile and so cannot agglomerate to condense out. The oxygen is held in solid solution in a supersaturated (i.e., greater than would be allowed if the wafer were in thermodynamic equilibrium) condition. If the wafer is annealed at about 650°C, the precipitates are shaped like rods and lie along the ⟨110⟩ directions in the (100) planes [13]. When the wafer is annealed at about 800°C, square precipitates form on the (100) planes with [111] rounded edges [14]. For wafers annealed at 1000°C, the precipitates are shaped like octahedra [15]. The amount of precipitated oxygen varies approximately as the seventh power of the initial oxygen concentration [16].

To use an intrinsic gettering process, the wafer should have an oxygen concentration of 15 to 20 ppm (Figure 2.11). If the concentration is much smaller than that, the oxygen impurities are too far apart to form agglomerates. Larger oxygen concentrations will precipitate, but they will also lead to wafer warpage, and other extended defects such as slip lines may thread through the active regions. As wafer size has increased, tolerance for these undesirable effects has decreased, leading to the need for low temperature oxygen precipitation. A typical intrinsic gettering process consists of three steps: outdiffusion, nucleation, and precipitation [17]. The purpose of the outdiffusion step is to reduce the concentration of dissolved oxygen in a denuded zone near the surface of the wafer. The depth of the denuded zone must exceed the deepest junction depth in the technology, plus the depletion width of the junction at maximum reverse bias. It is important not to use too deep a denuded zone, since this would reduce the effectiveness of the gettering. However, if the denuded zone is too close to the active device region, it can degrade device performance. Typical denuded zone depths are 20–30 μm. The denuded zone is formed by annealing the wafer at high temperature [18, 19] in an inert ambient. The temperature of the step must be high enough to allow the oxygen to diffuse out of the surface, but low enough to reduce the concentration of the oxygen to less than 15 ppm. The width of the denuded zone in silicon can be approximated by [20]:

$$L_d = \sqrt{0.091 \frac{cm^2}{sec} t \, e^{\frac{-1.2\,eV}{kT}}} \qquad (2.9)$$

By comparing Equations 2.8 and 2.9, one can find process times and temperatures that satisfy both conditions: $C_{ox} < 15$ ppm and $L_d > 10$ μm. For example, an anneal at 1200°C for 3 hr gives an oxygen concentration of 6×10^{17} cm^{-3} (12 ppm) and an approximate denuded zone depth of 25 μm. Finally, one needs to observe that the carbon concentration must also be controlled in wafers intended for intrinsic gettering use, since the oxygen can precipitate at carbon impurities [21]. The concentration of carbon in the wafer, therefore, is a factor in determining the size and shape of the oxygen precipitates [22]. It is desirable to keep C concentration less than 0.2 ppm.

Recently it has been shown that adding low concentrations of nitrogen to the wafer allows oxygen precipitation at lower temperature, reduces the size of voids in silicon, and increases the mechanical strength of the wafer [23]. These changes help reduce warpage in large-diameter silicon wafers [24, 25]. Nitrogen doping is easily done by replacing the argon gas ambient in typical Czochralski growth (see Section 2.4), with an A_F/N_2 mixture. Typical nitrogen concentrations are 1 to 10 ppb [26].

2.4 Czochralski Growth

The technique used to produce most of the crystals from which semiconductor wafers are cut is called *Czochralski growth*. The process was developed by Teal [27] in the early 1950s, although it was first developed by Jan Czochralski, who used it to draw thin metal filaments from the melt as early as 1918. Since silicon is a single component system, it is easiest to start by studying its growth. Once this is complete, some of the complications associated with compound semiconductor growth will be discussed. The production of the high purity polycrystalline materials that are melted in the Czochralski furnace will not be discussed. While this is an interesting exercise in distillation, it is not very relevant to the IC fabrication process.

Czochralski growth involves the solidification of a crystal from a melt. The material used in single crystal silicon growth is electronic grade polycrystalline silicon (polysilicon, or poly), which has been refined from quartzite (SiO_2) until it is 99.999999999% pure. The poly is loaded into a fused silica crucible that is contained in an evacuated chamber (Figure 2.12) [28]. The chamber is back-filled with an inert gas, and the crucible is heated to approximately 1500°C. Next, a small chemically etched seed crystal (about 0.5 cm in diameter and 10 cm long) is lowered into contact with the melt. This crystal must be carefully oriented since it will serve as the template for the growth of the much larger crystal, called the *boule*. Modern boules of silicon can reach a diameter of over 300 mm and are 1 to 2 m long.

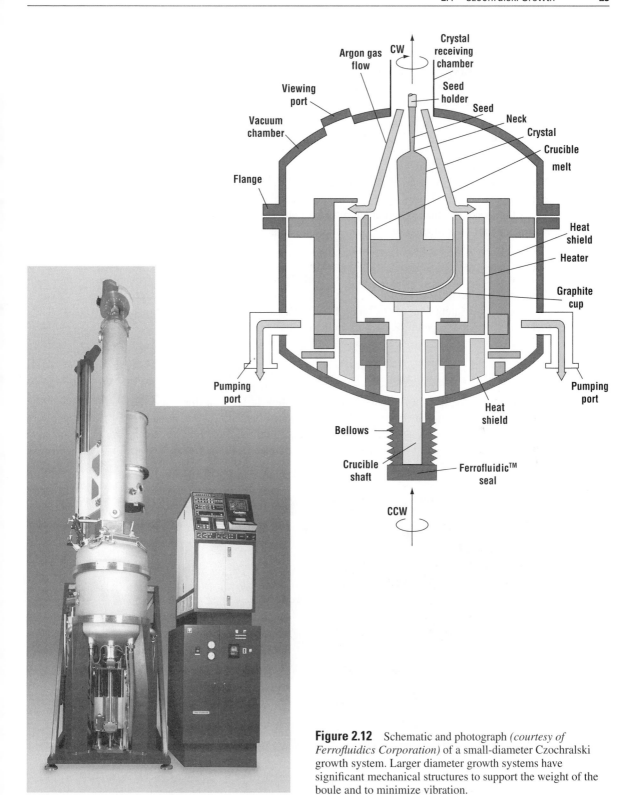

Figure 2.12   Schematic and photograph *(courtesy of Ferrofluidics Corporation)* of a small-diameter Czochralski growth system. Larger diameter growth systems have significant mechanical structures to support the weight of the boule and to minimize vibration.

Since both the liquid and the solid are at about the same pressure and have approximately the same composition, solidification must be accomplished by a reduction in temperature. As shown in Figure 2.12, temperature is lost by the increased surface area of the solid. Both natural convection and gray body radiation will cause the crystal to give off substantial heat and will give rise to a large thermal gradient across the liquid/solid interface. At the interface, additional energy must be lost to accommodate the latent heat of fusion. Balancing the energy flow in a unit volume at the interface with a simple 1-D analysis,

$$\left(-k_l A \frac{dT}{dx} \Big|_l \right) - \left(-k_s A \frac{dT}{dx} \Big|_s \right) = L \frac{dm}{dt} \tag{2.10}$$

where the ks are the thermal conductivities of liquid and solid silicon at the melting point, A is the cross-sectional area of the boule, T is the temperature, and L is the latent heat of fusion (approximately 340 cal/gm for silicon).

Both of the two thermal diffusion terms are positive under normal Czochralski growth conditions, with the first term larger than the second. This implies that there is a maximum rate at which the crystal can be pulled (Figure 2.13). This would occur if all of the heat diffusing up the solid is produced by the latent heat of fusion at the interface (i.e., the first term in Equation 2.10 is 0). Then there would be no temperature gradient in the liquid and

$$V_{max} = \frac{dx}{dt} = \frac{kA}{L} \frac{dT}{dm} = \frac{k}{\rho L} \frac{dT}{dx} \Big|_s \tag{2.11}$$

If an attempt is made to pull the crystal from the melt faster than this, the solid cannot conduct the heat away and the material will not solidify in a single crystal [30]. Typical values for the temperature gradient in silicon Czochralski are about 100°C/cm, although for pull rates near the maximum, the temperature gradient will vary inversely with the diameter of the crystal (Figure 2.14). To minimize

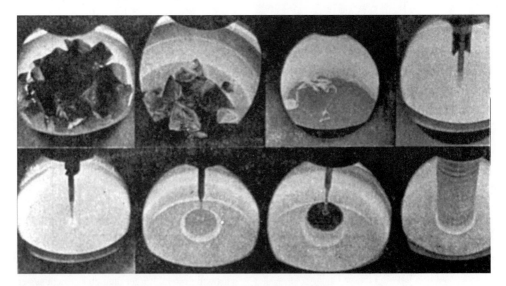

Figure 2.13 Time lapse sequence of boule being pulled from the melt in a Czochralski growth *(reprinted with permission of Lattice Press).*

the temperature gradient in the melt, the boule and melt are typically rotated in opposite directions during the growth.

In reality, the maximum pull rate is not normally used. It has been found that the crystalline quality is a sensitive function of the pull rate. The material near the melt has a very high density of point defects. It would be desirable to cool the solid quickly enough to prevent these defects from agglomerating. On the other hand, such rapid cooling means that large thermal gradients (and therefore large stresses) will occur in the crystal, particularly for large-diameter wafers. Czochralski growth processes use this effect to minimize dislocations in the boule by rapidly beginning the pull. This produces a narrow, highly perfect region [31–33] just below the seed crystal. Any dislocations in the seed crystal, whether there initially or caused by contact with the molten silicon, can be prevented from propagating into the boule in this manner [34]. An example of such an edge-terminated dislocation can be seen in Figure 2.15. The melt temperature is then lowered and the pull rate reduced to shoulder out the boule to the desired diameter. Finally, the pull rate and furnace temperature are stabilized using feedback control from an optical detector set to measure the width of the boule. The design of the heat shields is critically important in controlling the temperature distribution near the solid/melt interface and therefore in determining the defect density [36].

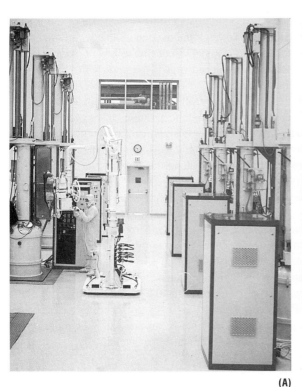

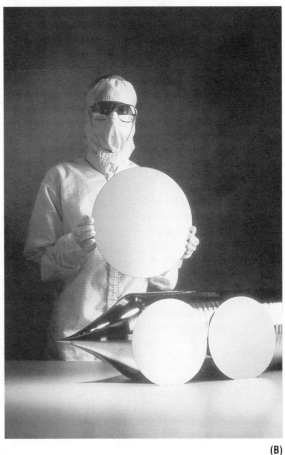

(A)

(B)

Figure 2.14   (A) A 200-mm silicon growth facility, (B) prototype 300-mm wafers *(photographs courtesy of MEMC Electronic Materials, Inc.).*

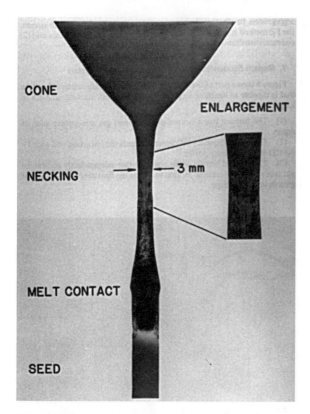

Figure 2.15 X-ray topograph of seed neck showing edge terminated dislocations *(reprinted by permission, Academic Press)*.

Generally, as the diameter of the crystal is increased, the pull rate must be decreased. This is because the heat loss is proportional to the surface area, which is proportional to the diameter of the crystal, while the power produced by fusion is proportional to the product of the pull speed and the cross-sectional area, which in turn is proportional to the square of the diameter. If the pull rate is too low, however, point defects will agglomerate. The type of defects formed most commonly are called *dislocation loops*. In semiconductor substrates these loops are called swirl because they are often distributed in swirls about the center of the wafer.

During the process of Czochralski growth several impurities will incorporate into the crystal. We have already discussed the importance of oxygen in the substrate. We are now in a position to understand its origin. Crucibles used to hold the molten silicon during the Czochralski process are usually fused silica (amorphous SiO_2, often mistakenly called quartz). At 1500°C, silica will release a considerable amount of oxygen into the molten silicon. Over 95% of the dissolved oxygen escapes from the surface of the melt as SiO [36]. Some of the oxygen will be incorporated into the growing crystal. Since this supply of oxygen is constantly replenished, it will be approximately constant along the length of the boule. The oxygen concentration can be controlled using the temperature of the melt. For reduced concentrations of oxygen, the boule can be grown under magnetic confinement. The first commercially significant magnetically confined Czochralski growths were reported in the early 1980s [37, 38]. Figure 2.16 shows a magnetically confined crystal growth system. The magnetic field may be directed along the length of the boule (axial); however, modern systems are nearly all perpendicular to the growth direction (transverse). Typical fields are 0.3 T (3 kG). In either case, the purpose of the field is to create a Lorentz force ($qv \times B$), which will change the motion of the ionized impurities in the melt in such a manner as to keep them away from the liquid/solid interface and therefore decrease the impurity incorporation in the crystal. In this arrangement, oxygen concentrations as low as 2 ppm have been reported [39]. This process also has the effect of minimizing resistivity variations across the wafer [40].

It is also common to introduce dopant atoms into the melt so that a wafer of a particular resistivity can be made. To do this, one can simply weigh the melt, determine the number of impurity atoms that would be needed, and add that weight of impurity. The process is complicated, however, by the fact that impurities tend to segregate at solid/liquid interfaces. That is, the solid may be more or less likely to contain an impurity than the liquid. A segregation coefficient k can be defined as

$$k = \frac{C_s}{C_l} \qquad\qquad (2.12)$$

where C_s and C_l are the impurity concentrations at the solid and liquid sides of the solid/liquid interface. Table 2.1 summarizes the segregation coefficients of common impurities in silicon [41].

Figure 2.16 Photographs of a commercial magnetically confined Czochralski system *(Thomas et al.)*.

Table 2.1 Segregation coefficients for common impurities in silicon

Al	As	B	O	P	Sb
0.002	0.3	0.8	0.25*	0.35	0.023

From Sze [41].
*The segregation coefficient of oxygen is somewhat controversial. Values closer to 1.0 are sometimes quoted.

To understand how dopant segregation impacts crystal uniformity, consider what happens if $k > 1$. In that case, the concentration of impurity in the solid is greater than that in the melt. To achieve this, a higher proportion of dopant is pulled from the melt than is contained in the liquid. Consequently, the impurity concentration in the melt must decrease as the boule is pulled. Referring to the phase diagrams or the solid solubility curves, the concentration shift of the melt also shifts the concentration of the solid. If we define X as the fraction of the melt that has solidified, and assume that the solution is well mixed, it can be shown that

$$C_s = kC_0(1-X)^{k-1} \qquad (2.13)$$

where C_0 is the initial melt concentration.

The well-mixed approximation is not very good due to the existence of thermal gradients in the melt. The basic effect is shown in Figure 2.17. The hot walls of the crucible cause the melt near the walls

---

Example 2.4

An initial melt has a concentration of 1 ppm boron.

(a) Find the concentration of boron at the top of the boule. (b) Find the position along the boule where the solid concentration is 1 ppm.

Solution

(a)   Equation 2.13: $C_s = kC_o (1-X)^{k-1}$,

for boron, $k = 0.8$

At the top of the boule $X = 0$, so

$$C_s = 0.8 \times 1 \text{ ppm} (1 - 0)^{-0.2} = 0.8 \text{ ppm}$$

$$C_s = 0.8 \times 10^{-6} = 5 \times 10^{22} \frac{\text{atoms}}{\text{cm}^3} = 4 \times 10^{16} \text{cm}^{-3}$$

(b)   Using Equation 2.13,

$$C_s = 1 \text{ ppm} = 0.8 \times 1 \text{ ppm} (1-x)^{-0.2}$$

$$1.25 = (1 - X)^{-0.2}$$

Solving, $x = 0.67$

---

to expand. The lower density of this heated material will cause it to rise. The cooler region of the melt near the boule will tend to sink. Collectively, the process is known as *natural convection,* and the flow patterns that it produces are known as *buoyancy-driven recirculation cells.* This effect is present to some extent whenever temperature gradients exist in a system that contains a liquid or gas with a finite, nonzero viscosity. The rotation of the crucible and the boule, along with the pulling action of the boule, the heat released by solidification at the solid–melt interface, and the surface tension of the melt, all contribute to the flow patterns in the melt [42]. For larger boules the flow in the melt is unstable. This instability occurs because the cooler surface layer tends to increase in density and so would sink in the melt. The instability gives rise to temperature fluctuations at the interface that can create small voids in the silicon [43]. Large crucibles and magnetic fields can be used to control this instability [44].

At the liquid/solid interface, the melt cannot flow, in the same way the water at the banks of a river cannot flow. In fact, because of the finite viscosity of the melt, there will be a region near the liquid/solid interface over which little flow of material will be present. This region of the melt is called the *boundary layer.* The impurities that are taken up by the solid must diffuse across this region. To take this effect into account, the segregation coefficient k can be replaced with an effective segregation coefficient k_e, where

$$k_e = \frac{k}{k + (1 - k)e^{-Vb/D}} \tag{2.14}$$

b is an effective boundary layer thickness, V is the pull velocity, and D is the impurity diffusivity in the molten semiconductor.

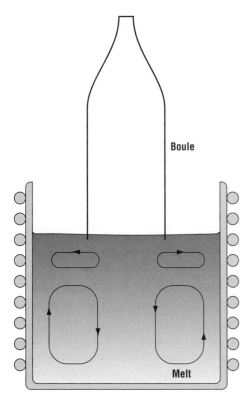

Boule

Melt

Figure 2.17   Idealized schematization of the formation of recirculation cells in a melt.

The growth of GaAs from the melt is significantly more difficult than the growth of silicon. One reason is the difference in the vapor pressure of the two materials. Stoichiometric GaAs melts at 1238°C (see Figure 2.2). At that temperature, the vapor pressure of gallium is less than 0.001 atm, while the vapor pressure of arsenic is about 10^4 times larger. It is obvious that maintaining stoichiometry through the boule will be challenging. A variety of systems have been designed to overcome this obstacle. This and the next section will present the two most popular choices: *liquid encapsulated Czochralski growth*, commonly called LEC [45], and the *Bridgman growth technique*. Bridgman wafers have the lowest dislocation densities (of order $10^3 \, \text{cm}^{-2}$) and are commonly used for fabricating optoelectronic devices such as lasers. LEC-grown wafers can be made with larger diameters; they are round, and can be made semiinsulating with resistivities of nearly 100 MΩ-cm. A disadvantage of LEC wafers is that they typically have defect densities greater than $10^4 \, \text{cm}^{-2}$. Many of these defects are due to thermoplastic stress arising from vertical temperature gradients of 60–80°C/cm [46]. The production and use of these semi-insulating substrates will be discussed in Chapter 15. As a result of this property, nearly all electronic GaAs devices are fabricated on LEC material.

Pyrolytic boron nitride (pBN) crucibles are used instead of quartz in LEC growth to avoid silicon doping of the GaAs boule from the quartz. To prevent outdiffusion of As from the melt, the LEC process uses a tightly fitting disk as shown in Figure 2.18 [47]. The most common sealant material is boric oxide (B_2O_3). A slight excess of arsenic is added to the charge to compensate for the arsenic loss that occurs until the cap becomes molten at about 400°C and seals the melt. Once the charge is molten, the seed crystal can be lowered through the boric oxide until it contacts the charge. During synthesis the pressure reaches 60 atm. Crystal growth is carried out at 20 atm [48]. For that reason, the process is sometimes called *high pressure LEC* or HP LEC. Typical pull rates are about 1 cm/hr [40].

The second problem encountered in LEC growth relates to differences in the material properties of silicon and GaAs. Table 2.2 summarizes the relevant data. The thermal conductivity of GaAs is about a third that of silicon. As a result, a GaAs boule is not able to dissipate the latent heat of

Table 2.2   Thermal and mechanical properties of semiconductors

	Melting Point (°C)	Thermal Conductivity (W/cm-K)	Critical Resolved Shear Stress at MP (MPa)
Si	1420	0.21	1.85
Ge	960	0.17	0.70
GaAs	1238	0.07	0.40

Thomas et al. [40].

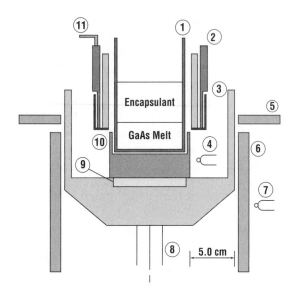

Figure 2.18   Schematic of a liquid-encapsulated Czochralski growth system. The labeled parts include (1) quartz crucible, (2) heat flux control system, (3) graphite shield, (4) cavity thermocouple, (5) radiation shield, (6) heater, (7) temperature control thermocouple, (8) water cooled support, (9) insulation support, (10) graphite crucible support, and (11) tubing support *(after Kelly et al.).*

fusion as rapidly as a silicon boule. Furthermore, the shear stress required to nucleate a dislocation at the melting point is about a fourth that of silicon. Not only is the material less able to dissipate heat, a smaller thermoplastic strain will induce defects. As shown in Figure 2.18, heaters may be used around the boule to limit the heat flux, and therefore the temperature gradient. This region must not be too hot or arsenic will desorb from the surface [49]. It is not surprising, therefore, that the Czochralski growth of GaAs is primarily limited to much smaller wafers than silicon and that defect densities of Czochralski grown material are many orders of magnitude larger than comparable silicon wafers.

If the dislocation density is kept low enough, it does not present an insurmountable barrier to the fabrication of ICs with moderate to high levels of integration; however, when it exceeds 10^4 cm^{-2}, the dislocations can have a significant impact on transistor performance [50]. It has been found that thermally annealing the GaAs ingot after Czochralski growth somewhat reduces the dislocation density [51]. Alloying about 0.1 atomic percent indium into the wafer has been found to minimize the impact of dislocations in Czochralski-grown GaAs. Ehrenreich and Hirth [52] have suggested that the replacement of a gallium atom with a much larger indium atom creates a strain field that traps the dislocations, effectively gettering them from any active layers grown on top of the wafer. It is now generally believed that the use of indium increases the critical resolved shear stress through a process known as *solid solution hardening* [53]. As a result, the dislocation density itself can be 10^3 cm^{-2} or less in indium-doped wafers. Because of the hardening, indium-doped wafers are even more brittle than pure GaAs and so more prone to chipping and breakage. This fact, combined with concerns of indium diffusion during the process and the material improvements made possible by boule annealing has caused the popularity of indium doped GaAs to decline in recent years. Initial wafer resistivity has also been found to have a significant effect on transistor performance. Very high resistivity semi-insulating material leads to less efficient activation (compensation) of implanted impurities and lower current drive [54]. Control of the exact concentration of defects that drive the wafer semi-insulating (see Section 15.5) is extremely important in order to get controlled transistor operation, particularly in implanted devices. A variation of LEC, called vapor pressure controlled Czochralski (VCZ), keeps the boule hot and uses an arsenic-containing ambient to suround the boule [55]. This provides better control of composition and therefore resistivity.

2.5 Bridgman Growth of GaAs

More than half of the market for growing GaAs is controlled by the horizontal Bridgman method and its variants [56]. The basic process is shown in Figure 2.19 [57]. The solid Ga and As components are loaded into a fused silica ampoule, which is then sealed shut. In many cases, the ampoule includes a separate solid arsenic chamber with a restricted orifice to the main chamber. The arsenic-only chamber

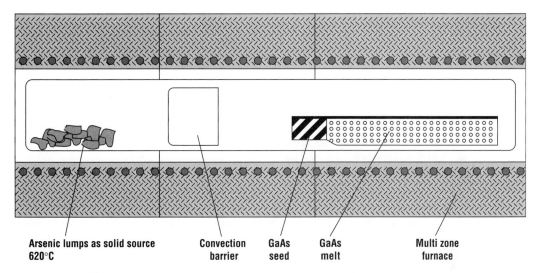

Arsenic lumps as solid source   **Convection**   **GaAs**   **GaAs**   **Multi zone**
620°C   **barrier**   **seed**   **melt**   **furnace**

Figure 2.19   Schematic of a horizontal Bridgman growth system *(after Sell).*

provides the arsenic overpressure necessary to maintain stoichiometry. The ampoule is loaded into a SiC tube, which rests on a semicircular trough, typically also made from SiC. A tube furnace is then slowly rolled past the charge. Although the charge may be passed through the furnace, the opposite is normally done to minimize any disturbance of the crystal as it solidifies. The temperature of the furnace is set to melt the charge when it is completely inside. As the furnace rolls past the ampoule, the molten GaAs charge in the bottom of the ampoule recrystallizes in a characteristic "D" shape. If desired, a seed crystal can be mounted to contact the melt. Typical crystal diameters produced this way are 1 to 2 in. The growth of larger crystals requires very accurate control of the stoichiometry of the axial and radial temperature gradients to control the dislocation density [58]. The important feature of the Bridgman approach is that the molten charge is held by the ampoule. This allows the process to be run with very small thermal gradients and, therefore, routinely produces wafers with dislocation densities of less than $10^3 \, \text{cm}^{-2}$, with reports as low as $10^2 \, \text{cm}^{-2}$.

Difficulties with standard Bridgman growth include its inability to produce high resistivity substrates because of the large intimate contact between the ampoule and the melt and the difficulty of making wafers with diameters greater than 2 in. A variety of techniques have been proposed to overcome this obstacle. Two of the most commonly cited are the vertical Bridgman [59] and vertical gradient freeze [60] methods. The basic idea of each approach is to turn a horizontal Bridgman apparatus on its side. The charge is held in a boat that typically is boron nitride, which is then sealed in a silica ampoule in such a way as to minimize the dead volume. A small additional arsenic charge may be added to maintain stoichiometry. The ampoule is loaded into a furnace, where it is raised to a temperature just above the melting point. The furnace is then slowly raised, cooling the ampoule and freezing the charge. Temperature gradients are less than 10°C/cm, and the growth rate is limited to a few millimeters per hour [61]. The vertical gradient freeze method, a variation of vertical Bridgman, adds multiple heaters to control the temperature gradients outside the furnace [62]. A 1997 report suggests that materials having resistivities of 42–67 MΩ-cm material can be grown by the vertical gradient freeze technique [63]. The vertical Bridgman technique has been extended to 4-in. and even 6-in. wafers with encouraging results [64].

2.6 Float Zone Growth

When extremely high purity silicon is required, the growth technique of choice is *float zone refining*. This technique is not generally used for GaAs. Figure 2.20 shows a plot of the crystal purity reported with both Czochralski growth and float zone refining. Carrier concentrations as low as 10^{11} cm^{-3} have been achieved with the float zone method. The basic feature of this growth technique is that the molten part of the sample is supported entirely by the solid part. There is no need for a crucible. The concept of the float zone method is shown in Figure 2.21. A rod of high purity polycrystalline material is held in a chuck while a metal coil driven by a high power radio frequency (RF) signal is slowly passed along its length. The field set up by the RF power induces eddy currents in the rod that lead to joule heating. The coil power is set such that the part of the rod closest to the coil is melted. Alternatively a focused electron beam can be used to locally melt the rod. The entire apparatus can be encased in a vacuum system or in an enclosed chamber with an inert gas purge.

To ensure growth along a preferred orientation, a seed crystal may also be used. In a method very similar to that of Czochralski growth, Dash injected a very thin seed crystal into the top of the molten rod [33]. In this technique, a thin neck about 3 mm in diameter and about 10 to 20 mm long is pulled, and the pull rate and temperature lowered to shoulder the crystal out to a larger diameter. The difficulty of the top-seeded technique is that the molten region must be able to support the weight of the entire rod, unlike Czochralski, where the boule has not yet been formed. As a result, this technique is limited to boules of no more than a few kilograms. Figure 2.22 shows an arrangement that can be used for the float zone growth of large diameter crystals [65]. In this case the crystal is bottom seeded. After a sufficient length of dislocation free material has been produced, a funnel filled with small spheres is slid upward until it supports the weight of the boule.

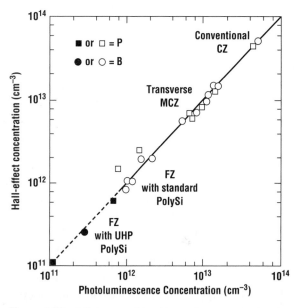

Figure 2.20   Minimum achievable carrier concentration for various growth technologies *(Thomas et al.)*.

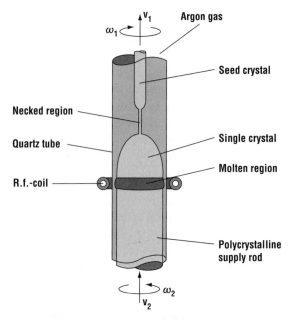

Figure 2.21   Schematic of a float zone refining system.

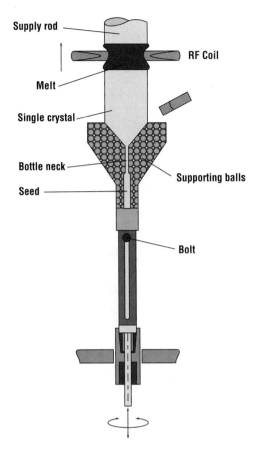

Supply rod

RF Coil

Melt

Single crystal

Bottle neck

Supporting balls

Seed

Bolt

Figure 2.22   Schematic of a float zone system for growing large diameter boules *(reprinted from Keller and Mühlbauer by courtesy of Marcel Dekker).*

A disadvantage of float zone growth is the difficulty of introducing a uniform concentration of dopants. There are four methods that can be used: *core doping, pill doping, gas doping,* and *neutron transmutation.* Core doping refers to the use of a doped polysilicon rod for the starting material. On top of this rod, additional undoped polysilicon is deposited until the average desired concentration is reached. The process can be repeated through several generations if necessary. Core doping is the preferred process for boron because its diffusivity is high and because it does not tend to evaporate from the surface of the rod. Neglecting the first few melt lengths, the concentration of boron in a boule is quite uniform. Gas doping is accomplished through the use of gases such as PH_3, $AsCl_3$, or BCl_3. The gas may be injected as the polysilicon rod is deposited, or it may be injected at the molten ring during the float zone refining. Pill doping is accomplished by drilling a small hole in the top of the rod and inserting the dopant in the hole. If the dopant has a small segregation coefficient, most of it will be carried with the melt as it passes the length of the boule, resulting in only a modest nonuniformity. Gallium and indium doping work well in this manner. Finally, for light n-type doping, float zone silicon can be doped through a process known as transmutation doping. In this process, the boule is exposed to a high brightness neutron source. Approximately 3.1% of the silicon is the mass 30 isotope. Under neutron flux, that isotope can be changed through the nuclear process [66]

$$.^{30}Si(n^{\cdot}, \gamma^{\prime}) \rightarrow .^{31}Si \overset{3.6\,h}{\rightarrow} -\,^{31}P + \beta^{\prime} \tag{2.15}$$

Of course, a disadvantage of this process is that it is not suitable for forming p-type silicon.

2.7 Wafer Preparation and Specifications

After the boule has been grown, the wafers must be made. The boule is first characterized for resistivity and crystal perfection. Then the seed and tail are cut off and the boule is mechanically trimmed to the proper diameter. The diameter at this point is slightly larger than the final wafer diameter, since additional etching will be done. For wafers 150 mm and less, flats are ground the entire length of the boule to denote the crystal orientation and the doping type and to provide a method for coarsely aligning the rotation of the wafer during subsequent photolithography steps. The largest flat, called the *primary*, is oriented perpendicular to the $\langle 110 \rangle$ direction. One or more minor flats will also be ground. Figure 2.23 shows the flat orientations for different wafer types. For larger wafers, a notch is ground into the edge.

After the flats have been ground, the wafer is dipped in a chemical etchant to remove the damage caused by the mechanical grinding. Each manufacturer has a proprietary mixture, but usually the mix is based on the $HF–HNO_3$ system. After etching, the boule is sliced into wafers. This is a critical step in the process as it will determine the wafer bow and flatness. Typically, a wire impregnated with diamond particles is used. The wafers may then be edge rounded in another mechanical grinding

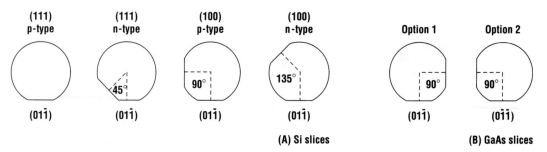

Figure 2.23   Standard flat orientations for different semiconductor wafers.

Table 2.3   Typical specification for 2-in. GaAs wafers

	LEC	Vertical Gradient Freeze
Resistivity (Ω-cm)	$>2 \times 10^7$	$>10^7$
Mobility (cm^2-V/s)	>4000	>5000
Carrier concentration (cm^{-3})	$<10^8$	—
Etch pit density (cm^{-2})	7×10^4(typical)	$<10^3$
Flatness (μm)	<4	<5
Thickness (μm)	450	500
Cost ($/cm^2)	10	15

Table 2.4   Typical specifications for state-of-the-art silicon wafer

Cleanliness (particle/cm^2)	<0.03
Oxygen concentration (cm^{-3})	Specified $+-3\%$
Carbon concentration (cm^{-3})	$<1.5\times10^{17}$
Metal contaminants bulk (ppb)	<0.001
Grown in dislocation (cm^{-2})	<0.1
Oxidation induced stacking faults (cm^{-3})	<3
Diameter (mm)	≥150
Thickness (μm)	625 or 675
Bow (μm)	10
Global flatness (μm)	3
Cost ($/cm^2)	0.2

From Shimura [7].

process. It has been found that edge-rounded wafers are less susceptible to defects created by mechanical handling during the subsequent processing. It also prevents the pile-up of liquids that will coat the wafer when using spin on processes.

Next a series of steps is performed to remove any residual mechanical damage and to prepare the wafers for device fabrication. First, the wafers are mechanically lapped in a slurry of alumina and glycerine, then etched as before to reduce the damage. Finally, one or both sides receive a polish in an electrochemical process involving a slurry of NaOH and very fine silica particles followed by a chemical clean to remove any residual contaminants. Tables 2.3 and 2.4 show typical wafer specifications.

Both silicon and GaAs technologies are migrating to larger wafer sizes, which allow more die per wafer and hence greater efficiency. The standard GaAs wafer is 100 mm, but new facilities use 150-mm wafers. The standard wafer for silicon IC manufacturing has been 200 mm, but most new fabs use 300 mm wafers. Plans to extend wafer diameters to larger sizes exist, but are unlikely to be implemented for at least another decade.

2.8 Summary and Future Trends

This chapter reviewed some of the most basic properties of semiconductor materials. An introduction to phase diagrams was given and the idea of solid solubility was presented. Next, a basic description of point, line, area, and volume defects was given. In the second half of the chapter, crystal growth methods were presented. Czochralski growth is the most common method for preparing silicon wafers. Liquid encapsulated Czochralski is the dominant method for growing GaAs for electronic applications, but high defect densities are a problem. For optoelectronic applications the preferred method is the Bridgman technique. Vertical gradient freeze shows promise and is expected to increase its market share.

Problems

1. A GaAs crystal is on a coordinate system such that an arsenic atom sits at the position 0,0,0 and a gallium atom sits at $a/4$, $a/4$, $a/4$. Find the x, y, z coordinates of the other three nearest neighbor gallium atoms to the arsenic atom at 0,0,0. What is the distance between these atoms?
2. For the crystal in Problem 1, find the three other nearest arsenic atoms to the cited gallium atom.
3. A mixture of 30% silicon and 70% germanium is heated to 1100°C. If the material is in thermal equilibrium, what is the concentration of silicon in the melt? At what temperature will the entire charge melt? The sample temperature is raised to 1300°C, then slowly cooled back down to 1100°C. What is the concentration of silicon in the solid?
4. During the molecular beam epitaxial growth (MBE) of GaAs layers, there is a strong tendency to form droplets of gallium on the surface of the wafer. These oval defects are a serious problem for MBE fabrication. Avoiding them requires a large arsenic-to-gallium flux ratio. Referring to the phase diagram (Figure 2.2), explain why thermodynamics would favor the formation of these droplets.
5. A process technology that has gained a great deal of interest in the last few years is rapid thermal annealing (see Chapter 6). The process allows wafers that contain high concentrations of dopant atoms to be heated to high temperatures very rapidly, minimizing dopant diffusion. Explain the desirability of such a process based on the discussion of phase diagrams and solid solubility.
6. A silicon wafer that has 10^{16} cm^{-3} of boron is found to have a neutral vacancy concentration of 2×10^{10} cm^{-3} at some processing temperature and a singly ionized vacancy concentration of 10^{9} cm^{-3} at the same temperature. Determine the temperature and the activation energy of the charged vacancy with respect to the intrinsic level, E_i.
7. Repeat Problem 6 if the wafer is doped with 2×10^{19} cm^{-3} of boron.
8. It is desired to form a denuded zone 10 μm thick using an 1100°C anneal. How long will the anneal need to be? What will the oxygen concentration be in the denuded zone?
9. If the temperature gradient in Czochralski silicon is 100°C/cm, calculate the maximum pull rate.
10. Assuming that the melt is at a uniform temperature and loses no heat except to the boule and that the boule is a perfect blackbody, set up the differential equations and boundary conditions in two dimensions that one would need to solve to find $T(r, z)$ in the boule.

11. A boule of single crystal silicon is pulled from the melt in a Czochralski process. The silicon is boron doped. After the boule is pulled, it is sliced into wafers. The wafer taken from the top of the boule has a boron concentration of $3 \times 10^{15}\,\mathrm{cm}^{-3}$. What would you expect for doping concentration of the wafer taken from the position corresponding to 90% of the initial charge solidified?

12. A Czochralski growth process is begun by inserting 1000 moles of pure silicon and 0.01 mole of pure arsenic in a crucible. For this boule, the maximum permissible doping concentration is $10^{18}\,\mathrm{cm}^{-3}$. What fraction ($x$) of the boule is usable?

13. In an attempt to form a heavily doped p-type boule of silicon, a CZ process is run in which the initial melt concentration of boron was 0.5%. Assume that the solidification temperature was 1400°C, and that the boule cooled rapidly after solidification.
     (a) What is the solid solubility of boron at this temperature?
     (b) What fraction of the boule must be pulled (i.e., solidified) before the concentration of the boron in the solid will begin to exceed the solid solubility of boron in silicon?
     (c) What is the concentration (in percent) of boron in the liquid at this point, assuming that the liquid has a uniform concentration.

14. A melt contains 0.1 atomic percent phosphorus in silicon. Assume the well-mixed approximation and calculate the dopant concentration when 10% of the crystal is pulled, when 50% of the crystal is pulled, and when 90% of the crystal is pulled.

15. A boule of silicon is pulled from a melt that contains 0.01% phosphorus (P) in the melt.
     (a) What concentration of phosphorus (P) would you expect at the top of the boule ($X = 0$)?
     (b) If the boule is 1 m long and has a uniform cross section, at what position (or X value) would you expect the concentration of phosphorus to be twice as large as it is at the top?
     (c) Now consider the melt to contain gallium as well. (Gallium is a p-type dopant for silicon, but it is not commonly used.) The concentration of gallium in the melt is such that at the top of the boule ($X = 0$), the concentrations of gallium and phosphorus are exactly equal. If the concentration of gallium near the bottom of the boule ($X = 0.9$) is twice that of the phosphorus, what is the segregation coefficient (k) for gallium?

16. Why does Bridgman growth tend to have higher impurity concentrations than LEC?

References

1. S. Mahajan and K. S. Harsha, *Principles of Growth and Processing of Semiconductors*, McGraw-Hill, Boston, 1999.
2. *Binary Alloy Phase Diagrams*, 2nd ed., vol. 2, ASM Int., Materials Park, OH, 1990, p. 2001.
3. *Binary Alloy Phase Diagrams*, 2nd ed., vol. 1, ASM Int., Materials Park, OH, 1990, p. 283.
4. *Binary Alloy Phase Diagrams*, 2nd ed., vol. 1, p. 319.
5. F. A. Trumbore, "Solid Solubilities of Impurities in Germanium and Silicon," *Bell Systems Tech. J.* **39**:210 (1960).
6. S. Ghandhi, *VLSI Fabrication Principles*, Wiley, New York, 1983.
7. F. Shimura, *Semiconductor Silicon Crystal Technology*, Academic Press, San Diego, 1989.
8. S. M. Hu, *Mater. Res. Soc. Symp. Proc.* **59**:249 (1986).
9. S. Mahajan, G. A. Rozgonyi, and D. Brasen, "A Model for the Formation of Stacking Faults in Silicon," *Appl. Phys. Lett.* **30**:73 (1977).
10. R. A. Craven, *Semiconductor Silicon 1981*, p. 254.
11. J. C. Mikkelson, Jr., S. J. Pearton, J. W. Corbett, and S. J. Pennycook, eds., *Oxygen, Carbon, Hydrogen and Nitrogen in Crystalline Silicone*, MRS, Pittsburgh, 1986.

12. H. R. Huff, "Silicon Materials for the Mega-IC Era," Sematech Technical Report 93071746A-XFR (1993).

13. A. Bourret, J. Thibault-Desseaux, and D. N. Seidmann, "Early Stages of Oxygen Segregation and Precipitation in Silicon," *J. Appl. Phys.* **55**:825 (1985).

14. K. Wada, N. Inoue, and K. Kohra, "Diffusion Limited Growth of Oxygen Precipitation in Czochralski Silicon," *J. Cryst. Growth* **49**:749 (1980).

15. K. H. Yang, H. F. Kappert, and G. H. Schwuttke, "Minority Carrier Lifetime in Annealed Silicon Crystals Containing Oxygen," *Phys. Stat. Sol.* **A50**:221 (1978).

16. H. R. Huff, H. F. Schaake, J. T. Robinson, S. C. Baber, and D. Wong, "Some Observations on Oxygen Precipitation/Gettering in Device Processed Czochralski Silicon," *J. Electrochem. Soc.* **130**:1551 (1983).

17. D. C. Gupta and R. B. Swaroop, "Effects of Oxygen and Internal Gettering on Donor Formation," *Solid State Technol.* **27**:113 (August 1984).

18. R. B. Swaroop, "Advances in Silicon Technology for the Semiconductor Industry," *Solid State Technol.* **26**:101 (July 1983).

19. D. Huber and J. Reffle, "Precipitation Process Design for Denuded Zone Formation in Czochralski Silicon Wafers," *Solid State Technol.* **26**:137 (August 1983).

20. M. Stavola, J. R. Patel, L. C. Kimerling, and P. E. Freeland, "Diffusivity of Oxygen in Silicon at the Donor Formation Temperature," *Appl. Phys. Lett.* **42**:73 (1983).

21. W. J. Taylor, T. Y. Tan, and U. Gosele, "Carbon Precipitation in Silicon: Why Is It So Difficult?" *Appl. Phys. Lett.* **62**:3336 (1993).

22. T. Fukuda, *Appl. Phys. Lett.* **65**:1376 (1994).

23. X. Yu, D. Yang, X. Ma, J. Yang, Y. Li, and D. Que, "Grown-in Defects in Nitrogen doped Czochralski Silicon," *J. Appl. Phy.* **92**:188 (2002).

24. K. Sumino, I. Yonenaga, and M. Imai, "Effects of Nitrogen on Dislocation Behavior and Mechanical Strength in Silicon Crystals," *Appl. Phys. Lett.* **59**:5016 (1983).

25. D. Li, D. Yang, and D. Que, "Effects of Nitrogen on Dislocations in Silicon During Heat Treatment." *Physica B* **273–74**:553 (1999).

26. D. Tian, D. Yang, X. Ma, L. Li, and D. Que, "Crystal Growth and Oxygen Precipitation Behavior of 300 mm Nitrogen-doped Czochralski Silicon," *J. Cryst. Growth* **292**:257 (2006).

27. G. K. Teal, "Single Crystals of Germanium and Silicon—Basic to the Transistor and the Integrated Circuit," *IEEE Trans. Electron Dev.* **ED-23**:621 (1976).

28. W. Zuhlehner and D. Huber, "Czochralski Grown Silicon," in *Crystals 8*, Springer-Verlag, Berlin, 1982.

29. S. Wolf and R. Tauber, *Silicon Processing for the VLSI Era, Vol. 1,* Lattice Press, Sunset Beach, CA, 1986.

30. S. N. Rea, "Czochralski Silicon Pull Rates," *J. Cryst. Growth* **54**:267 (1981).

31. W. C. Dash, "Evidence of Dislocation Jogs in Deformed Silicon," *J. Appl. Phys.* **29**:705 (1958).

32. W. C. Dash, "Silicon Crystals Free of Dislocations," *J. Appl. Phys.* **29**:736 (1958).

33. W. C. Dash, "Growth of Silicon Crystals Free from Dislocations," *J. Appl. Phys.* **30**:459 (1959).

34. T. Abe, "Crystal Fabrication," in *VLSI Electron—Microstructure Sci.* **12**, N. G. Einspruch and H. Huff, eds., Academic Press, Orlando, F2, 1985.

35. W. von Ammon, "Dependence of Bulk Defects on the Axial Temperature Gradient of Silicon Crystals During Czochralski Growth," *J. Cryst. Growth* **151**:273 (1995).

36. K. M. Kim and E. W. Langlois, "Computer Simulation of Oxygen Separation in CZ/MCZ Silicon Crystals and Comparison with Experimental Results," *J. Electrochem. Soc.* **138**:1851 (1991).

37. K. Hoshi, T. Suzuki, Y. Okubo, and N. Isawa, *Electrochem. Soc. Ext. Abstr.* St. Louis Meet., May 1980, p. 811.

38. K. Hoshi, N. Isawa, T. Suzuki, and Y. Okubo, "Czochralski Silicon Crystals Grown in a Transverse Magnetic Field," *J. Electrochem. Soc.* **132**:693 (1985).

39. T. Suzuki, N. Izawa, Y. Okubo, and K. Hoshi, *Semiconductor Silicon 1981*, 1981, p. 90.

40. R. N. Thomas, H. M. Hobgood, P. S. Ravishankar, and T. T. Braggins, "Melt Growth of Large Diameter Semiconductors: Part I," *Solid State Technol.* **33**:163 (April 1990).

41. S. Sze, *VLSI Technology*, McGraw-Hill, New York, 1988.

42. N. Kobayashi, "Convection in Melt Growth—Theory and Experiments," *Proc. 84th Meet. Cryst. Eng.* (Jpn. Soc. Appl. Phys.), 1984, p. 1.

43. M. Itsumi, H. Akiya, and T. Ueki, "The Composition of Octohedron Structures Thtat Act as an Origin of Defects in Thermal SiO_2 on Czochralski Silicon," *J. Appl. Phys.* **78**:5984 (1995).

44. H. Ozoe "Effect of a Magnetic Field in Czochralski Silicon Crystal Growth," in *Modelling of Transport Phenomena in Crystal Growth*, J. S. Szymd and K. Suzuki, eds MIT Press, Cambridge, MA, 2000.

45. J. B. Mullin, B. W. Straughan, and W. S. Brickell, J. Phys. Chem. Solid, **26**:782 (1965).

46. I. M. Grant, D. Rumsby, R. M. Ware, M. R. Brozea, and B. Tuck, "Etch Pit Density, Resistivity and Chromium Distribution in Chromium Doped LEC GaAs," in *Semi-Insulating III–V Materials*, Shiva Publishing, Nantwick, U.K., 1984, p. 98.

47. K. W. Kelly, S. Motakes, and K. Koai," Model-Based Control of Thermal Stresses During LEC Growth of GaAs. II: Crystal Growth Experiments," *J. Cryst. Growth* **113(1–2)**:265 (1991).

48. R. M. Ware, W. Higgins, K. O. O'Hearn, and M. Tiernan, "Growth and Properties of Very Large Crystals of Semi-Insulating Gallium Arsenide," *GaAs IC Symp.*, 1996, p. 54.

49. P. Rudolph and M. Jurisch, "Bulk Growth of GaAs: An Overview," *J. Cryst. Growth* **198–199**:325 (1999).

50. S. Miyazawa, and F. Hyuga, "Proximity Effects of Dislocations on GaAs MESFET V_t," *IEEE Trans. Electron. Dev.* **ED-3**:227 (1986).

51. R. Rumsby, R. M. Ware, B. Smith, M. Tyjberg, M. R. Brozel, and E. J. Foulkes, *Tech. Dig. GaAs IC Symp.*, Phoenix, 2, 1983, p. 34.

52. H. Ehrenreich and J. P. Hirth, "Mechanism for Dislocation Density Reduction in GaAs Crystals by Indium Addition," *Appl. Phys. Lett.* **46**:668 (1985).

53. G. Jacob, *Proc. Semi-Insulating III–V Materials*, Shiva Publishing, Nantwick, U.K., 1982, p. 2.

54. C. Miner, J. Zorzi, S. Campbell, M. Young, K. Ozard, and K. Borg, "The Relationship Between the Resistivity of Semi-Insulating GaAs and MESFET Properties," *Mat. Sci. Eng B.* **44**:188 (1997).

55. P. Rudolph, M. Newbert, S. Arulkumaran, and M. Seifert, *J. Cryst. Res. Technol.* **32**:35 (1997).

56. O. G. Folberth and H. Weiss, *Z. Naturforsch.*, **100**:615 (1955).

57. H.-J. Sell, "Melt Growth Processes for Semiconductors," *Key Eng. Mater.* **58**:169 (1991).

58. T. P. Chen, T. S. Huang, L. J. Chen, and Y. D. Guo, "The Growth and Characterization of GaAs Single Crystals by a Modified Horizontal Bridgman Technique," *J. Cryst. Growth* **106**:367 (1990).

59. R. E. Kremer, D. Francomano, G. H. Beckhart, K. M. Burke, and T. Miller, *Mater. Res. Soc. Symp. Proc.* **144**:15 (1989).

60. C. E. Chang, V. F. S. Kip, and W. R. Wilcox, "Vertical Gradient Freeze Growth of GaAs and Naphthalene: Theory and Practice," *J. Cryst. Growth* **22**:247 (1974).

61. R. E. Kremer, D. Francomano, B. Freidenreich, H. Marshall, and K. M. Burke, in *Semi-Insulating Materials 1990*, A. G. Milnes and C. J. Miner, eds., Adam-Hilger, London, 1990.

62. W. Gault, E. Monberg, and J. Clemans, "A Novel Application of the Vertical Gradient Freeze Method to the Growth of High Quality III–V Crystals," *J. Cryst. Growth*, **74**:491 (1986).

63. E. Buhrig, C. Frank, C. Hannis, and B. Hoffmann, "Growth and Properties of Semi-Insulating VGF-GaAs," *Mat Sci. Eng. B.* **44**:248 (1997).

64. R. Nakai, Y. Hagi, S. Kawarabayashi, H. Migajima, N. Toyoda, M. Kiyama, S. Sawada, N. Kuwata, and S. Nakajima, "Manufacturing Large Diameter GaAs Substrates for Epitaxial Devices by VB Method," *GaAs IC Symp.*, 1998, p. 243.

65. W. Keller and A. Mühlbauer, *Float-Zone Silicon*, Dekker, New York, 1981.

66. C. N. Klahr and M. S. Cohen, *Nucleonics* **22**:62 (1964).

Part II

Unit Processes I: Hot Processing and Ion Implantation

Once a new technology rolls over you, if you are not part of the steamroller you're part of the road.[1]

Until now, only the semiconductor substrate itself has been discussed. This section will begin a discussion of unit processes. These are the individual process steps carried out in typical fabrication technologies. A later section will discuss how these unit processes are put together to form functional blocks (known as process modules) and ultimately, a technology. This first section on unit processes will discuss those processes related to dopant introduction and movement as well as the growth of thermal oxides. Since dopants are necessary for all types of devices, they are some of the first processes developed for fabrication. For the device to operate properly, the doped regions must have the right concentrations and sizes. This section will therefore first discuss the movement of dopant impurity atoms through diffusion. Early technologies used gaseous or liquid vapor sources in high temperature ovens to introduce the impurities into the wafer. As device size was reduced, however, ion implantation was developed to better control the position and amount of impurity in the wafer. As standard implantation and high temperature annealing steps began to prove inadequate, special methods were developed to allow the formation of very shallow, heavily doped regions. One of the most important of these is rapid thermal processing, which will be discussed in Chapter 6. Chapter 4 will also cover the thermal oxidation of silicon. Unlike the other chapters, only silicon oxidation will be discussed, since this process is not used in compound semiconductor technologies.

---

[1]Stuart Brand, founder, Whole Earth Catalog.

Chapter 3

Diffusion

Every semiconductor device relies on the ability to fabricate well-controlled, locally doped regions of the wafer. The chemical impurities must therefore first be introduced into some sections of the wafer; they must be active so that they contribute the desired carrier; and they must be the concentration desired by the device designer. Frequently, concentration profiles will be described. As shown in Figure 3.1, the impurity concentration or the carrier concentration is plotted on the vertical axis. The depth into the wafer is plotted on the horizontal axis. Typically, the y variable will vary over many orders of magnitude. For that reason, the concentration is normally given on a logarithmic scale. Recall that the number density of silicon is 5×10^{22} atoms/cm^3, so that typical impurity concentrations (10^{17} atoms/cm^3) for active device regions are doped as lightly as a few parts per million.

After the impurities are introduced they may redistribute in the wafer. This may be intentional or it may be a parasitic effect of a thermal process. In either event, it must be controlled and monitored. The motion of impurity atoms in the wafer occurs primarily by diffusion, the net movement of a material that occurs near a concentration gradient as a result of random thermal motion. This chapter will introduce the differential equations that describe diffusion, solve the equations in closed form for two sets of boundary conditions, describe the physics involved in the diffusion coefficient, present models that describe the diffusion behavior of typical impurities in silicon and GaAs, and introduce software that calculates diffusion profiles under a wide variety of conditions.

3.1 Fick's Diffusion Equation in One Dimension

Any material that is free to move will experience a net redistribution in response to a concentration gradient. The movement will tend to reduce the size of the gradient. The source of this movement is the random motion of the material. Since the high concentration region has more impurity atoms, there is a net movement of impurities away from the concentration maximum. This is an effect not limited to impurities in semiconductors by any means. The basic laws of diffusion introduced here are used to describe heat transfer, the motion of electrons, gaseous impurities such as air pollution, and even animal population statistics.

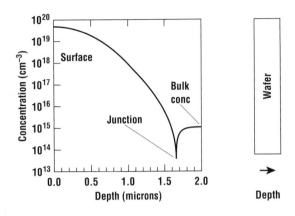

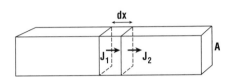

Figure 3.1 Typical concentration plot of impurities or carriers as a function of depth into the wafer. Note that these profiles are typically much less than 1% of the total wafer thickness.

Figure 3.2 A differential volume element in a bar of cross-sectional area A, where J_1 and J_2 are the flux of an impurity into and out of the volume element.

The basic equation that describes diffusion is Fick's first law,

$$J = -D\frac{\partial C(x, t)}{\partial x}$$

(3.1)

where C is the impurity concentration, D is the coefficient of diffusion, and J is the net flux of material. The units of J are number per unit time per unit area. The negative sign expresses the fact that there is net movement in the direction of decreasing concentration.

While Fick's first law accurately describes the diffusion process, in this application there is no convenient way to measure the current density of the impurity. Unlike electrical current, the diffusing material is usually poorly confined and not easily detected. Therefore, a second expression for Fick's law has been developed that describes the same concept, but with more readily measurable quantities. In developing this expression it is easiest to start with a long bar of material with a uniform cross section A (Figure 3.2). Consider a small volume of length dx, then

$$\frac{J_2 - J_1}{dx} = \frac{\partial J}{\partial x}$$

(3.2)

where J_2 is the flux leaving the volume and J_1 is the flux entering the volume. If these two fluxes are not the same, the concentration of the diffusing species in the volume must change. Recall that the number of impurities in this volume element is just the product of the concentration and the differential volume element ($A \cdot dx$). Then the continuity equation is expressed as

$$\frac{dN}{dt} = A\, dx\, \frac{\partial C}{\partial t} = -A(J_2 - J_1) = -A\, dx\, \frac{\partial J}{\partial x}$$

where N is the number of impurities in the volume element, or

(3.3)

$$\frac{\partial C(x, t)}{\partial t} = -\frac{\partial J}{\partial x}$$

From Fick's first law this can be written

$$\frac{\partial C(x, t)}{\partial t} = \frac{\partial}{\partial x}\left(D\frac{\partial C}{\partial x}\right)$$

(3.4)

Equation 3.4 is the most general representation of Fick's second law. If the diffusion coefficient is assumed to be independent of position, this reduces to the simpler form

$$\frac{\partial C(z, t)}{\partial t} = D\frac{\partial^2 C(z, t)}{\partial z^2}$$

(3.5)

where the position variable has been changed to z to suggest that the direction into the wafer (depth) is the one of primary interest. Finally, in three dimensions for an isotropic medium, Fick's second law is expressed as

$$\frac{\partial C}{\partial t} = D\nabla^2 C$$

(3.6)

One is left, then, with the solution of a differential equation that is second order in position and first order in time. This requires the knowledge of at least two independent boundary conditions. The solution of the differential equation will be discussed later in the chapter. First we will focus on the application of Fick's second law to the problem of diffusion in semiconductors and discuss the factors that determine the diffusion coefficient, D.

3.2 Atomistic Models of Diffusion

This section will discuss the physical mechanisms that determine the diffusion coefficient D. The argument begins by assuming that the crystal is isotropic. Without this approximation Fick's second law cannot be applied. Although it allows us to derive solutions to Equation 3.6, it breaks down when the concentration of the dopant is large. Then the diffusivity becomes a function of the doping concentration and therefore depth.

In a crystal, the lattice sites are represented as the minima of parabolic potential wells. Each atom is at rest only in the limit of 0 K. At nonzero temperatures the atoms oscillate about their equilibrium position. Now insert an impurity atom into this crystal. The atom may sit between lattice sites in an interstitial position. Typically, atoms that do not bond readily with the matrix material are interstitial impurities. These impurities diffuse rapidly, but they do not directly contribute to doping. A second type of impurity is one that replaces the silicon atom on the lattice site. These substitutional impurities will be the primary focus of the chapter. Table 3.1 lists a number of silicon impurities and categorizes them into primarily substitutional and primarily interstitial.

Assume that the impurity atom in Figure 3.3A moves one lattice site to the right. By symmetry, no net energy was expended. Yet, for a substitutional atom to move in the crystal, it must have

Table 3.1   Silicon impurities

Substitutional	P, B, As, Al, Ga, Sb, Ge
Interstitial	O, Au, Fe, Cu, Ni, Zn, Mg

Impurities in silicon tend to reside primarily on lattice sites (substitutional impurities) or primarily in the spaces between the lattice sites (interstitial impurities).

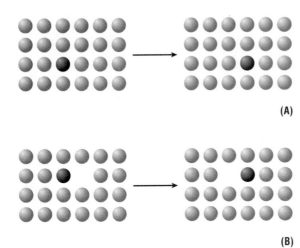

Figure 3.3   Diffusion of an impurity atom by direct exchange (A) and by vacancy exchange (B). The latter is much more likely owing to the lower energy required.

sufficient energy to surmount the potential well in which it rests. For the direct exchange shown in Figure 3.3A, at least six bonds must be broken for the host atom and the impurity to exchange positions. This is considerably easier, however, if the adjacent lattice site is occupied by a vacancy (Section 2.4). Then only three bonds must be broken. Vacancy exchange (Figure 3.3B) is therefore one of the important diffusion mechanisms for substitutional impurities.

Fair's vacancy model has been used to successfully describe the diffusion of many impurities in low and moderate concentrations at temperatures below 1000°C. It takes the simple picture from Figure 3.3B and adds one additional detail: vacancy charge. Recall that to fill its valence shell, each atom in the silicon matrix must form a covalent bond with its four nearest neighbors. In the presence of a neutral vacancy, these four atoms are left with an unsatisfied shell. If the vacancy captures an electron, it satisfies one atom's valence, but becomes negatively charged. Similarly, an adjacent atom can lose an electron, and the vacancy appears to be positively charged.

Since vacancies are very dilute in a semiconductor at typical processing conditions, each of the possible charged states can be treated as an independent entity. The diffusion coefficient then becomes the sum of all possible diffusion coefficients, weighted by their probability of existence. If we assume that the probability of charge capture is a constant, then the number of charged vacancies is proportional to the ratio $[C(z)/n_i]^j$, where $C(z)$ is the carrier concentration, n_i is the intrinsic carrier concentration, and j is the order of the charge state. Then, the most general expression for the total diffusion coefficient in the vacancy model is given by

$$D = D^o + \frac{n}{n_i} D^- + \left[\frac{n}{n_i}\right]^2 D^{2-} + \left[\frac{n}{n_i}\right]^3 D^{3-} + \left[\frac{n}{n_i}\right]^4 D^{4-}$$

$$+ \frac{p}{n_i} D^+ + \left[\frac{p}{n_i}\right]^2 D^{2+} + \left[\frac{p}{n_i}\right]^3 D^{3+} + \left[\frac{p}{n_i}\right]^4 D^{4+} \tag{3.7}$$

The intrinsic carrier concentration for silicon can be found from [1]

$$n_i(\text{cm}^{-3}) = n_{io}T(\text{K})^{3/2}e^{-E_g/2kT} \tag{3.8}$$

where $n_{io} = 7.3 \times 10^{15}$ cm^{-3} for silicon and $n_{io} = 4.2 \times 10^{14}$ cm^{-3} for GaAs. The bandgap can be determined by

$$E_g = E_{g0} - \frac{\alpha T(\text{K})^2}{\beta + T(\text{K})} \tag{3.9}$$

where E_{g0}, α, and β are 1.17 eV, 0.000473 eV/K, and 636 K for silicon and 1.52 eV, 0.000541 eV/K, and 204 K for GaAs. The intrinsic carrier concentrations for silicon and GaAs are shown in

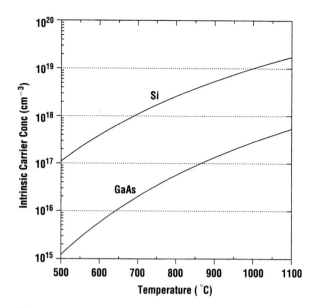

Figure 3.4 Intrinsic carrier concentration of silicon and GaAs as a function of temperature.

Figure 3.4. In heavily doped silicon the bandgap is also reduced by the bandgap narrowing effect

$$\Delta E_g = -7.1 \times 10^{-10} \text{ eV} \sqrt{\frac{n(\text{cm}^{-3})}{T(\text{K})}} \qquad (3.10)$$

For heavily doped diffusions ($C \gg n_i$) the electron or hole concentration is just the impurity concentration. For low concentration diffusions [$C(z) \ll n_i$] $p \approx n \approx n_i$. For substrates with excess free electrons (n-type), the positive charge terms in Equation 3.7 can be neglected, and for substrates with excess free holes (p-type), the negative charge terms can be neglected. Furthermore, the contributions of the third- and fourth-power terms are typically very small and are almost universally neglected. If charged vacancies must be considered, the electron or hole concentration, and therefore the diffusivity, is a function of position. In that case the simple form of Equation 3.5 cannot be used. Instead Equation 3.4 must be solved numerically.

If very dilute impurity profiles are measured before and after diffusion, a diffusion coefficient can be determined. If this procedure is repeated for several temperatures and the logarithm of the diffusivity is plotted against reciprocal temperature in kelvin, an Arrhenius plot will result. The neutral vacancy diffusivity is of the form

$$D^o = D^o_o e^{-E_a/kT} \qquad (3.11)$$

where E_a is the activation energy of the neutral vacancy, and D^o_o is a nearly temperature-independent term that depends on the vibrational frequency and geometry of the lattice. Table 3.2 summarizes the activation energies and preexponentials for some common dopants. Notice that the activation energies of all of the neutral vacancies shown in Table 3.2 diffusivities in silicon lie between 3.39 and

Table 3.2 Diffusion coefficients of common impurities in silicon and gallium arsenide

		Donors						Acceptors	
		$D_o^=$	$E_a^=$	D_o^-	E_a^-	D_o	E_a	D_o^+	E_a^+
As in Si	D			12.0	4.05	0.066	3.44		
P in Si	D	44.0	4.37	4.4	4.0	3.9	3.66		
Sb in Si	D			15.0	4.08	0.21	3.65		
B in Si	A					0.037	3.46	0.41	3.46
Al in Si	A					1.39	3.41	2480	4.2
Ga in Si	A					0.37	3.39	28.5	3.92
S in GaAs	D					0.019	2.6		
Se in GaAs	D					3000	4.16		
Be in GaAs	A					7e − 6	1.2		
Ga in GaAs	I					0.1	3.2		
As in GaAs	I					0.7	5.6		

From Runyan and Bean [2] and references quoted therein. Donors are labeled with a "D," acceptors with an "A," and self-interstitials with an "I." All preexponentials are in centimeters squared per second, and the activation energies are in electron-volts.

Example 3.1

Calculate the diffusivity of arsenic in silicon at 1000°C if the concentration of arsenic is much less than the intrinsic carrier concentration and again if the arsenic concentration is 1×10^{19} cm^{-3}. For $T = 1273$ K, $kT = 0.110$ eV. Then

$$D_i = 0.066e^{-3.44/0.110} = 1.6 \times 10^{-15} \text{ cm}^2/\text{sec}$$

Solution

According to Figure 3.4 and Table 3.2 the negative one vacancy must be considered for arsenic diffusion. Then

$$D_- = 12e^{-4.05/0.110} = 1.2 \times 10^{-15} \text{ cm}^2/\text{sec}$$

Recall from elementary semiconductor physics,

$$n = \frac{N_D}{2} + \sqrt{\left[\frac{N_D}{2}\right]^2 + n_i^2}$$

This equation has as its two limits, $n = N_D$ for $N_D \gg n_i$ and $n = n_i$ for $N_D \ll n_i$. In this case, $N_D \ll n_i$, so

$$D = D_i + D_- = 2.8 \times 10^{-15} \text{ cm}^2/\text{sec}$$

Referring to Figure 3.4, at 1000°C, $n_i = 10^{19}$ cm^{-3}. Then if N_D is 1×10^{19} cm^{-3}, $n = 1.61 \times 10^{19}$ cm^{-3}, and

$$D = 1.6 \times 10^{-15} + \frac{1.41 \times 10^{19}}{10^{19}} 1.2 \times 10^{-15} = 3.5 \times 10^{-15} \text{ cm}^2/\text{sec}$$

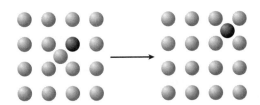

Figure 3.5 In interstitialcy diffusion, an interstitial silicon atom displaces a substitutional impurity, driving it to an interstitial site, where it diffuses some distance before it returns to a substitutional site.

3.66 eV. This is about an electron-volt greater than the activation energy of neutral vacancy creation. The additional energy represents the effective barrier to the exchange shown in Figure 3.3B.

A second important mechanism for diffusion in silicon relies on the presence of silicon self-interstitials. It is referred to as the interstitialcy method (Figure 3.5). In this case, an interstitial silicon atom displaces the impurity, driving it into an interstitial site. From there it moves rapidly to another lattice site, where the silicon atom is removed and becomes an interstitial. Interstitialcy is not believed to occur unless vacancy diffusion does as well. Boron and phosphorus are two impurities that tend to diffuse by both mechanisms. Either may be dominant depending on the process conditions. In principle, to find the effective diffusivity for these impurities one must add the contributions from both methods.

Some impurities tend to diffuse rapidly through interstitial spaces. There are two mechanisms by which these impurities may return to the lattice; they are summarized in Figure 3.6. In the

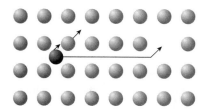

Figure 3.6   The kick-out (left) and Frank–Turnbull mechanisms (right).

Frank–Turnbull method the interstitial impurity is captured by a vacancy. In the kick-out mechanism the impurity replaces a lattice atom. These two mechanisms are distinct from the interstitialcy method in that they do not require the presence of self-interstitials to drive the process. Impurities that tend to diffuse through this mechanism are characterized by a low interstitial solubility to substitutional solubility ratio.

The analytic examples and problems in this chapter will, for the most part, ignore these effects even though they may be important. The parameters of Table 3.2 were derived by fitting Equation 3.7 to measured data. This does not mean that vacancy diffusion is the only, or even the most important mechanism for all impurities. It does, however, provide a simple way to estimate diffused profiles for many situations.

By comparing inert, oxidizing, and nitridizing dopant diffusion experiments, some light has been shed on the processes by which various impurities diffuse [3]. The diffusivity of impurities in a semiconductor depends on the concentration of vacancies. When a semiconductor is oxidized, a high concentration of excess interstitials is generated near the oxide/semiconductor interface [4, 5]. The excess concentration decays with depth due to vacancy/interstitial recombination. Near the surface these interstitials increase the diffusivity for boron and phosphorus. It is believed, therefore, that these impurities diffuse primarily by the interstitialcy process. Arsenic diffusivity is found to decrease under oxidizing conditions. An excess interstitial concentration is expected to depress the local vacancy concentration. Arsenic is therefore believed to diffuse primarily via the vacancy mechanism, at least when oxidizing conditions obtain. Experiments can also be done during thermal nitridization. Although this process is rarely done in practice, it injects a high concentration of vacancies into the substrate. These results have confirmed the conclusions of the earlier results by showing the opposite trends of oxidation.

To be able to predict a dopant profile after a diffusion in an oxidizing ambient, we need to find the diffusivity under these conditions. Note that, as with high concentration diffusion, the diffusivity is now a function of position. Strictly speaking, therefore, Equation 3.5 is no longer valid. To first order, however, dD/dz is small compared to dC/dz and can be ignored. Since the concentration of excess interstitials depends on the oxidation and recombination rates, the diffusivity must be a function of the oxidation rate. It has been shown that for diffusion under oxidizing conditions,

$$D = D_i + \Delta D \tag{3.12}$$

where [6]

$$\Delta D = \alpha \left[\frac{dt_{ox}}{dt} \right]^n \tag{3.13}$$

is the diffusivity enhancement or retardation due to oxidation. The exponent n has been found experimentally to be between 0.3 and 0.6. The term α may be positive (for oxidation-enhanced diffusion) or negative (for oxidation-retarded diffusion).

Thus far only diffusion in one dimension has been discussed. It is assumed that the dopant concentration is uniform across the wafer and therefore no net diffusion occurs laterally. This is not true near the edge of diffused patterns, such as under the gate of a MOSFET when the source and drain are being diffused. Diffusion will occur both vertically and laterally (i.e., under the gate). The lateral diffusion is generally assumed to proceed uniformly. This leads to an effective depletion of dopant near the edge of the feature, which reduces the junction depth. Experimental verification of lateral diffused profiles is extremely difficult, since dilute concentrations must be detected in very small volumes. For that reason detailed models of lateral diffusion are less well developed.

3.3 Analytic Solutions of Fick's Law

Returning to the assumption of a constant diffusivity, Fick's second law is a simple differential equation that can be solved subject to various boundary conditions. In practice the dopant profiles that are of interest are sufficiently complex, and the assumption that the coefficient of diffusion is constant is sufficiently questionable that Equation 3.5 must be solved numerically. There are two sets of boundary conditions, however, for which exact solutions can be derived. These solutions can be used to develop a basic understanding of diffusion processes and as rough approximations of actual profiles.

The first type of solution of Fick's law occurs when the source is fixed at the surface for all times greater than zero. Called a predeposition diffusion, the boundary conditions (one for time and two for position) are

$$C(z, 0) = 0$$
$$C(0, t) = C_s \qquad \text{(3.14)}$$
$$C(\infty, t) = 0$$

The solution for these conditions is given by

$$C(z, t) = C_s \, \text{erfc}\left(\frac{z}{2\sqrt{Dt}}\right), \quad t > 0 \qquad \text{(3.15)}$$

In this equation, C_s is the fixed surface concentration and erfc is a function known as the complementary error function. The complementary error function is tabulated in Appendix V and in many math handbooks for various values; $\sqrt{Dt}$ is a common feature in the solution of diffusion problems and is known as the characteristic diffusion length.

The dose of the predeposition diffusion varies with the time of the diffusion. To obtain the dose, the profile can be integrated as

$$Q_T(t) = \int_0^\infty C(z, t) \, dz$$
$$= \frac{2}{\sqrt{\pi}} C(0, t) \sqrt{Dt} \qquad \text{(3.16)}$$

This dose is measured in units of impurities per unit area, typically per square centimeter. Since the depth of the profile is typically less than 1 μm (10^{-4} cm), a dose of 10^{15} cm^{-2} will produce a large volume concentration ($>10^{19}$ cm^{-3}). Since the surface concentration is fixed for a predeposition diffusion, the total dose increases as the square root of the time.

The second type of solution of Fick's law is called the drive-in diffusion. In this case, an initial amount of impurity Q_T is introduced into the wafer and diffused subject to the boundary condition that Q_T be fixed. If the diffusion length is much larger than the width of the initial profile, the initial profile can be approximated as a delta function. Then the boundary conditions are

$$C(z, 0) = 0, \quad z \neq 0$$
$$\frac{dC(0, t)}{dz} = 0$$
$$C(\infty, t) = 0 \qquad \text{(3.17)}$$
$$\int_0^\infty C(z, t) dz = Q_T = \text{constant}$$

The solution to Fick's second law for these conditions is a Gaussian centered at $z = 0$:

$$C(z, t) = \frac{Q_T}{\sqrt{\pi Dt}} e^{-z^2/4Dt}, \quad t > 0 \tag{3.18}$$

The surface concentration C_s, decreases with time as

$$C_s = C(0, t) = \frac{Q_T}{\sqrt{\pi Dt}} \tag{3.19}$$

The reader can easily demonstrate that at $x = 0$, dC/dx is zero for all $t \neq 0$. Figure 3.7 shows a plot of the predeposition and the drive-in diffusions with $\sqrt{Dt}$ as a parameter.

One classic type of problem that uses these two diffusions is a predeposition followed by a drive-in. Recall that one of the boundary conditions for the drive-in was that the initial impurity concentration was zero everywhere except at the surface. In practice, the drive-in is a good approximation as long as

$$\sqrt{Dt}_{\text{predep}} \ll \sqrt{Dt}_{\text{drive-in}} \tag{3.20}$$

Now assume that boron is diffusing in a silicon wafer that has a uniform concentration of phosphorus, C_B. Also assume that $C_s \gg C_B$. Then, a depth will exist at which the concentration of boron exactly equals that of the background. Since boron is a p-type dopant and phosphorus is an n-type dopant in silicon, a p–n junction will exist at this depth, which is called the junction depth x_j. If the diffusion is a drive-in diffusion, one can show from Equation 3.18:

$$x_j = \sqrt{4Dt \ln\left[\frac{Q_T}{C_B\sqrt{\pi Dt}}\right]} \tag{3.21}$$

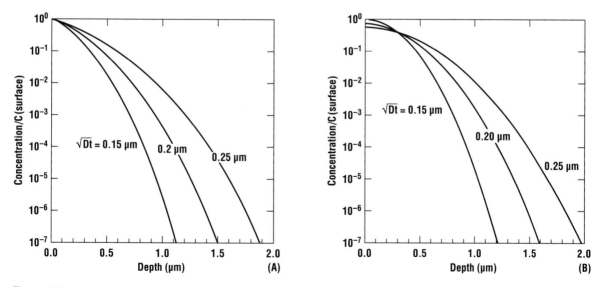

Figure 3.7   Concentration as a function of depth for (A) predeposition and (B) drive-in diffusions for several values of the characteristic diffusion length.

If the diffusion was a predeposition diffusion, from Equation 3.15,

$$x_j = 2\sqrt{Dt}\, \text{erfc}^{-1}\left[\frac{C_B}{C_s}\right]$$ (3.22)

Example 3.2

A wafer is heated to 1100°C and exposed to a high concentration source of arsenic. After 5 minutes, the wafer is removed from this source, the surface is sealed, and the wafer is annealed at 1200°C for 6 hr. Assume intrinsic diffusion. Find: (a) Q_T, (b) The final profile, and (c) The junction depth if the wafer initially was doped 1×10^{15} cm^{-3} p-type.

Solution

The initial process is a predeposition. Lacking more information, it is reasonable to set the surface concentration at the solid solubility. From Figure 2.4, $C_s \approx 2 \times 10^{21}$ cm^{-3}. From Table 3.2,

$$D = 0.066\, e^{-3.44/kT} + 12.0 \times 1 \times e^{-4.05/kT}$$

since we are assuming intrinsic ($n = n_i$) diffusion.
Solving,

$$D = 3.2 \times 10^{-14}\, \text{cm}^2/\text{sec at } 1100°C$$

$$D = 2.8 \times 10^{-13}\, \text{cm}^2/\text{sec at } 1200°C$$

Then

$$\left(\sqrt{Dt}\right)_{1100°C} = 0.092\ \mu\text{m} \ll \left(\sqrt{Dt}\right)_{1200°C} = 0.78\ \mu\text{m}$$

So we can use the δ approximation. For the predeposition,

(a) $Q_T = \dfrac{2}{\sqrt{\pi}} C_s \sqrt{Dt} = 2.1 \times 10^{16}$ cm^{-2}

(b)   The same dose can be used for the drive-in, so

$$C = \frac{Q_T}{\sqrt{\pi}\sqrt{Dt}}\, e^{-z^2/(2\sqrt{Dt})^2}$$

$$= 1.5 \times 10^{20}\ \text{cm}^{-3}\, e^{-(z/1.6\ \mu m)^2}$$

(c)   To find the junction depth

$$X_J = \sqrt{4\,Dt \ln\left(\frac{Q_T}{C_B\sqrt{\pi}}\sqrt{Dt}\right)}$$

$$X_J = 5.4\ \mu\text{m}$$

3.4 Diffusion Coefficients for Common Dopants

In this section the vacancy and interstitialcy models of diffusion will be applied to describe the diffusion coefficients that are appropriate for each of the commonly used impurities. These have been summarized in *Defect and Diffusion Forum* [7, 8], with hundreds of references to the diffusion of common impurities. The diffusivity of boron in silicon has been measured over a wide range of concentrations and temperatures [9]. Referring to Fair's vacancy model, the data can be fit with the intrinsic diffusivity and only the first positive vacancy term up to concentrations of 10^{20} cm^{-3}. Above this concentration not all of the boron can be accommodated into the lattice, and some must reside on interstitial sites or in cluster precipitates. The diffusivity of boron in this concentration range is sharply reduced in crystalline silicon [10], but highly mobile in amorphous silicon [11]. Interstitials play a key role in diffusion in this range [12]. Figure 3.8 shows a typical diffused high concentration boron profile.

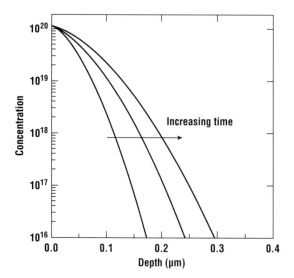

Figure 3.8   Typical profile for a high concentration boron diffusion.

Arsenic diffuses in silicon via neutral and single negatively charged vacancies. Since the diffusivity of arsenic in silicon is relatively low, arsenic is often chosen as the n-type dopant when minimal dopant redistribution is desired. Typical examples are the source/drain diffusion of submicron NMOS transistors and the emitter of many bipolar transistors. At low and medium concentrations the diffusivity is well described by simple intrinsic diffusion.

At high concentrations the diffusivity of As is clearly dependent on concentration. Some evidence exists to suggest that As combines with vacancies, perhaps as a VAs$_2$ cluster, and that these effects are time dependent [13]. In any case, the resultant diffusion profiles are significantly different from simple constant diffusivity predictions.

Arsenic also tends to form interstitial clusters at concentrations in excess of 10^{20} atoms/cm^3 that resist electrical thermal activation. This effect also tends to flatten the top of high concentration carrier profiles. During a high temperature anneal the cluster concentration moves toward a thermal equilibrium with substitutional arsenic. The maximum carrier concentration is given by

$$C_{\max} = 1.9 \times 10^{22} \text{ cm}^{-3} \, e^{-0.453/kT} \tag{3.23}$$

The relative concentrations are therefore determined by the anneal temperature. The clusters are not believed to be mobile. Instead, the arsenic atoms move individually.

Phosphorus diffuses much more rapidly than arsenic. Its application to the technology of ultra large scale integration (ULSI) is limited primarily to wells (or tubs) and isolation, although its higher diffusivity is helpful in reducing the peak electric field in MOS transistors (see Chapter 16). For completeness however, high concentration phosphorus profiles will be described. Figure 3.9 shows a high concentration phosphorus diffusion. The profile consists of three regions: the high concentration region, the low concentration region, and the transition or "kink" region [14]. Near the surface the concentration is nearly constant. In an early, influential model, the diffusivity in this region was believed to have two components: D_i, the normal phosphorus atom neutral vacancy exchange, and

$D_i^=$, which corresponds to positively charged phosphorus ions paired with double negatively charged vacancies to form single negatively charged pairs $(PV)^-$:

$$D_{Ph} = D_i + D_i^{2-} \left[\frac{n}{n_i}\right]^2 \tag{3.24}$$

Near the kink region the electron concentration falls sharply. Most of the ion vacancy pairs dissociate, and the diffusion of unpaired phosphorus ions continues into the substrate. The dissociation of the $(PV)^-$ pairs causes an excess vacancy concentration, which also increases the diffusivity in the tail region. More recent work suggests that, like boron, phophorus has a significant interstial-dependent contribution to the diffusivity [15], with self-interstitials predominating in the kink region and P interstitials dominating in the tail. The changeover from vacancy-dominant to kick-out dominant diffusion is responsible for the kink.

Gallium arsenide must be heavily doped for certain optoelectronic devices and for forming ohmic contacts. Diffusion in GaAs is significantly more complicated than in silicon. It depends not only on the charge state of the vacancy and interstitial, but whether it is a gallium vacancy (V_{Ga}) or an arsenic vacancy (V_{As}). The neutral vacancy diffusion coefficients for several common GaAs dopants are given in Table 3.1. This section will present the diffusion mechanisms of two of the primary GaAs dopants: zinc and silicon.

Zinc is a commonly used p-type dopant in GaAs technology. Figure 3.10 shows a series of predeposition diffusions of zinc into GaAs at 600°C for various times [16]. For low concentrations, the diffusion can be fit to a simple diffusivity. Values in the literature vary significantly, with valves of D disagreeing by 6 orders of magnitude at the same temperature [8]. At high concentrations, the diffusions show a broad, flat plateau and a sharp exponential tail. The plateau is limited by the solid solubility. An early model for the diffusion of zinc developed by Weisberg and Blanc [17] has two components. The first is the standard vacancy method of Fair, as described earlier in this chapter. The second, which proceeds much faster, is the Frank–Turnbull or substitutional interstitial (SI) process

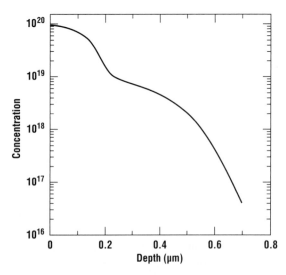

Figure 3.9 Typical profile for a high concentration phosphorus diffusion.

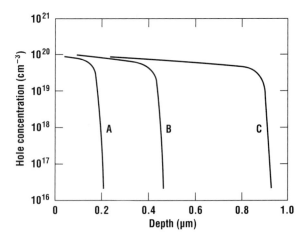

Figure 3.10 Predeposition diffusions of zinc into GaAs at 600°C for 5, 20, and 80 min *(after Field and Ghandi, used with permission, Electrochemical Society).*

(Figure 3.6). In this process, zinc atoms exist in two forms: a small concentration of positively charged interstitial ions Zn^+, which diffuses rapidly, and a substitutional ion Zn^- that diffuses much more slowly by neutral vacancy exchange. While multiple positive states may participate in the interstitial diffusion, it is believed to be dominated by the singly charged ion.

In this model the Zn^+ diffuses rapidly until it encounters a V_{Ga}. It becomes trapped at this location, gives up two holes, and becomes Zn^-, which markedly reduces its diffusivity. At high concentrations this process results in a concentration dependent diffusivity of the form

$$D_{Zn} \approx AC_s^2 \qquad\qquad (3.25)$$

where C_s is the concentration of substitutional zinc and A is a constant [18]. Thus the diffusion coefficient remains large until the concentration begins to drop. At this point, the diffusivity also drops, further sharpening the profile edge.

This model fails to describe a kink behavior often seen in the tail of the zinc profile. Kahen [19] modified the model to include the possibility of multiple charge states associated with the gallium vacancy. The author also included the possibility of the formation of a paired state between the substitutional negatively charged zinc ion and the interstitial positively charged zinc ion. The mechanism greatly reduces the diffusivity of the zinc. The model closely follows the kink behavior over a wide range of process conditions (see Figure 3.11).

One of the most common n-type dopants in GaAs is silicon (Si). Silicon is a group IV element and may be either a p-type or an n-type dopant, depending on the lattice site that it occupies. The carrier concentration then is the difference between the concentrations of silicon that reside on the two types of lattice sites. When this difference is small compared to the individual components, the semiconductor is said to be highly compensated. At high concentrations the diffusivity of silicon is concentration dependent [20, 21]. Greiner and Gibbons [20] proposed an Si_{Ga}–Si_{As} pair diffusion model in which a pair of impurities on adjacent (opposite type) lattice sites exchanges position with a pair of adjacent vacancies. This is believed to occur in a two-step process. In this type of process the semiconductor can be highly compensated, since the dopant atoms occupy both sites. The diffusivity then increases linearly with the dopant concentration.

The pair diffusion model has not been able to explain the diffusion results observed under rapid thermal annealing conditions (see Chapter 6). Furthermore, the model has been unable to reproduce the effects of substrate doping on the diffused profile. A more complete steady state model that takes into account charge effect, has been proposed by Yu et al. [22]. In this model, Si_{Ga^+} diffuses by exchange with an uncharged or triply negatively charged gallium vacancy. The diffusion mechanism of the Si_{As^-} atom is unknown but is assumed to be independent of the gallium atom. Coupled pairs of Si_{Ga}–Si_{As} are assumed to be immobile. Kahen et al. [23] have applied an Si_{Ga}/V_{Ga^-} model to fit experimental data, also with good results,

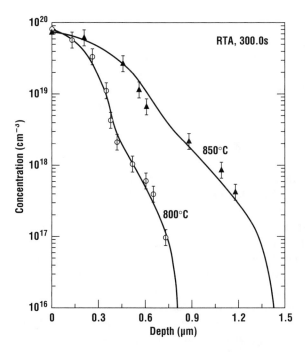

Figure 3.11 Comparison of the multiple charge model for zinc diffusion and experimental results *(after Kahen, used with permission, Materials Research Society).*

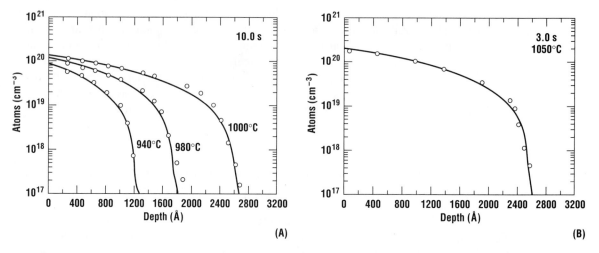

Figure 3.12   Diffusion of silicon in GaAs *(after Kahen et al., used with permission, Materials Research Society).*

although background doping was not considered. Other models suggest that the double negative Ga vacancy is the diffusion partner and $D_{si} \propto (n/n_i)^2$ [24]. Figure 3.12 shows the results of Kahen et al. In agreement with this model, Sudandi and Matsumoto showed that the diffusivity of silicon decreases sharply in Ga-rich LEC material where the concentration of gallium vacancies is expected to be low [15].

3.5 Analysis of Diffused Profiles

Once an impurity has diffused, it is desirable to be able to measure the impurity concentration as a function of depth and position. There are many techniques for obtaining depth profiles, but lateral profiles are much more difficult to obtain. The simplest technique for obtaining information about the profile is to measure its sheet resistance. Instead of a concentration profile however, only a single number

$$R_s = [q \int \mu(C) C_e(z) \, dz]^{-1}$$ (3.26)

is obtained, where $C_e(z)$ is the carrier concentration, $\mu(C)$ is the concentration-dependent mobility, and R_s is called the sheet resistance. As discussed in Chapter 1, R_s is quoted in units of ohms per square ($\Omega/\square$). Sheet resistance measurements are quick and easy to perform and give the process engineer useful information, particularly if a standard or target sheet resistance is known.

The sheet resistance can be measured in a variety of ways. The simplest is the use of a four-point probe (Figure 3.13A). Four-point probes are available in several geometries; the most common is collinear. In this case, current is passed between the two outer probes and the voltage is measured across the inner pair. The sheet resistance is found by measuring the ratio of the voltage drop to the forced current. The result is multiplied by a geometric correction factor that depends on the probe geometry and the ratio of the probe spacing to the thickness of the diffusion [26, 27]. For collinear probes, where the probe spacing is much larger than the junction depth, the factor is 4.5325 [28]. For this method to be useful in characterizing diffused profiles in semiconductors, the underlying substrate must be insulating or must be of much higher resistivity than the layer to be measured, or the layer to be measured must form a reverse-biased diode with the substrate. In the latter case,

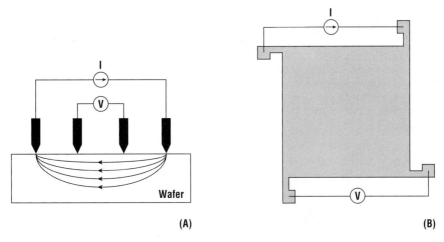

Figure 3.13   The four-point probe (A) and Van der Pauw (B) methods for determining the resistivity of a sample.

very shallow junctions can be penetrated with the probe if excessive force is used. Furthermore, the sheet resistance measurement will include the effect of the depleted region near the junction.

A second technique for measuring the sheet resistance is the Van der Pauw method [29]. The measurement is again done by contacting the edge of a sample in four places. A current is forced between one pair of adjacent contacts and the voltage is measured across the other pair (Figure 3.13B). To improve the accuracy, the probe connections are rotated 90° and the measurement repeated three times. Then the average resistance is calculated:

$$R = \frac{1}{4}\left[\frac{V_{12}}{I_{34}} + \frac{V_{23}}{I_{41}} + \frac{V_{34}}{I_{12}} + \frac{V_{41}}{I_{23}}\right] \qquad (3.27)$$

and

$$R_s = \frac{\pi}{\ln(2)} F(Q)R \qquad (3.28)$$

where $F(Q)$ is a correction factor that depends on the probe geometry. For a square, $F(Q) = 1$. In this technique, one must be careful to correctly measure the geometry. If a square sample is assumed, the contacts must be made on the sides of the sample [30]. This can be done by breaking off a piece of the wafer in the shape of a square and making ohmic contacts; however, it is more commonly done by photolithographically patterning a Van der Pauw structure, using oxide or junction isolation to restrict the diffusion geometry.

The sheet carrier concentration can also be combined with a measurement of the junction depth to provide a more complete description of the diffused profile. For deep junctions, this can be accomplished by beveling the wafer (see spreading resistance measurements) or by mechanically abrading a groove of known diameter in the surface of the wafer. The wafer is then immersed in a stain solution (Figure 3.14). The etch rate of the solution depends on the carrier type and concentration. P-type silicon etched in a $1:3:10$ mixture of hydrofluoric acid (HF), nitric acid (HNO_3), and acetic acid ($C_2H_4O_2$) will turn dark. A stain used for GaAs is a $1:1:10$ mixture of HF, hydrogen peroxide

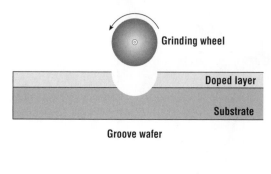

Groove wafer

After selective etch

Figure 3.14 In junction staining, a cylinder is used to groove the wafer. A doping-sensitive etch then removes part of the top layer. The junction depth can be found from the known diameter of the cylinder and the measured width of the lower abraded groove.

(H_2O_2), and water. In this case, the sample must be exposed to a bright light. The width of the stained region is measured after staining using an optical microscope with calibrated eyepieces. The junction depth can be determined from the known geometry of the bevel or groove. Limitations of accuracy and reproducibility prevent staining from being useful for junctions less than 1 μm deep. As a result, this method is much less popular than it once was.

A limitation of sheet resistance methods is that some knowledge of mobility is required to obtain even an integrated carrier concentration. The Hall effect (Figure 3.15) can be used to directly measure the integrated carrier concentration. In this measurement, a current flow in the diffused layer is also subjected to a magnetic field that is perpendicular to the flow. If one assumes that only holes are present in the diffusion, there will be a Lorentz force on each hole

$$\bar{F} = q\bar{v}x\bar{B} \tag{3.29}$$

Holes will be deflected by this force until the component of the field perpendicular to both the current flow and the field is large enough to equal the Lorentz force:

$$\mathscr{E}_y = v_x B_z \tag{3.30}$$

The establishment of this electric field is known as the Hall effect, and the resultant voltage is the Hall voltage,

$$V_h = v_x B_z w \tag{3.31}$$

where w is the width of the diffusion. The drift velocity of the hole can be related to the current by

$$v_x = \frac{I_x}{qwx_j\bar{C}_e} \tag{3.32}$$

where

$$\bar{C}_e = \frac{1}{x_j}\int_0^{x_j} C_e\,dx \tag{3.33}$$

Solving for the integrated carrier concentration,

$$\int_0^{x_j} C_e\,dx = x_j\bar{C}_e = \frac{I_x B_x}{qV_h} \tag{3.34}$$

This result can also be used to find the average Hall mobility of the sample if the four contacts are also used for a Van der Pauw measurement. Then the average Hall mobility is given by

$$\bar{\mu} = \frac{1}{qx_j\bar{C}_e R_s} \tag{3.35}$$

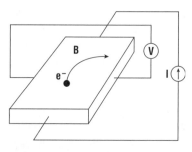

Figure 3.15 The Hall effect is able to simultaneously measure the carrier type, mobility, and sheet concentration.

For a diffused profile, this average mobility is of little interest; however, the Hall mobility is an often-quoted figure of merit for the quality of an epitaxial layer with a nominally uniform concentration.

All of the preceding techniques have a serious limitation: they provide information only about the integral of the profile. Several methods can be used to measure the carrier concentration as a function of depth. The first uses the capacitance–voltage characteristic of a diode (pn junction or Schottky) or an MOS capacitor. Although the MOS technique is widely used, it is more difficult to derive and requires a low interface state density at the Si/SiO$_2$ interface to be reliable. The diode method will be assumed here, but the techniques are very similar.

Assume that the structure can be described in the depletion approximation. For a one-sided step junction or a Schottky contact, the depletion width is given by

$$W = \sqrt{\frac{2\varepsilon(V_{bi} + V)}{qN_{sub}}} \tag{3.36}$$

where ε is the dielectric constant of the semiconductor, V_{bi} is the built in voltage of the diode, N_{sub} is the substrate doping concentration, and V is the externally applied voltage. The capacitance of the diode is

$$C = \frac{A\varepsilon}{W} = \sqrt{\frac{A^2 q \varepsilon N_{sub}}{2(V + V_{bi})}} \tag{3.37}$$

Differentiating with respect to voltage and solving for the impurity concentration,

$$N_{sub}(z) = \frac{8(V + V_{bi})^3}{A^2 q \varepsilon} \left[\frac{dC(z)}{dV} \right]^2 \tag{3.38}$$

To measure the substrate doping then, one needs to measure the capacitance in depletion as a function of the applied voltage and find the first derivative. The doping concentration as a function of voltage for each data point can be determined using Equation 3.39, and the depth corresponding to that point may be found using Equation 3.37.

The C–V method has several significant limitations. The first is that impurity concentrations in silicon above 1×10^{18} cm^{-3} cannot be measured. At these concentrations, the semiconductor becomes degenerate and acts more like a metal than a semiconductor. The second is that the depletion edges are not abrupt. Instead, they are graded over a few Debye lengths, where

$$L_D = \sqrt{\frac{\varepsilon kT}{q^2 C_{sub}}} \tag{3.39}$$

Consequently, abrupt doping profiles are not well described by their carrier profiles. Finally the C–V technique can profile only to the depth corresponding to breakdown voltage in Schottky diodes or inversion in MOS capacitors.

Several quantitative 2-D dopant profiling techniques are being developed, including nanospreading resistance and advanced dopant-sensitive etch systems. Perhaps the most promising is scanning capacitance microscopy (SCM) [31]. The SCM technique uses an atomic force microscope to scan a conducting tip over a sample. Typically the sample is cleaved and measured edge-on. The conductive tip is used to measure the capacitance in inversion. This can be readily converted to dopant concentration under the tip. Typical capacitances are less than 1 pF [32]. Although calibration is challenging, excellent quantitative agreement is possible.

The next approach to electrically profiling the carrier concentration, called *spreading resistance profilometry*, uses the dependence of the current crowding near a point contact on the local carrier concentration. Figure 3.16 shows a typical spreading resistance measurement. The sample is first beveled at a shallow angle by grinding and lapping. It is mounted in a chuck, and a pair of probes are placed in contact with the surface with a predetermined force. Thin asperities of the probe are believed to

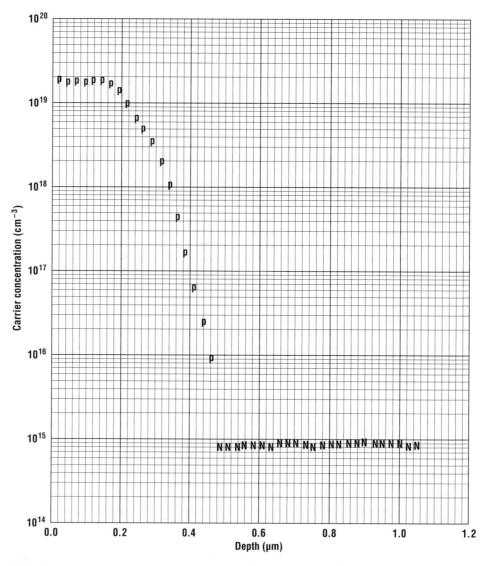

Figure 3.16   Typical spreading resistance profile showing measured carrier concentration as a function of depth *(used with permission, Solecon Labs).*

penetrate the semiconductor surface to a depth of order tens of angstroms. The current is crowded into the asperity, leading to a finite resistance between the probes. If this resistance is compared to a calibration standard of known concentration, methods exist [33] to deconvolute the resistivity to give carrier profiles. The technique can be used to measure profiles ranging from 10^{13} to 10^{21} cm^{-3}.

There are three primary limitations to spreading resistance measurements. The measurement depends critically on the reproducibility of the point contact. The probe tips must be carefully conditioned [34], and calibration standards must be run often. The measurement accuracy therefore depends on the level of experience of the operator. Commercial laboratories claim an accuracy of a factor of 2 over most ranges of doping. The second limitation of spreading resistance measurements is near-surface measurement. It is difficult to accurately produce a very flat shallow angle ($<0.5°$). Furthermore, it is often a matter of judgment as to where the surface begins unless there is an insulator on the surface. Measurements below 500 Å are not generally regarded as reliable. The final limitation to spreading resistance measurements is the materials to which it can be applied. The sample is assumed to be similar (i.e., having the same dependence of mobility on doping) to the calibration standards. This is not always the case, particularly for compound semiconductors. Furthermore, GaAs undergoes significant band bending near the surface. As a result, spreading resistance is not generally done in this system.

Various approaches to electrochemical profiling have also been established [35]. These involve electrochemically etching the wafer, measuring the capacitance and/or resistivity, and then repeating the process. By measuring the change in the capacitance or the conductivity as a function of etch time, a profile of the carrier concentration can be obtained. These methods are used widely for III–V electrical depth profiling.

A variety of techniques exist to measure chemical concentrations in thin films. To be useful in evaluating diffused impurities in semiconductors, however, the technique must be sensitive to concentrations of at least 1 ppm and ideally 1 ppb. This requirement leaves only one popular approach: secondary ion mass spectroscopy (SIMS).

Figure 3.17 shows a typical SIMS arrangement. The sample to be tested is loaded into the instrument and the system is pumped to ultrahigh vacuum (usually about 1×10^{-9} torr). The sample is then exposed to a beam of ions with an energy of between 1 and 5 keV. The energetic ions strike the surface of the sample, where they destroy the crystal lattice and eject material through a process known as sputtering. (The use of sputtering as a thin film deposition process will be discussed in Chapter 12.) A fraction of the ejected material is ionized, collected, and accelerated toward a mass spectrometer. Either negative or positive ions can be collected. The instrument can be run in a static mode with a slow sputter rate to survey the elements in the layer. The sputter rate can also be increased, and the concentration of several impurities can be measured as a function of depth.

If the sputter rate for the substrate is known for the incident ion and ion energy, or if it can be measured after the sputter is complete, this raw data of counts versus sputter time can be transformed into counts versus depth. Frequently this is done by measuring the depth of the sputtered crater after the SIMS data have been collected. Converting the counts data into chemical species concentrations is much more

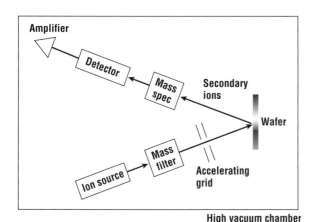

Figure 3.17 A typical SIMS arrangement. The sample is bombarded by high energy ions. The sputtered material is mass-analyzed to determine the composition of the substrate.

Table 3.3   Typical commercially available SIMS detection limits in single-crystal semiconductor samples

Impurity	Beam	Type	Detection Limit
Boron in silicon	O_2	+	1 to 5×10^{14} cm^{-3}
Arsenic in silicon	Cs	−	5×10^{14} cm^{-3}
Phosphorus in silicon	Cs	−	1×10^{15} cm^{-3}
Oxygen in silicon	Cs	−	1×10^{17} cm^{-3}
Carbon in silicon	Cs	−	1×10^{16} cm^{-3}
Silicon in GaAs	Cs	−	1×10^{15} cm^{-3}
Magnesium in GaAs	O_2	+	1×10^{14} cm^{-3}
Beryllium in GaAs	O_2	+	5×10^{14} cm^{-3}
Oxygen in GaAs	Cs	−	1×10^{17} cm^{-3}
Carbon in GaAs	Cs	−	5×10^{15} cm^{-3}

Data courtesy Mr. Charles McGee, Charles Evans and Associates.

difficult. The sputtering yield, the collection efficiency, the ionization efficiency, and the detector sensitivity are only approximately known and vary from day to day and run to run. The current best practice for reducing SIMS data is to run a calibration sample for each impurity immediately before and/or after the sample being measured. Implanted samples (see Chapter 5) are often used, since both the concentration and depth of the profile are well known. Even with this procedure, SIMS accuracy is at the very best about a factor of 2.

SIMS sensitivities vary widely according to the instrument and the technique being used. Early SIMS systems used Ar$^+$ beams. These beams have low ionization efficiencies for most impurities of interest for microelectronics applications. Reactive beams such as cesium and oxygen can produce sensitivities that are several orders of magnitude better. Furthermore, conditions used to optimize positive ion production generally will not be sensitive to impurities that tend to form negative ions. To obtain a complete inventory of the impurities present, at least two runs must be made. Under optimum conditions the sensitivities that can be expected from commercial laboratories are listed in Table 3.3.

Two problems with SIMS analysis are the time and cost of the profile. The sample must be pumped to high vacuum. Since the sample must be sputtered away during the analysis, the data collection is limited by the erosion rate. Samples thicker than 1 μm may take 4 to 8 hr to profile. Abrupt interfaces are also difficult to measure, particularly if they are buried. The sample erosion does not occur one atomic layer at a time; instead, the sample becomes progressively less planar. Combined with "knock on," the tendency of the incident ions to redistribute the impurities, the analysis of sharp concentration profiles can be significantly comprised. SIMS is nevertheless, a powerful and widely used tool.

3.6 Diffusion in SiO$_2$

In early fabrication technologies silicon dioxide (SiO$_2$) was frequently used as a mask for ion implantation. In current technologies, SiO$_2$ is usually used for its insulating properties. Silicon dioxide is a wide bandgap insulator and not a semiconductor, and thus unlike in silicon, the presence of impurities at concentrations in the ppm range in SiO$_2$ is generally believed to have little effect on the properties of the film. As a result, diffusion in SiO$_2$ has not been as extensively studied. When it has been studied, the work usually involves implantation of an impurity into a thick layer of SiO$_2$, followed by a thermal anneal and a measurement of the chemical profile of the impurity, and finally a derivation

Table 3.4   Diffusivity of various impurities in SiO₂

Element	D_o (cm²/sec)	E_a (eV)	C_s (cm^{-3})	Source
Boron	3×10^{-4}	3.53	$<3 \times 10^{20}$	Borosilicate
Phosphorus	0.19	4.03	8×10^{17} to 8×10^{19}	Phosphosilicate
Arsenic	250	4.90	1 to 6×10^{19}	Arsenosilicate
Antimony	1.31×10^{16}	8.75	5×10^{19}	Sb₂O₅ vapor

Data for boron and phosphorus from Ghezzo and Brown [36].

of a diffusion coefficient. Typically the data are taken at several temperatures to extract an activation energy. The classic reference for this type of work is Ghezzo and Brown [36], who summarized a variety of earlier references that are primarily from the 1960s. There is little agreement in the literature, particularly for values of D_o. Representative values from their summary are given in Table 3.4.

More recently, the diffusion of impurities in SiO₂ has become a topic of extreme interest. This is due to the use of heavily doped polycrystalline silicon (poly) as the gate electrode for MOS transistors, as will be described in Chapter 16. The poly is doped with phosphorus for NMOS devices and boron for PMOS devices. To get the best possible performance in the device, these gate electrodes are doped as heavily as possible. Furthermore, the grain boundaries in the poly allow for rapid diffusion to the poly/SiO₂ interface. Since the thickness of the gate oxide in current-generation MOS devices is 2 nm or less, this represents an enormous concentration gradient. Diffusion of these impurities through the gate oxide will shift the device threshold voltage and may affect the properties of the oxide, particularly as related to charge trapping and reliability [37]. This effect is commonly known as boron (or phosphorus) penetration.

Boron penetration has been heavily studied and is a major concern for small CMOS devices. Boron is believed to diffuse substitutionally for the silicon atom [38]. It has been found that when impurities diffuse through a thin layer such as a gate oxide, they move with a diffusion coefficient that is much larger than the values found from implanting boron into thick films of SiO₂. This effect has been found to be related to the high concentrations of boron used in the gate electrode, which leads to a 10× increase in diffusivity under realistic processing conditions. Values of 0.18 cm²/sec and 3.82 eV for the prefactor and activation energy have been suggested [39]. The effect of the segregation coefficient (see Chapter 4) further increases the concentration of the boron in the oxide, which may be as much as several percent. The high boron concentration leads to a softening of the oxide, which increases the diffusivity [40]. The increase in diffusivity with boron concentration leads to a loading of boron near the gate electrode and a profile qualitatively similar to that of As in Si. Uematsu has also proposed that the silicon interface is a source of dissolved SiO, which diffuses slowly through the oxide and increases the boron diffusivity [41].

The diffusion of phosphorus in SiO₂ is considered to be a less severe problem than the diffusion of boron, but, has also been studied [42]. Evidence exists that phosphorus dissolves into the SiO₂ interstitial locations as P₂, becomes substitutional by replacing Si, and thereafter diffuses on the Si sites [43].

The diffusion of these impurities in silicon dioxide has been found to be affected by the presence of high concentrations (>1%) of impurities. In particular, the incorporation of fluorine is known to increase the diffusivity of both boron and phosphorus. For boron the effect can be as much as an order of magnitude [44]. Hydrogen is also known to increase boron diffusivity. Of particular technological interest, however, are impurities that can reduce impurity diffusion. Nitrogen has been found to be particularly effective at this [45]. It is believed that substitutional diffusion of B and P requires a rearrangement in the local bonding structure from silicon bonding covalently to four nearest-neighbor oxygen atoms, to boron bonding covalently to three nearest-neighbor oxygen atoms or phosphorus bonding covalently to three nearest-neighbor oxygen atoms and forming a double bond to the fourth oxygen atom. The presence of nitrogen is believed to block this rearrangement [46]. The concentration of nitrogen required increases as the gate oxide thickness decreases. For deeply scaled (<100 nm) devices with gate oxide thicknesses of about 1.5 nm, one would like nitrogen concentrations of about 10%. Furthermore, this nitrogen should be near the poly gate, both to block the diffusion of boron into the gate oxide and to keep it as far away from the lower interface as possible, since it degrades the electrical properties of the Si/SiO$_2$ interface.

Example 3.3

Assume that the values given in the text are a reasonable approximation for the diffusivity of boron in SiO$_2$. If the concentration of boron at the surface of a 2nm oxide is 10^{21} cm^{-3}, find the flux of boron into the substrate for a one-minute anneal at 1000°C.

Solution

From Fick's first law

$$J = -D\frac{dc}{dx} \approx D\frac{10^{21}-0}{2 \text{ nm}}$$

From the text,

$$D = 0.18 \text{ cm}^2/\text{sec } e^{-3.82/kT} = 1.4 \times 10^{-16} \text{ cm}^2/\text{sec}$$

For a one-minute anneal,

$$Q_T = J \times 60 \text{ sec} = 1.4 \times 10^{-16} \text{ cm}^2/\text{sec} \times 5 \times 10^{27} \text{ cm}^{-4} \times 60 \text{ sec}$$

$$Q_T = 4 \times 10^{13} \text{ cm}^{-2}$$

This is a very large dose that will readily shift the device operation.

3.7 Simulations of Diffusion Profiles

It has probably become apparent to the reader that the various complications that arise in the calculation of diffused profiles, such as concentration-dependent diffusivities, preclude analytic calculations for all but the simplest examples. For that reason, numerical methods have been developed for the prediction of profiles in 1-, 2-, and 3-D. Often the results can be linked to device simulators so that the effect that a change in an impurity profile would have on the device characteristics can be predicted in a straightforward manner. If the simulation results are accurate, the device designer can optimize performance and test process sensitivity with far fewer runs through the fabrication facility, resulting in tremendous savings in cost and time.

While some corporations have developed proprietary software, one of the most popular packages for calculating impurity profiles is the Stanford University PRocess Engineering Module (SUPREM). SUPREM III performs detailed calculations in 1-D. SUPREM IV performs calculations in 2-D.

The outputs of these programs are the chemical, carrier, and vacancy concentrations as functions of depth into the semiconductor.

Before proceeding, however, a word of caution is in order. Students tend to regard the output of these programs as being infallibly correct. This is, of course, not true. The predictions of this code are only as good as the models and numerical techniques that are employed. In practice these model parameters must be rigorously checked to ensure accuracy. These programs should be regarded as calculation tools that allow the process engineer to access more complicated diffusion models. These models are also more realistic and usually, but not always, give fairly accurate results.

All diffusion process simulators are built on three basic equations. In 1-D these are the flux equation

$$J_i = -D_i \frac{dC_i}{dx} + Z_i \mu_i C_i \mathscr{E} \tag{3.40}$$

where Z_i is the charge state and μ_i is the mobility of the impurity; the continuity equation

$$\frac{dC_i}{dt} + \frac{dJ_i}{dx} = G_i \tag{3.41}$$

where G_i is the generation recombination rate of the impurity; and Poisson's equation

$$\frac{d}{dx}[\varepsilon \mathscr{E}] = q(p - n + N_D^+ - N_A^-) \tag{3.42}$$

where ε is the dielectric constant, n and p are the electron and hole concentrations, and N_D^+ and N_A^- are the concentrations of the ionized donors and acceptors. These equations are solved simultaneously over a 1-D grid that the user defines.

The diffusivity used for SUPREM is based on the vacancy model of Fair. The diffusivity is calculated using Equation 3.7. The values of E_a and D_o are included in a software look up table for boron, antimony, and arsenic in silicon. Finally, empirical models are added to take into account field-aided, oxidation-enhanced, and oxidation-retarded diffusion.

The examples in the book will use a set of software programs collectively called ATHENA© and marketed by Silvaco© (www.silvaco.com). This choice was made for three reasons: (1) The Silvaco code is built, in part, on Stanford's SUPREM IV software, which has a long and rich history; (2) the Silvaco code is a widely used TCAD (technology computer-automated design) suite in industry; and (3) Silvaco offers deep discounts for educational use and the educational software runs well on conventional PCs. For students who do not have access to this software, a parallel set of examples are planned for use on the NSF-sponsored Nanohub website (http://www.nanohub.org). These tools are available over the Web free of charge. See this book's website (http://www.oup-usa.org/isbn/01951360SS.html) to access this Nanohub-compatible examples.

To run the software, an input deck must be provided (see examples below). This file contains a series of comments, the initialize statement, materials statements, process statements, and output statements. It is constructed with the program deckbuild. The deck starts with the title card, which is simply a comment repeated on each page of the output. Several comment cards may follow. The user is encouraged to use these cards to document the process flow.

The next set of lines determines the simulation grid. Since we are only interested in the vertical direction, a very rudimentary grid can be set in the horizontal direction (line $\times$ commands). This will save computation time. Next the background concentration of the wafer is set to be able to determine junction depth. The adapt command allows the software to adapt the grid according to the doping profile. Next the diffuse command is invoked. The concentration of phosphorus at the surface of the wafer is given by c.phos. Since the diffusion ambient is not specified, it is assumed to be inert. Once this is done, the software can extract physical and electrical information about the impurity profile, including the junction depth and sheet resistance. For these commands, we want the first occurrence of the material silicon (there is only one), and we want the sheet resistance of the uppermost region, which was doped n-type by the diffusion. Finally, the solution is saved, and the postprocessing software, Tonyplot is invoked.

Example 3.4

Use Silvaco's Athena software to create a quasi-one-dimensional grid and do a subsequent solid source phosphorus predeposition diffusion.

Solution

```
go athena
#TITLE: Solid Source Phosphorus Diffusion Example 3.2

line x loc=0.0          spacing=0.02
line x loc=0.2          spacing=0.02
line y loc=0.0          spacing=0.02
line y loc=0.1          spacing=0.02
line y loc=0.4          spacing=0.04
line y loc=0.8          spacing=0.06
line y loc=1.5          spacing=0.10

init c.boron=3e14
method adapt

# Diffuse Phosphorus
diffuse time=60 temp=1000 c.phos=1e20

# extract junction depth
extract name="xj" xj material="Silicon" mat.occno=1 x.val=0.1
junc.occno=1

#extract 1D electrical parameters
extract name="sheet_rho" n.sheet.res material="Silicon"
mat.occno=1 x.val=0.1 region.occno=1

#  Save and plot the final structure
structure outfile=ex3_4.str
tonyplot
```

Upon running, the solution given is:

```
EXTRACT> extract name="xj" xj material="Silicon" mat.occno=1
x.val=0.1 junc.occno=1
xj=0.566734 um from top of first Silicon layer X.val=0.1
EXTRACT> #extract 1D electrical parameters
EXTRACT> extract name="sheet_rho" n.sheet.res
material="Silicon" mat.occno=1 x.val=0.1 region.occno=1
sheet_rho=36.7657 ohm/square X.val=0.1
```

and the Tonyplot result is shown in Figure 3.18. The software can be used to zoom in or to show a 3-D contour plot, which for this structure is rather uninteresting.

Example 3.5   **Predeopsition diffusion, followed by a drive-in**

```
go athena
#TITLE: Phosphorus Predep Followed by Drive-in Diffusion
Example 3.3

line x loc=0.0          spacing=0.02
line x loc=0.2          spacing=0.02
line y loc=0.0          spacing=0.02
line y loc=0.1          spacing=0.02
line y loc=0.4          spacing=0.04
line y loc=0.8          spacing=0.06
line y loc=1.5          spacing=0.10

init c.boron=3e14
method adapt

# Diffuse Phosphorus
diffuse time=60 temp=1000 c.phos=1e20

#              Grow pad oxide, 400A.
Diffuse        Temperature=1000  Time=17  DryO2

#              Deposit 800A of CVD Nitride.
Deposit        Nitride  Thickness=.0800  Spaces=15

#              Perform drive-in diffusion
Diffuse        Temperature=1000  Time=180

# extract junction depth and nitride and oxide thickness
extract name="xj" xj material="Silicon" mat.occno=1 x.val=0.1
junc.occno=1
```

```
extract name="tox" thickness material="oxide" mat.occno=1
y.val=0.1

#extract 1D electrical parameters
extract name="sheet_rho" n.sheet.res material="Silicon"
mat.occno=1 x.val=0.1 region.occno=1

# Save and plot the final structure
structure outfile=ex3_3.str
tonyplot
quit
```

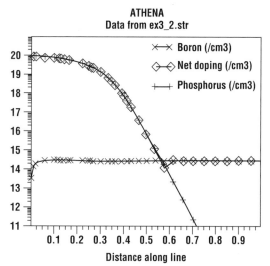

Figure 3.18   Plot of impurity as a function of depth for phosphorus drive-in.

This simple case (Example 3.4) is followed by a slightly more advanced example in Example 3.5. Here the predeposition is followed by a drive-in diffusion. Note that the role of sealing the interface, which is necessary for a drive-in, has been done by the addition of a thin layer of silicon nitride. As will be discussed later in the book, nitride is a good diffusion barrier; that is, many materials including phosphorus, have a low diffusivity in silicon nitride. The oxide, which is grown before the nitride deposition, is a practical matter, since the mismatch in the thermal expansion coefficients of nitride and silicon would lead to defect formation upon thermal cycling.

The results are given as follows:

```
EXTRACT> extract name="xj" xj material=
"Silicon" mat.occno=1 x.val=0.1
junc.occno=1
xj=1.15156 um from top of first Silicon
layer X.val=0.1
EXTRACT> extract name="tox" thickness
material="oxide" mat.occno=1 y.val=0.1
tox=402.545 angstroms (0.0402545 um) Y.val=0.1
EXTRACT> #extract 1D electrical parameters
EXTRACT> extract name="sheet_rho" n.sheet.res
material="Silicon" mat.occno=1 x.val=0.1 region.occno=1
sheet_rho=35.1746 ohm/square X.val=0.1
```

One can see that the profile is more similar to that of a complementary error function than the Gaussian and that the junction is much deeper because of the large thermal cycle. The sheet resistance does not change dramatically. Most of the impurity was activated during the predeposition.

3.8 Summary

This chapter reviewed the physics of diffusion and presented Fick's laws, the relations that govern diffusion. Two particular solutions were presented corresponding to drive-in and predeposition diffusion. The atomistic mechanisms of diffusion were presented along with heavy doping effects. The details of diffusion for a variety of popular dopants were also discussed. At high doping concentrations the diffusion coefficient is no longer constant, but frequently depends on the local doping concentration and concentration gradient. A numerical tool, was introduced that allows the student to calculate dopant profiles in the presence of these nonlinear effects.

Problems

(For all problems assume that the diffusivity can be approximated by the intrinsic diffusivity, unless specified otherwise.)

1. Assume that you have been asked to measure the diffusivity of a donor impurity in a new elemental semiconductor. What constants would you need to measure? What experiments would you attempt? Discuss the measurement techniques that you would use to measure the chemical and carrier profiles. What problems are likely to arise?

2. Construct a semilog plot using Table 3.2 of the three contributions (D_i, D_-, and $D^=$) to the diffusivity of phosphorus as a function of temperature from 700 to 1100°C. Assume that the phosphorus concentration is 10^{19} cm^{-3}.

3. Using Fair's vacancy model, including charge effects, calculate the diffusivity of arsenic (As) in silicon at 1000°C for the following arsenic doping concentrations:
    (a) 1×10^{15} cm^{-3}
    (b) 1×10^{21} cm^{-3}
    Hint: For both parts, the carrier concentration (n) is **not** equal to the doping concentration (C).

4. Stress is a high concentration effect that was not discussed. Arsenic, for example, is a much larger atom than silicon. When high concentrations of arsenic are incorporated in the lattice it creates strain. Qualitatively discuss how this strain may affect the diffusivity.

5. Delta doping is a process used in advanced GaAs fabrication to increase the Schottky barrier height of the gate electrode. This reduces the gate electrode leakage. Delta doping is done by depositing a monolayer of a p-type dopant material directly between the gate electrode and the GaAs. Assume that the atomic surface coverage is 1.5×10^{15} cm^{-2} and that the dopant is beryllium. After the gate patterning, the source/drain is annealed at 800°C for 10 min to activate the impurity. (a) If the gate material prevents any outdiffusion from the wafer, use first-order diffusion theory to calculate the junction depth if the channel is doped 1×10^{17} cm^{-3} n-type. (b) What surface concentration of Be will result? (c) Sketch the profile that you calculated using this simple theory, and the profile that might actually be expected. Briefly list two reasons for the difference.

6. The surface of a silicon wafer has a region that is uniformly doped with boron at a concentration of 10^{18} cm^{-3}. This layer is 20 Å thick (1 Å $= 10^{-4}$ μm $= 10^{-8}$ cm). The entire wafer, including this region, is uniformly doped with arsenic at a concentration of 10^{15} cm^{-3}. The surface of the wafer is sealed and it is heated at 1000°C for 30 min. Assume intrinsic diffusion.
    (a)   Find the concentration of boron at the surface after the anneal.
    (b)   Find the junction depth (boron concentration equal to arsenic concentration) after the anneal.

7. There is strong interest in forming very shallow pn junctions in silicon to build deeply scaled CMOS. Assume that you have access to an extremely low energy boron implanter and use it to implant a dose of 10^{15} cm^{-2} of boron to a negligible depth ($R_p \approx 0$). Next, the surface of the wafer is sealed and the wafer is annealed at 1000°C for 10 sec. Assuming intrinsic diffusion, find:
   (a) The final junction depth if the background concentration of the wafer is 10^{17} cm^{-3}.
   (b) The final concentration at the surface of the wafer.

8. A 10-Å-thick, uniformly sulfur (S)-doped layer is grown on top of a GaAs wafer. The doping concentration of this layer is 10^{18} cm^{-3}. The wafer is scaled with a layer of Si$_3$N$_4$ to prevent any outdiffusion, and it is annealed for 60 min at 950°C. Ignore all heavy doping effects.
   (a) Find the sulfur concentration at the surface after the anneal.
   (b) At what depth would the concentration be 10^{14} cm^{-3}?

9. A silicon wafer was doped in a 1000°C predeposition diffusion with phosphorus to its solid solubility limit. The process time was 20 min. After the predeposition, the surface of the silicon was sealed and an 1100°C drive-in was done. Find the drive-in time necessary to obtain a junction depth of 4.0 μm. Assume a substrate concentration of 10^{17} cm^{-3}. What is the surface concentration after the drive-in?

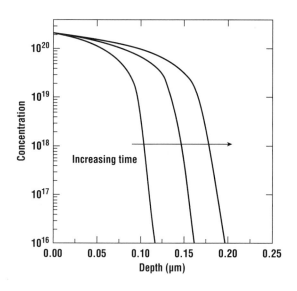

10. Refer to the accompanying plot, which shows the diffusion of arsenic into silicon. Assume that this was done at 1000°C.
    (a) Taking into account charged vacancy effects, calculate the diffusion coefficient of As at the surface of the wafer.
    (b) Calculate the intrinsic diffusivity
    (c) Is the diffusion shown in the figure a predeposition or a drive-in? Justify your answer. (Note that owing to heavy doping effects, you cannot get this from the shape of the curve.)

11. For deeply scaled MOSFETs, it is necessary to make very shallow source/drain junctions. Assume that one needs a P+/n junction that is 0.05 μm deep. The wafer is implanted with boron at extremely low energy ($R_p \ll 0.05$ μm) to a dose of 5×10^{15} cm^{-2}. An anneal must be done at 1000°C to repair the implant damage and to activate the impurity. Ignoring heavy doping effects (i.e., assume simple intrinsic diffusion) and transient diffusion effects, and assuming that none of the implanted boron leaves the wafer, how long should the diffusion be done? The (n-type) substrate concentration is 2×10^{17} cm^{-3}.

12. A 10-min 1100°C predeposition is done using arsenic. Assume that the surface concentration reaches the solid solubility limit. Next, the wafer is annealed for 24 hr at 1000°C. Use Equation 3.23 to predict the maximum carrier concentration. How does this compare to the solid solubility limit at 1000°C? Explain the distinction between solid solubility and Equation 3.23.

13. Assume that a wafer is uniformly doped. If a Schottky contact is formed on the surface, what would the C–V curve look like?

14. Boron-doped gate electrodes are desirable for certain highly scaled MOS transistors. Assume that the gate oxide thickness is 100 Å. Also assume that the concentration of boron in the oxide very close to the polysilicon gate is a constant of 10^{21} cm^{-3}. The maximum amount of boron

that can be diffused into the substrate without shifting the threshold voltage outside-allowable limits is 3×10^{11} cm^{-2}. Using Table 3.4, estimate the dose of impurities that enter the semiconductor after a 4-hr anneal at 1000°C. Will this anneal produce an unacceptable threshold voltage shift?

References

1. F. J. Morin and J. P. Maita, "Electrical Properties of Silicon Containing Arsenic and Boron," *Phys. Rev.* **96**:28 (1954).
2. W. R. Runyan and K. E. Bean, *Semiconductor Integrated Circuit Processing Technology*, Addison-Wesley, Reading, MA, 1990.
3. P. M. Fahey, P. B. Griffin, and J. D. Plummer, "Point Defects and Dopant Diffusion in Silicon," *Rev. Mod. Phys.* **61**:289 (1989).
4. T. Y. Yan and U. Gosele, "Oxidation-Enhanced or Retarded Diffusion and the Growth or Shrinkage of Oxidation-Induced Stacking Faults in Silicon," *Appl. Phys. Lett.* **40**:616 (1982).
5. S. Mizuo and H. Higuchi, "Retardation of Sb Diffusion in Si During Thermal Oxidation," *J. Appl. Phys. Jpn.* **20**:739 (1981).
6. A. M. R. Lin, D. A. Antoniadis, and R. W. Dutton, "The Oxidation Rate Dependence of Oxidation-Enhanced Diffusion of Boron and Phosphorus in Silicon," *J. Electrochem. Soc.* **128**:1131 (1981).
7. D. J. Fisher, ed., "Diffusion in Silicon—A Seven-year Retrospective," *Defect Diffusion Forum* **241**:1 (2005).
8. "Diffusion in Ga-As and other III–V Semiconductors," *Defect Diffusion Forum* **157–159**:223 (1998).
9. R. B. Fair, "Concentration Profiles of Diffused Dopants in Silicon," in *Impurity Doping Processes in Silicon*, F. F. Y. Wang, ed., North-Holland, New York, 1981.
10. H. Ryssel, K. Muller, K. Harberger, R. Henkelmann, and F. Jahael, "High Concentration Effects of Ion Implanted Boron in Silicon," *J. Appl. Phys.* **22**:35 (1980).
11. R. Duffy, V. C. Venezia, A. Heringa, B. J. Pawlak, M. J. P. Hopstaken, G. C. J. Maas, Y. Tamminga, T. Dao, F. Roozeboom, and L. Pelaz, "Boron Diffusion in Amorphous Silicon and the Role of Fluorine," *Appl. Phys. Lett.* **84**(21):4283 (2004).
12. A. Ural, P. B. Griffin, and J. D. Plummer, "Fractional Contributions of Microscopic Diffusion Mechanisms for Common Dopants and Self-Diffusion in Silicon," *J. Appl. Phys.* **85**(9):6440 (1999).
13. J. Xie and S. P. Chen, "Diffusion and Clustering in Heavily Arsenic-Doped Silicon— Discrepancies and Explanation," *Phys. Rev. Lett.* **83**(9):1795 (1999).
14. R. B. Fair and J. C. C. Tsai, "A Quantitative Model for the Diffusion of Phosphorus in Silicon and the Emitter Dip Effect," *J. Electrochem. Soc.* **124**:1107 (1978).
15. M. UeMatsu, "Simulation of Boron, Phosphorus, and Arsenic Diffusion in Silicon Based on an Integrated Diffusion Model, and the Anomalous Phosphorus Diffusion Mechanism," *J. Appl, Phys.* **82**(5): 2228 (1997).
16. R. J. Field and S. K. Ghandhi, "An Open Tube Method for the Diffusion of Zinc in GaAs," *J. Electrochem. Soc.* **129**:1567 (1982).
17. L. R. Weisberg and J. Blanc, "Diffusion with Interstitial-Substitutional Equilibrium. Zinc in Gallium Arsenide," *Phys. Rev.* **131**:1548 (1963).
18. S. Reynolds, D. W. Vook, and J. F. Gibbons, "Open-Tube Zn Diffusion in GaAs Using Diethylzinc and Trimethylarsenic: Experiment and Model," *J. Appl. Phys.* **63**:1052 (1988).

19. K. B. Kahen, "Mechanism for the Diffusion of Zinc in Gallium Arsenide," in *Mater. Res. Soc. Symp. Proc.*, Vol. 163, D. J. Wolford, J. Bernholc, and E. F. Haller, eds., MRS, Pittsburgh, 1990, p. 681.

20. M. E. Greiner and J. F. Gibbons, "Diffusion of Silicon in Gallium Arsenide Using Rapid Thermal Processing: Experiment and Model," *Appl. Phys. Lett.* **44**:740 (1984).

21. K. L. Kavanaugh, C. W. Magee, J. Sheets, and J. W. Mayer, "The Interdiffusion of Si, P, and In at Polysilicon Interfaces," *J. Appl. Phys.* **64**:1845 (1988).

22. S. Yu, U. M. Gosele, and T. Y. Tan, "An Examination of the Mechanism of Silicon Diffusion in Gallium Arsenide," in *Mater. Res. Soc. Symp. Proc.*, Vol. 163, D. J. Wolford, J. Bernholc, and E. F. Haller, eds., MRS, Pittsburgh, 1990, p. 671.

23. K. B. Kahen, D. J. Lawrence, D. L. Peterson, and G. Rajeswaren, "Diffusion of Ga Vacancies and Si in GaAs," in *Mater. Res. Soc. Symp. Proc.*, Vol. 163, D. J. Wolford, J. Bernholc, and E. F. Haller, eds., MRS, Pittsburgh, 1990, p. 677.

24. J. J. Murray, M. D. Deal, E. L. Allen, D. A. Stevenson, and S. Nozaki, *J. Electrochem. Soc.* **137**(7):2037 (1992).

25. D. Sudandi and S. Matsumoto, "Effect of Melt Stoichiometry on Carrier Concentration Profiles of Silicon Diffusion in Undoped LEC Sl-GaAs," *J. Electrochem. Soc.* **136**:1165 (1989).

26. L. B. Valdes, "Resistivity Measurements on Germanium for Transistors," *Proc. IRE* **42**:420 (1954).

27. M. Yamashita and M. Agu, "Geometrical Correction Factor of Semiconductor Resistivity Measurement by Four Point Probe Method," *Jpn. J. Appl. Phys.* **23**:1499 (1984).

28. D. K. Schroder, *Semiconductor Material and Device Characterization*, Wiley-Interscience, New York, 1990.

29. L. J. Van der Pauw, "A Method for Measuring the Specific Resistivity and Hall Effect of Discs of Arbitrary Shape," *Phillips Res. Rep.* **13**:1 (1958).

30. D. S. Perloff, "Four-point Probe Correction Factors for Use in Measuring Large Diameter Doped Semiconductor Wafers," *J. Electrochem. Soc.* **123**:1745 (1976).

31. A. Diebold, M. R. Kump, J. J. Kopanski, and D. G. Seiler, *J. Vacuum Sci. Technol. B* **14**:196 (1996).

32. J. S. McMurray, J. Kim, and C. C. Williams, "Direct Comparison of Two-Dimensional Dopant Profiles by Scanning Capacitance Microscopy with TSUPRE4 Process Simulation," *J. Vacuum Sci. Technol. B.* **16**:344 (1998).

33. M. Pawlik, "Spreading Resistance: A Comparison of Sampling Volume Correction Factors in High Resolution Quantitative Spreading Resistance," in *Emerging Semiconductor Technology*, D. C. Gupta and R. P. Langer, eds., STP 960, American Society for Testing and Materials, Philadelphia, 1987.

34. R. G. Mazur and G. A. Gruber, "Dopant Profiles in Thin Layer Silicon Structures with the Spreading Resistance Profiling Technique," *Solid State Technol.* **24**:64 (1981).

35. P. Blood, "Capacitance–Voltage Profiling and the Characterization of III–V Semiconductors Using Electrolyte Barriers," *Semicond. Sci. Technol.* **1**:7 (1986).

36. M. Ghezzo and D. M. Brown, "Diffusivity Summary of B, Ga, P, As, and Sb in SiO_2," *J. Electrochem. Soc.* **120**:146 (1973).

37. Z. Zhou and D. K. Schroder, "Boron Penetration in Dual Gate Technology," *Semicond. Int.* **21**:6 (1998).

38. K. A. Ellis and R. A. Buhrman, "Boron Diffusion in Silicon Oxides and Oxynitrides," *J. Electrochem. Soc.* **145**:2068 (1998).

39. T. Aoyama, H. Arimoto, and K. Horiuchi, "Boron Diffusion in SiO_2 Involving High Concentration Effects," *Jpn. J. Appl. Phys.* **40**:2685 (2001).

40. S. Sze, *VLSI Technology*, McGraw-Hill, New York, 1988.
41. M. Uematsu, "Unified Simulation of Diffusion in Silicon and Silicon Dioxide," *Defect Diffusion Forum*, **237–240**:38 (2005).
42. T. Aoyama, H. Tashiro, and K. Suzuki, "Diffusion of Boron, Phosphorus, Arsenic, and Antimony in Thermally Grown Silicon Dioxide," *J. Electrochem. Soc.* **146**(5):1879 (1999).
43. M. Susa, K. Kawagishi, N. Tanaka, and K. Nagata, "Diffusion Mechanism of Phosphorus from Phosphorus Vapor in Amorphous Silicon Dioxide Film Prepared by Thermal Oxidation," *J. Electrochem. Soc.* **144**(7):2552 (1997).
44. T. Aoyama, K. Suzuki, H. Tashiro, Y. Toda, T. Yamazaki, K. Takasaki, and T. Ito, *J. Appl. Phys.* **77**:417 (1995).
45. T. Aoyama, K. Suzuki, H. Tashiro, Y. Tada, and K. Horiuchi, "Nitrogen Concentration Dependence on Boron Diffusion in Thin Silicon Oxynitrides Used for Metal-Oxide-Semiconductor Devices," *J. Electrochem. Soc.* **145**:689 (1998).
46. K. A. Ellis and R. A. Buhrman, "Phosphorus Diffusion in Silicon Oxide and Oxynitride Gate Dielectrics," *Electrochem. Solid State Lett.* **2**(10):516 (1999).

Chapter 4

Thermal Oxidation

One of the major reasons for the popularity of silicon ICs is the ease with which silicon forms an excellent oxide, SiO_2. This oxide is widely used as an insulator both in active devices such as MOSFETs and in the region between the active devices, known as the field. Various methods are used for forming SiO_2. Chapter 4 will introduce thermal oxidation. This method produces oxides with the fewest defects both in the bulk and at the interface with the underlying silicon. Unfortunately, most other semiconductors do not form oxides of sufficient quality to be used in device fabrication. For that reason, this chapter will focus on silicon processing exclusively.

4.1 The Deal–Grove Model of Oxidation

The oxidation of silicon in molecular oxygen proceeds according to the overall reaction

$$Si(solid) + O_2(gas) \rightarrow SiO_2 \tag{4.1}$$

This process is called *dry oxidation* since it uses molecular oxygen rather than water vapor as an oxidant. The Deal–Grove model works well for predicting oxide thicknesses for thermal oxides larger than about 300 Å [1]. The growth of thin oxides will be discussed later in this chapter. High temperature is not necessary to grow an oxide. Silicon will oxidize in air at room temperature. Once an oxide forms, however, silicon atoms must travel through the oxide layer to react with the oxygen present at the surface of the wafer, or else oxygen molecules must travel through the oxide to reach the silicon surface where a reaction can occur. The process that drives this motion is the one discussed in the last chapter: diffusion. The diffusivity of Si in SiO_2 is several orders of magnitude smaller than the diffusivity of O_2. As a result, the chemical reaction occurs at the Si–SiO_2 interface. This has a very important effect. The interface produced by thermal oxidation has not seen the atmosphere. As a result, it is relatively free of impurities. The amount of silicon consumed due to the chemical reaction given by Equation 4.1 is about 44% of the thickness of the final oxide.

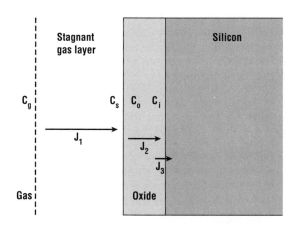

Figure 4.1   Schematic diagram of the oxidant flows during oxidation.

At room temperature, neither the silicon nor the oxygen molecules are sufficiently mobile to diffuse through the native oxide. After a while, the reaction effectively stops and the oxide will not get much thicker than 25 Å. For a sustained reaction to occur, the silicon wafer must be heated in the presence of an oxidizing ambient [2]. For now, assume that the ambient is O_2. Consider the diagram of a growing oxide (Figure 4.1). The four oxygen concentrations in this figure are C_g, the concentration of oxygen in the gas stream far from the wafer, C_s, the concentration of oxygen in the gas at the surface of the wafer, C_o, the concentration of oxygen in the oxide at the surface of the wafer, and C_i, the concentration of oxygen at the Si–SiO$_2$ interface. Define J as an oxygen flux, which is the number of oxygen molecules that crosses a plane of a certain area in a certain time. One can then define three oxygen fluxes of interest. The first is the amount of oxygen that moves from the bulk of the gas to the surface of the growing oxide film. Recall that Chapter 2 briefly mentioned stagnant layer formation in the molten silicon of a Czochralski crucible. This layer arises due to the finite viscosity of the melt. Similarly, if a gas stream of oxygen is made to flow past the surface of the wafer, a boundary layer will exist near the surface. The gas velocity in this boundary layer ranges from zero at the surface of the wafer to the bulk gas velocity at the opposite side of the boundary layer. To a first approximation, the oxygen molecules cannot be transported across this region by the gas flow: there is none. Instead, they must diffuse in a manner described by Fick's first law

$$J_1 \approx D_{O_2} \frac{C_g - C_s}{t_{sl}} \tag{4.2}$$

where t_{sl} is the thickness of the stagnant layer. C_g can be calculated using the ideal gas law:

$$C_g = \frac{n}{V} = \frac{P_g}{kT} \tag{4.3}$$

where k is Boltzmann's constant and P_g is the partial pressure of the oxygen in the furnace. Although this formulation is workable, it tends to underestimate the flux. It is more common to explicitly take into account the fact that some flow remains through much of what is called the stagnant layer by writing

$$J_1 = J_{\text{gas}} = h_g(C_g - C_s) \tag{4.4}$$

where h_g is called the *mass transport coefficient*.

The second oxygen flux is the diffusion of molecular oxygen across the growing oxide film. The concentration gradient needed to drive diffusion arises because the gas ambient acts as an oxygen source while the reacting surface acts as a sink. Then, assuming no sources or sinks of oxygen in the growing oxide, the concentration varies linearly and

$$J_2 \approx D_{O_2} \frac{C_o - C_i}{t_{ox}} \tag{4.5}$$

where the diffusivity is now that of oxygen in SiO_2. The third flux is the flow of oxygen reacting with the silicon to form SiO_2. This rate is determined by chemical reaction kinetics. Since there is an abundant supply of silicon at the reacting surface, the reaction rate and the flux are proportional to the oxygen concentration:

$$J_3 = k_s C_i \tag{4.6}$$

where the constant of proportionality, k_s, is the chemical rate constant for the overall reaction described in Equation 4.1. In equilibrium these three flows must balance:

$$J_1 = J_2 = J_3 \tag{4.7}$$

Combining Equations 4.4–4.6 leaves two equations and three unknown concentration: C_s, C_o, and C_i. Solving to find a growth rate requires one more equation. That equation is Henry's law, which says that the concentration of an adsorbed species at the surface of a solid is proportional to the partial pressure of that species in the gas just above the solid

$$C_o = H p_g = H k T C_s \tag{4.8}$$

where H is Henry's gas constant, and the ideal gas law has been used to replace p_g. Now there are three equations and three unknowns. After some algebra one can show that

$$C_i = \frac{H p_g}{1 + \dfrac{k_s}{h} + \dfrac{k_s t_{ox}}{D}} \tag{4.9}$$

where $h = h_g / H k T$. Finally, to obtain the growth rate, just divide the flux at the interface by the number of molecules of oxygen per unit volume of SiO_2, which is commonly labeled N_1. For oxidation by molecular oxygen, N_1 is half the number density of oxygen atoms in SiO_2, or 2.2×10^{22} cm^{-3}. The result is

$$R = \frac{J}{N_1} = \frac{dt_{ox}}{dt} = \frac{H k_s p_g}{N_1 \left[1 + \dfrac{k_s}{h} + \dfrac{k_s t_{ox}}{D} \right]} \tag{4.10}$$

Assuming that at time 0 the oxide thickness is t_0, the solution of this differential equation is of the form

$$t_{ox}^2 + A t_{ox} = B(t + \tau) \tag{4.11}$$

where

$$A = 2D \left(\frac{1}{k_s} + \frac{1}{h} \right)$$

$$B = \frac{2DH p_g}{N_1} \tag{4.12}$$

$$\tau = \frac{t_0^2 + A t_0}{B}$$

The parameters A and B are well known for a variety of process conditions. The more fundamental parameters such as diffusivity that go into A and B are rarely quoted. Furthermore, most silicon oxidation is done at atmospheric pressure. As a result, $k_s \ll h$, and the growth rate is nearly independent of the gas phase mass transport and, therefore, of the reactor geometry. When the oxidizing species is H_2O rather than O_2 (see Section 4.2), the same equations apply, but with different values for the diffusivity, mass transport properties, reactivity, pressure of the gas, and number of molecules per unit volume.

Since A and B are proportional to diffusivity both parameters will follow an Arrenhius function. The activation energies of A and B can be calculated from the activation energy of the diffusivities and reaction rates as shown by Equation 4.12. There is a reasonably good agreement (about 10% discrepancy) between the activation energies of the diffusivities of oxygen and water in fused silica and the activation energy of B. Furthermore, since the ratio B/A eliminates the diffusivity, its activation energy should depend primarily on k_s. As expected, the activation energy of B/A is in reasonable agreement with the Si–Si bond strength.

Finally, it is worth pointing out the significance of τ, since this is a common source of confusion. It arises due to the boundary condition for the differential equation at zero time. Notice that for sufficiently thick oxides, the rate of oxidation varies with the oxide thickness. If an oxidation is begun with an initial oxide thickness of t_0, it is not accurate therefore to calculate the thickness grown in the process and simply add it to t_0. Instead, the initial thickness must be used to determine τ, and it must be added to t to obtain a total effective oxidation time. It is as though the oxidation process started at a time $-\tau$. Then at $t = 0$ the thickness of the oxide is exactly t_0.

4.2 The Linear and Parabolic Rate Coefficients

Equation 4.11 has two important limiting forms. For sufficiently thin oxides, one can neglect the quadratic term. In that case

$$t_{ox} \approx \frac{B}{A}(t + \tau) \qquad (4.13)$$

On the other hand, if the oxide is sufficiently thick, the term that is linear in the thickness can be ignored and

$$t_{ox}^2 \approx B(t + \tau) \qquad (4.14)$$

Because of these two limiting forms, B/A is called the *linear rate coefficient* and B is called the *parabolic rate coefficient*. These are the numbers that are usually quoted for oxidation. Figure 4.2 shows the coefficients for oxidations in pure oxygen. This type of process is called a *dry oxidation*. The plots are summarized at typical process conditions in Table 4.1.

Oxygen is not the only oxidant that can be used in this process. Another very popular one is a mixture of O_2 and H_2O. Called wet oxidation, these processes have the advantage of oxidizing at a much higher rate than dry oxidations. The fundamental reasons for the higher rate are the higher diffusivity of H_2O compared with O_2 and the much larger solubility (Henry's constant) for H_2O compared with O_2. A disadvantage is that oxides grown wet are less dense. Thus wet oxidation is typically used when a thick oxide is required that will not be subjected to any significant electrical stress. Figure 4.3 shows the linear and parabolic rate coefficients for wet oxidation. Note that this process is typically run with excess O_2 to reduce the chance for explosion. As a result, the partial pressure of H_2O is less than one atmosphere. The B coefficient therefore depends on this flow ratio. Section 4.4 will discuss the structure of the oxide in more detail.

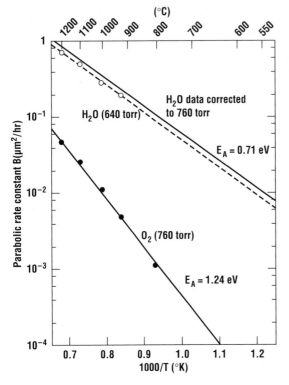

Figure 4.2   Arrhenius plot of the *B* oxidation coefficient. The wet parameters depend on the H_2O concentration and therefore on the gas flows and pyrolysis conditions *(after Deal and Grove)*.

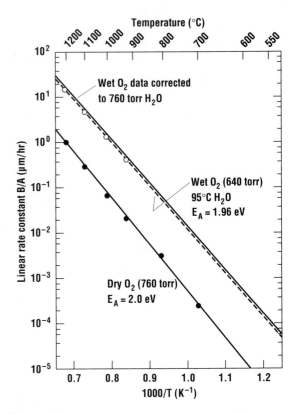

Figure 4.3   Arrhenius plot of the ratio *(B/A)* of the oxidation parameters *(after Deal and Grove)*.

Another gas ambient used in thermal oxidation is dry molecular oxygen with a small (1 to 3%) halogen concentration. The most commonly used halogen is chlorine [3]. There are several reasons for using this mixture. Most heavy metal atoms react with Cl to form volatile (i.e., gaseous) metal chlorides. Metallic contaminants are believed to originate from the heating elements and insulation around the fused silica flow tube in which the oxidation is done. The impurities diffuse through the

Table 4.1   Oxidation coefficients for silicon

Temperature (°C)	Dry			Wet (640 torr)	
	A (μm)	*B* (μm²/hr)	*τ* (hr)	*A* (μm)	*B* (μm²/hr)
800	0.370	0.0011	9	—	—
920	0.235	0.0049	1.4	0.50	0.203
1000	0.165	0.0117	0.37	0.226	0.287
1100	0.090	0.027	0.076	0.11	0.510
1200	0.040	0.045	0.027	0.05	0.720

The *τ* parameter is used to compensate for the rapid growth regime for thin oxides. *(After Deal and Grove.)*

wall of the furnace and may be incorporated into the growing oxide. Chlorine has the effect of constantly cleaning the gas ambient of these impurities. Oxides grown in a chlorine ambient are found to have not only fewer impurities, but also a better interface with the underlying silicon. The oxidation rate in O_2/Cl mixtures is higher than in pure O_2. For a 3% concentration of HCl in O_2 the linear rate coefficient doubles [4] (Figure 4.4).

The most commonly used Cl source has been HCl. Trichloroethylene (TCE) and trichloroethane (TCA) are sometimes used, since they are much less corrosive than HCl. TCE has the disadvantage of being carcinogenic. TCA has the potential of forming phosgene ($COCl_2$) at high temperature. Phosgene is a highly toxic substance also known as mustard gas. Stringent safety precautions must therefore be employed in a TCA furnace to prevent the conditions under which phosgene may form.

The Deal–Grove model accurately predicts that the oxidation rate slows as the oxide thickens. Oxides thicker than 1 μm require long exposures to very high temperatures. This will lead to dopant diffusion in the wafer, which is often undesired. In this regime the oxidation rate depends on the parabolic rate coefficient, which in turn depends on C_g, the equilibrium concentration of oxygen in the gas. As shown in the derivation of the model, C_g depends on the partial pressure of the oxygen in the gas phase. Increasing the oxygen pressure in the furnace will increase the parabolic rate coefficient B and can therefore decrease the time or the temperature required for growing thick oxides. Production equipment is available for growing oxides at high pressure. They are most commonly used in bipolar technologies for isolation oxide growth and are almost always done in a wet ambient. In this application the lower oxidation temperature is required to avoid excessive diffusion into the device structure of the heavily doped buried collector. This will be covered in more detail in Chapter 18. Figure 4.5 shows the

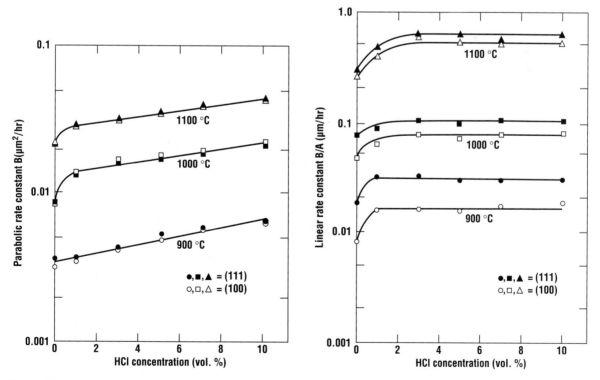

Figure 4.4   The effect of chlorine on the oxidation rate *(after Hess and Deal, reprinted by permission, The Electrochemical Society).*

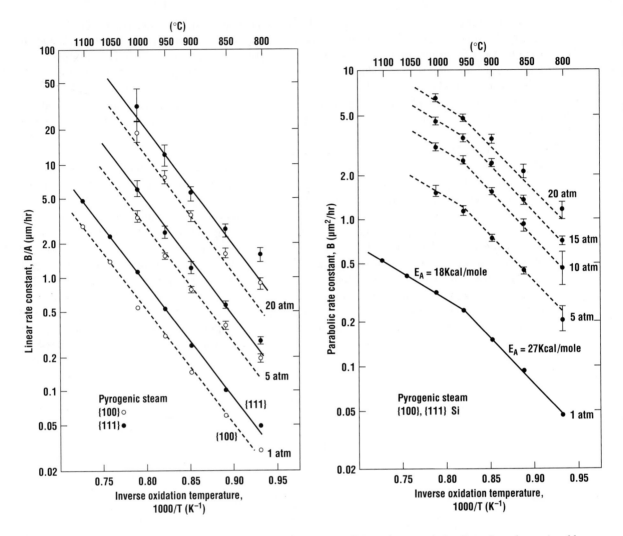

Figure 4.5   High pressure studies of the parabolic and linear rate coefficients in steam *(after Razouk et al., reprinted by permission, The Electrochemical Society).*

| **Example 4.1** | **A simple oxidation** |

Calculate the amount of silicon dioxide grown during a 120-min 920°C steam oxidation. Assume that the wafer initially had 1000 Å of oxide.

Solution

According to Table 4.1, at 920°C, $A = 0.50$ μm and $B = 0.203$ μm²/hr. Inserting these values into Equation 4.12,

$$\tau = \frac{0.1\ \mu m \times 0.1\ \mu m + 0.5\ \mu m \times 0.1\ \mu m}{0.203\ \mu m^2/hr} = 0.295\ hr$$

Using Equation 4.11 and the quadratic formula,

$$t_{ox} = \frac{-A + \sqrt{A^2 + 4B(t + \tau)}}{2} = 0.48 \ \mu m$$

In this case $A \approx t_{ox}$, so that neither linear nor quadratic approximations would be valid.

dependence of the parabolic rate coefficient on temperature with pressure as a parameter. As expected it is proportional to the oxygen pressure [5].

4.3 The Initial Oxidation Regime

The Deal–Grove model fits the oxidation rate over a broad range of parameters, if the linear and parabolic rate coefficients are known. By plotting the experimental thickness versus time for dry oxidation, however, one finds that the curve does not go through zero thickness or even through the native oxide thickness at zero time. Instead, it will typically intersect the zero time axis at a thickness of a few hundred angstroms. Figure 4.6 shows the experimentally measured oxidation rate [6] as a function of oxide thickness for a variety of temperatures. According to the Deal–Grove model, the oxidation rate should approach a constant value

$$\lim_{t \to 0} \frac{dt_{ox}}{dt} = \frac{B}{A} \qquad (4.15)$$

Instead, the oxidation rate increases by a factor of 4 or more. As a result, the Deal–Grove model severely underestimates the thickness of thin oxides. Interestingly, these thin oxides are widely used in one of the most critical applications: the gate dielectric of MOS transistors. The values of τ given in Table 4.1 can be used to correct the Deal–Grove model for dry oxidation to compensate for the excess growth that occurs in this initial growth regime. They should be used for dry oxidations done on bare silicon. With this correction the model accurately predicts thicknesses when $t_{ox} > 300$ Å, but predicts oxides that are thicker than reality for $t_{ox} < 300$ Å. There is no simple model that accurately takes into account rapid growth and an initial oxide.

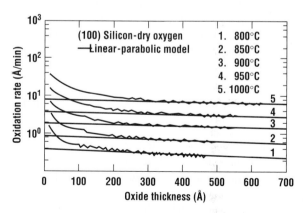

Figure 4.6 Oxide thickness dependence of the growth rate for very thin oxides oxidized in dry O_2. The substrates are lightly doped (100) silicon (*from Massoud et al., reprinted by permission, The Electrochemical Society*).

Several models have been put forward to explain the thin oxidation results. Typically, they postulate one of three mechanisms for the enhanced oxidation rate. The first involves some enhancement of the arrival rate of the oxidant at the interface. Deal and Grove suggested that an electric field exists across the oxide that enhances the diffusion during the early stages of oxidation. In this model, electrons from the substrate charge surface states create a field across the oxide. The diffusing species must be O_2^-. Other authors proposed variations on this basic mechanism [7, 8]. The problem with these models was that the oxidant must be ionized. Later studies found that the dominant diffusing species is neutral molecular oxygen [9]. Earlier, Revesz and Evans [10] had proposed that thin worm holes or microchannels with diameters of order 50 Å exist in the oxide. These holes aid in the movement of

Example 4.2

A bare Si wafer is oxidized in dry O_2 at 1000°C. The desired final thickness is 40 nm. (a) Find the time ignoring rapid growth. (b) Find the time taking rapid growth into account.

Solution

Using Table 3.1,

$$(0.04 \, \mu m)^2 + 0.165 \, \mu m \times 0.04 \, \mu m = 0.0117 \, \mu m^2/hr \, (t + \tau)$$

Solving

$$t + \tau = 0.7 \, hr$$

for (a) $\tau = 0$, so $t = 0.7 \, hr = 42 \, min$.
for (b) $\tau = 0.37$, so $t = 0.33 \, hr = 20 \, min$.

oxygen to the surface of the silicon. A difficulty with this model is that it cannot account for the increase in oxide thickness away from the point at which the microchannel intersects the silicon surface. Finally, it was suggested that the mismatch in the thermal expansion coefficients of the oxide and silicon causes stress in the oxide, and this stress may enhance the diffusivity of the oxidizing species [11]. All of these models share a fundamental flaw. The oxidation rate of silicon in the thin oxide regime is limited by the chemical reaction rate at the surface, not the arrival rate of the molecular oxygen. It is easy to show that the linear rate constant is independent of the diffusivity of the oxygen. It has since been suggested that the reaction rate of ionized molecules may be larger than that of neutral molecules [8] and that both ionized and neutral species may participate in the process [12].

A second class of models attempts to explain the thin oxide growth regime by appealing to an increase in the oxygen solubility in the oxide. Henry's law is valid only in the absence of dissociation or recombination of the adsorbed gas in the solid. This is not true when the oxide is very thin. Unfortunately this model has had very limited success as well. A third class of model suggests that the oxidation reaction occurs over some finite thickness. That is, the interface is not atomically abrupt [13]. The physical origin of the finite thickness is unknown, but it is suggested that some interfacial stress is relieved through an increase in defects near the silicon surface. Oxygen atoms or molecules may diffuse into the silicon slightly and react over these finite widths. The oxidation rate then follows the function

$$\frac{dt_{ox}}{dt} = \frac{B}{2t_{ox} + A} + C_1 e^{-t_{ox}/L_1} + C_2 e^{-t_{ox}/L_2} \tag{4.16}$$

where L_1 and L_2 are characteristic distances over which the reaction occurs and C_1 and C_2 are constants of proportionality. The distance L_1 is weakly temperature dependent and is typically of order 10 Å. The distance L_2 is temperature independent with typical values of about 70 Å. This model has been shown to accurately fit experimental data even for very low temperature oxidations [14]. One can integrate Equation 4.16 by ignoring the C_1 term. Then one can show that

$$t_{ox} = \frac{B}{A}(t + \tau) + L_2 \ln \left[1 + \frac{C_2}{B/A} \left(1 - e^{-B/A \, (t + \tau/L_2)} \right) \right] \tag{4.17}$$

where

$$\tau = L_2/B/A \ \ln \left[\frac{(B/A) \ e^{t_0/L_2} + C_2}{B/A + C_2} \right] \qquad \textbf{(4.18)}$$

and t_0 is the initial oxide thickness. If C_2 and L_2 are known, one can now more accurately predict the growth kinetics of thin oxides.

Example 4.3

One can use the data of Figure 4.6 to estimate C_2 and L_2. Find C_2 and L_2 for 1000°C (100) dry silicon oxidation and estimate the amount of SiO$_2$ grown after 2 min.

Solution

From Equation 4.16, if $L_1 = 0$

$$\lim_{t_{ox} \to 0} \left(\frac{dt_{ox}}{dt} \right) = \frac{B}{A} + C_2$$

From Figure 4.6, for $T = 1000°C$

$$50 \ \text{Å/min} = 9 \ \text{Å/min} + C_2$$

$$C_2 = 41 \ \text{Å/min}$$

To find L_2, we can use the fact that the e^{-t_{ox}/L_2} term decreases by a factor of 2.7 when $t_{ox} = L_2$. Again using Figure 4.6, $L_2 \approx 50 \ \text{Å}$.

Then, using Equation 4.17, and $B/A = 0.071 \ \mu\text{m/hr} = 11.8 \ \text{Å/min}$

$$t_{ox} = 11.8 \ \text{Å/min} \times 2 \ \text{min} + 50 \ \text{Å} \times \ln \left[1 + \frac{41}{11.8} e^{-11.8 \left(\frac{2}{50} \right)} \right]$$

$$= 81 \ \text{Å, compared to 23 Å for simple Deal–Grove.}$$

The excess oxidation shown in Figure 4.6 may actually be occurring at the surface of the oxide, not at the interface [15, 16]. This is believed to be due to the diffusion of defects such as oxygen vacancies from the Si–SiO$_2$ interface to the surface where they react. Delarious et al. [17] developed a detailed chemical model of the reaction, including the production of atomic oxygen at the interface by the chemical reaction of O$_2$. This explanation was supported by isotope tracer experiments by Gusev et al. [18]. In contrast with the Deal–Grove model, the authors found that for very thin oxides the reaction occurs over a range of distance near the silicon and at the top of the oxide.

4.4 The Structure of SiO$_2$

The form of SiO$_2$ that is important for microelectronic fabrication is called *fused silica*. It is amorphous and is thermodynamically unstable below 1710°C. Since, however, SiO$_2$ is a glass, it therefore has a glass transition temperature at which the viscosity of the glass drops sharply. Although still a solid and

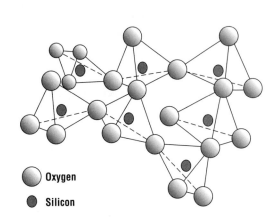

Figure 4.7 The physical structure of SiO$_2$ consists of silicon atoms sitting at the center of oxygen polyhedra.

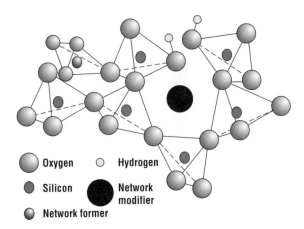

Figure 4.8 Schematic of impurities and imperfections in SiO$_2$.

still SiO$_2$, the material will reflow at these temperatures. In the temperature range of interest to most microelectronic processing, the rate of crystallization is so slow that it can be neglected. Although fused silica does not have long-range order, it does exhibit a short-range structure (Figure 4.7). The structure can be thought of as four oxygen atoms sitting at the corners of a triangular polyhedron. In the center of the polyhedron is a single silicon atom. Thus, each of the four oxygen atoms is approximately covalently bonded to the silicon atom, satisfying the silicon valence shell. If each oxygen atom is part of two polyhedra, the valence of the oxygen would also be satisfied, and a regular crystal structure called quartz would be the result. In fused silica some oxygen atoms, called *bridging oxygen sites*, are bonded to two silicon atoms. Some oxygen atoms are nonbridging, bonded to only one silicon. Thermally grown SiO$_2$ then can be considered to consist primarily of a randomly oriented network of polyhedra. The larger the fraction of bridging to nonbridging sites, the most cohesive and less prone to damage the oxide is. Dry oxides have a much larger ratio of bridging to nonbridging sites compared to wet oxides.

A variety of impurities (Figure 4.8) can also exist in thermal oxides. Some of the most common are water-related complexes. If H$_2$O is present during the growth, one reaction that can occur is the reduction of a bridging oxygen site into two hydroxyls:

$$Si:O:Si \rightarrow Si:O:H + H:O:Si \qquad \textbf{(4.19)}$$

These hydrogen atoms are only weakly bonded and can be removed under electrical stress or ionizing radiation, leaving a trap or potential charged state in the oxide. Other impurities are intentionally incorporated into thermally deposited SiO$_2$ to change its physical and electrical properties. Substitutional impurities that replace the silicon atom are called *network formers*. The most common network formers are boron and phosphorus. These impurities tend to reduce the bridging-to-nonbridging ratio, which allows the glass to flow at lower temperature. These impurities, however, are normally used in deposited oxides rather than in thermal oxides.

4.5 Oxide Characterization

The oxide thickness is an important parameter of the oxidation process, and thus many ways have been developed to measure it. This section will describe several methods for estimating the thickness of an oxide. Each has its inherent advantages and disadvantages. Most make some assumption about the oxide that may be valid only under certain circumstances. Aside from the techniques described here there are a number of thin film measurement techniques that can be used.

The first class of measurements involves a physical determination of the oxide thickness. To do this requires the production of a step in the oxide. Typically this is done with a mask followed by an etch. This type of step was introduced in Chapter 1 and will be reviewed in more detail in later chapters. Hydrofluoric acid (HF) etches oxide at a much higher rate than it etches silicon. Therefore, if a mask is applied to the wafer, the wafer immersed in HF, and then the mask is removed, a step nearly equal to the oxide thickness will be left. This step can be measured using a scanning electron microscope (SEM) if it is larger than 200 Å, or with a transmission electron microscope (TEM) if it is not. An easier approach is to use a surface profilometer, an instrument that measures surface topology by mechanically scanning a needle stylus while it is in contact with the wafer. The deflection of the needle is measured, amplified, and displayed as a function of position. Resolution of these instruments down to 2 Å is claimed by the manufacturers. Similarly, atomic force microscopy (AFM) can be used to measure the step. Profilometry has the advantage that it makes no assumptions other than the relative etch rates of the oxide and the silicon. Since part of the oxide must be etched to determine the thickness, this test is destructive and generally requires the use of a dedicated test wafer.

There are also several optical techniques that can be used to measure oxide thickness. The simplest is to partially immerse the unmasked wafer in dilute HF until the oxide on the submerged portion of the wafer is completely removed. Near the line between the etched and unetched oxide, a slow grading of the thickness will be found. If this edge is examined under a microscope a variety of colors will be seen starting from light brown (Table 4.2). These colors are due to interference between the incident and reflected light. By following the colors up to the top of the oxide an approximate thickness can be found.

There are at least two other, more sophisticated optical techniques for measuring the oxide thickness. The first is *ellipsometry*. In this technique, a polarized coherent beam of light is

Table 4.2   Color chart for silicon dioxide (refractive index of 1.48) and silicon nitride (refractive index of 1.97)

Color	SiO_2 Thickness (Å)	Si_3N_4 Thickness (Å)
Silver	<270	<200
Brown	<530	<400
Yellow-brown	<730	<550
Red	<970	<730
Deep blue	<1000	<770
Blue	<1200	<930
Pale blue	<1300	<1000
Very pale blue	<1500	<1100
Silver	<1600	<1200
Light yellow	<1700	<1300
Yellow	<2000	<1500
Orange-red	<2400	<1800
Red	<2500	<1900
Dark red	<2800	<2100
Blue	<3100	<2300
Blue-green	<3300	<2500
Light green	<3700	<2800
Orange-yellow	<4000	<3000
Red	<4400	<3300

Note that multiple orders exist. An SiO_2 film that appears red may be 730 to 970 Å, 2400 to 2500 Å, or 4000 to 4400 Å.

reflected off the oxide surface at some angle. Helium–Neon lasers are commonly used as a source. The reflected light intensity is measured as a function of the polarization angle. By comparing the incident and reflected intensity and the change in the polarization angle, the film thickness and the index of refraction can be found. To do this definitively requires measurement at more than one incidence angle or for more than one wavelength, since there is more than one thickness that will produce the same change in the light at any given angle or wavelength. Variable angle spectroscopic ellipsometers systematically vary both angle and wavelength and fit the data to a model to extract thickness and index. Ellipsometry has the advantage of being nondestructive, although it often requires that the oxide be grown on bare silicon. Since the ellipsometer's beam is quite large, it is also normally done on unpatterned wafers. It is also common to measure many points across the wafer and map the film thickness.

A second optical technique for measuring oxide thickness is *interference*. Commercially marked under the name of Nanospec®, this system uses light at a nearly normal incidence to the film. The reflected light intensity is measured as a function of the wavelength. When the wavelength of the light is such that the incoming and outgoing waves constructively interfere, an optical maximum will result. If the waves interfere destructively, a minimum will be seen. By measuring the difference in wavelength $\Delta\lambda$ between maxima or minima, the thickness can be determined as

$$t_{\text{ox}} = \frac{\Delta\lambda}{2n_{\text{ox}}}$$

(4.20)

where n_{ox} is the real part of the index of refraction of the oxide; n_{ox} is assumed to be wavelength independent in this model. The technique can measure the thickness of transparent films down to several hundred angstroms. The upper limit for thickness measurement is determined by optical losses in the film and by the ability to resolve high order peaks.

Electrical techniques are the most useful way to characterize oxide layers. The simplest electrical measurement is the breakdown voltage. In this test, a large number of metal electrodes are formed on top of the oxide. The voltage on the capacitor is increased while the electrical current through the oxide film is measured. The leakage current is too small to measure until a high electric field is reached. Eventually, for thin oxides, a current will be detected that will rise exponentially with voltage. Within a small voltage range the current increases discontinuously, signaling an irreversible rupture of the oxide. The dielectric field strength of thermal oxide is about 12 MV/cm. Thus, by knowing the breakdown voltage, one can estimate the oxide thickness. If, as is usually the case, the thickness is known, the breakdown field can be measured. A breakdown histogram is often presented as a first-order indication of oxide quality and defect density (Figure 4.9). Three groups of breakdown regions are evident. The low voltage group is known as extrinsic breakdown. These are killing defects such as pinholes in the growth process. The high voltage group is intrinsic breakdown. They typically cluster near the ultimate breakdown field of the oxide. The intermediate group is usually associated with

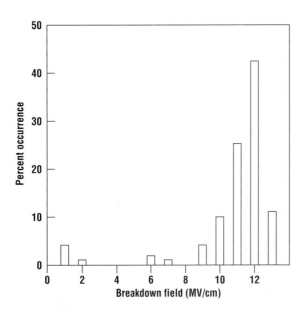

Figure 4.9   Typical breakdown histogram for an oxide.

weak spots in the oxide. The larger the fraction of breakdown events contained in the intrinsic breakdown group, the better the oxide quality. To be significant, the area of the capacitor tested should be similar to the active area on a chip.

A popular variant of the simple breakdown histograms is the charge to breakdown test. In this technique, the oxide is stressed to a point just below the breakdown field. This may be done in constant current, constant voltage, or ramped modes. This test is also known as the time-dependent dielectric breakdown (TDDB) test. For oxides thicker than 50 Å, the current in a TDDB test typically is of the Fowler–Nordheim type and follows the equation

$$J_{FN} = AE_c^2 e^{-B/E_c} \tag{4.21}$$

where E_c is the electric field at the point of tunneling into the oxide and A and B are constants that depend on the height of the barrier and the electron (or hole) effective mass [19]. A typical plot of the resultant curve for a constant voltage test is shown in Figure 4.10. The current through the oxide decreases due to trapping of electrons in the bulk of the oxide. The breakdown is believed to result from an accumulation of trapped positive charge near the interface. The charge to breakdown is measured by integrating the current from the start of the test until just before breakdown. It has been proposed [20] that the oxide is made up of robust and weak areas. The weak areas may be physically thinner than the rest of the oxide or perhaps have defects that easily trap charge.

An interesting aspect of this work is the application of TDDB tests to very thin oxides. A constant current TDDB test for a 62-Å oxide is shown in Figure 4.11. There is very little evidence of bulk electron trapping. This is believed to be due to the ability of the electrons to tunnel from trapped states in the oxide to either the substrate or the gate electrode [21] in a process called *tunnel annealing*. These thin oxides also show very large charge-to-breakdown values. This is believed to result from the small voltages required for stress conditions. For example, a 50-Å oxide requires only 5 V to develop a 10-MV/cm field. At these voltages, electrons in the oxide cannot gain sufficient energy in the oxide to suffer impact ionization and, therefore, to create the holes that eventually trap at the interface. Other less efficient hole generation mechanisms must be used. The area of reliability projections for very thin oxides has evolved considerably in the past decade, however.

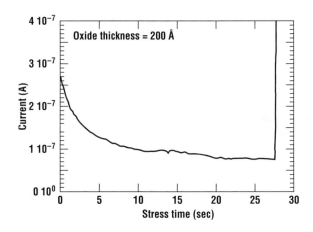

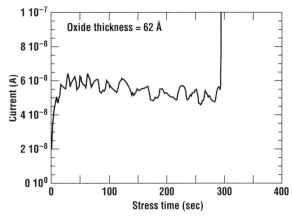

Figure 4.10   Constant voltage stress measurements of a 200-Å oxide film. The sharp increase in current near the end of the trace indicates that irreversible breakdown has occurred.

Figure 4.11   Plot similar to Figure 4.10 for a 62-Å oxide. Note the lack of charge trapping and the low voltage required for stressing.

More sensitive methods of evaluating oxides use capacitance–voltage measurements. Again, a metal film must be used as an upper electrode. The wafer serves as a lower electrode. Assume for the moment that the substrate is doped p-type. If a negative voltage is applied to the gate, additional holes are drawn to the Si–SiO$_2$ interface. This condition is known as accumulation. Now assume that a small ac signal is added to the dc bias and the ac current is measured. The out-of-phase magnitude of this current is proportional to the capacitance. If the diameter of the capacitor is large compared with the oxide thickness, the device can be considered to be a parallel plate capacitor. Then

$$C_{ox} = \frac{\varepsilon_{ox} \text{Area}}{t_{ox}} = \frac{k_{ox}\, \varepsilon_0 \, \text{Area}}{t_{ox}} \tag{4.22}$$

where k_{ox} is the relative permittivity or dielectric constant of the oxide and ε_0 is the permittivity of free space.

Again, this is a valuable method of measuring the oxide thickness. For very thin oxides however, one must also take into account the finite width of the accumulation layer in the semiconductor [14].

If the bias voltage is ramped from the negative set point to positive values, the measured capacitance will decrease. This occurs because the field changes sign, repeling the charge directly below the gate. This effectively increases the width of the dielectric, reducing the capacitance. Another way of viewing the situation is to imagine a second capacitor, which extends from the Si–SiO$_2$ interface to the edge of the depleted region of the substrate, has been formed in series with the oxide capacitor. The voltage at which this depletion layer forms depends on the doping concentration in the substrate and the oxide thickness. If these are known, an ideal C–V curve can be calculated and compared to the predicted value.

All SiO$_2$ layers have a thin layer of positive charge near the SiO$_2$–Si interface. This is known as the "fixed" or "built in" oxide charge. It is believed that the fixed charge is related to a thin (about 30 Å) transition region that lies near the interface and has excess silicon ions. This excess silicon broke away from the lattice during the oxidation process, but had not yet fully reacted with an oxygen molecule. The fixed charge density decreases if an inert anneal is done after the oxidation. A classic demonstration of this effect is the Deal triangle shown in Figure 4.12. Although the data shown are for $\langle 111 \rangle$ silicon, the same effect is seen in $\langle 100 \rangle$ material. The fixed charge density is dramatically decreased if the oxidation is followed by a high temperature inert ambient anneal [22]. For that reason, most oxidation processes include a short nitrogen or argon anneal before the wafers are pulled from the furnace. It has been found that the roughness of the Si–SiO$_2$ interface depends on the oxidation temperature. Higher temperatures produce more abrupt interfaces [23]. This effect has been attributed to the reduction in SiO$_2$ viscosity at high temperature [24].

The effect of the fixed charge is to shift the C–V curve laterally. This will in turn change the threshold voltage at which the MOS transistor "turns on" (see Section 16.1). If the shift is reproducible and is a small fraction of the threshold voltage, this does not represent a serious problem. The fixed charge density can be calculated from this voltage shift using

$$\Delta V_g = -\frac{Q_f}{C_{ox}} \tag{4.23}$$

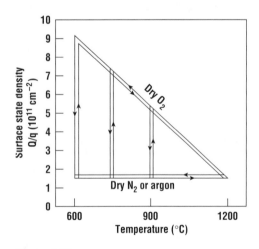

Figure 4.12 The Deal triangle showing the effects of a high-temperature inert (nitrogen or argon) postoxidation anneal on interface states and fixed charge density *(after Deal et al., reprinted by permission, The Electrochemical Society).*

Often the capacitance is normalized to the area of the device. Then the fixed charge density is quoted in units of coulombs per square centimeter. This effect can be difficult to measure, however, since several other effects also serve to shift the C–V trace along the voltage axis.

One of these other sources of lateral shift of the C–V trace is mobile ionic charge, most typically sodium ions. A trace contaminant, sodium has a large mobility in silicon dioxide. These charged ions have the same effect as the oxide fixed charge, except that one must also consider the charge distribution through the oxide. Then the lateral voltage shift is given by

$$\Delta V_g = -\frac{1}{\varepsilon_{SiO_2}} \int_0^{t_{ox}} \rho(x)x \, dx \qquad (4.24)$$

where $\rho(x)$ is the mobile ion density. In the limit that the charge is a delta function at the interface, it is easy to show that this reverts to the formula given for the fixed charge.

The presence of mobile ionic charge represents a serious reliability problem for field effect devices. At 100°C, the diffusion coefficient for Na in SiO_2 is about 3×10^{-15} cm²/sec. If the device is left at temperature for an hour, the characteristic diffusion length is about 3000 Å. Since gate oxides of most MOSFETs are less than a few hundred angstroms, the mobile ionic charge will continuously redistribute itself in the oxide in response to changes in the gate bias. This changes the C–V curve and, consequently, the threshold voltage of the device. A circuit may work initially, but fail after sufficient ionic drift.

Mobile ionic charges are easily detected by temperature bias stressing [25]. An MOS capacitor is first prepared and a C–V trace is taken. The sample is then heated above 100°C and a positive field of 2 to 5 MV/cm is applied to the gate. This bias and temperature are held for 10 to 20 min. The voltage is removed, and the capacitor is cooled to room temperature. Next the C–V trace is retaken. Finally the process is repeated, this time with a negative bias on the gate. Figure 4.13 shows typical results for a sample with positive ionic charge. Initially the ions are randomly distributed through the oxide. Under the influence of a positive gate potential, the ions drift to the interface, where they strongly shift the C–V trace. After a negative drift is applied, the ions move to the gate–SiO_2 interface where they do not affect the C–V trace. Then the mobile ionic sheet charge can be calculated from the voltage shift using Equation 4.24.

The ionic contamination can come from a variety of sources, including the furnaces themselves, wet chemicals used for processing and precleaning, and certain photoresists. Manufacturers have largely solved this problem, and VLSI or ULSI or "semiconductor grade" chemicals are widely available with very low ionic concentrations. The processes that have the most stringent requirements, such as gate oxidation, may use additional precautions such as a purged, double-walled furnace tube. A tube preclean consisting of 1 to 3% of HCl in an inert ambient has also been found to lower contamination levels by reacting the ions to form compounds more readily swept from the tube. Halogen-containing oxidations (see Section 4.2) also reduce mobile ionic charge.

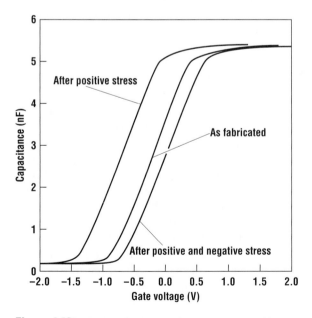

Figure 4.13   Typical C–V traces from temperature bias stress measurements of an oxide contaminated with a positive ion impurity.

Example 4.4

The capacitor used for Figure 4.13 is a square, 1000 μm on a side. (a) Find t_{ox}, (b) Find Q_{MI}, the mobile ionic charge per unit area.

Solution

(a) From Equation 4.22,

$$t_{ox} = \frac{k_{ox}\varepsilon_0 \text{ Area}}{C_{ox}} = \frac{3.9 \times 8.84 \times 10^{-14} \text{ F/cm} \times 10^{-2} \text{ cm}^2}{5.2 \times 10^{-9} \text{ F}}$$

$$= 6.6 \times 10^{-7} \text{ cm}^2 = 66 \text{ Å}$$

(b) From Figure 4.13, $\Delta V_g \approx -0.8$ V between +stress and −stress. From Equation 4.23,

$$Q_{MI} = C_{ox}\, \Delta V_g = \frac{5.2 \times 10^{-9} \text{ F}}{10^{-2} \text{ cm}^2} \times 0.8 \text{ V}$$

$$= 4.2 \times 10^{-7} \text{ C/cm}^2$$

$$Q_{MI} = 2.6 \times 10^{12} \text{ ions/cm}^2$$

where we used the fact that each ion has 1.6×10^{-19} C of charge.

Consider the diagram of the Si–SiO$_2$ interface shown in Figure 4.14 [26]. In the ideal picture the interface is atomically abrupt with perfect purity; however, a more realistic figure also includes oxide fixed charge and mobile ionic impurities. Another common imperfection is interface states. Due to the change from a perfectly ordered crystal to an amorphous solid, not all of the bonds at the interface will be satisfied. Since the surface atoms have incomplete valences and unsatisfied bonds, electrons and holes can readily be captured. These sites are therefore traps, whose energy depends on the local bond configuration. Since this is not fixed, there will be a distribution of trap energies throughout the bandgap. Because these traps are low in energy and are located right at the interface, it is also easy for carriers to detrap. As a result, the traps can fill and empty every time the gate is biased. To understand the magnitude of the effect, consider the fact that the unreconstructed (100) silicon surface has 6.8×10^{14} atoms/cm^2. Assume that the gate oxide is 150 Å thick. Finally, assume that the allowable voltage shift is 0.1 V and that the effect of a charged interface state is the same as that of an oxide fixed charge. Then the sum of the interface trap density and oxide fixed charge density must be less than 2.3×10^{-8} C/cm^2, or less than 1.4×10^{11} defect/cm^2, or about 1 in 5000 sites. The interface therefore, must be highly perfect for typical MOS transistors.

Since the charge state of the interface traps depends on the voltage, their effects on the C–V trace are more

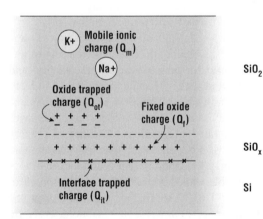

Figure 4.14   Silicon–silicon dioxide structure with mobile, fixed charge, and interface states (© 1980, IEEE, after Deal).

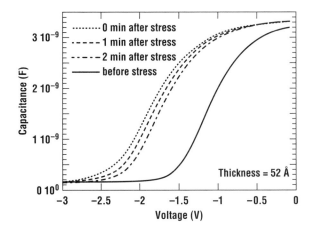

Figure 4.15   High frequency C–V traces showing the effects of interface states and fixed charge.

complicated than those of fixed or mobile ion charge. Figure 4.15 shows the effect of interface states and bulk charge on the C–V trace. After a thin (~50 Å) oxide has been stressed by passing a small current through it, both fixed charge and interface states are visible on the high frequency C–V trace. The fixed charge moves the trace to the left, while the interface states reduce the slope of the C–V trace in depletion. The fixed charge in these particular devices decreases spontaneously through *tunnel annealing*. The most popular techniques of measuring the interface state density all involve taking C–V measurements. In the Terman [27] method, the measured high frequency C–V curve is compared to the ideal curve calculated from the oxide thickness, the metal semiconductor work function, the fixed charge density, and semiconductor doping concentration. Alternatively, one can measure the difference between the high and low frequency C–V traces [28] or the difference between the low frequency and theoretical low frequency curve [29]. None of these calculations is trivial. To extract the interface state density, the doping concentration must also be assumed constant, which is generally not true. The reader is referred to Nicollian and Brews [30] for further information.

4.6 The Effects of Dopants During Oxidation and Polysilicon Oxidation

In nearly all oxidation processes the substrate has some doping concentration. This concentration will change during oxidation. The profile depends on the rate of diffusion of the impurity in the oxide and the segregation coefficient m, where

$$m = \frac{\text{concentration of impurity in Si}}{\text{concentration of impurity in SiO}_2} \qquad \textbf{(4.25)}$$

Redistribution during oxidation occurs in a manner analogous to that during crystal growth.

Grove et al. [31] developed a widely adopted pictorial representation of four classes of impurities shown graphically in Figure 4.16. If $m > 1$, the oxide rejects the impurity. As a result, it accumulates in the silicon under the growing film, reaching a maximum at the interface. This effect can be offset, however, if the impurity diffuses rapidly in the SiO_2. In this case, the dopant is rapidly removed from the interface. The impurity concentration in the substrate at the interface, although larger than in the oxide, is still less than the concentration in the bulk. If $m < 1$, the oxide is said to take up the dopant. Here the impurity concentration in the substrate decreases near the interface.

Figure 4.17 shows the segregation coefficient of boron in Si for a variety of oxidizing conditions [32]. In these experiments, near dry oxidations means an oxidizing furnace that uses O_2 as the feed gas; however, no special provisions have been made to remove water from the ambient. Obviously, the segregation coefficient can vary, but is well described by an Arrhenius function. Phosphorus, arsenic, and antimony all have segregation coefficients of about 10 [33].

The common silicon dopants all tend to enhance the oxidation rate of silicon when present in the substrate in high concentrations. Boron, which segregates into the oxide, enters and weakens the

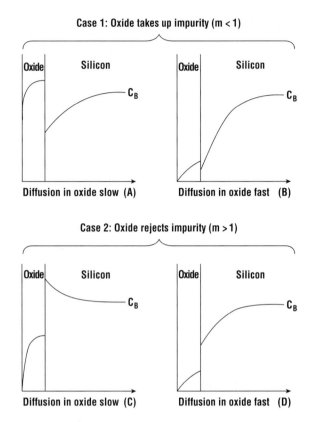

Figure 4.16   The effect of thermal oxidation on the impurity distribution in silicon and silicon dioxide. (A) Slow diffusion in oxide, $m < 1$ (boron in neutral or oxidizing ambient); (B) fast diffusion in oxide, $m < 1$ (boron in hydrogen ambient); (C) slow diffuser in oxide, $m > 1$ (phosphorus, arsenic); (D) fast diffuser in oxide, $m > 1$ (gallium) *(after Grove et al.).*

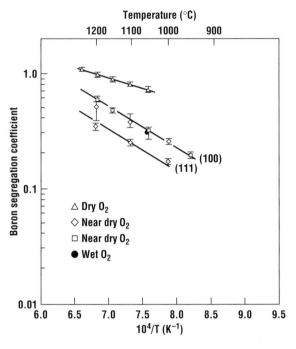

Figure 4.17   Temperature dependence of boron segregation coefficient for various types of oxidations *(reprinted by permission, McGraw-Hill, after Katz).*

glassy structure, reducing its viscosity. When the boron surface concentration exceeds 10^{20} cm^{-3}, this also has the effect of increasing the diffusivity of molecular oxygen. As a result the parabolic rate coefficient increases [34] when the surface is heavily doped with boron (Figure 4.18).

With heavy phosphorus doping the parabolic rate coefficient shows only modest increases, but the linear rate coefficient increases rapidly for surface doping levels greater than 10^{20} cm^{-3} (Figure 4.19) [35, 36]. This is believed to occur because the segregation coefficient causes the phosphorus to accumulate at the surface. A model to explain the increased reactivity of the surface suggests that this high concentration of phosphorus shifts the Fermi energy and thereby increases the surface vacancy concentration [37]. These vacancies provide additional oxidation sites and so increase the reactivity.

The oxidation of polysilicon is of considerable interest for applications such as the production of thin oxides between poly layers in electrically erasable, programmable read-only memories (EEPROMs), the oxidation of polysilicon plugs for DRAMs, and the oxidation of poly gates that occurs during reoxidation steps. It has been found that an enhanced oxidation rate occurs due to stress at the grain boundaries [38]. Figure 4.20 shows the oxide thickness for wet and dry undoped polyoxidations for various temperatures [39]. For oxidation temperatures less than 1000°C, the oxidation rates for undoped polysilicon are larger than either (100) or (111) silicon at short times. For longer oxidations the oxide thickness falls to a level intermediate between (100) and (111) silicon. When the polysilicon is heavily doped with phosphorus, the polyoxidation rate is less than that of either (100) or (111) single crystal silicon.

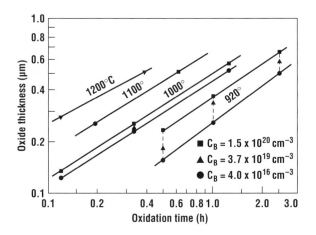

Figure 4.18   Silicon dioxide thickness versus wet oxidation time for three different surface concentrations of boron *(after Deal et al., reprinted by permission, The Electrochemical Society).*

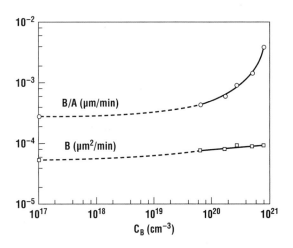

Figure 4.19   Oxidation rate coefficients for dry oxygen at 900°C as functions of the surface concentration of phosphorus *(after Ho et al., reprinted by permission, The Electrochemical Society).*

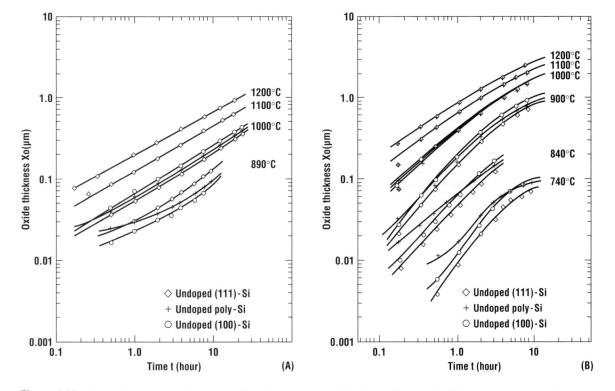

Figure 4.20   The oxidation of undoped polysilicon in (A) wet and (B) dry ambients *(after Wang et al., reprinted by permission, The Electrochemical Society).*

4.7 Silicon Oxynitrides

The last chapter highlighted a very significant problem for small MOS transistors, the diffusion of impurities from the heavily doped polysilicon gate electrode through the very thin gate oxide and into the channel region of the device. The problem is most pronounced for boron, and therefore it is typically called boron penetration. As noted in Chapter 3, the incorporation of nitrogen into silicon dioxides reduces boron diffusivity and therefore helps alleviate this problem. As a result, deeply scaled MOS transistors almost always use an oxynitride rather than pure SiO_2 as the gate dielectric. Nitrogen may be incorporated during the growth process, or after the growth, or both.

In addition to reducing boron penetration, nitrogen can have a few other residual benefits. At high concentration it can increase the dielectric constant. This is beneficial to the performance of the transistor (see Section 4.8). It has also been shown to have the potential to improve oxide reliability. One of the breakdown mechanisms for SiO_2 is the release of atomic hydrogen from bonds within the film. These defects form a percolation path that ultimately leads to a filamentary path through the oxide and ultimately irreversible breakdown. It has been suggested that by decreasing diffusivity, nitrogen makes it less likely that the debonded hydrogen will be lost from the film and more likely that it will be reincorporated in the oxide [40, 41].

Unfortunately, nitrogen also presents significant problems, particularly in the area of charge trapping. Early oxides that were nitrided in ammonia (NH_3) showed a high concentration of electron traps in the bulk of the oxide, even under low field operation [42]. These traps appear to be associated with the formation of hydroxyl (OH) radicals in the film. It has been shown that a reoxidation of the ammonia-nitrided oxides leads to a dramatic reduction in bulk charge trapping [43, 44]. The gates are called reoxidized nitrided oxides (RNO). The films can be grown in a single rapid thermal chamber through a sequence of steps and are attractive for modest concentrations of nitrogen [45]. The RNO process accomplishes another desirable task. Nitrogen near the silicon interface increases the interface state density leading to low carrier mobility in the channel and degraded turn-off characteristics for the transistor [46]. It is desirable, therefore, to have a nitrogen profile in the gate insulator that peaks near the gate electrode and approaches zero near the substrate. A result of reoxidizing in pure O_2, is to grow additional SiO_2, pushing the nitrided oxide away from the channel.

Ito et al. [47] first introduced the idea of nitriding a thermal oxide as an attractive alternative to pure SiO_2. When a thermal oxide is exposed to a high temperature NH_3 anneal, the oxide reacts to form an oxynitride, SiO_xN_y, where x and y depend on the process conditions, but typically $y \ll x$. Ammonia anneals can produce concentrations as high as $y = x$ at the surface that tails off into the bulk of the film, but rises again near the Si/SiO_2 interface. Although N_2 could be used instead of ammonia to avoid this problem, the anneal temperatures are much higher, and very little nitrogen is actually incorporated into the film. More recently there has been work on increasing nitrogen concentration near the surface of the film by very low energy implantation processes (see Chapter 5) such as plasma immersion doping [48]. In this process nitrogen ions are electrostatically accelerated toward the surface of pure SiO_2. As they impact, they become incorporated into the film. Since this is a nonequilibrium process, very high concentrations of nitrogen are possible and, if the ion energy is very low, the nitrogen is peaked near the upper surface of the oxide. The key problem with such a process in ensuring that the film so created does not have too many defects, which will trap charge, and that its properties will survive the various thermal processes needed to build the transistor.

The alternative is the use of nitrous (N_2O) or nitric (NO) oxide instead of, or in addition to, O_2 during the growth process. It is believed that N_2O rapidly decomposes into NO, O_2, and N_2 once

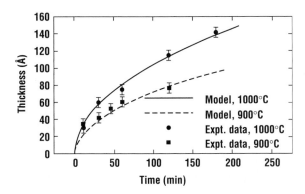

Figure 4.21   Growth of oxynitride in nitric oxide. Reprinted by permission from Journal of Applied Physics. Copyright 2003, American Institute of Physics.

introduced into a furnace [49]. The growth rate is far lower for these films than for pure SiO_2. For thermally grown oxynitrides, there always appears to be a nitrogen-deficient region near the top of the oxynitride. It has been shown that a higher concentration nitrogen region, typically about two-thirds of the film thickness and starting at the silicon interface, forms for growth in pure NO [50, 51], while it forms in the bulk of the film for an NO/O_2 growth. Several authors have attempted to explain the oxidation kinetics. Several have simply modified the Deal–Grove model by noting the reduced diffusivity of species in oxynitride [52, 53]. Dasgupta and Takoudis [54], as well as de Almeida and coworkers [55], developed a set of differential equations to describe the process. They found that a reaction–diffusion model fit the data well and that the growth kinetics and film composition resulted from dynamic effects. Examples of growth rate data and model predictions are shown in Figure 4.21.

4.8   Alternative Gate Insulators[+]

The search for alternative dielectric systems for the MOS gate oxide has a long and until recently not particularly fruitful history. The impetus for an alternative dielectric is usually either a higher dielectric constant for increased current drive at a given dielectric thickness, an increase in the ability of the gate dielectric to withstand electrical stress, or a reduced diffusivity for gate electrode impurities. The penalty that is often paid is an increase in the interface state density that leads to a reduced mobility due to Coulomb scattering and a poorly controlled threshold voltage.

Most of the gate insulators that one might consider are simple dielectrics. The polarization is due to a distortion of the valence shell electrons. A relationship can be developed for the permittivity of these materials [56]:

$$E_g = 20 \left[\frac{3}{\varepsilon + 2} \right]^2 \tag{4.26}$$

Thus increasing the permittivity reduces the bandgap. As the bandgap decreases, it becomes easier to inject charge into the gate insulator, and so the gate leakage current rises sharply. As a result only modest increases in permittivity should be considered for simple dielectrics.

A number of groups are investigating high permittivity layers that might someday be used to replace SiO_2. The reason for this interest is the observation that as the gate insulator is scaled below 3 nm, the gate leakage current density rises sharply. As shown in the measurements of Brar et al. [57] and Buchanan and Ho [58] (Figure 4.22), the current density rises about 2 orders or magnitude for every 0.5 nm decrease in thickness. Current projections are that power limitations will prevent scaling of the SiO_2 below 1.5 nm for mobile applications and about 1 nm for desktop applications. To overcome the thickness-versus-bandgap relationship (Equation 4.26), one must use lattice-polarizable materials. In these films, one or more atoms will shift position in response to an externally applied electric field. The resultant bond stretch produces a sizable dipole that leads to large permittivities.

Replacing SiO₂ with another material is an extremely daunting proposition. This is the heart of the MOSFET, and the semiconductor industry has invested billions of dollars and decades of research to perfect this material and its interface with silicon. A partial list of the requirements includes: chemical stability on silicon, thermal stability up to at least 1000°C, large band offsets for both electrons and holes, a low interface state density, formation of very thin interfacial layer on silicon, low charge trapping, low ionic impurities, good resistance to charge trapping and breakdown, compatibility with chemical vapor deposited polysilicon, low diffusivity of boron and phosphorus at typical processing conditions, and the ability to be etched. Some of these properties depend not only on the material, but on its physical form, how it is deposited, and how it is processed after the deposition. For example, a process that leaves a high concentration of chemical impurities is likely to lead to excessive charge trapping, but it may be cleaned up by a high temperature postdeposition anneal in oxygen. Polycrystalline films are generally considered less desirable than amorphous films, since grain boundaries are likely to be high diffusivity paths for dopants and because oxygen may be depleted from the grain boundaries, leading to trap formation. Amorphous films generally have lower permittivity that polycrystalline films, however, and so may not provide a sufficient leakage advantage.

Initial work in these high-k gate insulators was done by the author of this book using TiO₂ [59, 60]. It was quickly found that only certain materials met the criteria of chemical stability on silicon [61], among them ZrO₂ and HfO₂. The latter, with a permittivity of about 20, has become a leading candidate for gate dielectric applications, but is polycrystalline. Adding silicon dioxide to the film, making it Hf$_x$Si$_{1-x}$O₂, will make the film amorphous, with higher concentrations of silicon required as the maximum processing temperature increases. If one also wants to reduce impurity diffusion, nitrogen can be added, making HfO$_x$N$_y$ [62]. In general these amorphous films have permittivities in the range of 8 to 12.

Many transistors have been demonstrated using these high permittivity films, and in general, the carriers in the channel are found to be less mobile than channels formed under an SiO₂ gate insulator. Careful engineering [63] and the use of extremely thin SiO₂ as a buffer can reduce this effect with mobilities reported to be better than 90% of the ideal value. Figure 4.23 shows a typical performance comparison

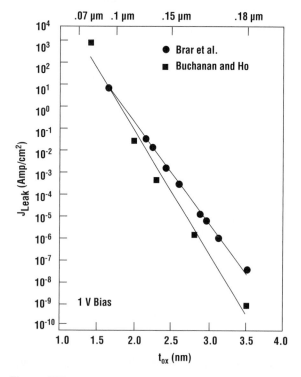

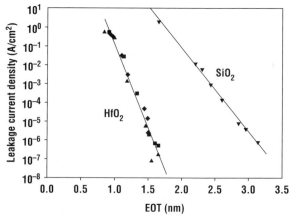

Figure 4.22   Measured leakage current at 1 V supply for different physical oxide thicknesses.

Figure 4.23   Leakage current density as a function of equivalent oxide thickness (EOT) HfO₂ and SiO₂.

[64] for pure HfO_2. The equivalent oxide thickness (EOT) is the thickness of SiO_2 that would give the same capacitance per unit area as the observed value. At this writing HfO_2 can produce usable (i.e., acceptable mobility) films with an EOT of slightly less than 1.0 nm, and $Hf_xSi_{1-x}O_2$ can produce films with an EOT of less than 1.5 nm. First commercial products using HfO_2 will probably appear in 2008 and will be focused on the mobile market where power constraints are the most severe. Metal gates may have to be developed simultaneously to avoid boron penetration problems [65]. Scaling these films to provide an EOT of 0.5 nm, which is the ultimate goal of the CMOS industry, will require new materials beyond HfO_2 such as $HfTiO_4$ [66] or $SrHfO_3$ [67].

4.9 Oxidation Systems

The horizontal diffusion furnace has been the mainstay of the semiconductor industry for decades. Although many innovations have been introduced, the basic idea remains unchanged. As shown in Figure 4.24, the furnace consists of four components: the gas source cabinets; the furnace cabinet, including power supplies, tubes, thermocouples, and elements; the load station; and the computer controller. As the name implies, the gas cabinet contains the source gases, typically in high pressure cylinders, and means for producing a controlled flow of these gases into the furnace. Typically, sets of pressure regulators, pneumatic valves, and gas filters are used in this application. The gases are plumbed to a shelf at the end of the furnace tube. This shelf contains additional valves and mass flow controllers (see Chapter 10) that control the flow of gases into the furnace tube.

The furnace cabinet houses long tubes, typically four, that are stacked vertically. Horizontal furnaces designed for 200-mm wafers typically have three tubes. Furnace heating elements are wound on high purity ceramic forms large enough to accept a 150- to 300-mm-diameter fused silica tube. The windings are often separated into three heating zones and fed by three-phase power through a high current power controller. The furnace temperature is monitored in at least three locations with thermocouples. The thermocouple voltages are read and compared to the desired temperatures, and the error signal is amplified and fed back to the power controllers. It is not unusual for the flat zone in the center of the furnace to exhibit 0.5°C uniformity over temperature ranges of 400 to 1200°C. If wet oxidations are to be done, additional equipment is also required to ensure that the hydrogen gas completely combusts upon injection.

The wafers are loaded into fused silica holders called *boats* in the load station. The boats typically hold 25 wafers in a vertical rack. The boats are loaded onto carriers that can hold up to 200 wafers. Early carriers were actually fused silica carriages, with skids or fused silica wheels that were slowly rolled into the furnace. Concerns with particulation, however, caused the replacement of these carriers by cantilevered load systems in which the boats are supported on two long rods. In these systems, the wafers never touch the furnace walls. Due to difficulty maintaining the rod linearity, many of these systems have now been replaced by soft landing systems. In these

Figure 4.24 Horizontal tube oxidation/diffusion system includes computer controller, load station, and four tubes *(photo courtesy of ASM International).*

stations, the loading system carries the boats into the furnace, deposits them, and then withdraws, minimizing the time at temperature for the boat support system. In each system care must be taken to avoid jostling the wafers during this process to avoid defect formation.

A microcomputer controls all of the furnace operations. Specific recipes are programmed for each process step. Disc drives are provided to store these recipes for future recall. For example, a gate oxidation step might include a preoxidation tube clean in a mixture of O_2 and HCl at 1100°C for 60 min, followed by a nitrogen purge and cool down to 800°C, followed by a slow push, or loading of the wafers in a mixture of O_2 and N_2, followed by a controlled furnace ramp to 1000°C, followed by an oxidation in O_2 and HCl, followed by a short N_2 anneal to 1050°C to reduce the oxide fixed charge, followed by a controlled cool down to 800°C, followed by a slow pull, or removal of the wafers. The microcomputer allows this to be a "one button" operation for highly reproducible results. It can also be programmed to anticipate thermal loads such as the paddle insertion by ramping the temperature just before loading to minimize temperature fluctuations. The temperature control systems are now completely digital and interface well with the furnace controller. The furnace controller also can be interfaced to a facility management computer. With the use of robotic load stations, this allows the diffusion furnaces to be completely automated.

Vertical furnaces are often used for large diameter wafers. As shown in Figures 4.25 and 4.26, these furnaces are similar to a horizontal furnace turned on end. Wafers are pushed from below the

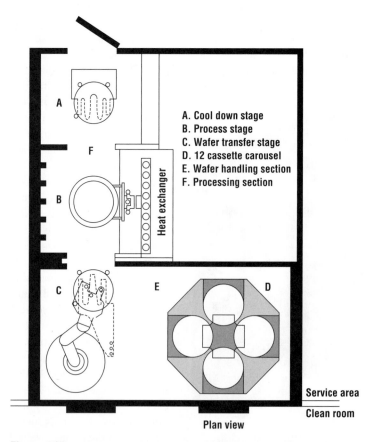

A. Cool down stage
B. Process stage
C. Wafer transfer stage
D. 12 cassette carousel
E. Wafer handling section
F. Processing section

Plan view

Figure 4.25   Plan view of vertical furnace. Cassettes with 25 wafers are loaded into the carousel and transferred to the process stage, where they are raised into the furnace (*courtesy of ASM International*).

Figure 4.26 Vertical oxidation system showing carousel, robot arm, and wafers about to be raised into the furnace (*photo courtesy of ASM International*).

furnace up into the tube. These systems have four main advantages. The vertical orientation means that the wafers remain horizontal. As a result, these furnaces are easier to automate. Robot wafer handlers can easily load and unload the tubes. Furthermore, the uniform spacing improves process uniformity across the wafer, particularly for large-diameter substrates. Also, there is no need to deposit the wafers in the tube. There is no net force on the cantilevers except down, so they do not tend to warp with usage and can remain in the furnace. Finally, vertical furnaces tend to have a smaller clean room footprint than horizontal furnaces.

A third type of system finding some acceptance is called minibatch of fast ramp furnaces. These systems typically hold 20 to 50 wafers and can ramp up at 100°C/min and down at 50°C/min without slip [68]. This compares to approximately 10°C/min in a conventional furnace.

4.10 Numeric Oxidations[+]

The last chapter introduced numerical calculation to simulate the diffusion profiles of impurities in silicon. The model can also be used to perform oxidations based on the Deal–Grove model. The tool has incorporated Arrhenius functions to describe the linear and parabolic rate coefficients for wet and dry silicon coefficients, as well as a rudimentary model for chlorinated oxidation.

Oxidation processes are accessed by the same command as diffusion processes: DIFFUSION. For oxidation to occur, simply add the parameter DRYO2 or WETO2 for dry and wet oxidations, respectively. The parameters for these processes can be adjusted before invoking the DIFFUSION statement, with the DRYO2 and WETO2 commands. High-pressure oxidations can be accomplished by also adding the parameter PRESSURE $= x$, where x is the oxygen or steam pressure in atmospheres. The temperature can be ramped by using the parameter T.RATE $= x$, where x is the time rate of change of the pressure in atmospheres per minute. For chlorinated oxidations the parameter is HCl $= x$, where x is the percentage of HCl in the ambient. For diluted flows one can also use the commands F.H2 $= x$, F.H2O $= x$, F.HCl $= x$, F.N2 $= x$, and F.O2 $= x$, where the x values correspond to the gas flows in standard liters per minute of the corresponding gases.

In the thin oxide regime, the program uses an empirical oxidation model [69]

$$\frac{dt_{ox}}{dt} = \frac{B}{2t_{ox} + A} + Ce^{-t_{ox}/L} \tag{4.27}$$

where B and A are the oxidation rate coefficients from the Deal–Grove model, and C and L are empirical constants. Notice that this equation closely follows the model of Massoud for $t_{ox} > L_1$, which is about 10 Å. Impurity segregation coefficients and oxidation-enhanced diffusion models are also incorporated in the code so that impurity redistribution during oxidation can be modeled.

Example 4.5

In the program that follows, find the oxide thickness after each step and the boron segregation coefficient. Which of the four categories in Figure 4.16 applies to boron? Ignoring the temperature ramps, how does the final thickness compare to the prediction of the Deal–Grove model?

```
go athena
#TITLE: Furnace Gate Oxidation (Thick)
line x loc=0.0          spacing=0.02
line x loc=0.2          spacing=0.02
line y loc=0.0          spacing=0.02
line y loc=0.1          spacing=0.02
line y loc=0.3          spacing=0.04
line y loc=0.6          spacing=0.04
line y loc=0.8          spacing=0.06
line y loc=1.5          spacing=0.10
init c.boron=1e17
method adapt

#              Push the wafers at 800 C
Diffuse      Temperature=27 T.rate=25.7667 Time=30
#extract oxide thickness
extract name="tox1" thickness material="oxide" mat.occno=1
y.val=0.1
#              Ramp the furnace to 1000 C
Diffuse      Temperature=800 T.rate=20 Time=10 F.N2=1.8 F.O2=0.2
#extract oxide thickness
extract name="tox2" thickness material="oxide" mat.occno=1
y.val=0.1
#              Grow gate oxide
Diffuse      Temperature=1000 DryO2 HCl=3.0 Time=30
#extract oxide thickness
extract name="tox3" thickness material="oxide" mat.occno=1
y.val=0.1
#              Ramp the furnace to 800 C
Diffuse      Temperature=1000 T.rate=-20 Time=10 F.N2=1.8
F.O2=0.2
#extract oxide thickness
extract name="tox4" thickness material="oxide" mat.occno=1
y.val=0.1
#              Pull the wafers at 800 C
Diffuse      Temperature=800 T.rate=-25.7667 Time=30
#extract oxide thickness
```

```
extract name="tox" thickness material="oxide" mat.occno=1
y.val=0.1

#  Save and plot the final structure
structure outfile=ex4_5
tonyplot
quit
```

Solution

The oxide thickness increases from 0 after the first ramp (which was done in an inert ambient) to 2.95 nm after the ramp to 1000°C, to 40.1 nm after the 1000°C cycle in dry O_2 and HCl, and finished at 40.3 nm. For Deal–Grove at 1000°C, $A = 0.165$ μm, $B = 0.0117$ μm²/hr, and $\tau = 0.37$. The presence of chlorine increases B slightly; however, for pure O_2, Deal–Grove predicts a thickness of 47.8 nm. From the impurity profile shown in Tonyplot, the concentration of boron on the oxide side of the Si/SiO$_2$ interface is about 3×10^{17} cm^{-3} while on the silicon side it is about 9×10^{15} cm^{-3}, giving a segregation coefficient of 0.03. Since the concentration in the silicon decreases only modestly as one approaches the interface, case (A) is the correct diagram.

4.11 Summary

This chapter introduced the topic of the thermal oxidation of silicon, presenting the Deal–Grove model. This model accurately predicts the oxide thickness of a wide range of oxidation parameters. Enhanced growth rates are seen for thin oxides. Although their origin is uncertain, several models that attempt to explain the results were presented. Oxidation is also known to induce defects in the bulk silicon because of the high concentration of self-interstitials that the process produces. Techniques for avoiding these defects are described. Finally, typical oxidation systems are described and the application of numerical simulation to oxidation is presented.

Problems

1. A 1000-Å gate oxide is required for some technology. It has been decided that the oxidation will be carried out at 1000°C, in dry oxygen. If there is no initial oxide, for how long should the oxidation be done? Is the oxidation in the linear regime, the parabolic regime, or between the two?

2. Repeat Problem 1 if the oxidation is done in a wet O_2 ambient.

3. It has been decided to grow the oxide in Problem 1 in two steps. A 500-Å oxide will be grown first, then the wafers will be reoxidized to a total thickness of 1000 Å. If the oxidations are carried out at 1000°C in dry O_2, calculate the time required for each of the oxidations.

4. It is necessary to grow a 1-μm field oxide to isolate the transistors in a certain bipolar technology. Due to concerns with dopant diffusion and stacking fault formation, the oxidation must be carried at 1050°C. If the process is carried out in a wet ambient at atmospheric pressure, calculate the required oxidation time. Assume that the parabolic rate coefficient is proportional to the oxidation pressure. Calculate the oxidation time required at 5 and at 20 atm.

5. It has been shown that when some technologies are used to grow very thin gate oxides, SiO$_2$ grown with water vapor may have better reliability than films grown in dry oxygen.

(a) Estimate the time (in seconds) that it would take to grow 30 Å of SiO_2 on a bare silicon wafer at 920°C with 640 torr of water vapor. (1 angstrom $= 10^{-4}$ μm).

(b) In an effort to make the process more controllable, the pressure of the water in the chamber is reduced to 76 torr (0.1 atm). Now how long would it take (in minutes) for the same oxidation?

6. Silicon oxidation is carried out using a new oxidant that shows no rapid growth regime but follows the Deal–Grove equation for all times. For short times the oxidation rate is constant at 1.0 nm/sec for 1000°C and at 0.5 nm/sec at 900°C.

(a) From the Deal–Grove model, find an expression for the initial oxidation rate in terms of some of the following: D, k_s, H, P_g, N_1. You can assume $k_s < h$.

(b) What would you expect the oxidation rate to be at 800°C?

7. A gate oxide is grown in 640 torr of water vapor at 1000°C. The oxide is grown on a bare (100) silicon wafer. The oxidation time is 2 mins.

(a) Assuming the tau (τ) is zero, use the Deal–Grove model to predict the oxide thickness that would be grown.

(b) When this process is run, the oxide thickness is actually found to be 600 Å (60 nm), but the oxidation rate after 2 mins agrees with the B/A ratio. Find the value of τ that should be used for these process conditions to take into account the rapid growth regime.

8. An oxidation of a bare silicon wafer is to be done in one atmosphere of dry O_2 at 1000°C. The desired thickness for this oxide is 400 Å (40 nm or 0.04 μm).

(a) Find the time needed for this oxidation if one ignores rapid growth effects.

(b) Find the time needed for this oxidation taking into account rapid growth effects.

9. For a particular CMOS process, you need to grow a gate oxide that is 2 nm thick (0.002 micron). You must do this in a furnace at 1000°C with an oxidation time of 10 mins. Assume that the wafer is initially bare and that rapid growth effects are not important. What dry oxygen partial pressure should you use to do the oxidation?

Use Figure 4.6 to estimate the time it would take with one atmosphere of dry oxygen taking into account rapid growth effects (The minor tick marks on the vertical axis represent 2, 4, 6, and 8.) You can assume that the oxidation rate does not change during the process since the oxide is so thin. Note that 1 nm $= 10$ Å.

10. A silicon wafer, which is a doped 10^{15} p-type, is heated to 1100°C for 1 hr. If the wafer is in a dry oxygen atmosphere, how much oxide will grow? Be sure to include rapid growth effects.

11. Assume that negatively charged O_2^- has exactly twice the diffusion coefficient in SiO_2 as neutral O_2 (due to the field-aided term), but 10 times the reactivity at the surface, with exactly the same activation energy for the reaction rate coefficient. Repeat Problem 1 in a source of 1 atm of O_2^-.

12. For submicron MOSFETs, it is often necessary to grow gate oxides of order 100 Å. Although process control is very difficult due to the short oxidation times involved, it is preferable to grow these oxides at high temperatures. Explain why.

13. One technique that has been used with some success to grow very thin oxides is to grow the oxides in a dilute mixture of oxygen and an inert species such as argon. Assume that we have a mixture containing 10% O_2 and 90% Ar. If we ignore the rapid oxidation regime normally associated with thin oxides and assume, as in Problem 4, that the parabolic rate coefficient B depends on the pressure of the oxidant, calculate the growth time for a 1000°C oxidation of 100 Å. Assume no initial oxide.

14. Derive Equations 4.17 and 4.18.

15. Assume that L_2 is 70 Å and C_2 is 50 Å/min. If the 100-Å oxide described in Problem 13 is grown at 1000°C in a dry oxygen ambient, and there is no initial oxide, determine the oxidation time both with and without the thin oxide rate enhancement term from the equation derived in

Problem 14. Are both of the limits required for the derivation in Problem 14 strictly satisfied? If not, qualitatively describe the effect.

16. Repeat Problem 15 for a growth done in an ambient of 10% oxygen and 90% argon.

17. A thermal oxide thickness is measured both by use of a Nanospec and by measuring the accumulated capacitance. The results are found to differ by 20%, even though the same wafer was used for both measurements. Give three possible errors that might account for the discrepancy.

18. A 250-Å gate oxide is found to have a 15-mV temperature bias stress shift. Calculate the number of mobile ions per unit area in the oxide.

19. Modify Example 4.5 to simulate 30-min wet oxidations at 920, 1000, and 1100°C. Compare the calculated oxide thickness with the Deal–Grove predictions. Calculate the segregation coefficient from your results and compare it to the graph in Figure 4.17. Change the substrate concentration to 1.5×10^{20} cm^{-3} of boron and rerun the 920°C simulation. Compare the results to those of Figure 4.18. What does this tell you about the model used by your software?

20. Use your software to grow several dry oxides at 1000°C for 1, 2, 5, and 10 min. Calculate the oxide thickness as a function of time. Does your code attempt to model the initial oxidation regime? If not, how can a program user increase the accuracy of moderate thickness (200 Å < t_{ox} < 1000 Å) dry oxidation runs?

References

1. B. E. Deal and A. S. Grove, "General Relationship for the Thermal Oxidation of Silicon," *J. Appl. Phys.* **36**:3770 (1965).

2. M. M. Aptyalia, in *Properties of Elemental and Compound Semiconductors*, H. Gates, ed., Interscience, New York, 1960, p. 163.

3. R. S. Ronen and P. H. Robinson, "Hydrogen Chloride and Chlorine Gettering: An Effective Technique for Improving Performance of Silicon Devices," *J. Electrochem. Soc.* **119**:747 (1972).

4. D. W. Hess and B. E. Deal, "Kinetics of Thermal Oxidation of Silicon in O$_2$/HCl Mixtures," *J. Electrochem. Soc.* **124**:735 (1977).

5. R. R. Razouk, L. N. Lie, and B. E. Deal, "Kinetics of High Pressure Oxidation of Silicon in Pyrogenic Steam," *J. Electrochem. Soc.* **128**:2214 (1981).

6. H. Z. Massoud, J. D. Plummer, and E. A. Irene, "Thermal Oxidation of Silicon in Dry Oxygen: Growth Rate Enhancement in the Thin Regime," *J. Electrochem. Soc.* **132**:2685 (1985).

7. M. Hamasaki, "Effect of Oxidation-Induced Oxide Charges on the Kinetics of Silicon Oxidation," *Solid State Electron.* **25**:479 (1982).

8. S. M. Hu, "Thermal Oxidation of Silicon," *J. Appl. Phys.* **55**:4095 (1984).

9. D. N. Modlin, Ph.D. Dissertation, Stanford University, Stanford, CA, 1983.

10. A. G. Revesz and R. J. Evans, "Kinetics and Mechanism of Thermal Oxidation of Silicon with Special Emphasis on Impurity Effects," *J. Phys. Chem. Solids* **30**:551 (1969).

11. A. Fargeix, G. Ghibaudo, and G. Kamarinos, "A Revised Analysis of Dry Oxidation of Silicon," *J. Appl. Phys.* **54**:2878 (1983).

12. R. B. Beck and B. Majkusiak, "The Initial Growth Rate of Thermal Silicon Dioxide," *Phys. Stat. Sol.* **A116**:313 (1989).

13. H. Z. Massoud, J. D. Plummer, and E. A. Irene, "Thermal Oxidation of Silicon in Dry Oxygen Growth-rate Enhancement in the Thin Oxide Regime—II. Physical Mechanisms," *J. Electrochem. Soc.* **132**:2693 (1985).

14. K. H. Lee, W. H. Liu, and S. A. Campbell, "Growth Kinetics and Electrical Characteristics of Thermal Silicon Dioxide Grown at Low Temperature," *J. Electrochem. Soc.* **140**:501 (1993).

15. C. J. Han and C. R. Helms, *J. Electrochem. Soc.* **134**:1299 (1987).

16. F. Rochet, B. Agius, and S. Rigo, "An ^{18}O Study of the Oxidation Mechanism of Silicon in Dry Oxygen," *J. Electrochem. Soc.* **131**:914 (1984).

17. J. M. Delarious, C. R. Helms, D. B. Kao, and B. E. Deal, "Parallel Oxidation Model for Si Including Both Molecular and Atomic Oxygen Concentrations," *Appl. Surf. Sci.* **39**:89 (1989).

18. E. P. Gusev, H. C. Lu, T. Gustafsson, and E. Garfunkel, "The Initial Oxidation of Silicon: New Ion Scattering Results in the Ultra-thin Regime," *Appl. Surf. Sci.* **104/105**:329 (1996).

19. R. H. Fowler and L. W. Nordheim, "Electron Emission in Intense Electric Fields," *Proc. R. Soc.* **A119**:173 (1928).

20. C. Hu, S. C. Tam, F.-C. Hsu, P.-K. Ko, T.-Y. Chan, and K. W. Terrill, "Hot-Electron-Induced MOSFET Degradation—Model, Monitor, and Improvement," *IEEE Trans. Electron Dev* **ED-32**:375 (1985).

21. K. H. Lee and S. A. Campbell, "The Kinetics of Charge Trapping and Oxide Trap Recombination in Ultrathin Silicon Dioxide," *J. Appl. Phys.* **73**:4434 (1993).

22. B. E. Deal, M. Sklar, A. S. Grove, and E. H. Snow, "Characteristics of the Surface State Charge of Thermally Oxidized Silicon," *J. Electrochem. Soc.* **114**:266 (1967).

23. M. T. Tang, K. W. Evans-Lutterodt, M. L. Green, D. Brasen, K. Krisch, L. Manchanda, G. S. Higashi, and T. Boone, "Growth Temperature Dependence of the Si(001)/SiO$_2$ Interface Width," *Appl. Phys. Lett.* **64**:748 (1994).

24. E. P. EerNisse, "Stress in Thermal SiO$_2$ During Growth," *Appl. Phys. Lett.* **35**:8 (1979).

25. E. H. Snow, A. S. Grove, B. E. Deal, and C. T. Sah, "Ion Transport Phenomena in Insulating Films," *J. Appl. Phys.* **36**:1664 (1965).

26. B. E. Deal, "Standardized Terminology for Oxide Charges Associated with Thermally Oxidized Silicon," *IEEE Trans. Electron Dev.* **ED-27**:606 (1980).

27. L. M. Terman, "An Investigation of Surface States at a Silicon/Silicon Dioxide Interface Employing Metal-Oxide-Silicon Diodes," *Solid State Electron.* **5**:285 (1962).

28. R. Castagne and A. Vapaille, "Description of the SiO$_2$-Si Interface Properties by Means of Very Low Frequency MOS Capacitance Measurements," *Surf. Sci.* **28**:157 (1971).

29. C. N. Berglund, "Surface States at Steam-Grown Silicon-Silicon Dioxide Interfaces," *IEEE Trans. Electron Dev.* **ED-31**:701 (1966).

30. E. H. Nicollian and J. R. Brews, *Metal Oxide Semiconductor Physics and Technology*, Wiley, New York, 1982.

31. A. S. Grove, O. Leistiko, and C. T. Sah, "Redistribution of Acceptor and Donor Impurities During Thermal Oxidation of Silicon," *J. Appl. Phys.* **35**:2695 (1964).

32. R. B. Fair and J. C. C. Tsai, "Theory and Direct Measurement of Boron Segregation in SiO$_2$ in Dry, Near Dry, and Wet O$_2$ Oxidation," *J. Electrochem. Soc.* **125**:2050 (1978).

33. A. S. Grove, *Physics and Technology of Semiconductor Devices*, Wiley, New York, 1967.

34. B. E. Deal and M. Sklar, "Thermal Oxidation of Heavily Doped Silicon," *J. Electrochem. Soc.* **112**:430 (1965).

35. L. E. Katz, "Oxidation," in *VLSI Technology*, S. M. Sze, ed., McGraw-Hill, New York, 1988.

36. C. P. Ho, J. D. Plummer, J. D. Meindl, and B. E. Deal, "Thermal Oxidation of Heavily Phosphorus Doped Silicon," *J. Electrochem. Soc.* **125**:665 (1978).

37. C. P. Ho and J. D. Plummer, "Si–SiO$_2$ Interface Oxidation Kinetics: A Physical Model for the Influence of High Substrate Doping Levels. I. Theory," *J. Electrochem. Soc.* **126**:1516 (1979).

38. J. C. Bravman and R. Sinclair, "Transmission Electron Microscopy Studies of the Polycrystalline Silicon-SiO$_2$ Interface," *Thin Solid Films* **104**:153 (1983).

39. Y. Wang, J. Tao, S. Tong, T. Sun, A. Zhang, and S. Feng, "The Oxidation Kinetics of Thin Polycrystalline Silicon Films," *J. Electrochem. Soc.* **138**:214 (1991).

40. T. Ito, H. Arakawa, T. Nozaki, and H. Ishikawa, "Retardation of Destructive Breakdown of SiO$_2$ Films Annealed in Ammonia Gas," *J. Electrochem. Soc.* **127**:2248 (1980).

41. F. L. Terry, R. J. Aucoin, M. L. Naiman, and S. D. Senturia, "Radiation Effects in Nitrided Oxides," *IEEE Electron Dev. Lett.* **EDL-4**:191 (1983).

42. S. K. Lai, D. W. Dong, and A. Hartenstein, "Effects of Ammonia Anneal on Electron Trapping in Silicon Dioxide," *J. Electrochem. Soc.* **129**:2042 (1982).

43. S. K. Lai, J. Lee, and V. K. Dham, "Electrical Properties of Nitrided-oxide Systems for Use in Gate Dielectrics and EEPROM," *IEDM Tech. Dig.*, 1983, p. 190.

44. T. Hori and H. Iwasaki, "Ultra-thin Re-oxidized Nitrided-oxides Prepared by Rapid Thermal Processing," *IEDM Tech. Dig.*, 1987, p. 570.

45. T. Hori, H. Iwasaki, and K. Tsuji, "Electrical and Thermal Properties of Ultrathin Reoxidized Nitrided Samples Prepared by Rapid Thermal Processing," *IEEE Trans. Electron Dev.* **36**:340 (1989).

46. M. L. Green, D. Brasen, L. Feldman, E. Garfunkel, E. P. Gusev, T. Gustafsson, W. L. Lennard, H. C. Lu, and T. Sorsch, "Thermal Routes to Ultrathin Oxynitrides," in *Fundamental Aspects of Ultrathin Dielectrics in Si-Based Devices*, E. Garfunkel, E. P. Gusev, and A. Y. Vul', eds., Kluwer, Dordrecht, Netherlands, 1998.

47. T. Ito, T. Nozaki, and H. Ishikawa, "Direct Thermal Nitridization of Silicon Dioxide Films in Anhydrous Ammonia Gas," *J. Electrochem. Soc.* **127**:2053 (1980).

48. S. Fukuda, Y. Suzuki, T. Hirano, T. Kato, A. Kashiwagi, M. Saito, S. Kadomura, Y. Minemura, and S. Samukawa, "Ultra Shallow Incorporation of Nitrogen into Gate Dielectrics by Pulse Time Modulated Plasma," Materials Research Society Symposium Proceedings, *Fundamentals of Novel Oxide/Semiconductor Interfaces Symposium*, **786**:239–244 (2003).

49. K. A. Ellis and R. A. Buhrman, "Furnace Gas-Phase Chemistry of Silicon Oxynitridation in N$_2$O," *Appl. Phys. Lett.* **68**:1696 (1996).

50. S. S. Dang and C. G. Takoudis, "Optimization of Bimodal Nitrogen Concentration Profiles in Silicon Oxynitrides," *J. Appl. Phys.* **86**:1326 (1999).

51. M. L. Green, E. P. Gusev, R. DeGraeve, and E. Garfunkel, "Ultrathin (less than or equal to 4 nm) SiO$_2$ and Si-O-N Gate Dielectric Layers for Silicon Microelectronics: Understanding the Processing, Structure, and Physical and Electrical Limits," *J. Appl. Phys.* **90**:2057 (2001).

52. W. Ting, H. Hwang, J. Lee, and D. L. Kwong, "Growth Kinetics of Ultrathin SiO$_2$ Films Fabricated by Rapid Thermal Oxidation of Si Substrates in N$_2$O," *J. Appl. Phys.* **70**:1072 (1991).

53. Z. Q. Yao, H. B. Harrison, S. Dimitrijev, D. Sweatman, and Y. T. Yeow, "High Quality Ultrathin Dielectric Films Grown on Silicon in a Nitric Oxide Ambient," *Appl. Phys. Lett.* **64**:3584 (1994).

54. A. Dasgupta and C. G. Takoudis, Growth Kinetics of Thermal Silicon Oxynitride in Nitric Oxide Ambient," *J. Appl. Phys.* **93**(6):3615 (2003).

55. R. M. C. de Almeida, I. J. R. Baumvol, J. J. Ganem, I. Trimaille, and S. Rigo, "Thermal Growth of Silicon Oxynitride Films on Si: A Reaction–diffusion Approach," *J. Appl. Phys.* **95**(4):1770 (2004).

56. S. A. Campbell, D. C. Gilmer, X. Wang, M. T. Hsich, H. S. Kim, W. L. Gladfelter, and J. H. Yan, "MOSFET Transistors Fabricated with High Permitivity TiO$_2$ Dielectrics," *IEEE Trans. Electron Dev.* **44**:104 (1997).

57. B. Brar, G. D. Wilk, and A. C. Seaburgh, "Direct Extraction of the Electron Tunneling Effective Mass in Ultrathin SiO$_2$," *Appl. Phys. Lett.* **69**:2728 (1996).

58. D. A. Buchanan and S.-H. Lo, "Growth, Characterization and the Limits of Ultrathin SiO$_2$ Based Dielectrics for Future CMOS Applications," in *The Physics and Chemistry of SiO$_2$ and the Si–SiO$_2$ Interface—III*, H. Z. Massoud, E. H. Poindexter, and C. R. Helms, eds., Electrochemical Society, Pennington, NJ, 1996, p. 3.

59. H. S. Kim, S. A. Campbell, D. C. Gilmer, and D. L. Polla, "Leakage Current and Electrical Breakdown in TiO$_2$ Deposited on Silicon by Metallorganic Chemical Vapor Deposition," *Appl. Phys. Lett.* **69**:3860 (1996).

60. S. A. Campbell, D. C. Gilmer, X. Wang, M. T. Hsieh, H. S. Kim, W. L. Gladfelter, and J. H. Yan, "MOSFET Transistors Fabricated with High Permittivity TiO$_2$ Dielectrics," *IEEE Trans. Electron Dev.* **44**:104 (1997).

61. K. J. Hubbard and D. G. Schlom, "Thermodynamic Stability of Binary Oxides in Contact with Silicon," in *Epitaxial Oxide Thin Films II*, Vol. 401, J. S. Speck, D. K. Fork, R. M. Wolf, and T. Shiosaki, eds. MRS, Pittsburgh, 1996, pp. 33–38.

62. T. Ino, Y. Kamimuta, M. Suzuki, M. Koyama, and A. Nishiyama, "Dielectric Constant Behavior of Hf–O–N system," *Jpn. J. Appl. Phys.*, Part 1: Regular Papers and Short Notes and Review Papers **45**(4 B):2908–2913 (Apr. 25, 2006).

63. S. A. Campbell, T. Z. Ma, R. Smith, W. L. Gladfelter, and F. Chen, "High Mobility HfO$_2$ N- and P-Channel Transistors," *Microelectron. Eng.* **59**(1–4):361–366 (2001).

64. Z. Zhang, B. Xia, W. L. Gladfelter, and S. A. Campbell, "The Deposition of Hafnium Oxide from Hf *t*-butoxide and Nitric Oxide," *J. Vacuum Sci. Technol.* A **24**(3):418–423 (May/June 2006).

65. J. H. Lee, Y. S. Suh, H. Lazar, R. Jha, J. Gurganus, Y. Lin, and V. Misra, "Compatibility of Dual Metal Gate Electrodes with High-*p* Dielectrics for CMOS," *IEDM, Tech. Dig.*, 2003, p. 323–326.

66. F. Chen, B. Xia, C. Hella, X. Shi, W. L. Gladfelter, and S. A. Campbell, "A Study of Mixtures of HfO$_2$ and TiO$_2$ as High-*k* Gate Dielectrics," *Microelectron. Eng.* **72**(1–4):263–266 (2004).

67. I. McCarthy, M. P. Agustin, S. Shamuilia, S. Stemmer, V. V. Afanas'ev, and S. A. Campbell, "Strontium Hafnate Films Deposited by Physical Vapor Deposition," *Thin Solid Films,* **515**(4):2527–2530 (2006).

68. P. Singer, "Furnaces Evolving to Meet Diverse Thermal Processing Needs," *Semicond. Int.* **40**:84 (1997).

69. H. Z. Massoud, C. P. Ho, and J. D. Plummer, in *Computer-Aided Design of Integrated Circuit Fabrication Processes for VLSI Devices*, J. D. Plummer, ed., Stanford Univ. Tech. Rep., 1982.

Chapter 5

Ion Implantation

The introduction of dopant impurities by predeposition diffusion was described in Chapter 3. In this process, dopant was diffused into the semiconductor from an infinite source at the surface of the wafer. The surface concentration was limited by the solid solubility, and the depth of the profile was determined by the time and diffusivity of the dopant. In principle, it seems that a more lightly doped profile could be achieved if the supply of dopant at the surface of the wafer is appropriately limited. For example, a very dilute mixture of dopant in an inert carrier gas can reduce the surface concentration. This process was used in early microelectronic technologies; however it was found very difficult to control. It was also found that the lightly doped profiles were often the most critical. The base of a bipolar transistor and the channel of a MOSFET are two examples of moderately doped profiles that must be very well controlled, since they determine the gain and threshold voltage, respectively.

In ion implantation, ionized impurity atoms accelerated through an electrostatic field strike the surface of the wafer. The dose can be tightly controlled by measuring the ion current. Doses for the process range from 10^{11} cm^{-2} for very light implants to 10^{16} cm^{-2} for low resistance regions such as source/drain contacts, emitters, and buried collectors. Some specialized applications require doses of more than 10^{18} cm^{-2}. By controlling the electrostatic field, the penetration depth of the impurity ions can also be controlled. Ion implantation therefore, provides the capability to tailor the dopant profile in the substrate to some extent. Typical ion energies range from 1 to 200 keV. Certain special applications, including forming deep structures such as retrograde wells (see Chapter 16), can require energies up to several megaelectron-volts.

After considerable research demonstrations during most of the 1960s, the first commercial implanters were introduced in 1973. Despite initial reluctance, the new method of dopant introduction soon became ubiquitous. By 1980, most processes were fully implanted. Although now widely used, ion implantation also has several drawbacks. The incident ions damage the semiconductor lattice. This damage must be repaired and in some cases complete repair cannot be done. Very shallow and very deep profiles are difficult or impossible. The throughput of ion implanters is limited for high dose implants, particularly compared to diffusion processes in which 200 wafers can often be run simultaneously. Finally, ion implantation equipment is expensive. A state-of-the-art system costs more than $4,000,000.

The process that this chapter describes provides a blanket dose that is essentially uniform across the wafer. To selectively dope regions of the wafer, an implant mask must be used. A variation of standard ion implantation is to focus the ion beam to a small spot and use this spot to provide a localized process capability. These processes are called *focused ion beam techniques*. For example, the ions can be used directly to provide lateral variations in the doping profile across a device. This method is too expensive and too slow to find widespread application in manufacturing. Ion beams are used, however, to repair mask defects and to selectively remove layers for diagnostic work.

5.1 Idealized Ion Implantation Systems

Ion implantation systems can be divided into three components (Figure 5.1): the ion source, the acceleration tube, and the end station [1]. The ion source (Figure 5.2) starts with a feed gas that contains the desired implant species. Common feed gases for use in silicon technologies are BF_3, AsH_3, and PH_3. For GaAs technologies common gases are SiH_4 and H_2. Most implanters set up for gaseous sources will allow any of several gases to be selected by opening the proper valve. The flow of gas can be controlled with a variable orifice. If the desired implant species is not available in gaseous

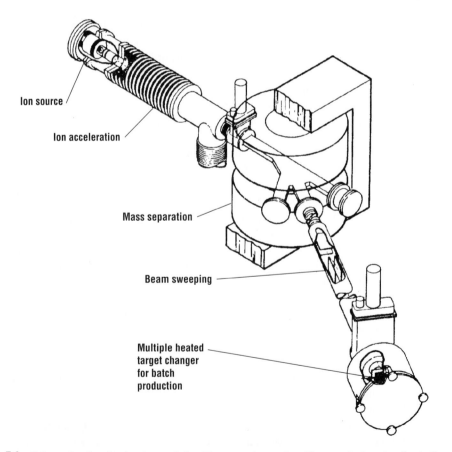

Ion source

Ion acceleration

Mass separation

Beam sweeping

Multiple heated
target changer
for batch
production

Figure 5.1   Schematic of an ion implanter *(after Mayer et al., reprinted by permission, Academic Press).*

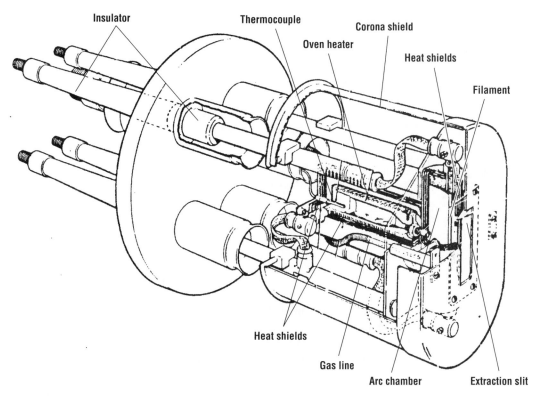

Figure 5.2 Schematic for a Freeman ion source. Solids can be vaporized in the oven heater, while gaseous sources may be injected directly into the arc chamber, newer system's use a Bernas ion source *(reprinted by permission, Elsevier Science, after Freeman in Mayer et al.).*

form, as shown in Figure 5.2, a solid charge can be heated and the resultant vapor used as the source. The material is heated in the oven and the vapor flows past the filament. For gaseous precursors, the oven is replaced with a simple gas feed. At high pressures, the electron current is often sufficient to maintain a glow discharge (see Chapter 10).

The gas flows into an arc chamber. This chamber has two purposes: to break up the feed gas into a variety of atomic and molecular species and to ionize some of these species. In the simplest such systems the feed gas flows through an orifice into the low pressure source chamber where it passes between a hot filament and a metal plate. The filament is maintained at a large negative potential with respect to the plate. Electrons boil off the filament and are accelerated toward the plate. As they do so they collide with feed gas molecules, transferring some of their energy. If the transferred energy is large enough, molecular dissociation can occur. For example, BF_3, breaks up into B, B^+, BF_2, BF_2^+, F^+, and a variety of other species in varying quantities. Negative ions may also be produced, but are less abundant. To improve the ionization efficiency, a magnetic field is often imposed in the region of the electron current. This produces a spiral path for the electrons, dramatically increasing the ionization probability. The positive ions are attracted to the exit side of the source chamber, which is biased at a large negative potential with respect to the filament. The positive ions then exit the source chamber through a slit. The resulting ion beam is often a few millimeters by 1 to 2 cm. The gas pressure in this part of the source is typically 10^{-5} to 10^{-7} torr, resulting in a stable arc between the filament and the anode. Maximum ion currents are typically a few milliamperes.

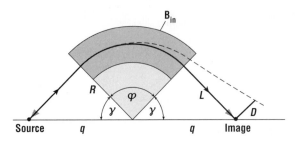

Figure 5.3 Mass separation stage of an ion implanter showing perpendicular magnetic field and ion trajectory: D corresponds to the displacement for an ion of $M+\delta M$.

The beam now consists of a variety of species, most of them ionized. The next task is to select the desired implant species. In the previous example, we may want to select only the B^+ from the beam and prevent the other species from continuing down the implanter. This is normally done with an analyzing magnet (Figure 5.3). The beam enters a large chamber that is also maintained at low pressure. A magnetic field perpendicular to the beam velocity exists in this chamber. Balancing force and acceleration,

$$\frac{Mv^2}{r} = qvB \qquad (5.1)$$

where v is the magnitude of the ion velocity, q is the charge on the ion, M is the ion mass, B is the magnetic field intensity, and r is the radius of curvature. If the ion obeys classical mechanics,

$$v = \sqrt{\frac{2E}{M}} = \sqrt{\frac{2qV_{ext}}{M}} \qquad (5.2)$$

where V_{ext} is the extraction potential. The derivation of Equation 5.2 ignores the energy imparted to the ions by collisions with electrons in the source. The spread of ion energies due to this effect is of order 10 eV. Since the extraction potential is typically 2 or 3 orders of magnitude larger than this, Equation 5.2 provides a good approximation of the true energy.

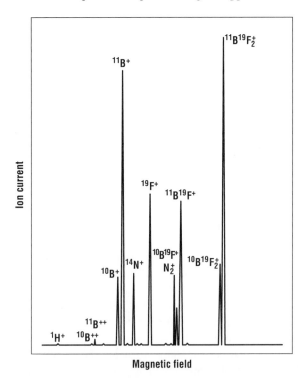

Another slit can be made in the mass analysis chamber such that only one mass (or more accurately one charge-to-mass ratio) will have exactly the correct radius of curvature to exit the source (Figure 5.4). The primary limitation to the resolution of such a system is the fact that the beam has some small divergence. This arises due to the finite slit size and small variations in the ion energy. As demonstrated in Figure 5.4 however, such a system can readily distinguish between the isotopes of boron, ^{11}B and ^{10}B[2].

Combining Equations 5.1 and 5.2,

$$r = \frac{Mv}{qB} = \frac{1}{B}\sqrt{2\frac{M}{q}V_{ext}} \qquad (5.3)$$

Normally, the analyzing field is everywhere perpendicular to the ion velocity, and the inlet and outlets are symmetric. Now assume that the field has been tuned in such a way as to allow an ion of mass M to exactly follow a circle of radius R. If an ion of mass $M + \delta M$ enters the filter the beam will be displaced by a distance [3]:

$$D = \frac{R}{2}\frac{\delta M}{M}\left[1 - \cos\phi + \frac{L}{R}\sin\phi\right] \qquad (5.4)$$

Figure 5.4 Typical mass spectrum for a BF$_3$ source gas (after Ryssel and Ruge, reprinted by permission Wiley).

Example 5.1   **Mass resolution**

If an analyzing magnet bends the ion beam through 45° and $L = R = 50$ cm, find the displacement D that would be seen if ^{10}B is sent through the system when it is tuned for ^{11}B. If the extraction potential is 20 kV, find the required field.

Solution

Since the difference in mass is 1 amu,

$$\frac{\delta M}{M} = \frac{1}{10} = 0.1$$

and

$$D = \frac{1}{2} \, 50 \text{ cm} \, \frac{1}{10} \, [1 - \cos 45° + \sin 45°] = 2.5 \text{ cm}$$

Since slit widths are typically a few millimeters, this analyzer could easily resolve a one-amu difference. Even so, heavier mass species will be more difficult to resolve and may require a larger radius magnet. According to Equation 5.3,

$$B = \frac{1}{0.50 \text{ m}} \sqrt{2 \, \frac{10 \times 1.67 \times 10^{-27} \text{ kg}}{1.6 \times 10^{-19} \text{ C}} \, 2 \times 10^4 \text{ V}} = 0.13 \text{ T} = 1.3 \text{ kG}$$

where the units of field are tesla (T) and kilogauss (kG).

Two masses are said to be resolved when D is larger than the width of the beam plus the width of the exit slit. The best resolution occurs when R is large and M is small. (Due to beam divergence, L has less effect than this simple analysis would suggest as long as it is of order a meter or more.) Most mass filters used for IC fabrication have filters with a radius of a meter or less.

Acceleration of the ions can be done either before or after the mass analysis. Accelerating first reduces the likelihood that ions will lose their charge before reaching the surface of the wafer, but requires a much larger magnet. The following discussion will assume analysis before acceleration. This tube can be several meters long and must be maintained at a relatively high vacuum ($<10^{-6}$ torr). This is necessary to avoid collisions during acceleration. The beam is first focused into either a spot or a ribbon using a set of electrostatic lenses. It then enters a linear electrostatic accelerator. The accelerator creates an electric field along the length of the tube to change the ion energy. In the event that the desired energy is less than the extraction potential, an opposing or bucking potential can be applied. This reduces the stability and spatial coherence of the beam however, and so is not often used.

The beam at this point is primarily composed of ions. Some neutrals may have reappeared. Commonly, the neutrals are ions that combined with thermal electrons:

$$^{11}\text{B}^+ + \text{e}^- \rightarrow \, ^{11}\text{B} \tag{5.5}$$

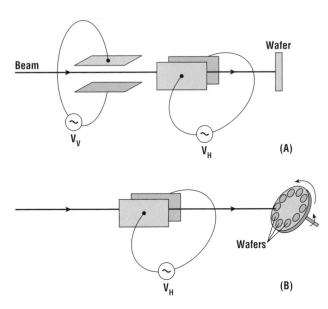

Figure 5.5   Typical scanning systems for ion implanters. (A) Electrostatic rastering commonly used in medium current machines. (B) Semielectrostatic scanning used on some high current machines.

They may also be ions that collided with other ions in the beam undergoing a charge exchange. Neutrals are highly undesirable, since they will not be deflected in electrostatic end station scanning mechanisms. They will travel in a beam to the wafer where they will continuously implant near the center of the wafer. To avoid this problem, most ion implantation systems are equipped with a bend. The beam passes between the parallel plates of an electrostatic deflection system. The neutrals are not deflected and so do not follow the bend, but instead strike a beam stop. The ions are sufficiently deflected by the plates to continue to travel down the tube.

Near the end of the tube are additional sets of deflection plates (Figure 5.5). In many implanters both horizontal and vertical pairs of plates are used in a manner analogous to a television set. The beam is rastered back and forth and up and down, writing uniformly across the wafer. In systems designed for high dose implantation, the beam is rastered in only one direction. Horizontal rastering is done by mechanically moving the wafer past the beam. This is done by placing a number of wafers on the perimeter of a spinning disc. The wafers may be clamped, or as is now common, held by centrifugal force. There are a number of advantages to such a system. The individual wafer does not have as large a thermal load, since a batch of them are implanted at the same time. The angle of the beam with respect to the wafer remains unchanged in the mechanically scanned direction. The total pump downtime is reduced by loading a batch of wafers simultaneously. In many such systems, two wheels are mounted in a side-by-side configuration. This allows one wheel to load and pump down while the other is being implanted.

In some modern high current machines, the wafers are also scanned mechanically in the radial direction as well. Electrostatic scanning requires the formation of an intense narrow beam. Because the beam current in high current machines is large, the increased electron density will cause the beam to expand or bloom. These space-charge effects lead to poorly controlled electrostatic scanning.

Finally, one needs to control the dose of the implant. This is done in the end station by placing the wafer in a Faraday cup. The cup is simply a cage that captures all of the charge that enters it. The ion current into the wafer is measured directly by connecting an ammeter between the Faraday cup and ground. The dose is just the time integral of the current divided by the wafer area. Electrical current is easy to measure over a broad range of values. For a 200-mm. wafer, typical values for implant current range from 1 μA to tens of milliamperes. To accurately measure the dose, one must guard against errors due to secondary electron ejection. This process will be described in more detail in Chapter 10. It involves the creation of large numbers of electrons, many of which have sufficient energy to escape the wafer, when a high energy ion strikes the surface of the wafer. To prevent secondary electron dose errors, the wafer is biased with a small positive voltage. This bias, typically tens of volts, is sufficient to attract all of the secondary electrons back to the surface of the wafer where they are reabsorbed.

The ion current may represent a considerable source of energy to the wafer

$$\text{energy} = \int \text{power } dt = \int IV \, dt = V \int I \, dt = VQ \qquad \textbf{(5.6)}$$

where Q is the charge delivered. As an example, consider a typical source/drain implantation for an FET. This would involve 2×10^{15} atoms/cm^2 at an energy of 10 keV. Photoresist is commonly used as the implant mask. The energy deposited in a 200-mm wafer is about 1 kJ. The energy is not deposited uniformly through the wafer, but rather in the top 200 Å or so. Since photoresist is a poor thermal conductor, it can get quite hot during an implant. The photoresist may flow or be baked to such an extent that it is difficult to remove after the implant. This is particularly true at the edge of the wafer, where a thicker bead of photoresist can form. To avoid these problems, the end stations often come with provisions to cool the wafer and thereby control the temperature. The other problem often seen in implanting through a resist mask is outgassing. The ions striking the surface break apart the organic molecules in the resist leading to the formation of gaseous H_2 that evolves from the surface, leaving behind involatile carbon [4]. Efficient removal of hydrogen by the vacuum pump is an important consideration. Heavily implanted resist layers often have a hardened carbonized layer near the surface that is difficult to remove later. The outgassing can raise the pressure in the end station sufficiently to cause neutralization of the beam through impact with the H_2 molecules, resulting in significant dose rate errors [5].

5.2 Coulomb Scattering°

Coulomb scattering is usually discussed during the study of classical mechanics. This topic is very relevant to any discussion of ion implantation. The typical scattering experiment is shown for the laboratory frame of reference in Figure 5.6. One usually thinks of the atoms in the wafer as having a neutral charge. The incident ions, however, are traveling at energies that allow them to penetrate the electron cloud around the nucleus of the atoms in the wafer. As a result, the target ions are more appropriately treated as charged ions with a screening cloud of electrons. The target ion is initially at rest in the laboratory frame of reference. An ion approaches with an incident velocity v and an impact parameter b (the distance of closest approach between the centers of the two atoms if no scattering were to occur). The target ion is assumed to be free. Once again, this is a good approximation for the high incident ion energies commonly used in microelectronic fabrication.

The objective is to calculate the amount of energy transferred to the target atom during the collision. The solution of this problem requires the conservation of energy, momentum, and angular momentum. The calculation is more involved than is warranted for our purposes and normally involves a translation into center-of-mass coordinates. It is covered in many undergraduate classical mechanics texts. Let us simplify the problem by considering two hard spheres that scatter elastically. If p_i and p_t are the final momenta of the incident and target spheres and p_o is the initial momentum of the incident sphere,

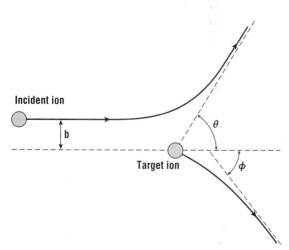

Figure 5.6   Typical scattering problem. Figure inset shows electrostatic potential as a function of distance between the nuclei. The impact parameter is labeled b.

$$\bar{p}_i + \bar{p}_t = \bar{p}_o \qquad \textbf{(5.7)}$$

Conserving angular momentum,

$$L_i + L_t = L_o = p_o b \tag{5.8}$$

Conserving energy,

$$\frac{p_i^2}{2m_i} + \frac{p_t^2}{2m_t} = \frac{p_o^2}{2m_o} \tag{5.9}$$

Then one can show that the energy lost by the incident sphere is

$$\Delta E = E_o \left[1 - \frac{\sin^2 \phi}{\cos \theta \sin \phi + \cos \phi \sin \theta} \right] \tag{5.10}$$

Thus, the energy loss is proportional to the incident energy and depends on the scattering angles. These angles depend on the ion masses and the impact parameter.

5.3 Vertical Projected Range

This section will use the scattering results to discuss what happens to the energetic ions in an implanter once they reach the surface of the wafer. Some models that have been developed to describe this process will be discussed qualitatively, and the difficulties of ion implantation in modern devices will be pointed out. When an energetic ion enters a solid, it will begin to lose energy. The distance that the ion travels in the semiconductor is its *range*, R (Figure 5.7). As discussed in the previous section, the energy loss depends on the impact parameter. Since the ion is entering a solid, the ions in the beam will have a range of impact parameters. As a result, the nature of the energy loss mechanism can, to a good approximation, be considered probabilistic. For a given flux of ions a range of distributions will result. For a uniform beam the quantity of interest is not the total distance traveled, but rather the average depth. This quantity is the projected range R_p. The energy loss in the target material is the result of two mechanisms [6]. The first is ion–electron interactions. This involves not only the valence electrons, but also the core electrons of the host material. Since a great deal of the space in the crystal is made up of the electron clouds from the atoms, many of these interactions will occur. Even if the electron is not in the path of the ion, energy may be transferred through Coulomb interaction. For a typical semiconductor implant, hundreds of thousands of these interactions occur. Furthermore, the mass ratio between the ion and an electron is of order 10^5. Any single electron ion interaction will not dramatically alter the momentum of the incident ion.

Due to the large numbers and small individual effects involved, these discrete interactions can be approximated by a continuum mechanism. That is, the effect of the electrons on the ion is very much like a

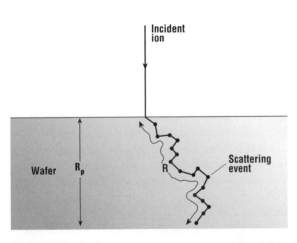

Figure 5.7   The total distance that an ion travels in the solid is the range. The projection of this distance along the depth axis is the projected range, R_p.

particle moving through a fluid. A viscosity due to the electrons can be assigned to the crystal media. In that model, there is a viscous drag force assumed to be proportional to the velocity for energies of interest to typical microelectronic applications:

$$F_D \propto v \propto \sqrt{E} \tag{5.11}$$

Since the ion must do work to move through the media, this force can also be expressed as an energy gradient. The energy loss per unit length due to electronic stopping is given the symbol S_e and

$$S_e = \left.\frac{dE}{dx}\right|_e = k_e\sqrt{E} \tag{5.12}$$

where k_e is a constant of proportionality that depends on the ion and target species as [7]

$$k_e \propto \sqrt{\frac{Z_iZ_t}{M_i^3M_t}\frac{(M_i + M_t)^{3/2}}{(Z_i^{2/3} + Z_t^{2/3})}} \tag{5.13}$$

In Equation 5.13, Z is the charge number (number of protons) and M is the mass (neutrons plus protons) for the incident and target ions. At very high energies, the viscous fluid model can no longer be invoked. Instead, S_e peaks and decreases as the incident ion energy is raised further.

The ion interaction with the lattice ions is very different from the interaction with the electrons. Consider for the time being an amorphous solid. It is known experimentally that ions penetrate semiconductor crystals to depths of thousands of angstroms. Since the atomic spacing is of the order of angstroms, about a thousand interactions can occur. Furthermore, the incident and target ions have masses of the same order of magnitude. It is possible for the incident ion to be scattered at a large angle relative to its incident velocity. As a result, nuclear interactions cannot be treated as a continuum. Instead, they must be treated as a series of discrete events. As described in the previous section, the angle at which the ion is scattered will depend on the impact parameter and on the masses and relative positions of the two ions. This means that the result of any interaction depends on all of the interactions that occurred previously, back to the first atomic layer of the solid. Since the ions are uniformly distributed over the surface of the wafer as they enter, a statistical distribution of depths will result.

To first order, Gaussian distributions can be used to model the range of depths that an ion might reach. Thus, the impurity concentration as a function of depth in an amorphous solid will be given by

$$N(x) = \frac{\phi}{\sqrt{2\pi}\Delta R_p}e^{-(x - R_p)^2/2\Delta R_p^2} \tag{5.14}$$

where R_p is the projected range, ΔR_p is the standard deviation of the projected range, and ϕ is the dose. The normalizing factors are chosen such that the integral of the profile from $x = 0$ to ∞ gives the dose. Due to the statistical nature of the ion implantation process, the ions will also be scattered laterally, penetrating past the edges of the mask. The profile must therefore be considered in two dimensions, with both lateral and vertical standard deviations.

To find R_p one must determine the S_n, the energy loss of the incident ion per unit length of travel due to nuclear stopping. The theory of nuclear stopping is involved. The reader, if interested, is referred to Dearnaley *et al.* [3] for an excellent treatment. It can be qualitatively described by the charged sphere model described in Section 5.2. The energy transfer during any collision will be a sensitive function of the impact parameter. The lower the impact parameter, the greater the energy loss. The average energy loss will also be a function of the ratio of the mass of the ion to that of the target atom. The smaller this ratio, the larger the average energy loss per collision will be. Finally, the average energy loss will be a function of the energy itself. The solid atom is chemically bonded. It is sitting in an approximately parabolic potential minimum. At ion low energies the average collision does not transfer enough energy to break this bond. The result is an almost elastic collision. The ion may change direction, but it will not lose much energy. Thus, S_n is expected to increase with ion energy at low energy. The momentum transfer is given by

$$\Delta p = \int F \, dt \tag{5.15}$$

At high velocities the collision time becomes so short that the energy loss decreases. Thus, S_n has a maximum at some energy. The maximum value of S_n can be approximated by the relation

$$S_n^o \approx 2.8 \times 10^{-15} \text{ eV-cm}^2 \frac{Z_i Z_t}{Z^{1/3}} \frac{M_i}{M_i + M_t} \tag{5.16}$$

where

$$Z = [Z_i^{2/3} + Z_t^{2/3}]^{3/2} \tag{5.17}$$

and S_n has a weak energy dependence. Figure 5.8 shows the nuclear and electronic components of $S(E)$ for several common silicon dopants as a function of energy [8, 9].

Once $S_n(E)$ and $S_o(E)$ are known, one can perform the integral

$$R_p = \int_0^{R_p} dx = \int_{E_o}^0 \frac{dE}{dE/dx} = \int_{E_o}^0 \frac{dE}{S_n + S_e} \tag{5.18}$$

to get the projected range. Once known, the projected range can be used to estimate [10] ΔR_p:

$$\Delta R_p \cong \frac{2}{3} R_p \left[\frac{\sqrt{M_i M_t}}{M_i + M_t} \right] \tag{5.19}$$

To avoid these complications, the projected range and its standard deviation are often obtained from LSS (Lindhard, Scharff, and Schiøtt) tables [11] or Monte Carlo simulations. These tables represent calculated values that assume a particular model for the electron density distributions of the ion and the solid atoms. The range values predicted by this model agree well with experimental results for concentrations close to the maximum. Figure 5.9 shows the projected range and implant straggle for several commonly implanted impurities [12].

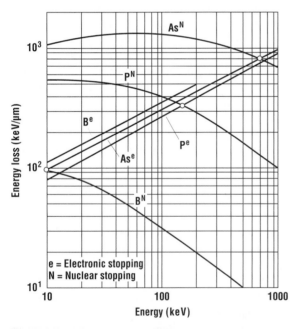

Figure 5.8   Nuclear and electronic components of $S(E)$ for several common silicon dopants as a function of energy (after Smith as redrawn by Seidel, "Ion Implantation," reproduced by permission, McGraw-Hill, 1983).

Example 5.2

Use Figure 5.8 to find $S_n(E)$ and $S_e(E)$ for a 100-keV As implant and calculate the projected range for $E = 100$ keV.

Solution

From Figure 5.8, $S_n \approx 1.2 \times 10^3$ keV/μm, while

$$S_e(E) \approx 30 \frac{\text{keV}^{1/2}}{\mu m}\sqrt{E}$$

Then

$$R_p = \int_{E_0}^{0} \frac{dE}{1.2 \times 10^3 + 30\sqrt{E}}$$

$$= 2\left[\frac{\sqrt{E}}{30} - \frac{1.2 \times 10^3}{(30)^2}\ln(30\sqrt{E} + 1.2 \times 10^3)\right]\Bigg|_{E_0}^{0}$$

$$= z\left\{1.33\ln(1.2 \times 10^3) - \left[\frac{\sqrt{100}}{30} - 1.33\ln(300 + 1.2 \times 10^3)\right]\right\}$$

$$R_p = 0.036 \ \mu m$$

The actual value is larger, presumably due to the rolloff in S_n at low energies.

Additional moments can be added to the profile to help describe the behavior at lower concentrations (Figure 5.10). The ith moment of a distribution is defined by

$$m_i = \int_0^{\infty} (x - R_p)^i N(x)\, dx \tag{5.20}$$

The first moment is just the normalized dose. The second moment is the product of the dose and ΔR_p^2. The third moment is related to the asymmetry of the distribution. This asymmetry is usually expressed in terms of the skewness, γ, where

$$\gamma = \frac{m_3}{\Delta R_p^3} \tag{5.21}$$

Negative values for the skewness represent an increased concentration on the surface side of the distribution, that is, for $x < R_p$. As an example, when implanting boron into silicon, higher concentrations are found near the surface than would be predicted by Equation 5.14. This occurs because boron is much lighter than silicon and so will suffer significant backscattering. As a result, boron has large negative values of skewness, particularly at high energy, where the effect is most pronounced. The fourth moment of the distribution is related to a distortion of the Gaussian peak. The distortion is expressed by the kurtosis β, where

$$\beta = \frac{m_4}{\Delta R_p^4} \tag{5.22}$$

The larger the kurtosis, the flatter the top of the Gaussian (Figure 5.10). Normal Gaussians have a kurtosis of 3. The values of γ and β can be found by performing Monte Carlo simulations [13], or more directly, by measuring actual profiles and fitting the results.

Boron in silicon is a rather special case. This distribution is most often described not by a modified Gaussian, but instead by a Pearson type IV distribution. Expressing this distribution in terms of the first four moments of the Gaussian,

$$n(x) = n(R_p)\exp\frac{\ln\left[b_0 + b_1(x - R_p) + b_2(x - R_p)^2\right]}{2b_2}$$

$$-\frac{b_1 + 2b_1b_2}{\sqrt{4b_0b_2^3 - b_1^2b_2^2}}\tan^{-1}\left[\frac{2b_2(x - R_p) + b_1}{\sqrt{4b_0b_2 - b_1^2}}\right] \tag{5.23}$$

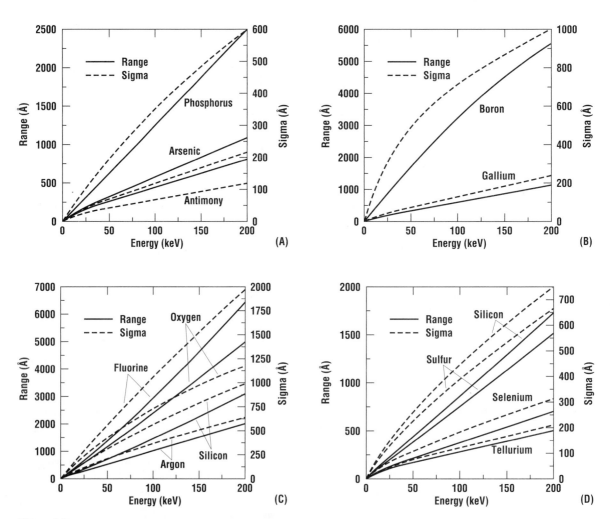

Figure 5.9 Projected range (solid lines and left axis) and standard deviation (dashed lines and right axis) for (A) n-type, (B) p-type, and (C) other species into a silicon substrate, and (D) n-type and (E) p-type dopants into a GaAs substrate, and several implants into (F) SiO$_2$ and (G) AZ111 photoresist (*data from Gibbons et al.*).

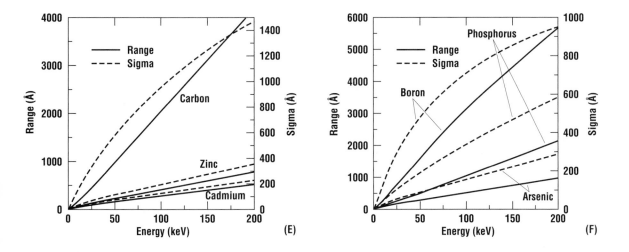

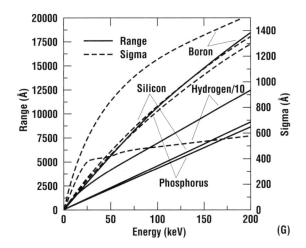

Figure 5.9 cont'd.   For legend see previous page.

where

$$b_0 = -\frac{\Delta R_p^2 (4\beta - 3\gamma^2)}{10\beta - 12\gamma^2 - 18}$$ (5.24)

$$b_1 = -\gamma \Delta R_p \frac{\beta + 3}{10\beta - 12\gamma^2 - 18}$$ (5.25)

and

$$b_2 = -\frac{2\beta - 3\gamma^2 - 6}{10\beta - 12\gamma^2 - 18}$$ (5.26)

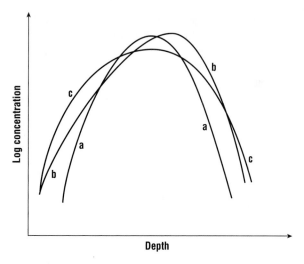

Figure 5.10   Curve (a) shows a standard Gaussian. (b) A negative skewness, which moves the peak slightly deeper and adds a tail extending toward the surface. (c) A large kurtosis flattens the peak of the distribution.

5.4 Channeling and Lateral Projected Range

When one is implanting into single-crystal materials, another complication can arise. Channeling can occur when the ion velocity is parallel to a major crystal orientation. In this situation, some ions may travel considerable distances with little energy loss (Figure 5.11), since nuclear stopping is not very effective and the electron density in a channel is low. Once in a channel, the ion will continue in that direction, making many glancing internal collisions that are nearly elastic until it comes to rest or finally dechannels. The latter may be the result of a crystal defect or impurity. Channeling is characterized by a critical angle Ψ (Figure 5.12)

$$\Psi = 9.73° \sqrt{\frac{Z_i Z_t}{E_o d}} \qquad (5.27)$$

where E is the incident energy in kiloelectron-volts and d is the atomic spacing along the ion direction in angstroms. For ions with velocity vectors much larger than Ψ away from a major crystallographic orientation, little channeling will occur [14]. The channeling direction need not be close to the initial ion velocity. A scattering event inside the target can redirect the incident ion along a crystallographic

Example 5.3

Use Equation 5.27 to estimate the critical angle for implanting boron into (100) silicon at 100 keV.

Solution

For (100) silicon, the closest atoms in the same plane are (0, 0, 0,) and ($a/2$, $a/2$, 0). The d value is

$$d = \sqrt{\left(\frac{a}{2}\right)^2 + \left(\frac{a}{2}\right)^2}$$

$$d = \frac{a}{\sqrt{2}} = 1.8 \text{ Å}$$

$$z_i = 5 \qquad z_t = 14 \quad \text{Then}$$

$$\Psi = 9.73° \sqrt{\frac{5 \times 14}{100 \times 1.8}}$$

$$\Psi = 6°$$

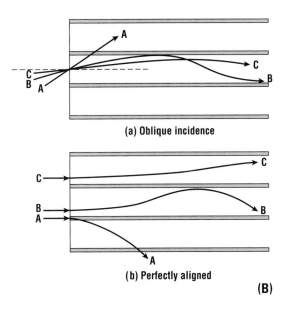

(a) Oblique incidence

(b) Perfectly aligned

(B)

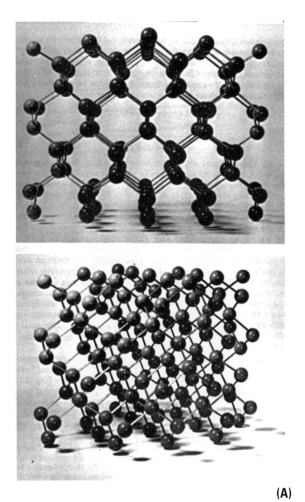

(A)

Figure 5.11   (A) Models of the diamond structure along a major crystal axis ⟨110⟩ and along a random direction. (B) Schematics of channeling *(reprinted by permission, Academic Press, after Mayer et al.).*

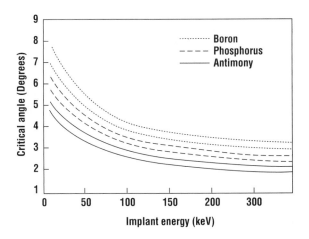

Figure 5.12   Critical angle at which channeling begins for common impurities in silicon. For each impurity, the upper curve is implanting into ⟨111⟩ substrates. The lower curves correspond to ⟨100⟩ substrates.

orientation, but the probability of this occurring is small, and so the effect is unlikely to produce a substantial distortion near the peak of the implant profile.

Channeling can produce a significant tail on the implant distributions. The effect is particularly pronounced in the implanting of light atoms on axis into a heavy matrix, since the ion's atomic radius is much less than the crystal spacing.

To avoid this tail, most IC implantation is done off axis. A typical tilt angle is 7°. To reduce the probability of an inadvertent line-up of the crystal planes, a twist angle of about 30° is also commonly used [15]. Some ions will still scatter along the crystal axis, and channeling effects will still occur. Another way to minimize channeling is to destroy the lattice before implantation. Preamorphization of silicon can be done with high doses of Si, F, or Ar before the dopant implantation. This will be covered in more detail in Section 5.6. Some channeling reduction has also been reported by implanting through a thin screen oxide to randomize the ion velocities before entering the crystal. This has the disadvantage of unintentionally implanting oxygen due to recoil or knock-on effects.

5.5 Implantation Damage

One component of energy transfer when a high energy ion enters a wafer is collision with lattice nuclei. Many of these atoms are ejected from the lattice during the process. Some displaced substrate atoms have sufficient energy to collide with other substrate atoms to produce additional displaced atoms. As a result, the implantation process produces considerable substrate damage that must be repaired during subsequent processing. Furthermore, if the implanted species is intended to act as a dopant, it must occupy lattice sites. This process of moving a large fraction of the implanted impurities onto lattice sites is known as impurity activation. Both damage repair and implant activation are normally done by heating the wafer (annealing) after implant. Often the anneal accomplishes both tasks simultaneously. For that reason, these two processes will be treated simultaneously. The activation of implants into GaAs will not be treated in this chapter since this is now most commonly done by rapid thermal annealing, the subject of the next chapter.

Since the energy loss per collision by nuclear energy transfer is typically much larger than the binding energy of the atom in the lattice, the crystal is damaged when it is implanted if the energy transfer is larger than some displacement energy [16]. There exists a threshold dose ϕ_{th} above which the damage is complete [17]. That is, after the implant no evidence of long-range order exists, and the surface of the substrate is rendered amorphous. The critical dose depends on implant energy, implant species, target material, and substrate temperature during implantation. Figure 5.13 shows the critical dose for several impurities in silicon as functions of the wafer temperature. At high temperatures, the substrate self-anneals, and the threshold dose becomes very large. Threshold doses for light ions are also large since a greater fraction of the energy loss is electronic.

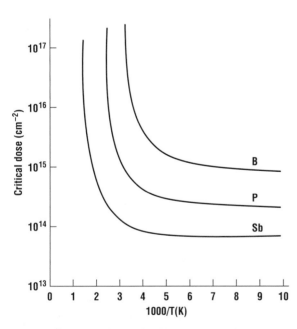

Figure 5.13   Critical implant dose required to amorphize a silicon substrate as a function of substrate temperature for several common silicon dopants *(after Morehead and Crowder).*

As an ion passes through the crystal, point defects consisting of interstitials and vacancies are created by direct interaction or through collisions with recoiled target atoms. The defects created by the implantation process are called *primary defects* [18]. Figure 5.14 shows Rutherford backscattering spectroscopy (RBS) spectra for an unimplanted wafer and a wafer implanted with $2 \times 10^{15} \text{ cm}^{-2}$ boron at 200 keV. RBS uses the scattering of He ions during an implant to determine the species present in the wafer. The dashed line represents the contribution of the random component of the beam to the spectra. The shaded area represents the signal from the displaced silicon atoms. From the area of the shaded region, it is estimated that the interstitial silicon concentration is $7 \times 10^{16} \text{ cm}^{-2}$, roughly 35 times the implant dose.

Secondary defects occur when an implanted wafer is annealed. As previously mentioned, point defects have a high energy in the crystal. This energy can be reduced by recombination or by agglomeration into extended defects [19]. Typically, these defects take the form of small point defect clusters such as divacancies or condense into higher dimensional defects like dislocation loops. For boron implanted into silicon, it appears that secondary defects occur when the Si interstitial concentration produced by the implant is $\geq 2 \times 10^{16} \text{ cm}^{-2}$. Ions such as P or Si that have masses comparable to that of silicon have larger critical interstitial concentrations, typically about $5 \times 10^{16} \text{ cm}^{-2}$. It has been suggested that implant light atoms like B form isolated defects, while heavier ions form more extended defects [20]. The interstitials created by moderate mass ions are bound in these defect clusters, and few are free to agglomerate in larger extended defects. Schreutelkamp et al. [18] have shown that secondary defects do not form for heavy ions $(Z > 69)$ except for megaelectron-volt implants. Even then, the critical interstitial concentration for a heavy ion like Sb exceeds 10^{17} cm^{-2}. Heavy ions, therefore, tend to amorphize the substrate before secondary damage is generated.

Annealing processes that minimize secondary defects are of considerable technological interest. In these applications, high temperatures are required to ensure that all of the dopant is activated and few residual extended defects remain. Isochronal annealing curves display the active carrier concentration normalized to the dose as a function of the annealing temperature for a fixed annealing time. Typical times are 30 or 60 min. Unless otherwise stated, the anneals are done in nitrogen. Figure 5.15 shows the result of an isochronal anneal experiment for boron [21]. At low temperature the carrier concentration is dominated by point defects. As the anneal temperature is raised, point defect repair begins, with some vacancies capturing nearby interstitials. This reduces the net trap concentration in the substrate, raising the free carrier concentration. In the 500 to 600°C range, the diffusivity of the defects is raised sufficiently to cause agglomeration and the formation of extended defects. This is particularly true at higher boron fluences, where the damage is large. Finally, at high temperature, these extended defects are annealed out and the active carrier concentration approaches the implant dose. This requires temperatures of 850°C to over 1000°C, with high fluence implants requiring the highest anneal temperatures.

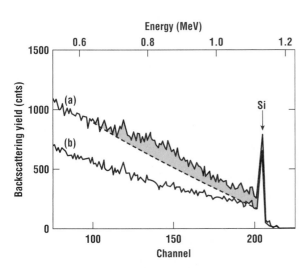

Figure 5.14   RBS spectra showing the backscattered He intensity as a function of energy. Curve (b) is an unimplanted wafer. The curve labeled (a) is a wafer after a 2×10^{15} B implant at 200 keV. The dashed line represents the component of the spectrum corresponding to the remaining crystalline silicon. The shaded area represents the contribution of displaced atoms *(reprinted by permission of Elsevier Science, after Schreutelkamp et al.).*

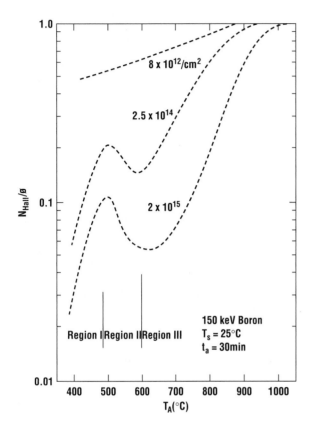

Figure 5.15 Fraction of implanted boron activated in silicon for several isochronal anneals *(after Seidel and MacRae, reprinted by permission, Elsevier Science).*

When the substrate has been rendered amorphous, either due to the implant itself or because of a preamorphization step (Section 5.6), the crystallinity is repaired by solid phase epitaxy (SPE). In theory, the crystal reforms, using the underlying undamaged substrate as a template. This causes most of the impurities to be incorporated into the growing lattice on nearly equal footing with the displaced substrate material. In discussing the annealing behavior of these amorphous layers, one must first recognize that amorphous layers need not extend to the surface. Figure 5.16 shows the damage induced in the substrate versus depth for a variety of arsenic implants [22]. Obviously, the degree of damage near the surface is much less than that further into the substrate. Buried amorphous layers are therefore possible by using high energy implants. Defects must penetrate the crystalline region before reaching the surface and being annihilated. Furthermore, SPE regrowth fronts will begin at both sides of the amorphous region. The plane where these fronts meet contains defects that can degrade device performance [23].

Generally, the annealing process for an amorphous layer involves an SPE regrowth of the layer. This regrowth can be done at temperatures as low as 600°C, since the regrowth velocity of silicon at this temperature is greater than 300 Å/min for $\langle 100 \rangle$ and about one order of magnitude less for $\langle 111 \rangle$ [24]. (For high dose implants, the SPE rate depends on the implanted species.) Thus, a 30-min anneal at 600°C will regrow almost 1 μm of material, which is well in excess of the amorphization depth of any reasonable implant. The activation of impurities occurs at much lower temperatures because of SPE. Figure 5.17 shows isochronal curves for the activation of phosphorus in single crystal and amorphous materials. A large percentage of the implant is activated even at 550°C.

The major concern with SPE regrowth of amorphous layers is residual defects. These layers include not only simple 1-D defects like dislocations, but also 2-D and 3-D defects like twins and stacking faults. The defects may originate from microislands of single-crystal material that are slightly displaced from their original position or become displaced during the SPE. These islands also serve as nucleation centers for regrowth. Defects will be formed when these disparate growth fronts meet. To reduce the concentration of the defects to an acceptable level, high temperature anneals are usually required (of order 1000°C) [25]; however, even these high temperature anneals may not be sufficient to remove all of the damage. Figure 5.18 shows the annealing characteristics of high-dose arsenic implants [26]. Even after 1000°C anneals, residual damage remains. The damage is concentrated in the regions near the edge of the implant. High temperature anneals also allow the activation of impurities in the implant tail regions, outside the amorphous layer. To avoid excessive diffusion during the high temperature anneal, the point defect concentration is first lowered by a low temperature SPE step.

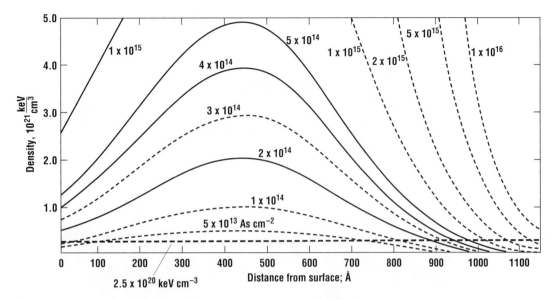

Figure 5.16   Damage density distribution for 100 keV arsenic implanted into silicon with dose as a parameter *(reprinted by permission, AIP)*.

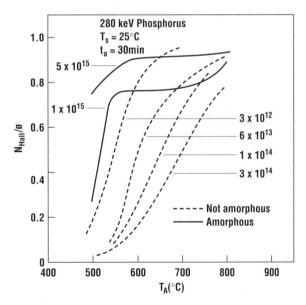

Figure 5.17   Isochronal annealing of phosphorus in silicon, with dose as a parameter. The solid lines correspond to implants that amorphized the substrate *(reprinted by permission, AIP, after Crowder and Morehead)*.

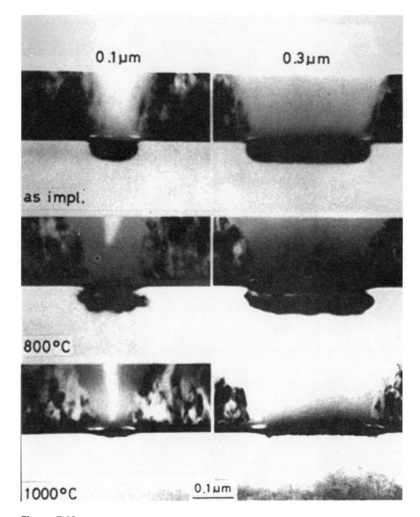

Figure 5.18   Cross-sectional TEMs image of implant damage in submicron areas for a high dose arsenic implant (5×10^{15} cm^{-2} at 25 keV) immediately after implant, after an 800°C anneal, and after a 1000°C anneal. Notice that even at the highest temperature residual damage remains at the edges of the implanted regions *(reprinted by permission of Elsevier Science, after Gyulai [26]).*

5.6 Shallow Junction Formation$^+$

In the production of very small devices, it is necessary to reduce junction depths (see Chapter 16). Most of the work on shallow junctions has focused on forming very shallow source/drain junctions for submicron and deep submicron or nanoscale CMOS. As a general rule of thumb, the source/drain junction depth should be no more than about 30% of the channel length. By that criteria, 65-nm CMOS transistors should have junction depths of 20 nm or less. Since boron has both a larger projected range (at the same energy) and a larger diffusivity (at the same temperature) than arsenic, the formation of ultrashallow P+/n junctions has drawn the most attention. This section will therefore focus on the formation of shallow P+/n junctions, but many of the techniques have been applied to forming N+/p junctions as well.

The task of forming shallow junctions can easily be subdivided into two tasks: forming a very shallow as-implanted profile, and limiting diffusion while achieving a high degree of dopant activation and damage repair. The first of these requires, at a minimum, a way of introducing boron to the silicon wafer at very low energies. Of course, one can apply traditional techniques (such as pictured in Figure 5.1) to such an application. The primary problem is the stability of the beam. At very low energies and high implant current densities, the ions in the beam repel each other and the beam broadens as it travels [27]. Lower implant energy, lighter mass ions, and a lengthy distance from analyzer to wafer, all exacerbate the effect. To combat blooming, ion implant tools may add electrons to the beam space in a controlled manner. This can be done by adding an easily ionized gas like xenon to the vacuum. As xenon is ionized, the electrons given off help keep the space charge of the beam low, thereby helping transport the beam to the target [28].

The first class of solutions to this problem is the use of molecular implants. The simplest example for boron implantation is BF_2, since it is a by-product of the decomposition of the typical boron source gas, BF_3. If one accelerates the BF_2 molecule to an energy E and it does not dissociate, all of the atoms in the molecule must be traveling at the same speed. The energy of any atom then is given by the molecule energy divided by the mass ratio:

$$E_B = E_{BF_2} \frac{m_B}{m_{BF_2}}$$

The smaller the mass ratio, the smaller the energy of the individual atom that can be implanted (however, the implanter loses mass discrimination at sufficiently high mass). Fluorine may also reduce the junction depth by combining with point defects in the silicon and therefore reducing the diffusivity [29, 30.] This has been demonstrated for B, As, and P implants into Si. Fluorine, however, appears to cluster in the damaged region of the implant and can result in microvoid formation in these areas.

For boron implantation there has been considerable interest in decaborane ($B_{10}H_{14}$) [31]. The energy ratio is now 11/124 or 0.089, allowing very low energy implants. Of course the atomic boron dose is 10 times larger than the molecule dose. A key problem with this is the development of a source that produces decaborane ions with very limited dissociation [32]. At these very low energies, incident ions may sputter (see Chapter 12) the surface, removing not only silicon atoms, but previously implanted boron atoms as well [33]. Some way of accounting for this is required to get an accurate final dose. Curiously, however, the use of clusters like decaborane appears to lead to a smoother surface than is obtained for monomer implant species [34]. Systems using boron clusters for very shallow junctions are now commercially available [35]. One major supplier is providing ClusterBoron material ($B_{18}H_{22}$) [36] as an implant source for very low energy boron implantation.

Another approach to this problem is to rethink how implantation is done. If one has a very short acceleration system, or perhaps none at all, beam spreading can be reduced. One can also use a much larger beam area to reduce the charge density. Very low energy ion implanters have been built around these concepts. Applied Materials' Quantum X Plus system, for example, allows implants down to 200 eV [37]. The ultimate version of this concept is generally called plasma immersion ion implantation (PIII). In this process the wafer to be implanted is placed in a plasma (see Chapter 10), using a gas that carries the desired implant species. The bias voltage of the plasma is adjusted to the desired value, and ions are implanted into the substrate. Such a technique can be extremely cost effective and can be used for surface treatments of structural materials [38]. Because of the broad area of immersion (the entire wafer surface), PIII has the potential to implant high concentrations of species without the stability problems of conventional implantation. For doping, the range of operation is for doses of order 10^{15} cm^{-2}, and ion energies between 1 and 30 eV [39]. Very shallow P+/n junctions suitable for MOS source/drains have been demonstrated this way [40]. PIII has several important drawbacks, however, at least as far semiconductor applications. There is no mass separation, and therefore all ionized fragments may be implanted into the wafer. It is extremely difficult to accurately control both dose and energy. Although high density SRAMs have been demonstrated [41] by using PIII, it is not clear that this form of ion implantation will replace conventional implantation in the foreseeable future. Other potential replacements for conventional high dose implantation include gas immersion laser doping or GILD [42], and projection gas immersion laser doping or P-GILD [43].

The second problem associated with shallow junction formation is channeling. Even in randomly oriented wafers channeling occurs [44] due to boron's high probability for deflection into a major crystallographic direction. This effect can be minimized by preamorphizing the wafer with a high-dose implant of a heavier atom before boron implantation. One popular choice is germanium. Ozturk and Wortman [45] have demonstrated about a 20% reduction in the junction depth (assuming $N_{sub} = 10^{16}$ cm^{-3}) when the wafers were preamorphized with a low energy Ge implant at a dosage of 3×10^{14} cm^{-2} before BF$_2$ implantation. No difference in diode characteristics was seen [45]. Junction depth reductions in excess of 40% have been reported for atomic B implants that are preamorphization with silicon implantation [46]. One disadvantage of these techniques is that the amorphous region must be recrystallized, and this recrystallization may leave residual defects.

The third problem associated with shallow junction formation is the anomalous diffusion that occurs during the high temperature anneal [47]. The effect is most pronounced in the tail region [48]. This effect is sometimes called the *transient annealing effect*. It has also been observed for phosphorus implants [49]. The cause is still uncertain. One model suggests that since only about 20% of the boron is substitutional as implanted [50], the anomalous diffusion occurs interstitially [51]. This model fails to explain the anomalous diffusion effects when the implant is done on axis. This latter result has reinforced theories that the transient diffusion effect is associated with damage and, in particular, with extended defects formed during annealing [52, 53]. More recently, Schreutelkamp has proven that the transient diffusion mechanism for boron in silicon is the addition of the kick-out–diffusion mechanism (see Chapter 3). Due to implant damage at the beginning of the anneal, a high concentration of self-interstitials is present. These interstitials kick out the substitutional boron to interstitial sites, where it rapidly diffuses. As the damage is annealed, the excess self-interstitial concentration drops, and so diffusivity decreases until late in the anneal, when extended defects are annihilated and give up excess interstitials that again drive the transient diffusion process.

5.7 Buried Dielectrics[+]

Thus far only one application of implantation, the introduction of dopants into the semiconductor substrate, has been discussed. A second application of the process is in device isolation. In GaAs technologies, this is a widely applied technique used to provide a surface isolation between adjacent devices fabricated in a conducting layer on an insulating substrate. Protons (H$^+$) are used most commonly, although other materials have been investigated. From an implantation process standpoint, there is little difference between this application and doping. We will therefore defer any further discussion of this isolation technique until Chapter 17. This section will cover the use of implantation to form silicon on insulator (SOI). This is done by performing a high-energy implant of N$^+$ to form Si$_3$N$_4$ in a process called SIMNI [54] or by performing a high-energy implant of O$^+$ to form SiO$_2$ in a process called SIMOX. The latter is much more popular.

Separation by IMplanted OXygen (SIMOX) was first reported by Izumi et al. [55]. The advantages of forming a buried insulator include increased radiation hardness, increased circuit speed, and increased packing density. The typical SIMOX process involves implanting the wafer with 150 to 300 keV O^+ at doses of about 2×10^{18} cm^{-2}. For most implanters, this involves a very long implantation time. Often the implant is done on axis to intentionally channel and minimize the

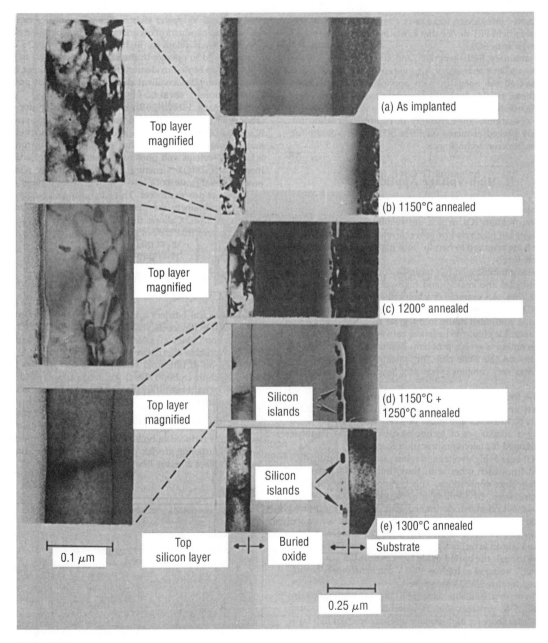

Figure 5.19 Cross-sectional TEMs of SIMOX layers for different anneal conditions. Samples were implanted at 150 keV with an oxygen dose of 2.25×10^{18} cm^{-2} *(after Lam, reprinted by permission, © 1987 IEEE).*

near-surface damage. The implant step produces a nearly amorphous layer with about twice as many oxygen atoms as silicon near the projected range. To minimize surface damage the wafer must be held at temperatures of at least 400°C during the implant [56]. This is very important because the amorphous oxide layer that will later be formed under the silicon prevents a recrystallization from the undamaged portions of the substrate. The implanted oxygen will also tend to diffuse from the projected range to form a broad flat concentration plateau if the wafer is held at about 500°C during the implant. The oxygen concentration in this region is the atomic concentration required to form SiO_2 [57].

Following the implant a very high-temperature step must be done to form the oxide. Typically the anneal is done under a deposited oxide cap. The anneal conditions are at least 1300°C for several hours. Some authors recommend annealing at temperatures over 1400°C [58]. Figure 5.19 shows a set of cross-sectional transmission electron micrographs of SIMOX wafers after various thermal anneals. Obviously, the crystal quality of the surface layer improves dramatically with anneal temperature. Residual oxygen atoms left in the implant tail also serve as thermal donors that can affect the carrier concentration in the active device regions. Fortunately, thermal anneals of 1300°C also reduce the thermal donor density to less than 10^{15} cm^{-3} [59].

Potential problems with SIMOX include heavy-metal impurities that lead to enhanced junction leakage, pinhole density, material quality, and thickness uniformity. Metallic contamination is present in all implants; however, typical SIMOX implant doses are nearly a thousand times those of conventional source/drain implants. The implanter components must be carefully designed for SIMOX use to avoid this problem. For example, silicon apertures and beam stops may be used rather than stainless steel or tungsten. Pinholes in the buried oxide occur when a particle is on the surface of the wafer during the SIMOX implant. The particle screens or partially screens the O_2 implant. The result is a singularity in the dielectric as it rises to the surface. Careful wafer cleaning and attention to particulate control can minimize this problem. Another concern is the dislocation density in the surface silicon. Although the material quality has improved dramatically through various process improvements, it still has of order 10^3 dislocations/cm². CMOS ICs can be fabricated quite well in SIMOX material, and 256K SIMOX SRAMs have been reported. Even bipolar for BiCMOS applications have been fabricated in this technology. Minority carrier lifetimes of about 10^{-7} sec have been observed.

One of the primary concerns for SIMOX is still cost, particularly in commodity components. Implanters have been specifically developed for the SIMOX process to address these concerns. Beam currents in these systems typically exceed 100 mA, allowing higher throughput and therefore reducing cost. With the use of these implanters, metallic contamination has been held to less than 10^{11} cm^{-2} and pinhole density to less than 0.2 cm². Thickness uniformity of the single-crystal layer is approaching 50 Å over 6-in. wafers.

5.8 Ion Implantation Systems: Problems and Concerns

Modern implanters have several technical problems that the idealized system described at the beginning of the chapter did not address. This section will review some problems besides damage annealing, that make the life of implant engineers interesting. Figure 5.20 shows a common problem for implantation of wafers with topology: shadowing. Any nonplanar masking layer will cast an implant shadow because of the tilt and twist angles of the implant. Within the decade, the decrease in the allowable thermal cycles and resultant diffusion, along with the increasing reliance on self-aligned structures to minimize parasitic capacitance, has increased the importance of shadowing effects. In a typical example, the *I–V* characteristics of an MOS transistor will be asymmetric with respect to the choice of source and drain contacts. The newest generation of implanters allows both

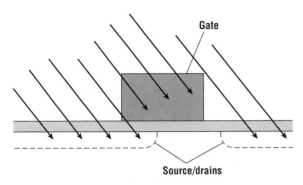

Figure 5.20 Simple shadowing example for deeply scaled MOSFET. As a result of the tilt angle one of the source/drain diffusions does not extend to the channel leading to poor *I–V* characteristics.

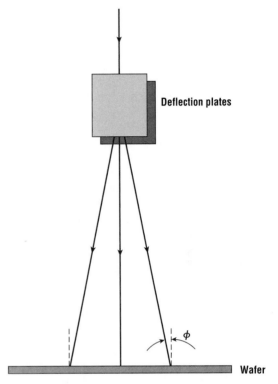

Figure 5.21 Variability of the angle between the incident beam and the surface normal in simple electrostatically scanned systems.

the implant angle and the twist to be varied while the wafers are under high vacuum. This allows multiple implant angles to be run efficiently. Often four angles are run in a "quad" arrangement.

The problem of shadowing can be further exaggerated by using simple electrostatic scanning as shown in Figure 5.21. The scanning plate deflection can cause the implantation angle to vary from 5° to 9° across the wafer unless a corrector magnet is added to compensate. Another problem with the tilt angle in electrostatically scanned implantation systems is that the dosage of the wafer will vary from one side to the other. The problem becomes more pronounced with large wafers, where the electrostatic angle can be 2° or more. High current implanters, in which the wafers are mechanically scanned in both directions, do not suffer from this problem; however, the wafer in these systems heats during the implant unless carefully cooled. Because of this heating and the edge-clamping process used, the wafers frequently flex, causing further implant angle uncertainty.

The increasing use of large diameter silicon wafers makes the task of producing highly uniform implants more difficult. Current and Keenan have also demonstrated that production implanters can have considerable variability from machine to machine, particularly at low dosages [60]. Besides the problems discussed in the previous section, dose rate monitoring errors can be present due to several effects. Any neutral species that strike the wafer will not produce an electric current and so will not be counted. Neutrals can be produced in the beam by recombination with residual gas atoms or thermal electrons [61]. Dosage errors can also be the result of incomplete secondary electron recapture. To prevent this problem, the wafer is held in a *Faraday cup* and the stage is typically held at a small positive bias. Electron flood guns are also commonly used to maintain charge neutrality and compensate for secondary electron loss.

Ion implantation is often assumed to be an extremely clean process. It is done in high vacuum, and the implant species is selected to correspond to the desired isotope of the desired species. Implanters have four types of commonly seen contamination. As with all processing equipment, one type of contamination is particles. If the particles fall on the wafer before implantation, they will screen the wafer from the implant. A later clean may remove the particle, but its unseen shadow remains behind as a virtually undetectable killing defect. Most particles are caused by improper wafer handling, improper pump-down procedures, clamping procedures, the use of unfiltered gas for venting, and in high current systems, the spinning disk [62]. The use of clampless wafer mounting in particular, has been found to have a significant effect on particle generation in ion implanters.

The effects of implanted charge on the MOS gate oxide integrity is also a very serious problem. This requires charge neutralization, either in the beam or on the surface of the wafer, by adding compensatory electrons. This effect is exacerbated by antenna effects. The charge will flow from parts of the gate electrode that reside on thicker oxides to the regions on top of thinner oxides to equalize the voltage.

The other three types of contamination are related to much smaller impurities that may fall on the wafer during implantation. To guard against them, implants are often done through a thin screen oxide that is removed after implantation. Hydrocarbons are produced in high current machines when vapors from diffusion pumps backstream into the beamline and are polymerized during implantation. These compounds are extremely resistant to subsequent chemical cleans. This problem can be avoided by carefully baffling the diffusion pumps, or, as is often done, by using cryo or turbo pumps. Heavy-metal impurities such as Fe, Cr, and Ni are sometimes seen when stainless steel apertures are used [63]. New machines commonly use graphite or copper-free aluminum for these components to avoid this effect. Finally, it is common to see some cross-contamination, particularly on low dose implants that follow a high dose implant of a different species. It is therefore important that lifetime killing impurities never be implanted with semiconductor implanters. Many larger fabrication facilities will often dedicate an implanter to a particular impurity to avoid any cross-contamination concerns.

5.9 Numerical Implanted Profiles[+]

The process simulator introduced in Chapter 3 to estimate diffusion profiles, and later used in Chapter 4 for oxidation, may also be used to calculate implant profiles. This feature is very useful, since in practice most impurities are implanted. Simulated impurities can be implanted, activated, and diffused to compare to real profiles. The software contains data for the implant parameters for most common dopants. The program also handles the implantation of films through multiple layers. Typically, the program will predict the profile based on simple Gaussian, two-sided Gaussian, Pearson type IV, or dual Pearson type IV distributions. Some versions of the program are also able to predict the profile based on Boltzmann transport and Monte Carlo methods.

A typical run that would be used to simulate a P^+ source/drain region is given in Example 5.4.

Example 5.4

Use Silvaco's Athena software to create a quasi-one-dimensional grid and do a 2-keV, $10^{15}\,cm^{-2}$ PMOS source/drain implant into a uniformly doped ($10^{17}\,cm^{-3}$) channel. Measure the junction depth and sheet resistance. How does the profile compare to a Gaussian? What if BF_2 is used instead of boron?

Solution

```
go athena
  #TITLE: Boron Implant Example 5.4

  line x loc=0.0    spacing=0.02
  line x loc=0.2    spacing=0.02
  line y loc=0.0    spacing=0.002
  line y loc=0.1    spacing=0.002
  line y loc=0.2    spacing=0.005
  line y loc=0.4    spacing=0.01
  line y loc=0.6    spacing=0.02

  init c.phos=1e17
  method adapt

  # implant Boron
  implant boron energy=2 dose=1e+15
  # extract junction depth
  extract name="xj" xj material="Silicon" mat.occno=1
  x.val=0.1 junc.occno=1

  #extract 1D electrical parameters
  extract name="sheet_rho" p.sheet.res material="Silicon"
  mat.occno=1 x.val=0.1 region.occno=1
  # Save and plot the final structure
  structure outfile=ex5_4.str
  tonyplot
```

Results

For boron, Athena predicts a junction depth of 58 nm and a sheet resistance of 233 $\Omega/\square$. In reality, the sheet resistance would be quite a bit larger, since no activation step was specified in this sequence. If one looks at the profile produced (Figure 5.22), the chemical boron concentration near the surface is quite a bit larger than the electrically active (net doping) concentration. Also notice that the profile near the peak is fairly distorted, in part due to the high concentration effects of this implant. For BF_2, replace the word "Boron" with "bf2." The junction depth is 25 nm, and the sheet resistance is 300 $\Omega/\square$, since less of the dopant is electrically active.

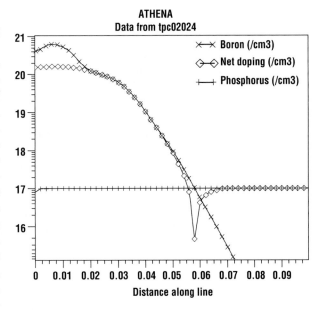

Figure 5.22

5.10 Summary

This chapter introduced the technology of ion implantation. The components of a modern ion implanter were described, and some limitations of ion implanting impurities were presented. Most implant profiles can be described by a Gaussian distribution. Additional moments of the Gaussian including skewness and kurtosis are sometimes used to better approximate experimentally observed profiles. After implantation the impurities must be annealed. The anneal step activates the impurities and repairs the implant damage. Different annealing recipes are called for depending on the amount of damage in the substrate. In silicon technologies, implantation can also be used to form buried insulators through high dose oxygen implantation. Finally the use of software to simulate implantation was covered.

Problems

1. A 30-keV implant of ^{11}B is done into bare silicon. The dose is 10^{12} cm^{-2}.
    (a)   What is the depth of the peak of the implanted profile?
    (b)   What is the concentration at this depth?
    (c)   What is the concentration at a depth of 3000 Å (0.3 µm)?
    (d)   The measured concentration is found to be an order of magnitude larger than the value predicted in part (c), although the profile agrees with answers (a) and (b). Give a possible explanation, assuming that the measured value is correct.
2. A particular silicon device needs to have an implant of boron with a peak at a depth of 0.3 µm (3000 Å) and a peak concentration of 10^{17} cm^{-3}. Determine the implant energy and dose that should be used for this process. Find the as-implanted junction depth if the substrate is n-type with a concentration of 10^{15} cm^{-3}.
3. Phosphorus is implanted into silicon. The implant parameters are a dose of 10^{15} cm^{-2} and an energy of 150 keV.
    (a)   Find the depth of the peak of the implant profile and its value at that depth.
    (b)   If the wafer originally had 10^{16} cm^{-3} of boron uniformly distributed throughout, find the depth(s) at which the concentration of phosphorus is equal to the concentration of boron.
4. A silicon wafer is implanted with both boron and arsenic at an energy of 100 keV and each with a dose of 10^{15} cm^{-2}.
    (a)   What are the projected range R_p and standard deviation ΔR_p for each implant?
    (b)   Draw a rough sketch of both of the implant concentrations versus depth on a single plot. Use a log scale for concentrations. Do not do any calculations for this; just show the relative positions and distributions for the two curves.
    (c)   Find the position(s) where the two implanted profiles have the same concentrations.
5. Your objective is to select an implant recipe for putting boron into silicon. The peak of the implant profile (i.e., maximum) should occur at 0.2 µm. The boron concentration at the peak should be 2×10^{17} cm^{-3}.
    (a)   What implant energy should be used?
    (b)   What implant dose (in (reciprocal centimeters squared)) should be used?
6. A MOSFET threshold voltage adjust implant is done through a gate oxide of 150 Å. The implant species is boron at 30 keV. Estimate the fraction of boron implanted in the oxide. (You may have to use the approximation that $x \ll R_p$ for the Gaussian.)
7. A mass spectrometer as described in the chapter is used for element extraction in an implanter. Calculate the magnetic field necessary to extract silicon (mass 28) if the extraction potential is 20 keV and the radius of curvature for the analyzer is 30 cm. Explain why one might also see some N_2 in the implanted profile if the source cabinet has a small vacuum leak.

8. A wafer is implanted with sulfur (S) at an energy of 100 keV and a dose of 1×10^{13} cm^{-2}. The wafer has a 500-Å-thick layer of AlGaAs on top of a very thick GaAs bulk, as shown in the accompanying figure. Assume that the AlGaAs layer behaves just like conventional GaAs.

AlGaAs

GaAs Wafer

   (a) At what depth is the implanted sulfur concentration the highest?
   (b) What is the concentration at this point?
   (c) Sketch a plot of the log of the concentration versus depth, indicating the AlGaAs and GaAs regions. Will most of the implanted S be in the AlGaAs or in the GaAs? Why?

9. A silicon wafer is implanted with phosphorus. The dose is 10^{15} cm^{-2}, and the energy is such that the projected range for the implant is 0.2 μm and the standard deviation of the implant is 0.05 μm. After the implant, the impurity is activated with a 1000°C rapid thermal anneal. Ignoring the transient enhanced diffusion effect, but taking into account doping effects on the diffusion, answer the following.
   (a) What is the net diffusion flux (J) of the phosphorus at the beginning of the anneal at a depth of 0.2 μm? Is it toward $x = 0$ or the other way?
   (b) What is the net diffusion flux (J) of the phosphorus at the beginning of the anneal at a depth of 0.3 μm? Is it toward $x = 0$ or the other way?

10. An implanted profile of arsenic in silicon has a range of 0.1 μm and an implant straggle (standard deviation) of 0.03 μm. The implant dose is 4×10^{14} cm^{-3}. The wafer is then heated to 900°C. You can ignore all transient enhanced diffusion effects.
   (a) Find the impurity concentration at 0.13 μm before the wafer is heated.
   (b) Find the diffusion coefficient D at this position before any diffusion begins.
   (c) Find the initial diffusion flux J in cm^{-2}-sec^{-1}.

11. A typical high-current implanter operates with an ion beam of 2 mA. How long would it take to implant a 150-mm-diameter wafer with O^{+} to a dose of 1×10^{18} cm^{-2}?

12. We can approximate the mass of an atom as twice the atomic charge number. Based on this calculate k_e as a function of Z_i assuming that the target is silicon. Repeat it for a germanium substrate and plot both curves. Discuss the significance.

13. The depth of the junction of the source/drain region in a MOSFET must be reduced as the gate length is scaled. It is highly desirable to produce low resistivity junctions thinner than 0.1 μm. Is this a significant problem for ion implantation? Justify your answer for both N^{+}/p and P^{+}/n junctions. What are the major problems in forming these structures?

14. One way to achieve very shallow implants is to use the molecule decaborane (B$_{10}$H$_{14}$) as the implant species. Assume that all of the boron is mass 11 and that hydrogen is mass 1. Since Figure 5.9 is not readable for very low energies, assume that for low energy implants the projected range for boron atoms is given by 32 Å/keV × atomic implant energy and the standard deviation of the range (straggle) is given by 12 Å/keV × atomic implant energy. The implant is done at a molecular energy of 2 keV and a molecular dose of 10^{14} cm^{-2}.
   (a) Find the position (depth) of the peak of the boron concentration.

(b)   Find the concentration of boron atoms at this peak.

(c)   Find the as-implanted junction depth if the wafer before implant is n-type with a dopant concentration of $10^{17}\,\text{cm}^{-3}$.

15.   Use your software to implant boron in to silicon with a dose of $10^{13}\,\text{cm}^{-2}$ and an energy of 100 keV. Record the chemical concentration. Plot this profile and a Gaussian distribution using the implant parameters found in Figure 5.9. Discuss the differences.

16.   In an earlier chapter you used your software to model a double-diffused NPN bipolar transistor. Repeat the exercise now using ion implantation. Assume that the wafer is a uniformly doped n-type at $10^{16}\,\text{cm}^{-3}$, a base width of 0.2 μm, and a Gummel number (the total amount of net dopant in the base) of about $10^{13}\,\text{cm}^{-2}$. Which impurities would you use for the base and the emitter? Use your software to develop an implant and activation cycle that would produce this profile. Plot the net active concentration to demonstrate your solution. Why is such a profile more controlled than the corresponding double-diffused profile?

References

1.   J. W. Mayer, L. Erickson, and J. A. Davies, *Ion Implantation in Semiconductors, Silicon and Germanium*, Academic Press, New York, 1970.

2.   H. Ryssel and I. Ruge, *Handbook of Ion Implantation Technology*, Wiley, New York, 1986.

3.   G. Dearnaley, J. H. Freeman, R. S. Nelson, and J. Stephen, *Ion Implantation*, New Holland Amsterdam, 1973.

4.   T. C. Smith, "Wafer Cooling and Photoresist Masking Problems in Ion Implantation," in *Ion Implantation Equipment and Techniques*, H. Ryssel and H. Glawischnig, eds., vol. 11, Springer Series in Electrophysics, Springer-Verlag, New York, 1983, p. 196.

5.   P. Burggraaf, "Resist Implant Problems: Some Solved, Others Not," *Semicond. Int.* **15**:66 (1992).

6.   J. F. Gibbons, "Ion Implantation in Semiconductors—Part I, Range Distribution Theory and Experiments," *Proc. IEEE* **56**:295 (1968).

7.   Y. Xia and C. Tan, "Four-parameter Formulae for the Electronic Stopping Cross-section of Low Energy Ions in Solids," *Nucl. Instrums, Methods* **B13**:100 (1986).

8.   B. Smith, "Ion Implantation Range Data for Silicon and Germanium Device Technologies," Research Studies, Forest Grove, OR, 1977.

9.   T. E. Seidel, "Ion Implantation," in *VLSI Technology*, S. M. Sze, ed., McGraw-Hill, New York, 1983.

10.   J. Lindhard and M. Scharff, *Phys. Rev.* **124**:128 (1961).

11.   J. Lindhard, M. Scharff, and H. Schiøtt, "Range Concepts and Heavy Ion Ranges," *Mat. Fys. Med. Dan. Vidensk. Selsk* **33**:14 (1963).

12.   J. F. Gibbons, W. S. Johnson, and S. W. Mylroie, *Projected Range Statistics*, Dowden, Hutchinson, and Ross, Stroudsburg, PA, 1975.

13.   W. P. Petersen, W. Fitchner, and E. H. Grosse, "Vectorized Monte Carlo Calculations for the Transport of Ions in Amorphous Targets," *IEEE Trans. Electron Dev.* **30**:1011 (1983).

14.   D. S. Gemmell, "Channeling and Related Effects in the Motion of Charged Particles Through Crystals," *Rev. Mod. Phys.* **46**:129 (1974).

15.   N. L. Turner, "Effects of Planar Channeling Using Modern Ion Implant Equipment," *Solid State Technol.* **28**:163 (February 1985).

16.   G. H. Kinchi and R. S. Pease, *Rep. Prog. Phys*, **18**:1 (1955).

17. F. F. Morehead and B. L. Crowder, "A Model for the Formation of Amorphous Silicon by Ion Implantation," in *First International Conference on Ion Implantation*, F. Eisen and L. Chadderton, eds., Gordon and Breach, New York, 1971.

18. R. J. Schreutelkamp, J. S. Custer, J. R. Liefting, W. X. Lu, and F. W. Saris, "Preamorphization Damage in Ion-Implanted Silicon," *Mat. Sci. Rep.* **6**:275 (1991).

19. T. Y. Tan, "Dislocation Nucleation Models from Point Defect Condensations in Silicon and Germanium," *Materials Res. Soc. Symp. Proc.* **2**:163 (1981).

20. B. L. Crowder and R. S. Title, "The Distribution of Damage Produced by Ion Implantation of Silicon at Room Temperature," *Radiation Effects* **6**:63 (1970).

21. T. E. Seidel and A. U. MacRae, "The Isothermal Annealing of Boron Implanted Silicon," in *First International Conference on Ion Implantation*, F. Eisen and L. Chadderton, eds., Gordon and Breach, New York, 1971.

22. S. Wolf and R. N. Tauber, *Silicon Processing for the VLSI Era*, vol. 1, Lattice Press, Sunset Beach, CA, 1986.

23. M. I. Current and D. K. Sadana, "Materials Characterization for Ion Implantation," in *VLSI Electronics—Microstructure Science* 6, N. G. Einspruch, ed., Academic Press, New York, 1983.

24. L. Csepergi, E. F. Kennedy, J. W. Mayer, and T. W. Sigmon, "Substrate Orientation Dependence of the Epitaxial Growth Rate for Si-implanted Amorphous Silicon," *J. Appl. Phys.* **49**:3906 (1978).

25. J. A. Pals, S. D. Brotherton, A. H. van Ommen, and H. J. Ligthart, "Recent Developments in Ion Implantation in Silicon," *Mat. Sci. Eng.* **B4**:87 (1989).

26. J. Gyulai, "Annealing and Activation", in *Handbook of Ion Implantaion Technology*, J. Ziegler, ed., Elsevier Science, Amsterdam (1992).

27. S. Radovanov, G. Angel, J. Cummings, and J. Buff , "Transport of Low Energy Ion Beam with Space Charge Compensation," 54th Annual Gaseous Electronics Conference, State College, PA, American Physical Society, 2001.

28. B. Thompson and M. Eacobacci, "Maximizing Hydrogen Pumping Speed in Cryopumps Without Compromising Safety," *Micro Mag.* May 2003.

29. K. Ohyu and T. Itoga, "Advantage of Fluorine Introduction in Boron Implanted Shallow P+n Junction Formation." *Jpn. J. Appl. Phys.* **29**(3):457 (1990).

30. D. Lin and T. Rost, "The Impact of Fluorine on CMOS Channel Length and Shallow Junction Formation," *IEDM Technical Digest*, p. 843.

31. M. A. Foad, R. Webb, R. Smith, J. Matsuo, A. Al-Bayati, T-Sheng-Wang, and T. Cullis, "Shallow Junction Formation by Decaborane Molecular Ion Implantation," *J. Vacuum Sci. Technol. B: Microelectron Nanometer Struct.* **18**(1):445–449 (2000).

32. A. S. Perel, W. K. Loizides, and W. E. Reynolds, "A Decaborane Ion Source for High Current Implantation" *Rev. Sci. Instrum.* **73**(2 II):877 (February 2002).

33. M.A. Albano, V. Babaram, J. M. Poate, M. Sosnowski, and D. C. Jacobson, "Low Energy Implantation of Boron with Decaborane Ion," Materials Research Society Symposium, *Proceedings* **610**:B3.6.1–B3.6.6 (2000).

34. C. Li, M. A. Albano, L. Gladczuk, and M. Sosnowski, "Characteristics of Ultra Shallow B Implantation with Decaborane," Materials Research Society Symposium, *Proceedings* **745**: 235–240 (2002).

35. "Axcelis' Imax High Dose, Low Energy Boron Cluster Implant Technology Added to Optima Platform," *Semicond. Fabtech* Monday, 21 August 2006.

36. D. Jacobson, T. Horsky, W. Krull, and B. Milgate, "Ultra-high Resolution Mass Spectroscopy of Boron Cluster Ions," *Nucl. Instrum. Methods, Phys. Res. B: Beam Interactions Mater. Atoms*

237(1–2): August, 2005, *Ion Implantation Technology Proceedings of the 15th International Conference on Ion Implantation Technology,* p. 406–410, 2005.

37. http://www.amat.com/products/Quantum.html?menuID=1_9_1.

38. K. Sridharan, S. Anders, M. Nastasi, K. C. Walter, A. Anders, O. R. Monteiro, and W. Ensinger, "Nonsemiconductor Applications," in *Handbook of Plasma Immersion Ion Implantation and Deposition*, André Anders, ed., Wiley, New York, 2000.

39. P. K. Chu, N. W. Cheung, C. Chan, B. Mizumo, and O. R. Monteiro, "Semiconductor Applications," in *Handbook of Plasma Immersion Ion Implantation and Deposition*, André Anders ed., Wiley, New York, 2000.

40. X. Y. Qian, N. W. Cheung, M. A. Lieberman, M. I. Current, P. K. Chu, W. L. Harrington, C. W. Magee, and E. M. Botnick, "Sub-100 nm p +/n Junction Formation Using Plasma Immersion Ion Implantation," *Nucl. Instrum. Methods Phys. Res. B: Beam Interactions Mater. Atoms* **55**(1–4):821 (Apr. 2, 1991).

41. M. Takase and B. Mizuno, "New Doping Technology—Plasma Doping for Next Generation CMOS Process with Ultra Shallow Junction-LSI Yield and Surface Contamination Issues," IEEE International Symposium on Semiconductor Manufacturing Conference, *Proceedings*, 1997, pp. B9–B11.

42. K. H. Weiner, P. G. Carey, A. M. McCarthy, and T. W. Sigmon, "Low-Temperature Fabrication of p±n Diodes with 300-angstrom Junction Depth," *IEEE Electron Dev. Lett.* **13**(7):369–371 (1992).

43. K. H. Weiner, and A. M. McCarthy, "Fabrication of Sub-40-nm p-n Junctions for 0.18 um MOS Device Applications Using a Cluster-Tool-Compatible, Nanosecond Thermal Doping Technique," *Proc. SPIE* **2091**:63–70 (1994).

44. D. R. Myers and R. G. Wilson, "Alignment Effects on Implantation Profiles in Silicon," *Radiation Effects* **47**:91 (1980).

45. M. C. Ozturk and J. J. Wortman, "Electrical Properties of Shallow P+n Junctions Formed by BF_2 Ion Implantation in Germanium Preamorphized Silicon," *Appl. Phys. Lett.* **52**:281 (1988).

46. H. Ishiwara and S. Horita, "Formation of Shallow P+n Junctions by B-Implantation in Si Substrates with Amorphous Layers," *Jpn. J. Appl. Phys.* **24**:568 (1985).

47. T. E. Seidel, D. J. Linscher, C. S. Pai, R. V. Knoell, D. M. Mather, and D. C. Johnson, "A Review of Rapid Thermal Annealing (RTA) of B, BF_2 and As Implanted into Silicon," *Nucl. Instrum. Methods B* **7/8**:251 (1985).

48. T. O. Sedgwick, A. E. Michael, V. R. Deline, and S. A. Cohen, "Transient Boron Diffusion in Ion-implanted Crystalline and Amorphous Silicon," *J. Appl. Phys.* **63**:1452 (1988).

49. G. S. Oehrlein, S. A. Cohen, and T. O. Sedgwick, "Diffusion of Phosphorus During Rapid Thermal Annealing of Ion Implanted Silicon," *Appl. Phys. Lett.* **45**:417 (1984).

50. H. Metzner, G. Suzler, W. Seelinger, B. Ittermann, H.-P. Frank, B. Fischer, K.-H. Ergezinger, R. Dippel, E. Diehl, H.-J. Stöckmann, and H. Ackermann, "Bulk-Doping-Controlled Implant Site of Boron in Silicon," *Phys. Rev. B.* **42**:11419 (1990).

51. L. C. Hopkins, T. E. Seidel, J. S. Williams, and J. C. Bean, "Enhanced Diffusion in Boron Implanted Silicon," *J. Electrochem. Soc.* **132**:2035 (1985).

52. R. B. Fair, J. J. Wortman, and J. Liu, "Modeling Rapid Thermal Diffusion of Arsenic and Boron in Silicon," *J. Electrochem. Soc.* **131**:2387 (1984).

53. A. E. Michael, W. Rausch, P. A. Ronsheim, and R. H. Kastl, "Rapid Annealing and the Anomalous Diffusion of Ion Implanted Boron," *Appl. Phys. Lett.* **50**:416 (1987).

54. K. J. Reeson, "Fabrication of Buried Layers of SiO_2 and Si_3N_4 Using Ion Beam Synthesis," *Nucl. Instrum. Methods* **B19–20**:269 (1987).

55. K. Izumi, M. Doken, and H. Ariyoshi, "CMOS Devices Fabricated on Buried SiO_2 Layers Formed by Oxygen Implantation in Silicon," *Electron. Lett.* **14**:593 (1978).

56. H. W. Lam, "SIMOX SOI for Integrated Circuit Fabrication," *IEEE Circuits Devices* **3**:6 (1987).

57. P. L. F. Hemment, E. Maydell-Ondrusz, K. G. Stevens, J. A. Kilner, and J. Butcher, "Oxygen Distributions in Synthesized SiO_2 Layers Formed by High Dose O^+ Implantation into Silicon," *Vacuum* **34**:203 (1984).

58. G. F. Celler, P. L. F. Hemment, K. W. West, and J. M. Gibson, "Improved SOI Films by High Dose Oxygen Implantation and Lamp Annealing," in *Semiconductor-on-Insulator and Thin Film Transistor Technology*, A. Chiang, M. W. Geis, and L. Pfeiffer, eds., *Mater. Res. Soc. Symp. Proc.* **53**, Boston, 1986.

59. S. Cristoloveanu, S. Gardner, C. Jaussaud, J. Margail, A.-J. Auberton-Hervé, and M. Bruel, "Silicon on Insulator Material Formed by Oxygen Ion Implantation and High Temperature Annealing: Carrier Transport, Oxygen Activation, and Interface Properties," *J. Appl. Phys.* **62**:2793 (1987).

60. M. I. Current and W. A. Keenan, "A Performance Survey of Production Ion Implanters," *Solid State Technol.* **28**:139 (February 1985).

61. H. Glawischnig and K. Noack, "Ion Implantation System Concepts," in *Ion Implantation Science and Technology*, J. F. Ziegler, ed., Academic Press, Orlando, FL, 1984.

62. P. Burggraaf, "Equipment Generated Particles: Ion Implantation," *Semicond. Int.* **14**(10):78 (1991).

63. E. W. Haas, H. Glawischnig, G. Lichti, and A. Bleicher, "Activation Analytical Investigation of Contamination and Cross Contamination in Ion Implantation," *J. Electron. Mater.* **7**:525 (1978).

Chapter 6

Rapid Thermal Processing

The last few chapters have discussed the redistribution of impurities at high temperature. For small devices this redistribution is often very undesirable, and so there is a great deal of emphasis on low temperature processes that minimize diffusion. Some processes, such as implant annealing, however, are not as effective at low temperature. Certain types of implant damage cannot be annealed out unless high temperatures are achieved. Furthermore, some dopants require anneal temperatures of at least 1000°C to be completely activated. The other avenue to reducing diffusion is to reduce the time at temperature. Standard furnace annealing is ill suited to short-time anneals. The wafers in such a system heat from the edges inward. To avoid excessive temperature gradients that lead to warpage, the wafers must be heated and cooled slowly [1]. As a result, even though the annealing time could be short, the long temperature ramps result in significant diffusion. At the same time, there has been an increasing emphasis on processes that operate on one wafer at a time rather than large batches. These single wafer processes provide the best uniformity and reproducibility, particularly for large wafers rapid thermal processing (RTP) describes a family of single-wafer hot processes that has been developed to minimize the thermal budget of a process by reducing the time at temperature in addition to, or instead of, reducing the temperature.

RTP was originally developed for implant annealing. Although this application is still common, the use of rapid thermal approaches to processing has spread to oxidation, chemical vapor deposition, and epitaxial growth. Central to all rapid thermal processes is a set of common problems: heating and cooling the wafer uniformly, being able to maintain a uniform temperature during the process, and measuring the wafer temperature. The next few sections will describe the nature of these problems and the approaches that rapid thermal systems have adopted to address them. The chapter will then review some applications of the technology starting from the traditional implant annealing and the more recently developed oxidation and nitridization and concluding with the formation of silicides. The use of RTP for chemical vapor deposition (RTCVD) and epitaxial growth will be delayed until later in the book, since these topics also require an understanding of general chemical vapor deposition techniques.

Rapid thermal processes can be divided into three broad classes by the type of heating that is carried out: adiabatic, thermal flux, or isothermal [2]. The first demonstration of RTP used an adiabatic heat source [3]. In this approach, fast pulses of light in a broad beam heat only the front surface (of order a few microns) of the wafer as long as the pulse length is short compared with the thermal time constant of the substrate. Adiabatic systems are often powered by broad beam coherent sources such as excimer lasers. Although this type of annealing system allows the shortest time at temperature, it has several important drawbacks. These include difficulties in controlling temperatures and anneal times, large vertical temperature gradients, and large capital equipment costs. Thermal flux systems use an intense spot source such as an electron beam or a focused laser that is scanned across the wafer. To avoid large lateral thermal gradients, the scan period must be short compared to the thermal time constant. Although this type of system has been used for research, the defects caused by lateral thermal nonuniformity are usually large enough to prevent their use for IC fabrication. Isothermal heating uses a broad beam of radiation to heat the wafer for many seconds. These systems may have minimal temperature gradients across and through the wafer. Typically they are powered by incoherent sources such as an array of tungsten–halogen lamps. The wafer rests on quartz pins in isothermal systems. Quartz is selected because of its chemical stability and its low thermal conductivity. This arrangement is sometimes called *thermal isolation*. This chapter will concentrate on isothermal RTP systems since almost all current generation commercial systems use this design.

6.1 Gray Body Radiation, Heat Exchange, and Optical Absorption°

There are four types of heat transfer that may be of interest to semiconductor processing: conduction, convection, forced flow, and radiation. Thermal conduction is the diffusion of heat through a solid or gas. The heat flow through a cross-sectional area A of a solid or an immobile gas or liquid is given by

$$\dot{q}(T) = k_{th}(T)A\nabla T \qquad (6.1)$$

where $k_{th}(T)$ is the thermal conductivity of the material. Dividing both sides by the area produces Fick's first law in terms of heat transfer. Since most of the optical energy in rapid thermal processing is absorbed in the first few microns of the wafer, thermal conduction through the wafer plays an important role in the final temperature distribution. When considering thermal conduction in a gas, however, one must also take into account gas flow, since it can alter the rate of heat transfer. If the flow is caused by an externally applied pressure gradient it is said to be forced flow. Examples include flows caused by gas injection or pumping. Flows that are in response to temperature gradients in an otherwise closed system are called *natural flows*. The movement of water in a heated pot is a natural flow. An effective heat transfer can be defined as

$$\dot{q} = h(T - T_\infty) \qquad (6.2)$$

where T_∞ is the temperature of the gas far from the wafer and h is an effective heat transfer coefficient that depends on both free and forced flow. For most geometries h is a function of temperature and the position on the wafer.

Example 6.1

Some "fast ramp" furnaces use forced flow of a hot gas to heat a wafer. If $T_\infty = 1000°C$ and $h = 2 \times 10^{-4} \dfrac{W}{cm^2\text{-}°C}$, find the initial ramp rate for a silicon wafer 0.7 mm thick.

Solution

From Equation 6.2

$$\dot{q} = |h(T - T_\infty)|$$

Initially $T \approx 30°C$, 50

$$\dot{q} = |2 \times 10^{-4} \text{ W/cm}^2\text{-}°C \ (30 - 1000)|$$
$$= 0.19 \text{ W/cm}^2$$

Then the heating rate is

$$\frac{dT}{dt} = \frac{\dot{q}}{C_p \times \rho \times thickness}$$

where C_p is the specific heat and ρ is the mass density. From Appendix II,

$$\frac{dT}{dt} = \frac{0.19 \text{ W/cm}^2}{0.75 \text{ J/gm-}°C \ 2.33 \text{ gm/cm}^3 \times 0.07 \text{ cm}} = 1.7°C/sec = 102°C/min$$

The amount of power that can be delivered by gas flow is limited. As a result, most rapid thermal systems use radiative heat transfer as the primary method of heat exchange. One of the basic parameters of radiative heat transfer is the spectral radiant excitance $M_\lambda(\lambda, T)$, the amount of power radiated by a body into a perfectly absorbing environment (black body) per unit surface area of the emitting object and per unit wavelength of the radiation. Planck's radiation law gives the spectral radiant exitance as

$$M_\lambda(T) = \varepsilon(\lambda) \frac{c_1}{\lambda^5 (e^{c_2/\lambda T} - 1)} \tag{6.3}$$

where $\varepsilon(\lambda)$ is the wavelength-dependent emissivity of the emitting body and c_1 and c_2 are the first and second radiation constants given by 3.71×10^{-12} W-cm^2 and 1.44 cm-K, respectively. When $\varepsilon = 1$, the emitting source is said to be a black body.

When $M_\lambda(\lambda, T)$ is integrated over all wavelengths from 0 to ∞, and it is assumed that the emissivity is wavelength independent, the result is the total exitance $M(T)$, which is given by the Stefan–Boltzmann equation

$$M(T) = \varepsilon\sigma T^4 \tag{6.4}$$

where σ is the Stefan–Boltzmann constant, 5.6697×10^{-12} W/cm^2-K^4. Comparing Equations 6.2 and 6.4, one sees that the amount of power radiated by an object is proportional to the fourth power of the temperature, while the power conducted through thermal conduction is proportional to the temperature difference between the object and the background. As a result, radiation is the dominant heat transfer mechanism at high temperature while thermal conduction is more important at low temperature. Since most rapid thermal systems operate in the high temperature regime, radiation exchange is the dominant heat exchange mechanism.

Differentiating Equation 6.1 and setting the result equal to zero, one can define λ_{max} as the wavelength at which the emitted power is maximized. Then

$$\lambda_{max} = \frac{0.2898 \text{ cm-K}}{T} \tag{6.5}$$

This relationship can be used to translate the temperature of an emitting body into a corresponding color temperature. Many lamps are specified this way, since it would be difficult to measure their filament temperature directly.

Example 6.2

Consider the sun to act as a blackbody. The surface temperature is $6000°C$, and it has a radius of 6.95×10^8 m. Find the peak wavelength and the total power radiated.

Solution

The peak wavelength is

$$\lambda_{max} = \frac{0.2898 \text{ cm-K}}{6273 \text{ K}} = 462 \text{ nm}$$

where 462 nm corresponds to visible blue.
The power radiated is

$$P = M(T) \times \text{area}$$
$$= 4\pi \times (6.95 \times 10^8 \text{ m})^2 \times 5.67 \times 10^{-8} \text{ W/m}^2\text{-k}^4 \times (6273 \text{ k})^4$$
$$= 5.3 \times 10^{26} \text{ W}$$

When radiation is incident on the surface of the wafer it may be reflected, absorbed, or transmitted. $\rho(\lambda, T)$ is defined as the fraction of reflected radiation and $\tau(\lambda, T)$ as the fraction of the transmitted radiation. According to Kirchhoff's law

$$\varepsilon(\lambda, T) = 1 - \rho(\lambda, T) - \tau(\lambda, T) \tag{6.6}$$

For opaque materials $\tau(\lambda, T) = 0$, and

$$\varepsilon(\lambda, T) = 1 - \rho(\lambda, T) \tag{6.7}$$

Once the emissivities of two bodies are known, the net power transfer between them can be calculated. We might take these two bodies to be the wafer and the lamp array, for example. Let ε_1 and ε_2 be the average emissivities for all wavelengths being considered for the two bodies. Then the net power transfer from body 1 to body 2 is

$$s = \dot{q}_{1\rightarrow2} - \dot{q}_{2\rightarrow1} = \sigma(\varepsilon_1 T_1^4 - \varepsilon_2 T_2^4)A_1 F_{A1\rightarrow A2} \tag{6.8}$$

where A_1 is the area of body 1, and $F_{A2\rightarrow A1}$ is a geometric constant called the *view factor* or configuration factor. $F_{A1\rightarrow A2}$ is the fraction of the total solid that the area A_2 subtends

$$F_{A1\rightarrow A2} = \frac{1}{A_1} \int_{A_2}\int_{A_1} \frac{\cos\beta_1 \cos\beta_2}{\pi r^2} dA_1 dA_2 \tag{6.9}$$

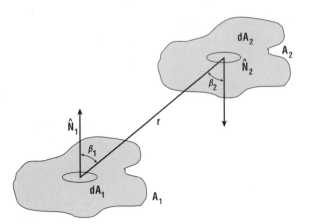

Figure 6.1   Geometry for calculating the view factors between two surfaces, A_1 and A_2.

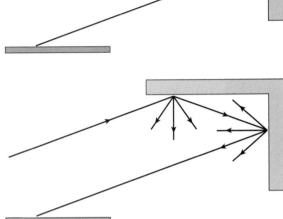

Figure 6.2   Possible optical paths among three surfaces including only single reflections. This diagram assumes diffuse reflections.

from the surface A_1; β_1 and β_2 are the angles from the surface normals as shown in Figure 6.1 [4]. The view factor can be calculated directly. Alternatively, the view factors for many simple geometries have been tabulated [5].

Consider what happens if a third surface such as a reflector is added (Figure 6.2). One might first consider three pairwise interactions, but that would include only the power radiated by the heated surfaces. The reflector in particular may be water cooled, keeping the surface temperature low. Yet the reflector surface may transmit a great deal of power because it increases the effective view factor between the wafer and the lamp. The reflector also allows the wafer to radiate to itself. If all possible reflections are included, the complexity of any pairwise scheme increases substantially. Alternatively, one can treat all of the surfaces simultaneously, including reflections, by using matrix methods [6, 7].

6.2 High Intensity Optical Sources and Chamber Design

Most isothermal systems use either tungsten–halogen lamps or long arc noble gas discharge lamps as power sources (Figure 6.3). The tungsten–halogen lamp consists of a tightly wound spiral tungsten filament encased in a fused silica envelope. The spiral winding increases the surface area of the lamp, thereby increasing the radiation efficiency. The spiral filament may be linear with a connector on both sides of the bulb or the filament may be shaped for single-ended termination. The quartz envelope of these lamps contains a halogenated gas. One common component is $PNBr_2$ [8]. Tungsten evaporates from the heated filament and deposits on the walls of the envelope. As the walls heat, the halide gas reacts with the tungsten to form volatile tungsten halides that diffuse to the much hotter filament, where they decompose and redeposit the tungsten. This process has an intrinsic feedback mechanism. As more tungsten deposits on the envelope, the increased quartz heating increases the etching reaction rate. This feedback prevents excessive buildup of tungsten.

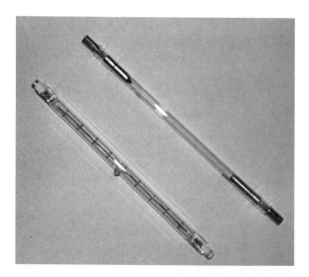

Figure 6.3 Tungsten–halogen lamps (left) and noble gas arc lamps (right) for use in rapid thermal systems.

Tungsten–halogen lamps emit as blackbodies. Most of the radiation is in the 0.8 to 4.0 μm range. Assume that it is necessary to be able to raise the temperature of a 150-mm wafer to 1100°C. Since the emissivity of silicon is approximately 0.7, the total exitance at this temperature (Equation 6.2) is about 2.5 kW. Each tungsten lamp consumes approximately 100 W of electrical energy per centimeter of lighted length. About 40% of that energy is converted into radiation in a typical filament lamp; however, that number varies significantly from lamp to lamp. An undoped, room temperature silicon wafer absorbs only photons with energy greater than the bandgap (1.1 eV). This means $\lambda \leq 1.1$ μm. Thus only about 15% of the incident energy is captured by the wafer initially. Since only a fraction of the emitted optical power is collected, it would be necessary to use an array of 15 to 30 8-in. lamps to heat the wafer.

Long arc noble gas discharge lamps contain two refractory metal electrodes sealed in a fused silica envelope with a noble gas such as krypton or xenon. The electrical properties of the discharge are determined by the electrode spacing, the inner diameter of the envelope, and the gas composition and pressure. When ignited with a high voltage pulse (roughly 2 kV per centimeter of lighted length) the gas is ionized and a dc path is established. These lamps emit strongly in the visible part of the spectrum. Their spectra include a very high temperature electron plasma that looks like a gray body with $\lambda_{max} \approx 200$ nm, along with discrete line spectra corresponding to the electronic transitions of the fill gas.

Discharge lamps consume up to 700 W/cm of lighted length. Optical conversion efficiencies are about 45%, and much more of the light is captured by the wafer. High power discharge lamps must be water cooled to be able to withstand these high power densities. Particular care must be taken to ensure that the quartz-to-metal seals and the metal electrodes remain cool. Using the same criteria to heating a 150-mm wafer to 1100°C, one would need only two to four 8-in. lamps; however, this advantage in power density is partially offset by the need to water-cool and electrically isolate the lamps for igniting them. Furthermore the lamps must be run by a more costly regulated dc source, while the filament lamps may operate directly from a simple line (i.e., 60-Hz) source.

Various chamber geometries have been used to optimize the power collection efficiency (Figure 6.4). At the same time, one must also try to design the chamber so that the wafer can achieve and maintain a uniform temperature. Many early RTP system designs use the reflecting cavity approach. In this design the wafer resides in a quartz flow tube. Linear tungsten–halogen lamps are above and below the flow tube. An N_2 purge may be used to cool the lamps and the outer surface of the quartz flow tube. The entire assembly is encased in a gold-coated box. The surface of the gold is roughened to ensure diffuse reflection. The purpose of this arrangement is to randomize the optical path so as to distribute the radiation uniformly across the wafer.

The use of reflecting cavities has been only partially successful in achieving a uniform temperature across the wafer. The wafer edge tends to be more cool than the center due to three effects (Figure 6.5). For a finite-sized array of lamps, the view factor from the outside of the wafer is less than that of the center. The edge of the wafer is almost completely blocked from direct radiation

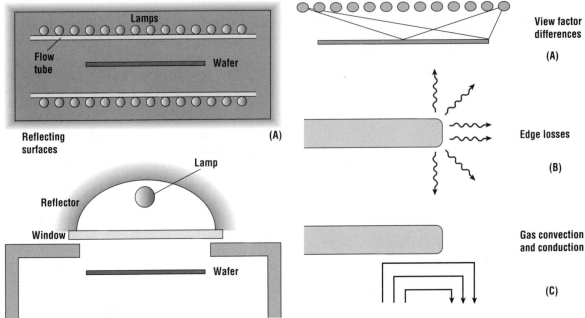

Figure 6.4   Various chamber designs include (A) the reflecting cavity and (B) a windowed system using an intense source and a shaped reflector.

Figure 6.5   Causes of thermal nonuniformity include (A) a reduced view factor to the lamp array for large *r*, (B) very small view factors along the wafer edge, and (C) nonuniform gas phase heat transfer *(after Campbell, 1994).*

exchange with the lamps. Finally, for common geometries, the edge of the wafer is more effectively cooled by the gas flow than the center. Collectively, these edge effects may lead to temperature gradients of several tens of degrees (Figure 6.6). These temperature gradients lead to process nonuniformity, and if severe enough, lead to slip and wafer warpage.

Various approaches have been used to minimize thermal nonuniformity in RTP systems. To compensate for the increased edge losses, radiant power to the edge of the wafer must be increased. Older RTP systems do this by shaping the reflectors or lamp spacings, or changing the transmissive properties of the fused silica flow tube. These approaches suffer from the fact that the amount of excess radiation required depends on the process temperature. Thus, the power distribution can be optimized for only one temperature and then only if the lamps do not age. Furthermore, temperature ramps require nearly uniform radiant distributions. As a result, the wafer edges overheat during temperature ramp up [9].

The solution to this problem is to divide the lamps into heating zones that can be independently controlled. For example, one set of lamps is arranged geometrically in such a way that it heats the outermost part of the wafer more than the center. A second set might heat the center of the wafer more effectively than the edges. A third might provide roughly uniform illumination. By choosing the proper combination of power settings, more uniform temperature can be obtained for a broad range of process conditions. This type of arrangement is called a *multizone heater.*

Most modern RTP systems are designed around this zone heating concept [10, 11]. At present the most common arrangement uses single-ended filament lamps housed in individual parabolic reflectors. Typical dimensions are 2.5-cm-long lighted filaments and approximately 2.5-cm-diameter

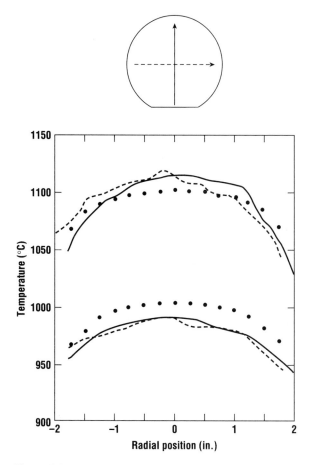

Figure 6.6 Typical wafer temperature distributions across the wafer in an early-generation rapid thermal system (*after Lord, © 1988 IEEE*).

reflecting cavities. The reflecting cavities are milled into an aluminum block and gold plated to reduce corrosion. The reflector can then be water cooled. The lamps and reflecting cavities are often laid out in a planar array with hexagonal symmetry, and grouped into banks corresponding to their distance from the center of the array. Over 100 1-kW lamps are typically used to heat a 200-mm wafer. By varying the power distribution between the heating rings, one can control the temperature profile across the wafer. This control can be done in response to actual wafer temperature profile measurements, or the system may be calibrated with a wafer using a large number of embedded thermocouples. This information is used to construct a semiempirical model of the wafer temperature under a wide variety of static and dynamic conditions. The power to the lamps is then distributed in such a way as to minimize thermal nonuniformity across the wafer. An interesting side note is that uniform temperature might not be desired in all processes. A zone-heated system allows the process designer to establish a semicontrolled temperature ramp across the wafer to compensate for gas phase depletion of other effects inside the chamber.

6.3 Temperature Measurement

One of the most difficult tasks associated with rapid thermal processing is accurate and reproducible temperature measurement. The wafer temperature is used in a feedback loop to control the lamp power output. In most pieces of process equipment, the wafer temperature is measured using a thermocouple embedded in the wafer holder or susceptor. This is not possible in RTP since there is no susceptor. The thermocouple can be placed as a point contact on the surface of the wafer, but the thermal impedance associated with such a contact will result in a substantial temperature difference between the wafer and the thermocouple. Furthermore, heat loss through the lead will cool the wafer locally, leading to stress and process nonuniformity in the vicinity of the contact. Finally, direct contact thermocouple measurements may allow chemical reactions with the silicon wafer. These reactions contaminate the wafer and shift the characteristics of the thermocouple.

For these reasons all temperature measurements in RTP systems are done indirectly. The most popular techniques are pyrometry (optical) or thermoelectric detectors. The most common thermoelectric detector for RTP is the thermopile. This device operates by the Seebeck effect. A bimetal junction is suspended on a thin membrane. When the junction is heated, it develops a small voltage. A second junction shielded from the radiation serves as a reference. The magnitude of the voltage difference between the two junctions varies linearly with the temperature difference between the junctions [12].

The second widely used method is pyrometry. Most pyrometers operate by measuring the radiant energy received in some band of energies, assuming the source to be a gray body of known emissivity, and converting the power to a source temperature using the Stefan–Boltzmann relationship. Care must be taken in selecting the operating wavelength to ensure that any lamp output at that wavelength is completely filtered before entering the chamber. Special-purpose glasses or other optical filters can be used for this purpose. Most commercial systems monitor some band in the mid-infrared (3–6 μm). Arc lamps produce insignificant radiation in this band. If tungsten–halogen lamps are used, the flow tube itself can be used to filter out most of the radiation from the lamps in this band. In that case, a hole or thinned window in the flow tube must be provided to allow the pyrometer a line of sight to the wafer. This type of system is called a spot-type radiation thermometer. Alternatively, are can use a fiber-optic element close to the surface of the wafer in an approach called a light pipe radiation thermometer. Both approaches have significant problems that must be over come [13]. The window thickness must be well known, or temperature errors as large as 100°C can result [14]. Another concern with pyrometry is that some process gases can absorb energy in the IR and so reduce the apparent wafer temperature [15]. Depending on the gas composition, pressure, and temperature, strong absorption bands can exist anywhere from 2 to 10 μm [16].

The major caveat in optical pyrometry is that the effective emissivity must be accurately determined. The effective emissivity includes both intrinsic and extrinsic contributions. The intrinsic emissivity is a function of the material, the surface finish, the temperature, and the wavelength at which it is measured. Fortunately, the intrinsic emissivity of bare silicon wafers is reasonably well characterized [17] as a function of temperature (Figure 6.7). Timans provided a comprehensive review of the optical properties of semiconductors [18]. At low temperature the emissivity depends on the doping concentration for energies less than the bandgap. As the temperature increases above 600°C, the semiconductor becomes intrinsic. Free carriers are present in sufficient numbers to produce a nearly wavelength-independent emissivity. The extrinsic emissivity relates to the amount of radiant energy from other sources that is reflected back onto the spot being measured and so increases the

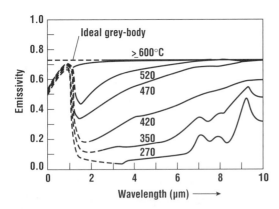

Figure 6.7   The emissivity of silicon as a function of wavelength with temperature as a parameter. Above 600°C the wafer is intrinsic *(after Sato, reprinted by permission, Japan. J. Appl. Phys.).*

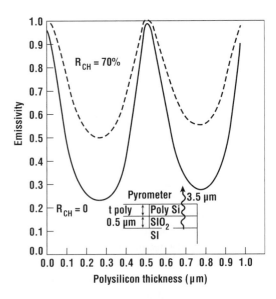

Figure 6.8   The effect of polysilicon thickness on effective emissivity for a polysilicon/SiO₂/silicon structure in nonreflective and 70% reflective cavities *(after Hill and Boys, reprinted by permission, Plenum Publishing).*

apparent temperature. Shields are often used to minimize this effect [19], which therefore must be measured in situ and may change as the chamber reflectivity changes. Furthermore, the presence of layers such as polysilicon and silicon dioxide can dramatically alter the apparent emissivity (Figure 6.8) at the measurement wavelength due to interference effects [20].

Example 6.3

The major source of uncertainty in pyrometry is an uncertainty in the emissivity. If the wafer temperature is 1000°C, what wavelength is most desirable to minimize the effect of this uncertainty?

Solution

According to Equation 6.3,

$$M_\lambda(T) = \varepsilon(\lambda)\frac{c_1}{\lambda^5(e^{c_2/\lambda T}-1)}$$

Differentiating and assuming that M and λ are fixed and that $e^{c_2/\lambda T} \gg 1$,

$$\frac{dT}{T} = -\frac{\lambda T}{c_2}\frac{d\varepsilon}{\varepsilon}$$

If the pyrometer is operating at 5 μm, a 5% uncertainty in the emissivity produces a 22°C error in the temperature. On the other hand, a pyrometer operating in the 0.94–0.96 μm range produces only a 4°C temperature error for the same 5% error in emissivity.

Commercial RTP systems can control the wafer center temperature for a bare wafer process such as an implant anneal to about 2°C; however, the absolute temperature is generally less well known due to emissivity uncertainties. Two or more color pyrometers measure the emitted power at multiple wavelengths to attempt to correct for emissivity changes. Implicit in their operation is the assumption that the ratio of the emissivity at these wavelengths is fixed. This is not always a good assumption, particularly when the process involves a film growth or deposition. Of course, the emissivity will change from that of the substrate material to the emissivity of the material deposited during the process. More problematic, however, is the situation that occurs when the film is at least partially transmissive. When the thickness of the film becomes an integral multiple of one-fourth of the measurement wavelength, destructive interference takes place that dramatically changes the apparent emissivity. For these processes, temperature reproducibility is severely degraded unless this effect can be corrected in software. It is possible to measure the emissivity directly by measuring the reflectivity of the wafer and using Kirchoff's law (Equation 6.7) [21]. For this scheme to work, the measurement must be done in the visible where the wafer is opaque [22]. The pyrometry should be done at wavelengths as close as possible to that of the reflection measurement; however, in practice this is difficult to achieve. Most commonly, the pyrometer is calibrated against an instrumented wafer, that is, with thermocouples embedded in it [23].

Another variation on this is to use the fact that the lamps are being powered by an ac source and so the light intensity will oscillate. Since the temperature of (and so radiation emitted from) the wafer is essentially constant, measuring the ac light intensity is largely measuring the reflectivity of

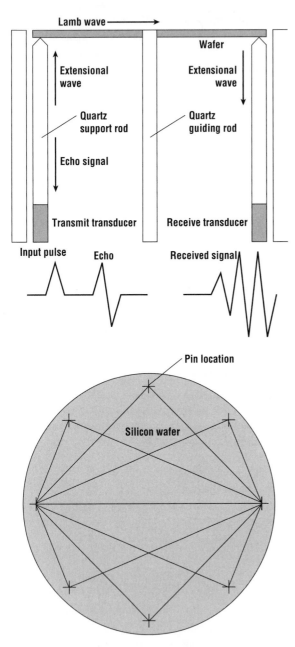

Figure 6.9   Acoustic temperature measurement side and top views *(after Degertekin et al., used by permission, Materials Research Society).*

the wafer. This can, in turn, be used to calculate the emissivity. First developed by Lucent, this method, called ripple pyrometry, has seen substantial use [24]. Improvements in the process have led to 1.5°C control of 300-mm wafers [25].

Various nonpyrometric techniques have been investigated to avoid the problems associated with assuming an emissivity. None of these, however, have seen much application in commercial RTP systems. Several authors have demonstrated the use of thermal expansion as a direct measure of the temperature. Diffraction gratings patterned on the surface of the wafer can be used with projection moiré interferometry [26] or in a diffraction order measurement [27]. In either case, the movement of the grating via thermal expansion can be measured and used to infer temperature changes. Of course, these methods require an appropriate pattern on the surface of the wafer that must be properly aligned with the optical probe beam and detector. Furthermore, noise can be a problem at high temperature due to strong gas flow conditions caused by the large thermal gradients in the chamber. Alternatively, the wafer diameter itself can be measured optically [28]. The demonstrated repeatability of this technique is 1%.

Another interesting approach to measuring temperature is the use of acoustic waves. It has been found that the velocity of sound in silicon is a linear function of the substrate temperature [29]. Acoustic waves can be launched through one of the quartz support pins and the wave detected using another pin (Figure 6.9). Since the transducers can act as both acoustic sources and sensors, there are $\frac{1}{2}(n^2 - n)$ possible paths. Because the measured velocity is averaged over the path, the temperature profile across the wafer can be extracted by taking multiple measurements along different paths [30]. This information is extremely valuable for multizone heating arrangements.

Finally, either reflectance or transmission can be used to measure the optical properties of the wafer if it is not metallized. At low temperatures one can use the wavelength dependence to the reflectance to infer a temperature. When the energy of the photon is equal to or greater than the bandgap of silicon, the reflectance is much lower. Since the bandgap is temperature dependent, one can accurately infer the temperature of the silicon up to about 600°C. At higher temperatures (and for heavily doped wafers), the carrier concentration is high enough that carrier absorption dominates the reflectance. The second technique measures the intrinsic carrier concentration by measuring the transmission of IR radiation through the wafer. Assuming $n_i \gg N_D$, N_A, one can infer the temperature from the measured inverse absorption length, which can be related to the intrinsic carrier concentration. Neither technique has seen widespread commercial use due to the restrictions on doping concentrations and metallizations.

6.4 Thermoplastic Stress°

The existence of the thermal gradients in the wafer gives rise to thermoplastic stress. If the stress becomes too large, defects such as dislocations and slip can result. Assuming that the substrates are isotropic, that dislocations do not occur, and that the temperature gradients vertically through the wafer can be neglected, one can estimate the thermal stresses on the wafer. Due to radial symmetry in $T(r)$, the shear stress will be zero. The radial and angular stress components are given by

$$\sigma_r(r) = \alpha E \left[\frac{1}{R^2} \int_0^R T(r')r'\,dr' - \frac{1}{r^2} \int_0^r T(r)r'\,dr' \right] \tag{6.10}$$

and

$$\sigma_s(r) = \alpha E \left[\frac{1}{R^2} \int_0^R T(r')r'\,dr' - \frac{1}{r^2} \int_0^r T(r)r'\,dr' - T(r) \right] \tag{6.11}$$

where α is the linear thermal expansion coefficient, E is Young's modulus, and R is the radius of the wafer.

Figure 6.10 shows a typical plot of the radial dependence of these stresses for a simple process consisting of a linear ramp to temperature, an anneal at temperature, and a linear ramp down. Under steady state conditions the wafer edges are slightly cooler than the center. Then the radial stress has a maximum near the center of the wafer and goes to zero at the edge. The tangential stress rises from zero at the center and is typically much larger than the radial stress component. Since this stress is so large and since defects will nucleate more readily at the wafer edge, most RTP-induced slip is seen at the edge of the wafer. The diamond structure of silicon tends to yield along the $\langle 110 \rangle$ directions in the (111) planes and so for p-type (100) wafers, the slip lines are parallel and perpendicular to the major flat orientation.

The yield strength is the limit at which a material will plastically deform. For silicon the yield strength is described by the Haassan formula

$$\sigma_{yield} = A \left[\frac{\dot{e}}{\dot{e}_o} \right]^{1/n} e^{E_a/kT} \tag{6.12}$$

where $\dot{e}$ is the strain rate and $\dot{e}_o$ is a reference strain rate normally taken to be 10^{-3} sec^{-1} [31].

The exact value of σ_{yield} depends on the oxygen and dopant concentrations in the wafer and the previous processing. Typical values include $A = 3630$ Pa, $E_a = 1.073$ eV, and $n = 2.45$ [32]. At high strain rates the yield strength is limited to 60 MPa, while at low strain rates σ_{yield} does not decrease below about 3.5 MPa [33]. Thus, the wafer can tolerate higher strains during transients, but

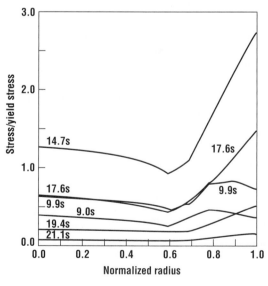

Figure 6.10 Normalized stress versus position on a wafer during a heating transient *(after Lord, © 1988, IEEE).*

the strain generated during transients can be much larger than that generated in the steady state, since simple single-zone RTP systems are optimized for steady state uniformity. Of course this is much less an issue in newer multizone-heated systems.

6.5 Rapid Thermal Activation of Impurities

The last chapter introduced ion implantation. Because of its ability to produce well-controlled impurity doses whose concentration may in fact exceed the solid solubility, the technique has become almost universally accepted. As device sizes have been reduced, the concentration gradients have increased, and the maximum allowable dopant redistribution has decreased. The implant damage in these wafers must still be removed by annealing. Depending on the implant species energy and dose, this may require temperatures as high as 1100°C [34]. The basic reason that rapid thermal processing was developed was to access these high temperatures while minimizing the thermal budget by reducing the time at temperature. The time required to anneal out defect clusters drops as the temperature rises, and reaches microseconds at 1000°C [35].

One of the most attractive features of RTP is that the wafer may not reach thermal equilibrium. This means that the electrically active doping profiles can actually exceed the solid solubility. Arsenic, in particular, is found to require only very short anneals to achieve a high level of activation [36]. Arsenic, in particular, can be activated to about 3×10^{21}, about 10 times its solid solubility if annealed for a few milliseconds [37]. The arsenic atoms have insufficient time to form clusters and condense out into inactive defects [38]. If the activation is too incomplete, however, the excess arsenic atoms seem to contribute a deep level [39]. These levels can be efficient generation/recombination centers leading to electrical leakage if they are close to p–n junctions.

It has generally been observed that low temperature and reduced time anneals of implanted species produce chemical junctions that are deeper than those predicted by simple diffusion theory [40, 41]. Diffusivity enhancements for boron in silicon have been reported to be as much as a factor of 1000. The origin of this enhancement is believed to be residual implant damage. The wafer is believed to have a high concentration of vacancies and self-interstitials after an implant [42]. This effect is sometimes called a transient effect or transient-enhanced diffusion. It has been shown that the increase in the junction depth of low dose implants is proportional to the square root of the implant energy. Furnace anneal processes often attempt to annihilate excess point defects through an extended treatment at temperatures of 500 to 650°C before heating to the activation temperature. RTP anneals may also include a brief low temperature step for the same reason.

The activation energies for transient-enhanced diffusion for the three most common silicon dopants are shown in Table 6.1 [43]. Although the result was initially somewhat controversial, it is now generally agreed that transient effects during the diffusion of arsenic are observed for very high dose implants but are much less pronounced than for boron. These transient effects decay with some characteristic time constant that is related to the rate of defect annihilation in the substrate.

Table 6.1   The activation energies (eV) for steady state intrinsic diffusion and transient diffusion in silicon

	Steady State	Transient
Boron	3.5	1.8
Arsenic	3.4	1.8
Phosphorus	3.6	2.2

Data taken from Fair [43].

It has also been shown that for boron and BF$_2$ not all of the chemical impurities are activated in RTP. The peak concentrations activate more fully at low temperature or BF$_2$ due to the increased degree of amorphization [44]. Furnace anneals always activate the low concentration implant tails, but may not fully activate the region near the peak concentrations due to thermodynamic considerations such as the solid solubility. It has been observed, however, that after RTP anneals the low concentration boron tails may not be fully activated, giving rise to an electrical junction that is more shallow than the chemical one [45, 46]. Figure 6.11 shows the difference between the chemical profile (solid line) and electrically active profile (closed points) after a 30-sec anneal at 1000°C for both boron and BF$_2$ implants. Both peak and tail concentrations are not fully activated. The nonactivated impurities are believed to be due to the formation of inactive boron interstitial pairs [47]. Figure 6.11 also shows that the tail of the B profile diffuses faster than the peak region. It is found that in the peak region the implant leads to dislocations and other extended defects. For $X > R_p + 1.5 \, \Delta R_p$ or $< R_p - 0.4 \Delta R_p$, the point defect density is high, leading to transient-enhanced diffusion [48].

Most implant activation in GaAs requires the use of a capping layer, usually Si$_3$N$_4$ or SiO$_x$N$_y$, to prevent the outdiffusion of arsenic from the wafer [49]: Si$_3$N$_4$ protects GaAs up to about 900°C, while SiO$_2$ is effective up to about 850°C [50]. The cap layer is typically deposited by plasma-enhanced chemical vapor deposition. As with silicon, the wafer is first annealed at close to 650°C for about 60 sec to reduce the point defect density. Activating the n-type impurity silicon in GaAs requires temperatures of 800 to 1000°C. Capping layers may be difficult to remove and, due to the thermal expansion mismatch, can lead to slip formation in the wafer. It has also been shown that GaAs implant activation can be done without capping layers using a specially designed graphite susceptor (Figure 6.12) [51].

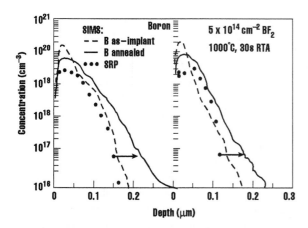

Figure 6.11 Chemical and active boron profiles after incomplete activation in RTP *(after Kinoshita et al., used by permission, Materials Research Society)*.

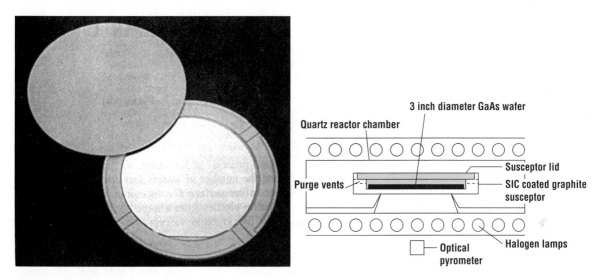

Figure 6.12 Photograph and schematic for capless arrangement for annealing GaAs *(after Kazior et al., © 1991 IEEE)*.

The wafer is placed inside a graphite cavity that is two or three times thicker than a typical wafer [52]. It has been shown that this type of annealing leads to an improved surface quality and a reduced particle count [53] compared with standard capped RTA. Finally, capless anneals can also be done by laying another GaAs wafer on top of the wafer to be annealed in the RTP. Care must be taken to ensure that the surface oxides have been removed from the upper wafer.

6.6 Rapid Thermal Processing of Dielectrics

Deep submicron devices require the production of very thin oxides of silicon. One technique for growing these oxides is to slow the oxidation rate by reducing the oxidation temperature. The difficulty with this approach is that the fixed charge and interface state density both tend to increase as the growth temperature is reduced [54]. As a result, rapid thermal oxidation (RTO) appears to be an attractive alternative, allowing short-time oxidations at suitably high temperatures in a well–controlled ambient. This indeed has become one of RTP's areas of application. Alternatively, it is possible to slow the oxidation rate by diluting the oxygen with argon or some other inert gas. Although this approach is capable of growing high quality oxides, the diffusion of impurities in the substrate is still a concern.

Virtually all RTO is done with dry oxygen. Early results showed that oxide thickness increased linearly with time, with an oxidation rate of approximately 3 Å/sec at 1150°C [55]. The oxides were electrically robust with good breakdown characteristics [56]. Rapid thermal oxidation uniformity is affected by the thermoplastic stress in the wafer caused by nonuniform temperatures and can result in a growth rate enhancement near the wafer edge where the stress is a maximum [57]. This effect is most pronounced for low temperature and short-time oxidations.

Numerous papers have been published regarding the oxidation rate of silicon during RTP. A representative set of data [58] is shown in Figure 6.13. A cohesive model of the oxidation rate during RTO has been difficult to develop for several reasons. The first is that RTO normally occurs completely within the initial oxidation regime. Furthermore, a wide variability in the experimentally measured oxidation rates has been reported in the literature. Part of this can be related to the difficulty in measuring the wafer temperature during growth. It is generally believed that as much as a 50°C temperature variability can be deduced from the literature [59] for many of the early RTO papers. Furthermore, the oxidation rate is known to increase with photon exposure [60]. The amount of this effect depends on the light source, with arc lamps having a more intense spectrum in the UV and therefore a greater effect. The enhancement is due to the photodissociation of O_2 molecules to form O^- ions, which drift to the Si interface under the influence of a space charge field. The effect is strongest for P^+ substrates and goes to zero for N^+ substrates [61].

As mentioned in an earlier chapter, very thin gate oxides have reliability concerns. It has been found that nitridizing an oxide can increase the ability of impurities to diffuse through the oxide and can reduce carrier trapping under electrical stress [62]. Heavy nitridization reduces the mobility of the inversion layer by 20 to 50% [63] and leads to an increase in electron trapping. This is believed to be due to an increase in hydrogen incorporation in the oxide [64]. Using RTP, a light nitridization can be done [65, 66] that avoids most of this mobility reduction. It has also been shown

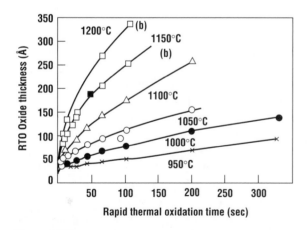

Figure 6.13   Typical data for oxide thickness as a function of time for a rapid thermal oxidation process *(after Moslehi et al., 1985).*

that using RTP to introduce small amounts (parts per million) of fluorine in an oxide produces excellent breakdown characteristics. This fluorine can be incorporated through the use of NF_3 in a rapid thermal processor [67].

It is also possible to produce nitrided oxides by growing the oxide directly in nitric (NO) or nitrous (N_2O) oxygen ambients. The N_2O-grown films have better reliability and are more resistant to boron penetration than O_2-grown films [68]. The peak nitrogen concentration occurs near the silicon interface and varies from about 1% (at 900°C) to about 1.7% (at 1100°C). Nitrogen incorporation in these films is known to slow the oxidation rate, presumably since it reduces the diffusivity of O_2 in the dielectric [69]. The NO-grown oxynitrides are known to incorporate even more nitrogen. Once again, the concentration depends on the temperature of the growth. The profile of nitrogen in NO-grown films is strongly affected by processing conditions, with much higher nitrogen concentrations in the bulk of the films [70]. NO can also be used to incorporate nitrogen in SiO_2 layers grown in O_2.

One significant potential advantage of RTP is its potential to be used as a hot wafer/cold wall process if the chamber walls are properly cooled. The importance of this observation is that multiple processes can be run in a single chamber, since the walls do not contaminate sequential process steps. The wafer temperature can be used to start and stop these reactions. The gas flowing in the chamber can be changed while the wafer is cool, preventing any chemical reaction. The advantage of this technique, of course, is the minimization of wafer handling and resultant contamination. Multiple layers can be deposited in this way with well-controlled interfaces. This capability was exploited very early in the development of RTP [71] by thermally growing a thin oxide and then depositing a layer of polycrystalline silicon using chemical vapor deposition to fabricate MOS capacitors. The capacitors were found to have excellent interfacial properties. Later work by the same group demonstrated that the lower silicon electrode could be grown selectively in silicon dioxide windows followed by oxidation of the epitaxial layer and finally polysilicon deposition so that both interfaces of the dielectric were formed in the RTP reactor [72].

6.7 Silicidation and Contact Formation

As will be described in some detail in Chapters 15, 16, and 18, for some device applications the resistivity of even heavily doped silicon is too large. In those cases, it is common to form metal silicides on top of the exposed silicon to reduce the resistivity. Metal silicides are often formed by depositing a thin metal layer on top of the wafer and then heating it to drive a silicidation reaction. Common metals for silicide formation include Ti, Pt, W, Ta, Mo, and Co. RTP allows the careful control of the silicidation temperature and ambient to minimize impurities and to promote the growth of the most desirable stoichiometry and phase of the silicide. In some cases the wafer has a partial coverage of SiO_2 before metal deposition. The metal does not react with the oxide and so can be easily removed wet chemically after the silicide reaction.

For many years $TiSi_2$ was been one of the most desirable films due to its low resistivity (13 $\mu\Omega$-cm). During the silicide formation anneal, however, silicon can diffuse along the grain boundaries of the $TiSi_2$, leading to an overgrowth of the silicide on top of the edges of the oxide (Figure 6.14) [73]. This overgrowth can be minimized by first annealing at a low temperature (450–600°C) to form TiSi. The unreacted metal is then stripped and a high temperature RTA is done to form the desired phase (C54) of the silicide. This phase of $TiSi_2$ requires an anneal of at least 750°C [74, 75], with some reports indicating that as high as 900°C is required [76]. When the initial silicidation is carried out in a pure N_2 ambient, lateral growth of the silicide is further impeded [77]. The titanium silicidation reaction is sensitive to ppm levels of H_2O or O_2 in the ambient, since oxidation and silicidation will compete at the interface. Oxygen will also dramatically increase the sheet resistance of the silicide. Silicides formed in N_2 containing 10 ppm of O_2, for example, have nearly

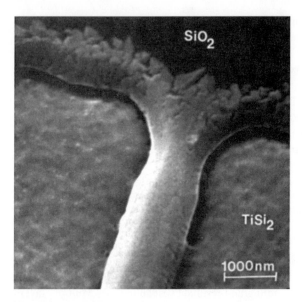

Figure 6.14   Results of a 900°C 30-sec formation of TiSi$_2$ in an oxide window. The ragged edges indicate silicide formed over the edges of the window *(after Brat et al., used by permission, The Electrochemical Society).*

40 times the resistivity of silicides formed in 0.1 ppm O$_2$ [78]. RTP has a significant advantage over standard furnace anneals with regard to oxygen contamination, because the smaller RTP flow tubes reduce backstreaming from the air [79].

As will be discussed in a later chapter, cobalt silicide is very attractive for small MOS devices. The silicide forms as Co$_2$Si at 300–370°C, shifts toward CoSi at about 500°C, and forms CoSi$_2$ at 700°C or higher. Higher final anneal temperatures lead to lower junction leakage, presumably due to a smoothing of the silicide/silicon interface [80]. Cobalt is the dominant diffuser, although in the CoSi$_2$ phase silicon diffusion dominates [81].

Another application of RTP for silicon technologies is the formation of barrier metals. These conducting films are used to prevent the interdiffusion of the silicon substrate and the aluminum-based alloys used for interconnecting devices. One common diffusion barrier is TiN. The film can be formed by reacting titanium with N$_2$ or NH$_3$ during a rapid thermal anneal, but this method produces high resistance contacts when small geometries are used [82]. It is preferred instead to deposit the TiN with close to perfect stoichiometry and then anneal the films in an RTA with a nitridizing ambient. This is sometimes called *grain boundary stuffing*, since the residual nitrogen is believed to occupy surface sites along the grain boundaries of the polycrystalline TiN. This nitrogen passivates the grain boundary and reduces the grain boundary diffusivity.

Rapid thermal processing has also been used for contact formation in GaAs technologies. As will be discussed in more detail in Chapter 15, low resistance ohmic contacts to n-type GaAs are formed by depositing and thermally annealing a layer containing a mixture of gold and germanium. Although early technologies used conventional ovens for this annealing, it has been shown that very desirable contacts can be formed reproducibly using a short anneal at about 450°C [83]. The interfacial layer between the metal and semiconductor is considerably broader in conventional anneals [84], and the surface of the contact is more smooth, which improves circuit yield. Since temperature is a critical parameter in forming these contacts, a backside metallization on the GaAs wafer is sometimes used to ensure a uniform temperature across the wafer [85].

6.8 Alternative Rapid Thermal Processing Systems

As discussed earlier in the chapter, one of the principal problems with RTP is thermal uniformity. more recently developed RTP systems use nontraditional chamber and/or heating designs to improve the thermal uniformity in RTP and to extend benefits of rapid thermal processing to small batch systems.

One of the more radical approaches to solving thermal uniformity in RTP is the resistively heated or quartz lamp–heated bell jar (Figure 6.15). The wall is either made from SiC [86] or quartz [87]. The chamber walls are typically kept several hundred degrees above the maximum wafer temperature. The lower part of the chamber is kept cool. The wafer is rapidly heated in the chamber. The heating rate is controlled by the velocity of the wafer elevator and is typically limited to about 100°C/sec [88].

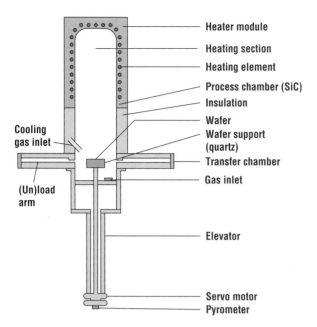

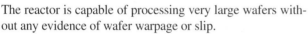

Figure 6.15 New designs for high uniformity RTP include the hot wall system *(after Roozeboom and Parekh).*

Figure 6.16 The Applied Materials Centura RTP Honeycomb source in the lifted position. The wafer is in the lower part of the photograph *(courtesy Applied Materials).*

The reactor is capable of processing very large wafers without any evidence of wafer warpage or slip.

An alternative to RTA is the fast ramp furnace. Similar in design to a small load, vertical tube furnace, these systems can heat up to 50 200-mm wafers at rates as high as 75°C/min. This is about an order of magnitude larger than conventional furnaces, but 100 times slower than true RTA systems. The systems use a model-based control of multiple heating elements to avoid excess thermal stress [89].

6.9 Summary

Rapid thermal processing was developed to enable short-time, high temperature implant annealing. Generally, the wafer rests on quartz pins in a flow tube and is heated using a bank of high intensity filament lamps (Figure 6.16). Problems with RTP include temperature measurement and thermal uniformity of the wafer. Excessive temperature gradients across the wafer cause thermoplastic stress that may lead to wafer warpage and/or slip. Rapid thermal processing has been extended to include silicide and barrier metal formation, thermal oxidation, chemical vapor deposition, and epitaxial growth.

Problems

1. Undoped silicon is nearly transparent for photons with energies close to or less than the bandgap (1.1 eV). Assume that a wafer is transparent for $\lambda > 1$ μm. If the wafer is heated with a tungsten–halogen lamp operating at 2000 K, what fraction of the incident energy is transmitted through the wafer? Would GaAs absorb more or less of the energy? Why?

2. The primary mechanism for absorbing this infrared radiation is by free carriers (electrons and holes). As the wafer heats up, the intrinsic carrier concentration increases (see Chapter 3). What effect will this have on RT processes? Roughly sketch a plot of temperature versus time for an undoped silicon wafer heated with a constant power tungsten–halogen lamp.

3. Large-diameter wafers are difficult for RTP at high temperature without the occurrence of thermoplastic-induced slip at the edge of the wafer. Describe the reasons for this slip. If the slip is more pronounced after high temperature ramps, what does this tell you about the radiation pattern at the surface of the wafer?

4. What problems might you expect if you tried to use your software to model rapid thermal annealing?

5. A rapid thermal processing system is used to heat a 200-mm-diameter silicon wafer to 950°C.
    (a) If the effective emissivity of the wafer is 0.7, use Equation 6.4 to calculate the amount of power necessary to maintain this temperature.
    (b) If one wants to ramp the temperature at 100°C/sec, what additional amount of power is required? You can assume that the wafer is 700 μm thick and that you can use the room temperature specific heat and mass density for silicon given in Appendix II.

6. Why has little work been done on rapid thermal oxidation in a wet ambient? (*Hint*: What are RT oxides used for? Are wet oxides suitable for this application?)

7. Explain the reasons for multizone heating in RTP.

References

1. S. M. Hu, "Temperature Distribution and Stresses in Circular Wafers in a Row During Radiative Cooling," *J. Appl. Phys.* **40**:4413 (1969).

2. C. Hill, in *Laser and Electron Beam Solid Interactions and Materials Processing*, J. F. Gibbons, L. D. Hess, and T. W. Sigmon, eds., Elsevier-North Holland, New York, 1981.

3. T. O. Sedgewick, "Short Time Annealing," *J. Electrochem. Soc.* **130**:484 (1983).

4. S. A. Campbell, "Rapid Thermal Processing," in *Computational Modeling in Semiconductor Processing*, M. Meyyappan, ed., Artech, New York, 1994.

5. J. R. Howell, *A Catalog of Radiation Configuration Factors*, McGraw-Hill, New York, 1982.

6. K. Knutson, S. A. Campbell, and F. Dunn, "Modeling of Three-Dimensional Effects on Temperature Uniformity in Rapid Thermal Processing of Eight Inch Wafers," *IEEE Trans. Semicond. Manuf.* **7**:68 (1994).

7. K. L. Knutson, *Theoretical and Experimental Investigations of Thermal Uniformity in Rapid Thermal Processing and Reaction Kinetics in Chemical Vapor Deposition for a Dichlorosilane/ Hydrogen System*, Ph.D. Dissertation, University of Minnesota, 1993.

8. J. R. Coaton and J. R. Fitzpatrick, "Tungsten-Halogen Lamps and Regenerative Mechanisms," *IEE Proc.* **127A**:142 (1980).

9. S. A. Campbell and K. L. Knutson, "Transient Effects in Rapid Thermal Processing," *IEEE Trans. Semicond. Manuf.* **5**:302 (1992).

10. P. P. Apte, S. Wood, L. Booth, K. C. Saraswat, and M. M. Moslehi, *Mater. Sci. Symp. Proc.* **224**:209 (1991).

11. M. M. Moslehi, J. Kuehne, R. Yeakley, L. Velo, H. Najm, B. Dostalik, D. Yin, and C. J. Davis, "In-Situ Fabrication and Process Control Techniques in Rapid Thermal Processing," *Mater. Sci. Symp. Proc.* **224**:143 (1991).

12. F. Roozeboom, "Temperature Control and System Design Aspects in Rapid Thermal Processing," *Mater. Res. Soc. Symp. Proc.* **224**:9 (1991).

13. B. K. Tsai, "A Summary of Light pipe Radiation Thermometry Research at NIST," *J. Res. NIST* **111**(1):9 (2006).

14. B. Brown, *Proc. 9th European RTP Users Group Meeting*, Harlow, U.K., 1992.

15. A. J. LaRocca, in *The Infrared Handbook*, W. L. Wolfe and G. J. Zissis, eds., Environmental Research Institute of Michigan, Ann Arbor, 1989.

16. F. Roozeboom, "Rapid Thermal Processing Status, Problems and Options After the First 25 Years," *Mater. Res. Soc. Symp. Proc.* **303**:149 (1993).

17. T. Sato, "Spectral Emissivity of Silicon," *Jpn. J. Appl. Phys.* **6**:339 (1967).

18. P. J. Timans, "The Radiative Properties of Semiconductors," in *Advances in Rapid Thermal and Integrated Processing*, F. Roozeboom, ed., NATO ASI Series E, Vol. 318, Kluwer, Amsterdam, 1996, p. 35.

19. D. P. Dewitt, F. Y. Sorrell, and J. K. Elliott, "Temperature Measurements Issues in Rapid Thermal Processing," in *Integrated Processing VI*, Vol. 470, MRS, Pittsburgh, 1997, p. 3.

20. C. S. Hill and D. Boys, "Rapid Thermal Annealing Theory and Practice," in *Reduced Thermal Processing for ULSI*, R. A. Levy, ed., Plenum, New York, 1989.

21. A. T. Fiory, C. Schietinger, B. Adams, and F. G. Tinsley, "Optical Fiber Pyrometry with In-Situ Detection of Wafer Radiance and Emittance—Accufiber's Ripple Method," *Mat. Res. Soc. Symp. Proc.* **303**:139 (1993).

22. J.-M. Dihlac, C. Ganibal, and N. Nolhier, "In-Situ Wafer Emissivity Variation Measurement in a Rapid Thermal Processor," *Mat. Res. Soc. Symp. Proc.* **224**:3 (1991).

23. R. Vandena Beele and W. Renken, "Study of Repeatability, Relative Accuracy and Lifetime of Thermocouple Instruments Calibration Wafers for RTP," in *Rapid Thermal Integrated Processing VI*, Vol. 470, MRS, Pittsburgh, 1997, p. 17.

24. B. Nguyemphu, M. Oh, and A. Fiory, "Temperature Monitoring by Ripple Pyrometry in Rapid Thermal Processing," in *Rapid Thermal and Integrated Processing V*, Vol. 42, MRS, Pittsburgh, 1996, p. 291.

25. M. Glück, W. Lerch, D Löffelmacher, M. Hauf, and U. Kreiser, "Challenges and Consent Status in 300 mm Rapid Thermal Processing" *Microelectronic Eng.* **45**:237 (1999).

26. S. H. Zaidi, S. R. J. Brueck, and J. R. McNeil, "Non Contact 1°C Resolution Temperature Measurement by Projection Moiré Interferometry," *J. Vacuum Sci. Technol.* **B10**:166 (1992).

27. S. R. J. Brueck, S. H. Zaidi, and M. K. Lang, "Temperature Measurement for RTP," *Mat. Res. Soc. Symp. Proc.* **303**:117 (1993).

28. B. Peuse and A. Rosekrans, "In-Situ Temperature Control for RTP Via Thermal Expansion Measurement," *Mat. Res. Soc. Symp. Proc.* **303**:125 (1993).

29. B. A. Auld, *Acoustic Fields and Waves in Solids*, Vol. 1, Wiley, New York, 1973.

30. F. L. Degertekin, J. Pei, Y. J. Iee, B. T. Khuri-Yakub, and K. C. Saraswat, "In-Situ Temperature Monitoring in RTP by Acoustical Techniques," *Mat. Res. Soc. Symp. Proc.* **303**:133 (1993).

31. A. E. Widmer and W. Rehwald, "Thermoplastic Deformation of Silicon Wafers," *J. Electrochem. Soc.* **133**:2405 (1986).

32. J. R. Patel and A. R. Chaudhuri, "Macroscopic Plastic Properties of Dislocation Free Germanium and Other Semiconductor Crystals. I. Yield Behavior," *J. Appl. Phys.* **34**:2788 (1963).

33. H. A. Lord, "Thermal and Stress Analysis of Semiconductor Wafers in a Rapid Thermal Processing Oven," *IEEE Trans. Semicond. Manuf.* **1**:105 (1988).

34. K. S. Jones and G. A. Rozgonyi, "Extended Defects from Ion Implantation and Annealing," in *Rapid Thermal Processing Science and Technology*, R. B. Fair, ed., Academic Press, Boston, 1993.

35. V. E. Borisenko and P. J. Hesketh, *Rapid Thermal Processing of Semiconductors*, Plenum, New York, 1997.

36. A. Leitoila, J. F. Gibbons, T. J. McGee, J. Peng, and J. D. Hong, "Solid Solubility of As in Si Measured by CW Laser Annealing," *Appl. Phys. Lett.* **35**:532 (1979).

37. A. Kagner, F. A. Baiocchi, and T. T. Sheng, "Kinetics of As Activation and Clustering in High Dose Implanted Si," *Appl. Phys. Lett.* **48**(16):1090 (1986).

38. J. L. Altrip, A. G. R. Evans, J. R. Logan, and C. Jeynes, "High Temperature Millisecond Annealing of Arsenic Implanted Silicon," *Solid State Electron.* **33**:659 (1990).

39. J. L. Altrip, A. G. Evans, N. D. Young, and J. R. Logan, "The Nature of Electrically Inactive Implanted Arsenic in Silicon After Rapid Thermal Processing," *Mater. Res. Soc. Symp. Proc.* **224**:49 (1991).

40. J. R. Marchiando, P. Roitman, and J. Albers, "Boron Diffusion in Silicon," *IEEE Trans. Electron Dev.* **TED-32**:2322 (1985).

41. R. B. Fair, "Low-Thermal-Budget Process Modeling with the PREDICT™ Computer Program," *IEEE Trans. Electron Dev.* **ED-35**:285 (1988).

42. Y. Kim, H. Z. Massoud, and R. B. Fair, "The Effect of Ion Implantation Damage on Dopant Diffusion in Silicon During Shallow Junction Formation," *J. Electron. Mater.* **18**:143 (1989).

43. R. B. Fair, "Junction Formation in Silicon by Rapid Thermal Annealing," in *Rapid Thermal Processing Science and Technology*, R. B. Fair, ed., Academic Press, Boston, 1993.

44. T. E. Seidel, R. Knoell, G. Poli, B. Schwartz, F. A. Stevie, and P. Chu, "Rapid Thermal Annealing of Dopants Implanted into Preamorphized Si," *J. Appl. Phys.* **58**(2):683 (1985).

45. H. Kinoshita, T. H. Huang, and D. L. Kwong, "Modeling of Boron Diffusion and Activation for Non-Equilibrium Rapid Thermal Annealing Application," *Mater. Res. Soc. Symp. Proc.* **303**:259 (1993).

46. S. R. Weinzierl, J. M. Heddleson, R. J. Hillard, P. Rai-Choudhury, R. G. Mazur, C. M. Ozburn, and P. Potyraj, "Ultrashallow Dopant Profiling via Spreading Resistance Measurements with Integrated Modeling," *Solid State Technol.* **36**:31 (January 1993).

47. I. W. Wu and L. J. Chen, "Characterization of Microstructural Defects in BF_2^+ Implanted Silicon," *J. Appl. Phys.* **58**:3032 (1985).

48. R. B. Fair, J. J. Wortman, and J. Lin, "Modelling of Rapid Thermal Diffusion of As and B in Si," *J. Electrochem. Soc.* **135**(10):2387 (1984).

49. S. S. Gill and B. J. Sealy, "Review of Rapid Thermal Annealing in Ion Implanted GaAs," *J. Electrochem. Soc.* **133**:2590 (1986).

50. T. E. Haynes, N. K. Chu, and S. T. Picraux, "Direct Measurement of Evaporation During Rapid Thermal Processing of Capped GaAs," *Appl. Phys. Lett.* **50**(16):1071 (1987).

51. S. J. Pearton and R. Caruso, "Rapid Thermal Annealing of GaAs in a Graphite Susceptor— Comparison with Proximity Annealing," *J. Appl. Phys.* **66**:2482 (1989).

52. T. E. Kazior, S. K. Brierley, and F. J. Piekarski, "Capless Rapid Thermal Annealing of GaAs Using a Graphite Susceptor," *IEEE Trans. Semicond. Manuf.* **4**:21 (1991).

53. S. K. Brierley, T. E. Kazior, and F. J. Piekarski, "Optimization of Capless Rapid Thermal Annealing for GaAs MESFETs," *Mater. Res. Soc. Symp. Proc.* **224**:451 (1991).

54. A. Joshi and D. L Kwang, *IEEE Trans. Electron Dev.* **12**(1):28 (1991).

55. J. Nulman, J. P. Krusius, and A. Gat, "Rapid Thermal Processing of Thin Gate Dielectrics— Oxidation of Silicon," *IEEE Electron Dev. Lett.* **EDL-6**:205 (1985).

56. M. M. Moslehi, K. C. Saraswat, and S. C. Shatas, *Proc. SPIE* **623**:92 (1986).

57. R. Deaton and H. Z. Massoud, "Effects of Thermally Induced Stresses on Rapid Thermal Oxidation of Silicon," *J. Appl. Phys.* **70**:3588 (1991).

58. M. M. Moslehi, S. C. Shatas, and K. C. Saraswat, "Thin SiO_2 Insulators Grown by Rapid Thermal Oxidation of Silicon," *Appl. Phys. Lett.* **47**:1353 (1985).

59. J.-M. Dilhac, *Mater. Res. Soc. Symp. Proc.* **146**:333 (1989).

60. R. Singh, S. Sinha, R. P. S. Thakur, and P. Chou, "Some Photoeffect Roles in Rapid Isothermal Processing," *Appl. Phys. Lett.* **58**(11):1217 (1991).

61. J. P. Pon Pon, J. J. Grob, A. Grob, and R. Stuck, "Formation of Thin Oxide Films by Rapid Thermal Heating," *J. Appl. Phys.* **59**(11):3921 (1986).

62. M. H. Moslehi and K. C. Saraswat, "Thermal Nitridation of Si and SiO_2 for VLSI," *IEEE Trans. Electron Dev.* **ED-32**:106 (1985).

63. M. A. Schmidt, F. L. Terry, Jr., B. P. Mather, and S. D. Senturia, "Inversion Layer Mobility of MOSFETs with Nitrided Oxide Gate Dielectrics," *IEEE Trans. Electron Dev.* **35**:1627 (1988).

64. T. Hori, H. Iwasaki, and K. Tsuji, "Electrical and Physical Properties of Ultrathin Renitrided Oxides Prepared by Rapid Thermal Processing," *IEEE Trans. Electron Dev.* **36**:340 (1989).

65. T. Hori and H. Iwasaki, "Improved Transconductance Under High Normal Field in Nitrided Oxide MOSFETs," *Electron Dev. Lett.* **10**:195 (1989).

66. T. Hori, "Inversion Layer Mobility Under High Normal Field in Nitrided-Oxide MOSFETs," *IEEE Trans. Electron Dev.* **37**:2058 (1990).

67. J. Kuehne, G. Q. Lo, and D. L. Kwong, "Chemical and Electrical Properties of Fluorinated Oxides Prepared by Rapid Thermal Oxidation of Si in O_2 with Diluted NF_3," *Mater. Res. Soc. Symp. Proc.* **224**:367 (1991).

68. G. L. Ling, D. Lopes, and G. E. Miner, "Uniform Ultra-Thin Oxides Grown by Rapid Thermal Oxidation of Silicon in N_2O Ambient," in *Rapid Thermal and Integrated Processing VI*, Vol. 470, MRS, Pittsburgh, 1997, p. 361.

69. J. M. Grant and Z. Karim, "Rapid Thermal Oxidation of Silicon in Mixtures of Oxygen and Nitrous Oxide," in *Rapid Thermal and Integrated Processing V*, Vol. 429, MRS, Pittsburgh, 1996, p. 257.

70. Z. Q. Yao, "The Nature and Distribution of Nitrogen in Silicon Oxynitrite Grown on Silicon in a Nitric Oxide Ambient," *J. Appl. Phys.* **78**(5):2906 (1995).

71. J. C. Sturm, C. M. Gronet, and J. F. Gibbons, "Limited Reaction Processing: *In-Situ* Metal-Oxide-Semiconductor Capacitors," *IEEE Electron Dev. Lett.* **EDL-7**:282 (1986).

72. J. C. Sturm, C. M. Gronet, C. A. King, S. D. Wilson, and J. F. Gibbons, "*In-Situ* Epitaxial Silicon-Oxide-Doped Polysilicon Structures for MOS Field-Effect Transistors," *IEEE Electron Dev. Lett.* **EDL-7**:577 (1986).

73. T. Brat, C. M. Osburn, T. Finsted, J. Liu, and B. Ellington, "Self Aligned Ti Silicide Formation by Rapid Thermal Annealing Effects," *J. Electrochem. Soc.* **133**:1451 (1986).

74. R. W. Mann, C. A. Racine, and R. S. Bass, "Nucleation, Transformation, and Agglomeration of C54 Phase Titanium Disilicide," *Mater. Res. Soc. Symp. Proc.* **224**:115 (1991).

75. R. Beyers and R. Sinclair, "Metastable Phase Transformation in Titanium-Silicon Thin Films," *J. Appl. Phys.* **57**:5240 (1985).

76. M. Bariatto, A. Fontes, J. Q. Quacchia, R. Furlan, and J. J. Santiago-Aviles, "Rapid Titanium Silicidation: A Comparative Study of Two RTA Reactors," *Mater. Res. Soc. Symp. Proc.* **303**:95 (1993).

77. S. S. Iyer, C.-Y. Ting, and P. M. Fryer, "Ambient Gas Effects on the Reaction of Titanium with Silicon," *J. Electrochem. Soc.* **302**:2240 (1985).

78. C. M. Osborn, H. Berger, R. P. Donovan, and G. W. Jones, "The Effects of Contamination on Semiconductor Manufacturing Yield," *J. Environ. Sci.* **31**:45 (1988).

79. C. M. Osborn, "Silicides," in *Rapid Thermal Processing, Science and Technology*, R. B. Fair, ed., Academic Press, Boston, 1993.

80. J. A. Kittle, Q. Z. Hong, H. Yang, N. Yu, et al. *Mater. Res. Soc. Symp. Proc.* **525**:331 (1998).

81. V. E. Borisenko and P. J. Hesketh, *Rapid Thermal Processing of Semiconductor*, Plenum, New York, 1997, p. 163.

82. G. D. Yao, Y. C. Lu, S. Prassad, W. Hata, F. S. Chen, and H. Zhang, "Electrical and Physical Characterization of TiN Diffusion Barrier for Sub-Micron Contact Structure," *Mater. Res. Soc. Symp. Proc.* **303**:103 (1993).

83. M. A. Crouch, S. S. Gill, J. Woodward, S. J. Courtney, G. M. Williams, and A. G. Cullis, "Structural and Electrical Properties of Ge/Au Ohmic Contacts to n-Type GaAs Formed by Rapid Thermal Annealing," *Solid State Electron.* **33**:1437 (1990).

84. J. B. B. Oliveira, C. A. Olivieri, J. C. Galzerani, A. A. Pasa, L. P. Cardoso, and F. C. dePrince, "Characterization of AuGeNi Ohmic Contacts on n-GaAs Using Electrical Measurements, Auger Spectroscopy, and X-ray Diffractometry," *Vacuum* **41**:807 (1990).

85. "Investigation of the Uniformity of Ohmic Contacts to N-Type GaAs Formed by Rapid Thermal Processing," *Solid State Electron.* **36**:295 (1993).

86. C. Lee and G. Chizinsky, "Rapid Thermal Processing Using a Continuous Heat Source," *Solid State Technol.* **32**(1):43 (1989).

87. C. Lee, U.S. Patent 4857689 (1989).

88. W. DeHart, *Microelectron. Manuf. Technol.* **14**(7):44 (1991).

89. K. Torres, D. Lam, R. Weaver, and G. Solomon, "The Performance of the Fast Ramp Furnace," in *Rapid Thermal and Integrated Processing VII*, Vol. 470, MRS, Pittsburgh, 1997, p. 193.

Part III

Unit Processes 2: Pattern Transfer

Any science or technology which is sufficiently advanced is indistinguishable from magic.[1]

The preceding part of the book presented processes required to introduce, activate, and diffuse dopants, and to grow oxides in silicon. A limitation of these processes is that they are done to the entire wafer. The essence of IC fabrication is the ability to transfer information from the IC designer's workstation to the semiconductor wafer. This third part of the book will deal with the processes involved in this transformation: lithography and etching. These are extremely important, since the lateral dimension of many layers and the proximity in which structures can be placed often determine the speed of the circuit. Pattern transfer is so important that a technology is often referred to in terms of its feature size (e.g., quarter-micron CMOS).

As discussed in Chapter 1, most lithography is done in a two-step process. First, the design is broken down into layers of information. Each layer is a map for the location of one film on the finished IC. Usually these layers are printed as photomasks. Once a photomask has been made, it can be used to pattern wafers by shining a light through it. The light falls onto a photosensitive material that is then developed to try to reconstruct the original layer image. The production of photomasks will not be explicitly discussed here. The process is very similar to that of lithography on wafers. Both optical lithography (Chapter 7) and electron beam lithography (Chapter 9) are used in photomask generation.

As the desired feature size continues to shrink, optical lithography is becoming much more difficult. As a result, new lithographic processes have been developed. Some, like excimer laser steppers and phase contrast masks are techniques that extend optical lithography to smaller features. Others use very short wavelength nonoptical radiation to reproduce fine lines. Chapter 9 will discuss some of the most popular processes: electron beam and extreme ultraviolet lithography. The continuing improvement of optical lithography and its extension to ever shorter wavelengths and the use of

---

[1]Arthur C. Clarke.

immersion optics have forestalled the widespread application of nonoptical techniques. It is presently anticipated that "optical" (in reality, deep ultraviolet) lithography will be capable of resolving features to 30 nm.

Once the photosensitive material has been patterned, it is often used as a stencil for reproducing the image in the underlying film. This is usually done subtractively by depositing the film everywhere and removing the unwanted material. The photosensitive material then must serve as a mask for this etching. Chapter 11 will cover both standard etching and an additive process called *liftoff* that is sometimes used for GaAs fabrication.

Chapter 7

Optical Lithography

7.1 Lithography Overview

Figure 7.1 shows a flowchart by which most ICs are designed. It begins with the identification of the chip function. If it is a complicated function, it may be broken down into several levels of subfunctions. The subfunctions are laid out on a floor plan that allocates space on a chip for the future design. At this point, the designer may construct a high level model of the chip to test functionality and to get an estimate of its performance. The designer then assembles the chip from predesigned pieces of circuitry called *cells*, using software to custom-design whatever is necessary. The layout of these cells is predicated on a set of design or layout rules. (See Figure 7.2 for an example of a few rules for a technology.) These rules are a contract between the fabrication facility and the designer. For each level, the layout rules will govern the smallest feature allowed, the smallest spacing allowed, the minimum overlap for features on this level to those on another level, the minimum spacing to underlying topology, and so on. If these rules are followed, the fabrication facility is obligated to ensure that the chip functions as designed. After the design is complete, it will be checked against the layout rules to ensure compliance. Finally, additional simulations may be run from the actual layout. If they do not meet specifications, the design is modified until they do. Depending on the sophistication of the computer-automated design (CAD) tools, some or all of these processes are automated. In the most advanced CAD systems, the tool is fed a high level description of the function to be implemented along with a file that contains the design rules. The program will generate a chip layout along with performance estimates. The designer then oversees the process and makes manual adjustments as necessary to improve performance.

The interface between the designers and the fabrication facility occurs through the production of photomasks. Each photomask contains an image of one layer of the process. Depending on the aligner to be used, the photomask may be the same size as the finished chip or an integral factor of that size. Figure 7.3 shows some typical photomasks. The masks may be the same size as the final image on the wafer (1×) or the image on the mask may be reduced during the exposure. Common reductions are 4×, 5×, and 10×. Most reticles used in manufacturing are 150 mm square, although larger microprocessor die

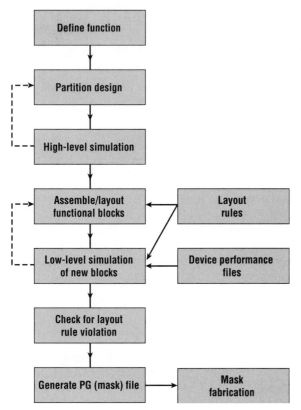

Figure 7.1 Simplified IC design process flow diagram.

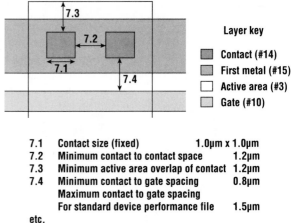

7.1	Contact size (fixed)	1.0µm x 1.0µm
7.2	Minimum contact to contact space	1.2µm
7.3	Minimum active area overlap of contact	1.2µm
7.4	Minimum contact to gate spacing	0.8µm
	Maximum contact to gate spacing	
	For standard device performance file	1.5µm
etc.		

Figure 7.2 Excerpt of typical design rule set. This portion deals with first metal rules for a particular technology.

sizes sometimes require a 225-nm reticle. Photomasks are fabricated on various types of fused silica. The most important properties for the mask include a high degree of optical transparency at the exposure wavelength, a small thermal expansion coefficient, and a flat, highly polished surface that reduces light scattering. On one surface of this glass is a patterned opaque layer. In most masks the opaque layer is chromium. After the masks have been patterned, the pattern may be verified by checking it against the database. Any unwanted chrome can be removed by laser ablation. Any pinholes in the chrome can be repaired with an additional deposition. This is a critical step, since the photomask may be used to pattern tens of thousands of wafers. Any defect in the mask large enough to be reproduced will be on every wafer made from the mask.

Lithography is the most complicated, expensive, and critical process in mainstream microelectronic fabrication. (Several good references exist on lithography for students who are interested in specializing in this area. Early classic books include Stevens [1] and Bowden et al. [2]. Elliott's book deals primarily with the topic of the next chapter but is also a reasonable reference [3]. A later book by Moreau [4] is one of the most comprehensive. Nanogaki et al. provide a more recent book on photoresists [5].) Some of the lithographic requirements to be met during the 1990s, can be found elsewhere [6, 7]. Figure 7.4 shows more recent projections. Lithography accounts for about one-third of the total fabrication cost, a percentage that is rising. A typical silicon technology will involve 15–20 different masks. For some BiCMOS processes, as many as 28 may be used. Although traditional GaAs technologies have required far fewer masking levels, the number is now increasing as well. Furthermore, the technology performance is often predicated on the ability to produce very fine lines. Even the evaluation of the performance of a lithographic process is difficult. A memory manufacturer might require very tight control on a critical feature size over hundreds of billions of transistors per wafer and over tens of thousands of wafers. A device researcher, on the other hand, might be perfectly satisfied if 50% of the features fall into some acceptable range on one wafer. Thus, it is all too easy to wind up comparing apples to oranges in an evaluation of photolithographic performance.

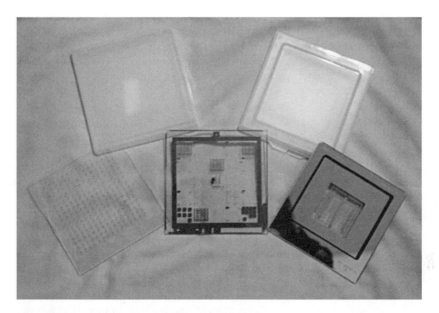

Figure 7.3 Typical photomasks including (from left) a 1× plate for contact or projection printing, a 10× plate for a reduction stepper, and a 10× plate with pellicles.

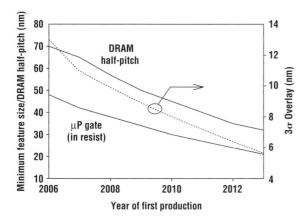

Figure 7.4 Projected lithography requirements showing overlay accuracy (right axis) and resolution requirements (left axis). Data taken from 2005 International Technology Roadmap for Semiconductors.

Figure 7.5 shows a cross-schematic of a simple system for optically exposing a wafer. An optical source on the top is used to shine light through the mask. The image is projected onto the wafer surface, which is coated with a thin layer of photosensitive material known as *photoresist*. Optical photolithography can therefore be divided into two parts. The design and operation of the exposure tool that makes the image of the photomask at the surface of the wafer are mainly problems in optical system design. The other half of the equation is the chemical processes that occur once the radiation of the image has been absorbed in the photoresist and the pattern developed. The dividing line is the pattern of radiation that strikes the surface of the wafer, also called the *areal image* of the mask. This chapter will review the basic physics of optical exposure. The tools discussed are all optical. They use either visible, ultraviolet (UV), deep ultraviolet (DUV) or extreme ultraviolet (EUV) light as the exposing radiation. They are called *aligners* because they have a dual purpose. Not only must they reproduce the image of a particular layer, they must also align that layer to the previous ones.

There are three primary measures of performance for an aligner (Table 7.1). The first is the resolution, the minimum feature size that can be exposed. Resolution is not a fixed number for a given aligner, although equipment manufacturers sometimes seem to lead one to believe otherwise. Resolution depends on the ability of the photoresist to reconstruct the pattern from the areal image.

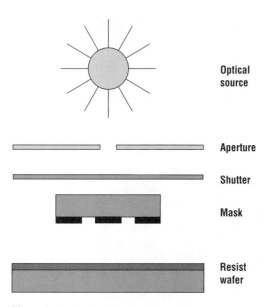

Optical
source

Aperture

Shutter

Mask

Resist
wafer

Figure 7.5   Schematic of a simple lithographic exposure system.

As already mentioned, very small features may be resolved with a particular optical tool and resist system, but the dimensional control is so poor that they cannot be reliably used. To be useful, resolution figures are often quoted as a minimum feature size that can be resolved and still maintain a certain feature tolerance. A typical number might be a six-standard-deviation (6σ) distribution of linewidths with no more than 10% variation.

The second performance metric for the aligner is registration, which is a measure of the overlay accuracy from layer to layer. Again, good measurements of registration are statistical in nature. If registration errors are completely random the mean error is zero. The width of the registration error distribution as measured by a figure like 6σ is a good indicator of overlay performance. This number depends on a variety of factors. Automated alignment systems are used to manufacture ICs. For these systems alignment tolerance depends strongly on the ability of the system to accurately locate the alignment marks. This, in turn, depends on the nature of the alignment marks used and on the films on the surface of the wafer. Simpler manual systems are typically used in research environments and are highly operator dependent. The third primary performance measure is throughput. Electron beam systems have excellent resolution, and the registration can be quite good. The throughput for a typical IC pattern that contains 10^8 transistors can be less than one wafer per hour, which is unacceptable in many applications.

With solid measures of these metrics, one must still bear in mind the intended technology. For most ULSI technologies the 6σ registration must be about one-third the minimum feature size for the lithographic tool to be fully useful. Throughput and process uniformity are also extremely important. Other technologies, such as GaAs metal semiconductor FET (MESFET), may have more lax requirements for registration and throughput, but may require excellent resolution. Even within a technology, different layers may have different requirements. Thus, some levels may be exposed on

Table 7.1   The effects (xx, strong; —, slight) of some of the resist parameters on process outcomes

	Resolution	Registration	Wafer-to-Wafer Control	Batch-to-Batch Control	Throughput
Exposure system	XX	XX	X	XX	XX
Substrate	X	X	XX	X	X
Mask	X	X	—	X	X
Photoresist	XX	X	XX	XX	XX
Developer	X	—	XX	XX	X
Wetting agent	—	—	XX	X	—
Process	X	X	XX	XX	XX
Operator[a]	X	XX	XX	X	XX

[a] Operator row entries are for manual aligners. Advanced tools use automated alignment and exposure systems that dramatically reduce the dependence of the results on the skill of the operator.

one type of tool and other levels on another. Using different types of processes for different layers is called *mix-and-match lithography*.

In discussing the optics of lithography, one must distinguish between those problems in which all dimensions are large compared with the wavelength of light and those in which this criterion is not satisfied. For example, in discussing an optical system that includes a light source, a reflector, and a lens, all of the geometries are of order 1 cm or larger. In such a system, the light can be treated as a particle traveling in straight lines between the components. The analysis used in this situation is called *ray tracing*. On the other hand, when the light passes through a mask where the feature sizes on the mask approach the optical wavelength, one must consider properties such as diffraction and interference. These phenomena require a description of light as an electromagnetic wave. We use three descriptions of the areal image. The first is the electric field (V/cm). The square of the electric field is the intensity (W/cm^2). Multiplying the intensity by the exposure time provides the dose, in joules per square centimeter (J/cm^2).

7.2 Diffraction°

This section will develop a basic understanding of the physics of lithographic exposure. This discussion can be found in any introductory optics text, in much more detail than can be presented here [8]. The integrals derived will be considerable simplifications, and even these can only be solved numerically. Nevertheless, without this background material, it is difficult to understand the lithographic limitations and the reason that one system is superior to another.

One starts with the understanding that light propagates as an electromagnetic wave. These waves can be described as

$$\mathcal{E}(\bar{r}, \nu) = E_o(\bar{r})e^{j\phi(\bar{r}, \nu)} \tag{7.1}$$

where E_o is the electric field intensity, j is the imaginary number, ϕ is the phase of the wave, $\bar{r}$ is the position, and ν is the frequency of the wave. Huygens's principle says that any local disturbance in an optical system, such as a mirror or a photomask, can be considered to generate a large number of spherical wavelets that propagate outward from the point of disturbance. To find the disturbed electromagnetic wave, one must sum over all of the wavelets.

Figure 7.6 shows the application of Huygens's principle to a system of interest. The upper figure shows the formation of point sources on the surface of the mask. The lower figure shows a point source of light being used to illuminate a mask. On the mask is a single feature: a long narrow aperture of width W and length L. Here the approximation will be made that the transparent region of the mask does not affect the incoming wavefront. The aperture can be divided into a large number of differential rectangular elements of width dx and length dy. The points P_w and P_m are on the wafer and mask, respectively. Each rectangular

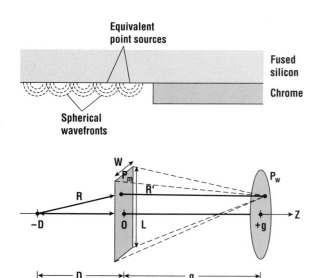

Figure 7.6 Huygens's principle applied to the optical system shown in Figure 7.5. A point source is used to expose an aperture in a dark field mask.

element generates a wavelet. The sum of the contributions to the exposure at a point on the surface of the wafer can be found using the integral

$$\mathcal{E}(R') = j\frac{A}{\lambda} \int \int_{\Sigma} \frac{e^{-jk(R,R')}}{RR'} \, d\sigma \tag{7.2}$$

where A is the amplitude of the sine wave at a unit distance from the source and Σ is the solid angle subtended by the aperture in the mask. Once one finds the electric field, the electromagnetic intensity at the surface of the wafer can be calculated by multiplying the field by its complex conjugate. If one lets W and $L \rightarrow \infty$, Equation 7.2 represents the summation of an infinite series of spherical waves that exactly reproduces the undisturbed wave. For such a wave the intensity

$$I = \mathcal{E}\,\mathcal{E}^* = E_o e^{j\phi} E_o e^{-j\phi} = E_o^2 \tag{7.3}$$

If, on the other hand, the aperture is of finite extent, the electric field is the superposition of plane waves with different phases. In the simplest case, where the aperture has been divided into only two elements,

$$I = [E_1 e^{j\phi_1} + E_2 e^{j\phi_2}][E_1 e^{j\phi_1} + E_2 e^{j\phi_2}]$$

$$= E_1^2 + E_2^2 + 2E_1 E_2 \cos(\phi_1 - \phi_2) \tag{7.4}$$

The interesting part of the equation, of course, is the cross-term due to interference between the wavelets. It gives rise to the oscillations that are a characteristic part of the diffracted image.

In practice, a realistic solution of Equation 7.2 is quite complicated, even for this very simple geometry. In photolithography one is only interested in two limiting cases. If Equation 7.2 is solved subject to the simplifying assumption that

$$W^2 \gg \lambda\sqrt{g^2 + r^2} \tag{7.5}$$

where r is the radial distance between the center of the diffraction pattern and the observation point (generally $r \approx W$), the result is near field or Fresnel diffraction. Actually, as $W \rightarrow \lambda$, even this approximation is no longer valid. Vector diffraction theory must be used, giving rise to polarization effects. The image of this type of diffraction is shown in Figure 7.7. The edges of the image rise gradually from zero, and the intensity of the image oscillates about the expected intensity. The oscillations decay as one approaches the center of the image. The oscillations are due to constructive and destructive interference of Huygens's wavelets from the aperture in the mask. The amplitude and period of these oscillations depend on the size of the aperture.

When W is small enough that the inequality in Equation 7.5 is no longer valid, the oscillations are large. When W is very large, however, the oscillations

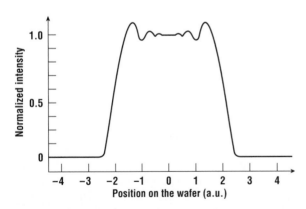

Figure 7.7   Typical near field (Fresnel) diffraction pattern.

rapidly die out, and one approaches simple ray tracing. Then, by geometric arguments, the width, of the image at the surface of the wafer is increased by an amount ΔW given by

$$\Delta W = W \frac{g}{D} \tag{7.6}$$

The other diffraction extreme occurs when

$$W^2 \ll \lambda \sqrt{g^2 + r^2} \tag{7.7}$$

This is called far field or Fraunhofer diffraction. Equation 7.7 is called the *Fraunhofer criterion*. Then Equation 7.2 can be simplified considerably. The intensity as a function of position on the surface of the wafer is given by

$$I(x, y) = I_e(0) \left[\frac{(2W)(2L)}{\lambda g} \right]^2 I_x^2 I_y^2 \tag{7.8}$$

where $I_e(0)$ is the flux density (typically expressed in W/cm^2) in the incident beam and

$$I_x = \frac{\sin\left[\frac{2\pi x W}{\lambda g} \right]}{\frac{2\pi x W}{\lambda g}} \tag{7.9}$$

$$I_y = \frac{\sin\left[\frac{2\pi y L}{\lambda g} \right]}{\frac{2\pi y L}{\lambda g}} \tag{7.10}$$

where L is the length of the lines and spaces. In Equations 7.9 and 7.10, x and y are the coordinates of the observation point on the surface of the wafer. A plot of this function and its square in 1-D is shown in Figure 7.8. The function has a sharp maximum at $x = 0$ and goes through 0 at integer multiples of one-half.

Real systems are much more complicated than these simple expansions, however. The light source is not a point, but a finite volume. It may also emit a number of wavelengths. The light is collected through a lens/mirror assembly. Each of the optical components will have some imperfections such as local distortions and aberrations. The mask itself will reflect, absorb, and phase-shift the incident radiation. Reflections on the surface of the wafer further complicate matters. As a result, the image produced on the surface of the wafer (the areal image) can only be approximated numerically. Even then sophisticated software is required.

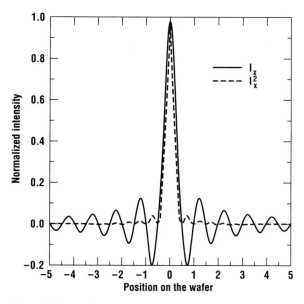

Figure 7.8 Typical far field (Fraunhofer) image.

Several commercial programs are available to do these calculations. In the following sections, some simple approximations will be presented that are frequently used to determine resolution in place of these more exact calculations of aerial intensity.

7.3 The Modulation Transfer Function and Optical Exposures

In discussing the resolution of a system, it is customary to discuss a series of lines and spaces called a *diffraction grating* rather than a single aperture. If the Fraunhofer criterion is met, one can roughly approximate the areal image by the superposition of the individual intensities. (Actually, the interference between peaks must be taken into account; but if the peaks are well separated, this is a modest effect.) Figure 7.9 shows a construct of the normalized intensity from such a grating as a function of position on the wafer. The minimum intensity is no longer 0. Instead, define I_{max} as the maximum intensity of the radiant pattern and I_{min} as the minimum intensity. In Figure 7.9, they are about 5.0 and 1.0, respectively.

The modulation transfer function (MTF) of an image can be defined as

$$\text{MTF} = \left(\frac{I_{max} - I_{min}}{I_{max} + I_{min}} \right) \tag{7.11}$$

The MTF is a strong function of the period of the diffraction grating. As the period of the grating decreases, MTF decreases. Physically, one can think of the MTF as a measure of the optical contrast in the areal image. The higher the MTF, the better the optical contrast. For the grating in Figure 7.9, the MTF is about 0.67.

Figure 7.10 shows a plot of areal intensity for a grating mask, based on simple Fresnel diffraction. Superimposed on this intensity plot are two simplified resist response indicators. In Figure 7.10A, the resist responds ideally: a single line exists at an exposure energy density D_{cr}. All regions of the photoresist that receive exposures greater than D_{cr} will completely dissolve during the develop process. All regions of the wafer that receive exposures below D_{cr} will not be attacked during the develop process. As the widths of the lines and spaces are decreased, diffraction makes only small changes in the widths of the lines and spaces until I_{min} multiplied by the exposure time $> D_{cr}$ or I_{max} multiplied by the exposure time $< D_{cr}$. Figure 7.10B shows a more realistic resist response model. The resist now has two critical exposure energy densities. For $D < D_0$, the resist will not dissolve in the developer. For $D > D_{100}$, the resist will completely dissolve in the developer. For the intermediate shaded regions ($D_0 < D < D_{100}$), the image will partially develop. As W decreases, the MTF goes down, and the areal intensity quickly enters a regime where the diffraction grating can no longer be perfectly reproduced on the wafer. The point at which this occurs depends on the values of D_0 and D_{100} and, therefore, on the resist being used. In common resist systems, when the MTF is less than about 0.4, the image can no longer be reproduced.

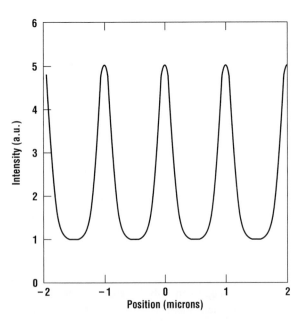

Figure 7.9   Far field image for a diffraction grating.

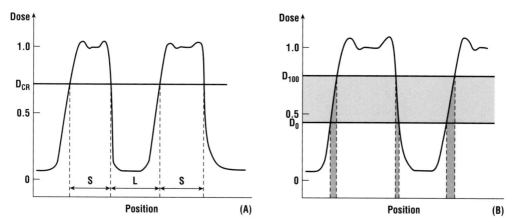

Figure 7.10   Plot of dose versus position on the wafer. Dose is given by the intensity of the light in the aereal image multiplied by the exposure time. Typical units are millijoules per centimeter squared (mJ/cm^2).

Example 7.1   Numerical simulation of diffraction gratings

Use a numerical tool to predict the areal images of long, equal-sized lines and spaces for a g-line optical system with a numerical aperture of 0.63. Use $W_{min} = 0.8 \, \mu m$ and $0.4 \, \mu m$.

Solution

In the Athena code, one uses Optolith™. The following simple deck can be used to run this situation.

```
go athena
# Example 7.1: Optolith execution of a simple diffraction
grating
#
# Set the illumination wavelength
illumination g.line
#
# Set the shape of the illuminating source
illum.filter clear.fil circle sigma=0.3
#
# The projection system numerical aperture
projection na=.43
#
# The shape of the pupil of the projection system
pupil.filter clear.fil circle
#
# Define the mask using rectangles for a diffraction
grating(cd=0.8 um)
layout x.low=-2.5 z.low=2.8 x.high=2.5 z.high=3.6
layout x.low=-2.5 z.low=1.2 x.high=2.5 z.high=2.0
layout x.low=-2.5 z.low=-0.4 x.high=2.5 z.high=0.4
layout x.low=-2.5 z.low=-2.0 x.high=2.5 z.high=-1.2
layout x.low=-2.5 z.low=-3.6 x.high=2.5 z.high=-2.8
```

```
#
# Define the grid for the areal image
image win.x.lo=-3 win.z.lo=-4 win.x.hi=3 win.z.hi=4 dx=.05
opaque
#
# Store the aereal image in a structure file
structure outfile=test1.str intensity
#
# Plot the aereal image during the run
tonyplot -st test1.str -set test1.set
#
quit
```

The result is shown in Figure 7.11 for a contour plot and a 1-D cut through the center:

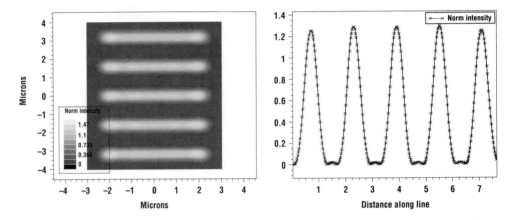

Figure 7.11 Contour plot (left) and 10 slice through the center of projected intensity as a function of position for a grating with 1.6 μm pitch, exposed at R = 436 nm and NA = 0.43.

In calculating the areal image of the smaller period grating, the area can be reduced or the number of lines increased. Here we did the latter and got the cut line in Figure 7.12 for the intensity plot. Compared to the larger period plot, we can see that the contrast of this image is much less. The best contrast comes from the outside lines, presumably due to a lack of interference from adjacent line pairs. This is an artifact of the limited size of the grating. Clearly it would be very difficult to resolve this image in a resist.

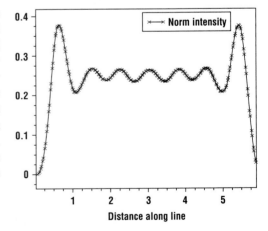

Figure 7.12 Same 1D plot for a grating with a pitch of 0.8 μm exposed on the same system.

7.4 Source Systems and Spatial Coherence

The next few sections will use the concepts of source systems and spatial coherence to discuss actual aligners. We will start with the component common to all aligners: the source system. This consists of the light source itself and any reflecting and/or refracting optics used to collect, collimate, filter, and focus the source. From the previous discussion, one should have a sense that the wavelength of the exposing radiation is a critical parameter of the lithography process. All things being equal, the shorter the wavelength, the smaller the feature size that can be exposed. Of course, the exposure requires that a certain amount of energy be deposited. Furthermore, this energy must be deposited uniformly across the wafer. To maintain reasonable exposure times, one requires an intense source at these short wavelengths. This section will review some of the most popular optical sources.

For many years the most common type of optical source for photolithography has been the high pressure arc lamp. Arc lamps are the brightest incoherent sources available. They are also used to pump lasers. Long gap arc lamps, sometimes used to heat wafers in rapid thermal processors, were mentioned in Chapter 6. Arc lamps for lithography have much shorter gaps, as shown in Figure 7.13. The lamp consists of two conducting electrodes sealed in a fused silica envelope. The electrodes, one pointed and the other round, end in a gap of about 5 mm. Most lamps contain a mercury vapor at a pressure close to 1 atm when the lamp is cold. To ignite the lamp, a high voltage spike is applied across the gap. The spike voltage must be sufficient to ionize the gas. Several kilovolts is a common strike potential. The mixture of partially ionized gas, electrons, and energetic neutral species inside the tube is called a *plasma* or a *glow discharge*. This topic will be discussed in Section 10.4. For now, appreciate that the plasma will conduct current. Most arc lamp power supplies also include a boost supply, typically a capacitor charged to several hundred volts, that ensures the plasma stability in the moments after the strike. The ionized gas in the lamp is very hot and the pressure in the bulb during operation may reach 40 atm. Typical electrical power dissipation by a photolithography arc lamp bulb is 500–1000 W. The emitted optical power is a little less than half of that.

During operation the lamp contains two optical sources. The high temperature electrons in the arc act as a very hot gray body source, radiating power as (Equation 6.3)

$$M_\lambda(T) = \frac{\epsilon(\lambda, T)\, C_1}{\lambda^5\,(e^{C_2/\lambda T} - 1)} \tag{7.12}$$

where $M_\lambda(T)$ is the energy density distribution, and C_1 and C_2 are the first and second optical constants as defined in Chapter 6. The electrons in the arc lamp plasma typically have temperatures of order 40,000 K. This corresponds to a peak emission at a wavelength of 75 nm, which is very deep in the ultraviolet. Since this energy is above the bandgap of the fused silica envelope, much of it will be absorbed before it leaves the lamp housing. Lamp manufacturers sometimes add impurities to the fused silica to enhance this absorption, since this highly energetic emission results in ozone production in the lamp assembly.

The second optical source in the lamp is the mercury atoms themselves. Collisions with the energetic electrons push the electrons of the mercury atoms into bound high energy states. When they decay into lower energy states, they emit optically at the wavelength corresponding to the energy transition. These line spectra

Figure 7.13   Typical high pressure, short arc mercury lamp *(courtesy Osram Sylvania).*

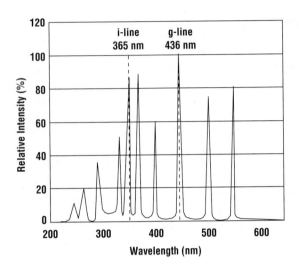

Figure 7.14 Line spectra of typical mercury arc lamp showing the positions of the two lines most commonly used in lithography.

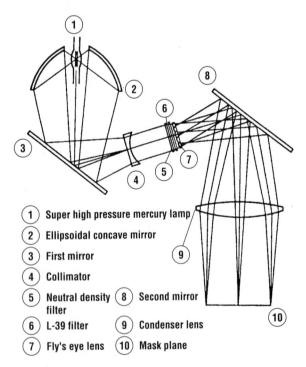

① Super high pressure mercury lamp
② Ellipsoidal concave mirror
③ First mirror
④ Collimator
⑤ Neutral density filter
⑥ L-39 filter
⑦ Fly's eye lens
⑧ Second mirror
⑨ Condenser lens
⑩ Mask plane

Figure 7.15 Schematic of a typical source assembly for a contact/proximity printer *(after Jain).*

are so sharp that they can be used to identify the primary species in the plasma. Figure 7.14 shows the line spectra of a typical mercury lamp [9]. The lines have been named according to their energy. Aligners often filter out all but a single line. At present, the older optical exposure equipment is g-line (436 nm), and i-line (365 nm) systems are very common. To extend the use of arc lamps deeper into the UV, xenon can be used as the fill gas. Xenon has a strong line at 290 nm, with much weaker lines at 280, 265, and 248 nm; however, excimer lasers have proven to be more popular sources for wavelengths less than 365 nm.

During operation, energetic mercury ions bombard the negative electrode. As they do so, they eject small amounts of the electrode material through a physical process known as *sputtering.* (See Chapter 12.) Some of the sputtered electrode coats the inside wall of the fused silica housing. Furthermore, the high temperature seen by the inside surface of the lamp may also cause a slow devitrification of the fused silica, leaving a cloudy white appearance. The combination of sputtering and devitrification reduces the output intensity. The additional absorbed energy also causes the lamp envelope to get hotter. Ultimately the lamp will fail, often by exploding, severely damaging the aligner's optics. For that reason, aligner lamps are usually cooled with fans during operation and are replaced after a set number of hours of use.

The simple optical design of the source in Figure 7.5 is, of course, not very practical. Only a very small fraction of the radiation from the lamp will reach the wafer. Unless the wafer is spherical (not usually a desired condition), the power density at the surface will be nonuniform. There are, therefore, four primary objectives for the design of the optical source system. The first is to collect as much of the radiation as possible. Without such collection, exposure times are impractically long. The second objective is to make the radiated intensity uniform over the field of exposure, preventing some parts of the wafer from being overexposed while others are underexposed. The third objective is to collimate and shape the radiation to the extent needed. Normally perfect collimation is not desired; instead a few degrees of divergence is often used. Finally, the source must select the exposure wavelength(s). Figure 7.15 shows a schematic of a source assembly for a simple but practical aligner [10].

Collection is most often done by the use of parabolic reflectors. By placing a point source at the focal point of the reflector, all of the radiation so captured is collimated. Arc lamps do not radiate uniformly. The plasma acts as a diffuse, semitransparent source. The electrodes and bulb shape the plasma so as to maximize the optical intensity perpendicular to the arc direction. For a properly designed reflector, little radiation is lost from the bottom of the lamp. The dark spot at the top of the lamp allows one to have a hole in the reflector to bring in the power leads and exhaust cooling air. The finite extent of the arc prevents such an arrangement from being perfectly collimated. If less collimation is desired, the lamp can be moved away from the focal point.

To increase the uniformity of the optical source, some type of optical integrator must be used. One common component is a fly's-eye lens. This is a large fused silica lens containing many small lenslets. The lenslets decollimate the incident light, and a second objective recollects the light, and reshapes it to the desired dimensions. A second approach that addresses both collection and integration is the use of fiber optic bundles. One end of the fibers either surrounds the arc source or is used immediately after the parabolic reflector. The fibers are mixed so as to provide a uniform source at the other end of the bundle. Optical fibers are sometimes used with arc lamps but are commonly used for excimer laser sources.

The wavelength selection is done through a set of filters. It is common to use a cold mirror to absorb the infrared radiation from the lamp. This prevents unintentional exposure and heating of the components further downstream. The wavelength selection can be done through the use of optical notch filters or a series combination of high pass and low pass filters. A mechanical shutter completes the source assembly.

Since most arc lamps are efficient emitters in the near-UV and visible wavelengths, they are inefficient in the deep UV. Excimer lasers are the brightest optical sources in this part of the spectrum. The word "excimer" is a concatenation of the words excited and dimer. True excimers, therefore, contain excited dimers (molecules with two atoms of the same element, like F_2). A molecule that has one or more electrons in excited energy levels will be designated using an asterisk, as F_2^*. Excimer lasers are more properly called *exciplex lasers,* since most modern excimers contain high pressure mixtures of two or more elements. These elements do not react when they are in the ground state, but if one or both are excited, a chemical reaction will proceed. In most excimers, one of the precursors is a halogen or halogen-containing compound such as NF_3 and the other is a noble gas. A common example is XeCl, where the reaction that leads to lasing is

$$Xe^* + Cl_2 \rightarrow XeCl^* + Cl \qquad \text{(7.13)}$$

The excited molecule emits in the deep UV, returning it to the ground state, where it immediately dissociates. If enough energy is supplied to maintain a large population of the noble species in the excited state, lasing will continue to occur. The energy is usually supplied by 10- to 20-kV arcs across two flat plate electrodes spaced from 1 to 2 cm apart. The arc can be strobed at rates up to several hundred hertz.

Some common excimer laser sources are given in Table 7.2 [11]. Of primary interest is the lasing wavelength and the power. A typical resist exposure dose is 10 to 50 mJ/cm^2. A laser must be able to produce about 1 J at the surface of the wafer so that a field several centimeters on a side can be exposed in no more than 1 sec. To achieve this, the laser should put out close to 20 W. Although

Table 7.2   Practical excimer sources that are potentially useful in semiconductor photolithography

Material	Wavelength (nm)	Max Output (mJ/pulse)	Frequency (pulses/sec)
F_2	157	40	500
ArF	193	10	2000
KrF	248	10	2000

Data taken from Patel [11].

XeCl puts out considerable power, it is not much further into the deep UV than arc lamps. The combination of high power and deep UV lines makes ArF and KrF attractive sources for advanced optical lithography. Although F_2 excimers have also been used at 157 nm for contact lithography [12], the low output power makes them impractical for production use.

Excimers emit strongly in a multimode fashion with a relatively poor spatial coherence. That is, excimer laser beams are more divergent than beams from argon ion lasers. This is a strong drawback to many laser applications, but it is actually an advantage for lithography. Lasers with high spatial coherence induce speckle, which occurs when phase variations are introduced into wavefronts by surface reflections [13]. For comparison, an Ar ion laser has a bandwidth of <0.0001 nm, while a free-running excimer has a bandwidth of about 1 nm. In lithography applications, excimer sources are often line narrowed to less than a picometer, but are still much broader than an argon ion laser. Figure 7.16 shows a typical speckle pattern for a narrow line laser exposure.

The containment of a high pressure halogen, along with high voltage sometimes used to pump the laser, makes safety a concern. Early excimer systems were considered exotic, dangerous, and unreliable. Furthermore, the very high energy density pulses (10 kW/cm^2) tended to degrade the optical quality of the lenses of early excimer systems by devitrifying the silica. The laser pulses also densify the fused silica optics. This gradually increases the distortion in the system. After a sufficient number of pulses, the optics must be replaced. This image of excimer lithography has improved considerably, and currently exposure tools based on KrF and ArF excimer sources are widely used in manufacturing lines at most IC manufacturers. Figure 7.17 shows the optical train of an excimer

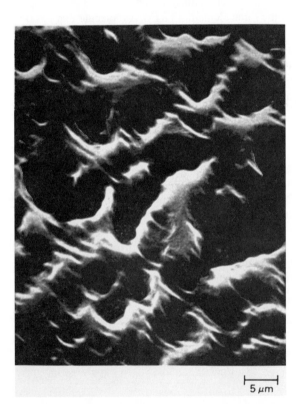

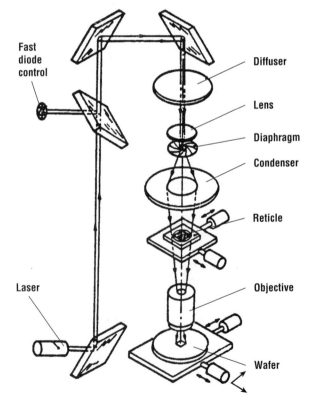

Figure 7.16   Speckle pattern obtained from exposing a pattern using a narrow linewidth laser as the optical source *(after Jain).*

Figure 7.17   Optical train for an excimer laser stepper *(after Jain).*

based 10:1 reduction stepper (Section 7.6). Excimers must still be serviced after about 10^8 pulses and replaced after about 10^9 pulses. For 157-nm exposures, it is generally believed that SiO_2 will not be a suitable lens material. Instead CaF_2 or MgF_2 may be required [14]. Needless to say, this will be an extremely challenging proposition. One of the primary remaining problems for deep UV excimer lithography is the development of suitable commercial photoresists. Typical optical resists are almost opaque at these wavelengths. All of the excimer energy is deposited in the top layers of conventional i-line resists [15], resulting in poor image formation. This topic will be discussed further in the next chapter.

Example 7.2   **Numerical simulation of diffraction gratings**

Redo Example 7.1 for a 0.4-μm grating using an ArF source (193 nm).

Solution

To do this, replace the g-line parameter in the illumination command with "lambda=0.193." The cutline result is shown in Figure 7.18. The MTF of the image is very close to one, suggesting that it would be easy to resolve such an image. Clearly the reduction in wavelength makes a tremendous difference in the ability of the system to resolve small features.

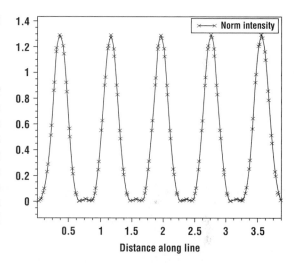

Figure 7.18   Intensity as a function of position for a 0.8 μm period grating (0.4 μm features) exposed at 193 nm with NA = 0.43.

7.5 Contact/Proximity Printers

The simplest type of aligner is a contact printer. In contact printing, the mask is pressed against the resist-coated wafer during exposure. The primary advantage of contact printing is that fairly small features can be made using comparatively inexpensive equipment. In the typical contact exposure system shown in Figure 7.19, the mask is held chrome-side down in a frame just below the microscope objectives. Vernier screws are then used to move the wafer with respect to the mask. Once the wafer is aligned to the mask, the two are clamped together, the microscope objectives are retracted, and the wafer/mask assembly is wheeled into the exposure station. Here radiation from a high intensity lamp (housed in the black box, Figure 7.19, upper right) is used to expose the wafer. After exposure, the carriage is returned to the inspection station for unloading. The figure insert shows the wafer ready for unloading.

Ideally, the entire wafer is in contact, with the mask. Because of this contact, the gap between the wafer and the optical disturbance (photomask) goes to zero and diffraction effects are minimized.

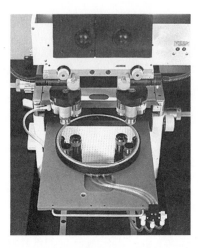

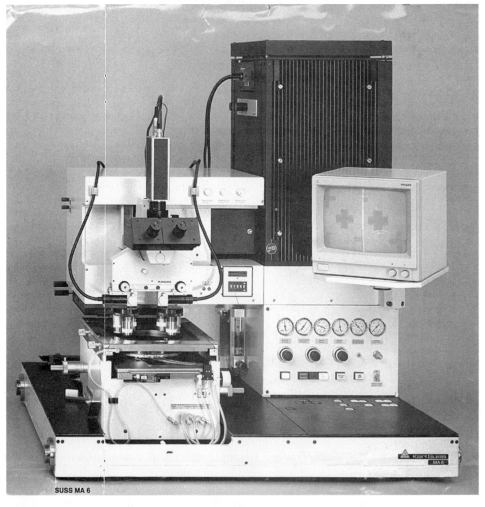

Figure 7.19   Typical contact exposure system *(courtesy of Karl Suss)*.

Thus the MTF should be ≈1.0 for essentially any feature size. Actually, due to the finite resist thickness, the gap cannot be zero. Furthermore in real contact printers, the mask contact, varies across the wafer surface. This occurs since neither the wafer nor the mask is perfectly flat. Pressures ranging from 0.05 to 0.3 atm are used to push the mask into more intimate contact with the wafer. This is called the *hard contact mode of exposure*. In the most extreme cases, thin film masks are sometimes used to help promote the contact. Then the resolution is limited primarily by light scattering in the resist. Features as small as 10 Å [16, 17] have been produced using this method, with extremely thin resists as demonstration vehicles. In more useful resists, features as small as 0.1 μm have been demonstrated using contact printing and deep submicron sources [18, 19]. It is easier to use these sources in contact printing, since the optics are so simple. Resolution using more common sources is about 0.5 μm.

The major disadvantage of hard contact lithography is defect generation due to the contact between the resist-coated wafer and the photomask. Defects are generated both on the wafer and on the mask on every contact cycle. For this reason, contact printers are typically limited to device research or other applications that can tolerate high defect levels. Proximity printing was developed to avoid defect generation. In this type of exposure tool, the mask floats off the surface of the wafer, typically on a cushion of nitrogen gas. The gap between the wafer and mask is controlled by the flow of nitrogen into this space. Separations of 10–50 μm are common. Since there is no longer any (intentional) contact between the wafer and the mask, defect generation is sharply reduced.

The problem with proximity printing is a reduction in the resolution. Consider a mask consisting of a single aperture of width W in a dark field. Assume that this mask is exposed using a monochromatic, nondivergent light source, such as a broad beam laser. Figure 7.20 shows the areal intensity of such an image as a function of the gap [20]. When g is small such that

$$\lambda < g < \frac{W^2}{\lambda} \qquad (7.14)$$

the system is in the near field region of Fresnel diffraction. The areal image produced by such an aperture is a well-known function of λ, g, and W, and closely approximates the ideal image. The small intensity oscillation near the edges of the pattern is called *optical ringing*. In a real optical system in fact, the beam will not have perfect spatial coherence and so even the modest ringing pictured here will not exist. As the gap is increased, however, eventually

$$g \geq \frac{W^2}{\lambda} \qquad (7.15)$$

Then the image approaches that of far field Fraunhofer diffraction, as shown in Figure 7.8.

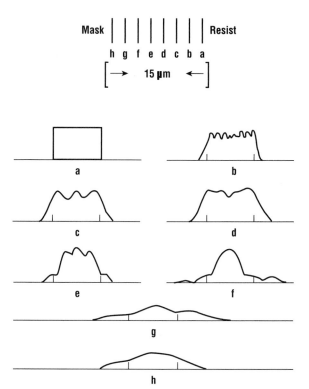

Figure 7.20 Intensity as a function of position on the wafer for a proximity printing system where the gap increases linearly from $g = 0$ to $g = 15$ μm *(after Geikas and Ables)*.

Table 7.3 Maximum allowable proximity gap for near and deep UV sources as a function of the feature size normalized to the gap required for 2.5-μm resolution with a deep UV source

Feature Size (μm)	Maximum Gap for Near UV Source	Maximum Gap for Deep UV Source
2.5	0.63	1.0
2.0	0.37	0.61
1.0	0.08	0.24
0.5	0.05	0.07

Data taken from Lin [19].

When the gap is very large, the image is severely degraded. As a rule, one can say that features less than

$$W_{min} \approx \sqrt{k\lambda g} \tag{7.16}$$

cannot be resolved in this type of optical printing, where k is a constant that depends on the resist process. Typical values of k are close to 1. For a gap of 20 μm and an exposure wavelength of 436 nm (g-line), the minimum feature size for proximity printing is about 3.0 μm. This, however, would require an excellent resist process. A minimum feature size with some process latitude would be about 50% larger than this. To further improve the resolution, it is necessary to reduce the wavelength or the gap. Reducing the gap increases the risk of contact. A second problem with small gaps is that the variations in the gap that may be caused by wafer or mask nonplanarity, dirt particles, resist beads, and unintentional tilt lead to linewidth variations across the wafer. A more attractive alternative is to reduce the wavelength. Systems used to expose fine features, therefore, typically operate deep in the UV. This process is limited primarily by high brightness sources and compatible optical materials for refractive optics. Table 7.3 shows the maximum allowable gap as a function of feature size for near and deep UV sources [21] for a 10% image distortion.

Example 7.3

An exposure system can be run in contact or proximity mode and uses an i-line source. If the resist thickness is 0.7 μm and $k = 0.8$, find W_{min} for hard contact, and air gaps of 10, 20, and 30 μm.

Solution

For hard contact

$$W_{min} = \sqrt{0.8 \times 0.7 \ \mu m \times 0.365 \ \mu m}$$
$$= 0.45 \ \mu m$$

For a 10-μm air gap, the actual gap $\approx$10.7 μm:

$$W_{min} = \sqrt{0.8 \times 10.7 \ \mu m \times 0.365}$$
$$= 1.77 \ \mu m$$

Similarly, W_{min} is 2.46 μm and 2.99 μm for gaps of 20 and 30 μm, respectively.

7.6 Projection Printers

Projection printers were developed to obtain the high resolution of contact printing without the defects. They have become by far the most widely used exposure tool for the manufacture of ICs. This section will introduce the basic equations and then will discuss common types of systems: scanners, steppers and scanning steppers. Figure 7.21 shows a simple example of a Kohler projection lithography system. The mask is held between the condenser and a second set of lenses called the projector or objective. The purpose of the projector is to refocus the light onto the wafer. In some cases, the light from the condenser is not collimated but instead is focused on the plane of the projector.

Part of the light from the mask in Figure 7.21 has been diffracted to a large angle. To try to reimage the pattern onto the wafer one must, at minimum, collect that diffracted light. The numerical aperture (NA) of the system is defined as

$$NA = n \sin(\alpha) \tag{7.17}$$

where α is one-half the angle of acceptance of the objective lens and n is the refractive index of the medium between the objective and the wafer. In the case of optical aligners, the exposure is typically done in air where $n = 1.0$. Typical values for NA range from 0.16 to 0.8.

The resolution of a projection system can be limited by imperfections in the optical train. These imperfections may take the form of lens irregularities such as aberrations, inclusions, or distortion, or the separation between the mask and the objective may be incorrect. In most aligners used in IC manufacturing, however, the optics are sufficiently well made that the resolution is limited by the ability of the optical train to collect and reimage the light. This limit is referred to as Rayleigh's criterion and is given by

$$W_{min} \approx k_1 \frac{\lambda}{NA} \tag{7.18}$$

where k_1 is a constant that again depends on the ability of the resist to distinguish between small changes in intensity. Typical values for k_1 range from 0.8 down to 0.4. The theoretical limit is believed to be 0.25. This means that an aligner with an NA of 0.6, together with a 365-nm source, can be used to image lines as small as 0.2 μm. Since the mask does not contact the wafer during exposure, defects are not created by projection aligners.

One route to finer lines is to develop higher NA lenses. Progress has been made in these area, but there has been a technical price. The depth of focus can be described as the distance along the optical train that the wafer can be moved and still keep the image in focus. For a projection system this is given by

$$\sigma = \frac{n\lambda}{NA^2} \tag{7.19}$$

Increasing the numerical aperture increases the resolution linearly, but decreases the depth of focus quadratically. Again taking $\lambda = 365$ nm with an NA of 0.4, $\sigma = 2.3$ μm. Increasing the NA to 0.6 reduces the depth of focus to 1.0 μm. Maintaining the depth of

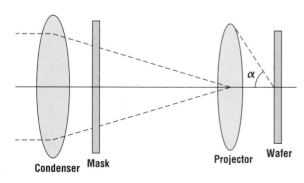

Figure 7.21   Schematic for the optical train of a simple projection printer.

focus across a 200-mm wafer is difficult, since the wafer topology alone may be as much as 2 μm unless some planarization is done. Added to this are the wafer bow and flatness at that stage of the technology and the finite resist thickness. Some compromise must therefore be achieved between resolution and depth of focus.

The other avenue to reducing the minimum feature size is using shorter wavelength sources. Most lithography tools now in use have moved from g-line to i-line, and many have begun working at the 248-nm emission of a KrF excimer laser. One of the difficulties in this transition is optics. Virtually all modern projection printers use diffractive rather than reflecting optics. The photon energy at 248 nm is rather large (4.9 eV). It becomes difficult to manufacture large-diameter lenses with a sufficiently high degree of perfection that are completely transparent at this wavelength. Small inclusions of hydroxyls, for example, may cause dark spots in the mid and deep UV. This problem is greatly exaggerated for excimer sources. The photon associated with the ArF excimer has an energy of 6.3 eV. Perfect fused silica has a band edge of about 8 eV, but since it is an amorphous material, the band edge is not sharp. Making optical components of sufficient quality and size for these wavelengths is extremely challenging. For an F_2 excimer, the photon energy is 7.7 eV, making the use of SiO_2 largely impossible, both for lenses and for photomasks. At 193 nm one can use fluorinated SiO_2, since the addition of fluorine extends the range of transparency deeper into the UV. Critical components may also be made from CaF_2. Using CaF_2 however, introduces a major problem. Its coefficient of thermal expansion (14 ppm) is 28 times larger than that of SiO_2. This great difference tremendously restricts the tolerance of the optical system to heating. Consider a 225-mm 4 $\times$ reticle. If the maximum allowable misalignment due to reticle heating is 5 nm the allowed ΔT for SiO_2 is 0.04°C. This is extremely difficult. For CaF, however, the allowed ΔT is only 0.0014°C.

Finally, the image resolution is also a function of the spatial coherence of the source. As a rough measure of the spatial coherence, first consider a circular source of some known diameter. The light from this source passes through an aperture known as the pupil, where a condenser lens attempts to collimate the beam. Then the spatial coherence S is approximately given by

$$S \approx \frac{\text{source image diameter}}{\text{pupil diameter}} \tag{7.20}$$

Figure 7.22 shows the modulation transfer function of a diffraction grating period on the mask [22]. Such a mask consists of equally sized lines and spaces, each of width W. The ordinate of the plot is the spatial frequency, given by

$$\nu_{ap} = \frac{1}{\Gamma} = \frac{1}{2W} \tag{7.21}$$

The spatial frequency has been normalized to the Rayleigh criterion,

$$\nu_o = \frac{1}{W_o} = \frac{NA}{0.61\lambda} \tag{7.22}$$

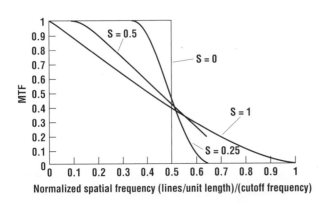

Figure 7.22 Modulation transfer function as a function of the normalized spatial frequency for a projection lithography system with spatial coherence as a parameter.

For a source with perfect spatial coherence ($S = 0$), the MTF drops abruptly at the Rayleigh criterion. It is not surprising that for a given grating frequency less than the highest resolvable, the MTF decreases as S increases. Interestingly, however, in grating periods less than that proscribed by the Rayleigh limit, the MTF increases with decreasing spatial coherence. Depending on the critical modulation transfer function of the resist, it may be preferable to use a radiation source that does not have perfect spatial coherence.

Example 7.4   **Resolution limits for optical projection lithography**

It has been determined that a resist is capable of resolving images with an MTF of 0.4. If the exposure tool has an NA of 0.35, an exposure wavelength of 436 nm, and an S of 0.5, what is the minimum feature size that this tool can resolve? What is the depth of focus? If this source were replaced with an i-line source (365 nm), how would these numbers change?

Solution

According to Figure 7.22, an MTF of 0.4 with $S = 0.5$ corresponds to a normalized spatial frequency of $0.52\nu_0$. Evaluating Equation 7.22 yields, $\nu_0 = 1.32~\mu m^{-1}$. Then, the resolution is 0.684 line pair per micron. This corresponds to a pitch (the sum of the minimum line and space widths) of 1.46 μm or a linewidth of 0.73 μm. If an i-line source is used instead and nothing else changes, the resolution improves to 0.61 μm. Evaluating Equation 7.19, one sees that the depth of focus is 3.56 μm for g-line and 2.98 μm for i-line.

During the 1970s, 1:1 scanning projection aligners revolutionized the microelectronics industry. Prominent among these are the scanning mirror projection aligners first developed by Perkin-Elmer [23, 24]. In these systems the wafer and mask are clamped into a cartridge that scans a narrow arc of radiation across the mask and the wafer. Figure 7.23 shows a schematic of the scanning system. During the scan, an arc of light that shines through the photomask reflects off two spherical mirrors. The scanning feature ensures that the light stays near the center of the optical system. A major advantage of this type of system is that the exposure optics can use reflective (catadioptric) elements. This makes the system less sensitive to chromatic aberrations. Aside from the source itself, no large expensive quartz lenses are required. The NA of a typical mirror projection system, however, is only about 0.16. This type of system (use) used to geometries of about 2.0 μm at very high throughput. Because of this high throughput, commercial scanning mirror projection systems are still used to fabricate some simple ICs, but they are less competitive. Excimer sources have also been applied to scanning systems using a cylindrical lens and mirror arrangement to transform the rectangular beam into the required crescent shape [25]. Using these sources one can obtain 1.0-μm features.

In the recent past, 1:1 scanning systems were replaced by step and repeat projection aligners. These systems are refractive, frequently with a reduction built in. Early steppers used 10:1 reduction lens; 5:1 and 4:1 lenses are now more common. This means that only a small region of the wafer, called a *field* (typically 0.5–3 cm^2), is exposed at a time [26]. This allows systems to be built with very high NAs and therefore, high resolution. Between exposures the wafer must be mechanically moved to the next field, hence the common name *steppers*. Although features as small as 0.1 μm have been demonstrated [27], they involved unacceptable constraints such as very small fields of view and very minimal focus/exposure latitude. The i-line steppers can repeatably pattern features below 0.5 μm over fields of more than 2.5 cm on a side [28]. Steppers based on KrF are capable of production at the 0.18-μm level [29] and even 0.13-μm IC production.

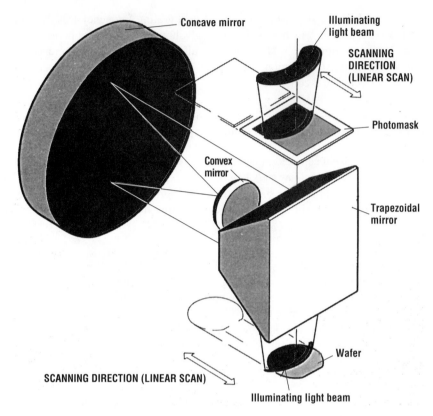

Figure 7.23   Schematic for the operation of a scanning mirror projection lithography system *(courtesy of Canon U.S.A.)*.

The past ten years have seen two very significant developments in exposure tools in the effort to continue scaling to small dimensions. Both have relied increasing the numerical aperture of the optical system. The first is to replace the conventional stepper with a step-and-scan system. In this approach the reticle and the wafer are simultaneously scanned through the optical column. Since a small area is exposed at any given time, this has two major advantages. The first is that a higher numerical aperture can be achieved. Scanners have obtained NAs of 0.7 and above. The second is that scanners can accommodate very large exposure areas. At this writing, the maximum field size is 26 mm by 33 mm. These machines are the mainstays of IC fabrication for features between 50 and 130 nm [30, 31].

The second major advance is called immersion lithography. Recall from Equation 7.17 that NA is proportional to the index of refraction of the media. Until recently, the medium was always air, so $n = 1.0$. Replacing the gap with a liquid allows n to be increased ($n = 1.43$ for water), lowering the minimum feature size.

The depth of focus (DOF) is another very important factor. Since the systems being considered have high NA values, a high NA version of the classical Rayleigh criterion is used rather than Equation 7.19 [32]:

$$\text{DOF} = \frac{\lambda}{4n \times \sin^2(\theta_p/2)} \tag{7.23}$$

where θ_p is the angle of propagation of the first order of a grating with pitch P ($2W$ for a grating with equal lines and spaces), and given by

$$\theta_p = \frac{\lambda}{n \times P} \tag{7.24}$$

It is common to show the ratio between a dry system and a wet system as [32]:

$$\text{Ratio} = \frac{\text{DOF}_{n(\text{liquid})}}{\text{DOF}_{n=1(\text{air})}} = \sqrt{\frac{n^2 - (\lambda/2P)^2}{1 - (\lambda/2P)^2}} \tag{7.25}$$

Using this equation, it can be seen that the ratio goes to n at large pitches. In the region in which the immersion systems will be used, the ratio can be considerably larger than n, giving a large DOF benefit to using immersion lithography instead of dry. Critical dimensions become more uniform due to the increased DOF [32].

Immersion is implemented by having the liquid dispensed locally onto the stage to be exposed. After exposure, the lens moves to the next exposure site with the liquid remaining under the lens due to surface tension [33, 34]. An example of this setup is shown in Figure 7.24 [35]. If the gap size between the lens and wafer warrants it, an air curtain from the nozzle is used to contain the liquid. Another method is to draw the liquid back into the nozzle after exposure and redispense it onto the next field [36]. (Figure 7.25). The main

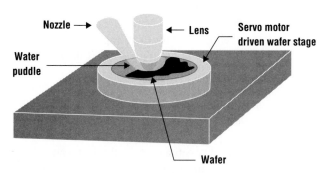

Figure 7.24   Setup of an immersion system using surface tension (*from Switkes et al., reprinted from the May 2003 edition of Microlithography World. Copyright 2003 by PennWell.*)

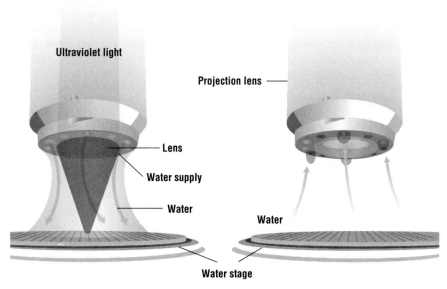

Figure 7.25   Nozzle system used by Nikon to put the water down and suction it up for each stage (*from Geppert, reprinted by permission IEEE.*)

advantage of this approach is that the liquid can be refreshed for each stage on the wafer, and tighter temperature control can be achieved.

First-generation immersion lithography systems have used water [36–38]. Water has many advantages. First, the optical absorbance is about 0.01/cm at 193 nm, which means that most of the light will be transmitted to the resist. Water has a refractive index of 1.436 at 193 nm, which corresponds to being able to image 65-nm nodes and approach the 45-nm node. Water is known to be compatible with the wafer; it is easy to purify and is in abundant supply. Second-generation immersion tools have looked at higher index fluids. Some of the liquids being studied for use in immersion lithography are acids such as phosphoric and sulfuric acid, surfactants, and quaternary ammonium salts [38].

One of the obvious concerns with immersion lithography is the potential for interaction between the immersion fluid and the resist. Typical effects in the resist include T-topping and microbridging [39]. The effect of photoresist compounds over the long term on the lens system is another major concern [40, 41]. Chemicals could diffuse through the fluid and deposit on the lens, causing the lens to degrade over time. Top coats for the resist are being tested for their impact on the system [41]. Another major concern associated with the change to immersion lithography is microbubbles in the fluid. The bubbles can cause scattering of the exposure light [33, 42, 43]. Bubble formation can occur when the gas-saturated water is shifted to a state of oversaturation due to pressure or temperature changes [43]. Suggested solutions, include the use of degassed water and carefully designed fill nozzles [33, 36]. Another source of bubbles is air trapped in the topography of the resist [44].

The first steppers used global alignment and focusing. In such a system, the focus and alignment are set at one time for the entire wafer. Newer systems are capable of site-by-site control of these variables. The system can automatically adjust the alignment and focus at every field on the wafer. Thus they are able to avoid many of the depth-of-focus problems normally associated with high NA systems, greatly reducing the effects of wafer warpage and distortion. This also makes the use of large diameter wafers much more feasible. To accommodate very large numerical apertures combined with short wavelengths in new i-line and excimer systems, the newest generation of machines is also beginning to use field by field-leveling systems [45] that automatically adjust the height of the wafer for each field to keep the image in focus.

The primary disadvantage of steppers is throughput. Although 1:1 scanning projection printers can achieve throughput of nearly 100 wafers and scanners per hour, steppers typically operate at 20–50 wafers/hr. The throughput of the system is determined by

$$T = \frac{1}{O + n \cdot [E + M + S + A + F]} \tag{7.26}$$

where n is the number of die per wafer, E is the exposure time, M is the stage movement time per exposure, S is the stage settling time, A is the site-by-site alignment time (if used), F is the autofocus time (if used), and O is the overhead associated with loading/unloading the wafer, prealigning it, then moving the wafer under the column and preforming the global alignment. Some of this can be done concurrent with the previous wafer exposure to reduce or eliminate the O term. Since it is common for n to be $\sim$100, the total of the times inside the bracket is critical to the commercial success of steppers. It must be 2 sec or less for the tool to be practically useful. Figure 7.26 shows a commercial lithography tool. Table 7.4 gives data for tools for manufacturing ICs from three current vendors.

Figure 7.26 Configuration of Step-and-Scan 193-nm system. The laser is at the left. The reticle is at the upper right, while the wafer is at the lower right (*photo courtesy ASML*).

Table 7.4 Tools commercially available as of mid-2006 for manufacturing of integrated circuits

	110–150 nm	**90 nm**	**65 nm**	**Immersion**
ASML	5500/1150C 248 nm 15-nm overlay $0.55 < NA < 0.8$	5500/1150C 193 nm 12-nm overlay $0.5 < NA < 0.7$		TS XT:1700Fi 45-nm resolution 7-nm overlay $0.75 < NA < 1.2$
Canon	FPA-6000EX6 248 nm 25 nm overlay $0.5 < NA < 0.65$	FPA-6000ES6a 248 nm 8-nm overlay $NA = 0.8$		
Nikon	NSR-S208D 248 nm $NA = 0.82$ 10 nm overlay		NSR-S308F 193 nm $NA = 0.92$ 8-nm overlay	NSR-S610C 193 nm $NA = 1.3$ 45-nm resolution

Data from company websites.

7.7 Advanced Mask Concepts[+]

As the quest for producing finer features at ever higher densities continues to challenge lithographers, new methods have emerged to improve the defect density and resolution by the method of mask fabrication. Several proposed methods such as predistorted reticles and the use of optical proximity correction are so difficult that they are unlikely to be implemented on a large scale in the near future. This section will discuss three current or potential mask improvements. Pellicles are already in widespread use. Antireflective coatings and phase contrast masks are in limited manufacturing applications at this writing.

The widespread acceptance of reduction stepper systems has made mask making much easier. Instead of making a mask with 0.5-μm lines, a 5× stepper requires a mask with only 2.5-μm minimum features. Furthermore, small defects such as particles probably would not be imaged on the surface of the wafer. If the particles are large enough to be printed, however, unlike a scanning photolithography system the stepper will repeat the defect in every field across the wafer. In some cases, the die size is large enough that only one die can be printed per exposure area. A defect on such a mask will make every die on the wafer nonfunctional. Very careful attention to potential defects is therefore essential in stepper plates. The types of defects to be detected are clear defects such as pinholes, notches, and missing geometries, and opaque defects such as bridges and particles. Other mask defects include scratches and chips from improper handling and run-out/run-in and magnification errors.

Two types of automated mask inspection systems have been developed. The easier of the two to implement is the die-to-die comparison technique. In this system, the mask is inspected at two nominally identical locations by illuminating the bottom while maintaining two inspection objectives on the top that are set at a fixed interval. The system then logs the locations of all positions for which the intensities measured at the two objectives are not sufficiently similar. Of course, such a system cannot work with stepper plates that have only one die on the mask. In that situation, the mask geometries must be directly compared to the database from which the mask was made. This method tends to generate many more false positives than a direct compare, but is still a good screen for a subsequent visual inspection.

The previous steps are often used to ensure that the stepper plate is free of defects as it leaves the mask shop. To minimize the impact of particles in the fab, stepper plates are often pellicalized. In this process, a thin coating of transparent material similar to Mylar is stretched over a cylindrical frame on either side of the mask. The frame stands off the membrane at a distance of about 1 cm from the surface of the mask. The purpose of the pellicle is to ensure that any particles that fall on the mask are kept outside the focal plane of the optical system [46]. Damage to pellicles when one is using high energy sources such as excimers is still a problem at this writing.

In any optical system, some fraction of the light will not follow the desired optical train, but instead will be lost. In enclosed systems, such as those used for photolithography, light rays may reflect off multiple surfaces and strike the surface of the wafer, degrading the areal image. One problem with pellicles is the loss of transmission and increased light scattering from the surface of the Mylar. For example, the light scatter in a Perkin-Elmer III scanning projection printer has been shown to degrade 3σ linewidth tolerances for a 2-μm line from ±0.31 μm to ±0.42 μm [47]. Even the light scatter from the chromium lines in the mask can be a significant source of linewidth variation, particularly in deep submicron patterns [48]. One new mask technique to reduce this problem is to use a 10% antireflective coating on the lens side of the mask.

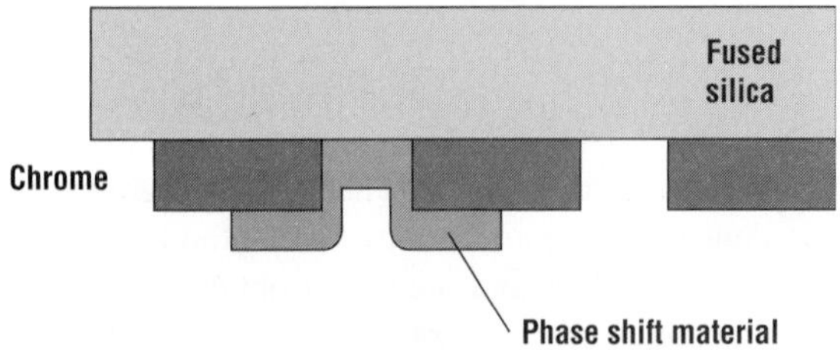

Perhaps the most dramatic improvement in resolution that masks can provide is through the use of phase shifting. The concept of phase shifting the optical image, which was first suggested by Levenson et al. [49], is illustrated in Figure 7.27. A mask containing a diffraction grating is overcoated with a phase-shifting material at twice the period of the grating in such a manner that every other aperture is covered by the material. The thickness and refractive index of the material are sufficient

Figure 7.27   Basic concept of phase shift masks as described by Levenson *et al.* [49].

to exactly shift the phase of the light by 180° with respect to the light that does not see the phase shift material. Typical shift materials have a thickness $t = \lambda/[2(n-1)]$ $n \approx 1.5$, so $t \sim \lambda$. Ideally the material would not attenuate, reflect, or scatter the incoming light. The result of the phase-shifting is that the tail of the diffracted distributions from adjacent features would destructively rather than constructively interfere. This dramatically improves the modulation transfer function at the surface of the wafer and, therefore, the resolution. Phase-shifted masks (PSM) have received an enormous amount of interest. Fujitsu Ltd., Toshiba Corp., and Matsushita Electric all market fully functional 64-Mb DRAMs fabricated using PSM [50].

Since Levenson's introduction of the concept, a myriad of phase-shifting techniques have been suggested. Some, such as Levenson's original proposal, add films to the surface. Others accomplish the phase-shifting by etching the quartz mask itself to a sufficient depth [51]. Although these techniques work very well on diffraction gratings, application to actual patterns is rather difficult. As an alternative, one can use self-aligned phase shifters. In this type of technology, only the region near the edge of the mask feature is exposed to the phase-shifted radiation [52]. This has the effect of sharpening the areal image, but in a straightforward manner that is easy to implement. Figure 7.28 shows a typical self-aligned method [53], sometimes called *rim phase shifting*. The chromium pattern is used to expose a layer of resist spun on top of the reticle. After development, a layer of resist remains that is perfectly aligned with the chromium pattern. The resist layer is used as a mask to etch the quartz plate. Finally, the resist is used to undercut the chromium a controlled distance, and the resist is stripped. What remains is a quartz ledge around each chromium line. This ledge acts as a local phase shifter that dramatically improves contrast and the repeatability of the linewidth of small features.

These simple phase-shifting techniques have been extended considerably into a more aggressive approach called optical proximity correction (OPC). An easy way to understand OPC is to first break up the exposure field of size X by Y into pixels of size Δx by Δy. Next consider a mask as an exposure matrix, M, consisting of 0s where the pixel on the mask is clear and 1s where the pixel on the mask is opaque. The matrix would have $X/\Delta x$ rows and $Y/\Delta y$ columns. If this mask is used to expose a wafer, one can construct an areal image matrix W consisting of the same number of pixels. Ideally, the two matrices would be identical except for a constant. In practice, the process of exposure causes a distortion of the areal image, and therefore of the matrix W. One can construct a matrix equation,

$$W = S \cdot M$$

where S is a matrix that represents the exposure. It contains all of the information about the optical system. Ideally S is a unity matrix. In practice it contains off-diagonal elements corresponding to the image distortion.

OPC then becomes, at least in principle, an effort to find S^{-1}. If one can identify S^{-1}, it can be applied to the mask to obtain a new mask M' where

$$M' = S^{-1} \cdot M$$

Then

$$W = S \cdot M' = S \cdot S^{-1} \cdot M = M$$

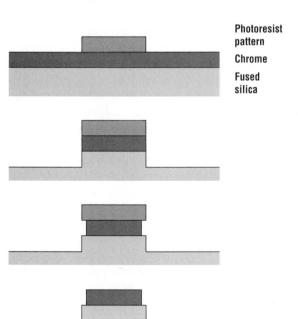

Photoresist
pattern

Chrome

Fused
silica

Figure 7.28 Self-aligned method of phase shifting the radiation at the edges of a pattern.

The mask compensates for the optical distortion of the system. In theory, once S^{-1} is known, it can be applied to any mask to correct for diffraction and other degradation of the optical signal. Although the matrix would seem to be huge ($\sim 10^{10}$ pixels), it is a very sparse matrix. Furthermore the distortion of simple features such as isolated lines is reproducible from one area of the mask to another. OPC therefore allows one to extend the capability of an optical tool at the expense of more complicated and therefore more expensive masks. It is limited by the pixel size and the digital nature of the mask data. Multiple layers of partially absorbant material can be used to obtain more detailed OPC masks, however, OPC masks can become quite expensive, approaching $100,000 for a state-of-the-art plate for a single level.

7.8 Surface Reflections and Standing Waves

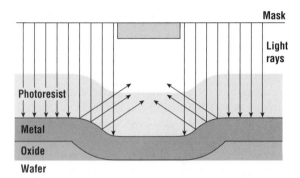

Figure 7.29   Light from the exposed regions can be reflected by wafer topology and be absorbed in the resist in nominally unexposed regions.

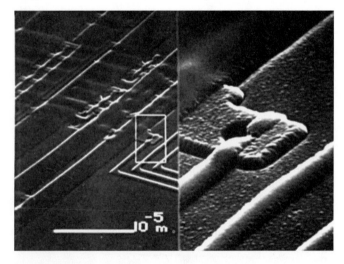

Figure 7.30   Resist lines for patterning the metal running diagonally from the lower left. On the right is an expanded view of the geometry described in Figure 7.31 *(after Listvan et al.).*

Until now, this chapter has considered only the optical train up to the wafer. To get an exposure through its full thickness, the photoresist must be partially transparent. The light that penetrates the resist may reflect back off the surface of the wafer and change the optical energy deposited in the resist. The required exposure time in a photolithography process, therefore, depends on the films on the wafer. A specular metal layer on the surface will require a shorter exposure than exposures over less reflective films.

One problem with surface reflections is image control. This is particularly a problem for late process steps where significant topology may have been developed. The problem is shown schematically in Figure 7.29. In this case, the metal line is running into an oversized contact. The contact reflects the light and exposes the pattern, resulting in poor definition of the line in the contact and possibly a hole in the metal. Figure 7.30 shows a real pattern with this problem. Resist lines running over patterned reflective sublayers have poorly controlled linewidths. An expanded view of one part of the pattern is shown at the right. The resist line that enters the contact hole is almost completely gone.

Several solutions have been developed to avoid this problem. It is possible to control the reflectivity of a layer by changing the deposition parameters, but this often requires some compromise of the deposited film characteristics. A better solution is to avoid topology and planarize the layers. This avoids not

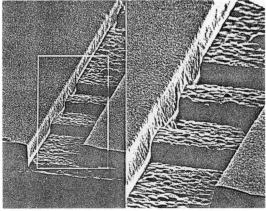

Figure 7.31   Micrographs of resist lines on reflective metal without and with a 2700-Å antireflecting layer under the resist *(after Listvan et al.).*

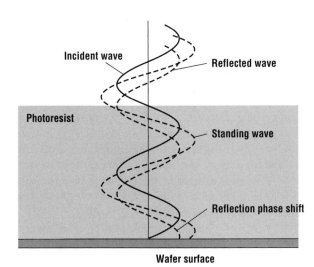

Figure 7.32   The formation of standing waves in the resist.

only unintentional side exposures, but a host of other problems, both for lithography and for metal deposition. The planarizing layer may be temporary, such as a thick absorbing resist layer applied under a thinner imaging resist. This type of process will be covered in the next chapter. The second solution is to planarize the layers through etching processes. This will be discussed in Chapter 11.

One solution to surface topology effects that has been applied with considerable success is the use of antireflective polymers under the photoresist [54]. Figure 7.31 shows resist lines patterned on specular metal films with the same topology as in Figure 7.30. In the image at the right, a 2700-Å antireflective layer was spun on before the photoresist. After resist develop, a short etch in an oxygen plasma was used to remove the exposed antireflective material. Side reflections are completely eliminated, and dimensional control is excellent. (For comparison, a micrograph of a standard resist line on a similar substrate is included in Figure 7.31.) The drawback of such a process is its complexity. The layer must be applied and baked before the resist is applied. It must not interact with the resist, and it must be noncontaminating. The layer must be etched after resist develop, and it must be stripped after resist removal.

An examination of micrographs of most resist lines will reveal the presence of ridges along the resist edges. These ridges are caused by standing waves in the aerial image. Standing waves are due to constructive and destructive interference between the incident and reflected light. Figure 7.32 illustrates the effect. As a result of this interference, the intensity of the light in the resist varies with thickness. With the change in the exposure comes a change in the resist dissolution rate (see Chapter 8). The use of an anti reflecting coating completely eliminates standing wave patterns.

7.9 Alignment

It was suggested at the outset of this chapter that photolithographic processes should be judged by three metrics: resolution, registration, and throughput. One should not underestimate the importance of registration in microelectronic fabrication. It limits packing density and, therefore, circuit performance. To understand why, consider the following analogy. A skyscraper is to be built floor by floor. Each floor is to be built on the ground. All plumbing and electrical work is done before assembly of the building. Connections are made automatically when each floor is placed on the previous one. If one hopes to get water and power on the top floor, each level must line up very closely to those above and below it. To accomplish this feat, the tradesmen decide to top each floor off with rubber connectors on each pipe and electrical contact pads on each wire. Needless to say, if 10^6 of the connections must be made at each level, the size of the pads and connectors could easily determine the amount of space left for the actual office. If the building is 200 feet, on a side and 50% of the space is allocated for these passthroughs and the spaces between them, they must be about 2 mm on a side. Assuming that the wires and pipes are 1 mm in diameter, the required building alignment for the construction is ± 0.5 mm.

A reasonable rule for ULSI lithography is that registration errors should be no more than one-fourth to one-third of the resolution. As already mentioned, steppers have the potential for aligning each exposure field individually. To utilize this site-by-site alignment economically, an automated alignment system must be used. Typically this is done by reflecting light off the surface of the wafer and back through the mask. A box-within-a-box arrangement is used, and the wafer is moved until the widths of the intensity peaks on either side of the center box are equal. Often a single global alignment is done manually first to ensure that each die will be properly aligned to minimize the search time.

Figure 7.33 demonstrates two possible types of misalignment. Simple x, y, and θ errors are called *misregistration* and are due to poorly aligning the mask and wafer. In many steppers, the wafer is not aligned directly to the mask. Instead, both the wafer and mask are aligned to the optical train of the exposure system in two separate steps. Misregistration errors can occur if these two alignment schemes do not exactly match. A periodic process of baseline correction is done to prevent these systematic errors. Run-out or run-in is a net difference in the distance between die. On $1\times$ systems such as contact/proximity and scanning projection printers, run-out is often caused by changes in the physical size of the wafer after it has gone through a series of high temperature steps. Steppers reduce this problem in a global alignment mode and almost completely eliminate it using a good site-by-site scheme. This type of error can easily be caused by small changes in the temperature of the mask. Fused silica is typically the material of choice for the mask, due in part to its low thermal expansion coefficient (5×10^{-7}°C^{-1}). Preventing an expansion of more than 20 nm across an 8-in. plate requires temperature control of about 0.15°C. This is not straightforward to obtain when large amounts of optical energy must be transmitted through the mask. Advanced exposure systems are used in submicron steppers to control the temperature of the mask to minimize this problem. Die rotation and misalignment due to alignment marker distortion are also possible. To find the total overlay tolerance one must sum the squares of the individual components:

Misalignment   **Runout**

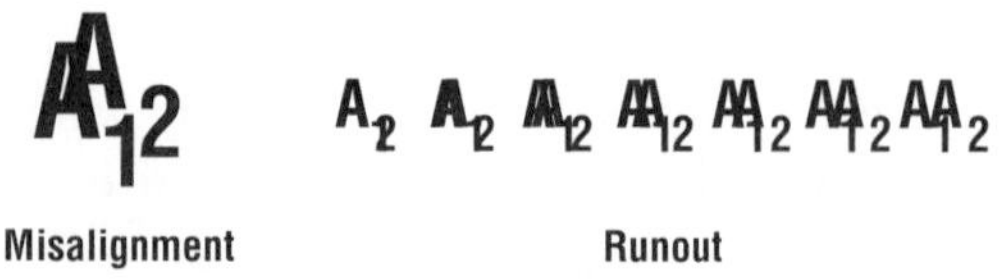

Figure 7.33   Two typical registration errors.

$$T = \left[\sum T_i^2 \right]^{1/2}$$

(7.27)

7.10 Summary

This chapter concentrated on the production of the areal image, the optical intensity as a function of position on the surface of the wafer. For the small features of interest in integrated circuit production, diffraction effects are extremely important. Simple contact printers can be used for pattern structures to less than 1 μm, but these systems are highly defect prone. To avoid this problem the mask can be floated above the wafer in a process known as proximity printing, but at a cost of degraded resolution. Projection lithography systems capable of submicron resolution were introduced. To achieve increased resolution in either type of optical system, it is desirable to use shorter wavelengths of exposing radiation. Although mercury arc lamps have historically been the most widely used source, excimer lasers are dominant in current-generation and advanced lithography tools. Finally, methods to increase resolution through mask making were introduced, primarily the use of phase-shifted masks and optical proximity correction.

The chapter began by the observation that lithography plays a critical role in determining the performance of a technology. As such, lithography has long been the gating process in technology development. As a result, it is natural to wonder how far optical lithography can be pushed. Nonoptical techniques, several of which will be reviewed in Chapter 9, suffer from severe drawbacks compared to optical lithography. There is a considerable amount of truth to the suggestion that the limit of optical lithography is roughly three generations beyond the current state of the art, and has been for the last 20 years. It is expected that 193-nm sources, immersion lithography, and OPC will extend optical lithography to at least 45 nm, and perhaps to 30 nm. If large-scale optics can be created for the F_2 laser, optical lithography will probably be extended to 20 nm. Resist improvements and mask refinements such as phase shifting and OPC are decreasing the minimum feature size faster than the exposing wavelength.

Problems

1. Some arc lamps produce a significant amount of energy in the deep UV because of the high energy electrons in the plasma. Ozone creation is therefore, a significant concern.
   (a) Calculate the plasma temperature required for the blackbody component of the radiation to be maximum at 200 nm.
   (b) If the volume of the lamp is 0.1 L and the gas is at 1 atmosphere at room temperature, use the approximation

   $$E \approx \frac{3}{2} kT$$

   where k is the Boltzmann constant, to determine the internal energy of the gas. This is a rough approximation that assumes the ionization of the gas to be negligible and that all species are at the same temperature. Note that 1 mole of gas occupies 22.8 L at room temperature and 1 atmosphere.
2. Show that for $x \ll \lambda g/w$ and w small enough to satisfy the Fraunhofer criterion of $I(x,y)$ independent of position in the x direction. Explain the significance of this result.
3. Use Equations 7.8, 7.9, and 7.10 in a 3-D graphing program to plot the areal intensity versus x and y for a 3 × 1-μm aperture. Assume that $\lambda = 436$ nm and $g = 25$ μm.
4. In an effort to make a relatively inexpensive aligner, which is capable of producing very small features, an engineer replaces the optical source of a simple contact printer with an ArF laser.
   (a) List two problems that the engineer is likely to encounter in trying to use this kind of aligner to make simple discrete devices. Assume that device yield is unimportant.

(b) Assume for this problem that the resist constant for the process is 0.8, and that, in hard contact, the gap is equal to the resist thickness. If the resist is 1.0 μm thick, what is the minimum feature size that can be achieved?

(c) How thin must the resist be made to achieve a 0.1-μm resolution?
If the exposure could be done successfully, and if the resulting resist image is to be used as a mask to etch some feature (for example, to etch a gate electrode), what problems might one have with this procedure?

5. Plot resolution and depth of field as a function of exposure wavelength for a projection aligner with 100 nm $\leq \lambda \leq$ 500 nm. Use $k = 0.75$ and NA = 0.26. On the same plots, recalculate these functions for NA = 0.41. Discuss the implications of these plots for the technologist that must manufacture transistors with 0.5-μm features.

6. A proximity aligner is used to expose 1-μm apertures. The gap is 25 μm. The separation between the mask and the g-line source is 0.5 m. Is the Fresnel criterion (Equation 7.5) satisfied? At what feature size would the inequality no longer be valid? How does this compare with the simple feature size prediction of Equation 7.16?

7. Repeat Problem 6, if the source is replaced with an i-line lamp and with an ArF laser.

8. A particular resist process is able to resolve features whose MTF is $\geq$ 0.3. Using Figure 7.21, calculate the minimum feature size for an i-line aligner with an NA = 0.4 and $S = 0.5$.

References

1. G. Stevens, *Microphotography*, Wiley, New York, 1967.
2. M. Bowden, L. Thompson, and C. Wilson, eds., *Introduction to Microlithography*, American Chemical Society, Washington, DC, 1983.
3. D. Elliott, *Integrated Circuit Fabrication Technology*, McGraw-Hill, New York, 1982.
4. W. M. Moreau, *Semiconductor Lithography, Principles, Practices, and Materials*, Plenum, New York, 1988.
5. S. Nanogaki, T. Heno, and T. Ho, *Microlithography Fundamentals in Semiconductor Devices and Fabrication Technology*, Dekker, New York, 1998.
6. P. Burggraaf, "Lithography's Leading Edge, Part 2: I-line and Beyond," *Semicond. Int.* **15**(3):52 CA, (1992).
7. *The National Technology Roadmap for Semiconductors*, 1997 ed., Semiconductor Industry Association, San Jose, CA, 1997.
8. For example, M. V. Klein, *Optics*, Wiley, New York, 1970.
9. M. Bowden and L. Thompson, in *Introduction to Microlithography*, M. Bowden, L. Thompson, and M. Lacombat, eds., American Chemical Society, Washington, DC, 1983.
10. K. Jain, *Excimer Laser Lithography*, SPIE Optical Engineering Press, Bellingham, WA, 1990.
11. R. Patel, "Excimer Lasers for Optical Lithography," *Vacuum Thin Film* 30 (March 1999).
12. H. Craighead, J. C. White, R. E. Howard, L. D. Jackel, R. E. Behringer, J. E. Sweeney, and R. W. Epworth, "Contact Lithography at 157 nm with an F$_2$ Excimer Laser," *J. Vacuum Sci. Technol. B* **1**:1186 (1983).
13. P. Concidine, "Effects of Coherence on Imaging Systems," *J. Opt. Soc. Am.* **56**:1001 (1966).
14. J. H. Bruning, "Optical Lithography Below 100 nm," *Solid State Technol.* **41**(11):59 (1998).
15. M. S. Hibbs, "Optical Lithography at 248 nm," *J. Electrochem. Soc.* **138**:199 (1991).
16. A. Voschenkov and H. Herrman, "Submicron Resolution Deep UV Photolithography," *Electron. Lett.* **17**:61 (1980).

17. A. Yoshikawa, S. Hirota, O. Ochi, A. Takeda, and Y. Mizushima, "Angstroms Resolution in Se–Ge Inorganic Resists," *Jpn J. Appl. Phys.* **20**:L81 (1981).

18. H. Smith, "Fabrication Techniques for Surface-Acoustic-Wave and Thin-Film Optical Devices," *Proc. IEEE* **62**:1361 (1974).

19. B. Lin, "Deep UV Lithography," *J. Vacuum Sci. Technol.* **12**:1317 (1975).

20. G. Geikas and B. Ables, *Kodak Photoresist Seminar*, 1968, p. 22.

21. B. Lin, in *Fine Line Lithography*, R. Newman, ed., North-Holland, Amsterdam, 1980, p. 141.

22. J. E. Roussel, "Submicron Optical Lithography?" *in Semiconductor Microlithography*, *Proc. SPIE* **275**:9 (1981).

23. D. A. Markle, *Solid State Technol*, **22**:50 (June 1979).

24. M. C. King, "New Generation of Optical 1:1 Projection Aligners," in *Developments in Semiconductor Microlithography IV, Proc. SPIE* **174**:70 (1979).

25. R. T. Kerth, K. Jain, and M. R. Latta, "Excimer Laser Projection Lithography on a Full-Field Scanning Projection System," *IEEE Electron Dev. Lett.* **EDL-7**:299 (1986).

26. P. Burggraaf, "Wafer Steppers and Lens Options," *in Semicond. Int.* 56 (March 1986).

27. K. Hennings and H. Schuetze, *SCP Solid State Technol.* 31 (July 1966).

28. M. A. van den Brink, B. A. Katz, and S. Wittekoek, "A New 0.54 Aperture i-line Wafer Stepper with Field by Field Leveling Combined with Global Alignment," in *Optical/Laser Microlithography IV*, V. Pol, ed., *Proc. SPIE* **1463**:709 (1991).

29. R. Unger, C. Sparkes, P. DiSessa, and D. J. Elliott, "Design and Performance of a Production-Worthy Excimer-Laser-Based Stepper," in *Optical/Laser Microlithography IV*, V. Pol, ed., *Proc. SPIE* **1674**:708 (1992).

30. Bert Vleeming, Barbra Heskamp, Hans Bakker, Leon Verstappen, Jo Finders, Jan Stoeten, Rainer Boerret, and Oliver Roempp, "ArF Step-and-Scan System with 0.75 NA for the 0.10 μm node," *Proc. SPIE* **4346**:634, *Optical Microlithography XIV*, Christopher J. Progler, ed., September 2001.

31. Bernard Fay "Advanced Optical Lithography Development, from UV to EUV," *Microelectron Eng*. **61–62**:11–24 (July 2002).

32. D. Gil, T. Brunner, C. Fonseca, and N. Seong, "Immersion Lithography: New Opportunities for Semiconductor Manufacturing," *J. Vacuum Sci. Technol. B* **22**(6): (November/December 2004).

33. Nikon, "Immersion Lithography: System Design and Its Impact on Defectivity," July 2005.

34. B. Smith, A. Bourov, Y. Fan, F. Cropanese, and P. Hammond, "Amphibian XIS: An Immersion Lithography Microstepper Platform," *Proc. SPIE* **5754** (2005).

35. M. Switkes, M. Rothschild, R. R. Kunz, S.-Y. Baek, and M. Yeung, "Immersion Lithography: Beyond the 65 nm Node with Optics," *Microlithography World* (May 2003); found at "Immersion Lithography," ICKnowledge.com (2003).

36. L. Geppert, "Chip Making's Wet New World," *IEEE Spectrum* (May 2004).

37. S. Owa, Y. Ishii, and K. Shiraishi, "Exposure Tool for Immersion Lithography," IEEE/SEMI Advanced Semiconductor Manufacturing Conference, 2005.

38. S. Peng, R. French, W. Qiu, R. Wheland, and M. Yang, "Second Generation Fluids for 193 nm Immersion Lithography," *Proc. SPIE* **5754** (2005).

39. J. Park, "The Interaction of Ultra-Pure Water and Photoresist in 193 nm Immersion Lithography," Microelectronic Engineering Conference, May 2004.

40. J. Taylor et al., "Experimental Techniques for Detection of Components Extracted from Model 193 nm Immersion Lithography Photoresists," *Chem. Mater.* **17**:4194 (2005).

41. M. Slezak, Z. Liu, and R. Hung, "Exploring the Needs and Tradeoffs for Immersion Resist Topcoating," *Solid State Technol.* (July 2004).

42. H. Sewell, D. McCafferty, L. Markoya, and M. Riggs, "Immersion Lithography, Next Step on the Roadmap," Brewer Science ARC Symposium, 2004.

43. B. Smith, A. Bourov, Y. Fan, F. Cropanese, and P. Hammond, "Air Bubble-Induced Light-Scattering Effect on Image Quality in 193 nm Immersion Lithography," *Appl. Opti.* **44**:3904 (2005).

44. A. Wei, M. El-Morsi, G. Nellis, A. Abdo, and R. Engelstad, "Predicting Air Entrainment Due to Topography During the Filling and Scanning Process for Immersion Lithography," *J. Vacuum Scie. Technol. B* **22**(6) (Nov/Dec 2004).

45. R. Unger and P. DiSessa, "New i-line and Deep-UV Optical Wafer Steppers," in *Optical/Laser Microlithography IV*, V. Pol, ed., *Proc. SPIE* **1463**:709 (1991).

46. R. Herschel, "Pellicle Protection of Integrated Circuit Masks," in *Semiconductor Microlithography VI, Proc. SPIE* **275**:23 (1981).

47. P. Frasch and K. Saremski, "Feature Size Control in IC Manufacturing," *IBM J. Res. Dev.* **26**:561 (1982).

48. B. J. Lin, "Phase-Shifting and Other Challenges in Optical Mask Technology," 10th Annu. Symp. Microlithography, *SPIE* **1496**:54 (1990).

49. M. D. Levenson, N. S. Viswnathan, and R. A. Simpson, "Improving Resolution in Photolithography with a Phase Shifting Mask," *IEEE Trans. Electron Dev.* **ED-26**:1828 (1982).

50. G. E. Flores and B. Kirkpatrick, "Optical Lithography Stalls X-rays," *IEEE Spectrum* **28**(10):24 (1991).

51. A. K. Pfau, W. G. Oldham, and A. R. Neureuther, "Exploration of Fabrication Techniques for Phase-Shifting Masks," in *Optical/Laser Microlithography IV*, V. Pol, ed., *Proc. SPIE* **1463**:124 (1991).

52. A. Nitayama, T. Sato, K. Hashimoto, F. Shigemitsu, and M. Nakase, "New Phase-Shifting Mask with Self-Aligned Phase-Shifters for a Quarter-Micron Photolithography," *Tech. Dig. IEDM*, 1989, p. 3.3.1.

53. Y. Yanagishita, N. Ishiwata, Y. Tabata, K. Nakagawa, and K. Shigematsu, "Phase-Shifting Photolithography Applicable to Real IC Patterns," in *Optical/Laser Microlithography IV*, V. Pol, ed., *Proc. SPIE* **1463**:124 (1991).

54. M. A. Listvan, M. Swanson, A. Wall, and S. A. Campbell, "Multiple Layer Techniques in Optical Lithography: Applications to Fine Line MOS Production," in *Optical Microlithography III: Technology for the Next Decade, Proc. SPIE* **470**:85 (1983).

Chapter 8

Photoresists

The last chapter discussed the production of the areal image, the pattern of radiation produced at the surface of the wafer during an optical exposure. To transfer a pattern, the radiation must strike a photosensitive material, and it must change the properties of that material in such a way that a replica of the mask is left on the surface of the wafer when the photolithography process is complete. The photosensitive compound used in microelectronics is called *photoresist* or simply, *resist*. This chapter will discuss the effects of radiation on the properties of photoresist. It will begin with novolac-based systems, since they have been commonly used in IC fabrication, and will proceed to resists used with short-wavelength exposures.

8.1 Photoresist Types

One of the most basic categories into which photoresist can be divided is its polarity. Following exposure of the photoresist to the radiation, it is immersed in a developer solution. Positive photoresist responds in a way that makes the exposed regions dissolve more quickly during the development process. Ideally, the unexposed regions will remain unchanged. Negative photoresist responds in the opposite manner. Unexposed regions of the resist will dissolve in the developer, while exposed regions remain behind. Positive resists tend to have the best resolution and are therefore much more popular for IC fabrication. They will be covered in more detail; however, negative resists will also be briefly discussed.

The photoresists used for IC fabrication normally have at least three components: a resin or base material, a photoactive compound (PAC), and a solvent that controls the mechanical properties, such as the viscosity of the base, keeping it in a liquid state. In positive resists, the PAC acts as an inhibitor before exposure, slowing the rate at which the resist will dissolve when placed in a developing solution. Upon exposure to light, a chemical process occurs by which the inhibitor becomes a sensitizer, increasing the dissolution rate of the matrix. Ideally, the inhibitor would completely prevent any dissolution of the photoresist, and the enhancer would produce an infinite dissolution rate. Of course, this is not achieved in practice.

The two most utilitarian metrics for the performance of a resist are sensitivity and resolution. Sensitivity refers to the amount of light energy (usually measured in millijoules per square centimeter) necessary to create the chemical change just described. The more sensitive the photoresist, the faster the process, since for a given exposure intensity, a shorter exposure time will be necessary. "Resolution" refers to the smallest feature that can be reproduced in a photoresist. As explained in the last chapter, this strongly depends on the exposure tool and the photoresist process itself. But even in a fixed exposure tool, this metric in particular carries with it a considerable uncertainty.

8.2 Organic Materials and Polymers°

Many of the compounds that will be considered in this chapter are carbon-based organic molecules. Since this is commonly outside the area of expertise of many of the students who might use this text, this section will introduce these materials and review a few of their relevant properties. Like silicon, carbon has four electrons in its valence shell. It requires four additional electrons to complete the shell. Unlike silicon, however, carbon readily bonds with itself in molecular form, allowing it to form complex chains and long repetitive molecules. Life on Earth is based on this ability. Carbon also readily bonds with hydrogen and materials to the right of it in the periodic table. The molecules discussed in this chapter are made up of carbon, oxygen, hydrogen, and nitrogen, with small amounts of additional elements. Over the past decades, methods have also been developed to combine hydrocarbons with heavy metals such as gallium to produce a class of materials known as organometallics. These saturated aliphatic (open-chain) compounds will be discussed separately in another chapter. This section will concentrate on two classes of carbon compounds: aromatic rings and long-chain polymers.

Aromatic rings consist of six carbon atoms arranged in a planar hexagonal structure. In the simplest such compound, benzene, a single hydrogen atom, is attached to each carbon atom (Figure 8.1). Often these hydrogens are omitted from the chemical symbol for brevity. The carbon atom gains two electrons by covalently bonding to two nearest neighbor carbon atoms, it gains one electron by bonding to the hydrogen. The last unpaired electron from each atom participates in a delocalized bond that takes

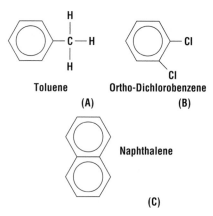

Figure 8.2 Some aromatic-based compounds based on (A) single-site substitution, (B) double-site substitution, where the first term defines the position (ortho, meta, or para) of the second substitution relative to the first, and (C) aromatic condensation.

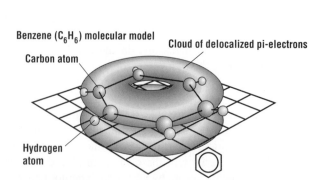

Figure 8.1 Diagram of simple benzene aromatic ring. The delocalized pi-bond electrons are in a ring that surrounds the nuclei. The symbol indicates the currently accepted ring notation.

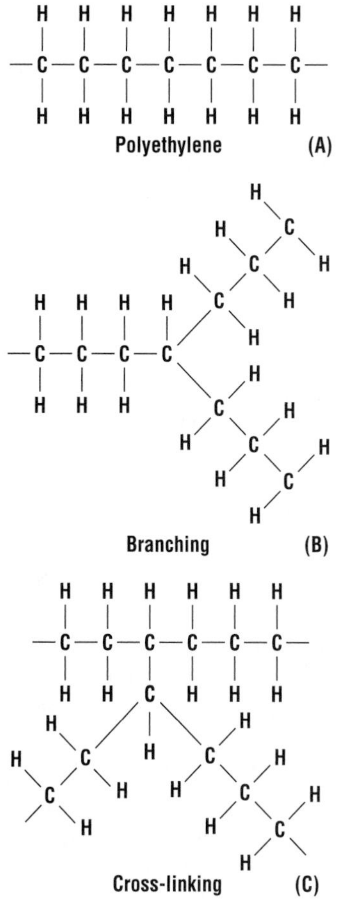

Figure 8.3

Figure 8.3 (A) Polyethylene, an example of a simple polymer. (B) Branched-chain polymers. (C) Cross-linking.

the shape of a doughnut around the benzene molecule. It is this highly mobile delocalized pi-electron that gives the aromatics their unique properties. According Hückel's rule, a monoclinic cyclic ring must have $4n + 2$ electrons, where $n = 0, 1, 2, \ldots$. Thus a ring with 5 electrons will not have this property.

A wide variety of compounds can be made by making simple variations of the benzene ring. As shown in Figure 8.2, toluene, a common solvent, is formed by replacing one of the hydrogens by a methyl (CH_3) group. Similarly, single hydrogen substitutions can produce phenol (OH), aniline (NH_2), chlorobenzene (Cl), and styrene ($HC=CH_2$). (A "=" sign in the molecular formula indicates that two electrons from each atom are participating in the bond.) An organic acid called carboxylic acid can be formed by adding a carboxyl group (COOH). This compound in particular will be important in the discussion of positive photoresists. The xylene class of solvents is formed by replacing two hydrogen atoms with methyl groups. Finally, aromatic rings can attach directly to each other. The simplest such combination, naphthalene, is shown in Figure 8.2. Much larger molecules can be formed in linear and 2-D arrays of aromatic rings. Graphite is an example of such a compound. Many of the suspected carcinogens in cigarette smoke are smaller condensed aromatics. Finally, not all rings have six carbon atoms. Rings can also form with five-atoms, although they are less common. Often these five-atom rings are attached to one or more benzene rings.

Polymers are very large molecules formed by linking together many smaller repeating units called *monomers*. A polymer can contain as few as five monomers, or it may contain thousands. Typical polymers include plastics, rubbers, and resins. Because of carbon's ease of bonding to itself, many polymers are carbon based. The simplest polymer is polyethylene (Figure 8.3A). It consists of a long chain of carbon atoms, each bonded to two hydrogen atoms. The monomer for polyethylene then is CH_2. A polymer may also be a branched chain as shown in Figure 8.3B. These molecules are stronger and have a higher density. Finally, polymers may cross-link, that is, bond with themselves or with other polymers (Figure 8.3C). Again, this increases the strength, and more importantly for this application, it reduces the ability of these molecules to dissolve in typical solvents, since, all other things equal, larger and heavier molecules dissolve more slowly than smaller, lighter ones. Resists, for example, are found [1] to have a dissolution rate proportional to MW^{-2}. On the other hand, if polymers are broken apart into a number of shorter chains, the molecules are more readily dissolved.

As with the benzene ring, many common compounds are simple variations of the basic polyethylene chain. If one hydrogen on every other carbon is replaced by a chlorine atom, the resultant material is polyvinyl chloride (PVC), which is commonly used to make plastic pipes for plumbing; PVC can also be softened to make simulated leather and raincoats.

One can now imagine a simple resist. If exposure of a long-chain polymer leads primarily to chain scission, the polymer dissolves more readily in the developer. Thus the polymer acts like a positive tone resist. If exposure of the polymer leads primarily to cross-linking, the exposed resist dissolves more slowly in the developer. The polymer acts as a negative tone resist. One of the problems with this simple resist is that it may be difficult to ensure that only one of these processes occurs. Chemists then can alter the backbone of the polymer to increase the likelihood of its dissolution or they can add reactive side chains to promote cross-linking. In the former case, however, it may be difficult to use the resist as an etch mask since the resist may also be attacked by a subsequent etch process.

8.3 Typical Reactions of DQN Positive Photoresist

For many years the most popular positive resists have been in a class of compounds called DQN, corresponding to the photoactive compound diazoquinone (DQ) and matrix material (N), respectively. As will be discussed later in the chapter, these resists cannot be used for very short wavelength exposures. As a result many alternatives have been developed. For i-line and g-line exposures, however, DQN is the dominant formulation. These resists evolved from materials used to make blueprints [2]. The matrix material in DQN resists is a thick resin called *novolac*. As shown in Figure 8.4, novolac is a polymer whose monomer is an aromatic ring with two methyl groups and an OH group Novolac comes in ortho-, meta-, and para-, forms. Most novolac is used as an adhesive when making plywood. By itself, it dissolves easily in an aqueous solution. Solvents are added to the resin to adjust the viscosity. This is an important parameter for the application of the resist to the wafer. Most of the solvent is evaporated from the resist before the exposure is done and so plays little part in the actual photochemistry. The solvents used in positive resists are generally combinations of aromatic compounds such as xylene and various acetates, although newer resists targeted toward deep UV applications are moving away from these compounds.

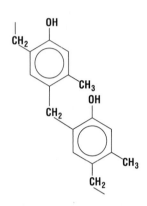

Figure 8.4 Meta-Cresol novolac, a commonly used resin material in g- and i-line applications. The basic ring structure may be repeated from 5 to 200 times.

The most commonly used PACs in these resists are the diazoquinones, such as the one shown in Figure 8.5. The part of the molecule from SO_2 down, which includes the two lower aromatic rings, is specific to particular resist manufacturers and plays only a secondary role in the exposure process. We will therefore simplify the DQ molecule by representing it with a generic R. In this state, the PAC acts as an inhibitor, reducing the dissolution rate of the resist in the developer by a factor of 10 or more. This occurs by a chemical bonding of the PAC and the novolac at the surface of the resist, where it is exposed to the developer [3], although the exact mechanism is still the subject of debate [4]. Some have also suggested that for the inhibitor mechanism to be operative, thermal cycles such as the softbake are necessary after application [5]. It is believed that the inhibitor mechanism is due to interactions between the resin and the photoactive compound. A popular early model, called the Stonewall model, suggested that the small molecules (PACs) help keep the large molecules in place in the way that small rocks help to keep a large boulders in place in a wall [6]. In this model, the carboxylic acid dissolves very quickly in the developer, increasing the effective area between the developer and the novolac matrix. In unexposed regions an azocoupling reaction between the matrix and the PAC retards the dissolution rate [7]. This inhibition depends on the chemical structure of the novolac matrix [8]. The dissolution mechanism is key to obtaining a high contrast resist. More recent models including the domain model [9] and the host-guest model [10], are somewhat more sophisticated, but still rely on PAC/matrix interactions to explain the inhibit/enhance process.

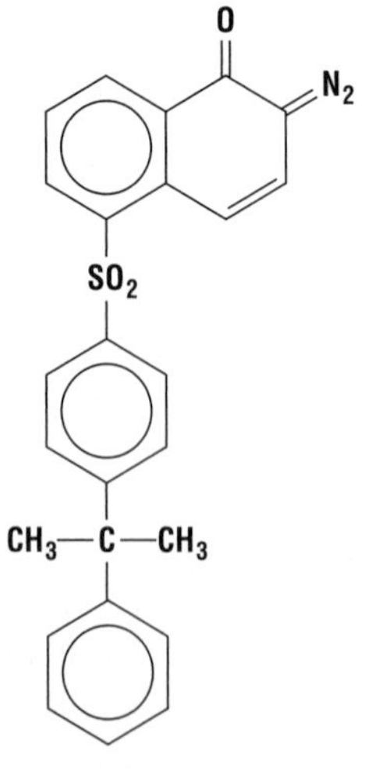

The nitrogen molecule (N_2) in the PAC is weakly bonded. As shown in Figure 8.6, the addition of UV light will free the nitrogen molecule from the carbon ring, leaving behind a highly reactive carbon site. One way to stabilize the structure is to move one of the carbons outside the ring. The oxygen atom is then covalently bonded to this external carbon atom. This process is known as a Wolff rearrangement. The resultant molecule shown in Figure 8.6 is called ketene. In the presence of water, a final rearrangement occurs in which the double bond to the external carbon atom is replaced with a single bond and an OH group. This final product is carboxylic acid.

Figure 8.5 Diazo quinone (DQ), the most commonly used photoactive compound for g- and i-line applications. The right-hand ring is not an aromatic but has a double bond.

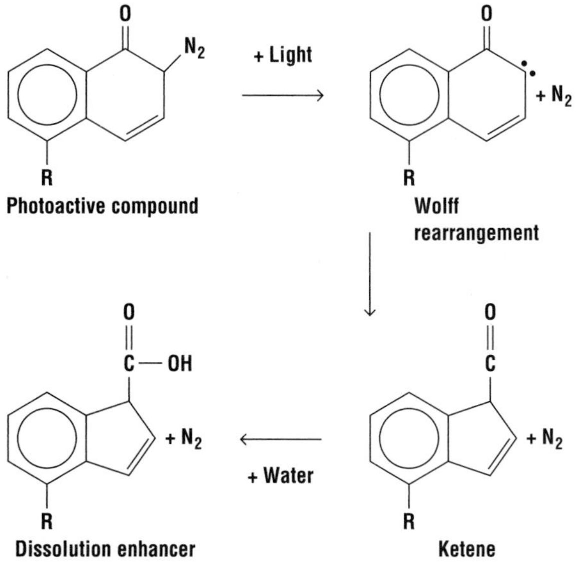

Figure 8.6 Photolysis and subsequent reactions of DQ upon UV exposure.

This process works as a PAC because the starting material will not dissolve in a base solution (one with a pH > 7). If the PAC is added to the matrix in about a 1:1 mixture, the photoresist is almost insoluble in a base solution. Carboxylic acid, on the other hand, readily reacts with and dissolves in base solutions [11]. This dissolution occurs for two reasons. The resin/carboxylic acid mixture will rapidly take up water. The nitrogen released in the reaction also foams the resist, further assisting the dissolution [12]. The chemical reaction that occurs during this dissolution is the breakdown of the carboxylic acid into water-soluble amines such as aniline and salts of K (or Na, depending on the developer). As already mentioned, the novolac resin itself is already water soluble and so dissolves easily. This process continues until all of the exposed resist is removed. It is not unusual for the dissolution rate of the photoresist to change by more than an order of magnitude under this process. Only light, water, and the ability to remove the nitrogen gas are required to drive the process. Typical developer solutions are KOH or NaOH diluted with water.

One of the great advantages of DQN photoresists is that the unexposed areas are essentially unchanged by the presence of the developer, since it does not penetrate the resist. Thus, a pattern of small lines on a clear field that is imaged onto a positive resist keeps its linewidth and shape. Another advantage is that novolac is a long-chain aromatic ring polymer that is fairly resistant to chemical attack. The photoresist pattern is therefore a good mask for subsequent plasma etching. Most negative resists work by cross-polymerization, a process in which the large resin molecules attach to each other to become less soluble. A typical negative resist is an azide-sensitized rubber such as cyclized polyisoprene. Negative photoresists have very high photospeeds and adhere well to the wafer without pretreatment. The primary disadvantage of these resists is swelling. Swelling refers to a broadening of the linewidth during the development phase, which takes place in organic solvents rather than aqueous solutions. An after-develop bake will typically cause the lines to return to their original dimension, but this swelling and shrinking process often causes the lines to be distorted. Closely spaced lines may come in contact during the swelling phase. As a result, negative resists are generally not suited to features less than 2.0 μm, unless very thin imaging layers are used in a multilayer process. Pinholes are also a serious concern in such an application.

8.4 Contrast Curves

As mentioned earlier, resolution is a very useful metric for resist performance, but it is highly dependent on the exposure tool. A function known as contrast (γ) is used to characterize the resist more directly. The contrast of a resist is measured by first coating a wafer with a layer of photoresist. Assume for now that a positive tone resist is used. The resist thickness is measured and the resist then given a uniform exposure of light for a small period of time. The exposure dose then is the light intensity (mW/cm^2) multiplied by the exposure time. Next, the wafer is immersed in a developer solution for a fixed period of time. Finally, the wafer is removed from the developer, rinsed off, and dried, and the remaining resist thickness is measured. If the light intensity was not too large, little of the PAC has changed from an inhibitor to an enhancer, so the thickness should be about the same as the original thickness. The experiment is then repeated for increasingly larger doses. If one normalizes the remaining resist thickness and plots it versus the logarithm of the incident dose, a γ or contrast curve as shown in Figure 8.7 will be obtained. The curve has three regions: low exposure where almost all of the resist remains; high exposure, where all of the resist is removed; and the transition region between these two extremes.

To derive a numerical value for the contrast of a photoresist, first approximate the steeply sloped portion of the curve by a straight line. The line extends from the lowest energy dose for which all of the resist is removed. This energy density is called D_{100}. The dose at which the line has a y-value of 1.0 is approximately the lowest energy needed to begin to drive the photochemistry. This energy is called D_0. Then the contrast is defined as

$$\gamma = \frac{1}{\log_{10}(D_{100}/D_0)} \tag{8.1}$$

which is simply the slope of the line.

Contrast can be thought of as a measure of the ability of a resist to distinguish between light and dark portions of the mask. For the moment, the significance of contrast may not be apparent.

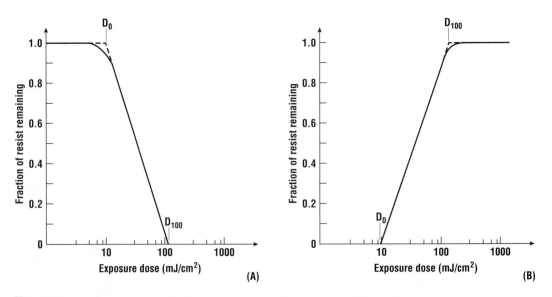

Figure 8.7   Contrast curves for idealized resists: (A) postitive tone and (B) negative tone.

Table 8.1   Contrasts for several commercial resists for different wavelengths, λ

λ (nm)	AZ-1350	AZ-1450	Hunt 204
248	0.7	0.7	0.85
313	3.4	3.4	1.9
365	3.6	3.6	2
436	3.6	3.6	2.1

Data from Leers [13]. The AZ formulations are products of Shipley.

Consider the exposure of a diffraction grating. As discussed in Chapter 7, the radiant intensity varies smoothly near the edges of the lines and spaces. The higher the contrast of the resist, the sharper the line edge. Typical optical photoresists have a contrast of 2 to 5. This means that D_{100} is $10^{1/5}$ to $10^{1/2}$ times larger than D_0. Furthermore, contrast curves are not fixed for a given resist. They depend on the development process, the softbake and postexposure bake processes, the wavelength of the exposing radiation, the surface reflectivity of the wafer, and several other factors. One task of the lithographer is to adjust the photoresist processing to maximize the contrast while maintaining an acceptable photospeed. Table 8.1 shows typical contrasts for several resists for various wavelengths.

Example 8.1

Find the contrast from Figure 8.7.

Solution

For either plot, $D_{100} = 100$ mJ/cm² and $D_0 = 10$ mJ/cm².
Then

$$\gamma = \frac{1}{\log_{10}(100/10)} = 1$$

Example 8.2

One can also use contrast curves to make a rough approximation of the final resist profile.

Solution

Consider the contrast curve obtained from a commercially available resist, shown in Figure 8.8A. Assume that the areal image can be approximated by two straight lines. The intensity is a constant from $x = 0$ to $x = 1.0$ μm and varies linearly from $x = 1.0$ to $x = 2.5$ μm (Figure 8.8B). Then, we can multiply this exposure energy by the time to obtain a profile of dose as a function of position (not shown). Finally, we can map the dose plot onto the contrast curve and determine the resist profile as shown in Figure 8.8C. Note that these curves are only very approximate profiles. They ignore 2-D and 3-D development of the resist and the vertical exposure variation. From Figure 8.8C one can extract a plot of feature size W versus exposure dose. These curves are often used to measure the sensitivity of the critical dimension to the exposure dose.

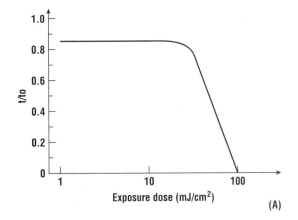

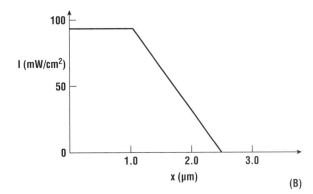

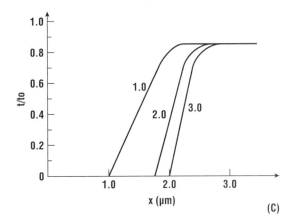

Figure 8.8 (A) Measured contrast curve for a commercial DQN resist. (B) Simple areal image. (C) Approximate profiles for 1-, 2-, and 3-sec exposures.

For typical resists, low exposure dosages are less than 50 mJ/cm². At these energies, the resist profile depends primarily on the low exposure and transition regions of the contrast curve. These exposures produce shallow angle resist profiles that are less dependent on the quality of the areal image. For doses greater than about 150 mJ/cm², the exposed regions are typically well above D_{100}. Then, the resist profiles depend primarily on the optical image and light scattering, and absorption in the resist and the profiles are fairly sharp. Although a sharper image is usually desirable, it comes at a cost of a longer exposure time and therefore slower throughput. Most exposures are done in the moderate or high exposure regimes. Subsequent comments shall be confined to these regimes.

Once the light begins to penetrate the resist its intensity will decrease as

$$I = I_0 e^{-\alpha z} \tag{8.2}$$

where α is the optical absorption coefficient in the photoresist, having units of inverse length. In general, D_0 is independent of the resist thickness. The energy density D_{100} is inversely proportional to the absorbance A, where

$$A \equiv \frac{\int_0^{T_R} [I_0 - I(z)]dz}{I_0 T_R} = 1 - \frac{1 - e^{-\alpha T_R}}{\alpha T_R} \tag{8.3}$$

where T_R is the resist thickness. Then one can show that [14]

$$\gamma \approx \frac{1}{\beta + \alpha T_R} \tag{8.4}$$

where β is a dimensionless constant. As one might reasonably expect, γ increases as the thickness of the resist decreases. If the resist is too thin, however, it may have poor step coverage as it covers the topography. It may also be unable to withstand the etch of the underlying layers. Some compromise must therefore be made between absolute resolution and more practical resist parameters.

8.5 The Critical Modulation Transfer Function

Another resist figure of merit that can be determined from the contrast is the critical modulation transfer function (CMTF). It is approximately the minimum optical modulation transfer function necessary to obtain a pattern. The critical modulation transfer function is defined by

$$\text{CMTF}_{\text{resist}} = \frac{D_{100} - D_0}{D_{100} + D_0} \qquad (8.5)$$

It can be found using the contrast as

$$\text{CMTF}_{\text{resist}} = \frac{10^{1/\gamma} - 1}{10^{1/\gamma} + 1} \qquad (8.6)$$

A typical value for the CMTF is about 0.3. The function of the CMTF is to provide a simple test for the resolution of a photoresist. If the MTF of an areal image is less than the CMTF, that image will not be resolved. If it is greater, it is possible to resolve the image. Like the contrast, it gives us a single number to determine the resolution.

Example 8.3

A new photoresist is being developed for use in a KrF stepper. If the stepper has NA = 0.6 and a spatial coherence of 0.5, what is the minimum feature size when $\gamma = 3.0$? If the resist has $\gamma = 4.0$, what will the minimum feature size be?

Solution

Using Equation 8.6, the CMTF of the two resists are found to be 0.366 and 0.206. Setting these two numbers equal to the MTF shown earlier in Figure 7.22, we find that the resolutions are 0.52 and 0.63 line per unit length normalized to the cutoff frequency. For a KrF stepper with a 0.6 NA lens, the cutoff frequency is 3.97 μm^{-1}. Then, the resolution of the exposure tool using these two resists would be 2.06 and 2.50 line pairs per micron or minimum feature sizes of 0.24 and 0.20 μm.

8.6 Applying and Developing Photoresist

The steps in a photolithographic process are listed in Figure 8.9. For positive resists, the wafer must typically be pretreated before resist application in order to obtain a smooth, uniform coverage of the photoresist with good adhesion of the resist to the wafer. The first step in the pretreatment is typically a dehydration bake. Done at 150–200°C, in either vacuum or dry nitrogen, this step is intended to drive off most of the adsorbed water on the surface of the wafer. At this temperature, about one monolayer of water remains on the surface. Dehydration bakes can be done at much higher temperatures to further remove all of the adsorbed water, but these high temperature bakes are much less common.

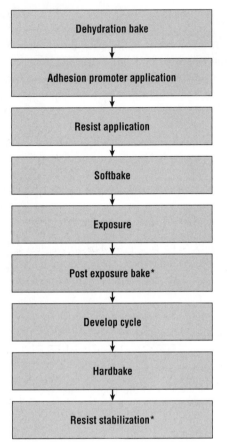

Dehydration bake

↓

Adhesion promoter application

↓

Resist application

↓

Softbake

↓

Exposure

↓

Post exposure bake*

↓

Develop cycle

↓

Hardbake

↓

Resist stabilization*

*Optional steps

Figure 8.9 Typical process flow in a photolithography step.

Immediately following the bake the wafer is often primed with hexamethyldisilazane (HMDS), which acts as an adhesion promoter. Vapor priming can be done by suspending the wafer above a container of the high vapor pressure HMDS liquid and allowing the vapor to coat the surface of the wafer. Liquid HMDS can also be applied directly on the wafer by dispensing a fixed volume and spinning the wafer to spread out the liquid to a very thin uniform coating. Using either method, one monolayer of HMDS bonds readily with the surface of the wafer, even if it is partially hydroxylated. The other side of the molecule bonds readily with the resist.

The wafer is coated with photoresist after it is primed. The most common application method is spin coating. The wafer is first mounted on a vacuum chuck, which is a flat, hollow metal disc connected to a vacuum line. The chuck has a number of small holes in the surface. When a wafer is placed on its surface, the vacuum draws the wafer into intimate contact with the chuck. A predetermined amount of resist is then dispensed on the surface of the wafer. Torque is applied to the chuck to rapidly accelerate it at a controlled rate up to a maximum rotational speed, usually 2000–6000 rpm. The acceleration stage is crucial to obtaining good uniformity, since the solvents begin evaporating from the resist as soon as it is dispensed. The wafer is spun at this speed for a fixed period of time, then decelerated in a controlled manner to a stop. A variation of this method, called a dynamic dispense, is to apply some or all of the resist while the wafer is spinning at low speed. This allows the resist to spread across the wafer before the high speed spin.

Resist thickness and thickness uniformity are critical parameters in developing a good photolithography process. Thickness is not a strong function of the dispense amount. Typically, less than 1% of the dispensed resist remains on the wafer after spinning. The rest flies off during the spin. To avoid a redeposition of this material, spinners have splash guards around the chuck. The thickness of the resist is primarily determined by its viscosity and the spin speed. Higher viscosities and slower spin speeds will produce thicker layers of photoresist. The resist thickness is found to vary with spin speed as [15]

$$T_R \propto \frac{1}{\sqrt{\omega}} \tag{8.7}$$

A typical process might be a 30-sec spin at 5000 rpm to produce a resist thickness of about 0.5 μm. At this point, the resist has a tacky consistency since less than one-third of the solvent remains.

After spinning, the wafers must undergo a softbake or prebake. The function of this step is to drive off most of the solvent in the resist and to establish the exposure characteristics [16]. The dissolution rate in the developer will be highly dependent on the solvent concentration in the final photoresist. Generally, shorter times or lower temperature softbakes lead to an increased dissolution rate in the developer and so to a higher sensitivity, but at the cost of lower contrast. High temperature softbakes can actually begin to drive the photochemistry of the PAC, leading to resist dissolution of unexposed regions in the developer. In practice, the softbake cycle is empirically determined through trial and error by optimizing the contrast while retaining an acceptable photosensitivity. Typical soft-bake temperatures are 90–100°C. Times range from 30 sec on a hotplate to 30 min in an oven. The solvent concentration remaining after softbake is usually about 5% of the original concentration.

After softbake the wafer is exposed. This process was described in some detail in the previous chapter. After exposure, the wafer must be developed. Nearly all positive resists use alkaline developers such as KOH dissolved in water. Because Na and K can affect transistor reliability, IC manufacturers have switched to nonalkaline developers such as tetramethyl ammonium hydroxide (TMAH). During the develop cycle, the carboxylic acid reacts with the developer to form amines and metallic salts. In doing so, it reduces the pH. Care must be taken to constantly maintain a pH of at least 12.5 in the developer if a consistent process is to be maintained [1]. In simple immersion developing, this is done by periodically dumping and refilling the tank, for example, after a certain number of wafers have been processed. To maintain a more consistent develop process, the develop time is often increased between tank refills. A method more commonly used in manufacturing is to ensure a supply of fresh developer for each wafer by dispensing it on track equipment in a process known as puddle develop [17] or by loading a batch of wafers into a spray developer system [18].

During the develop process the developer solution penetrates the surface of the exposed resist creating a gel. The depth of the gel, called the penetration depth, is negligibly small in novalac-based resins. This is not true for many negative tone resists, where the swelling of the penetration region can lead to a distortion of resist features.

Like many chemical reactions, the develop process is very temperature sensitive. It is important therefore, to control the develop temperature carefully if accurate control of linewidths is to be maintained. Developer temperature often must be controlled to better than 1°C. In the case of spray developers, the temperature drop associated with the adiabatic expansion as the developer is forced from the nozzle must be taken into account. To compensate for this effect, heated spray nozzles are sometimes used.

The develop process can affect the resist contrast and therefore the resist profiles. Figure 8.10 shows the contrast curves for Megaposit® SPR500™ [19] photoresist developed using MF CD-26. The film thickness for this curve was 1.085 µm, and the exposure was done with an i-line source. The soft-bake was done for 60 sec using a 90°C hotplate, and a 60 sec 110°C hotplate bake was done after exposure. When a 60-sec single spray puddle application of the developer was used, the contrast was 3.14. However, if two spray dispenses are done, each of which is 30 sec long, the contrast increases to 4.12. Obviously, the photolithography process occurs in a multidimensional space, and optimization of the process is an important and challenging task.

The surfactant HMDS was used to ensure a more uniform coating between the photoresist and the wafer. It also helps ensure that patterned features do not lift off the wafer during rinse steps. Surfactants may also be added to the developer solution. During the develop process it will migrate to the surface of the wafer, orienting itself with the hydrophobic portion of the surfactant toward the resist and the hydrophilic end toward the developer. This reduces the surface tension and improves the ability of the developer to wet the surface of the wafer [20]. In some cases, agitation, such as ultrasonic

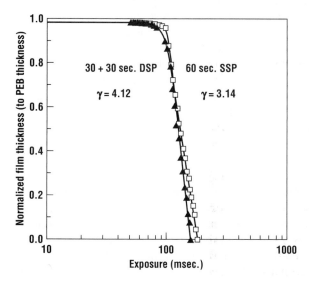

Figure 8.10 Contrast curves for Megaposit SPR500 resist using MF CD-26 developer *(courtesy Shipley).*

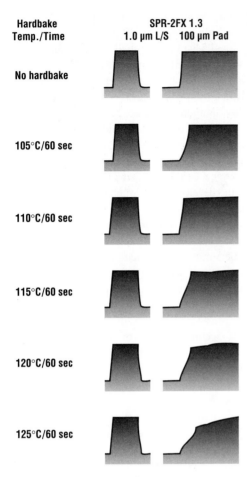

Hardbake Temp./Time	SPR-2FX 1.3 1.0 μm L/S 100 μm Pad
No hardbake	
105°C/60 sec	
110°C/60 sec	
115°C/60 sec	
120°C/60 sec	
125°C/60 sec	

Figure 8.11   Resist profiles of a 1-μm line and space and a large feature in SPR-2FX resist for different hardbake temperatures *(courtesy Shipley)*.

waves, may be used in conjunction with the surfactant to optimize the contrast [21, 22]. The surfactant may also serve as a dissolution inhibitor, blocking the developer from the unexposed regions. Mckean et al. used this effect to increase the contrast of a thick film resist by a factor of 3 [23].

High temperature bakes may also be done after develop. Sometimes called a hardbake, this process is used to cross-link the resist and so harden it against further energetic processes such as ion implantation and plasma etching. Figure 8.11 shows 1.0-μm line and space and large-feature resist profiles for 60-sec hotplate bakes of varying temperatures. The resist is Shipley Megaposit SPR-2FX, which is used for sub-micron g-line exposures. At sufficiently high temperatures, the resist profile begins to reflow. This can be used to produce a shallow profile so that during a subsequent etch the angle can be reproduced in the underlying film. The reflow temperature depends on the resist to be used. Some resists, formulated for use in energetic processes, do not flow below 200°C.

There are several types of systems available for resist processing. The simplest, and the one most likely to be encountered in a university laboratory, is a pair of convection ovens for processing batches of wafers through hard and softbakes and a single-wafer spinner. Often resist is applied using a syringe. Even this crude and inexpensive equipment can be used to fabricate submicron features, but uniformity and reproducibility leave much to be desired. At the other extreme, industrial fabrication facilities normally use automated photoresist processing systems (Figure 8.12) sometimes called tracks. In these systems, wafers move from a storage cassette to a hotplate or infrared lamp oven where a dehydration bake is done. Next, they are moved to a dispense station, where HMDS is applied and spun, and then photoresist is dispensed and spun. Next, they move to a second hotplate or infrared lamp for a softbake. Finally, they are sent to a second storage cassette from which they are taken for exposure. Cool plates are often used after the hotplates to ensure a repeatable temperature during dispense and to avoid contamination of the wafers by the receiving cassette.

Some tracks are linked to the exposure tool in a cluster arrangement for automatic exposure. These tools have the highest overall consistency and reproducibility. In this application, the automated resist processing equipment need not be integrated into a stand-alone unit. Instead, separate bake, coat, and develop modules may be arranged in a circular fashion along with the exposure tool. At the center of the cluster, a robot arm can pick and place the wafers at the appropriate station. Such systems can schedule the wafers for resist processing in a consistent manner to make the time between coat and expose and the time between expose and develop the same for each wafer. This can be a critical factor in the success of new resists designed for exposures at 248 nm and below. They can also reduce the particulation associated with the belt transport mechanisms typically found in tracks [24]. In either type of automated system, photoresist thickness variations of less than 50 Å (1.0%) across a wafer and less than 100 Å wafer to wafer are commonly attained. Due to the nature of photoresist and the number of mechanical parts in these systems however, fully automated systems require frequent attention.

Figure 8.12   Top view of a photoresist processing system including cassette load and unload, resist application, bake and develop stations, and a central robot; more modern systems are in a controlled environment and integrated into an exposure tool *(courtesy silicon Valley Group)*.

8.7   Second-Order Exposure Effects

An important parameter in the selection of a resist is its absorption spectrum. If the wafer is to be exposed using an aligner with a particular line source, one must know the α of the resist at that wavelength. If α is large (say more than one over the thickness of the resist), only the top of the resist will effectively be exposed. Upon develop, the lower part of the resist will be left behind, and the wafers will appear underdeveloped. If α is too large, little of the light will be absorbed during the exposure and long exposure times will be required. An advantage of the diazoquinones is that they absorb strongly at both g- and i-lines of mercury but do not absorb very well in the mid-UV and the visible [25]. This allows the use of room light in the lithography area, as long as the deep UV components have been filtered out. Typical DQ sensitizers also do not absorb well in the deep UV.

One must also take into account the absorbance of the resin portion of the resist. Light absorbed by the resin does not reach the PAC and so does not drive the photochemistry. Pure novolac is water-white. Due to an oxidation step in the resist fabrication process, it turns an orange-brown [26]. All novolac compounds absorb strongly in the deep UV. Novolac-based resists are therefore not appropriate for exposures in the deep UV, since the resin itself will absorb most of the light. Modified novolac, however, can be used down to about 250 nm [27]. As will be discussed later in this chapter, if it is

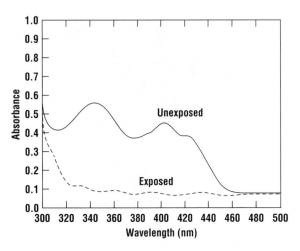

Figure 8.13 Total absorbance of a layer of SPR511-A resist before and after exposure. The difference between the two curves is the actinic absorbance *(courtesy Shipley).*

desired to use the DQN resist as an opaque layer for planarizing a subsequent resist layer, a high temperature postbake will increase the absorptivity of the layer in the mid-UV.

To further complicate matters, for most resists the absorbance changes on exposure. Typically it decreases. The actinic absorbance is defined as the difference between the unexposed and the exposed absorbance (see Figure 8.13). This effect is also known as bleaching. The advantage of bleaching is that it can provide a more uniform exposure. As the top layers of the resist are exposed, they become partially transparent, allowing a fuller exposure of the lower layers.

When photoresist is exposed over topology, a new set of problems arises. The last chapter described the effects of surface reflections on the areal image. A second problem that exists in the presence of topology is

---

Example 8.4

Using the data in Figure 8.13, find the inverse absorption length, α, at i-line before and after exposure. Assume $T_R = 0.5 \ \mu m$.

Solution

From Equation 8.3,

$$A = 1 - \frac{1 - e^{-\alpha T_R}}{\alpha T_R}$$

From Figure 8.13, $A = 0.4$ before exposure and 0.1 after exposure.

$$0.4 = 1 - \frac{1 - e^{-\alpha T_R}}{\alpha T_R}$$

$$\frac{1 - e^{-\alpha T_R}}{\alpha T_R} = 0.6$$

Solving $\alpha T_R \approx 1.1$, so $\alpha \approx 2.2 \ \mu m^{-1}$
After exposure $A = 0.1$, so $\alpha T_R \approx 0.2$ and $\alpha \approx 0.4 \ \mu m^{-1}$

---

linewidth variations due to resist thickness variation. Since the photoresist is a viscous film, it does not coat conformally. Instead, it will tend to smooth out the surface topology (Figure 8.14). On top of a step it will be thinner than the nominal thickness. Immediately next to a step it will be thicker than the nominal thickness. Since the step height is often about the same as the resist thickness, these variations can be substantial. The resist thickness changes will lead to linewidth variations: a bulging of the photoresist line just off the step and a necking at the step. These linewidth variations are caused

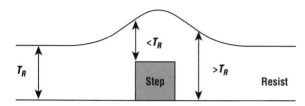

Figure 8.14 Cross-sectional view of resist as it covers a vertical step.

by an effective overexposure on the step and an effective underexposure just off the step. In the case of metal interconnect, these variations can be a serious reliability concern, since the linewidth necking near the step often occurs near a region of poor step coverage in the deposition and near a point where a large gradient exists in the thermal conductivity to the substrate. All of these effects tend to make it likely that a metal line will fail in operation.

One solution to this problem is to planarize the layers if possible. This will be discussed in Chapter 12, since planarization has many more impacts than just lithography and is being aggressively pursued as part of an advanced interconnect strategy. Another solution is the use of multilayer resists. In a typical application, a thick planarizing polymer is first applied. The polymer is called locally planarizing because it can smooth steps in the topology but does not affect changes in height that occur over long distances such as the wafer bow. The lower level may also be opaque to the exposing radiation to prevent surface reflections (see Section 7.8). The upper layer of photoresist may be quite thin to improve the contrast and minimize light scattering. In some cases a third material such as SiO_2 may be deposited between the two polymers to prevent their interaction and to act as an etch mask for subsequent pattern transfer. This is particularly useful if the lower level is PMMA or a similar polymer with a low etch resistance.

After application and softbake the upper resist is exposed and developed. The photoresist is then used as a mask to etch the underlying polymer, which in turn is used as the processing mask. This allows the upper or imaging photoresist to be quite thin and therefore have high contrast. It is how very fine features are produced in contact lithography. The drawback of such approaches, of course, is process complexity. In applications in which small features are required, however, planarization of the substrate in at least some of the process steps is almost mandatory.

Example 8.5

Use software to expose the 0.4-μm micron patterns used in the numerical examples in the last chapter. Assume i-line exposure and NA = 0.5. Adjust the exposure time to get a good image.

Solution

```
go athena
#
# OPTOLITH input file: Example 8_4.in
# Example of resist deposition, exposure, and develop
# Define the grid 2 um by 2 um
line x loc=0.00 spac=0.05 tag=left
line x loc=1.00 spac=0.025
line x loc=2.00 spac=0.05 tag=right
line y loc=1. spac=0.1 tag=top
line y loc=2.0 spac=0.1 tag=bottom
init silicon orientation=100
```

```
#
# Define exposure parameters for the Dill exposure model and
development
# rate parameters for the kim development rate model for the
photoresist.
rate.develop name.resist=PR i.line\
  r1.kim=0.085329 r2.kim=0.000002 r3.kim=11.74276 \
  r4.kim=0.0 r5.kim=0.0 r6.kim=0.0 r7.kim=0.0 \
  r8.kim=0.0 r9.kim=0.0 r10.kim=0.0 \
  a.dill=0.525 b.dill=0.0298 c.dill=0.02 \
  Dix.0=7.55e-13 Dix.E=3.34e-2
#
# Define index of refraction of user defined photoresist.
optical photo name.resist=PR lambda=0.365 refrac.real=1.6
refrac.imag=0.02
#
# Deposit user defined photoresist PR
deposit photoresist name.resist=PR thick=.5 divisions=20
#
# Use symmetry to reduce computation time (Note x.grid above
definition only for x>0)
structure mirror left
structure outfile=ex8_4a.str
#
# Run the imaging module for the diffraction grating; Note the
change in orientation.
illumination i.line
illum.filter clear.fil circle sigma=0.3
projection na=.5 flare=2
pupil.filter clear.fil circle
#
layout x.low=1.4 z.low=-2.5 x.high=1.8 z.high=2.5
layout x.low=0.6 z.low=-2.5 x.high=1.0 z.high=2.5
layout x.low=-0.2 z.low=-2.5 x.high=0.2 z.high=2.5
layout x.low=-1.0 z.low=-2.5 x.high=-0.6 z.high=2.5
layout x.low=-1.8 z.low=-2.5 x.high=-1.4 z.high=2.5
image clear win.x.l=-3 win.z.l=0 win.x.h=3 win.z.h=0 x.p=31
z.p=3 one.d
structure outfile=ex8_4b.str intensity
#
# Resist exposure, post exposure bake, and development
expose dose=100 na=0
structure outfile=ex8_4b.str
bake time=60 seconds temp=125
develop kim time=100 steps=5
structure outfile=ex8_4c.str
tonyplot -st ex8_4c.str -set ex8_4.se
```

Solutions for exposure doses of 60, 100, and 140 mJ/cm² appear in figure 8.15.

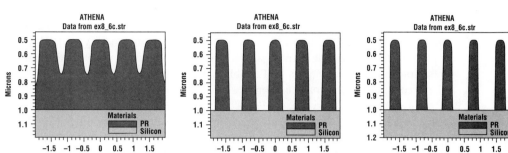

Figure 8.15   Resist profiles for 60, 100, and 140 mJ/cm². Notice the line narrowing ($W < 0.4$ μm) at the higher dose. Dose optimization will require a measurement of the line width as a function of the dose (swing curve).

8.8   Advanced Photoresists and Photoresist Processes[+]

This section will review new resist technologies that show promise for future deep submicron lithography. The categories are somewhat arbitrary. All of the topics covered are intended in some form for deep UV application, although they are not listed as such. Silicon-containing resists can be used as a contrast enhancement layer, or they can be a resist themselves.

The last chapter discussed the utility of deep UV sources such as excimer lasers and indicated that these exposure systems are now being used by a number of IC manufacturers. One of the major difficulties with this application is the lack of good deep UV resists. Novolac compounds begin to strongly absorb just below 250 nm, making KrF (248 nm) exposures marginally acceptable, but precluding the much more desirable ArF (193 nm). Furthermore DQ does not bleach very effectively at 248 nm. This leads to highly nonuniform exposures through the depth of the resist, and therefore sloping sidewalls for small features. Generally it is accepted that DQN cannot be used for features much below 0.18 μm, and that both a new matrix material and a new photoactive material must be found for 193-nm exposures.

When DQN resists are used at short wavelengths, there are typically chemically amplified. In chemically amplified resist (CAR) systems, an additional photoactive compound is added to the matrix and photosensitizer. Upon exposure to light, the chemical amplification agent acts in such a way as to greatly increase the primary photochemical process. A key to this process is that a single photoevent catalyzes many subsequent bond-breaking events. A typical example is the use of a photoacid generator (PAG). Once it absorbs a photon, the PAG becomes chemically reactive dissociating the matrix material. Early CARs contained onium salts [28] such as some early IBM resists [29]. Exposure dosages as low as 10 mJ/cm have been reported using CARs with conventional DQN resists. This resist system has been used in the early manufacture of 1-Mbit DRAMs [30] and in 16-Mbit DRAM pilot line operation [31]. One liability of these salts is the metallic contaminants left in the resist. Other workers have demonstrated CARs using nonmetallic components [32, 33]. Later CARs have used halogens [34], sulfonic acid esters [35], or sulfonyl compounds [36]. In many cases a protection agent is used to ensure that the unexposed PAG does not attack the resist before it is unzipped.

Although the use of chemical amplification in mid-UV applications can be thought of as a manufacturing enhancement, it plays a much more vital function in the deep UV resists currently being developed. The high energy of the photons makes matrix absorption almost unavoidable. All resists currently being developed for deep UV (248 nm and below) applications therefore use some type of chemical amplification [37] as the photoactive compound. Typical DUV resists consist of a matrix that is somewhat photosensitive, a PAG, a protective agent, and modifiers such as solvents. None of these materials is well established for deep UV applications. Table 8.2 shows potential 193-nm components and considerations for a single-layer resist.

Polymethyl methacrylate (PMMA) is a prototype deep UV resist or is a matrix in more sophisticated resists. This is a resist much favored by lithographers intent on displaying the ultimate in resolution, but has limited utility as an imaging layer otherwise. Since PMMA is most commonly used as an electron-beam resist, its structure will be discussed in more detail in the next chapter. Briefly, however, it is a long-chain polymer consisting of alternating segments of H–C–H and CH_3–C–$COOCH_3$. Often this chain is compressed or "rolled up." Under deep UV exposure the long chains can be broken and one or more of the carbon atoms left with an unsatisfied bond, or either the methyl (CH_3) group or the ester ($COOCH_3$) side chain can be affected. If the main chain breaks, the resulting short molecules are more easily dissolved in a developing solution. Microfoaming by the dissociated gaseous products (CH_3, CH_3OH, and $HCOOCH_3$) also increases the dissolution rate [39].

PMMA has two primary drawbacks. The first is that the plasma etch tolerance of the resist is very low, lower in fact, than most films to be etched. Therefore unless a very thick layer of PMMA is used to protect a very thin film, the resist will disappear during the etch before the film does. This is generally not feasible, however, since resist features with aspect ratios larger than 4 are not generally considered to be mechanically stable. Furthermore the dissociation of the PMMA changes the chemistry of the plasma etch and frequently leads to polymeric deposits on the surface of the wafer. The other primary limitation of PMMA is a low sensitivity. Typical optical exposure doses are greater than 200 mJ/cm^2, while required sensitivities are typically 5 to 10 mJ/cm^2. As described earlier, various PAGs have been added [40] to PMMA-like compounds to make more practical resists. In some cases these additives also can increase the etch resistance. A second approach to increase the sensitivity is to expose the wafer at an elevated temperature. This has the additional advantage of increasing the contrast. Contrasts of 7 have been reported for exposures at 140°C [41].

Although the sensitivity may be improved, the poor etch tolerance and short shelf life of PMMA-based resists have forced the development of alternative deep UV matrix materials. Although alternate materials such as poly(4-hydroxy-styrene) are much more transparent than more commonly used novolac compounds, they are extremely difficult to inhibit to a sufficient degree [42].

Table 8.2   Possible materials and considerations for a 193-nm resist

Polymer	PAG Considerations	Protection Agents	Additives
Acrylate	Efficiency	*t*-Butyl ester	Glass transition temperature modifier
Methacrylate	Acidity	Alicyclic ester	Resolution enhancer
Cyclic olefin	Volatility	Low bake temperature	Coating aid
Alternating copolymer	Size		
Hybrid			

From Allen [38].

The substitution of dimethyl groups has reduced the dissolution rate of this matrix material by a factor of 500, close to what is necessary for a high performance resist [43]. An alternative imaging scheme for deep UV resists is to use a completely different chemical path. Some of the most promising prototype 193-nm resists use a PAG, a protection agent, and various acrylic-based matrixes [44]. Although an improvement compared to simple PMMA, acrylic-based resists share the problem of etch resistance. To improve the etch resistance, various groups have been grafted onto the side chains of the monomer. The use of pendant groups such as norbonyl methacrylate or adamantyl methacrylate has been found to improve the etch resistance close to that of novalac, but degrade the sensitivity and resolution [45]. Acrylics also typically show poor adhesion during development.

Common problems with CARs include a deterioration of the image during the time between exposure and postexposure bake and a deterioration of the surface of the resist after coating when exposed to room air. In the former case, diffusion of the PAG from exposed to unexposed regions can lead to a degradation of the image [46]. In the latter case isolated lines appear in a characteristic "T" shape upon developing. Many CARs are also sensitive to shelf life. Demonstrations of 193-nm single-layer resists have been performed with CARs [47]. Table 8.3 lists delay-time effects for several resists.

Contrast enhancement layers (CELs) allow the use of an optical lithography exposure tool for smaller features than would otherwise be attainable. They have been demonstrated primarily with DQN resists. The basic process involves a material that is spun onto a resist-coated wafer after the softbake step [48]. This material must be nominally opaque to the exposing wavelength, but undergoes a bleaching reaction upon exposure that renders it transparent. Effectively the use of a CEL transfers the mask to this upper layer, which is in hard contact with the resist. After exposure, the contrast enhancement layer is stripped before develop. CEL is particularly important for deep UV resists, since the optical sources may be less intense and the matrix materials may tend to absorb the radiation.

Inorganic resists fall into a class of resists known as charge transfer compounds. In these systems insolubility is brought about by a change in the polarity of the resist molecule. The prototypical inorganic resist is Ag-doped Ge–Se. In this process, a 2000-Å layer of Ge–Se is first deposited by sputtering or evaporation. Next, a 1000-Å layer of Ag is plated on top of the Ge–Se layer in an $AgNO_3$-containing bath. Upon exposure to light from 200 to 460 nm [49], a photodoping process creates Ag_2Se. This can readily be dissolved in a potassium iodide acid solution [50]. After excess silver removal, the image can be dry-developed in a CF_4 or other fluorine-containing plasma [51]. As shown in Figure 8.16, both positive and negative tones can be achieved. Inorganic resists have the advantage of very high contrast ($\gamma \approx 7$), resulting in the ability to produce fine lines even in g-line exposure systems [52]. Due to their thin film nature, they require a thick planarizing underlayer. The deposited Ge–Se layer tends to have pinholes. Many more defects may be added during the plating process. As a result, these inorganic resists have not found much acceptance.

Table 8.3   Commercial available deep UV resists including delay-time effect

Source	Name	Delay-Time Effects
OCG	CAMP 6	After 5 min
IBM	APEX-E	After 15 min
	XP-2105	None up to 2 hr
Shipley	XP-2198	
	XP-3036	
BASF	ST2	Depends on softbake

After Paniez et al. [46].

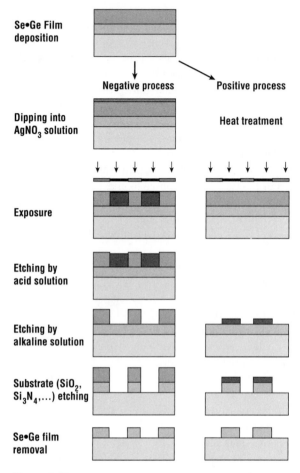

Se•Ge Film
deposition

Negative process   **Positive process**

Dipping into
AgNO₃ solution   **Heat treatment**

Exposure

Etching by
acid solution

Etching by
alkaline solution

Substrate (SiO₂,
Si₃N₄,...) etching

Se•Ge film
removal

Figure 8.16   Processing sequence for Ag/Se–Ge resists
(after Yoshikawa et al., reprinted by permission, AIP).

The second category of new resist materials is called dry developable. The most interesting class of these materials is the silicon-containing resists. Organosilanes were first proposed as resist sources by Hofer and coworkers [53]. Their motivation was as follows: since the matrix materials such as novolac tend to absorb in the deep UV, one must try to construct an imaging process that will occur in a very thin resist layer, but can be transferred into a thicker underlayer through an anisotropic plasma etch (see Chapter 11). The top layer must be highly resistant to such plasma processes. Silicon-containing resists are attractive, since they form SiO_2, which protects the underlying resist [54] in an oxygen plasma.

The field of silicon-containing resists gained considerable momentum when it was discovered that polymeric silicon materials called polysilynes could be fabricated [55]. Upon exposure to deep UV radiation, significant bleaching was observed due to the replacement of the Si–Si network with a cross-linked siloxane network [56]. The exposed resists can be developed as negative resists in nonpolar solvents such as toluene or xylene, which will dissolve the unexposed areas. Wet-developed contrasts greater than 7 are obtained in this system. Figure 8.17 shows 0.2- and 0.15-μm lines imaged in a 300-Å poly(n-butylsilyne) resist on top of a novolac resin. After 110 mJ/cm^2 exposure and 15-sec toluene development, the underlying novolac resist was etched in an oxygen plasma designed to etch vertically. The resists can also be developed in a selective plasma etch system. If the

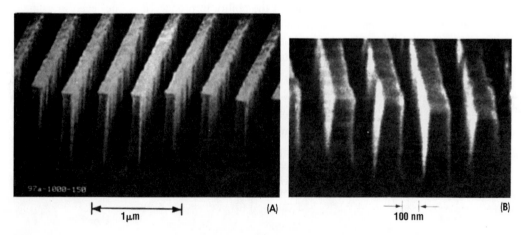

(A) 1 µm
(B) 100 nm

Figure 8.17   Imaging of (A) 0.2 µm and (B) 0.15 µm lines in a 300-Å poly(n-butylsilyne) resist over a novolac resin *(after Kunz et al., used by permission, SPIE).*

etch is selective for silicon to silicon dioxide, a positive image will result. If the etch is selective for silicon dioxide over silicon, a negative tone will develop. The most commonly used process, which etches silicon selectively to silicon dioxide, is an HBr plasma. At present, a contrast of 5 has been obtained in such a system. It has also been found that a separate polysilyne resist layer need not be applied. Instead, the surface of the resist can be silylated by gaseous [57] or liquid [58] exposures at elevated temperatures, or by exposure to a suitable plasma environment [59]. As with the bilayer process, high contrasts and deep submicron images can be obtained.

8.9 Summary

This chapter reviewed the basic chemistry of the most popular positive tone resists called DQN. These resists have good photospeeds for mid-UV applications, but the novolac-based resins are not appropriate for deep UV applications. One of the most important resist figures of merit, contrast, was discussed. Several methods were explored for determining the minimum feature size once the contrast curve and the optical system are known. An introduction to the resist processing sequence and typical resist processing equipment was given. Finally, new types of resists that have high contrast and are appropriate for deep UV applications were discussed. These resists will abandon the novolac matrix for exposures below 248 nm.

Problems

1. Calculate the CMTF for AZ-1450 at the four wavelengths listed in Table 8.1. Assuming $NA = 0.4$, use Figure 7.21 to determine the minimum feature size for an aligner with $S = 0.5$ using this resist at the various wavelengths.
2. A 0.6-μm-thick layer of a particular photoresist has $D_0 = 40$ mJ/cm^2 and $D_{100} = 85$ mJ/cm^2.
   (a) Calculate the resist contrast.
   (b) Calculate the CMTF.
   (c) It is found if the resist thickness is cut in half, D_{100} reduces to 70 mJ/cm^2 while D_0 is unchanged. What is the higher contrast possible in this resist without changing the resist processing?
3. Discuss why the resist contrast might decrease at short wavelengths (high photon energies). Include in this discussion the effects of these photons on both the matrix and the PAC.
4. Derive Equation 8.6.
5. It is pointed out in Chapter 8 that certain resists have contrasts as large as 7. If a positive tone resist has $D_0 = 10$ mJ/cm^2, and has $\gamma = 7$, find D_{100}. It is desired for some applications to make a tapered resist profile, that is, a resist edge that is not vertical, but rather rises more slowly from the exposed region. To do this, would you use a high dose illumination or a low dose illumination? Justify your answer. What limits the exposure?
6. Explain why resist bleaching is a desirable effect.
7. Would you expect the bilayer inorganic/novolac method discussed in the chapter to work well over moderate (<5000 Å) topography? Justify your answer. If surface reflections are a problem, how could that be addressed?
8. A state-of-the-art exposure tool uses a 248-nm exposing wavelength with a spatial coherence of 0.75. It has a numerical aperture of 0.6. If this tool is used with a resist with a contrast of 3.5, calculate in microns the minimum line size that can be resolved.

9. The accompanying plot of intensity versus position shows the optical intensity at the surface of the wafer when exposing a diffraction grating in a projection printing system. Calculate the modulation transfer function for this image. If the image is exposed in a resist that has a contrast of 3.0, would you expect it to be resolved? You must justify your answer. Estimate to the closest tenth the smallest contrast for which this image will be resolved.

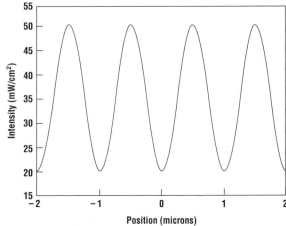

10. A university lab uses a g-line stepper with NA = 0.4 and a spatial coherence of 0.5. On a good day, with flat substrates, this machine can resolve 0.8-μm lines and 0.8-μm spaces, but nothing smaller. Find the contrast of the resist process being used.

11. An i-line stepper with a spatial coherence of 0.5 and NA = 0.5 is used to expose a resist with a contrast of 4.1. Find
    (a) The depth of field.
    (b) The minimum linewidth that can be resolved.

12. For –1 um < x < 1 μm, an image is found to follow the equation:

$$I = 20 \text{ mW/cm}^2 + 100 \text{ mW/cm}^2 \times [\sin(2\pi x/1 \text{ μm})]^2$$

    (a) What is the MTF for this image?
    (b) Using the gamma curve shown in Figure 8.7a, for a one-second exposure, estimate the resist thickness at x = 0, x = 0.1 μm, and x = 0.25 μm.
    (c) Roughly sketch the profile from –1 < x < +1 μm, indicating the position of the center of all lines that would appear in the resist.

13. The figure shows the areal image of a line that is being exposed in a positive tone resist. The resist has a D_0 of 30 mJ/cm² and a D_{100} of 80 mJ/cm². Assume that the width of the resist line is measured as shown, as the distance between the two bottom corners.
    (a) Find the MTF of the image
    (b) If the linewidth is measured from the points at which the resist thickness reaches zero, find the linewidth for exposure times of 1, 2, 3, and 4 sec. *Hint*: This corresponds to an exposure dose of D_{100}.

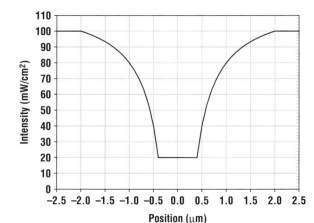

14. The minimum feature size using the g-line (436-nm) stepper in a university's laboratory is 0.8 μm. The stepper has an NA of 0.45.

(a) If the spatial coherence (S) is 0.5, calculate the CMTF of the resist process.

(b) If the laboratory needs to produce 0.5-μm lines and spaces, what CMTF is required? What contrast would that be?

15. Assume that a wafer is being exposed with a proximity printer. In the far field limit in one dimension,

$$I(x) \approx I_1(0) \left[\frac{2W}{\lambda g} \right]^2 I_x^2$$

where $I_1(0)$ is the linear flux density in the incident beam at the center of the aperture and I_x is given by Equation 7.9. Assume that we are using a 1.0-μm-thick positive tone resist with $D_0 = 30$ mJ/cm^2 and $D_{100} = 100$ mJ/cm^2. For $D < D_0$ assume that no resist is removed in the develop process. For $\lambda = 436$ nm (0.436 μm), $g = 10$ μm, and $I_1(0) = 100$ mW-μm^2/cm^2, calculate the resist profiles (thickness versus x) for $W = 1$ μm and exposure times of 1, 2, 4, and 7 sec. Using the distance over which the resist is 0.5 μm thick as the width of the developed space, plot the space width versus exposure time for these four points. You may wish to set up your coordinate system so that the center of the exposure is at $x = 0$. Then the spacing on the mask goes from -1 μm to $+1$ μm. The problem is symmetric about the origin, and you have to calculate and plot only half of the resist profile.

References

1. R. A. Arcus, *Proc. SPIE* **631**:124 (1986).
2. P. Hanson, "Unconventional Photographic Systems," *Photogr. Sci. Eng.* **14**:438 (1970).
3. J. Pacansky and J. Lyerla, "Photochemical Studies on a Substituted Naphthalene-2, 1-Diazooxide," *J. Electrochem. Soc.* **124**:862 (1977).
4. M. Furuta, S. Asaumi, and A. Yokota, "Mechanism of Dissolution of Novolac-diazoquinone Resist," in *Advances in Resist Technology and Processing VIII*, H. Ito, ed., *SPIE Proc.* **1466**:477 (1991).
5. V. Rao, L. L. Kosbar, C. W. Frank, and R. F. W. Pease, "The Effect of Sensitizer Spatial Distribution on Dissolution Inhibition in Novolac/Diazonaphthoquinone Resists," in *Advances in Resist Technology and Processing VIII*, H. Ito, ed., *SPIE Proc.* **1466**:309 (1991).
6. M. Hanabata, Y. Uetani, and A. Furuta, "Design Concepts for a High-performance Positive Photoresist," *J. Vacuum Sci. Technol. B* **7**:640 (1989).
7. S. Nonogaki, T. Ueno, and T. Ito, *Microlithography Fundamentals in Semiconductor Devices and Fabrication Technology*, Dekker, New York, 1998.
8. T. Kajita, T. Ota, H. Nemoto, Y. Yumoto, and T. Miura, "Novel Novolac Resins Using Substituted Phenols for High Performance Positive Photoresists," *Proc. SPIE* **1466**:161 (1991).
9. K. Honda, B. T. Beauchemin Jr., E. A. Fitzgerald, A. T. Jeffries III, S. P. Tadors, A. J. Blakeney, R. J. Hurditch, S. Tan, and S. Sakaguchi, *Proc. SPIE* **1466**:141 (1991).
10. T. Kajita, T. Ota, H. Nemoto, Y. Yumoto, and T. Miura, *Proc. SPIE* **1466**:161 (1991).
11. C. G. Willson, "Organic Resist Materials—Theory and Chemistry," in *Introduction to Microlithography*, L. F. Thompson, C. G. Willson, and M. J. Bowden, eds., *Advances in Chemistry Series* **219**, Americon Chemical Society, Washington, DC, 1983.
12. W. Hinsberg, C. Willson, and K. Kanazawa, "Use of a Quartz Crystal Microbalance Rate Monitor to Examine Photoproduct Effects on Resist Dissolution," in *Proc. SPIE Resist Technol.* **539**:6 (1985).

13. D. Leers, "Investigation of Different Resists for Deep UV-Exposure," *Solid State Technol.* 91 (March 1981).

14. W. M. Moreau, *Semiconductor Lithography, Principles, Practices, and Materials*, Plenum, New York, 1988, p. 31.

15. D. Meyerhofer, "Characteristics of Resist Films Produced by Spinning," *J. Appl. Phys.* **49**:3993 (1978).

16. B. D. Washo, "Rheology and Modeling of the Spin Coating Process," *IBM J. Res. Dev.* **21**:190 (1977).

17. R. F. Leonard and J. A. McFarland, "Puddle Development of Positive Resist," *SPIE Proc.* **275**, *Semiconductor Microlithography VI* (1981).

18. D. Burkman and A. Johnson, "Centrifugal On-Center, Flood Spray Development of Positive Resist," *Solid State Technol.* 125 (May 1983).

19. Megaposit and SPR500 are trademarks of Shipley Corporation.

20. G. E. Flores and J. E. Loftus, "Lithographic Performance and Dissolution Behavior of Novolac Resins for Various Developer Surfactant Systems," *Proc. SPIE* **1672**:317 (1992).

21. H. Shimada, I. Toshiyuki, and S. Shimomura, "High Accuracy Resist Development Process with Wide Margins by Quick Removal of Reaction Products," *Proc. SPIE* **2195**:813 (1994).

22. T. Iwamoto, H. Shimada, S. Shimomura, M. Omedera, and T. Ohmi, "High-Reliability Lithography Performed by Ultrasonic and Surfactant-Added Developing System," *Jpn. J. Appl. Phys.* **33**:491 (1994).

23. D. R. McKean, T. P. Russel, and A. F. Renaldo, "Thick Film Photoresist Resolution Enhancement with Surfactant Surface Treatment," *Proc. SPIE* **2438**:673 (1995).

24. S. Clifford, B. Hayes, and R. Brade, "Results of Photolithographic Cluster Cells in Actual Production," *Proc. SPIE* **1463**:551, *Optical/Laser Microlithography IV* (1991).

25. G. Willson, R. Miller, D. McKean, T. Thompkins, N. Clecak, and D. Hofer, *J. Am. Chem. Soc.* **40**:54 (1983).

26. A. Knop, in *Applications of Phenolic Resins*, Springer-Verlag, Berlin, 1979.

27. E. Gipstein, A. Duano, and T. Thompkins, "Evolution of Pure Novolac Cresol–Formaldehyde Resins for Deep U.V. Lithography," *J. Electrochem. Soc.* **129**:201 (1981).

28. J. V. Crivello, "Possibility of Photoimaging Using Onium Salts," *Polym. Eng. Sci.* **23**:953 (1983).

29. R. D. Allen, G. M. Wallraff, D. C. Hofer, and R. R. Kunz, "Photoresists for 193-nm Lithography," *IBM J. Res. Dev.* **41**:95 (1997).

30. J. G. Maltabes, S. J. Holmes, J. R. Morrow, R. L. Barr, M. Hakey, G. Reynolds, W. R. Brunsvold, C. G. Willson, N. J. Clecak, S. A. MacDonald, and H. Ito, in *Advances in Resist Technology and Processing VII*, M. P. C. Watts, ed., *Proc. SPIE* **1262**:2 (1990).

31. S. Holmes, R. Levy, A. Bergendahl, K. Holland, J. Maltabes, S. Knight, K. C. Korris, and D. Poley, in *Optical/Laser Microlithography III*, V. Pol, ed., *Proc. SPIE* **1264**:61 (1990).

32. C. Renner, U.S. Patent 4, 371, 605 (1983).

33. W. Brunsvold, W. Montgomery, and B. Hwang, "Non-metallic Acid Generators for i-Line and g-Line Chemically Amplified Resists," in *Advances in Resist Technology and Processing VIII*, H. Ito, ed., *SPIE Proc.* **1466**:368 (1991).

34. G. Buhr, R. Dammel, and C. Lindley, "Non-ionic Photoacid Generating Compounds," *ACS Polym. Mater. Sci. Eng.* **61**:269 (1989).

35. F. M. Houlihan, A. Schugard, R. Gooden, and E. Reichmanis, "Nitrobenzyl Ester Chemistry for Polymer Processes Involving Chemical Amplification," *Macromolecules* **21**:2001 (1988).

36. G. Pawlowski, R. Dammel, C. Lindley, H.-J. Merrem, H. Roschert, and J. Lingau, "Chemically Amplified DUV Photoresists Using a New Class of Photoacid Generating Compounds," *Proc. SPIE* **1262**:16 (1990).

37. P. Burggraaf, "What's Available in Deep-UV Resists," *Semicond. Int.* 56 (September 1994).

38. R. Allen, "Progress in 193 nm Photoresists," *Semicond. Int.* **20**(10):72 (1997).

39. P. Van Pelt, "Processing of Deep-Ultraviolet (UV) Resists," *SPIE Proc.* **275**:150 (1981).

40. U.S. Patent 4,405,708 (1983), U.S. Phillips.

41. K. Harada and S. Sugawara, "Temperature Effects on Positive Electron Resists Irradiated with Electron Beam and Deep-UV Light," *J. Appl. Polym. Sci.* **27**:1441 (1982).

42. G. Pawlowski, T. Sauer, R. Dammel, D. J. Gordon, W. Hinsberg, D. McKean, C. R. Vicari, and C. G. Willson, "Modified Polyhydroxystyrenes as Matrix Resins for Dissolution Inhibitor Type Photoresists," *Proc. SPIE* **1262**:391 (1990).

43. K. Przybilla, H. Röschert, W. Spiess, C. Eckes, S. Chaterjee, D. Khanna, G. Pawlowski, and R. Dammel, "Progress in Deep UV Resists," in *Advances in Resist Technology and Processing VIII*, H. Ito, ed., *SPIE Proc.* **1466**:144 (1991).

44. R. D. Allen, G. M. Wallraff, W. D. Hinsberg, and L. L. Simpson, "High Performance Acrylic Polymers for Chemically Amplified Photoresist Application," *J. Vacuum Sci. Technol. B* **9**:3357 (1991).

45. G. M. Wallraff, R. D. Allen, W. D. Hinsberg, C. F. Larson, R. D. Johnson, R. DiPietro, G. Breyta, and N. Hacker, "Single-Layer Chemically Amplified Photoresists for 193 nm Lithography," *J. Vacuum Sci. Technol. B* **11**(6):2783 (1993).

46. P. J. Paniez, C. Rosilio, B. Mouanda, and F. Vinet, "Origin of Delay Times in Chemically Amplified Positive DUV Resists," *Proc. SPIE* **2195**:14 (1994).

47. R. R. Kunz, R. D. Allen, W. D. Hinsberg, and G. M. Wallraff, "Acid Catalyzed Single-Layer Resists for ArF Lithography," *Opt. Eng.* **32**(10):2363 (1993).

48. B. F. Griffing and P. R. West, "Contrast Enhancement Lithography," *Solid State Technol.* 152 (May 1985).

49. G. Benedikt, U. S. Patent 4,571,375 (1986).

50. Y. Yoshikawa, O. Ochi, H. Nagai, and Y. Mizushima, "A Novel Inorganic Photoresist Utilizing Ag Photodoping in Se-Ge Glass Films," *Appl. Phys. Lett.* **29**:677 (1977).

51. Japanese Patent 82,50430; *Chem. Abstr.* **97**:48231 (1982).

52. E. Ong and E. L. Hu, "Multilayer Resists for Fine Line Optical Lithography," *Solid State Technol.* June 1984.

53. D. C. Hofer, R. D. Miller, and C. G. Willson, "Polysilane Bilayer UV Lithography," *SPIE Proc.* **469**:16 (1984).

54. G. N. Taylor, M. Y. Hellman, T. M. Wolf, and J. M. Zeigler, "Lithographic, Photochemical, and O_2 RIE Properties of Three Polysilane Copolymers," *SPIE Proc.* **920**:274 (1988).

55. P. A. Bianconi and T. W. Weidman, "Poly(*n*-hexylsilyne): Synthesis and Properties of the First Alkyl Silicon $[RSi]_n$ Network Polymer," *J. Am. Chem. Soc.* **110**:2342 (1988).

56. R. R. Kunz, P. A. Bianconi, M. W. Horn, R. R. Paladugu, D. C. Shaver, D. A. Smith, and C. A. Freed, "Polysilyne Resists for 193 nm Excimer Laser Lithography," in *Advances in Resist Technology and Processing VIII*, H. Ito, ed., *SPIE Proc.* **1466**:218 (1991).

57. Ki-Ho Baik, L. Van den Hove, A. M. Goethals, M. Op. de Beeck, and R. Borland, "Gas Phase Silylation in the Diffusion Enhanced Silylated Resist Process for Application to Sub-0.5 μm Optical Lines," *J. Vacuum Sci. Technol. B* **8**:1481 (1990).

58. Ki-Ho Baik, L. Van den Hove, and R. Borland, "Comparative Study Between Gas- and Liquid-Phase Silylation for the Diffusion-Enhanced Silylated Resist Process," *J. Vacuum Sci. Technol. B* **9**:3399 (1991).

59. K. Kato, K. Taira, T. Toshihiko, T. Takahashi, and K. Yanagihara, "Effective Parameters of DESIRE Process to Controlling Resist Performance at Sub-half to Quarter Micron Rule," in *Resist Technology and Processing IX*, H. Ito, ed., *SPIE Proc.* **1672**:415 (1992).

Chapter 9

Nonoptical Lithographic Techniques[+]

The last two chapters discussed optical lithography as the primary tool for imaging patterns on the wafer surface. It was pointed out that resolution limits are a serious concern for this process. Predicting the location of these limits has proven to be an exercise in frustration for those who work in alternate patterning technologies. Optical lithography has been extended through the use of steppers with high numerical aperture lens, extreme UV sources, scanning steppers with phase-contrast masks or other optical proximity correction techniques, and now immersion lithography. Optical lithography supporters believe that the technology can be extended to at least 30 nm, and perhaps as far as 22 nm if suitable optical materials can be developed for 157-nm sources. Optical lithography at these dimensions is expected to be extremely expensive, however. Even mask sets, which are typically used for 2000 to 3000 wafers in production, could cost more than $2 million each if full optical proximity correction were required on all levels. If optical lithography cannot be extended in a cost-effective manner, or if feature sizes are to be scaled below 22 nm, new lithographic techniques must be developed. These methods are collectively called nonoptical or next-generation lithographies (NGLs). A large number of NGLs have been developed; however, all face serious obstacles for widespread use in building integrated circuits. Recall from Chapter 7 that the Rayleigh limit of resolution is proportional to the wavelength. A general feature of many nonoptical lithographies, then, is the use of very short wavelength energy sources. The two sources considered in this chapter, x-ray and electron beam lithography, have wavelengths so small that diffraction no longer defines the lithographic resolution. The key problem with using very short wavelength, and therefore very high energy, sources is the mask. No material is known that will allow most of this energy to go through a thick, mechanically stable plate as one might need for a mask. The chapter will review four techniques that are being developed to overcome this problem: maskless e-beam direct writing, membrane masks used in proximity x-ray and in projection electron beam lithography, and reflection masks used for projection x-ray lithography. The Chapter will then review imprint lithography, an approach to patterning that does not use a radiation pattern.

9.1 Interactions of High Energy Beams with Matter°

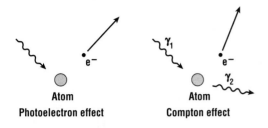

Figure 9.1   The two dominant interaction processes for high energy photons with matter: the photoelectric effect and Compton scattering.

The two most likely exposure sources for a nonoptical lithography system are short wavelength photons (x-ray) and high energy electrons (e-beam). The interactions that occur in the photoresist and in the underlying layers are somewhat similar for the two sources. This section will review the physics of these interactions before going on to the process itself.

Typical x-ray sources used for lithography emit photons with energies of 1 to 10 keV. When these photons are incident on a solid, there are many possible interactions, but the two most likely ones are photoelectric absorption and the Compton effect. As shown in Figure 9.1, both processes involve an interaction between photons and electrons. For energies much less than 10 keV, photoelectric emission dominates [1], and the ejected electron carries with it almost all of the incident photon energy. The capture cross section for the photoelectric process depends on the mass of the target material. At higher energies the Compton process dominates. The Compton process can be thought of as a collision between an electron initially at rest and a photon of energy hc/λ and momentum h/λ where h is Planck's constant and c is the speed of light. A portion of the incident photon energy is lost to the electron during this scattering event. Since the energy needed to liberate an electron from the solid (work function) is typically two to three orders of magnitude lower in energy than the incident photon energy, the electrons in the target can approximated as free, and therefore the cross section for the Compton effect depends only on the electron density. By conserving both momentum and energy, one can show that

$$\lambda_2 - \lambda_1 = \lambda_c(1 - \cos\theta) \tag{9.1}$$

where λ_c is the Compton wavelength (0.0243 Å) and θ is the angle between the incident and final photon momentum. For x-ray wavelengths of 1 Å, only a small fraction of the energy is lost during a single Compton scattering event. As a result high energy x-rays will penetrate a considerable distance into many solids.

Most x-ray lithography is done at energies well below 10 keV ($\lambda \gg 1$ Å) where photoelectric absorption dominates. As will be discussed in the next section, the majority of the incident photon energy will ultimately be dissipated by secondary electrons generated by impact ionization. In that respect x-ray lithography and e-beam lithography use similar exposure mechanisms in the resist, once the initial photoelectron event has occured. An important difference between the two processes is that the secondary electrons generated in the resist during x-ray lithography are usually about an order of magnitude lower in energy than the primary electrons in e-beam systems. As a result, the distance over which the energy is spread in x-ray lithography is much less.

Due to the finite size of the capture cross section, the incident photons are not absorbed at the surface of the resist. Instead they will penetrate to some depth until absorption occurs. A reasonable approximation is to consider the resist to be an amorphous solid, with a single capture cross section. Recall from Chapter 7 that $\alpha(\lambda)$ is defined as the wavelength-dependent absorption coefficient. Then

$$\alpha(\lambda) = \sigma(\lambda)\rho/m \tag{9.2}$$

where ρ/m is the number density of the target material. The absorption coefficient is a function of the photon energy. Desired values are of order 1 μm^{-1} for most resists. If α is too large, the resist exposure

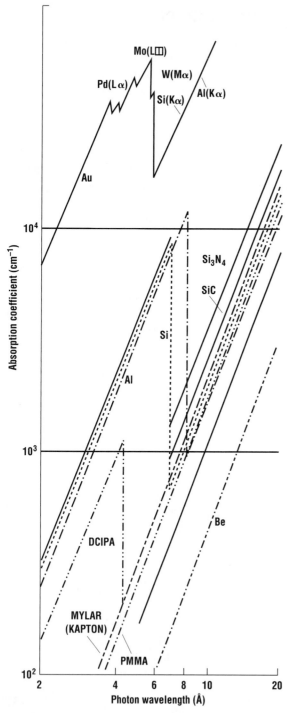

Figure 9.2 Absorption coefficients for some common materials as a function of photon energy *(after Glendenning and Cerrina [29], reprinted by permission, Noyes Publications).*

will be nonuniform. If it is too small, the photospeed will be low. The absorption coefficient also has discontinuities corresponding to the energies of the core levels of the atoms (Figure 9.2). Clearly the lower energy ($\lambda > 10$ nm) photons used in projection x-ray lithography will have much larger values of α. This will be a problem for resist exposures unless strong bleaching mechanisms are used. X-ray interaction with the underlying materials, and in particular x-ray-induced damage, is also a consideration for all NGLs because some fraction of the incident energy will pass through the resist and fall onto the wafer itself.

In electron beam lithography the interactions of electrons and solids are of primary importance. When energetic electrons enter matter, they can pass through undeflected, scatter elastically, or scatter inelastically. Elastic scattering events occur primarily through interaction with the atomic nucleus and can result in deflections larger than 90°. This type of interaction can be treated classically as Rutherford scattering from a screened Coulomb potential. The ratio of the scattered current to the incoming current is given by

$$R = \frac{1}{\sqrt{\theta^2 + [\lambda/2\,\pi a]^2}} \quad \textbf{(9.3)}$$

where λ is the wavelength of the electron and a is the atomic radius, which can be approximated by

$$a \approx 0.9\,a_o\,Z^{1/4} \quad \textbf{(9.4)}$$

where a_o is the Bohr radius (0.529 Å) and Z is the atomic number of the target.

There are a number of potential inelastic energy loss mechanisms including low energy (<50 eV) secondary electron generation, inner shell excitations leading to x-ray and Auger electron emission, electron–hole pair creation followed by recombination and photon emission (cathodoluminescence), and phonon excitation. Finding the path of electrons in the solid therefore is similar to finding implantation profiles (Chapter 5). One can do a Monte Carlo calculation where a fixed number of electrons are launched over some finite spot size. The path of each electron will be unique. In this technique each type of possible interaction must be modeled. One can build up a distribution after simulating

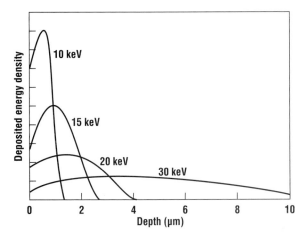

Figure 9.3 Deposited energy density for various energy electron beams incident on silicon as a function of depth.

the passage of a large number of electrons. Alternatively the process can be treated statistically using the Bethe equation [2]:

$$\frac{dE}{dx} = \left[\frac{N_A\,e^4}{2\pi\varepsilon_o^2}\right]\left[Z\frac{\rho}{A}\right]\left[\frac{1}{E}\ln\frac{E}{66J}\right] \qquad (9.5)$$

where x is the distance into the target, N_A is Avogadro's number, Z is the atomic number of the target, A is the atomic weight, ρ is the mass density of the target, and J is the mean ionization potential, which can be approximated by

$$J(eV) \approx 11.5 \cdot Z \qquad (9.6)$$

Then the projected range R_p can be found according to

$$\int_o^{R_p} dx = R_p = \int_{E_o}^0 \left[\frac{dE}{dx}\right]^{-1} dE \qquad (9.7)$$

Figure 9.3 shows a plot of deposited energy density for various energy electron beams incident on silicon. At typical direct-write e-beam energies (20 to 100 keV), most of the energy is deposited at depths greater than 1 μm. As a result, a great deal of the damage is in the substrate. When high energy electrons strike a surface, they may also induce chemical changes. This happens primarily by the cascade of relatively low energy secondary electrons. This will be discussed later in the chapter.

9.2 Direct-Write Electron Beam Lithography Systems

Electron beam lithography (EBL) systems may be used either for mask generation or to directly write patterns on the wafer. Currently EBL is the technology of choice for mask generation due to its ability to accurately define small features. Because of this book's focus on integrated circuit fabrication technologies, however, we will limit the discussion to direct-write EBL. Most direct-write systems use a small electron beam spot that is moved with respect to the wafer to expose the pattern, one pixel at a time. All will be discussed in the next section; several versions of projection and proximity electron beam lithography systems have also been developed [3]. The short penetration length of electrons precludes the use of a solid substrate such as quartz for the mask, however. Only a very thin membrane mask can be used. A stencil mask with cutouts through which the beam can pass [4] could also be used. Direct-write EBL systems can be classified as raster scan or vector scan, with either fixed or variable beam geometries. Each type of system has advantages. The selection depends on the type of writing that it is designed to do.

All electron beam systems have in common the need for an electron source with a high intensity (brightness), high uniformity, small spot size, good stability, and a long life. Brightness is measured in units of amperes per unit volume per steradian. Electrons can be removed from the cathode of the gun by heating the cathode (thermionic emission), applying a large electric field (field emission), a combination of the two (thermal-aided field emission), or with light (photoemission).

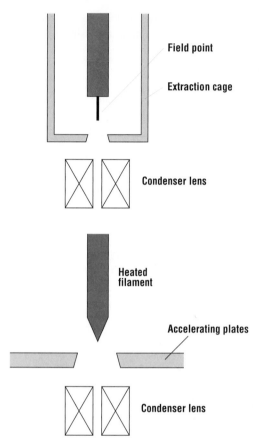

Figure 9.4   Simplified cross section schematics of field emission and thermionic emission electron guns.

Figure 9.4 shows cross sections of typical electron guns that might be used for EBL. The most common source is thermionic due to its high brightness. The filament material must be chosen to minimize its evaporation from the cathode and so produce the longest life possible. One of the primary figures of merit for the gun is the emitted electron current density, which is given by

$$J_c = AT^2 e^{-E_w/kT} \tag{9.8}$$

where A is Richardson's constant for the material (typically 10–100 A/cm²-K²) and E_w is the effective metal work function. These electrons are accelerated and some fraction of them are collected. The measure of the collected electron energy is the brightness, β. Although an increase in the emitted current density also generally increases β, if the increase in J_c decreases the collection efficiency, the percentage increase in the brightness is not as large as the percentage increase in the current density.

Most thermionic sources use either tungsten, thoriated tungsten, or lanthanum hexaboride (LaB$_6$). Tungsten filaments allow operation at pressures as high as 0.1 mtorr, but their current density is only about 0.5 A/cm². As a result their brightness is less than 2×0^4 A/cm³-sr. Thoriated tungsten cathodes have somewhat lower brightness at the same filament current and require higher vacuum (0.01 mtorr), but their maximum current density can be as large as 3 A/cm². Lanthanum hexaboride cathodes are the most popular [5], with a current density over 20 A/cm² and a brightness of nearly 10⁶ A/cm³-sr. The LaB$_6$ filaments require a vacuum of at least 10^{-6} torr, however, and must be well protected against sudden vacuum loss.

A concern with all filament sources is the apparent source size or crossover diameter. Because a large volume of wire is heated, the thermionic electron sources produce broad beams. The energy distribution from such a source is also quite broad. This leads to focusing problems similar to those found in large optical partially incoherent sources. For typical LaB$_6$ sources the crossover diameter is about 10 μm. To get to a 0.1 μm spot, a demagnification of 100× is required. Even though the source has a large current density, the inherent source brightness is greatly reduced in order to achieve deep submicron resolution.

It has been shown [6] that tips made from Zr/W/O annealed in forming gas (90% N$_2$ and 10% H$_2$) can be used in thermal-aided field emission guns to produce crossover diameters as small as 200 Å. As a result lithography systems using such a gun [7] can produce spots of 100 Å with current densities [8] as high as 1000 A/cm². To obtain acceptable field stability, however, this type of gun requires a constant vacuum of at least 1×10^{-8} torr. In spite of this limitation, this type of gun shows considerable promise for high resolution, high throughput EBL systems.

Once an electron stream is generated, it must be formed into a narrow beam. This is done in most practical EBL systems through a series of electrostatic lenses and a variety of apertures and knife edges. Figure 9.5 shows a typical lens arrangement. The final spot diameter on the surface of the wafer is given by

$$d^2 = d_o^2 + d_s^2 + d_c^2 \tag{9.9}$$

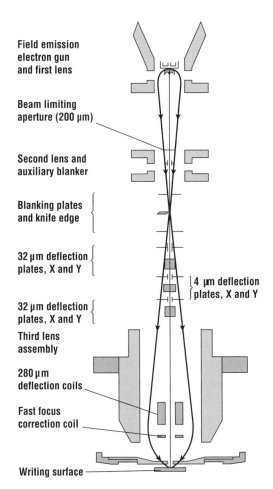

Field emission
electron gun
and first lens

Beam limiting
aperture (200 μm)

Second lens and
auxiliary blanker

Blanking plates
and knife edge

32 μm deflection
plates, X and Y

4 μm deflection
plates, X and Y

32 μm deflection
plates, X and Y

Third lens
assembly

280 μm
deflection coils

Fast focus
correction coil

Writing surface

Figure 9.5   System schematic of an early EBES
system. The basic column is similar in current
generation EBL systems *(after Herriot et al.,
reprinted by premission, © 1975 IEEE).*

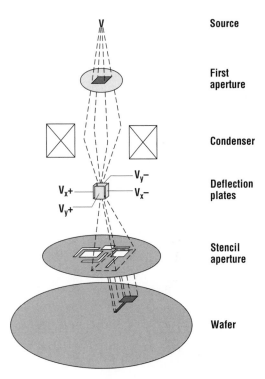

Source

First
aperture

Condenser

V_y-
V_x+ V_x-
V_y+

Deflection
plates

Stencil
aperture

Wafer

Figure 9.6   The use of stencil masks with EBL to
improve system throughput. This type of exposure is
useful only for exposing highly repetitive designs.

where d_s is the spherical aberration, d_c is the chromatic aberration due to the nonzero energy distribution, and d_o is the perfect lens diameter, which is limited by the finite source size and space charge effects. In commercial EBL systems d_s and d_c can be made negligibly small, and final spot sizes close to a few nanometers can be routinely achieved.

Most EBL systems use Gaussian beams, whose intensity varies approximately as a Gaussian with radius from the center of the beam. As will be described later in this chapter, the primary disadvantage of electron beam lithography is throughput. The process is too slow to economically manufacture most integrated circuits. To improve throughput, special-purpose EBL systems have been produced with shaped beams that combine direct-write and projection lithography [9]. In one variation a stencil mask is placed above the wafer (Figure 9.6) with a small number of predetermined geometries that are to be replicated many times (as in a DRAM or other memory). Also included is a blank area through which the beam can be directed for nonstandard geometries. When the beam is sent through this area, it can be rastered as in a standard EBL.

A second and much more popular method of producing a variable shaped beam is illustrated in Figure 9.7. In this case the beam is partially intercepted by a square aperture [10]. By changing the deflection, the width and length of the line can be varied. Alternatively, x and y shaping can be done separately. Furthermore an electrostatic rotation lens can be used to produce rectangles at 45° (or any other angle) to the primary axis. Since most images are made up of rectangles, this is an extremely

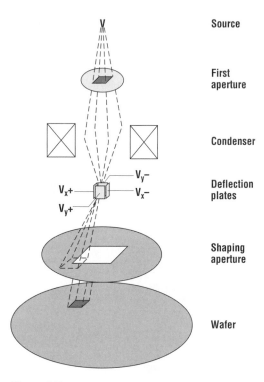

Figure 9.7   A variable shaped beam exposure system using mechanical beam stops for beam shaping. The broad beam exposes many pixels simultaneously, but dimensional control is not as reliable as in standard EBL.

efficient manner of performing EBL [11]. Hundreds or even thousands of pixels can be exposed simultaneously. Line-width is more difficult to control in this method, but for many applications the increase in throughput is sufficient to compensate for the reduction in linewidth control. Figure 9.8 shows a matrix of exposures ranging from 0.15×0.15 μm to 2.0 to 2.0 μm done in a single pass in a variable beam system [12].

In addition to beam shaping, another important factor to direct write throughput is the raster method. Raster scanning (Figure 9.9) was used in the first EBL systems including the early Bell Labs EBES [13]. It is a direct descendant of scanning electron microscopes. In a typical system the electron beam first passes through a pair of blanking plates that can deflect the beam into a beam stop. A second pair of plates is used to scan the beam in one direction, while the stage is mechanically scanned in the direction perpendicular to the beam scan. The area to be scanned is divided into fields from 100 μm to a few millimeters on a side. The scan trajectory is defined as the path that the beam would follow if every pixel were to be exposed. The data organized into a bit map that is organized serially along the scan trajectory. To improve edge resolution, the e-beam spot size is typically one-half to one-fifth of the minimum feature size, and several passes are taken through each exposure area.

If one is free to choose the polarity of the image, one can take advantage of the fact that much less than 50% of the

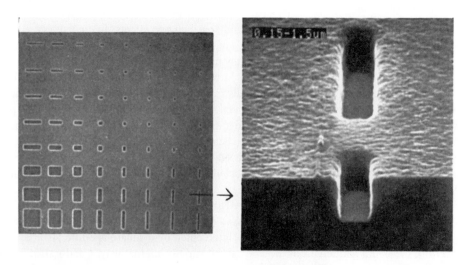

Figure 9.8   Exposure matrix in a variable shaped beam system *(after Hohn, reprinted by permission of SPIE).*

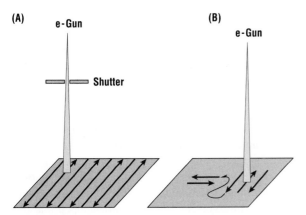

(A) e-Gun
Shutter

(B) e-Gun

Figure 9.9 A comparison of scanning methodologies: raster scan (A) and vector scan (B).

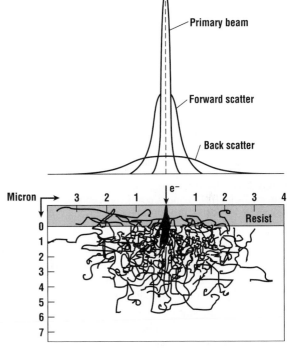

Figure 9.10 Monte Carlo simulation of electron trajectories during an EBL exposure. The upper curve indicates the forward- and backscattered components of the beam *(after Hohn, reprinted by permission, SPIE).*

die must normally be exposed in a typical pattern. In a raster scan method every pixel must be scanned serially, and so this the exposure time is nearly independent of the pattern. The pattern is written by opening and closing a shutter. Vector scan EBL systems have been developed to improve the throughput by directing the electron beam to only those parts of the chip to be exposed [14]. In this method the digital location of each area to be exposed is fed to x and y digital-to-analog converters (DACs). The beam is directed only to those pixels that must be exposed. Software must be used to sequence the pixels so as to minimize the time consumed by beam deflection.

Vector scan systems have a second important advantage over raster scan systems. Ignoring system aberrations and nonlinearities, the precision of the image address is determined simply by the width of the digital word. By using high speed, wide word DACs, each pixel can be placed on an extremely fine grid, without actually requiring the beam to access each pixel [15]. As an example, a 14-bit DAC provides 600-Å resolution over a 0.1-mm field. More commonly, multiple high speed 8- or 10-bit DACs are used in parallel to provide extremely high resolution (N/nm) over fields of order a millimeter. The speed of the DACs is a critical concern in determining system throughput for vector scan systems; however, this is a very active area of technology development.

One of the other major areas of concern for electron beam lithography is pattern distortion due to proximity effects. This refers for the tendency of scattered electrons to expose nearby areas that may not be intended for exposure. The effect is much more severe for electron beams than for optical exposures. The cause of these exposures can be readily seen by examining a Monte Carlo simulation of the electron trajectories from a typical e-beam exposure (Figure 9.10). We can divide the scattering types into forward and backscatter. Forward scatter occurs over a small range of angles with respect to the incident velocity and leads to a slight broadening of the image. Forward scatter can limit resolution unless high energy or very thin resists are used [16]. Backscatter contributes a large area fog of exposure. To minimize the effects of this broadening, most high resolution EBL is done in a thin resist layer.

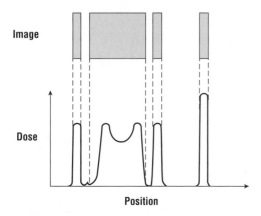

Image

Dose

Position

Figure 9.11   Small and large figures to be patterned with EBL requires position-dependent dosage to compensate for proximity effects.

The image is then transferred to a thicker masking layer that can be used to obtain the desired pattern. The incident beam energy has to be optimized for a given resist layer. The backscattering range, which depends on the square of the incident energy, can be quite large. For a high energy beam, this dispersal may be sufficient to reduce the exposure dose enough to prevent a significant change in the resist properties over the entire area unless the resist has a very high sensitivity. Otherwise the exposure dose must be corrected for backscatter as well.

If two pattern features are very close to each other, the overlapping trajectories, primarily from the forward scattering, lead to significant unintentional exposure [17] or proximity effect. A number of algorithms have been developed that can be used to modify the exposure time (and therefore dose) per pixel to compensate for proximity effects. The more comprehensive of these must be done offline by taking the fractured data for a design, running it through the proximity correction algorithm, and producing a dose-corrected fractured data file that can be used to run the exposure. Generally these algorithms calculate the concentration of surrounding features suitably weighted by their distance from the area to be exposed, and subtracts the amount of backscattered exposure that one expects to see [18]. A simple illustration of large and small features is shown in the upper part of Figure 9.11. The dose, corrected for proximity effect, taken along a line through the center of the features is shown in the lower part of the figure. The center of the large pad receives a lower dose than nominal, and the isolated line at the right receives a larger dose. Figure 9.12 shows a pattern consisting of lines and spaces, large pads with small rectangular holes, and isolated small rectangles. Figure 9.12A shows the desired intensity pattern, the energy distribution without proximity correction, and the corrected dose map. Figure 9.12B shows the results with and without correction when the smallest feature is 0.25 μm. Without proximity correction (right) the smallest features are not resolved and the longest line shows a significant linewidth variation that depends on the local image density [19].

Proximity correction algorithms usually begin by describing the scattered beam as a double Gaussian:

$$I = I_o\left[e^{-r^2/2\alpha^2} + \eta_E e^{-r^2/2\beta^2}\right]$$

(9.10)

To apply the correction, one needs to know the proximity parameters α (forward scattering), β (backscattering), and η_E (ratio of energy deposited due to backscattering to the energy deposited due to forward scattering) [20]. These parameters depend primarily on the resist used, the underlying substrate, and most importantly the accelerating voltage. Often these parameters are found by performing Monte Carlo simulations. Once the beam spreading is known, the deposited energy at each point can be determined and the dosage map can be adjusted until all of the points to be exposed receive a nearly uniform dose.

Proximity effects can be completely eliminated in an alternative approach to EBL: scanning tunneling microscopy EBL [21]. In this technique a field emission probe is brought into extremely close proximity to the surface of the wafer and kept there using a feedback control by monitoring the field emission current from the tip. Atomic force spectroscopic techniques can also be used. Incident electron energies in scanning tunneling microscopy are 4 to 50 eV. The technique has been used to expose resists and to do direct surface modification down to features of 200 Å. A major concern for

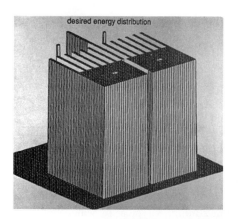

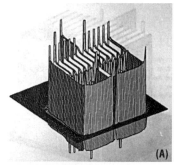
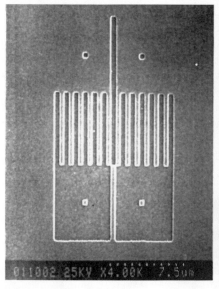
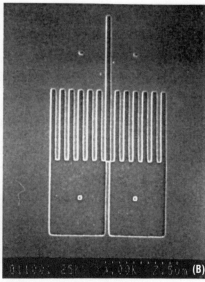

Figure 9.12 (A) Map of desired deposited energy, uniform dose deposited energy, and dose map needed to produce a uniform deposited energy; (B) electron micrograph of EBL exposures with and without proximity correction *(reprinted by permission, SPIE, after C.-Y. Chang et al., 1992).*

this technique is throughput, since the stage and/or tip must be mechanically scanned. Throughput of these systems is much less than that of traditional e-beam lithography. If a system consisting of a large array of these tips could be fabricated and controlled, however, STM lithography could see some application.

9.3 Direct-Write Electron Beam Lithography: Summary and Outlook

Like most "next-generation lithographies," EBL is being strongly pushed by the rapidly improving resolution of deep UV systems. By the mid-1970s electron beam lithography had demonstrated the capability to write lines and spaces less than 10 nm wide [22], while optical lithography was about

200× larger. Now the best commercial e-beam systems can write features about 5 nm, while deep UV is demonstrating features about 10× larger. With techniques such as 157-nm sources, immersion lenses, and optical proximity correction, "conventional" lithography may approach the 22-nm node, within a factor of 4 of e-beam; but at this writing, we do not foresee mask-based lithography surpassing the resolution of direct-write EBL.

The major concern with EBL is throughput. EBL can be thought of as a serial process, with the pattern information transferred to the wafer one pixel at a time. Exposures using masks, on the other hand, proceed in a massively parallel manner, with every pixel being exposed simultaneously. High brightness sources, vector scan systems, and low inductance deflection coils combined with large-bore lenses have all been developed over the past few years to improve throughput, but the technology remains at best an order of magnitude slower than optical lithography [23]. For example, the commercial system shown in Figure 9.13, which is intended for device production, can expose a wafer containing ten thousand 0.1×200-μm devices in about 20 min. A typical IC wafer, however, contains billions of transistors, making exposure times extremely long. EBL systems typically trade off resolution and throughput; that is, an EBL system designed to make sub-10-nm features will have very low throughput. (Table 9.1 lists specifications for a production e-beam system.) The shrinking difference between optical and e-beam performance has squeezed the market for high throughput, lower resolution e-beam systems designed for manufacturing. Due to throughput, direct-write EBL will never be seen as a production technology for the bulk of the IC industry as long as a viable alternative exists [24].

Aside from mask making, EBL has two primary applications. Variable beam systems have been used with high brightness sources to provide a fine feature prototyping or limited production capability.

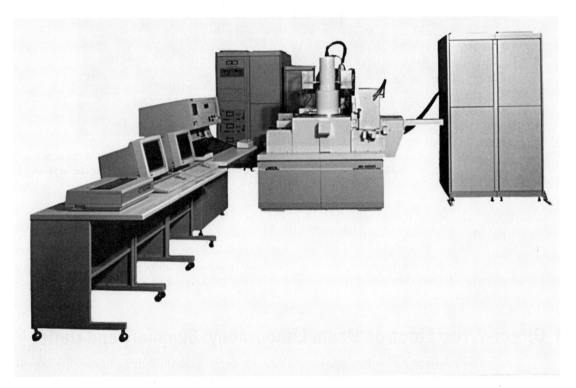

Figure 9.13   A JEOL-6000FS EBL system *(courtesy of JEOL Corporation)*.

Table 9.1   Critical specifications for a JEOL JBX-6000FS/SFE direct write electron beam lithography system designed for 0.1-μm device production

Maximum current density	2.0 A/cm^2
Beam diameter	0.005–0.2 μm
Cathode type	Thermionic field emission (Zr/O/W)
Accelerating voltage	25/50 kV
Writing accuracy	0.06 μm (2σ)
Minimum linewidth	0.02 μm

It is straightforward to use EBL to quickly modify patterns and produce a limited number of test die, since no mask-making step is required. Combined with its high resolution, this capability makes EBL an attractive technique for research and advanced prototyping. Typical applications in this market include the manufacture of high speed GaAs ICs such as millimeter microwave (MMIC) devices. The smaller wafer size, low device count, and the small volume of wafers, combined with the need for nanometer resolution, make EBL an ideal choice. The other primary application of EBL is device research. Again EBL is ideally suited to this task because of its excellent resolution, and its ability to rapidly image small quantities of demonstration vehicles. When throughput is not a serious concern, as is the case in discrete transistor research, reasonably high resolution (~50 nm) EBL systems can be made at relatively low cost by converting scanning electron microscopes.

9.4 X-ray Sources°

The second type of nonoptical lithography to be considered uses x-rays as the radiation. There are three x-ray sources that can be used for x-ray lithography (XRL). In order of increasing intensity (and complexity), they are electron impact, plasma, and storage rings. The ideal x-ray source should either be as small and as bright as possible (for proximity x-ray printing) or be uniform over a large area with as large an intensity as possible (for projection x-ray lithography). All of the x-ray sources must be operated under vacuum. Unlike EBL, however, most wafers are exposed at 1 atmosphere in x-ray lithography. This improves system throughput by avoiding the need to pump the wafers to high vacuum. Thin beryllium windows are used on the source to extract the x-rays. A film as thin as 25 μm can withstand a 1-atmosphere pressure differential across a small (<1 cm) diameter. Windows up to 6 cm in diameter have been developed for large area exposures [25]. Beryllium windows age with x-ray exposure and so must be replaced periodically.

The simplest type of x-ray source is the electron impact source. It uses a high energy electron beam incident on a metal target [26]. As discussed in Section 9.1, when energetic electrons strike a target, one of the primary energy loss mechanisms is through core-level electron excitation. When these excited electrons fall back to the core level, x-rays are emitted. These x-rays form a discrete line spectrum whose energies depend on the target material. In addition, a continuous spectrum of *Bremmsstrahlung* radiation is emitted due to the deceleration of the charged electrons [27].

One of the primary limitations of electron impact sources is power dissipation. If the target gets too hot, it will begin to evaporate. For this reason, refractory metals such as tungsten and molybdenum are often used for targets. The simplest sort of x-ray source, shown in Figure 9.14, is somewhat similar to the electron beam evaporator that will be discussed in a later chapter [28] except that in this case the hearth charge is water cooled to prevent evaporation of the target. To allow higher power densities, the water-cooled anode may be rotated at 7000 to 8000 rpm to dissipate the heat over a larger area [29]. In these systems an electrical power dissipation of up to 20 kW can be achieved.

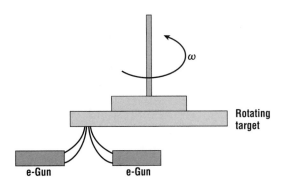

Figure 9.14   A simple rotating electron impact x-ray source uses electron beams focused on a rotating tungsten anode.

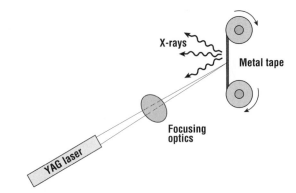

Figure 9.15   Laser plasma-heated x-ray source uses a focused high intensity pulsed laser to ablate a metal film. The superheated metal atoms radiate x-rays.

The next most intense class of sources use plasmas. There are two common types of plasma x-ray sources: laser-heated and electron discharge heated plasmas. Of these, laser-heated are the most popular. Although laser-heated systems have been built with excimers [30], most use a pulsed Nd glass slab laser. The laser energy is generated in 10 nsec, 20–25 J pulses of 1.053-μm radiation [31]. This energy is focused into a 200-μm-diameter spot at the surface of a thin metal film (Figure 9.15). The power density (of order 10^{13} W/cm^2) in each pulse is sufficient to vaporize the film. The superheated metal vapor radiates x-rays with wavelengths from 8 to 100 Å. Compact plasma sources have also been built for XRL [32].

The laser-heated plasma is well suited for lithography: it is an intense source with a very small diameter. The exposure energy can be metered using a photodiode, and the dose can be controlled digitally by changing the number of pulses. The metal film is frequently coated on a tape, allowing the target to be advanced as rapidly as the laser can be pulsed. State-of-the-art YAG lasers can be pulsed at 2 Hz. Typical conversion energies from the laser output to x-ray energy is about 10%, or about 2 J per pulse.

If the distance between the source and the wafer is 10 cm, and no reflectors or other optics are used, the energy density at the surface of the wafer is given by

$$E_d = \frac{E_o}{2\pi r^2} \qquad\qquad (9.11)$$

which is a little less than 4 mJ/cm^2 per pulse, or 8 mW/cm^2. As a comparison, rotating anodes in a similar geometry supply about 1.5 mW/cm^2 and stationary anodes supply about 0.2 mW/cm^2. If the resist requires an exposure of 40 mJ/cm^2, and power loss in the mask and exposure tool is ignored, as covered shortly, 10 pulses will be required. This will take about 5 sec per exposure, which because of the large field size may be adequate for IC production. Because of the short wavelength of the exposing radiation, these systems have excellent dimensional control. Proximity x-ray exposures for 130 nm features show total critical dimension variations (including mask errors) as low as 8.2 nm [33].

Synchrotrons are the brightest x-ray sources for lithography. Although the cost of the storage ring is high, the fact that a single ring can support a large number of exposure tools would make

Electron storage ring source

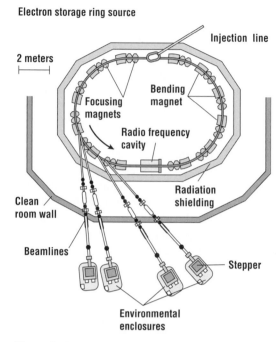

Figure 9.16 Basic schematic of an electron storage ring for XRL. Several exposure stations are indicated (*after Glendenning and Cerrina, reprinted by permission, Noyes Publications*).

synchrotrons attractive for production x-ray lithography. The basic electron storage ring is shown in Figure 9.16. Electrons are injected into the ring, where they are maintained at 10^6 to 10^9 eV. The energy of the electrons in the ring is given by

$$E \approx 0.3 \frac{\text{GeV}}{\text{T-m}} B \cdot r \qquad \textbf{(9.12)}$$

where B is the magnetic field in teslas of the bending magnet and r is the radius of curvature.

The energy lost in the emission process is made up in the RF cavity, where electrons are accelerated back to the storage ring energy. Electron injection must typically be repeated a few times per day, since the electron current will slowly decay due to leakage. The electrons may be injected as energies close to those of the storage ring, in which case they are first sent through a linear accelerator, or they may be injected at much lower energies and accelerated to speed in the synchrotron.

At each bending magnet location an intense x-ray beam is emitted. The simple storage ring shown in Figure 9.16 could support 12 exposure tools. When a nonrelativistic charged particle is accelerated in a circular path with angular frequency ω, it emits power as a dipole

$$\frac{dP}{d\Omega} = \frac{q^2 r^2 \omega^4}{4\pi c^3} \frac{1}{\sin^2 \theta} \qquad \textbf{(9.13)}$$

where c is the speed of light and Ω is the solid angle. In synchrotron sources the electrons are usually highly relativistic. This leads to a Doppler (blue) shift of the radiation energy and a narrowing of the angle of emission. The result is that the median wavelength of such a machine is given by

$$\lambda_C (\text{Å}) = 5.6 \frac{r(\text{m})}{E(\text{GeV})^3} = \frac{18.64}{B(\text{T})E(\text{GeV})^2} \qquad \textbf{(9.14)}$$

Most synchrotron sources that are currently operational were not designed as x-ray lithography sources and so have median wavelengths of 20 Å. New systems designed for proximity x-ray lithography have median wavelengths of about 10 Å.

Finally, the size of the storage ring is worth noting. Many current generation storage rings are quite large, with a radius of over 10 m [34]. The limitation is one of magnetic field. Large water-cooled electromagnets can generate about 20 kG (2 T) of field. For a 1-GeV beam, this will produce a turn radius of about 1 m, for a total ring diameter of perhaps 5 m. Such a ring, however, requires a great deal of power to operate. Large-bore superconducting magnets, on the other hand, can produce fields of 50 kG (5 T), resulting in turning radii of about 0.5 m. With this type of technology compact, storage rings with diameters of 2 m can be built [35]. The cost of operating these magnets is primarily the liquid helium refrigeration.

9.5 Proximity X-ray Exposure Systems

Unlike electrons and optical photons, it is difficult to construct any type of optics for x-rays. This difficulty increases as the wavelength decreases. Many experimental x-ray exposure tools therefore are 1-to-1 proximity printers or steppers. If the exposure tool is intended for IC manufacturing, it must be a stepper, since it is not currently possible to fabricate x-ray masks larger than a few inches on a side. The attractive feature of x-ray lithography is its potential for use in volume manufacturing. Figure 9.17 shows a commercial x-ray stepper that has been evaluated as an early prototype for manufacturing. This section will concentrate on a simple proximity exposure tool (Figure 9.18). A reflector may be used to collect and collimate the source if a rotating anode or plasma source rather than a synchrotron is used, delivering it to a narrow aperture; but unlike optical systems simple polished metal reflectors are extremely inefficient. A helium column or a high vacuum exists between the wafer and the beryllium window to avoid absorption in the gas. The wafer must be aligned to the mask optically and the mask clamped to the wafer carrier. The carrier assembly is then positioned in the beam and the x-ray shutter is opened. If the system is a stepper, the stage is moved under the mask before the next exposure is done. After all of the exposures are complete, the carrier is rolled out, and the wafer is removed for developing.

Figure 9.17   Commercial x-ray exposure tool *(courtesy Karl Suss).*

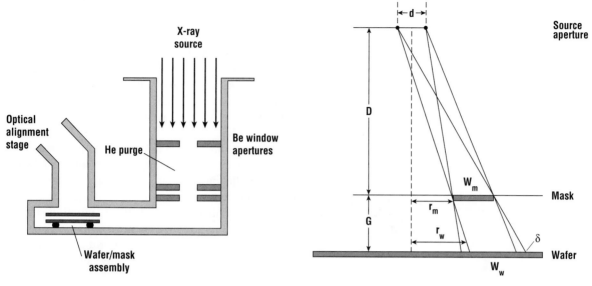

Figure 9.18 Simple proximity x-ray lithography aligner. The basic system is very similar to optical proximity systems.

Figure 9.19 Geometry of the exposure system shown in Figure 9.18.

Clearly with 10-Å photons, the features are $\gg \lambda$, so simple ray-tracing can be used. If one attempts to use the point source machine shown in Figure 9.18 to expose a pattern of lines and spaces, one can begin to appreciate the limits of x-ray lithography. Consider Figure 9.19. Let d denote the source aperture, D the aperture to mask spacing, G the mask-to-wafer gap, W_m the linewidth on the mask, W_w the linewidth on the wafer, and r_m and r_w the distance from the inner edge of the line on the mask and the wafer, respectively, to the centerline of the exposure system. As a result of the geometry, one can show that a pattern distortion given by

$$r_w = r_m + G \frac{r_m}{D}$$ (9.15)

occurs. For a gap of 25 μm, a D of 1 m, and an r_m of 2 cm, the shift is about 0.5 μm. This effect may be tolerable if all features are exposed on the same tool, or the masks are made to compensate for the distortion, as long as the mask-to-wafer spacing can be well controlled.

An even more serious limitation is penumbral blur. Referring to Figure 9.19, if one assumes the mask to be perfectly opaque, the finite source size will still lead to continuous shadow edges. The width of the outer shadow is considered the blur δ, given by

$$\delta = G \frac{d}{D}$$ (9.16)

The size of this blur is often considered to be a very coarse measure of the resolution of an x-ray exposure tool. For the dimensions that we considered before, using a 4 mm aperture, the resolution is of order 0.1 μm.

Of course penumbral blur is just a simple geometric effect. If one could produce a uniform wide-area beam, this problem could be avoided as well. Although synchrotrons naturally produce a large-area beam with a small divergence, it is not very uniform [36]. Either type of source could

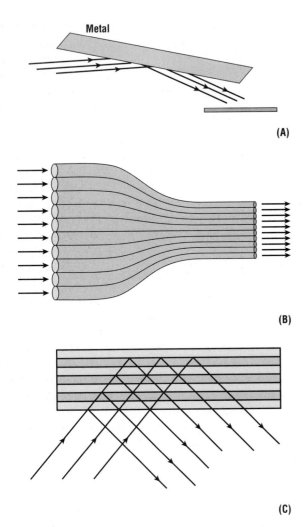

Figure 9.20   Possible choices for x-ray optics systems include glancing angle metal mirrors (A), Kumakhov lenses (B), and multilayer mirrors (C).

clearly benefit from x-ray lenses. There are three basic types of lenses currently being studied. The first and simplest is the glancing incidence mirror. As shown in Figure 9.20A, this is simply a highly polished metal surface that the beam strikes at a very low angle. Although such a system can be used in a simple application such as to raster the incident beam from a synchrotron, one cannot build up complicated optical systems with it.

The Kumakhov lens uses a large number of small-diameter glass capillary tubes to collect and redirect the radiation. Often the tubes may be drawn down to focus the radiation (Figure 9.20B). X-rays that enter the tube at small angles internally reflect off of smooth interior walls until they exit on the other side of the lens [37]. A current generation of Kumakhov lens is over a meter in length but has highly nonuniform areal intensity.

A third strategy for x-ray optics uses a multilayer mirror approach [38]. Originally developed for x-ray astronomy, these mirrors use alternating layers of two elements with widely different electron concentrations (Figure 9.20C). One material has a high mass (the scatterer) and the other has a light mass (the spacer). A common combination is molybdenum and silicon. At each layer only a small fraction of the incoming plane wave is reflected. If the spacer layer thickness is chosen properly, the reflected waves from the scattering layers constructively interfere. To have high efficiency reflection, the layer interfaces must be extremely smooth [39]. If the layer thicknesses are chosen properly, the reflected waves from each interface add constructively, and over half of the incident energy can be reflected for low energy x-rays [40]. Furthermore, if the mirror surface is properly shaped, not only can collimation be achieved, but so can reduction. If these mirrors allow a reduction stepper to be made with a very high intensity source, many of the mask distortion problems will be considerably reduced.

9.6 Membrane Masks

One of the most difficult aspects of x-ray lithographic technology is mask production. Since no material is highly transmissive for x-rays, the mask substrate must be a low Z thin film [41]. The mask information is transmitted using a patterned absorber layer on this membrane. Table 9.2 compares some of the films used most often for x-ray masks. The substrate materials are listed roughly in order of increasing desirability. For comparison, two common absorbers, tungsten and gold, have been included. Various alloys of Ta are also commonly used as absorbers. Although diamond has the most desirable properties for a substrate, it is also the most difficult to produce [42]. Due to the ease of

Table 9.2　Mechanical properties of typical x-ray mask materials

Material	Young's Modulus (GPa)	Thermal Coefficient of Expansion $(°C^{-1}) \times 10^{-6}$	Thickness for 50% Transmission (μm)
Silicon nitride	250	2.7	2.3
Silicon	47	2.6	5.5
Boron nitride	675	1.0	3.8
Silicon carbide	460	4.7	3.6
Diamond	11	1.0	4.6
Tungsten	397	4.5	0.8 for 10 dB
Gold	78	14.2	0.7 for 10 dB

fabrication, a common membrane material is nonstoichiometric Si_3N_4; however, SiC is preferred due to its increased stiffness and improved radiation tolerance [43]. The film is most commonly deposited using low pressure CVD. The temperature and gas compositions can be used to control the stress in the membrane [44].

All membrane mask processes begins with the fabrication of an x-ray mask blank. As shown in Figure 9.21, in the typical blank fabrication process a silicon wafer is first coated with the membrane material. In the case of silicon membranes, the surface of the wafer may be doped to be able to selectively etch the substrate [45]. The back of the wafer is patterned lithographically to protect the outer ring. The membrane material on the backside of the wafer is removed, then a long wet chemical silicon etch is done to remove most of the wafer. Once all of the wafer has been removed, the remaining ring is epoxy mounted to a Pyrex ring for additional strength and mechanical stability.

Once the wafer blank has been completed, the mask must be made. A typical additive sequence of membrane mask fabrication is shown in Figure 9.22. A thin tantalum/gold plating base is deposited on the completed blank and a thick stencil resist is spun on and baked. Thin layers of chromium and imaging resist are deposited on the stencil resist. The top resist is exposed, developed, and used to pattern the chromium layer. The chromium is used as a hard mask in a vertical etch of the stencil resist. The chromium is then stripped, and gold is plated up selectively in the resist trenches [46]. Finally the resist is stripped, and the plating base may be removed.

A subtractive process is shown in Figure 9.23. The mask blank is coated with tungsten and a resist is applied, patterned, and used as a mask to etch the tungsten. The resist is stripped to complete the process. Each technique has advantages. The subtractive process is simpler, but it requires the development of a vertical etch of tungsten that can stop on the membrane. This is much more difficult than achieving etch selectivity of resist to silicon, since the common etchant for both silicon and tungsten is fluorine. The lack of selectivity tends to limit the aspect ratio of tungsten absorbers. Gold can be used as an absorber in the subtractive process, but it suffers from redeposition

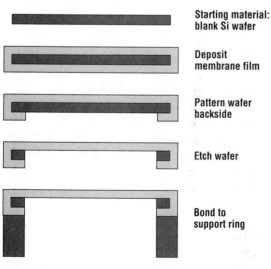

Starting material: blank Si wafer

Deposit membrane film

Pattern wafer backside

Etch wafer

Bond to support ring

Figure 9.21　X-ray mask blank fabrication process produces a membrane stretched across a mechanical support ring.

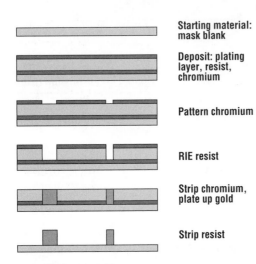

Starting material: mask blank

Deposit: plating layer, resist, chromium

Pattern chromium

RIE resist

Strip chromium, plate up gold

Strip resist

Figure 9.22   Additive process for x-ray mask fabrication.

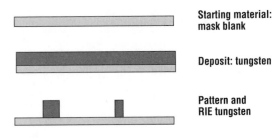

Starting material: mask blank

Deposit: tungsten

Pattern and RIE tungsten

Figure 9.23   Subtractive process for x-ray mask fabrication.

problems during the patterning step [47]. Redeposition can be avoided through the use of high energy ion milling [48], but this process can easily damage the membrane. Stress control is critical in the tungsten deposition process to avoid peeling and mask distortions. Stress is not a problem for a plating process; however, many silicon fabrication facilities are extremely reluctant to introduce gold because of possible contamination problems.

Pattern distortion is a serious problem in membrane masks. In addition to placement errors associated with the e-beam writing of the mask, serious distortion may arise from nonuniform clamping of the mask in the system [49]. Other sources of distortion include stress changes or differential thermal expansion of the mask materials. The stress on the mask blank may change during the addition of the various layers on the mask. After exposure and patterning the mask can relax back toward its previous condition. During exposure thermal expansion differences among the ring, the membrane, the absorber, and the resist can also cause local distortions. For that reason, choosing materials with matching expansion coefficients is very desirable. To provide optimal stability, the membrane stress should be large compared to the additional layers, but not large enough to lead to buckling and warpage. It should also be as thin as possible to minimize exposure times and mask heating due to absorption. The former can be overcome with high brightness sources, but the mask must also be optically transparent for most alignment techniques.

The major obstacles to implementation of simple proximity printing is the mask [50] and the continuing success of optical lithography. A number of groups are working on proximity x-ray lithography, and improvements are being made [51]; however, many issues are far from resolved. IBM has succeeded in fabricating masks with complex IC images [52]. The image placement needed for 50-nm technologies will be extremely challenging on a thin membrane mask [53]. Furthermore, very thin gaps (~10 μm ± 1 μm) will be needed. Mask distortion and aging at these small feature sizes are also extremely serious. Many question the ability of the mask to tolerate the handling implicit in day-to-day use in a production environment. Questions regarding mask cleaning and repair have not been resolved. Finally there is no commercial supplier of masks. These concerns have dropped proximity x-ray lithography from a leading contender as an NGL to more of a second-tier status.

9.7 Projection X-ray Lithography

Just as optical lithography overcame mask problems by moving from proximity/contact printing to projection lithography, XRL would benefit greatly if a projection technique could be developed. Optics for such a system would be mirrors, typically the reflecting high Z/low Z type discussed already. To avoid the problems associated with membranes, the masks can be made on a reflecting rather than a transmissive surface. This solves most of the current-generation x-ray mask problems [54]. These systems must use soft x-rays ($\lambda > 50$ Å) [55] where the reflectivity is large [56]. This radiation is sometimes called extreme ultraviolet (EUV). Normally a pulsed plasma source is used, since these can be quite rich in soft x-rays. This type of system has already demonstrated feature sizes of 0.1 μm [421], and scaling to smaller sizes appears to be feasible. Figure 9.24 shows a system designed around this principle. The mask is scanned from left to right, while the wafer is scanned from right to left.

Projection x-ray systems have several shortcomings, generally revolving around the mirror/mask structure. Since many layers are required, the defect density per layer must be extremely small. The spectral bandwidth is only 2 to 3%, limiting the power transmission unless a highly monochromatic source is used. The stability of the mirrors and coating after long-term x-ray exposures is a concern. Perhaps the most vexing problem for projection x-ray is the requirements for machining and polishing tolerances in the mirror fabrication process [57]. This has two parts: the atomic abruptness of the layer structure, which can severely impact the reflectivity of the mirrors, and the substrate on which the mirrors and masks are built.

The most widely used pair of materials for projection x-ray consists of Si as the low Z layer and Mo as the high Z layer. This combination can achieve about 60% reflectivity at wavelengths greater than 124 Å, which is the silicon L absorption edge [58]. This requires abrupt (<3 Å roughness) interfaces. These mirrors contain hundreds of repeats of the two materials. To minimize interdiffusion, the deposition of these layers must be done at low temperature. RF deposition systems put a lower heat load on the mirror and so are useful in this application [59]. Nickel/Chromium has been explored as a materials pair for shorter wavelength sources. Models predict that it should be useful down to 50 Å; however, obtaining smooth interfaces has been extremely difficult, with typical roughness values of 6 to 8 Å. Takenaka et al. have fabricated a mirror with a roughness of 2.5 Å by RF sputtering at $-20°C$, giving a reflectivity of about 20% [60].

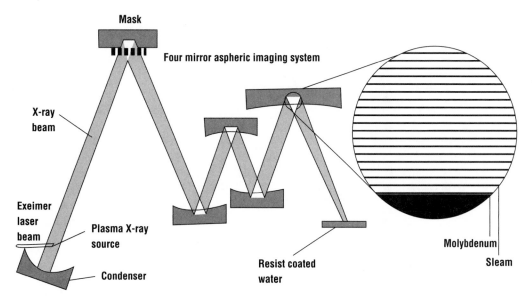

Figure 9.24   An x-ray projection lithography system using x-ray mirrors and a reflective mask *(after Zorpette, reprinted by permission, © 1992 IEEE).*

The second problem with x-ray projection optics is the substrate on which the mirror/mask is built. Early systems used 130-Å radiation with Schwarzschild microscope objectives consisting of two spherical mirrors [61]. To obtain a substantially larger exposure area, one must use aspheric mirrors. Several designs have been proposed, but all require mirror machining tolerances of less than 10 Å to achieve diffraction-limited results [62]. This is approximately two orders of magnitude better than state of the art in aspheric mirror fabrication techniques. There is some hope that diffraction gratings could be used on spherical mirrors [63]. This would significantly improve the manufacturability of a projection x-ray tool.

9.8 Projection Electron Beam Lithography (SCALPEL)

Direct-write e-beam lithography is too slow to be a viable production technique except in niche applications. If one could combine the high throughput of projection optical lithography with the resolution of e-beam, the result would be very attractive. The difficulty is finding a suitable masking material that would have adequate contrast and transparency to allow reasonable exposures, and still have the mechanical stability needed for repeated use. Typical stencil masks through thick plates have very limited applicability.

One of the most promising techniques to come out of e-beam lithography is called SCALPEL (SCattering with Angular Limitation Projection Electron-beam Lithography) [64, 65]. The basic idea behind SCALPEL was first introduced by Berger and Gibson at Bell Laboratories in 1989 [66]. It is the use of scattering contrast versus absorption contrast to create an image. That is, the system sends a broad collimated electron beam through a mask, whose "clear" areas consist of a thin membrane of a low atomic number (Z) material. The dark areas, which are patterned in a high-Z material, are not designed to absorb the electrons, rather they scatter them at a sufficiently large angle to prevent them from going through an orifice to reach the wafer (Figure 9.25). One of the primary problems with

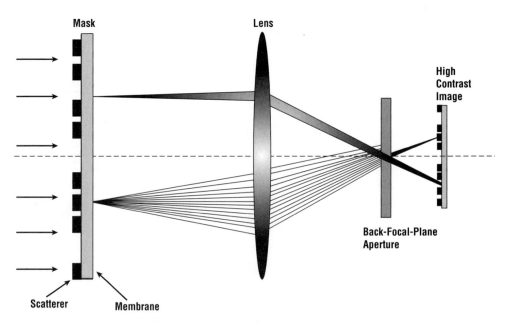

Figure 9.25   SCALPEL principle of operation.

membrane masks is the heating caused by beam absorption. The SCALPEL technique completely avoids the problem as the beam is only scattered, not absorbed in the mask. This allows one to use very high energies, typically about 100 keV, at which most membranes are almost completely transparent.

The electrons are created using a cathode in a temperature-limited (i.e., thermionic emission) mode. In early versions of the SCALPEL gun, a large flat single crystal of LaB_6 was heated to about 1250 K, producing a 10 µA current, which is highly uniform over a 1 mm diameter with a brightness of 1000 A/cm^2-sr [67]. Due to uniformity and stability issues associated with LaB_6, it has been replaced with a Ta disc cathode with a brightness of only 250 A/cm^2-sr. Notice that this is quite different from direct-write sources. For SCALPEL one wants a large-area, highly uniform, low-brightness source, with a very high total emittance. Magnetic lenses are then used to focus the electrons. Since low-aberration, large-field magnetic lenses are impractical, SCALPEL systems have small exposure areas, typically of order 1 mm on a side. The mask and wafer chambers and the beam path are kept at vacuums of order 10^{-7} torr, while the gun is maintained at pressures below 1×10^{-8} torr. The pattern is exposed by scanning the mask and wafer synchronously through the illumination, with stage velocities ratioed 4:1 to given a reduction to the exposure.

The SCALPEL mask comprises a ~0.1-µm low-Z membrane, often a silicon-rich silicon nitride, and a 0.05-µm high-Z pattern layer, typically W/Cr. Notice this is much less that the 1- to 2-µm thick layers typically used for dark regions in proximity x-ray masks. The high-Z material scatters the electron beam to high angles, while the scattering in the low-Z material produces a negligible angular change in the velocity. An aperture in the back-focal plane of the projection system blocks the scattered electrons, generating a high contrast arial image. To improve the mechanical stability of the membrane mask, the image is divided into ~1-mm rows that are stitched together in the final image. A skirt region containing the deflecting high-Z material surrounds each exposure area to form the edge of the image.

The SCALPEL process, although once a leading contender as an NGL, has several difficulties that must be resolved [68]. Naturally no source of commercial SCALPEL masks exists, nor will mask repair be trivial. After a 4× reduction, the image size on the wafer of one scan area is 0.25 mm × 3 mm. The small field size means that actual patterns must be stitched together. This is a serious concern, since maintaining critical dimension control over stitched features is extremely difficult. SCALPEL has adopted a semiblended approach in which features near the edge exist in a shared zone where partial exposures are done from each side. This leads to much less significant stitching errors. Perhaps the most difficult problem with SCALPEL is the throughput/image quality trade-off. High throughput in a SCALPEL system requires a high electron fluence. Just as was discussed for low energy ion implantation, space charge effects lead to a blooming or spreading of the beam. (In ion implantation, the energy at which this is important is ~1 keV, vs the 100 keV usually used here; but the ions are ~1000× more massive than the electrons used in SCALPEL) Excellent image quality can be obtained at low electron fluence, and reasonable throughput can be obtained at high fluence. It remains to be seen whether high throughput systems can be built for nanometer features.

9.9 E-beam and X-ray Resists

Image production in the resist in x-ray and e-beam lithography is very different from that for traditional optical lithography. In an optical process the energy of the absorbing photon is well defined. The photon energy is given by

$$E = h\nu = \frac{h\,c}{\lambda} \tag{9.17}$$

Table 9.3   Common e-beam lithography resists

Resist	Tone	Sensitivity (μC/cm^2)	Contrast	Etch Rate	Other
PMMA (typical)	+	120	Low	High	Good adhesion Good shelf life
ZEP 520-12	+	40	Medium	Low	Poor adhesion
APEX-E	+	3	Medium	Low	Poor adhesion, chem amp
SAL-601	−	10 to 20	High	Moderate	Poor adhesion, chem amp
EBR-9	+	1.2	Low	High	
Novolac	+/−	200–500	Medium	Low	

Chemically amplified (chem amp) resists are generally susceptible to atmospheric contamination and often have short shelf lives. Resists with poor adhesion require the use of an adhesion promoter such as hexamethyldioxysiloxane (HMDS).

For an i-line source the photon energy is 3.4 eV. In contrast, a high concentration of secondary electrons with a wide range of energies is produced upon exposure to either electron beams or x-rays. Rather than designing the resist so that a single chemical reaction is driven by the exposure, the resist must be designed so that the desired reaction occurs preferentially, but recognize that many reactions will occur. Because the energetic beams penetrate the resist well into the substrate, undesirable reactions (damage) in the substrate must also be considered. This section will discuss intentional chemical reactions in the resist. The next section will discuss exposure damage. The section will concentrate on x-ray damage, but the effects are similar for e-beam exposures as well.

In resists, two types of chemical changes are of particular interest. The photoresists that are often used in nonoptical lithography consist of long-chain carbon polymers (see Section 8.2). Upon irradiation, atoms on adjacent chains may be displaced and the carbon atoms will bond directly. This process is known as cross-linking. Highly cross-linked molecules dissolve more slowly in a developer solution. A material in which cross-linking is the dominant reaction upon exposure is a negative resist. As discussed in Chapter 8, radiation can also disrupt the polymer chains, rendering them more soluble in the developer. A material in which chain scission is the dominant reaction upon exposure is a positive resist.

The most important resist criteria are contrast and sensitivity for the exposure type and energy (Table 9.3) and resistance to damage during plasma etching. One of the most commonly used positive photoresists for high resolution work is polymethyl methacrylate (PMMA). The monomer for PMMA is

$$-[CH_2=CCH_3(COO(CH_3))]- \tag{9.18}$$

It has poor sensitivity, but good contrast (typically $\gamma = 2$–4). In PMMA both cross-linking and fragmenting of the polymeric chains occur (Figure 9.26 shows cross-linking in a modified PMMA); but the rate of scission is much larger than that of

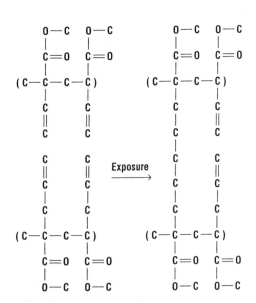

Figure 9.26   Cross-linking of an e-beam resist where the basic PMMA structure has been modified through the addition of a C=C side chain to promote cross-linking.

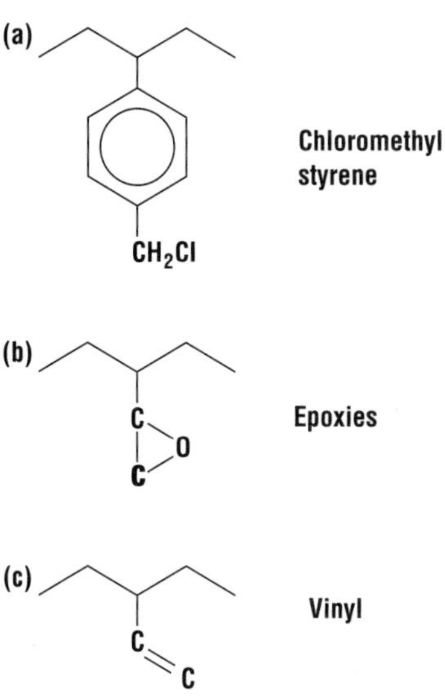

(a)

Chloromethyl styrene

CH_2Cl

(b)

Epoxies

(c)

Vinyl

Figure 9.27 Common groups used to promote cross-linking.

cross-linking. The fragments are about 100 Å long. PMMA is not exposed by white light; it has an indefinite shelf life and excellent resolution. Like many simple hydrocarbon polymers, however, it stands up very poorly in plasma etching environments. As a result, many PMMA-like resists are limited to mask making where only a thin Cr wet etch on a flat substrate is required. When these resists are used for direct writing, PMMA is typically used as the imaging resist for a liftoff process. A variety of derivatives of PMMA have also been demonstrated. One worthy of note is EBR-9, in which the polymer structure has been replaced by

$$-[CH_2 - CCl(CO_2(CH_2CF_3))]- \qquad (9.19)$$

Manufactured by Toray in Japan, EBR-9 has one of the highest sensitivities for positive resists that do not use chemical amplification, roughly 15 times that of PMMA.

A number of negative resists have also been fabricated. These resists have components on the polymer chain that enhance the likelihood of cross-linking. Typical cross-linking components include chloromethyl styrene, epoxies, and vinyl groups (see Figure 9.27). During exposure the polymers readily cross-link at these positions, reducing the solubility of the resist in developer. The negative resists generally have sensitivities as good as or better than the best positive resists; but they have low contrasts and are prone to swelling during the develop cycle.

X-ray exposures are typically done in thick resist layers and produce vertical sidewalls with well-controlled feature sizes. In contrast, thin layers of imaging resists are often used in e-beam lithography to minimize scattering effects. The pattern is transferred into a hard mask by liftoff or etching. When the resist must be directly used in an etching environment, either a novolac-based optical photoresist or a novolac/sulfone copolymer is used. These resists are even less sensitive than PMMA. For example, Shipley AZ-1350, a commonly used positive tone optical resist, acts as a negative resist under e-beam exposure with a contrast of 3.2 and requires a dose about 10 times that required for PMMA.

9.10 Radiation Damage in MOS Devices

One of the consequences of the use of energetic beams such as x-rays and e-beams for lithography is that the beam is not confined to the resist layer. Either the beam itself or the secondary electrons or high energy radiation generated by the beam will lead unintentional and often undesirable chemical changes in the underlying layers [69]. The most sensitive of the common IC structures to these effects are oxides and oxide/semiconductor interfaces. This section will review some of the primary results of radiation exposure and discuss the repair of these effects. For a comprehensive review the reader is referred to Ma and Dressendorfer [70].

It has been observed that irradiation causes increases in the fixed oxide charge density, the neutral trap density, and the interface state density of MOS devices (see Chapter 4 for a discussion of these defects). The fixed charge shifts the threshold voltage of the MOS transistor. The interface states shift the threshold and reduce the carrier mobility. Neutral traps in the oxide can cause the device to degrade more rapidly under electrical stress. Trapped charge changes the local field in the oxide and may ultimately lead to destruction. As shown in Figure 9.28, the process by which these three effects arise is quite complex. It may involve a direct bond-breaking mechanism during

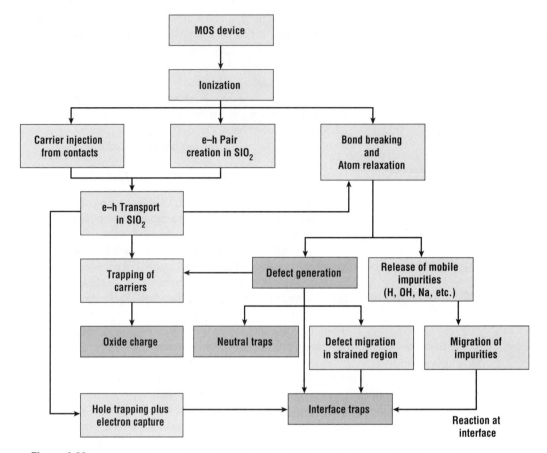

Figure 9.28 Radiation damage flow path *(after Ma and Dressendorfer, reprinted by permission, Wiley).*

the absorption event, or the damage may be caused by the absorption of carriers created in the oxide during the exposure. From Figure 9.28 it is apparent that the carriers in the oxide may fall into traps, or they may create traps through bond breaking. An example of the bond-breaking mechanisms believed to occur during the radiation of MOS devices is shown in Figure 9.29. In this example an oxygen vacancy leads to a strained Si–Si bond. Holes created during exposure are captured, causing the SiO_2 tetrahedra to relax back, leaving a permanently unsatisfied bond [71]. At the interface it is believed that hole capture releases atomic hydrogen, which had originally passivated defect sites. This released hydrogen atom can then cause additional defects.

Since the number of ionizing events in the oxide increases with oxide thickness, the damage is most pronounced in thick oxides. This dependence can be approximated by

$$N_{fc}, N_{it} \approx k t_{ox}^n \tag{9.20}$$

where k is a constant, N_{fc} and N_{it} are the fixed charge and interface state density, respectively, and n is approximately 2 [72, 73]. At a typical e-beam exposure dose (2×10^{-5} C/cm^2 or about 10^{14} electrons/ cm^2), a MOSFET with a 500-Å gate oxide will experience a fixed charge increase of about 10^{12} cm^{-2}.

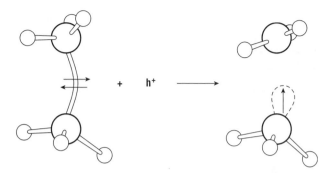

Figure 9.29   An example of a trap creation process believed to occur upon x-ray irradiation of MOS structures. The larger atoms are silicon, the smaller are oxygen. Due to an oxygen vacancy, the two silicon atoms are initially bonded together.

Most of the increase in fixed charge density and interface state density can be repaired by a 400°C anneal in forming gas [74]. This is a significant result, since the final anneal of most silicon technologies is done at temperatures at least that high. Neutral traps, on the other hand, require anneal temperatures of at least 700°C. There are indications [75] that these traps are not completely removed until the anneal temperature approaches 1000°C. As a result, any back-end energetic beam processes such as metal patterning may leave neutral traps that cannot be removed, and therefore these technologies will be more susceptible to later damage.

It has been shown that very thin gate oxides, grown with dry oxygen at low temperature are quite radiation hard when used with deep submicron devices [76], but some radiation effects remain. Irradiated unannealed n-channel devices with 50-Å gate oxides were found to degrade about twice as fast as unirradiated devices under electrical stressing [77]. The efficiency of neutral trap annealing of these thin oxides is found to depend on the annealing ambient [78], with pure hydrogen showing good effects even at 450°C. As a result, oxide damage is not expected to be a serious barrier to x-ray lithography of deep submicron devices.

9.11 Soft Lithography and Nanoimprint Lithography

The development of novel lithographic techniques may have applications both to mainstream IC fabrication, if the resolution is high enough, and to novel applications of micro and nanofabrication technologies, if they provide unique advantages compared to conventional lithography. A detailed review of these techniques is contained in Sotomayar Torres [79]. These novel areas include microelectromechanical systems (MEMS), biotechnology, microanalysis, sensors, and printable electronics. In some cases these nontraditional applications require the formation of patterns on curved, soft, or uneven substrates at the nanoscale [80]. The two classes of techniques to be covered in this section share a common two-step process: (a) fabrication of a mechanical stamp that replaces the photomask in conventional lithography and (b) using the stamp to form a pattern reproducibly on a substrate. The distinction between the two classes is the nature of the stamp, whether it is flexible and so capable of molding itself to the substrate, or if it is mechanically rigid and so better able to reproducibly form and position extremely small features.

Soft lithography is a collection of techniques that all involve the use of a soft stamp. These techniques include microcontact printing (μCP) [81], nanoimprint lithography (NIL), replica molding (REM) [82], microtransfer molding (μTM) [83], micromolding in capillaries (MIMIC) [84], and

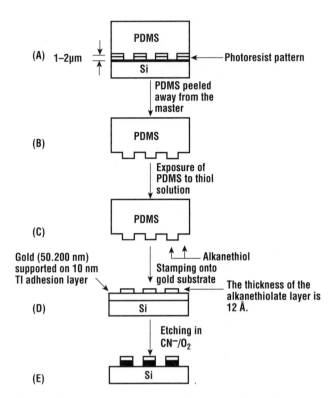

(A) 1–2μm
PDMS
Photoresist pattern
Si

PDMS peeled away from the master

(B)
PDMS

Exposure of PDMS to thiol solution

(C)
PDMS

Gold (50.200 nm) supported on 10 nm TI adhesion layer

Alkanethiol
Stamping onto gold substrate

The thickness of the alkanethiolate layer is 12 Å.

(D)
Si

Etching in CN⁻/O₂

(E)
Si

Figure 9.30 Schematic description for the μcp fabrication of Au patterns using an elastomer called PDMS (Sylgard 184, Corning): (A) PDMS stamp fabrication, (B) PDMS detach from the master, (C) exposure to the alkanethiol ink, (D) contacting with Au substrate, and (E) etching Au on the substrate *(reprinted with permission from Applied Physics Letters, from Ref. 88.) Copyright 2002, American Institute of Physics.*

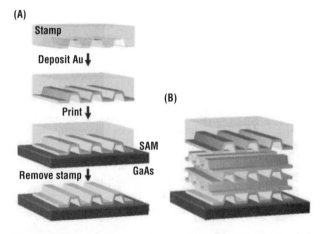

(A)

Stamp

Deposit Au ↓

Print ↓

(B)

SAM

Remove stamp ↓

GaAs

Figure 9.31 Schematic of a 3D micromolding process using a gold layer to form fluid flow channels (reprinted with permission from Nanoletters, from Ref. 92). *Copyright 2003, American Chemical Society.*

solvent-assisted micromolding (SAMIM) [85]. In the μCP process (Figure 9.30), an elastomeric stamp is prepared by a molding process and coated with a chemical ink capable of forming a self-assembled monolayer (SAM). The SAM is transferred to the target substrate. The resultant patterned SAMs can be used as ultrathin resists in selective wet etching [86] or as templates to control the wetting, dewetting [87], nucleation, growth [88], and deposition of other materials [89]. In this example, the transferred SAM was used as a masking layer and the underlying gold layer was etched. Once the master has been formed in silicon or some other hard material, the elastomeric element can be used as soft molds for imprinting [90] and conformable phase masks [91, 92]. The technique is able to generate submicron features with inexpensive laboratory apparatus, and patterns can be transferred onto a curved or uneven surface of target substrate owing to the flexibility of the elastomeric elements.

Soft lithography enables a fabrication of nanoscale three-dimensional structures. Zaumseil et al. reported 3D nanostructure formed by nanotransfer printing [93]. They showed complex 2-D and 3-D structures with nanometer resolution. Figure 9.31 illustrates a procedure for single and multilayer printing for three-dimensional structures. In the first step, a conformal layer of Au was deposited onto the surface of the stamp. Contacting this stamp with a substrate resulted in covalent bonding between the Au coating and the substrate. A monolayer of octanedithiol on a GaAs substrate provided exposed thiol groups that can bond to the Au coating when the stamp contacts the substrate. If the surface of the stamp is treated such that the coated film does not adhere well to it, then removing the stamp leaves a pattern with the geometry of the relief features onto a target substrate. An array of sealed nanochannels was formed from this printing procedure. Multiple prints enables stacking of such nanochannels to multilayer nanochannels.

Although submicron μCP has been demonstrated, the ability of this technique to reliably produce very small images is uncertain. Some important minimum feature size considerations include the fidelity of the contact printing process, the anisotropy of the etching process, and the deformation/distortion of an elastomer [94, 95]. Related processes include various types of molding steps in which a material is deposited on the elastomer mask in a way designed to fill the grooves left in the elastomer from the original hard pattern. The elastomer with the embedded material is then placed on a substrate, and the material in the spaces is released to the substrate. This technique has demonstrated features as small as 30 nm.

Nanoimprint lithography (NIL) is a new, low cost method for producing nanoscale features on hard, flat substrates. In this case, a rigid form is pressed into a polymer to form a relief image. Normally an anisotropic etch of the polymer follows the imprinting process to completely remove the polymer in the thinned areas, while leaving polymer in the raised areas. Differences in the process generally involve the removal of the stamp and the stabilization of the resist. As shown in Figure 9.32, there are two commonly accepted ways to do these. The thermal process, also called hot embossing, was developed first [96]. It involves heating the polymer so that it will flow readily into the mold features. The polymer is then cooled while the mold is in place to harden it. Finally, the mold is removed from the substrate. The second technique, also called step and flash, uses UV light to harden the polymer. In this case the mold must be made from a material such as fused silica that is reasonably transmissive to UV. Although this technique uses UV light and fused silica, neither technique is concerned with diffraction. The UV cure in UV-NIL is a flood exposure, not a patterned exposure. As a result, simple g-line systems consisting of a lamp, a parabolic reflector, and a shutter are sufficient to replicate nanoscale features once the stamp has been made. One can afford to use a slow process like electron beam lithography to make the stamp because it will serve to make many wafers.

Many polymers have been demonstrated as usable for NIL. During thermal NIL, the film is heated to facilitate flow of the polymer [97]. The most thoroughly studied thermal NIL material is PMMA. PMMA has glass transition temperature ($\sim100°C$) well above room temperature, but still easy to achieve. Conventional photoresists have been patterned with NIL [98]. For UV-NIL, one needs the ability to fully cross-link (i.e., cure) the polymer with acceptably short exposure times; however, cure-induced shrinkage needs to be minimized to limit distortion of the pattern transfer process [99]. Epoxies, vinyl ethers, and acrylate are popular materials for this purpose because of their commercial availability and ability to photocure rapidly [100–102]; as with photoresists, however,

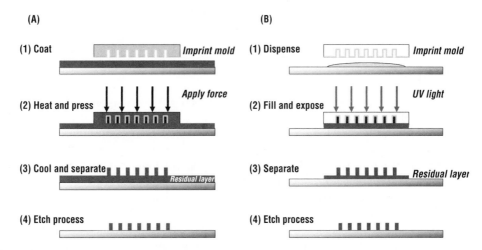

Figure 9.32 Process diagrams for (A) thermal nanoimprint lithography (NIL) and (B) ultraviolet-assisted NIL. In both processes, a mold is pressed into a soft material to form a physical relief image of the mold (*reproduced by permission of the MRS bulletin, from Ref. 103*).

multicomponent materials are most commonly used for commercial applications [103]. Multilayer films are also of interest because it is possible to mold in the top layer and transfer the image to a thicker base layer.

Although NIL is now considered a viable contender as a replacement for DUV lithography around 20 nm, and commercial systems are beginning to appear, there are many technical challenges that must be overcome. For any NIL process to be successful, one must ensure that all of the polymer is released by the mold. Any adhering material will propagate defects from substrate to substrate. An integrated cleaning step may prove to be a necessary part of the process. The second problem relates to flatness and particles. The surface must allow a uniform, intimate contact with the mold over a field, where the field size is likely to be ~5 cm on a side. There has been some work on UV-NIL using semicompliant molds to help compensate for flatness problems. In general, however, pattern stability on the scale of a few 10s of nanometers must remain a concern. For thermal NIL, one must be aware of thermal expansion problems as they relate to pattern shift and material stability.

9.12 Summary

Because of diffraction effects, optical lithography is not expected to be capable of being extended to feature sizes much less than 0.07 μm; however, costs for optical patterning at these dimensions will rise sharply. Some of the possible candidates for nanoscale lithography are direct-write and projection electron beam (EBL) and proximity and projection x-ray (XRL) lithography. Direct-write EBL has demonstrated the ability to produce feature sizes of about 10 nm, exceeding device requirements for the foreseeable future. The process is too slow, however, to make most integrated circuit manufacturing viable. Proximity x-ray lithography has the potential for volume production since, like optical lithography, it is a parallel rather than a serial process. The best x-ray sources include synchrotrons and laser-heated plasmas. Proximity x-ray lithography uses shorter wavelength photons. Masks must be fabricated on a thin membrane. Distortion of the mask due to mounting, thermal effects, and exposure aging are all serious concerns. Projection x-ray lithography uses reflecting masks and softer x-rays, avoiding some of the problems associated with proximity systems, but presents a new set of difficulties. SCALPEL, a projection e-beam system, was presented. SCALPEL also uses membrane masks, but avoids many of the problems found in the proximity x-ray technique. Finally, new techniques, including nanoimprint lithography, were reviewed.

Problems

1. Construct a table of the wavelength and energy of g-line, i-line, and ArF laser optical sources, 10-Å x-ray photons, and the de Broglie wavelength of 10-keV electrons. Use Equation 7.16 to predict the proximity printing diffraction-limited minimum feature size for each source if the proximity gap is 10 μm and $k = 1$.
2. If one examines the Monte Carlo results for an e-beam exposure shown in Figure 9.11, some of the scattering events involve scattering angles (the angle between the incident and outgoing velocity vectors) of greater than 90°. Explain briefly why this is seen far more often in e-beam lithography than in ion implantation.
3. Estimate the parameters in Equation 9.10 for the beam broadening indicated in the upper half of Figure 9.10. Use the lateral scale on the lower part of the diagram (I_o will be arbitrary).
4. Refer to Figure 9.5 for an e-beam column. Where might you expect to observe x-rays from this column? Where would they be particularly intense?

5. A proximity x-ray mask is made up of tungsten-absorbing patterns on a silicon nitride membrane. If the membrane heats 10°C during exposure and the field size is 2.5 cm on a side, how much can the field be distorted due to thermal expansion? Using approximation that the maximum allowable distortion is about one-fourth the minimum feature size, what is the maximum temperature rise allowable on the mask if the minimum feature size it 0.1 μm?

6. (a) Penumbral blur (see Equation 9.16) can be greatly reduced in x-ray lithography by using contact rather than proximity printing. Why isn't this commonly done?

    (b) What are the relative advantages of thermionic and field emission sources for high resolution direct-write electron beam lithography?

    (c) A typical proximity correction is applied to an electron beam exposure. Which pattern would receive a longer exposure time, a isolated line or a line in a dense pattern of features? Justify your answer.

7. How thin must a silicon nitride membrane be for 90% transmission at the wavelength used in Table 9.2.

8. According to Equation 9.16, the penumbral blur can be minimized if the aperture or the gap is reduced. Explain the trade-offs involved for each of the two parameters.

References

1. F. B. McLean, H. E. Boesch, Jr., and T. R. Oldham, "Electron-Hole Generation, Transport, and Trapping in SiO_2," in *Ionizing Radiation Effects in MOS Devices and Circuits*, T. P. Ma and P. V. Dessendorfer, eds., Wiley-Interscience, New York, 1989.

2. J. J. Muray, "Electron Beam Processing," in *Beam Processing Technologies*, N. G. Einspruch, S. S. Cohen, and R. N. Singh, eds., Academic Press, San Diego, CA, 1989.

3. P. Nehmiz, W. Zapka, U. Behringer, M. Kallmeyer, and H. Bohlen, "Electron Beam Proximity Printing," *J. Vacuum Sci. Technol. B* **3**:136 (1985).

4. J. Frosien, B. Lischke, and K. Anger, "Aligned Multilayer Structures Generated by Electron Microprojection," *Proc. 15th Int. Symp. Electron, Photon, Ion Beam Technol.*, 1980, p. 1827.

5. J. Frosien, B. Lischke, and K. Anger, "Aligned Multilayer Structures Generated by Electron Microprojection," *Proc. 15th Int. Symp. Electron, Photon, Ion Beam Technol.*, 1980, p. 1827.

6. Commercially available from FEI Co., Beaverton, Oregon.

7. M. Gesley, F. J. Hohn, R. G. Viswanathan, and A. D. Wilson, "A Vector Scan Thermal Field Emission Nanolithography System," *J. Vacuum Sci. Technol. B* **6**:2014 (1988).

8. H. Nakazawa, H. Takemura, and M. Isobe, "Thermally Assisted Field Emission Electron Beam Exposure System," *J. Vacuum Sci. Technol. B* **6**:2019 (1988).

9. H. C. Pfeiffer, "Variable Spot Shaping for Electron Beam Lithography," *J. Vacuum Sci. Technol.* **15**:887 (1978).

10. H. C. Pfeiffer, "Recent Advances in EBL for High Volume Production of VLSI Devices," *IEEE Trans. Electron Devi.* **ED-26**:663 (1979).

11. L. Veneklasen, "A High Speed EBL Column," *J. Vacuum Sci. Technol. B* **3**(1) (1985).

12. F. Hohn, "Electron Beam Lithography, Directions in Direct Write and Mask Making," in *Electron-Beam, X-Ray, and Ion-Beam Submicrometer Lithographies IX*, SPIE Proc. **1263**: 152 (1990).

13. P. Herriot, R. Collier, D. Alles, and J. Stafford, "EBES; A Practical Electron Lithography Systems," *IEEE Trans. Electron Dev.* **ED-22**:385 (1975).

14. T. H. P. Chang, M. Hatzakis, A. D. Wilson, A. Speth, A. Kern, and H. Luhn, "Scanning Electron Beam Lithography for Fabrication of Magnetic Bubble Circuits," *IBM J. Res. Dev.* **20**:376 (1976).

15. L. Veneklasen, "Electron Beam Patterning and Direct Write," in *Handbook of VLSI Microlithography*, Noyes, Park Ridge, NJ, 1991.

16. M. Bolorizadek and D. C. Joy, "Effects of Low Voltage on Electron Beam Lithography," *Proc. SPIE* **6151**:61512c (2006).

17. T. H. P. Chang and A. D. G. Stewart, "Proximity Effect in Electron Beam Lithography," *J. Vacuum Sci. Technol.* **12**:1271 (1975).

18. C.-Y. Chang, G. Owen, F. R. Pease, and T. Kailath, "A Computational Method for the Correction of Proximity Effect in Electron-beam Lithography," in *Electron-Beam, X-ray, and Ion-beam Submicrometer Lithographies*, SPIE Proc. **1671**:208 (1992).

19. P. Vermeulen, R. Jonckheere, and L. Van Den Hove, "Proximity Effect Correction in e-Beam Lithography," *J. Vacuum Sci. Technol. B* **7**:1556 (1989).

20. T. H. P. Chang and A. D. G. Stewart, *Electron-Beam, X-ray, and Ion-Beam submicrometer Lithographies*, **1089**:97 (1969).

21. C. R. K. Marrian and E. A. Dobisz, "Scanning Tunneling Microscope Lithography: A Viable Lithographic Technology?" in *Electron-Beam, X-ray, and Ion-Beam submicrometer Lithographies*, SPIE Proc. **1671**:166 (1992).

22. A. N. Broers, W. W. Molzen, J. J. Cuomo, and N. D. Wittels, "Electron Beam Fabrication of 80 Å Metal Structures," *Appl. Phys. Lett.* **29** (1976).

23. A. Gonzales, "Recent Results in the Application of Electron Beam Direct Write Lithography," in *Electron-Beam, X-ray, and Ion-Beam Submicrometer Lithographies VIII*, 374 (1989).

24. R. DeJule, "E-Beam Lithography, The Debate Continues," *Semicond. Int.* **19**:85 (1996).

25. K. Hara and T. Itoh, "Study of Large-field Beryllium Window for SR Lithography," in *Electron-Beam, X-ray, and Ion-beam Submicrometer Lithographies*, SPIE Proc. **1671**:391 (1992).

26. M. Lepselter, D. S. Alles, H. Y. Levinstein, G. E. Smith, and H. A. Watson, "A Systems Approach to 1 μm NMOS," *Proc. IEEE* **71**:640 (1983).

27. S. E. Bernacki and H. I. Smith, "Characteristic and Bremmsstrahlung X-ray Radiation Damage," *IEEE Trans. Electron Devi.* **22**:421 (1975).

28. T. Hayasaka, S. Ishihara, H. Kinoshita, and N. Takeuchi, "A Step-and-Repeat X-Ray Exposure System for 0.5 μm Pattern Replication," *J. Vacuum Sci. Technol. B* **3**:1581 (1985).

29. W. B. Glendenning and F. Cerrina, "X-ray Lithography," in *Handbook of VLSI Microlithography*, W. B. Glendinning and J. N. Helbert, eds., Noyes, Park Ridge, NJ, 1991.

30. R. Fedosejevs, R. Bobkowski, J. N. Broughton, and B. Harwood, "keV X-ray Source Based on High Repetition Rate Excimer Laser-produced Plasmas," in *Electron-beam, X-ray, and Ion-beam Submicrometer Lithographies II*, **1671**:373 (1992).

31. K. Fujii, Y. Tanaka, K. Suzuki, T. Iwamoto, S. Tsuboi, Y. Matsui, "Overlay and critical dimension control in proximity x-ray lithography" NEC Research and Development, v 42, n 1, January, 2001, p 27–31.

32. L. Malmqvist, A. L. Bogdanov, L. Montelius, and H. M. Hertz, "Nanometer Table-top Proximity X-ray Lithography with Liquid-target Laser-Plasma Source," *J. Vacuum Sci. Technol. B* **15**(4):814 (1997).

33. G. K. Celler, J. Frackoviak, R. R. Freeman, C. W. Jurgensen, R. R. Kola, A. S. Novembre, L. F. Thompson, L. E. Trimble, and D. N. Tomes, "Evaluation of a Laser-Based Proximity X-Ray Stepper," in *Electron-beam, X-ray, and Ion-beam Submicrometer Lithographies II*, **1671**:312 (1992).

34. J. B. Murphy, "Electron Storage Rinds as X-Ray Lithography Sources: An Overview," in *Electron-Beam, X-Ray, and Ion-Beam Submicrometer Lithographies IX*, **1263**:116 (1990).

35. D. E. Andrews, M. N. Wilson, A. I. Smith, V. C. Kempson, A. L. Purvis, R. J. Anderson, A. S. Bhutta, and A. R. Jorden, "Helios: A Compact Superconducting X-ray Source for Production Lithography," in *Electron-beam, X-ray, and Ion-beam Submicrometer Lithographies IX*, **1263**:124 (1990).

36. W. B. Glendinning and F. Cerrina, "X-ray Lithography," in *Handbook of VLSI Microlithography*, W. B. Glendinning and J. N. Helbert, eds., Noyes, Park Ridge, NJ, 1991.

37. "Soviet X-Ray 'Lens' Seen as Promising," *IEEE Institute*, May/June 1992, p. 1.

38. E. Spiller, "Reflective Multilayer Coatings for the Far UV Region," *Appl. Opt.* **15**:2333 (1975).

39. D. W. Kruger, D. E. Savage, and M. G. Lagally, "Diffraction Determination of the Size Distribution of Nanocrystalline Regions in a Crystalline Substrate," *Phys. Rev. Lett.* **63**:402 (1989).

40. D. G. Stearns, N. M. Ceglio, A. M. Hawryluk, and R. S. Rosen, "Multilayer Optics for Soft X-Ray Projection Lithography: Problems and Prospects," in *Electron-beam, X-ray, and Ion-beam Submicrometer Lithographies*, 80 (1991).

41. A. R. Shikunas, "Advances in X-ray Mask Technology," *Solid State Technol. J.* **27**:192 (1984).

42. H. Windischmann, "A 75 mm Diamond X-ray Membrane," in *Electron-beam, X-ray, and Ion-beam Submicrometer Lithographies IX*, **1263**:241 (1990).

43. P. Seese et al., *Proc. SPIE* **1924**:457 (1993).

44. R. Nachman, G. Chen, M. Reilly, G. Wells, H. H. Lee, A. Krasnoperova, P. Anderson, E. Brodsky, E. Ganin, S. A. Campbell, and F. Cerrina, "X-ray Lithography Processing at CXrL from Beamline to Quarter-micron NMOS Devices," *Proc. SPIE* (1994).

45. D. L. Spears and H. I. Smith, "X-ray Lithography—A New High Resolution Replication Process," *Solid State Technol. J.* **15**:21 (1972).

46. G. E. Georgiou, C. A. Janoski, and T. A. Palumbo, "DC Electroplating of Submicron Gold Patterns on X-ray Masks," *Proc. SPIE—Int. Soc. Opt. Eng.* **471**:96 (1984).

47. R. E. Acosta, J. R. Maldonado, L. K. Towart, and J. R. Warlaumont, "B-Si Masks for Storage Ring X-ray Lithography," *Proc. SPIE—Int Soc. Opt. Eng.* **448**:114 (1983).

48. J. L. Bartelt, C. W. Slayman, J. E. Wood, J. Y. Chen, C. M. McKenna, C. P. Minning, J. F. Coakley, R. E. Hollman, and C. M. Perrygo, "Mask Ion-Beam Lithography: A Feasibility Demonstration for Submicrometer Device Fabrication," *J. Vacuum Sci. Technol.* **19**:1166 (1981).

49. D. L. Laird and R. L. Engelstad, "Effects of Mounting Imperfections on an X-ray Lithography Mask," in *Electron-Beam, X-ray, and Ion-Beam Submicrometer Lithographies II*, **1671**:366 (1992).

50. S. Uchiyama, "Current Status and Issues with X-ray Mask," *IEEE Int. Conf. Microelectron Test Struct.* 1998, p. 61.

51. L. G. Lesoine and J. A. Leavey, "IBM Advanced Lithography Facility: The First Five Years," *Solid State Technol.* **41**:101 (July 1998).

52. R. Butsch, W. A. Enichen, M. S. Gordon, T. R. Groves, J. G. Hartley, J. W. Pavick, H. S. Pfeiffer, R. J. Quickle, J. D. Rockrohr, and W. Stickel, "Performance Enhancements on IBM's EL-4 Electron-beam Lithography System," *J. Vacuum Sci. Technol. B* **13**(6):2478 (1995).

53. S. Hector, "Status and Future of X-ray Lithography," *Microelectron. Eng.* **41/42**:25 (1998).

54. G. Zorpette, "Rethinking X-ray Lithography," *IEEE Spectrum* **29**:33 (June 1992).

55. H. Kinoshita, K. Kurihara, Y. Ishii, and Y. Torii, "Soft X-ray Reduction Lithography Using Multilayer Mirrors," *J. Vacuum Sci. Technol. B* **6**:1648 (1989).

56. D. G. Stearns, R. S. Rosen, and S. P. Vernon, "Multilayer Mirror Technology," in *Soft-X-Ray Projection Lithography Technical Digest*, Optical Society American, Washington, DC, 1992, p. 44.

57. B. E. Newnam and V. K. Viswanathan, "Development of XUV Projection Lithography at 60–80 nm," in *Electron-Beam, X-ray, and Ion-Beam Submicrometer Lithographies II*, **1671**:419 (1992).

58. D. G. Stearns, R. S. Rosen, and S. P. Vernon, "Fabrication of High-Reflectance Mo-Si Multilayer Mirrors by Planar-Magnetron Sputtering" *J. Vacuum Sci. Technol. A* **9**:2662 (1991).

59. H. Takenaka, H. Kinoshita, Y. Ishii, and M. Oshima, "Fabrication, Performance, and Applications of Multilayer Mirrors for Soft X-rays," *NTT R&D* **43**(1):39 (1994).

60. H. Takenaka, T. Kawamura, and H. Kinoshita, "Fabrication and Evaluation of Ni/C Multilayer Soft X-ray Mirrors," *Thin Solid Films* **288**:99 (1996).

61. H. Kinoshita, K. Kurihara, Y. Ishii, and Y. Torii, "Soft X-ray Reduction Lithography Using Multilayer Mirrors" *J. Vacuum Sci. Technol. B* **7**:1648 (1989).

62. K. Kurihara, *J. Photopolym. Sci. Technol.* **5**:173 (1992).

63. H. Fukuda and T. Terasawa, "New Optics Design Methodology Using Diffraction Grating on Spherical Mirrors for Soft X-ray Projection Lithography," *J. Vacuum Sci. Technol. B* **13**(2):366 (1995).

64. L. R. Harriott, S. D. Berger, C. Biddick, M. Blakey, S. Bowler, K. Brady, R. Camarda, W. Connelly, A. Crorken, J. Custy, R. DeMarco, R. C. Farrow, J. A. Felker, L. Fetter, L. C. Hopkins, H. A. Huggins, C. S. Knurek, J. S. Kraus, R. Freeman, J. A. Liddle, M. M. Mkrtchyan, A. E. Novembre, M. L. Peabody, R. G. Tarascon, H. H. Wade, W. K. Waskiewicz, G. P. Watson, K. S. Werder, and D. L. Windt, "Preliminary Results from a Prototype Projection Electron-beam Stepper SCALPEL Proof-of-Concept System," *J. Vacuum Sci. Technol. B* **14**:3825 (1996).

65. See also: http://www.bell-labs.com/project/SCALPEL/.

66. S. D. Berger and J. M. Gibson, "New Approach to Projection-Electron Lithography with Demonstrated 0.1 Micron Linewidth," *Appl. Phys. Lett.* **57**:153 (1990).

67. W. DeVore and S. D. Berger, "High Emittance Electron Gun for Projection Lithography," *J. Vacuum Sci. Technol. B* **14**(6):3764 (1996).

68. Stuart T. Stanton, J. Alexander Liddle, Warren K. Waskiewicz, Masis M. Mkrtchyan, Anthony E. Novembre, and Lloyd R. Harriott, "Critical Issues for Developing a High-Throughput SCALPEL System for Sub-0.18 Micron Lithography Generations," *Proc. SPIE* 3331 (1998) .

69. K. H. Zaininger and A. G. Holmes-Siedle, "A Survey of Radiation Effects in Metal-Insulator-Semiconductor Devices," *RCA Rev.* **28**:208 (1967).

70. T. P. Ma and P. V. Dressendorfer, eds., *Ionizing Radiation Effects in MOS Devices and Circuits*, Wiley-Interscience, New York, 1989.

71. F. B. McLean, H. E. Boesch, Jr., and T. R. Oldham, "Electron-Hole Generation and Trapping in SiO_2," in *Ionizing Radiation Effects in MOS Devices and Circuits*, T. P. Ma and P. V. Dressendorfer, eds., Wiley-Interscience, New York, 1989.

72. C. R. Viswanathan and J. Maserjian, "Model for the Thickness Dependence of Radiation Charging in MOS Structures," *IEEE Trans. Nucl. Sci.* **NS-23**:1540 (1976).

73. N. S. Saks, M. G. Ancona, and J. A. Modolo, "Radiation Effects in MOS Capacitors in Very Thin Oxides at 80 K," *IEEE Trans. Nucl. Sci.* **NS-31**:1249 (1984).

74. J. M. Aitken, "1 μm MOSFET VLSI Technology: Part III—Radiation Effects," *IEEE J. Solid State Circuits.* **SC-14**:294 (1979).

75. M. Shimaya, N. Shiono, O. Nakajima, C. Hashimoto, and Y. Sakakibara, "Electron-Beam Induced Damage in Poly-Si Gate MOS Structures and Its Effect on Long Term Stability," *J. Electrochem. Soc.* **130**:945 (1983).

76. K. H. Lee, S. A. Campbell, R. Nachman, M. Reilly, and F. Cerrina, "X-ray Damage in Low Temperature Ultrathin Silicon Dioxide," *Appl. Phys. Lett.* **61**:1635 (1992).

77. S. A. Campbell, K. H. Lee, H. H. Li, and F. Cerrina, "Charge Trapping and Device Degradation Induced by X-ray Irradiation in Metal Oxide Semiconductor Field Effect Transistors," *Appl. Phys. Lett.* **63**:1646 (1993).

78. C. C.-H. Hsu, L. K. Wang, D. Zicherman, and A. Acovic, "Effect of Hydrogen Annealing on Hot-Carrier Instability of X-ray Irradiated CMOS Devices," *J. Elect. Mater.* **21**:769 (1992).

79. C. M. Sotomayor Torres, *Alternative Lithography: Unleashing the Potentials of Nanotechnology*, Kluwer Academic/Plenum, New York, 2003.

80. Y. Xia, and G. M. Whitesides, *Angew. Chem. Int. Ed. Engl.* **37**:550 (1998).

81. A. Kumar, and G. M. Whitesides, *Appl. Phys. Lett.* 1993, **63**:2002 (1993).

82. Y. Xia, E. Kim, X. M. Zhao, J. A. Roger, M. Prentiss, and G. M. Whitesides, *Science*, **273**:347 (1996).

83. X. M. Zhao, Y. Xia, and G. M. Whitesides, *Adv. Mater.* **8**:837 (1996).

84. E. Kim, Y. Xia, and G. M. Whitesides, *Nature* **376**:581 (1995).

85. E. Kim, Y. Xia, X. M. Zhao, and G. M. Whitesides, *Adv. Mater.* **9**:651 (1997).

86. Y. Xia, X. M. Zhao, E. Kim, and G. M. Whitesides, *Chem. Mater.* **7**:2332 (1995).

87. A. Kumar, and G. M. Whitesides, *Science* **263**:60 (1994).

88. C. S. Chen, M. Mrksich, S. Huang, G. M. Whitesides, and D. E. Ingber, *Science* **276**:1245 (1997).

89. H. Yang, N. Coombs, and G. A. Ozin, *Adv. Mater.* **9**:811 (1997).

90. B. D. Aumiller, E. A. Chandross, W. J. Tomlinson, and H. P. Weber, *Appl. Phys. Lett.* **45**:4557 (1974).

91. J. A. Rogers, K. E. Paul, R. J. Jackman, and G. M. Whitesides, *Appl. Phys. Lett.* **70**:2658 (1997).

92. H. Schmid, H. Biebuych, and B. Michel, *Appl. Phys. Lett.* **72**:2379 (1998).

93. J. Zaumseil et al., *Nano Lett.* **3**:1223 (2003).

94. J. N. Lee, C. Park, and G. M. Whitesides, "Solvent Compatibility of Poly(dimethylsiloxane)-Based Microfluidic Devices," *Anal. Chem.* **75**:6544 (2003).

95. C. Y. Hui, A. Jagota, Y. Y. Lin, and E. J. Kramer, "Constraints on Microcontact Printing Imposed by Stamp Deformation," *Langmuir* **18**:1394 (2002).

96. S. Y. Chou, P. R. Krauss, and P. J. Renstrom, "Imprint Lithography with 25-nanometer Resolution," *Science* **272**:85, (1996).

97. H. D. Rowland, A. C. Sun, P. R. Schunk, and W. P. King, "Impact of Polymer Film Thickness and Cavity Size on Polymer Flow During Embossing: Toward Process Design Rules for Nanoimprint Lithography," *J. Micromech. Microeng.* **15**:2414 (2005).

98. C. Gourgon, C. Perret, and G. Micouin, "Electron Beam Photoresists for Nanoimprint Lithography," *Microelectron. Eng.* **61–62**:385 (2002).

99. S. Johnson, R. Burns, E. K. Kim, M. Dickey, G. Schmid, J. Meiring, S. Burns, C. G. Willson, D. Convey, Y. Wei, P. Fejes, K. Gehoski, D. Mancini, K. Nordquist, W. J. Dauksher, and D. J. Resnick, "Effects of Etch Barrier Densification on Step and Flash Imprint Lithography," *J. Vacuum Sci. & Technol. B: Microelectron. Nanometer Struct. Process. Meas. Phenom.* **23**:2553 (2005).

100. E. K. Kim, M. D. Stewart, K. Wu, F. L. Palmieri, M. D. Dickey, J. G. Ekerdt, and C. G. Willson, "Vinyl Ether Formulations for Step and Flash Imprint Lithography," *J. Vacuum Sci. Technol. B: Microelectron. Nanometer Struct. Process. Meas. Phenom.* **23**:2967 (2005).

101.   X. Cheng, L. J. Guo, and P.-F. Fu, "Room-Temperature, Low-Pressure Nanoimprinting Based on Cationic Photopolymerization of Novel Epoxysilicone Monomers," *Adv. Mater. (Weinheim, Germany)* **17**:1419 (2005).

102.   E. K. Kim, N. A. Stacey, B. J. Smith, M. D. Dickey, S. C. Johnson, B. C. Trinque, and C. G. Willson, "Vinyl Ethers in Ultraviolet Curable Formulations for Step and Flash Imprint Lithography," *J. Vacuum Sci. Technol. B. Microelectron. Nanometer Struct. Process. Meas. Phenom.* **22**:131 (2004).

103.   M. D. Stewart and C. G. Willson, "Imprint Materials for Nanoscale Devices," *MRS Bull.* **30**:947 (2005).

Chapter 10

Vacuum Science and Plasmas

U ntil now most of the unit processes that have been discussed are run at atmospheric pressure. The microelectronic processes that will be covered in the next four chapters, however, are run in vacuum chambers. The first half of this chapter will review vacuum science and technology. It will first discuss the fundamental physics of molecules and atoms in a vacuum chamber. It will then go through some of the basic equipment used to produce, contain, and measure vacuums. Further information can be obtained in several references [1–4]. The second half of the chapter is devoted to the physics and technology of plasmas or glow discharges. This type of system is used for the physical and chemical deposition of thin films and for etching. Subsequent chapters will discuss the details of each of these processes, but this chapter will lay the foundations for understanding plasma processes.

10.1 The Kinetic Theory of Gases°

One of the topics that will arise through the next section of the book is the behavior of gases in a vacuum chamber. To appreciate the behavior of gas in chemical reactions, gaseous beams, heat flow, and surface bombardment, we need to develop a model of gas molecules. One of the most common, the kinetic theory of gases, treats gaseous molecules as hard spheres. Then the probability distribution of velocities is given by the Maxwell speed distribution (Figure 10.1). For a simple monotonic gas the probability of a molecule having a certain speed is given by

$$P(v) = 4\pi \left[\frac{m}{2\pi kT} \right]^{3/2} v^2 e^{-mv^2/2kT} \tag{10.1}$$

where m is the mass of the molecule, k is Boltzmann's constant, v is the magnitude of the velocity, and T is the temperature in kelvins. The speed, or average magnitude of the velocity, is given by

$$|v| \equiv \bar{c} = \int_0^\infty v P(v) dv = \sqrt{\frac{8kT}{\pi m}} \tag{10.2}$$

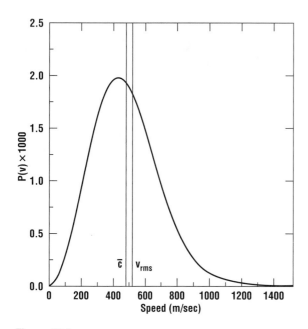

Figure 10.1   A Maxwellian speed distribution of particles; P(v) is the probability that a particular particle will have the target magnitude of velocity.

The average component of the velocity in any direction is given by

$$\bar{v}_x = \bar{v}_y = \bar{v}_z = \sqrt{\frac{2kT}{\pi m}} \qquad (10.3)$$

and the root mean square velocity is given by

$$v_{\text{rms}} = \sqrt{\frac{3kT}{m}} \qquad (10.4)$$

The thermal velocity is often large. For example, molecular nitrogen at room temperature has an average x-component of the thermal velocity of about 240 m/sec, or 530 mi/hr. The direction of the thermal velocity is random, and therefore if there are no externally applied forces, the average velocity is zero. If a small pressure gradient is imposed across the gas, there will be a net macroscopic flow from high to low pressure. On a microscopic level this induced flow velocity vector is added onto the much larger thermal velocity. Individual atoms may be traveling against the pressure gradient at any instant of time, but on average the flow will move from high to low pressure.

One of the primary mechanisms by which gases at atmospheric pressure change velocity is gas phase collisions. Consider a molecule of diameter d moving randomly in the gas. If another molecule of the same type is within a distance d of the projected path of the first atom, a collision occurs. The molecule then has a collision cross section of πd^2, which in the simple hard sphere approximation, is four times the actual cross section of a single molecule. The probability that a collision will occur over a distance L is given by

$$P = L\pi d^2 n \qquad (10.5)$$

where n is the number of gas molecules per unit volume. For simple diatomic molecules like N_2 and O_2, d is commonly taken as 3 Å. If we set $P \approx 1$, the average distance between collisions, commonly called the mean free path λ, can be approximated as

$$\lambda \approx \frac{1}{\pi d^2 n} \qquad (10.6)$$

A more rigorous statistical treatment will show that

$$\lambda \approx \frac{1}{\sqrt{2}\pi d^2 n} \qquad (10.7)$$

Since n is not commonly known, it must be calculated from a macroscopic equation of state. The ideal gas law

$$n = \frac{N}{V} = \frac{P}{kT} \qquad (10.8)$$

where P is the pressure of the chamber, is commonly used. Combining equations,

$$\lambda = \frac{kT}{\sqrt{2}\pi d^2 P} \qquad (10.9)$$

A host of useful equations can be derived from this simple model. Table 10.1 lists a few gas characteristics as derived from kinetic theory. In this table, $\overline{C}_v$ is the heat capacity per unit volume of the gas species and $\overline{c}$ is the average speed as defined in Equation 10.2. These formulas are applicable only when $\lambda \ll L$, where L is a characteristic length of the chamber (for example, the chamber diameter). This is called the viscous flow regime. We will, for the time being, confine our remarks to this pressure regime.

Finally, a topic must be introduced that may seem a bit abstract, but ultimately will prove quite useful. Recall from Chapter 3 that a flux density J_n can be defined as the net flux of molecules through a unit area per unit time. J_n can also be thought of as the number of molecules that strike a surface of unit area per unit time. It is given by

$$J_n = \frac{n\overline{v}_x}{2} = \sqrt{\frac{n^2 kT}{2\pi m}} = \sqrt{\frac{P^2}{2\pi kTm}} \qquad (10.10)$$

Once again this can be a very large number. For example, J_n for 1 atm of nitrogen at room temperature is about 3×10^{23} cm^{-2} sec^{-1}.

Before finishing this section, a word is required about units of pressure. Table 10.2 lists some of the most common units and their meanings. The unit of torr (the pressure that would raise a

Table 10.1   Various formulas that can be derived from the kinetic theory of gases

Parameter	Symbol	Equation
Diffusivity	D	$\overline{c}\lambda/3$
Viscosity	η	$mn\overline{c}\lambda/3$
Thermal conductivity	k_{th}	$\dfrac{nC_v\overline{c}\lambda}{3}$

Table 10.2   Conversion factors for various pressure units

Units	Conversion Factor
Standard atmospheres (atm)	1.333×10^{-3}
Pounds per square inch (psi)	1.933×10^{-2}
Torr or mmHg (torr or mm)	1000
Pascal or N t/m^2 (Pa)	133.3
Micrometers of Hg (μm)	1000×10^3
Millitorr (mtorr)	1.000×10^3

To change from torr to the desired unit, multiply by the conversion factor. To change to torr from the listed unit, divide by the conversion factor.

column of mercury 1 mm) has been commonly used in describing vacuum equipment. We have adopted this convention in this book. However, it is not an MKS unit. To use the equations developed in this chapter you must convert to a consistent unit set.

Example 10.1

Using Table 10.1, show why H_2 and He have the highest thermal conductivities of all gases. Calculate the thermal conductivity of He at room temperature if $C_v = 3/2\ k$ and $d = 1\text{Å}$. Compare your answer to the accepted value.

Solution

From Table 10.1

$$k_{th} = \frac{nC_v\bar{c}\lambda}{3}$$

$$= \frac{P}{kT}\frac{C_v}{3}\sqrt{\frac{8kT}{\pi m}}\frac{1}{\sqrt{2}\pi d^2}\frac{kT}{P}$$

$$= \frac{2C_v}{3\pi d^2}\sqrt{\frac{kT}{\pi m}} \text{ (independent of } P\text{)}$$

Note that smaller atoms/molecules will have lower mass and lower diameter. This increases k_{th}.

For He, $m = 4 \times 1.67 \times 10^{-27}$ kg, so

$$k_{th} = \frac{2 \times \frac{3}{2} \times 1.38 \times 10^{-23} \text{ J/K}}{3 \times \pi \times (10^{-10} \text{ m})^2} \times \sqrt{\frac{1.38 \times 10^{-23} \text{ J/K} \times 300 \text{ K}}{\pi \times 4 \times 1.67 \times 10^{-27} \text{ kg}}}$$

$$= 0.19 \text{ W/m-K}$$

The accepted value is 0.15 W/m-K.

10.2 Gas Flow and Conductance

This section will describe the equations used to calculate pumping speeds and gas flows. In measuring the flow of a simple liquid like water, a common unit might be the volumetric flow rate, say in gallons per hour. Although volumetric flow rate is sometimes used in describing gas flows, particularly in describing pumping speed, the problem of using the gas flow in this manner is that gases are much more compressible than liquids. To avoid this problem, throughput is used to describe the amount of gas flowing through a system. The mass of a gas in some volume V is

$$G = \rho V \qquad \qquad \textbf{(10.11)}$$

where ρ is the mass density ($m \times n$). Then the mass flow rate is

$$q_m = \frac{dG}{dt} \qquad \qquad \textbf{(10.12)}$$

The throughput of a gas Q, which has units of pressure–volume/time, is given by

$$Q = q_m \frac{P}{\rho} \qquad (10.13)$$

Gas flows are often measured in terms of a standard volume, that is, a volume that an equivalent amount of gas would occupy at 0°C and 1 atm of pressure. For example, one standard liter is the amount of gas that would occupy a liter at 1 atm at 273 K. Since 1 mole of gas occupies 22.4 L at standard conditions, one standard liter is 1/22.4 mole. Alternatively, one standard liter per minute is a throughput of 760 torr-liters per minute. A flow of gas, then, is often measured in standard liters or standard cubic centimeters per minute.

Now consider a simple vacuum system shown in Figure 10.2. Gas flows through the chamber, which is assumed to be at a uniform pressure P_1. A tube connects the chamber to a pump and the inlet of the pump is at a pressure P_2. The conductance C of a vacuum component is given by

$$C = \frac{Q}{P_1 - P_2} \qquad (10.14)$$

The conductances can be calculated for a variety of geometries or they can be found in tables. Like electrical conductances, conductances in parallel simply add, while conductances in series add as inverses:

$$\frac{1}{C_{series}} = \frac{1}{C_1} + \frac{1}{C_2} + \frac{1}{C_3} + \cdots \qquad (10.15)$$

Although detailed conductance calculations are beyond the scope of this book, we note in passing that the conductance of a tube of diameter D and length L in the viscous flow regime is

$$C = 1.8 \times 10^5 \text{ torr}^{-1} \text{ sec}^{-1} \frac{D^4}{L} P_{av} \qquad (10.16)$$

where P_{av} is the average of P_1 and P_2. Elbows, bends, and narrow sections all reduce the conductance of a system. Flowing a large amount of gas through a vacuum system while keeping the chamber pressure close to the pump pressure requires a vacuum system with a very large conductance. Such a system must therefore be designed with large-diameter tubes, and the pump must be placed close to the chamber.

Often it is desired to control the pressure in a vacuum chamber during a process. The volumetric flow rate of a typical pump cannot be independently controlled. Since, for most pumps, the pump speed (volumetric displacement) is nearly constant over a broad range of inlet pressures, the throughput rises with the inlet pressure. The pressure in the chamber can be set by adjusting the flow rate of the gas in the chamber, but this variable is normally reserved for optimizing other process parameters such as uniformity. The pressure in a vacuum chamber can be

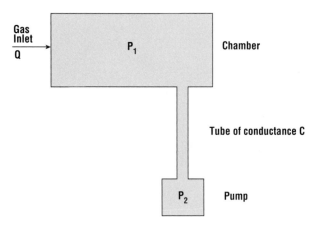

Figure 10.2   A simple vacuum system showing a uniform pressure chamber with inlet flow Q, a vacuum pump, and a tube of conductance C.

Butterfly valve   **Venetian blind valve**

Figure 10.3   Variable conductance valves used in
small- and large-diameter vacuum lines.

easily controlled, however, by inserting a variable conductance valve (Figure 10.3) into the pumping line. This
can be done by adding a simple vane that can be rotated
to partially block the tube, or in large-diameter vacuum
plumbing, a venetian blind arrangement can be used.
A pressure monitor attached to the chamber along with
a feedback loop can be used to maintain the pressure in
the chamber for a broad range of pumping rates and
inlet flows.

Pumps are usually specified in terms of pumping speed S_p

$$S_p = \frac{Q}{P_p} = \frac{dV_p}{dt}$$ **(10.17)**

Example 10.2

A chamber is attached through a line 2 long and 2.5 cm in diameter to a vacuum pump.
The process that we want to run would require a chamber pressure P_{ch} ($=P_1$ in Equation 10.14)
of 1 torr and a throughput of 1 standard liter per minute (slm). Calculate the required pumping
speed in liters per minute, using P_p ($=P_2$ in Equation 10.14) for the pump pressure.

Solution

From Equation 10.14,

$$P_{ch} = P_p + \left(\frac{Q}{C}\right)$$

$$P_{ch} - P_p = \frac{Q}{1.8 \times 10^5 \,\text{torr}^{-1}\,\text{sec}^{-1} \dfrac{D^4}{L} \dfrac{P_{ch} + P_p}{2}}$$

$$P_{ch}^2 - P_p^2 = \frac{Q}{9 \times 10^4 \,\text{torr}^{-1}\,\text{sec}^{-1} \dfrac{D^4}{L}}$$

$$P_{ch}^2 - P_p^2 = \frac{760 \,\text{torr} \;\; 1000 \,\text{cm}^3/\text{min} \times \dfrac{1}{60} \,\text{min/sec}}{9 \times 10^4 \,\text{torr}^{-1}\,\text{sec}^{-1} \,(2.5 \,\text{cm})^4/200 \,\text{cm}}$$

$$P_{ch}^2 - P_p^2 = 0.72 \,\text{torr}^2$$

$$P_p = \sqrt{1 \,\text{torr}^2 - 0.72 \,\text{torr}^2} = 0.53 \,\text{torr}$$

Then

$$S = \frac{Q}{P} = \frac{760 \,\text{torr} \;\; 1 \,\text{L/min}}{0.53 \,\text{torr}}$$

$$= 1440 \,\text{L/min}$$

where P_p is the inlet pump pressure. For example, a pump rated for 1000 L/min (slm), will pump 1000 slm at 1 atm inlet pressure. On the other hand, if the inlet pressure is 0.1 atm and the pumping speed at this pressure remains 1000 L/min, the maximum gas flow that this same pump could accept is 100 slm. Furthermore, the pumping speed generally is not a constant. It depends on the gas being pumped and the inlet pressure. The next section will discuss the pressure ranges commonly used for microelectronic fabrication and the types of pumps used for these systems.

10.3 Pressure Ranges and Vacuum Pumps

Two pressure ranges corresponding to the mean free path of the gas molecules have been previously identified. The division between viscous flow and molecular flow occurs at about 1 mtorr. The equations for the behavior of the gas in these two ranges are entirely different. This division is based on the physics of the molecule–molecule and molecule–wall interactions. As a practical matter, vacuum regions are also often defined by the technology required to achieve them. The divisions are rather flexible, but let us approximately define them as follows:

Rough vacuum	0.1 torr–760 torr	10 Pa–10^5 Pa
Medium vacuum	10^{-4} torr–10^{-1} torr	10^{-2} Pa–10 Pa
High vacuum	10^{-8} torr–10^{-4} torr	10^{-6} Pa–10^{-10} Pa
Ultrahigh vacuum	$<10^{-8}$ torr	$<10^{-10}$ Pa

Most of the processing equipment used in semiconductor fabrication operates in the rough or medium vacuum regime. To ensure a pure chamber, however, they are often pumped into the high or ultrahigh vacuum regime before introducing the process gases. For that reason, we will also discuss the production of high vacuum. The production of ultrahigh vacuum is a difficult task.

Rough vacuum pumps all involve the positive displacement of gas through the mechanical movement of a piston, vane, plunger, or diaphragm. All of these pumps involve three steps: capture of a volume of gas, compression of the captured volume, and gas expulsion. The simplest conceptual picture of such a pump is a piston pump (Figure 10.4). The gas to be pumped is drawn into the cylinder through a valve as the piston is drawn back into the cylinder. During the next part of the cycle both valves are closed and the gas is compressed. Near the end of the stroke the second valve is opened and the gas is expelled to the higher pressure region. Often these valves open automatically in response to a pressure difference. If an ideal gas is used, the pressure differential is just the ratio of the fully expanded to fully compressed volumes. If, for example, the exhaust pressure is 1 atm and the compression ratio is 100:1, the lowest pressure that can be achieved in this simple pump is 0.01 atm (7.6 torr). These steps may be run in a series of stages to obtain higher pressure differentials between inlet and outlet.

This simple piston pump is not widely used for microelectronic processing schematically. Instead, a very common pump for rough and medium vacuums is the rotary vane system shown schematically in Figure 10.5. A metal cylinder attached to an electric motor is rotated in a cylindrical chamber about an axis that is displaced from the chamber center. Spring-loaded vanes slide along the wall of the chamber, sealing off different

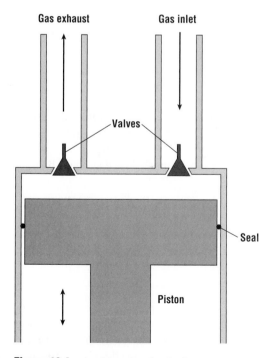

Figure 10.4   A schematic of a single-stage, two-valve piston pump.

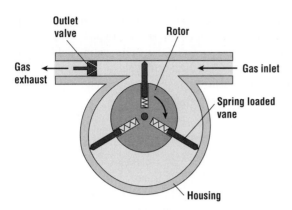

Figure 10.5 One of the most common types of pumps for microelectronic processing is the rotary vane vacuum pump.

areas of the pump. Oil is used to seal the vane and as a lubricant to help with this sliding action. The oil also helps to cool the pump, dissipating the heat generated by friction and by the compression of the gas. As with the piston pump, the rotary vane system works by compressing the gas as the pump rotates. The circular motion of the pump eliminates the need for crankshafts that translate the rotational motion of the motor into the up and down motion required for the piston pump. These pumps are commonly used in both single- and two-stage models, with a variety of throughputs. The ultimate vacuum of the single-stage rotary vane pump is about 20 mtorr, while a two-stage pump can produce a vacuum of less than 1 mtorr.

One of the problems associated with compression-type pumps is the potential for the condensation of vapors. As the gas is compressed, if the partial pressure of a gaseous vapor exceeds the vapor pressure of the corresponding liquid at the gas temperature, it will begin to condense, forming droplets of liquid. These liquids mix with the pump oil and may lead to corrosion. As a simple example, consider water. The vapor pressure of water at room temperature is about 20 torr. If the inlet gas is compressed by a factor of 10^4, water will condense if the partial pressure of water in the chamber is more than 2 mtorr. The problem is most acute when one is pumping corrosive condensibles such as Cl_2 and the chlorosilanes. To avoid this problem, a small flow (or bleed) of an inert gas such as N_2 can be injected into the chamber. The use of these gas ballasts limits the ultimate pressure of the pump. Figure 10.6 shows the pumping characteristics of some typical rotary vane pumps with and without the use of gas ballast. The pumping speed is nearly constant over a broad range of inlet pressures, eventually falling as the inlet pressure approaches the ultimate vacuum. Similarly, the throughput is nearly proportional to the inlet pressure over most of the range.

Example 10.3

Assume that the desired chamber for a particular process is 0.1 torr and that there is no pressure drop between the pump and the chamber. Also assume that a gas flow of 1 slm is needed to obtain suitable process results. Would the D65 pump whose characteristics are shown in Figure 10.6 be acceptable if no ballast is used? What is the maximum flow that could be used with this pump?

Solution

According to Figure 10.6, pumping speed of the D65 pump at 0.1 torr is about 40 cfm (1150 L/min). This would allow a maximum flow rate of about 1150 × 0.1/760 or 0.15 slm. Therefore this pump is not adequate for this application.

There are two ways to obtain higher throughputs: increase the volume pumped on each stroke or increase the rotational speed. The former is expensive. The latter is subject to limitations of heat dissipation in the vanes. This limits rotary vane pumps to about 2000 rpm. Alternatively, one can construct a pump without sliding seals. This allows very high rotation speeds. Compression is acheived by using a narrow clearance between the rotating pieces and the walls. In the simplest case,

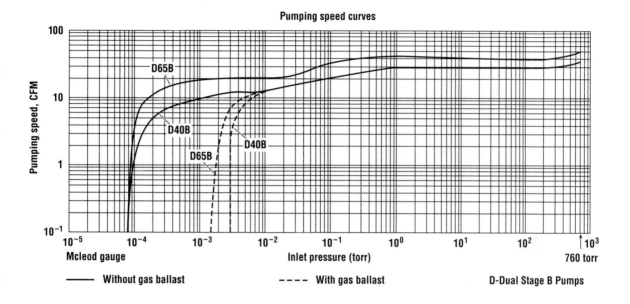

Pumping speed curves

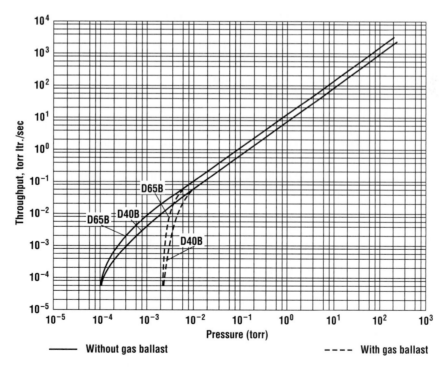

Figure 10.6   The pumping characteristics of Leybold D65® and D40® two-stage rotary vane vacuum pumps with and without gas ballasting *(courtesy Leybold)*.

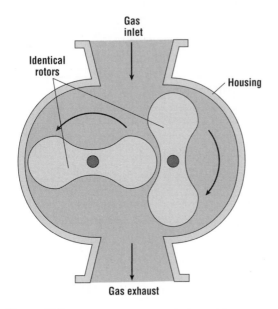

Figure 10.7   Schematic diagram of a Roots blower.

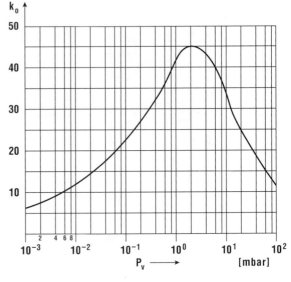

Figure 10.8   Compression ratio as a function of inlet pressure for a typical Roots blower (*courtesy Leybold*).

this can be used as precompressor for a conventional rotary vane pump. Pumps that have been designed for this application are called *blowers*. The most common blower in microelectronic fabrication is the Roots blower. As shown in Figure 10.7, the Roots blower is a positive displacement pump that consists of two figure-eight rotors that revolve at high speed in opposite directions and have very small clearances (<0.1 mm) with respect to each other and the pump walls. There is no mechanical seal between the surfaces. As a result, the compression ratio of these pumps is only about 30:1 and depends on the inlet pressure (Figure 10.8), but due to their rapid rotation the pumping speed is large. Allowances for thermal expansion to prevent scoring that reduces pump life preclude higher compression ratios. Two-stage blowers can also be used for larger compression ratios.

The inlet pressure of the roughing pump can be raised by placing a Roots blower before the rotary vane pump. The increase in the pumping speed of a Roots blower with a zero flow compression ratio of k_o and pumping speed S_R and a rotary vane pump with pumping speed S_{RV} can be calculated using the expression

$$S_{\text{eff}} = S_{RB}S_{RV}\frac{k_o}{S_{RB} + k_oS_{RV}} \tag{10.18}$$

Thus, the combination of a 40-cfm rotary vane pump and a 200-cfm blower with a compression ratio of 20 will provide an effective pumping speed of 160 cfm. Taking the previous example where the chamber pressure was 0.1 torr, the inlet pressure of the roughing pump after the blower now rises to 20×0.1 or 2 torr. According to Figure 10.6, if the blower is backed with a D65 rotary vane pump, this increases the maximum permissible chamber flow rate from 0.1 slm to about 3 slm.

The use of an inert purge or a filter before the roughing pump is also common practice. As the pump operates, the pump oil gets hot. The vapor pressure of pump oil at this temperature is quite high. Unless there is a flow of gas into the pump, the oil vapor flows from the pump into the chamber in a process known as backstreaming. A common example is the use of a roughing pump to evacuate a sealed chamber. As the pressure in the chamber falls, so does the flow of gases into the pump. Eventually the flow drops low enough to allow backstreaming. The pump oil vapor condenses on and contaminates the cold chamber walls. An inert purge flow in front of the pump will reduce backstreaming,

Figure 10.9 Cutaway view of a diffusion pump *(courtesy Varian).*

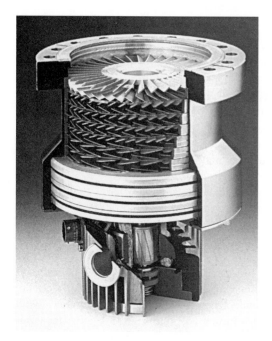

Figure 10.10 Cutaway view of a small turbomolecular pump. Notice the change in the blade angle and shape going from the high vacuum (top) to low vacuum (bottom) ends *(courtesy Varian).*

but will also increase the apparent ultimate pressure of the pump. Chemical traps are also available to solve this problem.

Alternatively one can use an oilless or "dry" pump. This type of pump uses a number of blower like stages in series to pump chambers to millitorr pressures and exhaust the gas to 1 atm. Dry pumps are more expensive and larger than a rotary vane pump of comparable throughput, but their noncontaminating nature make them popular for high purity and for corrosive applications.

High vacuum pumps for microelectronic fabrication fall into two categories: those that pump the gas by transferring momentum to gaseous molecules and those that trap gaseous molecules. Of these, the former are preferred for pumping corrosive and toxic gases, or high flows of gases, and the latter are preferred for pumping small flows of inert gases or when only the high vacuum pump is being used for pumping down the chamber before processing. Valves are almost always used to isolate high vacuum trapping pumps, allowing the chamber to be pumped by medium and rough vacuum pumps when first pumping down from an atmosphere and once the process gas begins to flow.

The two most popular types of momentum transfer pumps are diffusion pumps and turbomolecular pumps. Diffusion pumps (Figure 10.9) are extremely simple and robust. They operate by heating an oil at the bottom of the pump. The pump oil vapors rise through the center stack and are ejected through vents at very high speeds. They then strike cooled walls at the top of the pump, condense, and run down the walls. Gases are primarily pumped by momentum transfer between the vapor stream and the gas molecules. Gas molecules may also be transported by dissolving in the vapor droplets. As the oil is heated again at the bottom of the pump, the gas is given off and removed through a roughing pump, which is connected to the fitting shown at the right of Figure 10.9. It is possible to obtain compression ratios of 10^8 in these pumps.

Diffusion pumps have high pumping speeds as long as the inlet pressure is in the molecular flow regime. Most diffusion pumps may not be exposed to an atmosphere, because a chemical reaction called cracking may occur between the hot pump oil and the oxygen in the air. Even more serious are concerns that some of the pump oil vapor will not condense in the pump and will backstream into the vacuum system, leading to contamination. Diffusion-pumped systems may use baffles or cooled traps to remove most of the backstreamed pump oil. Because of this concern, diffusion pumps are generally not desirable when high purity is required.

A turbomolecular (or turbo) pump as shown in Figure 10.10 has a large number of stages in series. Each stage consists of a fan blade that rotates at extremely high speed (>20,000 rpm) and

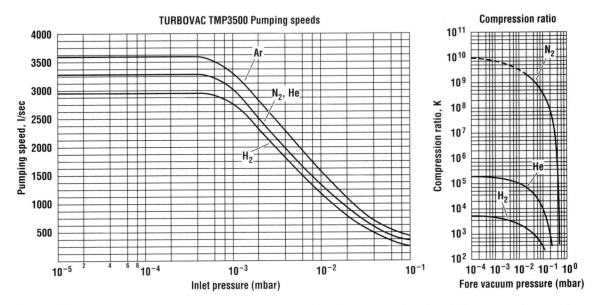

Figure 10.11 Pumping speed and compression ratio of a typical turbo pump as a function of inlet pressure for several representative gases (*courtesy Leybold*).

a stationary set of blades called a *stator*. The spacing between the stator and rotor is of order 1 mm. Each stage may have a modest compression ratio, but because of the large number of stages, the total pump has compression ratios as large as 10^9. The high pressure side of the pump must be attached to a roughing pump, since a low outlet pressure must be maintained. Since the momentum transfer depends on the mass of the gaseous molecule, the compression ratio depends strongly on gas being pumped. A typical pump that has a compression ratio of 10^9 for N_2 will have a compression ratio of less than 10^3 for H_2 (Figure 10.11). As a result, turbopumped chambers can have high concentrations of light gases such as H_2 and He. Because pump oil vapors have high mass, however, turbopumped systems are very clean.

When the ultimate in purity is required, gas entrainment pumps are used. One of the most common in use in microelectronic processing is the cryopump. As shown in Figure 10.12, a cryopump consists of a closed-cycle refrigerator. The cold head of the refrigerator, which is maintained at about 20 K, is contained in a pump body that attaches to the vacuum system. The cold head is generally constructed of copper or silver and may have coatings of active charcoal to further trap gases. A radiation shield is often used to minimize the thermal load on the head. All of the chamber gases except He will condense on the cold head, although gases with high boiling points are pumped more efficiently than those with low boiling points. Eventually, the cold head becomes saturated with adsorbed gases whose low thermal conductivity prevents further adsorption. At this point, the cryopump must be isolated from the chamber, heated and pumped to desorb the collected gases, and then cooled down to return it to service. Cryopumps may not be turned on until the chamber is at a rough vacuum, or their pumping capacity will be greatly diminished.

Another type of gas entrainment pump used for high vacuum is the sorption pump. Sorption pumps operate by chemically or physically absorbing the gas molecules. Early sorption pumps used carbon, but the material in modern pumps is commonly some form of activated Al_2O_3. These alumino-silicate zeolites are permeated by internal cavities interconnected by uniform-diameter

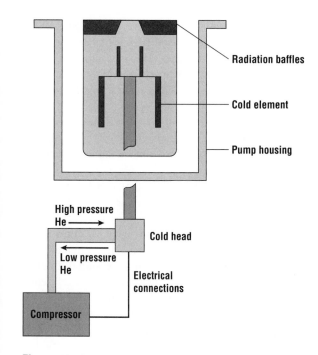

High pressure
He →

Low pressure
He ←

Figure 10.12   Schematic view of a typical cryopump.

pores. The pore size varies with the zeolite that is chosen but is typically between 3 and 15 Å. This type of pump is also called a *molecular sieve* due to the sorbate material. Sorption pumps are often operated by cooling them to liquid nitrogen temperatures and allowing the gas molecules to physically adsorb on the pore walls. The pumping efficiency depends on the size of the gas molecules relative to the size of the pores: N_2, CO_2, H_2O, and heavy hydrocarbons are pumped well by most sorption pumps, but light noble gases such as He are not. As with cryopumps, sorption pumps can be recycled by heating them while under a vacuum.

The third type of entrainment pump uses replenishable coatings of reactive metal to remove gaseous molecules. In titanium sublimation pumps a titanium-containing filament is heated to deposit a thin film of highly reactive metal on the inside surface of the pump. As the inner surfaces become coated with gas molecules, the pumping speed falls. To regenerate the pump, the filament is reheated and a fresh coating of Ti is put down. Sputter ion pumps operate in a very similar manner, but the thin reactive layers are replenished by sputtering (see Chapter 12) instead of sublimation. The inner surface of sputter ion pumps may be shaped to maximize the area being coated. As with sublimation pumps, the most common material deposited to pump the gas is titanium. Both sublimation and sputter ion pumps are extremely clean, robust, and simple to operate.

10.4 Vacuum Seals and Pressure Measurement

In the rough and medium vacuum region, elastomers, commonly in the form of O-rings, are used to seal the vacuum chamber. The O-rings have the advantages of low cost and ease of reuse. A common application is the wafer introduction and exit doors of the vacuum system. The choice of the elastomer depends primarily on the chemistry of the chamber and on the temperature that the elastomer must withstand. One of the most popular elastomer materials is viton, a vulcanized rubber. Some sort of mechanical support must be built into the joint to prevent the O-ring from being pulled into the chamber by the vacuum. One popular choice is to seat the O-ring in a groove. Common practice is to recess the O-ring groove to 70% of the diameter of the O-ring and to make the groove wide enough to prevent the compressed O-ring from filling the groove; 140% of the diameter of the O-ring is often used. This ensures a metal-to-metal contact between the joining surfaces in the sealed condition, reducing any vibration at the seal and thereby improving the reliability. Two common types of O-ring-sealed flanges are shown in Figure 10.13. Because of reliability concerns regarding O-ring seals, highly toxic gases may be contained by a double-O-ring arrangement, where the space between the O-rings is sampled for the presence of the toxic.

Elastomer seals begin to leak at pressures below 10^{-7} torr. For high and ultrahigh vacuum systems, metal-to-metal seals must be used. In this type of seal, the sealing material must be plastically deformed during the sealing process. Early high vacuum systems used a malleable wire such as gold in much the same manner as an elastomer O-ring. One of the most common high vacuum seals in

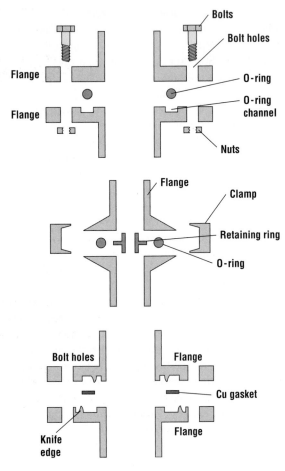

Figure 10.13 Two types of O-ring seals for medium vacuums and the Conflat® flange used for sealing high vacuum systems.

current use, the Conflat® flange, is shown in Figure 10.13. A flat, soft, 2-mm-thick metal ring is pinched between two stainless steel knife edges. The most common gasket material is copper or a copper/silver alloy. A uniform closing force must be applied to ensure a tight seal. This is usually provided by closely spaced bolts around the edge of the flange. To maintain high reliability, the sealing gaskets must be replaced after every use, and the knife edges must be protected against damage when the surface is exposed.

Pressure in a vacuum system can be measured with a number of different gauges, the most common of which in microelectronic applications are capacitance manometers, thermal conductivity gauges, and ionization gauges. Capacitance manometers fall into a echanical class of gauges that depend on a pressure difference between the chamber to be measured and a reference volume, to cause a mechanical deflection. Capacitance manometers detect the movement of a thin metal diaphragm. Although these gauges can be used to detect pressures as low as 1 mtorr, they are most often used at pressures above 1 torr. They are noncontaminating, and since they measure the pressure of the gas directly, they are independent of the gas being measured.

A family of popular vacuum gauges operates by measuring the thermal conductivity of the gas and from that, inferring a pressure. This type of gauge dates back to 1906 [5, 6]. They operate by passing a current through a wire and measuring the temperature of the wire. Wire-operating temperatures are kept low (a few hundred degrees centigrade) to prolong wire life, maximize sensitivity, and ensure that most of the heat transfer is thermal conduction through the gas rather than radiative heat loss. In Pirani gauges, the resistance of the wire is carefully measured by a Wheatstone bridge. The power to the filament is adjusted to balance the bridge and therefore maintain a constant wire temperature. In thermocouple gauges, a thermocouple is spot-welded to the heater wire to measure the wire temperature. Thermocouples consist of two dissimilar metals, and they produce a small temperature-dependent voltage. The measurement of this voltage is easier than balancing a Wheatstone bridge, but the technique has less sensitivity than Pirani gauges. Either type of gauge has a resolution of about 1 mtorr. Care must be taken in using this type of gauge, however, in that the thermal conductivity depends on the gas in the chamber.

Neither mechanical deflection nor thermal conductivity gauges are useful in measuring pressures much below 1 mtorr. The types of gauges most commonly used in high and ultrahigh vacuum applications are ionization gauges. They operate by using an electron stream to ionize the gas in the gauge and an electrical field to collect the ions. The ion current produced this way is a function of the pressure in the chamber. The electron stream can be created by heating a filament (hot cathode) or by a plasma (cold cathode). Of these, hot cathode gauges are much simpler to operate, but may present problems in highly corrosive ambients or in ultraclean chambers where filament outgassing is a concern. Furthermore, hot filament gauges usually cannot be operated at pressures above 10^{-4} torr due to short filament life. The ultimate pressure for either type of gauge is ultimately limited by the ability to measure very small ion currents, but some types of ionization gauges can detect pressures as low as 10^{-12} torr.

10.5 The DC Glow Discharge°

Many of the low pressure processes that are described in the next two chapters will involve the use of a glow discharge or plasma. These include various etch processes, chemical vapor deposition, and sputtering. Even the bulbs in many lithography tools are plasma sources. The term "glow discharge" refers to the light given off by the plasma. Plasmas can be used in place of high temperature to crack molecules and so drive some reaction chemistry. Plasma may also be used to create and accelerate ions. In some applications both effects may be important.

A plasma is a partially ionized gas. Assume that the inlet flow of gas contains molecule AB made from atoms A and B. The types of processes that may occur in the glow discharge can be characterized as [7]

Dissociation	$e^* + AB \rightleftarrows A + B + e$
Atomic ionization	$e^* + A \rightleftarrows A^+ + e + e$
Molecular ionization	$e^* + AB \rightleftarrows AB^+ + e + e$
Atomic excitation	$e^* + A \rightleftarrows A^* + e$
Molecular excitation	$e^* + AB \rightleftarrows AB^* + e$

where the superscript "*" refers to a species whose energy is much larger than the ground state. Dissociated atoms or molecular fragments are called *radicals*. Radicals have an incomplete bonding state and are extremely reactive. In Ar or other elemental plasmas, there are no radicals. Ions are charged atoms or molecules such as A^+ and AB^+. They may have more than one positive charge or may even be negatively charged. In simple capacitive discharge plasmas, the radicals may constitute 1% of the total plasma; charged species may be less than 0.01%. Although not specifically represented, energetic ionized species are possible as well.

Figure 10.14 shows a simple plasma reactor. Two parallel plates are contained in a vacuum system and attached to a dc supply through vacuum power feedthroughs. A typical pressure for a plasma process is 1 torr. A high voltage source (often a charged capacitor) is connected momentarily to the circuit to start the plasma. The inductor protects the dc supply from the high voltage arc. At 1 torr the required voltage for a 10-cm electrode spacing is about 800 V, while about 500 V is required for a 5-cm spacing.

Until the arc is struck, the gas will not conduct current since it acts as an insulator. Consider what happens as the plasma forms. If the voltage is high enough, the field in the reactor will exceed the breakdown field of the gas, and a high voltage arc will flash between the two electrodes. This arc will create a large number of ions and free electrons. Because of the electric field in the chamber, the electrons will be accelerated toward the positively charged anode, and the ions will be accelerated toward the negatively charged cathode. Due to their small mass, the electrons

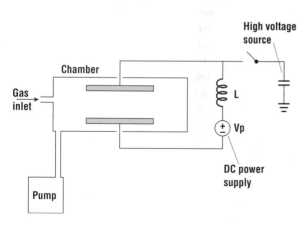

Figure 10.14   A simple parallel-plate plasma reactor.

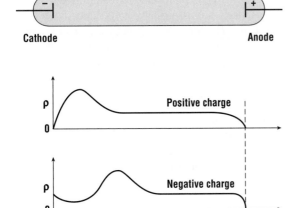

Figure 10.15 Positive and negative charge densities and electric field as a function of position in the plasma.

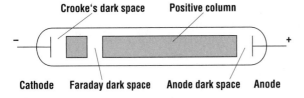

Figure 10.16   Structure of a dc plasma.

will be accelerated much more rapidly than the slowly moving ions. The ions travel across the tube and eventually strike the cathode. When they do so, they release a cloud of secondary electrons from the material in the cathode. These electrons are accelerated back toward the anode. If the voltage across the electrodes is large enough, when these high energy electrons collide inelastically with neutral atoms they create more ions. This process of secondary electron release and ion creation sustains the plasma.

Even in this simple 1-D case, the solution for the movement of ions in the plasma is beyond the scope of this text, since the differential equations describing the field and charge distribution are coupled and must be solved simultaneously. Figure 10.15 shows typical positive and negative charge densities and the electric field strength as functions of position in the plasma. In the bulk of the plasma, the densities of ions and electrons are equal. Since the electrons are rapidly accelerated from the cathode, the electron density near the cathode is much less than the ion density, and the region has a net positive charge. Near the edge of this positively charged region, the electrons have gained enough energy to create ions, and so the ion density increases with distance into the plasma. This positive charge shields the remainder of the plasma from the cathode, reducing the field and therefore the ionization rate. As a result, the ion density peaks and then falls to a constant value through the remainder of the plasma.

When a moderate energy electron scatters inelastically off of a neutral atom, it may excite a core level electron to a high energy state. This state is very short-lived (of order 10^{-11} sec), and so the excited atom or molecule does not move an appreciable distance between excitation and decay. When the electron decays back to its ground state, it gives off the energy in the form of visible radiation. The light from the glow discharge arises because of this optical emission process. For this process to occur, a high concentration of moderate energy electrons is required. Electrons with energies greater than 15 eV primarily ionize the gas molecules rather than excite them.

The requirement for moderate energy electrons precludes optical emission near each of the electrodes. These regions are called *dark spaces*. The region just above the cathode where most of the electrons have very low energies is called *Crooke's dark space* (Figure 10.16). The anode is a sink for electrons, and so the electron density just above the anode is too small for appreciable emission. This dark space is called the *anode dark space*. Finally, there is a region above the cathode where the electrons have been accelerated to very high energies, leading to ionization, and there are few electrons with energies appropriate for emission. This region is called the *Faraday dark space*.

One of the most important aspects of the glow discharge for fabrication purposes is the large electric field in Crooke's dark space. Ions that drift and diffuse to the edge of this region are accelerated rapidly toward the cathode. If the cathode is covered with wafers or other materials of interest, we can use this ion bombardment to drive various processes. The width of the dark space depends on the chamber pressure. At low pressures, the mean free path of electrons increases, and so the width of the dark space increases. By controlling the chamber pressure, one can control the energy with which ions strike the surface. This effect limits dc plasmas to pressures greater than about 1 mtorr.

10.6 RF Discharges

In many cases, the material on one or more of the electrodes is insulating. Take as an example an application that will be covered in the next chapter, the etching of silicon dioxide. In this process, wafers with an upper layer of SiO_2 are first patterned lithographically and then placed on a plasma electrode. The exposed materials on the top of the wafers, photoresist and oxide, are both insulators. As ions strike the surface of the wafer and secondary electrons are ejected, these layers become charged. The charge accumulates on the surface and the field is reduced until the plasma is ultimately extinguished. To solve this problem, the plasma can be driven by an ac signal. Sources are in the radio frequency (RF) range, commonly 13.56 MHz, which in the United States has been set aside by the Federal Communications Commission for this application. Figure 10.17 shows a typical RF plasma system. A tuning network is used to match the impedance between the plasma and the power source. A blocking capacitor is used to dc isolate the supply from the chamber.

At low frequencies, the plasma follows the excitation, and the width of the dark spaces pulse with the applied signal. When the rate of excitation is greater than 10 kHz, the slow ions in the plasma cannot follow the voltage change. Electrons, however, are rapidly accelerated. During alternate half-cycles, electrons strike the surface of each electrode, giving both a net negative charge with respect to the plasma. In this case, one dark space exists in the vicinity of each electrode. Figure 10.18 shows the dc voltage as a function of position across the chamber. Superimposed on this dc level is the RF signal. Since the plasma is conductive, the voltage drop across the glow discharge is small. Due to electron depletion, however, large dc voltage drops exist between the plasma and the electrodes.

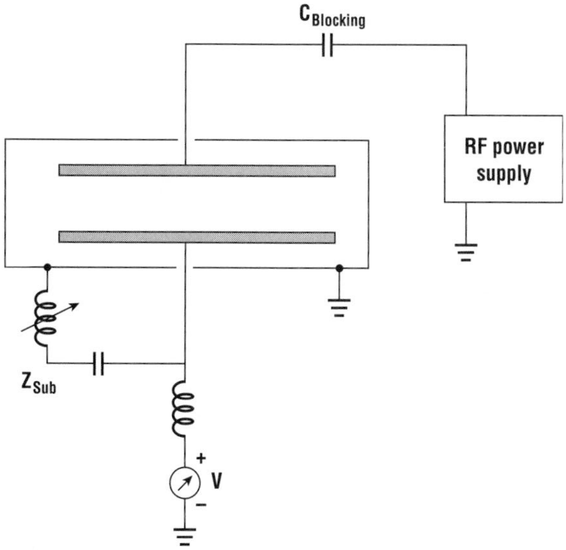

We can label the voltages as

$$V_1 \equiv V_{plasma} - V_{top}$$
$$V_2 \equiv V_{plasma} - V_{bottom} \qquad \textbf{(10.19)}$$

If the electrodes have the same area, by symmetry they must have the same potential difference. By conserving the current through the plasma, it can be shown [8] that for an asymmetric chamber

$$\frac{V_1}{V_2} \approx \left[\frac{A_2}{A_1}\right]^4 \qquad \textbf{(10.20)}$$

where A_1 and A_2 are the areas of the two electrodes. To maximize the voltage difference between the plasma and the lower electrode and therefore the ion bombardment energy on the lower electrode, it is desirable to

Figure 10.17   Schematic of an RF plasma system.

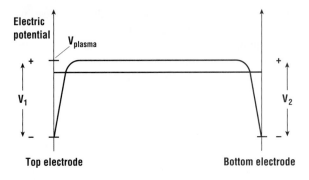

Figure 10.18   Typical plot of dc voltage as a function of position in an RF plasma.

increase the area of the upper electrode. This can be done by connecting the upper electrode to the walls of the chamber. For safety reasons this point is typically set to dc ground. Now the effective area of the upper electrode is the sum of the areas of the actual electrode and the chamber walls. In practice it is found that the exponent in Equation 10.20 is not a constant, but decreases with increasing area ratio [9]. It is difficult to obtain a voltage ratio much larger than 10:1. It is also possible, however, to adjust the dc bias voltage on the grounded chamber and the lower electrode by changing the value of the inductor in the *LC* circuit shown in Figure 10.17 [10].

Example 10.4

Consider the plasma etch chamber shown in Figure 10.19. It consists of a single powered platen, on which a wafer may be placed for etching. The platen is a 200-mm-diameter disc, which sits in the center of a 350-mm-diameter chamber. The chamber height is 150 mm. The chamber itself serves as the second electrode and is grounded. The chamber pressure is 10 mtorr.

(a) If the plasma is at +0.1 V, what is the dc voltage on the electrode?

(b) Would you expect a substantial amount of ion bombardment of the wafers?

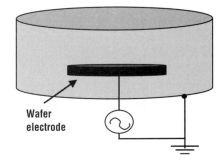

Wafer electrode

Figure 10.19

Solution

(a) The voltage can be found from Equation 10.20 after the areas have been calculated. The area inside the chamber is

$$A_1 = 2\pi(17.5 \text{ cm})^2 + 2\pi \times 17.5 \text{ cm} \times 15 \text{ cm}$$

$$= 3572 \text{ cm}^2$$

The electrode area is

$$A_2 = 2\pi(10 \text{ cm})^2 = 628 \text{ cm}^2$$

Then

$$\frac{V_2}{V_1} = \left(\frac{3572}{628}\right)^4 = 1047$$

If $V_{ch} = 0$ V, then $V_{Plasma} = 0.1$ V and

$$V_{Electrode} = -1047 \times 0.1 + 0.1 = -104.6 \text{ V}$$

(b) The wafers are on the smaller powered electrode with a large bias voltage, so one would expect extensive ion bombardment.

10.7 High Density Plasmas

As already mentioned, the concentrations of ions and radicals in a simple capacitive discharge plasma are a small fraction of the total gas. For many years, this type of system, commonly operated at 13.56 MHz, was the dominant plasma source for microelectronic processing. Since it is the ion bombardment and/or chemical reactions that will be of interest for driving the process, it would be desirable to find ways to increase their relative concentrations. In addition to increasing process throughput, other plasma parameters could be traded off to improve specific plasma properties. A variety of techniques have been developed for forming these enriched plasmas, including inductively coupled plasmas, magnetron plasmas, and electron cyclotron resonance plasmas. We will refer to them generically as high density plasmas (HDP). Very simple HDP systems have been used in sputtering applications for many years. These techniques spread to etchers in the late 1980s and early 1990s and are now being used very successfully for chemical vapor deposition.

    If a magnetic field is added to a plasma, the Lorentz force will deflect the motion of the electrons in a direction perpendicular to both the velocity and the magnetic field

$$F = q\overline{v} \times \overline{B} \tag{10.21}$$

If $|\overline{v}|$ is constant, such a field will induce a circular motion with a radius

$$r = mv/qB \tag{10.22}$$

Because of their large mass, ions will move through the field with only minor deflections, crossing the discharge long before traveling a full circle, while electrons will follow a helical path of much smaller radius. Effectively, then, the path of the electrons is many times larger in the presence of the field. For a fixed mean free path, the opportunity for impact ionization is much greater with the field on, and so the ion density and free radical density in such a system will be large. In one configuration called a magnetron, which is commonly used for sputtering, the secondary electrons being accelerated away from the cathode can actually be trapped by the magnetic field and return to the cathode, where the

Example 10.5

If electrons are ejected from a target at 30 eV, what magnetic field is needed to create a circular path of radius 5 mm?

Solution

From Equation 10.22

$$r = \frac{mv}{qB} = \frac{\sqrt{2mE}}{qB}$$

Solving,

$$B = \frac{[2(9.1 \times 10^{-31}\text{ kg}) \times 30\text{ eV} \times 1.6 \times 10^{-19}\text{ J/eV}}{1.6 \times 10^{-19}\text{C} \times 5 \times 10^{-3}\text{ m}}$$

$$= 3.7\text{ mT} = 0.37\text{ G}$$

Such a field is easily obtained.

Glow discharge

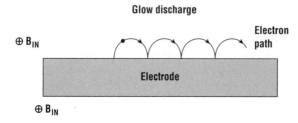

Figure 10.20   In a simple magnetically confined plasma, electrons ejected from the cathode are confined by the Lorentz force to stay in the cathode dark space.

cycle will repeat (Figure 10.20). Because most of the electrons remained trapped in this region, the ion density becomes quite large and ion bombardment of the cathode increases dramatically.

More recently, new configurations of these HDP reactors have been developed and applied to etching, chemical vapor deposition, and physical vapor deposition. The term "high density plasmas" refers to processes with ion concentrations in excess of 10^{11} cm^{-3}. Generically, all of these sources use transverse electric and magnetic fields to increase the distance that electrons travel in the plasma, allowing frequent collisions between electrons and atoms to boost the radical and ion densities in the plasma. In addition, HDP systems allow one to operate the plasma reactor at much lower pressures (typically below 10 mtorr) and lower dc biases (a few tens of volts). HDP sources include helicon plasma sources, inductively coupled plasma sources, and planar coil sources that all operate at traditional plasma frequencies. Another class of high frequency sources exists that operates at frequencies of more than 1 GHz. Of these we will review only the most well known, the electron cyclotron resonance (ECR) source. For further information on these and other HDP sources, the reader is referred to Popov [11].

An ECR plasma uses a perpendicular magnetic field along with an alternating electric field (Figure 10.21). The electric field increases the magnitude of the electron's velocity, and the magnetic field changes the direction of the velocity vector. Consider, for example, an electron on the surface of this page that is initially at rest. An electric field of $E_o \cos \omega t$, which is directed left to right at time $t = 0$, is switched on. The field will cause the electron to accelerate to the left, then slow down and reverse direction, as it oscillates back and forth. Now switch on a magnetic field that is directed into the page. As the electron moves to the left, the magnetic field will deflect the electron toward the top of the page. If the frequency of oscillation is set to the electron cyclotron resonance frequency

$$\omega = \omega_o = \frac{eB}{m} \qquad (10.23)$$

the amount of deflection caused by the magnetic field is just enough to turn the electron by 180° as the direction of the field changes sign. As a result, the electron will move in a circle. This condition is called electron resonance.

The electron in resonance gains energy throughout the circle if the mean free path is much larger than the circumference of the circle. This effect increases the power coupling efficiency and reduces the breakdown field [11] to as little as 10 V/cm. As with other magnet-ically enhanced plasmas, the use of an ECR dramatically increases the density of ions and free radicals in the discharge. If the process to be run depends on the ion flux from the plasma, an ECR will improve the rate at which the process occurs. For reasonable fields, the required cyclotron frequency is large. As an example, for a field of 900 G (0.09 T), ω is of order

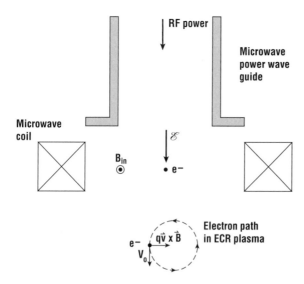

Figure 10.21   In an ECR plasma, an alternating field causes the electrons to move in circular orbits, dramatically increasing the ion density.

1.6×10^{10} rad/sec, or about 2.5 GHz. An ECR system, therefore, requires the generation of high power microwave signals that are coupled into the reactor. The microwave energy is typically carried by rectangular waveguides. The power may be transmitted into the chamber through a quartz window [11]. A limitation of ECR plasma is the scalability of the power supply. Generally supplies with outputs greater than 6 kW are not available. Such supplies would be highly desirable for 300-mm wafer systems.

Many other HDP systems make use of the fact that one can induce magnetic and/or electric fields in the reactor. One can see from Faraday's equation that an alternating electric field leads to an ac current, which in turn induces a magnetic field. Similarly, a changing magnetic field induces an electric field. This can be seen most easily in a planar coil source, as shown in Figure 10.22, when an RF current is applied to a coil above a chamber. Current flowing through the coil induces an alternating B field in the chamber. The coil may be inside the chamber [12] or it may be isolated from the discharge by a dielectric window [13]. The efficiency of the system decreases with the distance between the substrate and the plasma generation region. As a result, chambers using planar HDP sources tend to be rather squat. In this type of system the ion density is linearly proportional to the RF power.

The voltage of the coil is first raised with respect to the substrate until a plasma is established. Damage effects are serious concerns during this initial strike due to the high potential. Once the plasma is struck, the essential features of the coil are that it carries an RF current and produces a field that is nearly axially symmetric. An outer shield may be used to isolate the large RF fields from surrounding equipment. An inner sheath also may or may not be used. If an inner shield is used, slots are cut in the shield along the length of the coil. As with any conductor, the RF field will penetrate some distance into the plasma. This distance is called the skin depth. The skin depth is inversely proportional to the excitation frequency. The skin depth is also inversely proportional to the conductivity of the plasma. Thus the higher the power, the higher the ion concentration, and the smaller the skin depth. At high pressures (~1 torr), a current that opposes the coil current flows through a visible

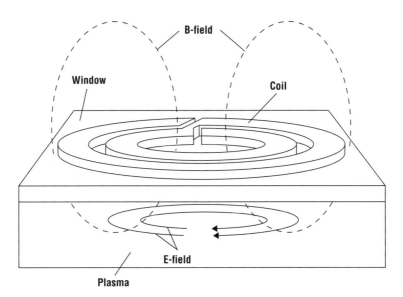

Figure 10.22   Schematic of ICP system showing the RF coil, the dielectric window, and the magnetic and electric fields *(From Hopwood [13], reprinted with permission from Plasma Sources Science and Technology. Copyright 1992, IOP Publishing).*

plasma ring. The field falls sharply inside the ring. Since the current is RF rather than dc, the current flowing through the plasma induces an RF magnetic field inside the reactor. If a shield is used, slits can be cut in the shield to allow the field to penetrate only in certain regions. When no shield is used, the RF coil also acts as a capacitive element, On alternate half-cycles, the coil and the substrate attract electrons, leading to electron bombardment of the substrate and chamber walls. This leads to a net negative charge on all surfaces and therefore a dc field.

Helical resonators operate with an antenna designed to have an electrical length of $\lambda/4 + n \cdot \lambda/2$ or $\lambda/2 + n \cdot \lambda/2$, where n is 0, 1, 2, etc., and λ is the wavelength of the excitation frequency, for quarter- and half-wavelength resonators, respectively. The antenna launches a helical electromagnetic wave and a plasma inside a dc magnetic field. Typical dc fields for helicon sources are 275 G, considerably lower than ECR fields [14]. Although various plasma modes can be excited, the mode of greatest interest for IC processing is the high density or H-mode. For example, a large-area H-mode helicon source can be created with a 13.56-MHz RF supply and a 24-cm helical antenna. This antenna consists of a pair of spiral lines that is rotated 180° in the azimuthal direction and terminated in two rings, producing a helix [15]. Although not yet in widespread use, helicon sources may be used in a variety of etching and plasma deposition processes for large wafers.

10.8 Summary

This chapter discussed some science and technology that will form the basis of many of the subsequent chapters. Vacuums can be well described by kinetic theory that treats gas molecules as hard spheres. Equations have been developed to describe macroscopic quantities in terms of microscopic variables such as velocity distributions. Rough and medium vacuums typically use rotary vane pumps, O-ring seals, and thermal conductivity gauges. When high flow throughputs are required, a Roots blower may be used before the rotary vane pump. For high vacuum systems, either momentum transfer (turbo or diffusion) or entrainment (cryo, sorption, or sublimation pumps) is used. These systems use metal-to-metal seals, and pressure is measured with ionization gauges. The second half of the chapter described plasmas or glow discharges. The basic dc plasma reactor can be used only when both electrodes are coated with conducting materials. When insulating layers are exposed, an ac source must be used. The voltage difference between the electrode and the plasma depends on the areas of the electrodes. High-density plasmas use both electric and magnetic fields. They produce high concentrations of radicals and low-energy ions.

Problems

1. Air pressure decreases as altitude increases. At 18,000 ft above mean sea level, the atmospheric pressure is about half of what it is at sea level (i.e., 760 torr). The temperature also decreases by about 70°C. Does this change the speed distribution of the air molecules? Does it change the mean free path? How much does each change?

2. If the transition between the molecular flow and viscous flow regimes occurs when the mean free path is 1 cm, at what pressure will this occur? (Assume $d = 3$ Å and room temperature.)

	Measured k_{th} at 0°C
Hydrogen	0.172
Nitrogen	0.024

3. The accompanying table summarizes the measured thermal conductivities (k_{th}) of several gases at 0°C and 1 atm pressure in units of joules per degree centigrade-meter-second (J/°C-m-sec). Table 10.1 provieds an equation for the thermal conductivity according to the kinetic theory of gases. Assuming that $C_v = 5/2\, n\, k$ for N_2 and H_2, (n is the number density of the gas; that is, $n = P/kT$ from the ideal gas law). If the diameters of the molecules/atoms are $d = 1.5$ Å for H_2, and 2.0 Å for N_2, calculate the theoretical thermal conductivities and compare them with the measured values. Would you expect the thermal conductivity to increase, decrease, or be constant if we increased the gas temperature? What about the gas pressure?

4. For a particular process to work well, the chamber pressure must be low enough for the mean free path of the neutral gas molecules to be at least 5 mm (~ the dark space width). If the mean free path is 3 cm or more (~ half the electrode spacing), the plasma is extinguished. Assume that the primary gas in the reactor is CF_4 and that the average molecular temperature is 200°C. You can take the diameter of the CF_4 to be 3 Å. Find the pressure range in millitorr over which this process will work. (*Hint*: Molecular masses are on the periodic table on the final page of your book.)

5. Thermocouple gauges measure the thermal conductivity of a gas to infer a temperature. Derive a relationship between pressure and thermal conductivity assuming a kinetic model for the gas. H_2 and He have very large thermal conductivities. Explain why.

6. For a certain process to be successful, it must be run at a chamber pressure of 1 torr, with a gas flow of 1 standard liter per minute (slm). Find the pumping speed necessary (in liters per minute) to do this if
   (a) The pump is directly attached to the chamber;
   (b) The pump is attached with a pipe that is 2 cm in diameter and 100 cm long.

7. Water in a pan is heated to the boiling point (100°C), where the water vapor pressure is 1 atmosphere. If the system maintains the equilibrium vapor pressure,
   (a) Find the flux of water molecules in reciprocal centimeter squared-seconds coming out of the pan.
   (b) Find the mean free path of water molecules under this condition in centimeters. Assume that the diameter of the water molecule is 2×10^{-10} m , there is no other molecule in the atmosphere except water, and $T = 100$°C.

8. To avoid vibration and improve serviceability, vacuum pumps must sometimes be located on a different surface than the process equipment. In some facilities, this is done by drilling holes through the floor and putting the pumps below the clean room on a separate floor. Discuss the drawback of this approach. If the distance between the pump and the system is 3.0 m, the tubing diameter is 5 cm, and the total gas flow is 1.0 slm, what is the minimum chamber pressure that could possibly be achieved?

9. If a D65 pump is used for the application in the previous problem and no gas ballast is used, estimate the pressures at the pump inlet and in the vacuum chamber.

10. A process pump has a pumping speed (S) of 2000 L/min, independent of inlet pressure. This pump is connected to a process chamber through 10 m of tubing that is 5 cm in diameter. If the desired chamber pressure is 1.0 torr, calculate the maximum gas throughput that can be introduced into the chamber in standard liters per minute. (*Hint*: Recall that $Q = P \cdot S$.)

11. The D65B pump shown in Figure 10.6 is used without ballast. It is connected to a process chamber through a 2-m-long straight pipe with a diameter of 5 cm. The process is run in the viscous flow regime and used 2.5 slm of gas.
   (a) For the conditions described, what will the chamber pressure be?

(b)   If you wanted to be able to easily control the pressure in this system for a variety of flows, how would you change the simple vacuum system?

12.   Explain why thermal conductivity type gauges will not work in an ultrahigh vacuum.

13.   Why are entrainment pumps generally not used in processing with toxic gases?

14.   It is possible to determine the chemical composition of a plasma by measuring the optical emission spectrum (i.e., optical intensity as a function of wavelength). Explain why.

15.   Referring to the material discussed in Section 10.7, if a process relies on ion bombardment of the wafers, would you want to put the wafers on the electrode connected to the chamber walls or on the electrode that is isolated from the chamber walls?

16.   Would you expect that adding magnetic confinement to a plasma would increase the x-ray emission? Justify your answer.

17.   For a certain process, the maximum allowable pressure is 0.20 torr. If a D65 rotary vane pump is used and conductance losses in the vacuum system can be ignored; what is the maximum permissible inlet flow in standard cubic centimeters per minute?

18.   Repeat the preceding Problem assuming that a blower with $k_o = 20$ and $s = 200$ cfm is inserted before the rotary vane pump.

References

1.   *Vacuum Technology: Its Foundations, Formulae and Tables*, Leybold AG, Export, PA, 1992.

2.   A. Roth, *Vacuum Sealing Techniques*, American Institute of Physics, New York, 1994.

3.   T. A. Delchar, *Vacuum Physics and Techniques*, Chapman and Hall, London, 1993.

4.   M. H. Hablanian, *High Vacuum Techniques: A Practical Guide*, m. Dekker, New York, 1990.

5.   M. Pirani, *Dtsch Phys. Ges. Verk.* **8**:686 (1906).

6.   W. Voege, *Phys. Z* **7**:498 (1906).

7.   B. Chapman, *Glow. Discharge Processes*, Wiley, New York, 1980.

8.   H. R. Koenig and L. I. Maissel, "Application of R. F. Discharges to Sputtering," *IBM J. Res. Dev.* **14**:168 (1970).

9.   C. M. Horwitz, "RF Sputtering-Voltage Division Between the Two Electrodes," *J. Vacuum Sci. Technol. A* **1**:60 (1983).

10.   J. S. Logan, "Control of RF Sputtered Films Properties Through Substrate Tuning," *IBM J. Res. Dev.* **14**:172 (1970).

11.   O. A. Popov, *High Density Plasma Sources: Design, Physics and Performance*, Noyes, Park Ridge, NJ, 1995.

12.   M. Yamashita, *J. Vacuum Sci. Technol. A* **7**:151 (1989).

13.   J. Hopwood, "Review of Inductively Compled Plasmas for Plasma Processing," *Plasma Sources Sci, Technol.* **1**:109 (1992).

14.   D. L. Flamm, D. E. Ibbotson, and W. L. Johnson, U.S. Patent 4, 918, 031 (1990).

15.   K. Suzuki, H. Sugai, K. Nakamura, T. H. Ahn, and M. Nagatsu, "Control of High-density Plasma Sources for CVD and Etching," *Vacuum* **48**(7–9):659 (1997).

Chapter 11

Etching

After a photoresist image has been formed on the surface of a wafer, the next process often involves transferring that image into a layer under the resist by etching. The chapter will begin with simple wet chemical etching processes where the wafer is immersed in a solution that reacts with the exposed film to form soluble by-products. Ideally, the photoresist mask is highly resistant to attack by the etching solution. Although still used for noncritical processes, wet chemical etching is difficult to control, is prone to high defect levels due to solution particulate contamination, cannot be used for small features, and produces large volumes of chemical waste. The chapter will therefore go on to discuss dry or plasma etch processes.

It is useful to begin the discussion of etching by identifying the appropriate figures of merit. The primary one is the etch rate, which has dimensions of thickness per unit time. A high etch rate is generally desirable in a manufacturing environment. Too high an etch rate, however, may render a process difficult to control. Common desired etch rates are hundreds or thousands of angstroms per minute. When a batch of wafers is etched simultaneously, the etch rate can be less than the rate required for a single-wafer etch process. Several related figures are equally important. Etch rate uniformity is measured in terms of percentage variation of the etch rate. It is quoted across a wafer and from wafer to wafer. Selectivity is the ratio of the etch rates of various materials, such as the photoresist or the underlying layer, referenced to the etch rate of the film being patterned. A particular process may be quoted as having a selectivity of 20 to 1 for polysilicon over oxide, meaning that polysilicon etches 20 times faster than oxide.

Undercut is the lateral extent of the etch under the photoresist mask. It can be quoted in two forms. The first is an undercut distance per side. For example, a particular etching process may be found to produce 0.8-μm lines when the patterned photoresist line is 1.0 μm wide. The process bias is 0.1 μm per side. As shown in Figure 11.1, the etched sidewalls are not always vertical. The amount of undercut therefore depends on how it is measured. Most electrical measurements of the line are sensitive to its cross-sectional area, providing an average undercut. If the etch process attacks the resist pattern, this attack will also produce an etch bias. Some processes are designed to intentionally do this. For now, however, assume that the resist mask does not change during the etching.

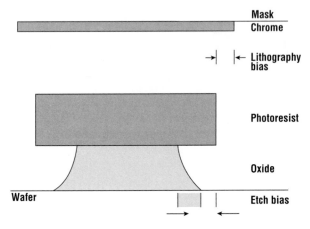

Figure 11.1 Typical isotropic etch process showing the etch bias.

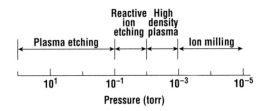

Figure 11.2 Types of etch processes on a chamber pressure scale.

The second method of describing undercut is to quote the etch rate anisotropy. Anisotropy is given by

$$A = 1 - \frac{R_L}{R_V} \qquad (11.1)$$

where R_L and R_V are the lateral and vertical etch rates. A process is said to be perfectly anisotropic ($A = 1$) if the lateral etch rate is zero. On the other hand, $A = 0$ would imply that the lateral and vertical etch rates are identical.

All of the preceding metrics can be expressed quantitatively. There are several other metrics that are somewhat more difficult to quantify. The first is the substrate damage. As an example, p–n junctions have been shown to degrade under certain types of plasma etching. The amount of degradation depends not only on the etch process, but on the depth and type of the junction. Finally, the process must be safe, both to the operator and to the environment. Chlorine plasmas, in particular, are known to produce hazardous by-products that must be neutralized before they are exhausted into the air. Many plasma etch processes are based on Freon chemistries that are known to degrade the environment. Replacing these etches is a continuing area of research.

Etching can be done by physical damage, chemical attack, or some combination of the two. One can define a scale of the etch mechanism that describes the range of etch processes (Figure 11.2). Ion milling, for example, is done with an energetic beam of inert atoms in a very low pressure chamber, such that the length of the mean free path of ions is much greater than the chamber diameter. It is at the extreme of purely physical etching. These processes are characterized by a high degree of anisotropy, but the etch rate is nearly independent of the substrate material. The selectivity of ion beam milling, therefore, is close to 1. Wet etching is at the other extreme of the spectrum with no physical attack. This type of process is usually characterized by low anisotropy, but it may have high selectivity.

11.1 Wet Etching

Wet etching is a purely chemical process that can have serious drawbacks: a lack of anisotropy, poor process control, and excessive particle contamination. However, wet etching can be highly selective and often does not damage the substrate. As a result, although the process is much less popular than it once was, it continues to be used for a wide range of "noncritical" tasks. Comprehensive reviews of wet etching can be found elsewhere [1].

Since the reactive species is normally present in the etchant solution, wet chemical etching consists of three processes: movement of the etchant species to the surface of the wafer, a chemical reaction with the exposed film that produces soluble by-products, and movement of the reaction products away from the surface of the wafer. Since all three steps must occur, the slowest one, called the *rate-*

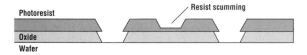

Figure 11.3 Resist scumming occurs when the photoresist is incompletely developed. The residual resist may serve as an etch mask to prevent a complete etch process.

limiting step, determines the etch rate. Since it is generally desirable to have a large, uniform, well-controlled etch rate, the wet etch solution is often agitated in some manner to assist in the movement of etchant to the surface and the removal of the etch product. Some wet etch processes use a continuous acid spray to ensure a fresh supply of etchant, but this comes at the cost of the production of significant amounts of chemical waste.

For most wet etch processes, the film to be etched is not directly soluble in the etchant solution. It is usually necessary to use a chemical reaction to change the material to be etched from a solid to a liquid or a gas. If the etching process produces a gas, this gas can form bubbles that can prevent the movement of fresh etchant to the surface. This is an extremely serious problem, since the occurrence of the bubbles cannot be predicted. The problem is most pronounced near pattern edges. In addition to assisting the movement of fresh etchant chemicals to the surface of the wafer, agitation in the wet chemical bath will reduce the ability of the bubbles to adhere to the wafer. Even in the absence of bubbles, however, small geometry features may etch more slowly, due to the difficulty in removing all of the etch products. This phenomenon has been shown to be related to microscopic bubbles of trapped gas [2]. Another common problem for wet etch processes is undetected resist scumming. This occurs when some of the exposed photoresist is not removed in the develop process (Figure 11.3). Common causes are incorrect or incomplete exposures and insufficient developing of the pattern. Due to the high selectivity of wet etch processes, even a very thin layer of resist residue is sufficient to completely block the wet etch process.

In the 1990s, wet etching enjoyed something of a resurgence. Automated wet etch benches were developed that allow the operator to precisely control the etch time, bath temperature, degree of agitation, bath composition, and the degree of misting in spray etches [3]. Increased use of filtration, even in hot, very aggressive, compounds, has helped control particle deposition concerns. Even with these improvements, however, wet etching is still not regarded as practical for most features smaller than 2 μm.

One of the most common etching processes is the wet etching of SiO_2 in dilute solutions of hydrofluoric acid (HF) [4]. Common etchants are 6:1, 10:1, and 50:1, meaning 6, 10, or 50 parts (by volume) of water to one part HF. A 6:1 HF solution will etch thermal silicon dioxide at about 1200 Å/min. Deposited oxides tend to etch much faster. The ratio of the deposited film etch rate in HF to that of thermal oxides is often taken as a measure of its density. Doped oxides such as phosphosilicate glass and borophosphosilicate glass etch faster yet, as the etch rate increases with impurity concentration. Solutions of HF are extremely selective of oxide over silicon [5]. Some etching of silicon does occur, since the water will slowly oxidize the surface of the silicon and HF will etch this oxide. Selectivities are commonly better than 100:1. Wet etching of oxide in HF solutions is, however, completely isotropic.

The exact reaction pathway is complex and depends on the ionic strength, the solution pH, and the etchant solution [6]. The overall reaction for etching SiO_2 is

$$SiO_2 + 6HF \rightarrow H_2 + SiF_6 + 2H_2O \tag{11.2}$$

Since the reaction consumes HF, the reaction rate will decrease with time. To avoid this, it is common to use HF with a buffering agent (BHF) such as ammonium fluoride (NH_4F), which maintains a constant concentration of HF through the dissolution reaction

$$NH_4F \rightleftharpoons NH_3 + HF \tag{11.3}$$

where NH$_3$ (ammonia) is a gas. Buffering also controls the pH of the etchant, which minimizes photoresist attack.

Silicon nitride is etched very slowly by HF solutions at room temperature. A 20:1 BHF solution at room temperature, for example, etches thermal oxide at about 300 Å/min, but the etch rate for Si$_3$N$_4$ is less than 10 Å/min [7]. More practical etch rates of silicon nitride can be obtained in H$_3$PO$_4$ at 140 to 200°C [8]. A 3:10 mixture of 49% HF (in H$_2$O) and 70% HNO$_3$ at 70°C can also be used, but is much less common. Typical selectivities in the phosphoric etch are 10:1 for nitride over oxide and 30:1 for nitride over silicon. If the nitride layer is exposed to a high temperature oxidizing ambient, therefore, a dip in BHF is often done before the nitride wet etch to strip any surface oxide that may have grown on top of the nitride.

Wet etching has also been widely used to pattern metal lines. Since the metal layers used in ICs are often polycrystalline, the lines produced by wet etching sometimes have ragged edges. A common aluminum etchant is (by volume) 20% acetic acid, 77% phosphoric acid, and 3% nitric acid. Most of the metal interconnect in silicon technologies is not elemental but rather dilute alloys. In many cases, these impurities are much less volatile in the bath than the base material. In particular, silicon and copper additions to aluminum are often difficult to completely remove in standard aluminum wet etch solutions.

Although several techniques have been used for wet etching silicon, most of them use a strong oxidant to chemically oxidize the silicon and HF to etch the oxide. A common etchant solution is a combination of HF and nitric acid (HNO$_3$) in water. The overall reaction is given by

$$Si + HNO_3 + 6HF \rightarrow H_2SiF_6 + HNO_2 + H_2 + H_2O \qquad (11.4)$$

Often acetic acid (CH$_3$COOH) is used as a diluent rather than water. Figure 11.4 shows diagrams the etch rate of silicon in HF and HNO$_3$ [9]. Notice that the three axes are not independent. To find the etch rate, draw lines from the percentages of HNO$_3$ and HF. They should intersect at a point on the line corresponding to the remaining percentage of diluent. At low HNO$_3$ concentrations,

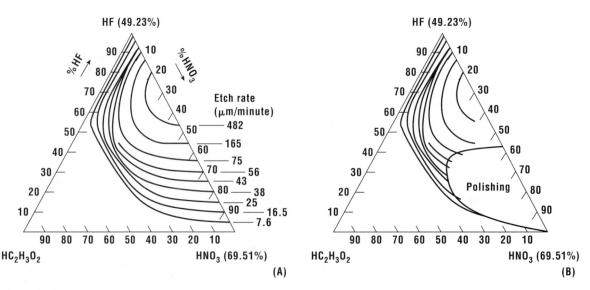

Figure 11.4 The etch rate of silicon in HF and HNO$_3$ *(after Schwarz and Robbins, reprinted by permission of the publisher, The Electrochemical Society Inc.)*.

the etch rate is controlled by the oxidant concentration. At low HF concentrations, the etch rate is controlled by the HF concentration. The maximum etch rate for this solution is 470 μm/min. A hole can be etched completely through a typical wafer at this rate in about 90 sec.

Example 11.1

A solution contains two parts of ~70% HNO_3, six parts of ~49% HF, and two parts of acetic acid. Find the etch rate of silicon in this solution.

Solution

Drawing the intersecting lines (as shown in Figure 11.5) produces an etch rate of about 165 μm/min. Note that all three lines intersect at a single point.

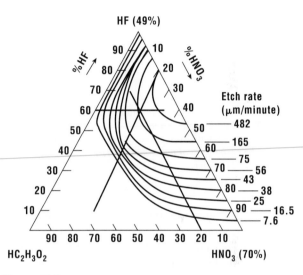

Figure 11.5   Intersection lines imposed on Figure 11.4A to show etch rate.

The most common wet etchants for GaAs [10, 11] are H_2SO_4–H_2O_2–H_2O [12], Br_2–CH_3OH [13], NaOH–H_2O_2 [14], and NH_4OH–H_2O_2–H_2O [15]. Figure 11.6 shows the etch contours for the H_2SO_4–H_2O_2–H_2O system [16]. The curves are qualitatively similar to those of Figure 11.4. At high sulfuric acid concentrations, the solution is quite viscous, and so the etch may be limited by diffusion of fresh etchant to the wafer. Etches are uneven in these solutions. As a result, the acid content is usually kept to less than 30%. As with all etching solutions using H_2O_2, the etch rate decreases with time.

Gallium acsenide etches that are selective to Al_xGa_{1-x} As are often required since many advanced devices consist of very thin layers of one material on top of the other. Selective etches are needed to make electrical contact to the various layers. The first reports of a selective etch used a 30% solution of H_2O_2 diluted with NH_4OH or H_3PO_4 to control the pH [17]. Selectivities as large as 30:1 have been reported in this system [18]. Redox solutions such as I_2–KI, $K_3Fe(CN)_6$–$K_4Fe(CN)_6$, and $C_6H_4O_2$–$C_4H_6O_2$ are now more commonly used due to their high etch rates and good selectivities. Tijburg and van Dongen [19] reported selective etching of GaAs over AlGaAs for $K_3Fe(CN)_6$–$K_4Fe(CN)_6$ solutions with pH values greater than 9 and selective etching of AlGaAs over GaAs for the same solutions with pH values between 5 and 9. In addition, NH_4OH–H_2O_2–H_2O and $K_3Fe(CN)_6$–$K_4Fe(CN)_6$ have been found to be highly selective etches for GaAs and AlGaAs over InGaAs [20].

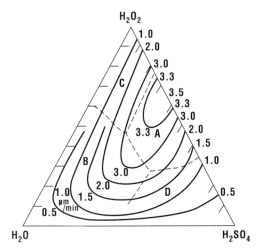

Figure 11.6   The etch rate of GaAs in H_2SO_4, H_2O_2, and H_2O. The right leg is the concentration of H_2SO_4, the bottom leg is H_2O, and the left leg is H_2O_2. All scales increase in the clockwise direction *(after Iida and Ito, reprinted by permission of the publisher, The Electrochemical Society Inc.).*

SEM TOP VIEW (100) DI ETCH

SEM CROSS-SECTIONAL VIEW (100) DI ETCH

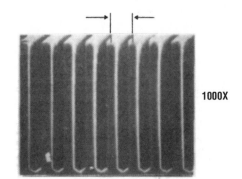

1000X

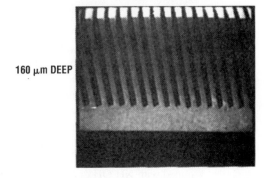

160 μm DEEP

Figure 11.7   (100) silicon wafers after directional etching in KOH, isopropyl alcohol, and water. The upper photo shows a 50-μm-deep etch. The lower photographs are of 80-μm-deep trenches etched at 10-μm pitch on (110) and 10° off (110) *(after Bean [21], ©1978 IEEE).*

It is sometimes useful to be able to etch deep patterns in semiconductor substrates with directional control. Wet etching can be directional if a cystalline material is etched. Although directional etching will be discussed in much more detail later in the chapter, this section will point out that some solutions will etch much faster along certain crystal directions. As a result, these etches produce sharp facets with well-controlled angles in the single-crystal substrates [21]. In the zincblende structure, the density of bonds in the (111) directions is much larger than in the (100) or (110) directions. Some etchants will therefore etch more slowly in the (111) directions.

Common directional wet etchants for silicon are mixtures of KOH, isopropyl alcohol, and water. A 23.4:13.5:63 mixture etches 100 times faster in the (100) directions than in the (111) [22], although some researchers have reported etch rate ratios up to 200:1 [23]. Since this etchant contains no HF, a simple thermal oxide can be used as the masking layer. Figure 11.7 shows a cross-sectional view of (100) wafers after directional etching. Closely spaced features will result in a characteristic V shape with sidewalls at 54.7° to the surface normal. More

Table 11.1   Wet chemical etch solutions used to delineate defects in (100) and (111) silicon

(100)	Schimmel [27]	2:1; HF/CrO_3(1 molar)	Works well on lightly doped layers (>0.6 Ω-cm)
	Modified Schimmel [27]	4:3:2; $HF/H_2O/CrO_3$ (1 molar)	Works well on heavily doped layers
	Secco d'Aragona [28]	2:1; $HF/K_2Cr_2O_7$ (0.15 molar)	Delineates oxidation-induced stacking faults
(111)	Sirtl and Adler [29]	1:1; HF/CrO_3 (5 molar)	Requires agitation
	Dash [30]	1:3:10; $HF/HNO_3/CH_3COOH$	Etch rates are slow and concentration dependent

widely spaced features will etch deeper into the substrate, exposing more of the (111) surface. If the wafers are etched long enough, these facets can extend completely through the wafer. A common directional etchant for GaAs is Br_2–CH_3OH [24].

A variety of special-purpose wet etches have also been developed for selectively etching certain regions of the wafers. This section will review two: doping selective and defect selective. Doping-selective etches were developed primarily for junction staining, providing steps that can be used to show the depth of a p–n junction (see Section 3.5). More recently, these techniques have been applied to the fabrication of etch stop layers for sensors and other special structures. One of the most common doping-selective etches is a 1:3:8 mixture of $HF/HNO_3/CH_3COOH$. The etch rate of either type of heavily doped ($>10^{19}$ cm^{-3}) layers of silicon is about 15 times larger than the etch rate of lightly doped layers [25]. An ethylenediamine/pyrocatechol/water mixture, however, etches lightly doped silicon but does not attack heavily doped p-type layers [26].

Defect-selective etching is done to decorate (highlight) defects in the wafer. Defects that may be invisible before etching are often readily observable by simple optical microscopy after preferential defect staining. By counting the defects after staining, a defect density can be derived. Although some defect stain etches have been developed for GaAs, the most common (summarized in Table 11.1) have been used in silicon.

11.2 Chemical Mechanical Polishing

One way to achieve global planarization is with chemical mechanical polishing (CMP). The CMP process is designed to produce a globally flat surface, free of scratches and contamination. Due to the need for high NA lens with a small depth of field and for many layers of metal interconnect, CMP usage has increased dramatically, with a total market exceeding $2 billion by 2008 [31]. Although originally developed for interconnected planarization, it is now also being applied to front-end processes such as device isolation as well. A simple process schematic is shown in Figure 11.8 . The wafer is mounted on a rotating carrier, which is placed into contact with a rotating pad. A slurry of chemicals and particles is put on the pad and is transported under the wafer. The pad has small asperities on the surface to provide fresh slurry to the wafer and to remove the by-products.

The simplest equation for the removal rate is Preston's equation [32]

$$RR = k_P P v$$

where k_P is Preston's coefficient, P is the downward pressure, and v is the relative velocity between the pad and the wafer. Typical removal rates are several thousand angstroms per minute [33]. For many processes, particularly for polishing hard materials, the removal mechanism is believed to be the formation of a hydrated surface layer that is abraded away by contact with the particles in the

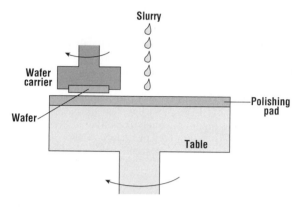

Figure 11.8 Chemical mechanical polishing of the surface of a partially processed wafer done to achieve a high degree of global planarization.

slurry. The formation and depth of this layer are increased by pressure. Some typical CMP process parameters are given in Table 11.2 [34]. Increasing the pad pressure increases the removal rate linearly, but often increases the step height ratio of the polished features [35] as well as increasing the residual oxide damage and metal contamination of the surface layer [36].

Gross mechanical damage is prevented by the fact that the particles chosen for the slurry are not harder than the film that is being polished [37]. Surface angles on CMP wafers are approximately 1° compared to 10° on reflow glass. Due to the flat surface that is formed, CMP wafers can have far fewer defects, such as metal stringers and opens, both of which commonly occur near the edges of severe topology. The smoothness that can be attained after CMP can be approximated by the Hertzian penetration depth

$$R_s = {}^3\!/_4 \, \phi \, P/[2K_p E]$$

where ϕ is the diameter of the abrasive particle, E is the Young's modulus of the material being polished, and K_p is a constant related to the density of the particle. For a closely packed material, $K_p = 1$. For the case of polishing silicon with silica particles with $\phi = 100$ nm, $K_p = 0.5$, and a pressure of 1.5 MPa, $R_s \sim 0.3$ nm, indicating that an essentially smooth surface can be attained for this system.

In the earliest and most common CMP processes, a thick spin-on or deposited layer of SiO_2, is first applied. The wafer is polished in an alkaline slurry containing colloidal silica (a suspension of abrasive SiO_2 particles) and an etching agent [38] such as a dilute HF see Figure 11.8 for a simplified schematic). Common matrix solutions for the suspension are KOH and NH_4OH. The pH, which is typically around 10, is maintained, to keep the silica particles negatively charged to repel each other and so avoid the formation of a large gel network. A pH buffering agent is sometimes used to ensure the stability of the process. The size of the particles that are used depends on the desired removal rate, with literature reports ranging from 0.03 to 0.14 µm. Typically sized particles (~0.05 µm) agglomeorate to form clusters about 0.25 µm in diameter [39]. The solids content in the slurry is kept to 12–30%.

CMP has also been extended to the planarization of metals such as copper [40, 41] and tungsten. For metal planarization, acidic (pH < 3) slurries are used. These slurries do not form colloidal suspensions, and so some agitation must be used to maintain uniformity [42]. Alumina is the most commonly used abrasive for tungsten CMP because it is closer in hardness to tungsten than most other abrasives. Tungsten is removed by continuous, self-limiting oxidation of the tungsten surface and subsequent mechanical abrasion [43, 44]. The slurry forms a hydrated tungsten oxide that is

Table 11.2 Typical CMP process parameters and results for oxide planarization

Thermal oxide removal rate	(Å/min)	600–800
Deposited oxide removal rate	(Å/min)	1000–1500
Polishing time	(min)	~10
Pad pressure	(psi)	6
Pad rotation	(rpm)	10
Wafer rotation	(rpm)	12

After Nanz and Camilletti [34].

selectively removed by alumina particles of order 200 nm. It has been shown that for typical CVD tungsten, the removal rate increases as the film gets thinner. This has been related to the change in tungsten grain size [45]. Generally the tungsten CMP process is optimized to obtain a large selectivity for W over SiO_2. Selectivities approaching 30 are commonly achieved.

Chemical mechanical polishing of copper is particularly interesting due to the low resistivity of copper and because copper is extremely difficult to etch in a plasma. Thus copper can be patterned by a CMP technique called damascene processing. This will be discussed in Chapter 15 . Copper is polished in an aqueous solution containing particles several hundred nanometers in diameter. The slurry must meet two basic requirements: any copper mechanically abraded must be dissolved by the slurry, and the polishing rate of the surfaces that come into contact with the pad must be higher than the removal rate of the low-lying copper. Typical slurries include ammonium hydroxide (NH_4OH), nitric acid (HNO_3), and hydrogen peroxide (H_2O_2) [46]. The oxidizing agents form copper oxide on the surface, which greatly inhibits removal unless the abrasives can remove the oxides. Polish rates up to 1600 nm/min have been achieved [47]. Unlike tungsten, copper is a soft metal. Mechanical effects have a significant effect on the polish process. The polish rate has been found to be proportional to applied pressure and relative linear velocity. The pad condition and pressure application mechanism are found to be particularly important for coppper CMP.

Since CMP processes do not generally have any end-of-process indicator, one must develop processes with a high selectivity or strive for very reproducible removal rates. Polish rate drift is a serious problem. The condition of the polishing pad is a key determinant of the removal rate, since the porosity of the pad determines the slurry arrival rate at the surface of the wafer. Glazing of the pad tends to occur after several runs and slows the polishing rate [48]. The solution to this problem is frequent conditioning of the pad to obtain a consistent roughness. Pad glazing is removed by contacting the pad with a diamond-impregnated workpiece. This can be done either during the wafer polish (*in situ*) or between wafer polishing (*ex situ*) [49]. This must be traded off against defect density, however, since post-CMP wafers often show greatly increased particle counts if polished immediately after pad conditioning [35].

Postpolish cleaning is an important step in the CMP unit process. Generally one must trade off polishing goals (uniformity planarity, throughput) against cleaning goals (particles, scratches, and other surface damage; residual ionic and metallic contaminants). Megasonic agitation may be used in combination with a soft pad scrubber or a cleaning solution to assist in the removal of the colloidal suspension from the wafer [50]. Usually the wafer is transferred to a second pad that is reserved for cleaning. This transfer must be timed to prevent the drying of the suspension on the surface of the wafer, after which residue removal is considerably more difficult. Furthermore, scratches left behind after CMP may collect metal that is hard to remove in standard plasma etches. These embedded metal-filled scratches are sometimes called rails and will short out subsequent metal lines. Rails are commonly produced in tungsten CMP processes in which the surface tungsten is removed down to an interlayer oxide, leaving tungsten only in the through holes. Such a process is done instead of the selective tungsten and the deposition and etchback processes described in Section 15.7 . The hard alumina particles often severely scratch the oxide surface once all of the tungsten has been removed. Cleanup after tungsten CMP may also be significantly more difficult due to the large electrostatic potential of tungsten particles under typical process conditions [51]. A dilute (100:1) HF step may be included to lift off many of the smaller metal particles and to reduce residual surface damage.

11.3 Basic Regimes of Plasma Etching

Etching in a plasma environment has several significant advantages when compared with wet etching. Plasmas are much easier to start and stop than simple immersion wet etching. Furthermore, plasma

Table 11.3   Typical etch chemistries

Si	CF_4/O_2, CF_2Cl_2, CF_3Cl, $SF_6/O_2/Cl_2$, $Cl_2/H_2/C_2F_6/CCl_4$, C_2ClF_5/O_2, Br_2, SiF_4/O_2, NF_3, ClF_3, CCl_4, CCl_3F_5, C_2ClF_5/SF_6, C_2F_6/CF_3Cl, CF_3Cl/Br_2
SiO_2	CF_4/H_2, C_2F_6, C_3F_8, CHF_3/O_2
Si_3N_4	$CF_4/O_2/H_2$, C_2F_6, C_3F_8, CHF_3
Organics	O_2, CF_4/O_2, SF_6/O_2
Al	BCl_3, BCl_3/Cl_2, $CCl_4/Cl_2/BCl_3$, $SiCl_4/Cl_2$
Silicides	CF_4/O_2, NF_3, SF_6/Cl_2, CF_4/Cl_2
Refractories	CF_4/O_2, NF_3/H_2, SF_6/O_2
GaAs	BCl_3/Ar, $Cl_2/O_2/H_2$, $CCl_2F_2/O_2/Ar/He$, H_2, CH_4/H_2, $CClH_3/H_2$
InP	CH_4/H_2, C_2H_6/H_2, Cl_2/Ar
Au	$C_2Cl_2F_4$, Cl_2, $CClF_3$

Cotler and Elta [52].

etch processes are much less sensitive to small changes in the temperature of the wafer. These two factors make plasma etching more repeatable than wet etching. Most important for small features, plasma etches may have high anisotropies. Plasma environments may also have far fewer particles than liquid media. Finally, a plasma etch process produces less chemical waste than wet etching.

As shown in Figure 11.2 Figure 11.2, there is a wide variety of dry etch processes with differing amounts of physical and chemical attack. Overlaid on this is the variety of etch chemistries that is used in each type of etch system. Table 11.3 lists some of the most common [52]. Clearly, a complete review of this topic would be complicated and lengthy. The remainder of the chapter will review a representative sample of these processes. A few of the typical etch chemistries for the most common etch process will also be presented. For further information of plasma etching, the reader is referred to any of several references [53–55].

The preceding chapter introduced the concept of a glow discharge. For a plasma etch process to proceed, six steps must occur. A feed gas introduced into the chamber must be broken down into chemically reactive species by the plasma. These species must diffuse to the surface of the wafer and be adsorbed. Once on the surface, they may move about (surface diffusion) until they react with the exposed film. The reaction product must be desorbed, diffused away from the wafer, and be transported by the gas stream out of the etch chamber. As with wet etching the etch rate is determined by the slowest of these steps.

In a typical plasma etch processes, the surface of the film to be etched is subjected to an incident flux of ions, radicals, electrons, and neutrals. Although the neutral flux is by far the largest, physical damage is related to the ion flux. Chemical attack depends an both ion flux and the radical flux. Often this bombardment sets up a modified surface layer that is several atomic layers thick.

11.4 High Pressure Plasma Etching

The earliest plasma etch equipment, introduced into IC fabrication facilities in the early 1970s, was based on high pressure, low power plasmas, where the mean free path of the species in the plasma is much less than the chamber size. The plasma in such a process is used to start and stop the chemical reaction of etching. It does this by producing a reactive species from an inert precursor. Since the energy of ions in the plasma is quite low, the etching process depends primarily on the chemistry of the plasma.

Plasma chemistry is extremely complex. To begin to understand the chemistry of the glow discharge, it is useful to start from a prototype system. The most widely studied plasma etch chemistry is that produced by carbon tetrafluoride (CF_4). Assume that a flow of CF_4 gas is established in a chamber, where a pressure of 500 mtorr (a high pressure plasma) is maintained. Silicon wafers with a photoresist mask are in contact with the plasma. The process intent is to etch the substrate silicon. This choice does not mean that only CF_4 or even fluorinated plasmas are run at high pressure. Chlorine and other species may also be used in high pressure plasmas. By the same token, fluorinated species are sometimes used for low pressure reactive ion etching.

One can begin with a simple energy balance argument as discussed in Morgan [54]. This model suggests that etching can occur if a chemical reaction that produces a gaseous or high vapor pressure liquid or solid is energetically favored. Applied to the etching of silicon, the basic idea is that volatilizing silicon with a halogen requires the replacement of Si–Si bonds with Si–halogen bonds. The amount of energy necessary to break a C–F bond in CF_4 is 105 kcal/mole. The amount of energy required to break an Si–Si bond is 42.2 kcal/mole. For CF_4 to etch silicon, the sum of these two energies (147 kcal/mole) must be less than the energy of the Si–F bond (130 kcal/mole)

$$C \veebar F + Si \veebar Si = Si - F + 17 \text{ kcal/mole} \qquad (11.5)$$

where $\veebar$ represents a bond-breaking event. Since there is a net positive energy required for the reaction, CF_4 will not etch silicon directly. Even if there was a net negative energy, the bond breaking imposes a large kinetic barrier that can make the process unworkably slow. In a plasma, however, collisions with high energy electrons will dissociate (crack) some of the CF_4 molecules, producing free fluorine atoms and molecular radicals. The first term of Equation 11.5 is now ignored from an energy balance point of view. The formation of SiF compounds is then energetically favored. Furthermore by choosing the feed gas, chamber pressure, and plasma power, one can increase the density of species that are energetically favored to etch the film.

In a typical plasma, most of the species are unreacted gas molecules. The density of these molecules at the prototype chamber pressure is about 3×10^{16} cm^{-3}. After the unreacted feed gas, the next most common (5 to 10%) species in the plasma are neutral radicals from the precursor; CF_3, CF_2, C, and F will be found in this CF_4 plasma. These radical species are extremely reactive. Assuming a simple kinetic theory gas, one can use Equation 10.10 to roughly estimate the flux of radicals striking the surface:

$$J_n = \sqrt{\frac{n^2 kT}{2\pi M}} \qquad (11.6)$$

Assuming a radical species temperature of 500 K, the radical bombardment rate due to simple diffusion is of order 10^{23} m^{-2} sec^{-1}. Also assuming that every radical that strikes the surface etches one silicon atom, and that the radical flux is the rate-limiting step, the resultant etch rate would be about 1000 Å/min. Thus, one could conceive of such a purely chemical process that would provide ample etch rates. In practice, not all of the radical species will etch all of the substrate materials. Some of the reaction by-products such as carbon do not react with silicon or do not form volatile species. As these materials coat the surface of the wafer, they slow the etching reaction. Furthermore, not all reactant species that strike the surface of the wafer will stick.

Typical plasma power densities in this prototype reactor create an ion concentration of order 10^{10} cm^{-3}. In a CF_4 plasma, the most abundant ion species is CF_3^+. Because of the small ion concentration, most of the etching will not be directly due to ions in the plasma. Rather, the constant bombardment of ions incident on the surface creates damage in the form of unsatisfied bonds that are exposed to the reactive radicals. The neutral radicals diffuse to the surface due to the concentration gradient in the plasma. On the surface they quickly react, form volatile products, and are pumped

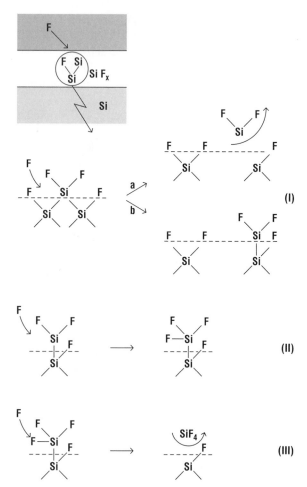

Figure 11.9 Proposed mechanism of plasma etching of silicon in CF$_4$. A 1- to 5-atom-thick SiF$_x$ layer forms on the surface. A silicon atom on the upper level is bonded to two fluorine atoms. An additional fluorine atom may remove the silicon as SiF$_2$. It is much more likely, however, that additional fluorine atoms bond to the silicon atom until SiF$_4$ forms and desorbs *(after Manos and Flamm, reprinted by permission, Academic Press).*

away. Referring back to Equation 11.5, ion bombardment of the surface eliminates most of the energy penalty associated with the Si–Si bond breaking event. This makes the forward reaction even more energetically favored. A classical page by Coburn and Winters [56] demonstrated vividly that a combination of chemical and physical attack etches much faster than either mechanism alone.

The CF$_4$ plasmas can be used to selectively etch silicon on silicon dioxide or silicon dioxide on silicon. To understand how, start with a picture of what happens at the surface of the wafer. It is believed that in a fluorine plasma, the surface silicon atoms are bonded to two F atoms, forming a fluorinated skin that may be several atoms thick. Both SiF$_2$ and SiF$_4$ are volatile species, but SiF$_2$ will not readily desorb since it is chemically bonded to the wafer. As shown in Figure 11.9, the arrival of additional F atoms will reduce the number of bonds between the surface silicon atom and the substrate until eventually the silicon fluoride molecule can be released at a minimum of energy. The primary sources of these atoms are atomic fluorine, F$_2$, and radicals of the form CF$_x$, where $x \leq 3$.

Without ion bombardment, F$_2$ radicals will etch SiO$_2$ very slowly, since there is a net energy loss in the reaction

$$F \underline{\vee} F + Si \underline{\vee} O = Si - F + -5 \text{ kcal/mole} \quad \textbf{(11.7)}$$

while CF$_3$ radicals will etch SiO$_2$ much more aggressively because the C—O bond energy compensates for the Si—O bond breaking. In a pure fluorine beam at low energy, the selectivity of Si of SiO$_2$ is almost 50:1 when the wafer is held at room temperature and about 100:1 at $-30°$C. As a result, plasma processes for etching silicon selectively over silicon dioxide are those that produce high atomic fluorine concentrations. Although it is possible to use F$_2$ as the feed gas, this is not desirable due to the particularly high toxicity of the gas. The preferred species are CF$_4$, C$_2$F$_6$, and SF$_6$, all of which produce very large concentrations of free F in a plasma. It is found experimentally that the addition of small concentrations of oxygen to a CF$_4$ feed gas will increase the etch rate of both silicon and silicon dioxide [57]. It is believed that the oxygen reacts with the carbon atoms to produce CO and CO$_2$. This removes (scavenges) some of the carbon from the plasma, thereby increasing the F concentration. These plasmas are called fluorine rich. Figure 11.10 shows plots of some of the species concentrations in a CF$_4$ plasma as a function of the amount of oxygen in the feed gas [58]. The F concentration can be increased by an order of magnitude by the addition of 12% O$_2$ to a CF$_4$ plasma producing about an order of magnitude increase in the Si etch rate, while the increase in the etch rate of SiO$_2$ is considerably less. At higher oxygen concentrations, the

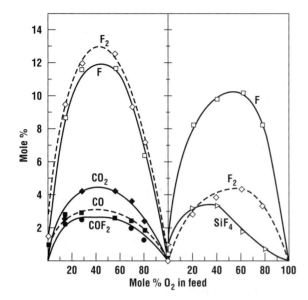

Figure 11.10   Species concentration in a CF_4 plasma as a function of the amount of oxygen in the feed gas *(after Smolinsky and Flamm, reprinted by permission, AIP).*

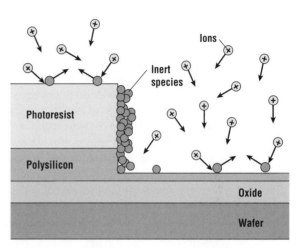

Figure 11.11   Schematic diagram of a high pressure anisotropic etch showing the formation of sidewall passivating films.

selectivity of silicon to oxide drops sharply as the molecular and atomic oxygen chemisorb on the silicon surface, making it appear more like SiO_2 [59].

A typical application of a selective etch of Si to SiO_2 is patterning a polysilicon gate for an MOS transistor over a thin gate oxide. Although the addition of O_2 to CF_4 can increase the selectivity of the etch, the processes are not necessarily anisotropic. It is possible to etch silicon anisotropically in a high pressure fluorine plasma by encouraging the formation of nonvolatile fluorocarbons that deposit on the surface of the wafer. The deposit can only be removed by physical collisions with incident ions. Figure 11.11 shows a schematic of an anisotropic high pressure plasma etch process. Fluorocarbon films deposit on all surfaces [60], but the ion velocity, which follows the electric field, is nearly vertical. As a result, as the etching proceeds there is little ion bombardment of the sidewalls and the fluorocarbon film accumulates. (Due to collisions with the ions, neutral species may also strike the surface of the wafer, but they are not shown in Figure 11.11 .) The nature of the film is a sensitive function of plasma conditions [61]. If it is not reactive, ion bombardment of the horizontal surfaces drives reactions with the underlying substrate [62, 63]. Under steady-state conditions an approximately 15-Å layer of fluorinated silicon is formed under the fluorocarbon [64]. The process of producing nonvolatile species that reduce the etch rate is known as *polymerization*. The film is said to passivate the sidewall, preventing lateral etching. This is one of two techniques for producing anisotropy. The composition of this film has been studied for a variety of chemistries [65–69].

One way to encourage the formation of these films is to add hydrogen to the plasma. The hydrogen scavenges fluorine, creating a carbon-rich plasma. The excess carbon can then form these nonvolatile products. The same thing happens if C_2F_6 is used as the feed gas instead of CF_4. In many cases, the etch products from the resist, which are also hydrocarbons, may also participate in the polymer formation process. Scavenging fluorine, however, does the opposite of what was done when oxygen was added to the plasma. As a result, the silicon-to-oxide selectivity drops in such an etch process. One can therefore obtain anisotropy or selectivity of Si to SiO_2 in such a plasma, but not both simultaneously.

The gases in a plasma reactor, however, can be changed during the etch process. In the case of MOS technologies, to obtain good dimensional control of polysilicon gate electrodes, one might start the etch process in a CF_4 plasma with H_2. When the etch is most of the way through the polysilicon, one could shut off the H_2 and instead add O_2. There may be some undercutting during this part of the process, but before undercutting can begin, the plasma must first erode the accumulated polymer. As a result, these hybrid etch processes, while using very simple equipment and relatively innocuous feed gases, can produce adequate etch profiles for features down to about 1 μm.

As just described, adding a small amount H_2 to a CF_4 plasma causes the etch rate of both silicon and silicon dioxide to decrease, in good agreement with our nonvolatile species model. At moderate H_2 concentrations, the etch rate of SiO_2 exceeds that of Si. The chemistry can therefore be used to selectively etch SiO_2 over Si. In this case, the hydrogen reacts with the fluorine radicals to form HF, which etches silicon dioxide but not silicon [70]. The plasma in this situation is said to be fluorine deficient. At moderate H_2 concentrations, the nonvolatile fluorocarbon film deposition process that increases anisotropy further increases the selectivity. On the surface of the SiO_2, the oxygen produced during the ion bombardment will react with the carbon to form CO and CO_2, both of which are volatile and so are pumped away. No such reaction occurs over the silicon, and so the etch selectivity of SiO_2 over Si increases sharply with the addition of H_2 to the plasma (Figure 11.12). Eventually, the carbon deposition rate swamps the ability of the plasma etch process to remove it, and instead of etching, there may be a net deposition.

Coburn and Kay recognized that for fluorocarbon systems the onset of polymerization depends on the fluorine to carbon ratio [71, 72]. After taking into account the formation of HF, CO, CO_2, and other scavenging compounds, polymerization proceeds when the remaining carbon concentration in the feed gas mixture is greater than half of the fluorine concentration. Examples of gases at the point of

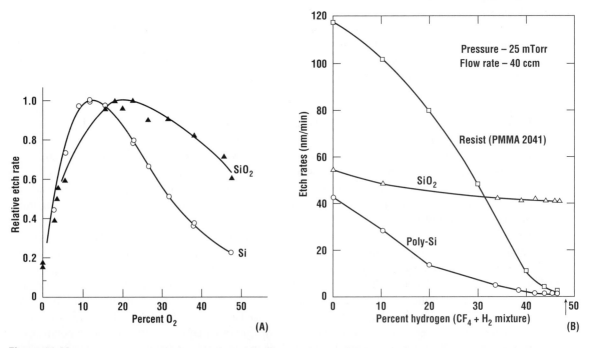

Figure 11.12   Etch rate of Si and SiO_2 in (A) CF_4/O_2 plasma *(after Mogab et al. [59], reprinted by permission, AIP)*, and (B) CF_4/H_2 plasma *(after Ephrath and Petrillo [100], reprinted by permission of the publisher, The Electrochemical Society Inc.)*.

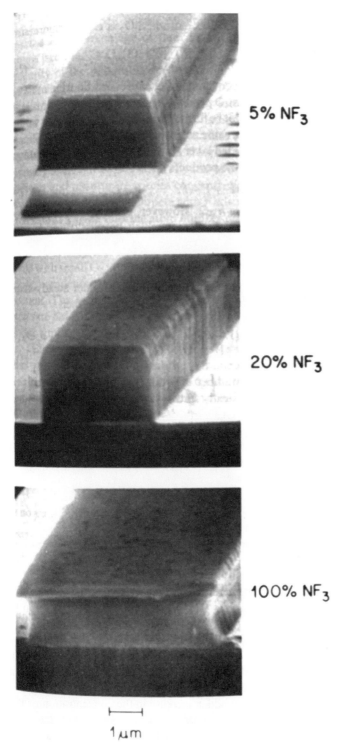

5% NF$_3$

20% NF$_3$

100% NF$_3$

⊢——⊣

1 μm

Figure 11.13 Etch profiles of SiO$_2$ with increasing concentration of NF$_3$ *(after Donnelly et al., reprinted by permission, AIP).*

polymerization include C$_2$F$_4$, CHF$_3$, and a 1:1 mixture of CF$_4$ and H$_2$. Highly selective plasma etch processes operate close to this point and can achieve etch selectivities of over 20:1 for etching SiO$_2$ over Si. The hydrogen can be supplied as H$_2$ gas. It can also be incorporated into the fluorinated precursor as CHF$_3$ or even CH$_2$F$_2$. It can also be added in the form of methane (CH$_4$), ethane (C$_2$H$_6$), or other simple hydrocarbons. In the latter case, one must take into account that this not only adds hydrogen to the plasma, but carbon as well.

As was just pointed out, the concept of depositing a nonvolatile polymeric film during the plasma etch process has an important application in producing etch selectivity. As was the case for silicon etching, sidewall passivation can produce highly anisotropic etch profiles when one is etching SiO$_2$; however, there is no trade-off of selectivity required. As a result, it is relatively easy to use fluorinated plasmas to produce nearly vertical etching of SiO$_2$ over Si. Figure 11.13 shows etch profiles of SiO$_2$ with increasing concentrations of NF$_3$ [74]. At the highest concentrations, there is a large etch rate, but the polymeric formation is suppressed and the etch is isotropic with substantial undercut of the mask.

Selective etching of silicon nitride to both silicon dioxide and silicon cannot be achieved simultaneously due to its intermediate position in terms of bond strength and electronegativity. For selectivity of nitride to oxide, silicon nitride can be etched in plasma processes similar to those used to etch silicon, that is, plasmas rich in atomic fluorine. A typical example is a CF$_4$–O$_2$ plasma [75]. Etching of silicon nitride selective to silicon can be done in a typical silicon dioxide etching process such as a CF$_4$–H$_2$ plasma.

In some applications, the plasma etch rate for a process is measured for a given set of process conditions, and the time to etch is inferred. Such etches provide poor reproducibility, since subtle changes in the etch process or in the film properties can result in unacceptable under or over etch. One such subtlety is loading effect, the tendency for the etch rate in many plasma etches to decrease with an increase in the area of the exposed film being etched. Usually, this is caused by a depletion of the reactive species in the plasma. Loading effect can be described by the equation

$$R = \frac{R_o}{1 + kA}$$ (11.8)

where R_o is the empty chamber etch rate, A is the area of the exposed film to be etched, and k is a constant [75]. Often k can be reduced by fixing the reactor pressure and increasing the flow of the etch gases, but this approach is subject to pumping limitations. A related problem is aspect ratio dependent etching. For small features with aspect ratios larger than one, the etch rate may slow down as small features are etched. This may be due to consumption of etch species or to gas transport limitations. The result is that small holes or spaces may be underetched, while large open areas are overetched.

Example 11.2

A plasma etch process is found to produce an etch rate of 300 Å/min when it is used to etch a single wafer. When a second wafer is added to the reactor, the etch rate falls to 240 Å/min. What etch rates would you expect for three and four wafers?

Solution

Denote the area to be etched on each wafer as A_o and the etch rates as R_n, where n is the number of wafers. Then from Equation 11.8,

$$\frac{R_1}{R_2} = 1.25 = \frac{1 + 2kA_o}{1 + kA_o}$$ (11.9)

Solving this equation for the kA_o product gives a value of 1/3. Then, inserting back into Equation 11.8 using either of the two known etch rates gives $R_o = 400$ Å/min. Finally, for three and four wafers, $R_3 = 200$ Å/min, and $R_4 = 171$ Å/min.

If a more repeatable process is required, some method of detecting either the thickness of the film remaining to be etched or the presence of etch by-product in the plasma can be used to determine the etch end point. If the film being etched is transparent, it is possible to use the same constructive and destructive interference effect described in Chapter 4 for measuring oxide thickness. The sum of the reflected and transmitted optical beams will oscillate with film thickness until the transparent layer is removed. This end point method, however, requires a substantial area of unpatterned film to focus and align the laser beam.

Alternatively, one can monitor the chemistry of the plasma either by using mass spectrometry of the etch effluent or by using the peaks of the optical emission spectra. In either case one can look for the extinction of the etch product of the desired film or the emergence of an etch product of the substrate. It is also possible to detect changes in the reactive species. If a substantial part of the gas is consumed by the reaction, the end point will be seen as a sudden increase in the concentration of the etchant. Table 11.4 lists some of the most common etch processes and the associated detection wavelengths. Figure 11.14 shows a commercial parallel-plate plasma etcher. The system contains two etch chambers fed by a central load lock and robot. Windows are provided at the front of each chamber for possible spectral analysis of the glow discharge.

Table 11.4   Characteristic optical emission wavelengths used for end point detection of plasma etch processes

Film	Etchant	Wavelength (Å)	Emitter
Al	CCl_4	2614	AlCl
		3962	Al
Resist	O_2	2977	CO
		3089	OH
		6563	H
		6156	O
Si	CF_4/O_2	7037	Fl
		7770	SiF
	Cl_2	2882	Si
Si_3N_4	CF_4/O_2	3370	N_2
		7037	F
		6740	N
SiO_2	CHF_3	1840	CO
W	CF_4/O_2	7037	F

After Manos and Flamm [53].

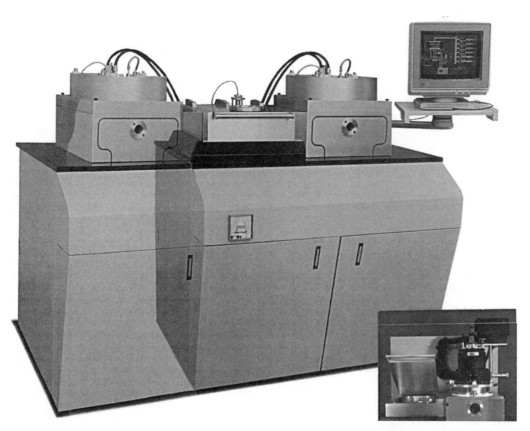

Figure 11.14   A computer-controlled, dual-chamber, parallel-plate plasma etch system *(courtesy Plasma Therm)*.

11.5 Ion Milling

Ion milling is at the opposite extreme of the spectrum of etch processes from high pressure plasma etches. Pure ion milling or ion-beam etching involves no chemical reactions with the etch species, since it uses noble gases such as argon. It is a strictly mechanical process, sometimes called the micromechanical analog of sandblasting. The physics of the etch process is analogous to sputtering. The reader is referred to Sections 12.7 and 12.8 for a review of the physical interactions between incident ions and solid targets. Ion milling has two significant advantages compared to high pressure plasmas: directionality and applicability. The directionality of the erosion is due to the fact that the ions in the beam are accelerated by a strong vertical electric field, and the chamber pressure is so low that atomic collisions are extremely unlikely. As a result, the ions' velocity as they impinge on the surface of the wafer is almost completely vertical. Anisotropic etching is possible for any material because it is chemistry independent. The second advantage of ion milling is that it can be used to pattern a wide variety of materials including compounds and alloys even if there are no suitable volatile etch products. The erosion rate of the target does not vary by much more than a factor of 3 from material to material. Thus, ion milling is popular for patterning YBaCuO, InAlGaAs, and other ternary and quaternary systems.

The ability of ion milling to pattern a wide range of materials is also one of its most serious liabilities. The process selectivity to photoresist and underlying layers is generally close to 1:1 unless a chemical component is added to the process. The other drawback with ion milling is throughput. Most ion sources are no more than 200 mm in diameter. Thus, for large silicon wafers, ion milling is a one-wafer-at-a-time process. Combined with the low erosion rates and the need for high vacuums, this makes ion milling impractical for high volume manufacturing in silicon-based technologies. For III–V technologies, however, the smaller wafer sizes and reduced number of wafers per lot make ion milling quite viable.

The most popular source for ion milling is the Kaufman source (Figure 11.15). This basic source was originally developed as a space-based rocket engine [76]. Unlike simple plasma processes, it has direct and independent control over both the ion bombardment energy and ion flux [77, 78]. A Kaufman source consists of an electron filament heated using the supply V_f. The filament is held at a voltage V_a below the anode potential. Electrons boil off the filament and are accelerated toward the anode grid. The voltage V_a must be large enough to ensure that the accelerated electrons impact the neutral gas atoms with enough energy to ionize them. Typically, this is about 40 V. Much higher voltages are not desirable, since high energy ions that strike the inside of the chamber walls will erode the chamber and so contaminate the system. To maintain the plasma, the source is held at about 10^{-3} torr.

The source body is perforated on one side. A grid held at a voltage V_g below the target just below this side of the source body accelerates the ejected ions toward the target. Typical accelerating potentials are 500 to 1000 V. Ions arrive at the target surface with an energy given by

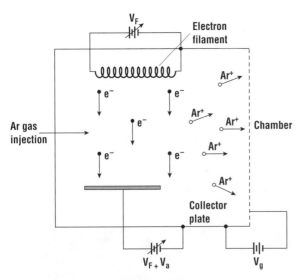

Figure 11.15 Cross section schematic of a Kaufman ion source.

$$E_{ion} = q|V_p - V_g| = q|V_a + V_{pa} - V_g| \quad (11.10)$$

where V_p is the plasma voltage with respect to ground, and V_{pa} is the plasma voltage with respect to the anode. This latter number is determined by the ion flux and the geometry of the discharge. As discussed in Chapter 10, V_{pa} is positive. Because of the large anode area, however, it is only a few volts. The extracted ion energy then is simply controlled by changing V_a or V_g.

To increase the ion density, a magnetic field is applied to the source as well. Field strengths are about 100 G [79]. (The use of magnetic fields to enhance the ion density in a plasma was discussed in Section 10.7.) Typical Kaufman sources can produce ion currents of almost an ampere over areas as large as 300 mm in diameter. This maximum current is limited by the electric field produced by a high density ion stream. The maximum current is approximately

$$j_{max} \approx K \sqrt{\frac{q}{m}} \frac{V_t^{3/2}}{I_g^2} \tag{11.11}$$

where K is a constant for a given chamber, q/m is the charge-to-mass ratio of the ions, V_t is the potential difference between the screen and accelerator grids, and I_g is the distance between the grids [80]. Typical values of I_g are 1 to 2 mm. These systems produce beams with divergences of 5° to 7°.

Example 11.3

Calculate the maximum flux from a Kaufman argon source when the potential difference between the grids is 500 V, the grid spacing is 1 mm, and $K = 2 \times 10^{-15}$ sec/F-m. Assume singly ionized argon.

Solution

Substituting $q = 1.6 \times 10^{-19}$ C and $m = 40 \times 1.67 \times 10^{-27}$ kg into Equation 11.11, j_{max} is directly calculated as 35 mA/m². This corresponds to 2.2×10^{17} ions/m²-sec.

Electron cyclotron resonance (ECR) and other advanced plasma sources have now been developed for etching. The basic operation of an ECR plasma is described in Section 10.7. An advantage of such a source for ion milling is the elimination of the filament [81]. This reduces source heating and therefore contamination, and improves system reliability. This advantage is most notable if a reactive species is used rather than argon (see CAIBE, later in this section). By using a large flux at a low enrgy, such a system may have less substrate damage than reactive ion etchers [82]. The disadvantages of these systems include increased cost and complexity for the source and the need for large and more expensive pumping systems. Since the flux density in an ion mill is limited by Equation 11.11, substantially higher erosion rates are not obtained with these sources [83] unless modified extraction methods are used.

The ions that are ejected from the source enter the target chamber. To maintain a highly directional etch, the target chamber is pumped to the lowest pressure the incident flux and the pumping speed will allow. This minimizes collisions between the ions and residual gas molecules. Cryopumps or turbomolecular pumps are typically used to evacuate the target chamber. It is normally maintained at a pressure 10 to 100 times lower than the source. Neglecting space charge effects, ions travel in a straight line to the target electrode, which contains the wafers to be milled. Since the ions are positively charged, a voltage will build up on the surface unless the wafer and the film being etched are conductive. To avoid charging effects, an electron flood gun is used. Exactly the same principle is used for maintaining accurate dosimetry in ion implantation (Chapter 5).

Figure 11.16 shows some of the problems that may occur during ion milling. Since the process erodes the mask, any taper in the masking layer will be transferred to the pattern. After the etch is done and the photoresist mask has been removed, the resultant pattern is broadened and appears very similar to an undercut. Since the eroded material from the target is not volatile, some of it will redeposit on the surface of the wafer. This can lead to an uneven etch and a considerable amount of organic residue from the photoresist mask. The latter problem is particularly pronounced on the sidewalls of the features. Finally, some milling cross sections show an enhanced erosion rate at the edges of the pattern. This problem, called *trenching,* most often occurs when mask erosion causes the sidewalls of the pattern to be tapered at a steep angle. Some of the low angle ions will reflect off the tapered surface toward the pattern edge, where they cause trenches. Commercial ion beam systems often allow the target to be tilted and rotated with respect to the incident beam during milling to minimize these effects [84].

It is necessary to use reactive species to increase selectivity in ion milling. The mixture of ions producing physical damage along with chemical attack provides the second approach to anisotropic etching since the physical damage occurs primarily on horizontal surfaces. It will be exploited in reactive ion etching, but the name Chemically Assisted Ion-Beam Etching (CAIBE, pronounced kay-bee) or reactive ion milling will be reserved for those processes in which the wafer is in a broad-area, well-collimated ion beam, but not in the plasma chamber itself. An inert species such as argon can

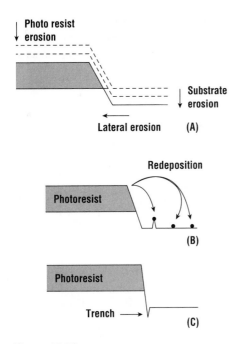

Figure 11.16   Problems that may occur during ion milling: (A) mask taper transfer, (B) redeposition from the mask, and (C) trenching.

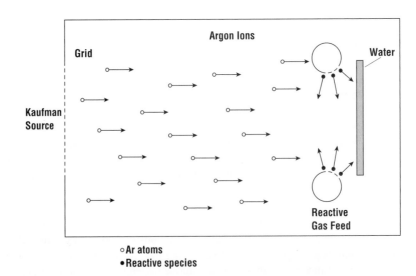

Figure 11.17   The use of a reactive bleed near the wafer surface in a conventional ion mill to introduce a chemical component to the etch process.

be mixed with small amounts of oxygen to reduce the erosion rate of many metals in an ion mill. This has particular applications in the etching of GaAs over $Al_xGa_{1-x}As$. It has been shown that a 2% O_2 dilution in Ar at 100 eV etches GaAs at hundreds of angstroms per minute, while the etch rate of $Al_{0.5}Ga_{0.5}As$ is not measurable [85].

The difficulty with introducing some reactive species, of course, is that they will tend to attack the source components. In particular, the hot filament is readily attacked. Oxygen is known to simply shorten filament life, but more aggressive precursors such as the halogens may be completely unusable for this reason. Some species may also have problems with gas phase polymerization that leads to coating of the inside of the source. It is therefore necessary to design ion source to accommodate a very reactive plasma. As previously discussed, ECR sources are desirable for this application.

Another way to avoid this problem is to add a reactive bleed near the wafer surface in a conventional ion mill (Figure 11.17). This type of etch, sometimes called an ion-assisted chemical etch, proceeds primarily by adsorption of the precursor gas on the wafer surface, simultaneously with ion bombardment. The ion stream leads to both adsorbed species decomposition and substrate damage to drive a chemical reaction. Since the chemical process depends on ion bombardment, however, this type of process cannot achieve the level of selectivity that can be obtained in a purely chemical process.

11.6 Reactive Ion Etching

Reactive ion etching (RIE) was developed because there was a very strong need for an anisotropic etch with much higher selectivity than can be achieved in ion milling. The name is something of a misnomer, since the ions are not the primary etching species in this process. A more appropriate name that is sometimes used is ion-assisted etching, although RIE (pronounced as either R I E or rye) is far more common.

Two common systems for RIE are shown in Figure 11.18. Unlike high pressure plasma etchers, the wafers rest on the powered electrode. In the parallel-plate reactor, the ground electrode is attached to the chamber wall to enlarge its effective area. In the hexode, the ground electrode is the chamber wall. As discussed in Chapter 10, the effect of this arrangement is to increase the potential difference from the plasma to the powered electrode and so the energy of the ion bombardment. For the parallel-plate arrangement to be effective, the plasma must contact the chamber walls. As pressures increase (>1 torr), the plasma contracts and loses contact with the walls. RIE, however, is done in low pressure plasmas, where the mean free path in the plasma is at least of order millimeters. In this region, the plasma remains in good contact with the walls, and a large potential appears between the plasma and the powered electrode.

Chlorine-based plasmas are commonly used to anisotropically etch silicon, GaAs, and aluminum-based metallizations. Although corrosive, chlorinated precursors such as CCl_4, BCl_3, and Cl_2 have high vapor pressures, and both the precursors and etch products are easier to deal with than the comparable bromides or iodides. Fluorine, iodine, and bromine plasmas can also be used under RIE conditions, but the chlorine system will serve as a model for this section.

It is relatively easy to develop a basic understanding of the anisotropic etching of silicon in a chlorine RIE. Undoped silicon etches very slowly in a Cl or Cl_2 ambient without the addition of ion bombardment. Heavily n-type doped silicon or polysilicon, however, spontaneously etches at high rates without bombardment in the presence of Cl, but not in the presence of Cl_2. The doping enhancement, which can be as much as a factor of 25, depends on the carrier concentration in the film rather than the chemical identity of the dopant [86, 87].

This very pronounced doping effect implies that the chlorine etching process involves electron transfer from the substrate. The model that has developed postulates that in a chlorine plasma, atomic

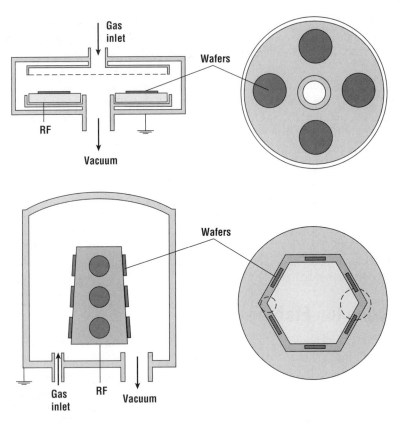

Figure 11.18 Top and side views of parallel-plate and hexode batch RIE systems. Typical conditions for either are 50 mtorr and 5 kW/m². For larger wafers the exhaust in the upper figure is drawn from the periphery rather than the center.

chlorine chemisorbs on the silicon but does not break the underlying Si–Si bonds. The absorption of more chlorine atoms is impeded by stearic hindrance, the repulsion of Cl once a monolayer coverage has been achieved (Figure 11.19). Once the surface chlorine becomes negatively charged, however, it can bond ionically with the substrate. This frees additional chemisorption sites and greatly increases the probability that the chlorine atoms will penetrate the surface and produce volatile silicon chlorides.

Chlorine penetration of the surface is also dramatically increased by ion bombardment. As a result, those surfaces subject to ion bombardment etch much more rapidly than those that are not. In particular, the vertical sidewalls receive very little ion bombardment. Because of these effects, the etch profile of undoped polysilicon or single-crystal silicon in a Cl_2 RIE is almost completely anisotropic. Unfortunately, the charge transfer mechanism described earlier produces isotropic etch profiles in heavily doped layers such as polysilicon gates [88] and aluminum metallizations. In these structures, it is necessary to obtain anisotropic etch profiles through sidewall polymerization schemes. This is commonly done by adjusting the relative concentrations of Cl_2 and an inhibitor-forming gas such as BCl_3, CCl_4, or $SiCl_4$. It can also be done by combining fluorinated precursors and Cl_2. A common mixture is 90% C_2F_6 with 10% Cl_2. By controlling the ratio of Cl to F, one can control the degree of undercut and obtain a range of etch profiles [89]. This is particularly useful in etching multiple layer films such as silicides on polysilicon with acceptable profiles [90]. As with many other passivation schemes, the etch rate must be traded off against the etch selectivity.

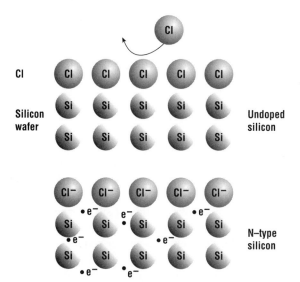

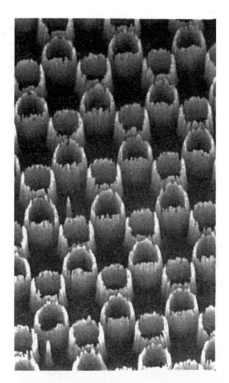

Figure 11.19 Once a monolayer of chlorine atoms has built on the surface, it impedes any chlorine addition.

The lower half of Figure 11.20 shows a silicon line etched with BCl_3, Cl_2, and O_2 in a sidewall passivation technique [91]. In this micrograph the sidewall film has begun to peel off the line. Notice that the top of the sidewall film is thicker than the bottom because of the difference in the deposition time. This produces a slight taper in the etch profile. This effect is a serious concern at very small dimensions, where the thickness of the sidewall coating becomes a substantial fraction of the feature size. The sidewall film, which is more clearly shown in the upper film, is almost pure SiO_2. The other features in the lower micrograph are sometimes called grass or black silicon [92]. They are due to the unintentional deposition of nonvolatile species and incomplete removal by ion bombardment. Grass appears in a microscope as a darkened area and often occurs when operating a sidewall passivation RIE process under heavy deposition conditions. Grass formation is suppressed by reducing the polymer formation by decreasing the flow of BCl_3,

Figure 11.20 Scanning electron micrographs of sidewall passivation films created in a $HCl/O_2/BCl_3$ plasma. In the upper micrograph, the photoresist and silicon posts have been removed by subsequent wet etching. The lower photo shows the sidewall film partially peeled from the etched pattern *(after Oehrlein et al., reprinted by permission, AIP).*

increasing the O_2 concentration, or by increasing the ion bombardment by reducing the pressure or increasing the power.

Reliable and repeatable etching of aluminum and aluminum-based compounds requires considerable attention to detail. Aluminum readily oxidizes to form Al_2O_3, which etches very slowly in most Cl plasmas. To avoid problems with etch uniformity and to maintain proper etch timing, a short sputter etch in Ar or an appropriate chemical RIE can be used before beginning the Cl-based RIE. At the end of the etch, it is also necessary to remove any adsorbed chlorinated species from the surface of the wafer. When the wafers are removed from the vacuum chamber, these compounds can react with moisture in the air to form hydrochloric acid, leading to metal corrosion. Several methods have been developed to avoid this problem. The simplest is a dip in deionized water immediately after the etch. Other postetch treatments can be done inside the vacuum chamber. These include an inert sputter etch, an oxygen plasma, or a fluorinated plasma to displace the adsorbed chlorine atoms [93]. If an insufficient clean is carried out, metal lines will have small voids, called mouse bites, along the line edges.

A problem related to etching aluminum compounds is the volatilization of silicon and/or copper additives to the metal. Copper residues are especially difficult to remove due to the low vapor pressures of these copper chlorides at room temperature. Furthermore, the copper residues are even more prone to corrosion than their aluminum counterparts. Etching compounds with greater than 2% copper generally require increased ion bombardment and/or heated substrates. Another way to etch these compounds is the use of sacrificial aluminum in the etch chamber to provide a sufficient amount of $AlCl_3$, an etch product that helps volatize copper chlorides.

Finally, a serious problem in chlorinated RIE is resist erosion due to Cl attack. This is particularly aggravated in aluminum RIE because of the thickness of typical interconnect layers, the topology that may exist, and the presence of the etch product $AlCl_3$, which also accelerates resist attack. Some resist vendors supply chlorine-resistant resists to address this problem, but even with these resists, erosion remains a concern.

Most anisotropic etching of GaAs is also done in chlorinated reactive ion etchers. Purely chemical, isotropic etching occurs in high pressure plasmas, while anisotropic etching occurs for chamber pressures of less than 10 mtorr under strong ion bombardment [94]. It is believed that the primary group III by-product, $GaCl_3$, accumulates on the surface of the wafer [95] even though the compound is reported to have a high vapor pressure. Due to differences in the etch rates of group III and group V halides, GaAs etches more rapidly along some crystal planes than in others. In low power, high pressure Cl_2 plasmas, significant faceting can occur. To avoid this problem, compounds may be added to form polymers in the plasma and passivate the sidewalls.

Another chemistry that can be used for anisotropically etching GaAs is hydrogen based. Arsenic, of course, forms several hydrides, the most stable of which is AsH_3. To volatize gallium, 5 to 25% methane must be added [96]. The maximum etch rate occurs at about 500 K [97]. As will be discussed in Section 14.9, both the long-lived radical monomethyl gallium and the stable compound trimethyl gallium [$Ga(CH_3)_3$] are volatile products. At high methane concentrations, however, excessive polymerization occurs and etching stops. Anisotropic etching with very high etch rates has also been demonstrated with chloromethanes [98].

In fluorinated RIE as in chlorinated RIE, etch anisotropy may be due to either sidewall passivation, physical damage of the horizontal surfaces, or some combination of the two. The physical damage mechanism is believed to be described by the chemical sputtering model [99]. In this model, the ion bombardment supplies energy, increasing the mobility and reactivity of surface species. The development of anisotropic etch profiles due to sidewall passivation was discussed in a previous section. In such a process, some ion bombardment of the surfaces is required to allow any etching at all [100]. In reactive ion etching, there is a substantial ion bombardment, suggesting that these tools may also be used with fluorinated chemistries. The high energy bombardment in RIE, however, damages the substrate, particularly when etching SiO_2 over Si. This will be discussed in the next section.

11.7 Damage in Reactive Ion Etching[+]

One of the limitations of reactive ion etching is residual damage left in the substrate after the etch. Typical ion fluences of 10^{15} ions/cm^2 are delivered at energies of 300 to 700 eV in a RIE. Both substrate damage [101] and chemical contamination are serious issues. The latter is a particular concern in polymerization etches, which are known to leave behind residual films. Gas phase particle deposition is also a significant problem [102]. Furthermore, one can often find metallic, impurities, including Fe, Ni, Na, Cr, K, and Zn, on the surface of the wafer after the etch [103] due to sputtering of the electrodes, chamber, and fixtures in contact with the plasma. Techniques developed for the removal of these impurities include O_2 plasma treatments followed by wet acid cleaning [104, 105] and H_2 plasma treatments [106]. The drawback of these postetch treatments is the increased complexity of the process.

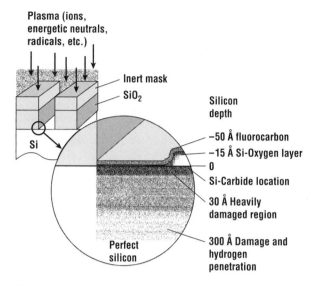

Figure 11.21 A cross section schematic of the results of a typical etch of SiO2 down to Si using CF_4/H_2 *(after Oehrlein, Rembetski, and Payne, reprinted by permission, AIP).*

Physical damage and drive-in of impurities are a second class of problem with RIE. After a typical etch in a carbon-containing RIE, the top 30 Å is heavily damaged with an extensive concentration of Si–C bonds, while significant damage is as deep as 300 Å [107]. RIE processes done in a hydrogen-containing ambient also have Si–H defects that can be observed electrically as deep as 400 Å [108] and are very difficult to remove [109]. Hydrogen may actually penetrate several microns into the surface, where it can deactivate dopants in the substrate [110]. Figure 11.21 shows a cross

---

Example 11.4   Etching a 0.5-micron-wide trench using ATHENA ELITE

Use the following deck:

```
Go Athena

#TITLE: Trench etch using rie model
#
line x loc = 0.00 spac = 0.05
line x loc = 2.00 spac = 0.05
line y loc = 0.00 spac = 0.05
line y loc = 0.00 spac = 0.05
init orientation = 100 space.m = 1
# Deposit oxide to serve as an etch mask
deposit oxide thick = 0.10 div = 5
# Remove oxide in 0.5 um area to be etched by defining the four
corners of the removal area
etch oxide start x = 0.75 y = -10
```

```
etch cont x = 0.75 y = 10
etch cont x = 1.25 y = 10
etch done x = 1.25 y = -10
# Define how the etch machine 11_3 etches silicon and oxide.
# Chemical is a uniform etch rate in nm/min directional is the
etch rate in the ion direction in nm/min
# Divergence is the angular divergence of the ions
rate.etch machine = 11_3 rie silicon n.m chemical = 35
directional = 90 divergence = 40
rate.etch machine = 11_3 rie oxide n.m chemical = 3.0
directional = 15 divergence = 40
#
etch mach = 11_3 time = 5.0 minutes dx.mult = .5
structure outfile = anelex 18.str
tonyplot anelex 18.str
quit
```

Solution

In this example, which uses the ELITE routine from Silvaco's Athena code, SiO_2 is used as a hard mask for the etch. Comparing the final profile (Figure 11.22) with the input deck, you can see that some of the oxide was lost during the etch. Also note that the beam in not perfectly normal to the surface. This leads to a shaped profile in the final trench. You can try adjusting the divergence parameter or the ratio of the nondirection (chemical) etch rate to the directional etch rate to get a flavor for what happens.

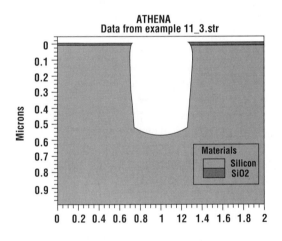

Figure 11.22   Etching a trench 0.5 μm wide.

section schematic of the results of a typical etch of SiO_2 down to Si [91]. Removal of this damage requires an initial clean followed by annealing at temperatures of over 800°C [111]. It is also possible to design RIE processes without hydrogenated species [112].

11.8 High Density Plasma (HDP) Etching

Some of the first uses of high density plasma systems, introduced in Chapter 10, were for etching. Perhaps this is because the benefits of using an HDP system were more obvious than were those for other plasma processes. High density sources use crossed magnetic and electric fields to dramatically increases the distance that free electrons in the plasma travel. This, in turn, increases the rate of dissociation and ionization compared with a simple diode plasma operating at the same pressure. This high density of ions and radicals can be used to increase the etch rate, or they can be traded off against other advantages. For examples, one can obtain acceptable ion and radical densities at very low pressures. This allows one to decouple the bias on the wafer-containing electrode from the ion density. Often this is done in HDP etch systems by attaching the wafer-containing electrode to a second RF source. The dc bias is set by the RF power of this second source. Because of the long mean free path in the system at low pressure, substrate biases of 10–30 V are often enough to produce anisotropic etching. This low energy means large selectivities and little residual damage. This is extremely useful for etching down to very thin layers (as in the gate etch for CMOS) or for etching contacts down to very thin junctions, or for etching down to active layers such as the poly etch process for a double-poly bipolar process. Furthermore, etching at lower pressures ensures more vertical incidence of the ions and therefore less reduction of the etch rate for high aspect ratio features [113]. This effect is sometimes called microloading [114]. The HDP source provides a high concentration of these low energy ions, ensuring acceptable etch rates. One disadvantage of HDP etching is that the high ion fluence can charge floating structures, like MOS gates, excessively. This can lead to excess leakage in the gate insulator due to residual etch damage [115]. An overview of conventional and HDP etching is given by Sugawara [116].

When sidewall passivation is used to produce an anistropic etch, the low pressure used in most HDP etchers severely restricts the production of the nonvolatile species. This may simply be a matter of increasing the flow ratio of the BCl_3 to Cl_2 or CH_2F_2 to CF_4, for example. At these high ratios of polymer formation, however, nonvolatile species will build up on the chamber walls. Eventually this buildup flakes off, depositing dust on the wafers. Commonly, O_2 plasma cleans along with wall heating are used to control polymeric buildup in HDP systems [117]. Wall temperature can also be used to fine-tune the etch selectivities and/or the etch rate uniformity. Alternatively, some geometries can be employed that prevent the plasma from reaching the walls of the chamber to minimize contamination.

The first HDP sources applied to etching were electron cyclotron resonance (ECR) systems [118]. Both magnetic coil and permanent magnets [119] have been used on the walls of the reactor to achieve the resonance condition. The critical concern is obtaining sufficient uniformity over a broad area to get iniform etch rates. Coil systems allow real-time changes in the magnetic field, while permanent magnets often provide more localized fields. Typical ECR etch systems use a 2.45-GHz source. This requires a field of 875 G to achieve resonance. The plasma products are transported to the surface of the wafer with a divergent magnetic field [120]. Typical ECR etch pressures are 0.1 to 10 mtorr. Typical plasma powers are 10 to 100 W.

As an example of the use of an ECR plasma, consider a gate etch. This requires etching 300 nm of polysilicon down to 1.5 nm of SiO_2. Obviously, a very high selectivity will be required for etching silicon over oxide. A high selectivity to the resist is also desirable to improve the linewidth control. From the previous discussion we know that Cl can be used to selectively etch silicon over SiO_2, but anisotropy can be achieved only in undoped films or with a sidewall passivation technique. Since silicon dioxide does not react chemically with Cl radicals, the etch rate is negligible if no sputtering of the film occurs. The Cl^+ sputter threshold for Si is about 20 eV, while that of SiO_2 is about 50 eV [121]. Thus an ECR operating

Figure 11.23 The Applied Materials high density plasma silicon etch system *(photo courtesy Applied Materials)*.

with chlorine ion energies of 20 to 50 eV should obtain extremely high selectivities for Si over SiO_2. Kanai et al. demonstrated a selectivity of 200:1 at low power in such a system [122].

Perhaps one of the most popular types of HDP etch systems is the inductively coupled reactor. HDP systems can be divided into the cylindrical coil or vertical barrel type of system, and the planar coil system. The former is shown in Figure 11.23. One of the challenges of designing a planar coil system is ensuring a uniform plasma density, since the reactor tends to have a peak under the center of the coil. This problem gets progressively more difficult as the wafer size increases. The typical coil-to-wafer spacing is 3 to 8 cm. Inductively coupled systems do not use an externally applied magnetic field. Instead, RF current through the coil produces an oscillating magnetic field that creates an electric field as described by Faraday's law. This induced magnetic field alters the path of the electrons in the plasma, increasing the plasma density.

ICP sytems are widely used for gate etching due to their ability to obtain high silicon-to-oxide selectivity at low power. Typical ICP power for gate etching is approximately 50 W. For silicon trench etching higher etch rates are desirable since trench depths can be of the order of 10 μm for DRAM technologies. At 200 to 800 W the selectivity of silicon over oxide decreases significantly [123]; however, selectivity is not a concern in this type of etch. At high power densities the anisotropy can decrease due to the high concentration of reactive radicals. This can be countered by increaing the substrate bias and/or using sidewall passivation.

All three types of high density sources have been applied to etching. All three produce similar etch rates and etch profiles for similar etch chemistries [124]. Often the choice comes down to parameters such as uniformity, repeatability, ease of use, starting effects, and source compatibility with the etcher geometry.

11.9 Liftoff

Most GaAs technologies were developed around liftoff rather than etching. The process is still popular as an alternative to ion milling for patterning difficult to etch materials. The sequence for liftoff is shown in Figure 11.24. A thick layer of resist is spun and patterned. Next, a thin layer of metal is deposited using evaporation (Chapter 12). One characteristic of evaporation is its difficulty in covering high aspect ratio features. If a reentrant profile is obtained in the resist, a break in the metal is

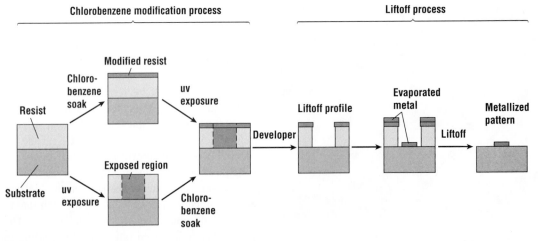

Figure 11.24   Process sequence for a liftoff operation *(after Hatzakis et al., © 1980 International Business Machines Corporation).*

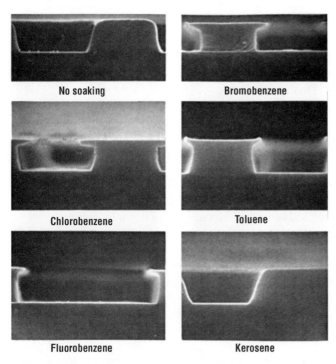

No soaking **Bromobenzene**

Chlorobenzene **Toluene**

Fluorobenzene **Kerosene**

Figure 11.25 Liftoff profiles after various resist treatments *(after Hatzakis et al., © 1980 International Business Machines Corporation).*

virtually assured. Next, the wafer is immersed in a solution capable of dissolving the photoresist. The metal lines that were deposited directly on the semiconductor remain, while the metal deposited on the resist lifts off of the wafer as the resist dissolves. Etch damage to the substrate is avoided and the lines patterned with infinite selectivity with no undercut. Since the process in its simplest form requires only a wet bench and perhaps ultrasonic agitation, it is widely used in research laboratories.

Methods that have been used to form a reentrant resist profile generally harden the surface of the resist. This can be done to some extent by promoting cross-linking by deep UV exposure, in a suitable plasma environment, or by ion implantation. Another solution is the use of multiple layers of resist such as a DQN resist on PMMA. After PMMA coating, the upper layer of resist is spun, baked, exposed, and developed as normal with a UV source. The patterned upper layer can then be used as a mask for a deep UV exposure of the PMMA, which is then overdeveloped using a solution that does not attack the upper resist. The result is a pronounced ledge that is very difficult to cover. Due to the complexity of multilayer resist processing, the most popular method of producing reentrant profiles is soaking a single-layer DQN resist after softbake in chlorobenzene [125, 126] or similar compounds. Typical soak times are 5 to 15 min. The soak process reduces the dissolution rate of the upper surface of the resist [127]. After developing the pattern, therefore, a ledge appears (Figure 11.25). The thickness of the ledge depends on the soak time, the temperature of the chlorobenzene bath, and the resist prebake cycle [128].

Liftoff processes have several shortcomings. The first is that the surface topology must be very smooth, since the metal deposition step is designed to have poor step coverage. Therefore, either the technology must be limited to one layer of metallization, or each layer must be planarized before liftoff patterning. This effectively prevents the use of sputtering. The other serious problem is that the metal lifted off remains solid and floats in the bath. Pieces of it are very likely to redeposit on the surface of the wafer. Unless the patterns are very simple, liftoff has serious yield impacts.

11.10 Summary

This chapter reviewed a variety of etch techniques that ranged from the purely physical (ion milling) to the purely chemical (wet etching). In between the two extremes are chemically assisted ion-beam etching, reactive ion etching, and high pressure plasma etching. Generally, the greater the degree of ion bombardment, the higher the etch anisotropy, but the lower the selectivity. Sidewall passivation was introduced as a technique for retaining high etch anisotropy with good selectivity. A wide variety of etches is used. Two of the most common, fluorine and chlorine plasma, have been described in some detail. Liftoff processes were developed to avoid the need for etching altogether. In these processes, the resist is patterned before metal deposition.

Problems

1. A solution consisting of four parts 70% HNO_3, four parts 49% HF, and two parts $HC_2H_3O_2$ is used to etch silicon. If the solution is held at room temperature, what etch rate would you expect? If the compound is to remain at two parts $HC_2H_3O_2$, what mixture of these same chemicals would be suitable to etch silicon at a rate of approximately 10 μm/min?

2. A hole must be wet-etched through a silicon wafer that is 700 μm thick. A mixture of two parts of $HC_2H_3O_2$, two parts of 49.2% HF and six parts of 69.5% HNO_3 is mixed together to do the etch.
    (a) How long should the etch take?
    (b) The etch is found to take nearly twice as long as predicted. Assuming that the initial concentrations of the proper chemicals are used, list three things that might have caused the reduction in the apparent etch rate and what you would do to solve each one.

3. Consider the GaAs etch diagram (Figure 11.6). The left axis is for H_2O_2, the right axis is for H_2SO_4, and the bottom axis is for water. A solution of 30% H_2O_2, 50% H_2SO_4, and 20% water is made up and used for etching a GaAs wafer.
    (a) Estimate the etch rate.
    (b) In practice it is found that the deposition rate slows with time. Suggest two possible mechanisms for this slowdown.
    (c) Would you expect this etch to be isotropic or anisotropic? Justify your answer. Remember that not all wet etches are isotropic.

4. One can etch SiO_2 using HF vapor. In this technique, the wafer is suspended above a bath of HF and the vapor comes off of the surface of the liquid, and then diffuses thrugh the air to the surface of the wafer where it reacts with the oxide.
    (a) Assume that the process is limited by the flux of HF molecules to the surface of the wafer, and that, for the conditions used, the partial pressure of HF near the wafer is 100 torr. If it takes four HF molecules to be incident on the surface to remove a single silicon atom from the SiO_2 and the density of silicon atoms in SiO_2 is 2×10^{22} cm^{-3}, determine the expected etch rate at room temperature. (You can assume that the oxygen immediately volatizes once the silicon is removed.)
    (b) The measured etch rate is many orders of magnitude less than the answer from part (a). It is found to be a sensitive function of the wafer temperature, with higher wafer temperatures resulting in higher etch rates. List two possible explanations for the lower etch rates that would be consistent with the observed temperature dependence.

5. An etch engineer has only a high pressure plasma etch and an ion mill available. Select which you would use in the following applications. Be sure to justify your answer.
    (a) Etching a 5000-Å polysilicon layer that serves as the upper electrode of a large square capacitor. The capacitor dielectric is 50 Å of SiO_2.
    (b) Recessing the channel of a GaAs FET. For this application, the residual etch damage must be minimized.
    (c) Anisotropic patterning of a thin layer of $YBa_2Cu_3O_7$ on a thick insulating layer.

6. An ion mill is used to etch a feature into a silicon wafer. The mill uses the ion source and conditions described in Example 11.2
    (a) What etch rate of silicon would you expect?
        Hints: (1) The sputter yield for Ar$^+$ is given in Figure 12.14; (2) the number density of a elemental solid is the mass density divided by the atomic weight.

(b) The process is found to have a poor selectivity to the photoresist. What could you do to increase the silicon removal rate to try to obtain a better selectivity?

(c) Would you expect the process to be nearly isotropic or nearly anisotropic? Assume vertical resist walls.

7. A cylindrical parallel-plate RF etch chamber is constructed with 12-in.-diameter electrodes. The chamber diameter is 18 in. and the height of the chamber is 6 in. One of the electrodes is attached to ground. The dc voltage between the electrodes is measured to be 20 V when a plasma is established. Assuming that the plasma is in contact with the chamber walls, calculate the potential difference between the plasma and each electrode. Explain the significance of the result.

8. Describe the differences between high pressure plasmas and reactive ion-etch systems. Explain when each is the preferred process.

9. Describe the difference between chemically assisted ion beam etching (CAIBE) and ion-assisted chemical etching.

References

1. W. A. Kern and C. A. Deckert, "Chemical Etching," in *Thin Film Processing*, J. L. Vossen, ed., Academic press, New York, 1978.

2. K. McAndrews and P. C. Subanek, "Nonuniform Wet Etching of Silicon Dioxide," *J. Electrochem. Soc.* **138**:863 (1991).

3. P. Burggraaf, "Wet Etching: Alive, Well, and Futuristic," *Semicond. Int.* **58** (July 1990).

4. W. Kern, "Chemical Etching of Dielectrics," in *Etching for Pattern Definition*, H.G.Hughes and M.J. Rand, eds., Electrochemical Society, Pennington, NJ, 1976.

5. S. M. Hu and D. R. Kerf, "Observation of Etching of n-Type Silicon in Aqueous HF Solutions," *J. Electrochem. Soc.* **114**:414 (1967).

6. J. S. Judge, in *Etching for Pattern Definition*, H. G. Hughes and M. J. Rand, eds., Electrochemical Society, Princeton, NJ, 1976.

7. L. M. Loewenstein and C. M. Tipton, "Chemical Etching of Thermally Oxidized Silicon Nitride: Comparison of Wet and Dry Etching Methods," *J. Electrochem. Soc.* **138**:1389 (1991).

8. J. T. Milek, *Silicon Nitride for Microelectronic Applications, Part 1—Preparation and Properties*, IFI/Plenum, New York, 1971, p. 1.

9. B. Schwartz and H. Robbins, "Chemical Etching of Silicon: Etching Technology," *J. Electrochem. Soc.* **123**:1903 (1976).

10. For a comprehensive listing of etching solutions for groups III–V, see Kern and Deckert [1] and references therein.

11. R. E. Williams, *Gallium Arsenide Processing Techniques*, Artech, Dedham, MA, 1984.

12. S. Adache and K. Oe, "Chemical Etching Characteristics of (001) GaAs," *J. Electrochem. Soc.* **130**:2427 (1983).

13. Y. Tarui, Y. Komiya, and Y. Harada, "Preferential Etching and Etched Profiles of GaAs," *J. Electrochem. Soc.* **118**:118 (1971).

14. D. W. Shaw, "Enhanced GaAs Etch Rates Near the Edges of a Patterned Mask," *J. Electrochem. Soc.* **113**:958 (1966).

15. J. J. Gannon and C. J. Nuese, "A Chemical Etchant for the Selective Removal of GaAs Through SiO_2 Masks," *J. Electrochem. Soc.* **121**:1215 (1974).

16. S. Iida and K. Ito, "Selective Etching of Gallium Arsenide Crystals in the H_2SO_4-H_2O_2-H_2O System," *J. Electrochem. Soc.* **118**:768 (1971).

17. R. A. Logan and F. K. Reinhart, "Optical Waveguides in GaAs-AlGaAs Epitaxial Layers," *J. Appl. Phys.* **44**:4172 (1973).

18. J. J. LePore, "Improved Technique for Selective Etching of GaAs and $Ga_{1-x}Al_xAs$," *J. Appl. Phys.* **51**:6441 (1980).

19. R. P. Tijburg and T. van Dongen, "Selective Etching of III–V Compounds with Redox Systems," *J. Electrochem. Soc.* **123**:687 (1976).

20. D. G. Hill, K. L. Lear, and J. S. Harris, Jr., "Two Selective Etching Solutions for GaAs on InGaAs and GaAs/AlGaAs on InGaAs," *J. Electrochem. Soc.* **137**:2912 (1990).

21. K. E. Bean, "Anisotropic Etching of Si," *IEEE Trans. Electron Dev.* **ED-25**:1185 (1978).

22. S. Wolf and R. N. Tauber, *Silicon Processing for the VLSI Era*, Vol. 1, Lattice Press, Sunset Beach, CA, 1986.

23. D. L. Kendall and G. R. de Guel, in *Orientation of the Third Kind: The Coming Age of (110) Silicon Micromachining and Micropackaging of Transducers*, C. D. Fung, P. W. Cheung, W. H. Ko, and D. G. Fleming, eds., Elsevier, Amsterdam, 1985, p. 107.

24. P. D. Greene, "Selective Etching of Semi-Insulating Gallium Arsenide," *Solid-State Electron.* **19**:815 (1976).

25. H. Muraoka et al., "Controlled Preferential Etching Technology," in H. R. Huff and R. R. Burgess, eds., *Semiconductor Silicon 73*, Electrochemical Society, Pennington, NJ, 1973, p. 327.

26. J. C. Greenwood, "Ethylene Diamine-Catechol-Water Mixture Shows Preferential Etching of p-n Junction," *J. Electrochem. Soc.* **116**:1325 (1969).

27. D. G. Schimmel, "Dry Etch for <100> Silicon Evaluation," *J. Electrochem. Soc.* **126**:479 (1979).

28. F. Secco d'Aragona, "Dislocation Etch for (100) Planes in Silicon," *J. Electrochem. Soc.* **119**:948 (1972).

29. E. Sirtl and A. Adler, *Z. Metallk.* **52**:529 (1961).

30. W. C. Dash, "Copper Precipitation on Dislocations in Silicon," *J. Appl. Phys.* **27**:1193 (1956).

31. T. Abraham, "GB-288 Chemical Mechanical Polishing Equipment and Materials: A Technical Market Analysis, 2004," Business Communications Company, Inc., www.bccresearch.com/advmat/GB288.html.

32. R. Jairath, D. Mukesh, M. Stell, and R. Tolles, "Role of Consumables in the Chemical-Mechanical Polishing (CMP) of Silicon Oxide Films," *Proc. 1993 ULSI Symp.* (October 1993).

33. M. A. Fury, "Emerging Developments in CMP for Semiconductor Planarization," *Solid State Technol.* 47 (April 1995).

34. G. Nanz and L. E. Camilletti, "Modeling of Chemical-Mechanical Polishing: A Review," *IEEE Trans. Semicond. Manuf.* **8**:382 (1995).

35. F. Malik and M. Hasan, "Manufacturability of the CMP Process," *Thin Solid Films* **270**:612 (1995).

36. F. Kaufman, S.Cohen, and M. Jaso, "Characterization of Defects Produced in TEOS Thin Films Due to Chemical Mechanical Polishing (CMP)," in *Ultraclean Semiconductor Processing Technology and Surface Chemical Cleaning and Passivation,* MRS 386, Materials Research Society, Pittsburgh, 1995, p. 85.

37. W. L. Patrick, W. L. Guthrie, C. L. Standley, and P. M. Schiable, "Application of Chemical-Mechanical Polishing to the Fabrication of VLSI Circuit Interconnections," *J. Electrochem. Soc.* **138**(6):1778 (1991).

38. B. Davari, C. W. Koburger, R. Schulz, J. D. Warnock, T. Furukawa, M. Jost, Y. Taur, W. G. Schwittek, J. K. DeBrosse, M. L. Kerbaugh, and J. L. Mauer, "A New Planarization Technique Using a Combination of RIE and Chemical Mechanical Polish (CMP)," *IEDM Tech. Digest*, 1989, p. 61.

39. P. Singer, "Chemical-Mechanical Polishing: A New Focus on Consumables," *Semiconductor Int.* 48 (February 1994).

40. J. M. Steigerwald, R. Zirpoli, S. P. Murarka, D. Price, and R. J. Gutmann, "Pattern Geometry Effects in the Chemical-Mechanical Polishing of Inlaid Copper Structures," *J. Electrochem. Soc.* **141**(10):2842 (1994).

41. R. Capio, J. Farkas, and R. Jairath, "Initial Study on Copper CMP Slurry Chemistries," *Thin Solid Films* **266**(2):238 (1995).

42. E. Ferri, "CMP Chemical Distribution Management," *Proc. Semicond. West: Planarization Technology: Chemical Mechanical Polishing (CMP)*, July 1994.

43. F. B. Kaufman, D. B. Thompson, R. E. Broadie, M. A. Jaso, W. L. Gutherie, D. J. Pearson, and M. B. Small, "Chemical-Mechanical Polishing for Fabricating Patterned W Metal Features as Chip Interconnects," *J. Electrochem. Soc.* **138**:3460 (1991).

44. C.-W. Liu, W.-T. Tseng, B.-T. Dai, C.-Y. Lee, and C.-F. Yeh, "Perspectives on the Wear Mechanism During CMP of Tungsten Thin Films," *Proc. CMP VLSI/ULSI Multilevel Interconnection Conf.*, Santa Clara, CA, 1996.

45. I. Kim, K. Murella, J. Schlueter, E. Nikkel, J. Traut, and G. Castleman, "Optimized Process for CMP," *Semicond. Int.*, **9**:119 (November 1996).

46. R. Capio, J. Farkas, and R. Jairath, "Initial Study on Copper CMP Slurry Chemistries," *Thin Solid Films* **266**:238 (1995).

47. C. Sainio, D. Duquette, J. Steigerwald, and S. Muraka, "Electrochemical Effects in the Chemical-Mechanical Polishing of Copper for Integrated Circuits," *J. Elect. Mater.* **25**(10):1593 (1996).

48. I. Ali, S. R. Roy, and G. Shinn, "Chemical-Mechanical Polishing of Interlayer Dielectric: A Review," *Solid State Technol.* **37**(10):63 (1994).

49. L. Borucki et al., "A Theory of Pad Conditioning for Chemical-Mechanical Polishing," *J. Eng. Math.* **50**(1):1–24 (2004).

50. J. M. de Larios, M. Ravkin, D. L. Hetherington, and J. D. Doyle, "Post-CMP Cleaning for Oxide and Tungsten Applications," *Semicond. Int.* 121 (May 1996).

51. I. Malik, J. Zhang, A. J. Jensen, J. J. Farber, W. C. Krusell, S. Raghavan, and C. Rajhunath, "Post-CMP Cleaning of W and SiO_2: A Model Study," in *Ultraclean Semiconductor Processing Technology and Surface Chemical Cleaning and Passivation*, MRS 386, Materials Research Society, Pittsburgh, 1995, p. 109.

52. T. J. Cotler and M. Elta, "Plasma-Etch Technology," *IEEE Circuits, Dev. Mag.* 38 (July 1990).

53. D. M. Manos and D. L. Flamm, *Plasma Etching, An Introduction*, Academic Press, Boston, 1989.

54. R. A. Morgan, *Plasma Etching in Semiconductor Fabrication*, Elsevier, Amsterdam, 1985.

55. A. J. van Roosmalen, J. A. G. Baggerman, and S. J. H. Brader, *Dry Etching for VLSI*, Plenum, New York, 1991.

56. J. W. Coburn and H. F. Winters, "Ion and Electron Assisted Gas Surface Chemistry," *J. Appl. Phys.* **50**(5): 3189 to 3196 (1979).

57. V. M. Donelly, D. I. Flamm, W. C. Dautremont-Smith, and D. J. Werder, "Anisotropic Etching of SiO_2 in Low-Frequency CF_4/O_2 and NF_3/Ar Plasmas," *J. Appl. Phys.* **55**:242 (1984).

58. G. Smolinsky and D. L. Flamm, "The Plasma Oxidation of CF_4 in a Tubular, Alumina, Fast-Flow Reactor," *J. Appl. Phys.* **50**:4982 (1979).

59. C. J. Mogab, A. C. Adams, and D. L. Flamm, "Plasma Etching of Si and SiO_2—The Effect of Oxygen Additions to CF_4 Plasmas," *J. Appl. Phys.* **49**:3796 (1978).

60. J. W. Coburn, "*In-situ* Auger Spectroscopy of Si and SiO_2 Surfaces Plasma Etched in CF_4-H_2 Glow Discharges," *J. Appl. Phys.* **50**:5210 (1979).

61. R. d'Agostino, F. Cramarossa, F. Fracassi, E. Desimoni, L. Sabbatini, P. G. Zambonin, and G. Caporiccio, "Polymer Film Formation in C_2F_6-H_2 Discharges," *Thin Solid Films* **143**:163 (1986).

62. M. Shima, "A Study of Dry-Etching Related Contaminations of Si and SiO_2," *Surf. Sci.* **86**:858 (1979).

63. S. Joyce, J. G. Langan, and J. I. Steinfeld, "Chemisorption of Fluorocarbon Free Radicals on Si and SiO_2," *J. Chem. Phys.* **88**:2027 (1988).

64. C. Cardinaud and G. Turban, "Mechanistic Studies of the Initial Stages of Si and SiO_2 in a CHF_3 Plasma," *Appl. Surf. Sci.* **45**:109 (1990).

65. G. S. Oehrlein, K. K. Chan, and G. W. Rubloff, "Surface Analysis of Realistic Semiconductor Microstructures," *J. Vacuum. Sci. Technol. A* **7**:1030 (1989).

66. G. S. Oehrlein and J. F. Rembetski, "Study of Sidewall Passivation and Microscopic Silicon Roughness Phenomena in Chlorine-based Reactive Ion Etching of Silicon Trenches," *J. Vacuum Sci. Technol. B* **8**:1199 (1990).

67. K. V. Guinn and C. C. Chang, "Quantitative Chemical Topography of Polycrystalline Si Anisotropically Etched in Cl_2/O_2 High Density Plasmas," *J. Vacuum Sci. Technol. B* **13**:214 (1995).

68. K. V. Guinn and V. M. Donnelly, "Chemical Topography of Anisotropic Etching of Polycrystalline Si Masked with Photoresist," *J. Appl. Phys.* **75**:2227 (1994).

69. F. H. Bell and O. Joubert, "Polycrystalline Gate Etching in High Density Plasmas. II. X-Ray Photoelectron Spectroscopy Investigation of Silicon Trenches Etched Using a Chlorine-based Chemistry," *J. Vacuum Sci. Technol. B* **14**:1796 (1996).

70. M. M. Millard and E. Kay, "Difluocarbene Emission Spectra from Fluorocarbon Plasmas and Its Relationship to Fluorocarbon Polymer Formation," *J. Electrochem. Soc.* **129**:160 (1982).

71. J. W. Coburn and E. Kay, "Some Chemical Aspects of Fluorocarbon Plasma Etching of Silicon and Its Compounds," *IBM J. Res. Dev.* **23**:33 (1979).

72. J. W. Coburn and E. Kay, "Some Chemical Aspects of Fluorocarbon Plasma Etching of Silicon and Its Compounds," *Solid State Technol.* **22**:117 (1979).

73. V. M. Donnelly, D. E. Ibbotson, and D. L. Flamm, in *Ion Bombardment Modification of Surfaces: Fundamentals and Applications*, O. Auciello and R. Kelly, eds., Elsevier, New York, 1984.

74. F. H. M. Sanders, J. Dieleman, H. J. B. Peters, and J. A. M. Sanders, "Selective Isotropic Dry Etching of Si_3N_4 over SiO_2," *J. Electrochem. Soc.* **129**:2559 (1982).

75. C. J. Mogab, "The Loading Effect in Plasma Etching," *J. Electrochem. Soc.* **124**:1262 (1977).

76. B. A. Heath and T. M. Mayer, in *VLSI Electronics Microstructure Science 8, Plasma Processing for VLSI*, N. G. Einspruch and D. M. Brown, eds., Academic Press, New York, 1984.

77. J. M. E. Harper, "Ion Beam Techniques in Thin Film Deposition," *Solid State Technol.* **30**:129 (1987).

78. R. E. Lee, "Ion-Beam Etching (Milling)," in *VLSI Electronics Microstructure Science 8, Plasma Processing for VLSI*, N. G. Einspruch and D. M. Brown, eds., Academic Press, New York, 1984.

79. H. R. Kaufman, J. J. Cuomo, and J. M. E. Harper, "Techniques and Applications of Broad-Beam Ion Sources Used in Sputtering—Part 1. Ion Source Technology," *J. Vacuum Sci. Technol.* **21**:725 (1982).

80. J. M. E. Harper, "Ion Beam Etching," in *Plasma Etching: An Introduction*, D. M. Manos and D. L. Flamm, eds., Academic Press, New York, 1989.

81. S. Matsup and Y. Adachi, "Reactive Ion Beam Etching Using a Broad Beam ECR Ion Source," *Jpn. J. Appl. Phys.* **21**:L4 (1982).

82. A. S. Yapsir, G. S. Oehrlein, F. Wiltshire, and J. C. Tsang, "X-ray Photoemission and Raman Scattering Spectroscopic Study of Surface Modifications of Silicon Induced by Electron Cyclotron Resonance Etching," *Appl. Phys. Lett.* **57**:590 (1990).

83. C. Keqiang, A. Erli, W. Jinfa, Z. Hansheng, G. Zuoyao, and Z. Bangwei, "Microwave Electron Cyclotron Resonance Plasma for Chemical Vapor Deposition and Etching," *J. Vacuum Sci. Technol. A* **4**:828 (1986).

84. R. E. Lee, "Microfabrication by Ion-Beam Etching," *J. Vacuum Sci. Technol.* **16**:164 (1979).

85. H. Kinoshita, T. Ishida, and K. Kaminishi, "Surface Oxidation of GaAs and AlGaAs in Low-Energy Ar/O$_2$ Reactive Ion Beam Etching," *Appl. Phys. Lett.* **49**:204 (1986).

86. S. Berg, N. Nender, R. Buchta, and H. Norstrom, "Dry Etching of n-Type and p-Type Polysilicon: Parameters Affecting the Etch Rate," *J. Vacuum Sci. Technol. A* **5**:1600 (1987).

87. H. Okano, Y. Horiike, and M. Sekine, "Photo-Excited Etching of Poly-Crystalline and Single-Crystalline Silicon in Cl$_2$ Atmospheres," *Jpn. J. Appl. Phys.* **24**:68 (1985).

88. G. C. Schwartz and P. M. Schaible, "Reactive Ion Etching of Silicon," *J. Vacuum Sci. Technol.* **16**:410 (1979).

89. D. L. Flamm, D. N. K. Wang, and D. Maydan, "Multiple-Etchant Loading Effects and Silicon Etching in CClF$_3$ and Related Mixtures," *J. Electrochem. Soc.* **129**:2755 (1982).

90. L. Peters, "Plasma Etch Chemistry: The Untold Story," *Semicond. Int.* 67 (May 1992).

91. G. S. Oehrlein, J. F. Rembetski, and E. H. Payne, "Study of Sidewall Passivated and Microscopic Silicon Roughness Phenomena in Chlorine-Based Reactive Ion Etching of Silicon Trenches," *J. Vacuum Sci. Technol. B* **8**:1199 (1990).

92. G. K. Herb, D. J. Rieger, and K. Shields, "Silicon Trench Etching in a Hex Reactor," *Solid State Technol.* **30**:109 (1987).

93. Y. T. Fok, *Electrochem. Soc. Proc.* **80**:301, Electrochemical Society, Pennington, NJ, 1980.

94. D. E. Ibbotson and D. L. Flamm, "Plasma Etching for III–V Compound Devices: Part 1," *Solid State Technol.* **31**:77 (October 1988).

95. M. Balooch, D. R. Orlander, and W. J. Siekhaus, "The Thermal and Ion-Assisted Reactions of GaAs (100) with Molecular Cl," *J. Vacuum. Sci. Technol. B* **4**:794 (1986).

96. R. Cheung, S. Thomas, S. P. Beamont, G. Doughty, V. Law, and C. D. W. Wilkinson, "Reactive Ion Etching of GaAs Using a Mixture of Methane and Hydrogen," *Electronics Lett.* **23**:16 (1987).

97. J. M. Villalvilla, C. Santos, and J. A. Vallés-Abarca, "Temperature Dependence of Reactive Ion Beam Etching of GaAs with CH$_4$/H$_2$," *Vacuum* **43**:591 (1992).

98. V. J. Law and G. A. C. Jones, "Chloromethane-Based Reactive Ion Etching of GaAs and InP," *Semicond. Sci. Technol.* **7**:281 (1992).

99. T. J. Tu, T. J. Chang, and H. F. Winters, "Chemical Sputtering of Fluorinated Silicon," *Phys. Rev. B* **23**:823 (1981).

100. L. M. Ephrath and E. J. Petrillo, "Parameter and Reactor Dependence of Selective Oxide RIE in CF$_4$ and H$_2$," *J. Electrochem. Soc.* **129**:2282 (1982).

101. G. S. Oehrlein, "Dry Etching Damage of Silicon: A Review," *Mater. Sci. Eng. B* **4**:441 (1989).

102. G. S. Selwyn, J. Singh, and R. S. Bennett, "*In-situ* Laser Diagnostic Studies of Plasma-generated Particulate Contamination," *J. Vacuum. Sci. Technol. A* **7**:2758 (1989).

103. S. J. Fonash, "Overview of Dry Etching Damage and Contamination Effects," *J. Electrochem. Soc.* **137**:3885 (1990).

104. X.-C. Mu, S. J. Fonash, G. S. Oehrlein, S. N. Chakravarti, C. Parks, and J. Keller, "A Study of CClF$_3$/H$_2$ Reactive Ion Etch Damage and Contamination Effects in Silicon," *J. Appl. Phys.* **59**:2958 (1986).

105. J. P. Gambino, M. D. Monkowski, J. F. Shepard, and C. C. Parks, "Junction Leakage Due to RIE—Induced Metallic Contamination," *J. Electrochem. Soc.* **137**:976 (1990).

106.   J. P. Simko, G. S. Oehrlein, and T. M. Mayer, "Removal of Fluorocarbon Residues on CF_4/H_2 Reactive Ion Etched Silicon Surfaces with a Hydrogen Plasma," *J. Electrochem. Soc.* **138**:277 (1991).

107.   G. J. Coyle and G. S. Oehrlein, "Formation of a Silicon-Carbide Layer During CF_4/H_2 Dry Etching of Silicon," *Appl. Phys. Lett.* **47**:604 (1985).

108.   G. S. Oehrlein, R. M. Tromp, Y. H. Lee, and E. J. Petrillo, "Study of Silicon Contamination and Near-surface Damage Caused by CF_4/H_2 Reactive Ion Etching," *Appl. Phys. Lett.* **45**:420 (1984).

109.   G. S. Oehrlein, J. G. Clabes, and P. Spirito, "Investigation of Reactive Ion Etching Related Fluorocarbon Film Deposition onto Silicon and a New Method for Surface Residue Removal," *J. Electrochem. Soc.* **133**:1002 (1986).

110.   J. M. Heddleson, M. W. Horn, and S. J. Fonash, "Evolution of Damage, Dopant Deactivation, and Hydrogen Related Effects in Dry Etched Silicon as a Function of Annealing History," *J. Electrochem. Soc.* **137**:1960 (1990).

111.   S. J. Fonash, X.-C. Mu, S. Chakravarti, and L. C. Rathbun, "Recovery of Silicon Surfaces Subjected to Reactive Ion Etching Using Rapid Thermal Annealing," *J. Electrochem. Soc.* **135**:1037 (1988).

112.   J. P. Simko and G. S. Oehrlein, "Reactive Ion Etching of Silicon and Silicon Dioxide in CF_4 Plasmas Containing H_2 and C_2F_4 Additives," *J. Electrochem. Soc.* **138**:2748 (1991).

113.   K. Norjiri, E. Iguchi, K. Kawamura, and K. Kadota, "Microwave Plasma Etching of Silicon Dioxide for Half-Micron ULSIs," *Ext. Abstr. 21st Conf. Solid State Devices*, 1989, p. 153.

114.   P. Singer, "New Frontiers in Plasma Etching," *Semicond. Int.* **19**:152 (July 1996).

115.   M. Okandan, S. J. Fonash, O. O. Awadelkarim, T. D. Chan, and F. Preuninger, "Soft-Breakdown Damage in MOSFET's Due to High-Density Plasma Etching Exposure," *IEEE Electron Dev. Lett.* **17**(8):388 (1996).

116.   M. Sugawara, *Plasma Etching Fundamentals and Applications*, Oxford Science, New York, 1998.

117.   S. Wantanabe, "Plasma Cleaning and Etching Using a Quartz Bell Jar with SnO_2 Transparent Thin Film Heater in a CHF_3-SiO_2 Microwave Etching System," *Jpn. J. Appl. Phys.* **33**:3608 (1994).

118.   K. Suzuki, S. Okudaira, N. Sakudo, and I. Kanomata, "Microwave Plasma Etching," *Jpn. J. Appl. Phys.* **16**:1979 (1977).

119.   A. Hatta, M. Kubo, Y. Yasaka, and R. Itatani, "Performance of Electron Cyclotron Resonance Plasma Produced by a New Microwave Launching System in a Multicusp Magnetic Field with Permanent Magnets," *Jpn. J. Appl. Phys.* **31**:1473 (1992).

120.   S. Wantanabe, "ECR Plasma Etchers," in M. Sugawara, ed., *Plasma Etching Fundamentals and Applications.* Oxford Science, New York, 1998.

121.   W. M. Holber and J. Forster, "The Effect of Operating Parameters and RF Bias on Ion Energies in an ECR Reactor," *Proc. 11th Symp. Dry Processes*, IEE Japan, 1989, p. 9.

122.   S. Kanai, K. Nojiri, and M. Nawata, "Microwave Plasma Etching System," *Hitachi Rev.* **40**:383 (1991).

123.   J. H. Lee, Yeom, J. W. Lee, and J. Y. Lee, "Study of Shallow Silicon Trench Etch Process Using Planar Inductively Coupled Plasmas," *J. Vacuum Sci. Technol. A* **15**(3):573 (1997).

124.   J. T. C. Lee, "A Comparison of HDP Sources for Polysilicon Etching," *Solid State Technol.* **39**:63 (August 1996).

125.   B. J. Canavello, M. Hatzakis, and J. M. Shaw, "Single Step Optical Lift-off Processes," *IBM Technol. Disc. Bull.* **19**:4048 (1977).

126. M. Hatzakis, B. J. Canavello, and J. M. Shaw, "Single-Step Optical Lift-off Process," *IBM J. Res. Dev.* **24**:452 (1980).

127. R. M. Halverson, M. W. MacIntyre, and W. T. Motsiff, "The Mechanism of Single-Step Lift-off with Chlorobenzene in a Diazo-Type Resist," *IBM J. Res. Dev.* **26**:590 (1982).

128. G. G. Collins and G. W. Halsted, "Process Control of the Chlorobenzene Single-step Liftoff Process with a Diazo-type Resist," *IBM J. Res. Dev.* **26**:596 (1982).

Part IV

Unit Processes 3: Thin Films

I was born not knowing and have had only a little time to change that here and there.[1]

The previous chapters have discussed the processes involved in growing oxides, diffusing impurities, and transferring a pattern through photolithography and etching. These are primarily processes used to fabricate the transistor. This section of the book will discuss processes that can be used for depositing thin films. These are a very important set of processes, since all of the layers above the surface of the wafer must be deposited. Generally the techniques used to deposit metals are physical; that is, they do not involve a chemical reaction. Processes used to deposit semiconducting and insulating layers often involve chemical reactions. This distinction, however, is changing.

This section begins with the physical process of evaporation, which has been applied primarily to III–V fabrication. Evaporated films are unable to cover sudden steps and other severe topology on the surface of the wafer. Since some III–V technologies use liftoff for patterning, evaporation is ideally suited for this process. Chapter 12 will also cover a second physical deposition process: sputtering. Sputtering has been used extensively in silicon technologies. It is able to deposit a wide range of alloys and compounds and has a good ability to cover surface topology. The reasons for the choice of materials to be used for metallization will not be emphasized here but will be covered in more detail in Chapter 15. Chapter 13 will cover chemical vapor deposition, which has the best ability to cover surface topology of the three methods and produces the least substrate damage. Because the chemistry of each process is unique, Chapter 13 will cover the deposition of several representative materials. The use of these materials is described in Chapter 15.

Chapter 14 will extend the deposition discussion of the first two chapters into the area of epitaxial growth. The chapter first covers chemical vapor deposition of silicon from silanes and chlorosilanes. These processes are widely used in silicon production.

---

[1]Richard P. Feynman.

The chapter will then cover processes that have been developed to grow epitaxial layers with atomic level control over the thickness. These processes have been applied primarily to III–V technologies. One of them, molecular beam epitaxy, is essentially an extension of simple evaporation. Another, MOCVD, is an extension of chemical vapor deposition to lower growth temperatures using organometallic sources.

Chapter 12

Physical Deposition: Evaporation and Sputtering

The metal layers for all of the early semiconductor technologies were deposited by evaporation. Although still widely used in research applications and for III–V technologies, evaporation has been displaced by sputtering and in some cases by electroplating in most silicon technologies for two reasons. The first is the ability to cover surface topology, also called the *step coverage*. As the lateral dimensions of transistors have decreased, the thickness of many of the layers has remained nearly constant. As a result, the topology that the metal must cover has become more severe. Evaporated films have very poor ability to cover for these structures, often becoming discontinuous on the vertical walls. It is also difficult to produce well-controlled alloys by evaporation. Since some technologies require alloys to form reliable contacts and/or metal lines, it would be very difficult to use evaporation for these technologies in a manufacturing environment. Some III–V technologies use the poor step coverage of evaporation to advantage. Rather than depositing and etching metal layers, the film is deposited on top of a patterned photoresist layer. The films naturally tend to break at the edges of the resist, so that when the resist is subsequently dissolved, the layer on top of the resist is easily lifted. Rather than trying to form alloys, these technologies use thin layers of different metals. Such a stack would be difficult to etch, but this problem is circumvented with liftoff.

A simple evaporator is shown in Figure 12.1. The wafers are loaded into a high vacuum chamber

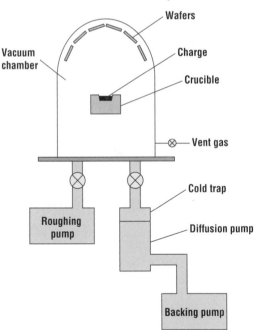

Figure 12.1   A simple diffusion-pumped evaporator showing vacuum plumbing and the location of the charge-containing crucible and the wafers.

that is commonly pumped with either a diffusion pump or a cryopump. Diffusion-pumped systems commonly have a cold trap to prevent the backstreaming of pump oil vapors into the chamber. The charge, or material to be deposited, is loaded into a heated container called the *crucible*. It can be heated very simply by means of an embedded resistance heater and an external power supply. As the material in the crucible becomes hot, the charge gives off a vapor. Since the pressure in the chamber is much less than 1 mtorr, the atoms of the vapor travel across the chamber in a straight line until they strike a surface, where they accumulate as a film. Evaporation systems may contain up to four crucibles to allow the deposition of multiple layers without breaking vacuum and may contain up to 24 wafers suspended in a frame above the crucibles. Furthermore, if an alloy is desired, multiple crucibles can be operated simultaneously. To help start and stop the deposition abruptly, mechanical shutters are used in front of the crucibles.

12.1 Phase Diagrams: Sublimation and Evaporation°

Section 2.1 discussed phase diagrams for multicomponent systems such as Al–Si. Phase diagrams are essentially maps in 2-D for binary mixtures and in 3-D for ternary mixtures showing regions of stability. For evaporation, one is usually interested in the behavior of single-component systems. This type of phase diagram is presented as a function of temperature and pressure. Evaporation only involves the pressure regime well below 1 torr. This is such a small range that we can ignore any pressure effects as well as the existence of multiple solid phases as they have little effect on evaporation.

   As the temperature of the sample is raised, the material typically goes through the solid, liquid, and gas phases. At every temperature there exists an equilibrium pressure P_e of vapor above the material. When the sample is below the melting temperature this process is called *sublimation*. When the sample is molten it is called *evaporation*. The semiconductor process called evaporation typically involves molten samples, but this is simply because this region of operation has high vapor pressures and so produces acceptable deposition rates. The distinction between molten and solid samples is of secondary importance for evaporation. Figure 12.2 shows the equilibrium vapor pressure as a function of temperature for a variety of elements [1]. A wide range of pressures is accessible simply by changing the temperature of the sample.

   When the charge is liquid, the vapor pressure is given by [2]

$$P_e = 3 \times 10^{12} \sigma^{3/2} T^{-1/2} e^{\Delta H_v / NkT} \tag{12.1}$$

where σ is the surface tension of the metal, N is Avogadro's number, and ΔH_v is enthalpy of evaporation. In practice, even small errors in the determination of the enthalpy will produce dramatic errors in the vapor pressure, so the data in Figure 12.2 are generally determined experimentally.

   To obtain reasonable deposition rates, the sample vapor pressure must be at least 10 mtorr. From Figure 12.2 it is obvious that some materials must be heated to much higher temperatures than others to obtain the same vapor pressure. Materials in the class known

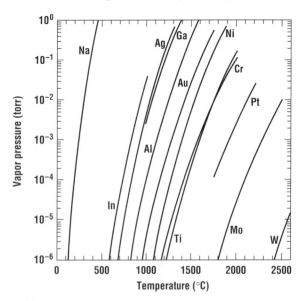

Figure 12.2   Vapor pressure curves for some commonly evaporated materials (*data adapted from Alcock et al.*).

as refractory metals, which includes Ta, W, Mo, and Ti, have very high melting temperatures and therefore have low vapor pressures at moderate temperatures. Tungsten requires a temperature in excess of 3000°C to obtain a vapor pressure of 10 mtorr. The evaporation of these refractory materials entails special equipment problems. Aluminum, on the other hand, has this same vapor pressure at 1250°C. Some ionic contaminants, such as K, have significant vapor pressure even at room temperature.

12.2 Deposition Rates

According to Equation 10.10, the number of gas molecules crossing a plane per unit time is given by

$$J_n = \sqrt{\frac{P^2}{2\pi kTm}} \tag{12.2}$$

where P is the pressure in the chamber and m is the atomic mass. Although this expression has been used so far to describe the flux of atoms striking the surface of a vacuum system, it can also be used to describe the loss rate of atoms from an evaporative source. Multiplying this flux rate by the mass of the molecule produces the Langmuir expression for the mass evaporation rate [3]

$$R_{ME} = \sqrt{\frac{m}{2\pi kT}}\, P_e \tag{12.3}$$

where P_e is the equilibrium vapor pressure of the crucible material.

Using this expression one can readily calculate the mass loss rate of the crucible as

$$R_{ML} = \int \sqrt{\frac{m}{2\pi kT}}\, P_e dA = \sqrt{\frac{m}{2\pi k}} \int \frac{P_e}{\sqrt{T}}\, dA \tag{12.4}$$

where the integral is done over the surface of the charge in the crucible. Equation 12.4 can be quite formidable unless two approximations are made. If the charge is completely molten, it is common to assume that natural convection and thermal conduction will keep the temperature of the charge nearly constant across the crucible. This is analogous to the rapid stirring approximation made when discussing Czochralski growth in Chapter 2. If it is also assumed that the opening of the crucible has a constant area A. Then

$$R_{ML} = \sqrt{\frac{m}{2\pi k}} \frac{P_e}{\sqrt{T}} A \tag{12.5}$$

Example 12.1

Consider a drop of water inside a room temperature vacuum chamber. If the drop forms a hemisphere of radius r_o, and if the drop remains at room temperature, calculate the time it will take to evaporate the drop. Assume $r_o = 1$ mm.

Solution

The number of water molecules in the drop is given by

$$N = \frac{1}{2} \frac{4}{3}\pi r^3 \cdot \rho \cdot \frac{1}{m}$$

The time rate of change of N is

$$\frac{dN}{dt} = 2\pi r^2 \frac{dr}{dt} \frac{\rho}{m}$$

Using Equation 12.2,

$$J = \frac{1}{A}\frac{dN}{dt} = \frac{2\pi r^2 (dr/dt)(\rho/m)}{1/2 \; 4\pi r^2} = \sqrt{\frac{P_e^2}{2\pi kTm}}$$

$$\frac{dr}{dt} = \sqrt{\frac{P_e^2 m}{2\pi kT\rho^2}} = \text{constant}$$

Since dr/dt is constant, $t = r_o/dr/dt$

$$t = \frac{r_o \rho}{P_e}\sqrt{\frac{2\pi kT}{m}}$$

At 27°C, P_e for H_2O is 27 torr = 3.6×10^3 Pa.
Then

$$t = \frac{10^{-3}m \times 2.33 \times 10^3 \text{ kg/m}^3}{3.6 \times 10^3 \text{ kg/m-sec}^2}\sqrt{\frac{2\pi \times 1.38 \times 10^{-23} \text{ J/K} \times 300 \text{ K}}{18 \times 1.67 \times 10^{-27} \text{ kg}}}$$

$$= 6.6 \times 10^{-4} \text{ sec}^2/\text{m} \times 53.7 \text{ m/sec}$$

$$= 35 \text{ msec}$$

In reality, water will be cooled during the evaporation, reducing P_e. Also, the last few monolayers of water will be held at the surface by electrostatic forces.

To find the deposition rate on the surface of the wafer, one needs to determine the fraction of that material leaving the crucible that accumulates on the surface of the wafer. The ultrahigh vacuum chamber makes the determination of this constant of proportionality relatively easy. The pressure is low enough to ensure that material ejected from the crucible travels in a straight line to the surface of the wafer. If it is assumed that all of the material that arrives at the wafer sticks and remains there, the arrival rate is governed by simple geometry. That is, the constant of proportionality is just the fraction of the total solid angle subtended by the wafer as seen from the crucible. This proportionality constant is the same view factor discussed in Chapter 6. It is given by [4]

$$k = \frac{\cos \theta \cos \phi}{\pi R^2} \tag{12.6}$$

where R is the distance between the surface of the crucible and the surface of the wafer and θ and ϕ are the angles between R and the surface normals of the crucible and wafer, respectively (see Figure 12.3A).

Equation 12.6 suggests that the deposition rate in the evaporator will depend on the location and orientation of the wafer in the chamber. Wafers directly above the crucible will be coated more heavily than wafers off to the side. Furthermore, θ, ϕ, and R actually vary across the surfaces of the crucible and the wafer so that film uniformity is a concern. One method commonly used to obtain good uniformity is to place the crucible and wafers on the surface of a sphere (Figure 12.3B). Then

$$\cos \theta = \cos \phi = \frac{R}{2r} \tag{12.7}$$

where r is the radius of the sphere. Combining Equations 12.5, 12.6, and 12.7, and recognizing that the deposition rate is just the mass arrival rate per unit area divided by the mass density of the film (ρ),

$$R_d = \sqrt{\frac{M}{2\pi k \rho^2}} \frac{P_e}{\sqrt{T}} \frac{A}{4\pi r^2} \tag{12.8}$$

In this equation, the first term depends only on the material to be evaporated, the second term depends on the temperature (and therefore the equilibrium vapor pressure), and the third term is determined by the geometry of the chamber.

To accommodate a spherical geometry, wafers are mounted in a hemispherical cage called a *planetary*. The wafers are flat, and so there will be small variations of the deposition rate across the surface of the wafer, but to a good approximation these can be ignored. Because all of the wafers are on the surface of the sphere, not only will the deposition occur at a uniform rate, the rate can be monitored at any single point on the sphere.

To obtain large deposition rates, evaporators are often operated with very high crucible temperatures. Then the vapor pressure in the region immediately above the crucible is high enough so that this region is in the viscous flow regime. Operation in this regime can lead to condensation of the evaporated materials into droplets. If these droplets travel to the surface of the wafer and adhere, the film can have very poor surface morphology. Operation in this regime also affects the deposition uniformity by creating a virtual source (Figure 12.4) at some distance above the crucible. To obtain the best uniformity, the evaporator must be run at low rates [5], although operation at these low rates requires extremely high vacuums to avoid contamination of the film.

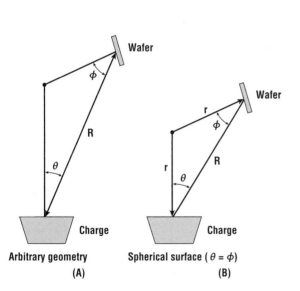

Figure 12.3   The geometry of deposition for a wafer (A) in an arbitrary position and (B) on the surface of a sphere.

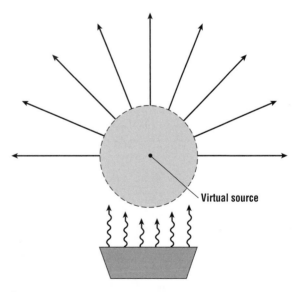

Figure 12.4   At high deposition rates, the equilibrium vapor pressure of the charge puts the region just above the crucible into viscous flow, creating a virtual source up to 10 cm above the top of the crucible.

The deposition rate is commonly measured using a quartz crystal rate monitor [6, 7]. This device is a resonator plate that is allowed to oscillate at the resonance frequency, which is then measured. The resonance frequency shifts due to the additional mass as material is deposited on top of the crystal. When enough material has been added, the resonance frequency will shift by several percent and the oscillator will no longer show a sharp resonance. The sensing elements are quite inexpensive, however, and are easily replaced. By linking the output of the frequency measurement system to the mechanical shutters, the thickness of the deposited layers can be well controlled over a wide range of deposition rates. Furthermore, the time rate of change of the deposited thickness can be fed back to the crucible temperature to maintain a constant deposition rate, if so desired. Crucibles must be constructed with tapered walls to avoid a progressive shadowing of part of the planetary as the charge is depleted. However, this decreases the surface area of the crucible as the charge is consumed. Again, this is not a serious concern if a crystal film thickness monitor is used to control the process.

Since the chamber does not represent a perfect vacuum, gases other than the vapor may also be incorporated in the deposited film. This deposition may be a simple physical mixing, or it may involve a chemical reaction. Because of the long mean free path in the chamber, if a chemical reaction occurs it will probably happen at the surface of the wafer. The incorporation of gaseous species may be intentional (reactive evaporation) or may be an unintentional incorporation of gases due to vacuum leaks or incomplete chamber evacuation.

Example 12.2

An evaporator is used to deposit aluminum. The aluminum charge is maintained at a uniform temperature of 1100°C. If the evaporator planetary has a radius of 40 cm and the diameter of the crucible is 5 cm, what is the deposition rate of aluminum? If the chamber also has a background pressure of 10^{-6} torr of water vapor and the water vapor is assumed to be at room temperature, determine the ratio of the arrival rates of the aluminum atoms and the water molecules. The mass density of solid aluminum is 2.7 gm/cm³ (2700 kg/m³).

Solution

According to Figure 12.2, at 1100°C the vapor pressure of aluminum is about 1×10^{-3} torr. The area of the crucible is 19.6 cm². The atomic mass of aluminum is 27. Using Equation 12.8 with MKS units,

$$R = 8.45 \times 10^{-6} \frac{\text{m}^2 \text{-sec}}{\text{kg}} \sqrt{K} \times 3.59 \times 10^{-3} \frac{\text{kg}}{\text{m-sec}^2 \text{-} \sqrt{K}} \times 9.8 \times 10^{-4}$$

$$R = 2.9 \times 10^{-11} \frac{\text{m}}{\text{sec}} = 17.8 \frac{\text{Å}}{\text{min}}$$

The arrival rate of aluminum atoms is just the growth rate times the number density of aluminum or

$$J_{Al} = R \times \frac{\rho}{M} = 1.78 \times 10^{18} \frac{\text{atoms}}{\text{m}^2 \text{-sec}}$$

According to Equation 10.10 and 12.2, the arrival rate of the water (atomic mass 18) is

$$J_{H_2O} = \sqrt{\frac{P_{H_2O}^2}{2\pi k T_{H_2O} m_{H_2O}}} = 4.8 \times 10^{18} \frac{\text{molecules}}{\text{m}^2 \text{-sec}}$$

Since the arrival rate of water is larger than that of aluminum, it would appear that the aluminum film will be heavily oxygen contaminated. In reality, the freshly evaporated aluminum film rapidly getters the residual oxygen and water vapor from the chamber. Most of this occurs before the shutter is opened. Such low deposition rates would still produce films with some oxygen contamination, however. This oxygen comes from the small amount of air that leaks past the chamber vacuum seals during the deposition. It is common practice, therefore, to deposit aluminum with crucible temperatures of about 1200°C where the vapor pressure is above 10^{-2} torr and the growth rate is several hundred angstroms per minnte.

12.3 Step Coverage

As already mentioned, one of the primary limitations of evaporation is step coverage. Figure 12.5 shows a schematic of the geometry of a film evaporated over a step. In this case, the step is the cross section of a contact etched through an insulating layer down to the substrate. Over this scale of distance (~1 μm), the incoming material beam can be considered nondivergent. Assuming that the incident atoms are immobile on the surface of the wafer, the topology will cast well-defined shadows, and the film will frequently be discontinuous on one side of the contact. As the deposition proceeds, the growing film on top of the insulator moves the edge of the shadow upward. The problem of covering steps is particularly acute for metallization layers because these are the last steps in the process, and unless some planarization technique is used, the accumulated topology may be quite severe. Some limited improvement can be made by optimizing the wafer orientation [8].

One frequently used method of improving the step coverage is to rotate the wafer in the evaporated beam. Thus, the hemispherical cages used to hold the wafers in the evaporator are designed to roll the wafer around the top of the evaporator. The deposition rate on the sidewalls is still less than the rate on the flat surface but is axially uniform. The aspect ratio of the contact is defined as

$$AR = \frac{\text{step height}}{\text{step diameter}} \tag{12.9}$$

Standard evaporation cannot be used to form a continuous film over features with aspect ratios greater than 1.0 and is very marginal for aspect ratios between 0.5 and 1.0.

A second method to improve step coverage is to heat the wafer. Many evaporators use banks of infrared lamps or low intensity refractory metal coils behind the planetaries to do this. Atoms that reach the wafer can diffuse across the surface before they chemically bond and become part of the growing film. When shadowing leads to a concentration gradient, this random motion will result in a net movement of material into the low rate deposition regions. In a manner similar to bulk

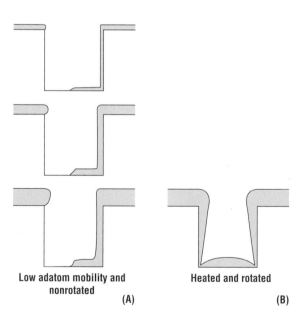

Low adatom mobility and nonrotated

(A)

Heated and rotated

(B)

Figure 12.5 (A) Time evolution of the evaporative coating of a feature with aspect ratio of 1.0, with little surface atom mobility (i.e., low substrate temperature) and no rotation. (B) Final profile of deposition on rotated and heated substrates.

diffusion discussed in Chapter 3, one can define a surface diffusivity which, to first order, follows a simple Arrhenius function

$$D_s = D_o e^{E_d/kT} \tag{12.10}$$

The surface activation energy is much smaller than that of the bulk diffusivity so that significant diffusion can occur at temperatures of a few hundred degrees centigrade. If the mean time before incorporation is τ, the characteristic surface diffusion length is

$$L_s = \sqrt{D_s \tau} \tag{12.11}$$

Because of the exponential dependence of D_s on temperature, heating the wafer above room temperature can dramatically increase L_s. It is not unusual for L_s to be much larger than the feature sizes of interest on the wafer. Many groups have demonstrated filling high aspect ratio structures using this technique [9]. One concern with heating the substrate when applied to alloy deposition is that the surface diffusivities of the constituent atoms may be very different. Thus, the composition of the film at the bottom of a contact may be different from the composition at the top of the structure. A second concern is that the increased substrate temperature can affect the film morphology, often leading to large grains. This can be avoided by using an ion beam after evaporation to redistribute the deposit [10]. The ions strike the surface and transfer energy to the evaporated film. The atoms in the film may redistribute by diffusion or by sputtering (to be discussed later in the chapter). Few evaporators are equipped with this capability, however. As a result, evaporation does not generally have any capability for independent control of morphology and step coverage.

Example 12.3   **Numerical simulation of deposited aluminum using Athena ELITE**

In this example aluminum is deposited into a contact with an aspect ratio of 1:1. An evaporator is used, both without and with planetary rotation.

```
go athena
# Example 12.3 - Effect of rotation in evaporated profiles
line x loc = 0.00  spac = 0.20
line x loc = 0.7   spac = 0.05
line x loc = 1.4   spac = 0.05
line x loc = 2.0   spac = 0.20
line y loc = 0.00  spac = 0.05
line y loc = 0.6   spac = 0.5
initialize
deposit oxide thick = .5 divis = 12
# Simple etch process to form a square contact hole.
etch oxide start x = 0.75 y = -10
etch cont x = 0.75 y = 10
etch cont x = 1.25 y = 10
etch done x = 1.25 y = -10
structure outfile = ex12_3_0.str
init infile = ex12_3_0.str
# Uni model refers to a unidirectional deposition with a small
(15°) angular spread
```

```
rate.depo machine = uni aluminum a.m sigma.dep = 0.20 uni
dep.rate = 300 angle1 = 15.0
# Run the deposition machine for five minutes
deposit machine = uni time = 5 minute divis = 5
structure outfile = ex12_3_1.str
# Now reload the initial structure and define a new machine
that includes planetary motion
init infile = ex12_3_0.str
rate.depo machine = planet1 aluminum a.m sigma.dep = 0.20 \
  planetar dep.rate = 300 angle1 = 15.0 angle2 = 40.0 angle3 = 6.0
c.axis = 20.0 p.axis = 10.0
deposit machine = planet1 time = 5 minute divis = 5
structure outfile = ex12_3_2.str
tonyplot -st ex12_3_1.str
tonyplot -st ex12_3_2.str
quit
```

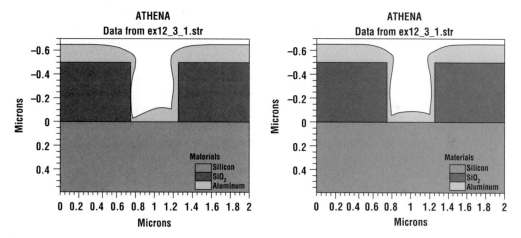

Figure 12.6 Athena simulation of the evaporation of aluminum in a 1:1 aspect ratio contact deposition conditions that reasonably approximate: simple evaporation (left) and evaporation onto a sample mounted in a rotating planetary (right).

Solutions:

The lack of step coverage is obvious for the deposition without rotation on the left. The profile on the right of Figure 12.6 has coverage on both sides, but the minimum thickness is less than 10% of the nominal (flat surface) thickness. Such a thin layer of metal could easily fail once sufficient current is passed through it.

12.4 Evaporator Systems: Crucible Heating Techniques

There are three types of crucible heating systems: resistive, inductive, and electron beam systems. A resistively heated system is the simplest type of source. Given a high vacuum chamber with power

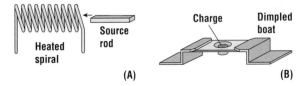

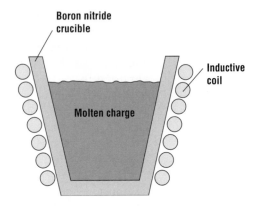

Figure 12.7   Resistive evaporator sources. (A) Simple sources including heating the charge itself and using a coil of refractory metal heater coil and a charge rod. (B) More standard thermal sources including a dimpled boat in resistive media.

Figure 12.8   Example of an inductively heated crucible used to create moderately charged temperatures.

feedthroughs, it is possible to construct a simple evaporator with only a small coil of wire and a simple variable transformer. The charge in such a system is a small solid bar laid over the heated element (Figure 12.7A). The input power is adjusted to prevent the charge from becoming molten and dripping through the coil. More practical arrangements are shown in Figure 12.7B. The charge is contained in a crucible that is heated resistively.

Since the filament wire must be at least as hot as the material to be evaporated, one of the problems with resistively heated crucibles is evaporation and outgassing from the wire. If a material such as aluminum is to be deposited, adequate vapor pressures can be obtained with only moderate power input. On the other hand, if a refractory metal must be deposited, there is often no suitable resistive heating element. One way to achieve at least moderate charge temperatures is the use of inductively heated crucibles. As shown in Figure 12.8, a solid charge is placed in a crucible, typically made of boron nitride (BN). A metal element is wound around the crucible, and RF power is run through the coil. The RF induces eddy currents in the charge, causing it to heat. The coil itself can be water cooled to keep its temperature below 100°C, effectively eliminating any loss of material from the coil.

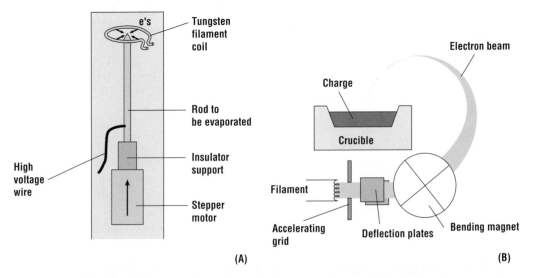

Figure 12.9   Electron beam evaporative sources. (A) A simple low flux source using a hot wire electron source and a thin movable rod. (B) A popular source using a 2707 source arc in which the beam can be rastered across the surface of the charge. The magnet must be much larger than shown to achieve the full 270° of arc.

While inductive heating can be used to raise the crucible temperature high enough to evaporate refractory materials, contamination of the charge from the crucible itself remains a serious problem. This effect can be avoided by heating only the charge and cooling the crucible. A common method for accomplishing this is electron beam (e-beam) evaporation. A simple low flux electron beam system (Figure 12.9A) consists of a loop of heated tungsten wire surrounding a thin rod of material held at a high bias with respect to the wire. Electrons boiling off the wire impact the rod, raising the temperature at the end of the rod and creating an atomic beam [11].

In most e-beam evaporators, an electron gun under the crucible ejects an intense, high energy beam. The location of the filament minimizes deposition of the filament material, typically W, on the surface of the wafer. A strong magnetic field bends the beam through 270° causing it to be incident on the surface of the charge (Figure 12.9B). The beam can be rastered across the charge to melt a significant fraction of the surface. The hot portion of the charge is then effectively self contained by the cooler portion of the charge.

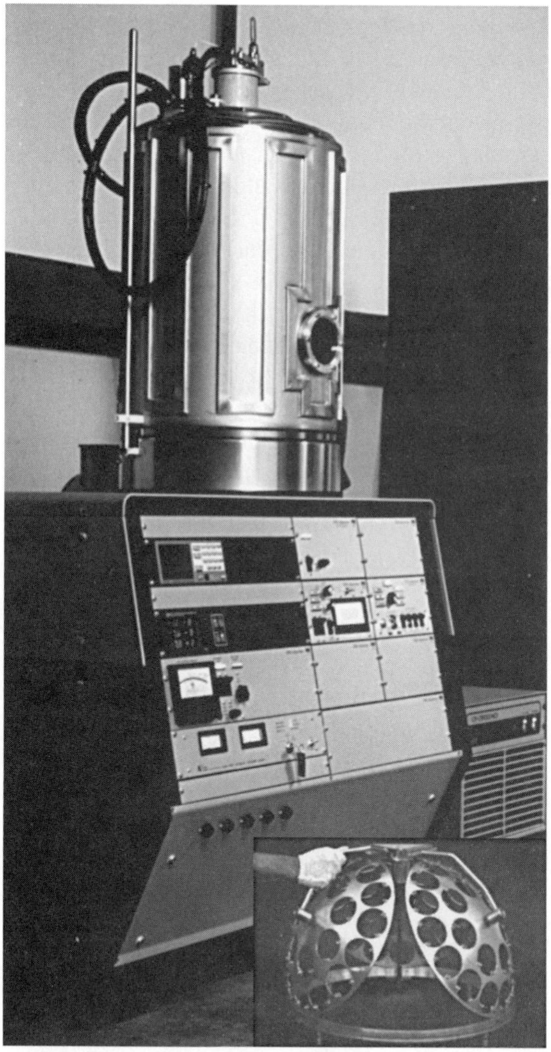

Because of their ability to easily deposit a wide range of materials, e-beam evaporators (Figure 12.10) are commonly used in GaAs technologies. When thermionic emission electron guns are used (see Section 9.2), the hot electron filaments remain a source of contamination in the chamber. Special care must be taken in e-gun design when these systems are operated at very high vacuums [12]. A more serious concern, particularly for silicon-based technologies, is radiation damage. The radiation is due to highly excited electrons in the material being evaporated decaying back to core levels. Since x-rays will damage the substrate and the dielectrics, e-beam evaporators cannot be used in MOS or other technologies that are sensitive to this type of damage unless later thermal annealing steps will be adequate to remove the damage. Even silicon bipolar technologies are susceptible to this type of damage, since transistors are typically isolated with MOS structures.

It is often convenient to have multiple sources in the evaporator, even if only one of them is to be used at a time. Such an arrangement allows the deposition of different materials without opening the high vacuum chamber. For a resistive evaporator, a high current switching box can be used, with each sample crucible having its own heating coil. E-beam evaporators are particularly well suited to this application, since the electron beam is easily steered between small charges using an electrostatic potential and perhaps a change in the magnetic field. Alternatively, a different charge can be moved into the beam mechanically by mounting the charges on a wheel that can be rotated into the irradiated position.

Figure 12.10 A commercial evaporator. Inset shows a planetary *(photographs courtesy of CHA Industries).*

12.5 Multicomponent Films

Often it is desirable to deposit alloys and compounds of materials. Figure 12.11 shows three possible methods for depositing alloy films using evaporation. It is possible to simply evaporate materials with very similar vapor pressures such as Al and Cu [13] from a properly prepared compound target. In some applications, such as the formation of ohmic contacts to GaAs, the vapor pressure of the constituent species is reasonably close and the variation in composition of the alloy is acceptable [14]. If, however, we place a solid sample of TiW with the desired composition in a crucible and evaporate it, the primary evaporated material may not be TiW, but rather some other combination of Ti and W. For example, at a crucible temperature of 2500°C, the vapor pressure of Ti is about 1 torr, while the vapor pressure of W is only 3×10^{-8} torr. The vapor that comes off initially is almost pure Ti. As the remaining melt composition changed due to this evaporation, the composition of the deposited film would slowly drift.

The basic problem with this process is the difference in vapor pressure of the various components. Controlling the composition with any reasonable accuracy is difficult; controlling it in many compounds is impossible. In coevaporation, multiple sources are run simultaneously in order to deposit an alloy structure. To deposit TiW, for example, two crucibles, one containing W and one containing Ti, are run at different temperatures. Although this is a significant improvement over a single-crucible process, the problem remains that the vapor pressure and ultimately the deposition rate are extremely sensitive functions of the temperature of the charge. In a single-component film, the absolute deposition rate is not that important, since a deposited film thickness monitor can be used to open and close the shutter. Unless multiple rate monitors are used in different regions of the evaporator [15], this type of system has no ability to control the flux of each component separately. During coevaporation therefore, even reasonable control of the film composition requires extraordinary temperature control.

An alternate technique for depositing multicomponent films is to preform a sequential deposition [16]. This is readily accomplished in a multiple source system by opening and closing shutters. After the deposition is complete, an alloy can be formed by elevating the sample temperature and allowing the components to interdiffuse. This can be assisted by depositing a large number of very thin layers with alternating compositions. The process requires the wafer to tolerate a subsequent high temperature step. The integrating time of the crystal rate monitor, however, prevents the use

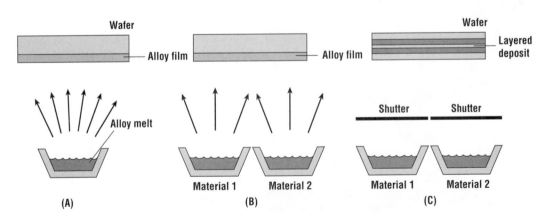

Figure 12.11   Methods for evaporating multicomponent films include (A) single-source evaporation, (B) multisource simultaneous evaporation, and (C) multisource sequential evaporation.

of this method for extremely thin layers without using very low deposition rates. Alternatively, the deposition can be controlled with the use of a mass spectrometer [17] or by measuring the mass of a deposited film on a sensor in the beam of each constituent with a microbalance [18].

12.6 An Introduction to Sputtering

Sputtering is the primary alternative to evaporation for metal film deposition in microelectronic fabrication. First discovered in 1852 [19], sputtering was developed as a thin film deposition technique by Langmuir in the 1920s [20]. It has better step coverage than evaporation, induces far less radiation damage than electron beam evaporation, and is much better at producing layers of compound materials and alloys. These advantages made sputtering the metal deposition technique of choice for most silicon-based technologies until the advent of copper interconnect.

A simple sputtering system, as shown in Figure 12.12, is very similar to a simple reactive ion etch system, a parallel-plate plasma reactor in a vacuum chamber. In a sputtering application, however, the plasma chamber must be arranged so that high energy ions strike a target containing the material to be deposited. (The requirements for obtaining a high energy ion flux were discussed in Chapter 11.) In sputtering, the target material, not the wafers, must be placed on the electrode with the maximum ion flux. To collect as many of these ejected atoms as possible, the cathode and anode in a simple sputtering system are closely spaced, often less than 10 cm. An inert gas is normally used to supply the chamber. The gas pressure in the chamber is held at about 0.1 torr. This results in a mean free path of order hundreds of micrometers.

Due to the physical nature of the process, sputtering can be used for depositing a wide variety of materials. In the case of elemental metals, simple dc sputtering is usually favored due to its large sputter rates. When depositing insulating materials such as SiO_2, an RF plasma must be used [21]. If the target material is an alloy or compound, the stoichiometry of the deposited material may be slightly different than the target material (Table 12.1). It has been shown, however, that the material with a lower sputter yield will accumulate on the surface of the target until the composition of the deposited film is approximately that of the bulk of the target [22]. (This is true only if the target temperature is kept sufficiently low to prevent solid state diffusion.) This makes sputtering very attractive not only for depositing elements, but a very wide range of materials.

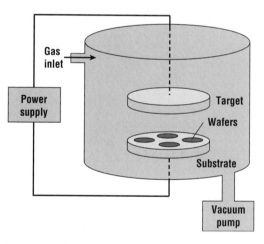

Figure 12.12 Chamber for a simple parallel-plate sputtering system.

Table 12.1 Composition of aluminum alloy films sputtered from composite targets

Aluminum Alloy Material	Target	Film
Si	0.5–1.0%	0.86%
Si	2%	2.8%
Cu	3.9–5.0%	3.81%
(Al + Si)Si	2%	2 ± 0.1%
(Al + Cu + Si)Si	4%	3.4%

See Wilson and Terry [23].

12.7 Physics of Sputtering°

The topic of glow discharges was covered in the latter half of Chapter 10. To briefly review, a plasma is initiated by applying a large voltage across a gap containing a low pressure gas. The required breakdown voltage is given by Paschen's law

$$V_{bd} \propto \frac{P \times L}{\log P \times L + b} \qquad (12.12)$$

where P is the chamber pressure, L is the electrode spacing, and b is a constant. Once a plasma is formed, ions in the plasma are accelerated toward the negatively charged cathode. When they strike the surface, they release secondary electrons, which are accelerated away from the cathode. They may collide with neutral species while crossing from cathode to anode. If the energy transfer is less than the ionization potential of the gaseous species, the atom can be excited to an energetic state (Table 12.2). The atom decays from this excited state through an optical transition, providing the characteristic glow. If the energy transfer is high enough, however, the atom will ionize and be accelerated toward the cathode. The bombardment of the cathode in this ion stream gives rise to the process of sputtering.

When an energetic ion strikes the surface of a material, four things can happen. Ions with very low energies may simply bounce off the surface. At energies of less than about 10 eV, the ion may also adsorb to the surface, giving up its energy to phonons (heat). At energies above about 10 keV, the ion penetrates into the material many atomic layer spacings, depositing most of its energy deep into the substrate, where it changes the physical structure. These high energies are typical for ion implantation. Between these two extremes, both energy transfer mechanisms occur. Part of the ion energy is deposited in the form of heat. The remainder goes into a physical rearrangement of the substrate. At this low energy, nuclear stopping at the surface is quite effective. Most of the energy transfer occurs within several atomic layers. When this happens, substrate atoms and clusters of atoms will be ejected from the surface of the substrate. The atoms and

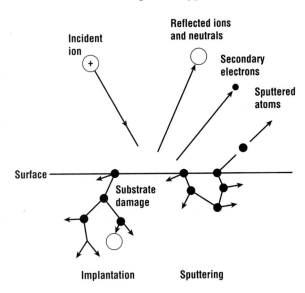

Figure 12.13   Possible outcomes for an ion incident on the surface of a wafer.

Table 12.2   First and second ionization potentials of some common gases

Atom	First Ionization (eV)	Second Ionization (eV)
Helium	24.586	54.416
Nitrogen	14.534	29.601
Oxygen	13.618	35.116
Argon	15.759	27.629

See Konuma [24].

atomic clusters ejected from the cathode escape with energies of 10 to 50 eV. This is about 100 times the energy of evaporated atoms. This additional energy provides sputtered atoms with additional surface mobility for improved step coverage relative to evaporation. At typical sputtering energies, about 95% of the ejected material is atomic. Most of the remainder is diatomic molecules [25].

At high energies such as those used in implantation, chemical bonding processes can be largely ignored and the target can be considered as simply a collection of atoms. At very low energies, no disruption of the target occurs, and a chemical model can be readily developed. At sputtering energies, however, the physics of the material removal is quite complicated, involving the coupled effects of bond breaking and physical displacement. Figure 12.13 shows some of the processes that may occur when an ion strikes a surface. A simple model developed by Wehner and Anderson [26], ignoring chemical effects and treating the substrate atoms as hard spheres, provides at least a qualitative picture of sputtering. An ion incident on the target surface may travel several atomic layers into the target until it strikes an atom with a small impact parameter and is deflected through a large-angle. This near head-on collision may also liberate a target atom that has a large momentum directed at a significant angle with respect to the surface normal. During this process, many of the bonds in the top layers of the target will be broken. If several of these large-angle collisions occur, the incident atom or the recoiled target atom may develop a significant velocity component parallel to the surface of the wafer. A subsequent collision can then eject an atom or small cluster of atoms.

12.8 Deposition Rate: Sputter Yield

The sputter deposition rate depends on the ion flux to the target, the probability that the impact of an incident ion will eject a target atom, and the transport of the sputtered material across the plasma to the substrate. The ion flux in a dc plasma can be approximated by the Langmuir–Child relationship

$$J_{ion} \propto \sqrt{\frac{1}{m_{ion}}} \frac{V^{3/2}}{d^2} \qquad \text{(12.13)}$$

where V is the voltage difference from the target electrode to the wafer, d is the dark space thickness, and m_{ion} is the mass of the ion. The value of the voltage exponent is not always found to be 1.5.

The sputter yield S is the ratio of the number of target atoms ejected from the target to the number of ions incident on the target. It depends on the ion mass, the ion energy, the target mass, and the target crystallinity. Figure 12.14 shows the sputter yield as a function of the ion energy for a variety of materials in an argon plasma [27]. For each target material, there exists a threshold energy, below which no sputtering

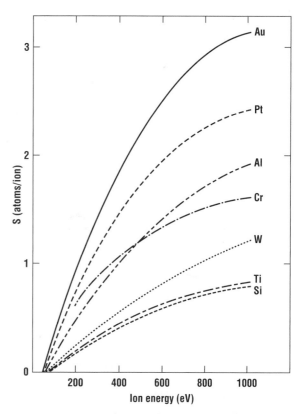

Figure 12.14 Sputter yield as a function of ion energy for normal incidence argon ions for a variety of materials *(after Anderson and Bay, reprinted by permission).*

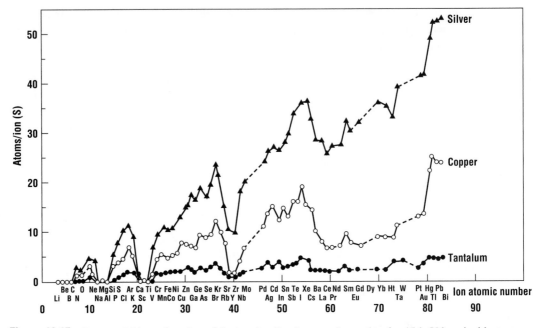

Figure 12.15   Sputter yield as a function of the bombarding ion atomic number for 45-keV ions incident on silver, copper, and tantalum targets *(after Wehner, reprinted by permission, AIP).*

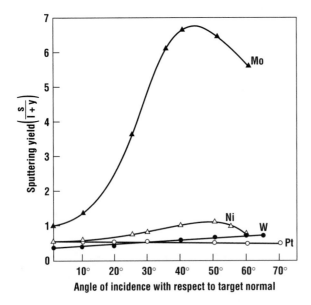

Figure 12.16   Typical angular dependence of the sputter yield for several different materials. The sputter profiles follow a cosine distribution *(after Wehner, reprinted by permission, AIP).*

occurs. This energy is typically in the range of 10 to 30 eV [28].

For ions whose energy is slightly greater than threshold, the sputter yield increases as the square of the energy up to about 100 eV, then linearly with energy up to about 750 eV. Above 750 eV, the yield increases only slightly until the onset of implantation [29]. The maximum sputter yield typically occurs at about 1 keV. The sputter yield is also a weak function of the plasma composition, generally increasing with ion mass (Figure 12.15) [30]. Notice from Figure 12.15 that the sputter yield is a maximum for bombarding ions with full or close to full valence shells. Noble gases such as Ar, Kr, and Xe have large yields.

The angular dependence of the sputter yield is a strong function of the target material and the incident ion energy. Materials such as Au, Pt, and Cu, which have high sputter yields, generally have little angular dependence. Materials such as Ta and Mo, which have low sputter yields, have a pronounced angular dependence at low ion energies. The maximum in the sputter yield occurs at about a 40° incidence angle with respect to the surface normal [31]. Figure 12.16 shows typical angular distributions for several materials [32]. At low energies a minimum exists at near normal incidence in a profile known as an undercosine. At high energies the sputter yield approaches [33]

$$S \propto \frac{M_{gas}}{M_{target}} \frac{\ln E}{E} \frac{1}{\cos \theta} \qquad (12.14)$$

where θ is the angle between the target normal and the velocity vector of the incident ion. At these energies the angular dependence of the net deposition rate approaches a simple cosine similar to that of evaporation. Crystalline targets, however, may have less regular distributions with maxima in the directions along the low Miller indices of the target.

The transport across the discharge is a complicated function involving computation fluid dynamics, including ion drift and diffusion. To improve the uniformity, many sputtering systems mechanically scan the wafers past a target. In some systems, the wafers may also be rotated.

Example 12.4

The system described in Example 10.4 is used to sputter aluminum by replacing the "wafer electrode" with a solid disc of aluminum that is the same size. Attached to the top and bottom of the chamber, facing the aluminum target, are 200-mm wafers. Assume that the gas in the chamber is argon, that 0.002% is ionized, and that all of the aluminum ejected from the target reaches the wafers. The plasma power is increased enough to obtain 0.5 V between the plasma and the grounded chamber. Find the deposition rate if the voltage between the target and the plasma is 104.6 V and the ions do not suffer any inelastic collisions as they cross the dark space.

Solution

If the power is increased to obtain a plasma potential of 0.5 V, the dc potential between the plasma and the target will be 523 V. A singly ionized argon atom will gain 523 eV of energy crossing the dark space. From Figure 12.14, the sputter yield for an aluminum target will be about 1.2 atoms/ion. The ion flux is given by Equation 10.10, where the ion partial pressure is 10 mtorr $\times$ 2 $\times$ 10^{-5}. Then J_{Ar+} is 4.8 $\times$ 10^{13} cm^{-2} sec^{-1}, which gives a J_{Al} of 5.8 $\times$ 10^{13} cm^{-2} sec^{-1}. Since the wafer area and the electrode area are the same, and since we assumed no transport losses, this is also the arrival rate of the aluminum. The number density of aluminum is the ratio of the mass density to the atomic mass, or 5.98 $\times$ 10^{22} cm^{-3}. Thus the deposition rate is 5.8 $\times$ 10^{13} cm^{-2} sec^{-1}/ 5.98 $\times$ 10^{22} cm^{-3} = 0.58 nm/min. Actually, somewhere between 20 and 60% of the aluminum is lost in most sputtering systems between the target and the walls. This is a rather slow rate; however, the ion density is low. Much higher rates can be obtained with dc magnetron systems (see next section) run at higher powers to get both higher ion concentrations and increased target bias, which gives larger sputter yield.

12.9 High Density Plasma Sputtering

As discussed in Section 10.7, the application of a magnetic field in a plasma causes the electrons to spiral around the direction of the magnetic field lines. Sputtering was the first process to make use of this effect. The systems use fixed bar magnets and the process is called magnetron sputtering. The radius of this orbital motion is given by

$$r = \frac{mv}{qB} \qquad (12.15)$$

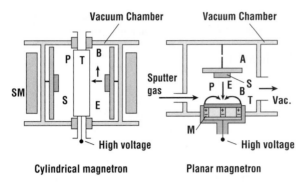

Cylindrical magnetron   **Planar magnetron**

Figure 12.17 Planar and cylindrical magnetron sputtering systems T: target; P: plasma; SM: solenoid; M: magnet; E: electric field; B: magnetic field *(after Wasa and Hayakawa, reprinted by permission, Noyes Publications).*

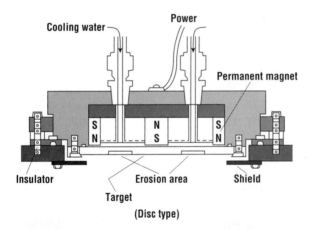

(Disc type)

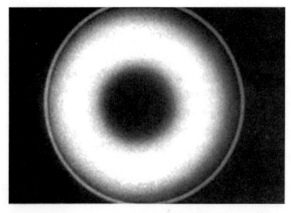

Figure 12.18 Detailed cross section of a rectangular planar magnetron target using permanent magnets to supply the field *(after Wasa and Hayakawa, reprinted by permission, Noyes Publications).*

The orbital motion of the electrons increases the probability that they will collide with neutral species and create ions. This increased ion density decreases the Crooke's dark space and increases the rate of ion bombardment of the target. Typical ion densities in normal sputtering systems are 0.001%, while in magnetron systems it often approaches 0.03%. The use of a magnetron also allows the formation of a plasma at lower chamber pressures, typically 10^{-5} to 10^{-3} torr.

Figure 12.17 shows two geometries that have been used for magnetron sputtering. A planar magnetron starts with the basic parallel-plate reactor shown in Figure 12.12. Either a solenoid coil [34] or a set of magnets is added behind the target in such a way as to create magnetic field lines parallel to the surface of the target. Circular planar magnetron targets can be made by arranging the bar magnets radially from the center of the target. A cylindrical magnetron starts with a cylindrical plasma chamber with a target electrode at the center. The wafer holding electrode is the vertical wall of the chamber. The magnetron is added by inserting the chamber in an electromagnetic winding that generates a vertical field [35]. This type of system is of less interest to microelectronic production due to the difficulty with placing large-diameter wafers on the walls of a cylindrical chamber. A typical field for either type of system is a few hundred gauss.

Figure 12.18 shows a detailed cross section of a planar magnetron target. The target material is typically hot pressed sintered discs. Most of the power in the plasma is dissipated in the form of heat in the target. To avoid excessive heating, the target is typically water cooled. Deionized water is often used to prevent an electrical short. The target material either is bonded to the backing plate using conductive epoxy or is held mechanically. To avoid sputtering the edge of the target or backing material, the edge of the target is surrounded by a ground shield. One of the concerns with the simple magnetron target, as shown in the lower half of Figure 12.18, is that the material in some parts of the target will erode much more rapidly than the material in other areas, resulting in a low utilization efficiency, commonly 25 to 35%. Other geometries have been shown to yield much better

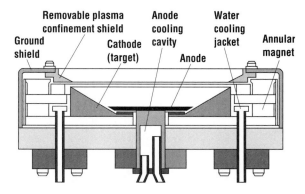

Figure 12.19 Cross-sectional diagram of an S-Gun® sputtering source *(courtesy Varian Associates)*.

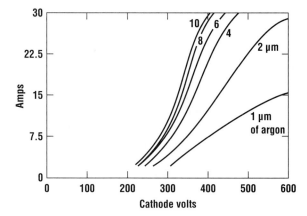

Figure 12.20 Current–voltage characteristics of a magnetron system with chamber pressure as a parameter. The lines are for 1, 2, 4, 6, 8, and 10 μm of Ar *(after Van Vorous, reprinted by permission, Solid State Technology).*

efficiencies [36]. Magnets may also be rotated to improve erosion uniformity. Magnet positions can also be used to tune deposition uniformity. Sputter utilization efficiency is especially a problem in modern systems where wafer load locks are used to avoid opening the chamber for long periods of time. Base vacuums of some systems are of order 10^{-10} torr. These systems may be unusable for several days after a target change while the main chamber is pumped to high vacuum.

One interesting target geometry [37], shown in Figure 12.19, is the S-Gun a registered trademark of Varian Associates. As in other magnetron systems, high deposition rates are achieved through crossed electric and magnetic fields. The anode and cathode are built into a single assembly. With an S-Gun, the wafers may then be mounted in a planetary cage, similar to the geometry of an evaporator. (One of the major applications of S-Guns is to upgrade an evaporator to have sputter deposition capability.) Typical deposition rates are hundreds of angstroms per minute. The *I–V* characteristics of a magnetron system are shown in Figure 12.20 [38, 39]. Such curves are characteristic of all magnetron sputtering and are of the form [40]

$$I = k(V - V_o)^n \tag{12.16}$$

where V_o is the minimum voltage necessary to maintain the plasma, and n is a positive number between 1.5 and 8, whose value depends on the geometry of the target and the pressure of the chamber.

12.10 Morphology and Step Coverage

Because the chamber pressure during most sputtering processes is high, ejected atoms will suffer many collisions before reaching the surface of the wafer. As a result, the deposition rate on the surface of the wafer will depend on the solid angle of the plasma at that point on the wafer. Once they reach the surface of the wafer, the adatoms will diffuse along the surface until they form nuclei of critical size. Once stable nuclei are formed, they capture more adatoms, forming islands. If the surface mobility is high, the islands will merge while still very thin, forming a smooth, continuous film.

The morphology of the deposited film has been described by a zone model first introduced by Movchan and Demchishin [41] and later modified by Thornton [42]. Figure 12.21 shows a schematic of the regions of film morphology as a function of substrate temperature and incident ion energy. The substrate temperature is normalized to the melting temperature of the film. Although the model is broadly applicable to materials that crystallize, the region boundaries vary somewhat from material to material. At the lowest temperature and ion energy, the film will be an amorphous, highly porous solid with a low

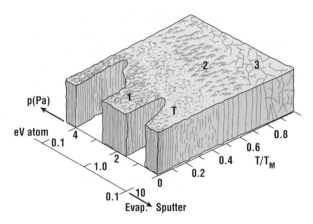

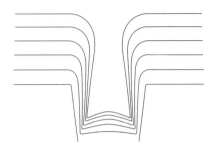

Figure 12.21   The three-zone model of film deposition as proposed by Movchan and Demchishin *(after Thornton, reprinted by permission, AIP)*.

Figure 12.22   Cross section of the time evolution of the typical step coverage for unheated sputter deposition in a high aspect ratio contact.

mass density. This is the first zone of the diagram. It is caused by the low adatom mobility of the growing film. Metal films deposited in this region can readily oxidize when exposed to air and so may also have high resistivities. If the chamber pressure is lowered or the substrate temperature is raised, the deposition process enters the "T" zone. Films deposited in this region are highly specular and have very small grains. For many microelectronic applications, this is the most desirable region of operation. Increasing the temperature and/or impinging energy further causes the grain size to increase. The second zone has tall narrow columnar grains that grow vertically from the surface. The grains end in facets. Finally, in zone 3, the film has large 3-D grains. The surfaces of the films in the second and third zones are moderately rough, and the films appear milky or hazy.

The most common microelectronic application of sputtering is the deposition of metal interconnect layers for ICs. Typically these layers are applied over thick insulating films such as SiO_2. To contact the device, holes are first etched in the SiO_2. As discussed earlier in the chapter, a crucial feature of the deposition process is the ability of the films to maintain an adequate thickness, even over high aspect ratio structures. The step coverage of films deposited by magnetron sputtering has been calculated by many authors [43, 44]. Figure 12.22 shows the development of a cross section over a typical high aspect ratio contact. On the top surface and near the upper corner the deposition rate is high. A more moderate deposition rate occurs on the sidewalls. The sidewall thickness tapers toward the bottom. At the bottom corner of the step a pronounced notch or crack may occur. This tendency increases with the aspect ratio of the feature unless substrate heating is done [45]. Compared to evaporated films, however, even low temperature deposited sputtered films have better step coverage, both because of the higher pressure and because of the incident energy of the deposited species. Sputtered films also tend to form a cusp, or protrusion, at the edge of an isolated step, due to the increased view factor at this position.

As with evaporation, the application of substrate heat will dramatically improve the step coverage due to surface diffusion. The wafer-containing electrode is often water cooled as well as resistively heated to provide a wide range of available temperatures. Very hot metal can completely fill contacts whose aspect ratio is greater than one, but if aluminum is used, a very robust barrier must be used between the silicon and the aluminum to avoid undesirable intermixing (see Chapter 15 on contact spiking). Furthermore, the temperature of the surface of the wafer during plasma deposition is difficult to control. Radiative heating from the target and bombardment by high energy secondary electrons present a large heat load to the wafer. RF sputtering systems are the most prone to uncontrolled substrate heating. Figure 12.23 shows a typical plot of substrate temperature versus plasma power for an SiO$_2$ target [46].

Step coverage remains a serious problem for high density interconnect, in purely sputtered interconnect technologies. Sufficient heating of the substrate during deposition in order to obtain the desired step coverage may produce unacceptably large grains or interdiffusion effects (see Al/Si/Cu deposition in section 12.12). A second technique for improving the step coverage in sputtering is to apply a bias to the wafers. If the bias is sufficiently large, the wafer will begin to be bombarded by energetic ions. This will tend to redeposit the sputtered material, somewhat improving step coverage.

New sputtering techniques have been developed in an effort to improve the step coverage of sputtered films into narrow, high aspect ratio contacts and vias. We have already discussed the use of hot aluminum metal as a technique for via filling and the concerns of reaction with the underlying silicon and the effects of the temperature on the aluminum grain structure. The next technique that was developed for this application was called force fill. The force fill technique intentionally sputters the film in a way that produces a pronounced cusp at the top corner of the contact. As the deposition continues, the deposition fronts at the top of the contact converge, pinching off any further deposition into the contact. This seals the low pressure sputtering gas (usually Ar) in a void inside the contact.

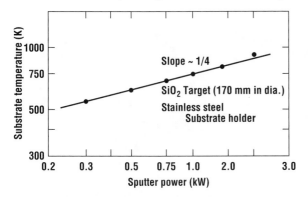

Figure 12.23   The temperature rise of substrates as a function of the plasma power for an argon RF plasma with an SiO$_2$ target (*after Wasa and Hayakawa, reprinted by permission, Noyes Publications*).

The step coverage of the film into the contact is extremely poor at this point. To rectify this situation, the wafers are put into an autoclave, where they are heated and pressurized to a number of atmospheres. This puts a force on the top of the metal bridges that seal the contacts. If the force at this point exceeds the yield strength of the heated metal, the seals collapse inward, pushing the metal down into the contacts. Force fill works well only for a certain range of contact sizes, and it does not improve the coverage of metal films over isolated steps.

The next level in improving the step of high aspect ratio contacts is collimated sputtering (Figure 12.24). This technique inserts a plate with high aspect ratio

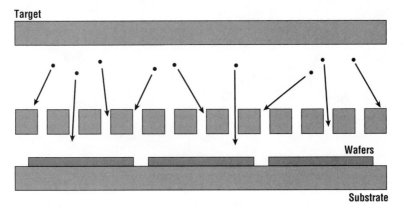

Figure 12.24   In collimated sputtering a disposable collimator is placed close to the wafers to increase directionality.

holes just above the wafer. If one sputters at low pressure (a few millitorr), the mean path will be long enough that few collisions will occur between the collimator and the wafer. Due to the high aspect ratio holes through the collimator, only species with velocities nearly perpendicular to the surface of the wafer will pass through the holes. This dramatically reduces the deposition rate (about $3\times$ for every 1:1 increase in the aspect ratio of the collimator, but acceptable rates are obtained for depositing diffusion barriers that are commonly 20 to 40 nm thick. Although early collimators were made from Al or Cu to reduce heating, most are now made from Ti to match the thermal expansion of the film being deposited (TiN). Although widely used, collimators have a number of problems. Higher aspect ratio contacts require higher aspect ratio collimators and correspondingly lower deposition rates. The material that deposits on the collimators builds up until it becomes so thick that it flakes off and may fall on the surface of the wafer. As a result, it is unusual to use collimators for contacts with aspect ratios larger than 3:1.

The state of the art for sputtering, or physical vapor deposition as it is now commonly called, is the ionized metal plasma (IMP) deposition (Figure 12.25). Ejected metal atoms pass through a second plasma that ionizes the plasma material. The angular distribution of material arriving on the surface of the wafer is controlled by the dc bias on the wafer and the ionization of the sputtered

Figure 12.25   The Endura system by Applied Materials uses a number of PVD or CVD chambers fed by a central robot. For conventional and IMP sputtering, targets are hinged to open upward. Two open chambers are shown, along with the load lock *(from Applied Materials)*.

species. The degree of ionization in the second plasma depends on its transit time, which in turn depends on the energy of the ejected materials. To gain sufficient ionization, IMP systems often operate close to 10 mtorr, where collisions slow the ejected materials. Because the IMP process produces near-vertical deposition, the coverage on horizontal surfaces is much better than on the sidewalls. Applied Materials has demonstrated contact bottom film thickness in 8:1 contacts that is 80% of the flat area deposition thickness, but sidewall coverage depends critically on the taper and is commonly less than 25%.

12.11 Sputtering Methods

All exposed surfaces in an RF plasma develop a negative potential with respect to the plasma due to the higher mobility of electrons than ions. In a typical sputtering system, most of the voltage drop is on the target electrode, but the bias on the substrate electrode leads to a bombardment on ions of the wafers as well. The bombardment leads to a removal of material from the surface of the wafer. This effect can be controlled by adjusting the dc bias on the electrode with respect to the plasma. This has two major applications in microelectronics: sputter cleaning and bias sputtering. The removal of all surface contaminants from the wafer before film deposition has been studied extensively for low temperature epitaxial growth. That topic will be covered in a later chapter. This section will restrict its attention to a purely physical method of cleaning: sputter etching of the substrates. A typical example of the use of a sputter deposition process is the deposition of a metal to form an ohmic contact to heavily doped silicon. The contacts are patterned and etched through a thick insulating layer of SiO_2. Immediately before loading into the sputtering system, the wafers are dipped in a dilute mixture of hydrofluoric acid and water (1:100) to remove any oxide that has regrown on the silicon after the contact etch. The wafers are rinsed in deionized water, spun dry, and loaded directly into the vacuum system for deposition. The brief exposure to a water rinse and the air, however, can allow a regrowth of a thin, patchy native oxide. To obtain a repeatable low resistance contact, it is desirable to remove this very thin oxide layer before metal deposition.

By reversing the electrical connections, it is possible to sputter from the substrates rather than from the target. This is frequently done for a short time before deposition to remove the native oxide and any residual contaminants from the surface of the wafer [47]. Sputter etching has serious problems, however. Sputtered material from the substrate electrode or from the oxide-coated regions of the wafer may deposit on the surface, leading to more contamination rather than less. The contamination may have serious consequences if it contains heavy metal impurities that cause junction leakage [48]. Organic contaminants, such as those obtained from condensed pump oil vapors, may be polymerized, making them very difficult to remove [49, 50].

Sputter-etched silicon layers show damage extending 40 to 110 Å into the wafer [51]. The surface of the silicon may contain up to 20 atomic percent Ar, depending on the bias condition [52]. Material removal from the surface of the wafer may also be nonuniform, leading to the formation of steep cones and etch valleys [53]; however, for a typical sputter etch the desired etch depth is less than 100 Å; more recently ionized metal plasma systems have been used with far less sputter damage. The development of a sputter preclean step, however, must be optimized experimentally for each sputtering system by measuring the effects of the sputter-cleaning step on the contact resistance, contact reliability, and the junction leakage.

For simple magnetron systems it is possible to adjust the bias on the substrate with respect to the plasma if the substrate and the deposited films are conductive. By placing a negative bias on the substrate, the ion bombardment of the substrate is increased. By controlling the bias, one can change the rate of deposition independent of the rate of sputter etching of the growing film. Since the sputter-etched film may redeposit on the wafer if sputtered at low bias, a net improvement in the step coverage

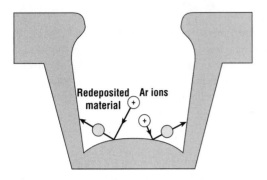

Figure 12.26   In bias sputtering, the ions incident on the surface of the wafer redistribute the deposited film to improve step coverage.

may be achieved (Figure 12.26). Vossen has shown that this technique can be used to actually produce a higher deposition rate on the sidewalls of a contact than on the surface [54]. High sputter etch rates, however, may lead to damage of the underlying substrate and faceting of the deposited layers. At lower bias voltages, the incident ion energy may also improve step coverage by increasing the adatom mobility [55].

It is also possible and often desirable to clean the target before beginning the deposition. To presputter, the plasma is ignited before opening the shutter so that the material on the top of the target is deposited on the backside of the shutter instead of on the substrates. A common application of presputtering is the removal of the native oxides formed on metal targets. For this application, the change in the glow discharge current can be used to determine the removal of the oxide [56]. The surface oxides generally have high secondary electron emission rates, and so the discharge current falls to a steady state value once the target is clean. Presputtering of these reactive species also getters the residual reactive gases such as O_2, H_2O, and N_2 from the chamber, both by gas phase reactions and by coating the back surface of the shutter with a highly reactive film.

12.12 Sputtering of Specific Materials

Many reports have been given regarding the utility of aluminum-based sputtered metallizations for silicon-based ICs [57, 58]. Pure aluminum has been replaced by aluminum silicon alloys to increase the reliability of ohmic contacts formed to shallow junctions. Typical silicon concentrations are 0.5 to 2.0 atomic percent. The addition of 0.5 to 1.0 atomic percent copper also reduces the tendency of the metal film to form hillocks [59] and dramatically improves the ability of lines formed from these films to pass high currents without electromigration degradation or stress-induced voiding. Each of these effects will be discussed in Chapter 15. The first part of this section will review the sputter deposition of aluminum alloy films.

To obtain large sputter deposition rates, most aluminum is deposited in planar dc magnetron systems. Since all of these films are alloys, control of the stoichiometry of the film is a primary concern. As discussed previously, the deposited film composition is usually close to the bulk target composition. At moderate substrate temperature where reevaporation of the deposited material can be neglected, the exact composition of the film is controlled by the transport properties of the constituents in the plasma [60]. At low chamber pressures, for example, sputtering from an AlCu target will result in slightly higher Cu concentrations in the film than in the target. The high copper composition is related to the ability of the Ar gas to thermalize the very light aluminum atoms, while collisions with the much heavier copper atoms have little effect [61]. More of the thermalized element is lost to the walls, and so less reaches the wafer. This result, called *selective thermalization* [62], is most pronounced with sputtering materials such as AlCu or TiW, in which one material has a much larger atomic mass than the other.

One way to obtain better control of the stoichiometry is to have multiple targets. By adjusting the power to each target, one can alter the composition of the deposited layer. A second method of controlling composition that does not require a second power supply is the use of composite targets that have different regions of concentration. In the simplest case, bits of one material can be attached to a target with an adhesive. The film composition will be determined by the ratios of the exposed areas. It is also possible, for example, to construct a target with cylindrically symmetric regions of two different materials. By changing the electrical properties of the plasma, one can control the composition of the deposited layer [63]. In most production sputtering, however, the target composition is simply chosen to give the desired film composition for a particular process.

Film resistivity is a sensitive function of the base pressure and gas composition in the chamber. Aluminum reacts readily with oxygen, nitrogen, and water. The resistivity rises sharply for nitrogen partial pressures above 10^{-6} torr [64]. If the primary source of nitrogen is an incomplete pumpout, much of the nitrogen will be gettered during the deposition. In this case, a presputter may substantially clean the chamber. More serious are small vacuum leaks that contribute nitrogen through the bulk of the deposit. To reduce the film resistivity, many production systems designed for aluminum deposition use a load lock to avoid the necessity of venting the main chamber to air. Wafers may be baked at temperatures greater than 400°C or may be exposed to deep UV sources to desorb most of the adsorbed species in the load lock prior to high vacuum chamber entry. A low temperature stage called a Meissner trap [65] is often used to condense residual gases and so reduce the pumpdown time.

Another variable of primary importance for IC fabrication is the deposited film reflectance. Films that have a low reflectance often appear hazy or milky. The large grain size of these films makes photolithography difficult, both because of an inability to see the alignment marks from previous layers and because of stray light scattered off the aluminum grains. The minimum reflectivity required for most UV photolithography is about 0.6. The factors that influence the specularity of Al-based metallizations include the substrate temperature [66], film thickness [67], and residual gases in the chamber [68]. It is found that the reflectivity obeys the relationship

$$R \propto e^{-[4\pi\sigma/\lambda]^2} \tag{12.17}$$

where λ is the wavelength of light and σ is the root mean square (RMS) surface roughness, as long as $\lambda \gg \sigma$ [69]. Since the material on which the aluminum is deposited is also known to affect the morphology of the deposited film [70], this may also affect the film specularity.

Figure 12.27 shows the percentage reflectance of Al–Si layers deposited at a constant rate as a function of the partial pressures of N_2, H_2, and H_2O. Generally, the higher the deposition rate, the less the effects of the residual gases. Once again, nitrogen has the most pronounced effect, seriously degrading the reflectivity if its partial pressure is greater than 10^{-6} torr. Although H_2 requires a partial pressure almost 10 times higher than N_2 for the same effect, hydrogen pumping is much less efficient. In a turbomolecular pump, hydrogen has the lowest compression ratio because of its light mass. In a cryopump, the pumping rate of hydrogen is low due to its low boiling point. Production systems may therefore use a getter pump, such as a titanium sublimation pump [71], to remove most of the hydrogen during the pumpdown. State-of-the-art sputter systems commonly operate with base pressures of about 1×10^{-9} torr.

When compounds whose components have very different sputter yields are to be deposited and one component can be obtained in a gaseous precursor, it may be preferable to deposit the film reactively. Reactive sputtering is a process in which the normally

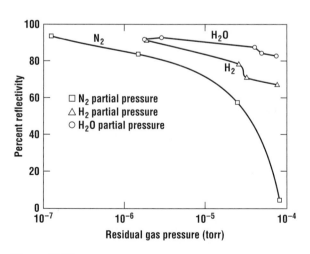

Figure 12.27   Reflectance of Al–Si layers at 400 nm as a function of the partial pressures of N_2, H_2, and H_2O *(after McLeod and Hartsough, reprinted by permission, AIP).*

inert sputter gas is replaced by a inert/reactive mixture. The composition of the deposited film can be controlled by varying the partial pressure of the reactive species in the plasma. The deposition rate is also affected by changing the partial pressure of the reactive species. Using reactive sputtering, the film composition can be smoothly controlled over a wide range of compositions. Because of the range of energies available to drive chemical reactions in the plasma, many compounds may be produced. The task is to select deposition and postdeposition annealing conditions so as to preferentially form the desired compound or phase.

As will be discussed in Section 15.7, there is a requirement in modern devices to form ohmic contacts to very thin junctions. To prevent an interdiffusion of the metal and silicon, a thin barrier metal is sometimes used. One popular barrier metal has been TiW; however, sputtering of TiW from a compound target produces high particle counts and so can be a serious problem for yielding dense ICs. Alternate diffusion barriers include TiN and WN films [72], produced both by chemical vapor deposition and by sputtering [73]. Unusual in its low resistivity, high thermal and chemical stability, and extreme hardness, TiN has also received considerable attention as a material for coating machining workpieces for reduced wear. The deposition of TiN will therefore be used as a vehicle to describe the process of reactive sputtering. The TiN layer must be free of pinholes and cracks. This is particularly difficult to achieve for very thin layers.

The primary variable for controlling the composition of the deposit is the nitrogen partial pressure in the sputtering system. Figure 12.28 shows the resistivity and composition of reactively sputtered TiN_x films as a function of nitrogen flow. A sharp jump in resistivity is seen at the onset of TiN formation [74]. Since it is difficult to maintain exact stoichiometry [75], the deposition is usually done with a nitrogen-rich ambient. Sharp corners or edges of the steps to be covered should be avoided to minimize stress cracking of the barrier metal (Figure 12.29) [76]. Films deposited in a nitrogen-rich ambient, however, have excess gas molecules that can passivate any microcracks that might develop in the film after deposition [77]. These films are known as stuffed barriers in the literature [78]. The films may also be annealed in a nitrogen or ammonia [79] ambient after deposition to increase the nitrogen packing and decrease the stress in the film. The residual gas partial pressures of H_2O and O_2 in the chamber should be less than 10^{-6} torr to avoid significant contamination of the

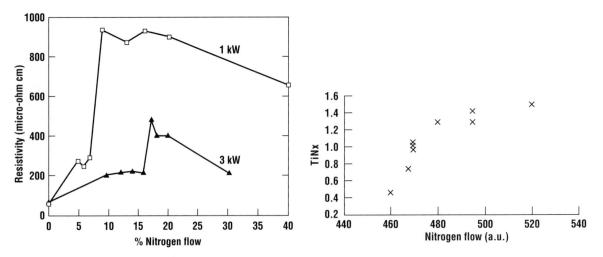

Figure 12.28   Resistivity and composition of reactively sputtered TiN as a function of the N_2 flow in the sputtering chamber (*after Tsai, Fair, and Hodul, reprinted by permission, The Electrochemical Society, and Molarius and Orpana, reprinted by permission, Kluwer Academic Publishing*).

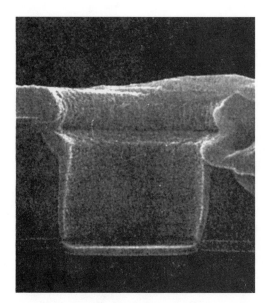

Figure 12.29 Cross section electron micrograph of a moderately high aspect ratio contact that has been sputter-deposited with TiN *(after Kohlhase, Mändl, and Pamler, reprinted by permission, AIP).*

Compressive stress   Tensile stress

Figure 12.30 The change in wafer deflection may be used to measure the stress in a deposited layer. This is typically measured using a reflected laser beam.

film [80]. Application of bias increases the film density for unheated depositions (zone 1 to zone T in Figure 12.21) [81]. Finally, it has been shown that a sputter cleaning of the substrate prior to reactive TiN sputtering may be necessary to obtain a low leakage diode when depositing on *n*-silicon [82].

12.13 Stress in Deposited Layers

A thin film deposited on a substrate can be either in tensile stress, in which the film would relax by contracting, or in compressive stress. If the stress in the film is too large, the film may peel from the surface of the wafer. The film stress and structure play important roles in the performance, and particularly in the reliability, of the metal lines. Large stress may give rise to void formation upon subsequent thermal cycling where the strain state provides a driving force for grain boundary diffusion.

One component of the stress is caused by thermal expansion mismatch of the film with the substrate. This stress is seen when the deposition is not done at room temperature. If E is Young's modulus for the material and ν is Poisson's ratio, and it is assumed that E and ν are temperature independent [83], this stress is given by

$$\sigma_{th} = \frac{E_{film}}{1 - \nu_{film}} \int_{T_o}^{T_{dep}} (\alpha_{film} - \alpha_{sub}) dT \qquad \textbf{(12.18)}$$

Intrinsic stress is less well understood. As already noted, films may form single-crystal grains during the deposition process. For polycrystalline films, stress can arise when the deposition is carried out at high temperature. Under these conditions, the larger grains may grow at the expense of the smaller grains. The increase in the crystalline order will lead to an overall stress. Thus, the intrinsic stress can depend on variables such as the substrate temperature, deposition rate, film thickness, and background chamber ambient. Generally, if the deposition is carried out at low temperature, the intrinsic stress is small. Taking into account both sources of stress,

$$\sigma = \sigma_{th} + \sigma_{bi} \qquad \textbf{(12.19)}$$

Stress in a film is commonly measured using the change in the bow of the wafer before and after film deposition (Figure 12.30). Then the film stress is given by

$$\sigma = \frac{\delta}{t} \frac{E}{1 - \nu} \frac{T^2}{3R^2} \qquad \textbf{(12.20)}$$

where δ is the change in the deflection of the center of the wafer, t is the film thickness, R is the radius of the wafer, and T is the thickness of the wafer.

12.14 Summary

Evaporation is an ultrahigh vacuum technique for depositing thin films. The material to be evaporated is heated in a crucible. The vapor of the material travels in a straight line to the substrates. Crucible heating can be done resistively, inductively, or with an electron beam. The latter technique is particularly useful for materials that require very high temperatures for deposition. The deposition rate depends on the vapor pressure of the charge and the geometry of the reactor. To improve the uniformity and reduce the effects of shadowing, the wafers are often mounted on a planetary that causes them to rotate during deposition. Substrate heating may also be used. The primary problems with evaporation include step coverage, alloy formation, and, in electron beam systems, radiation damage.

Sputtering is another purely physical process that uses a glow discharge to remove material from a target. The ejected material diffuses to and collects on the surface of the wafer. The primary advantages of sputtering compared to evaporation are improved step coverage and ease of deposition of alloys and compounds. Some compounds may also be deposited by reactively sputtering the target with a dilute mixture of argon and a reactive species. As a result of these advantages, sputtering has been used extensively for metallization in silicon technologies. High aspect ratio features require high substrate temperatures and/or substrate biases to obtain acceptable step coverage. These temperatures, however, promote the growth of large grains in the film. For the student who would like to investigate sputtering further, there are a variety of good texts. Stuart has written a short, qualitative introduction [84]. In addition, chapters on sputtering are included in Chapman and Mangano [85] and Vossen and Kern [86]. Excellent books that focus on sputtering include Konuma [24] and Wasa and Hayakawa [46].

Problems

1. In Example 12.2, the atomic arrival rate of aluminum is compared to the molecular flux of water vapor. For the same evaporator, what crucible temperature would be required to set the aluminum atomic flux equal to the water vapor flux calculated in the example? What would be the growth rate of the aluminum film?
2. The evaporator described in Example 12.2 is used to deposit nickel (Ni), which has a density of 8.9 gm/cm3 (8900 kg/m^3). If the crucible temperature is 1600°C, what is the deposition rate in angstroms (10^{-10} m) per minute?
3. An evaporator has a crucible with a 5-cm^2 surface area. The evaporator planetary has a 30-cm radius. Determine the crucible temperature needed to obtain a gold deposition rate of 1 Å/sec. The density and atomic mass of gold are 18,890 kg/m^3 and 197, respectively.
4. It is desired for a particular application to deposit a 50/50 mixture of gold (mass 197) and aluminum (mass 27) in an evaporator that has two independent thermal sources. To do this a researcher fixtures an evaporator with a large circular shutter, shown schematically in the accompanying illustration. The shutter spins at 10 rpm to allow the alternate deposition of each material. Assume that the opening is one-half of the total shutter area. The area of each crucible is

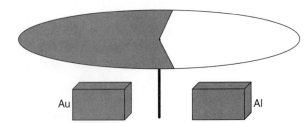

10 cm². The radius of the deposition sphere is 50 cm. Assume that the mass density of gold and aluminum are 18,900 and 2700 kg/m³, respectively. What temperature would you use for each of the sources if you want to deposit exactly one monolayer (assume 3 Å or 3×10^{-10} m) of each material on each cycle of the shutter rotation?

5. Problem 1 suggests that very high rate depositions may be desirable in reducing the film contamination. What drawbacks might exist for such a process?

6. It is desirable to deposit a mixture of Ga and Al using a single-source evaporator. If the deposition temperature is 1100°C, the mixture in the crucible is initially 50% molar, and both components have a sticking coefficient of one, what would the initial film composition be? How will this change with time?

7. The planetary concept used in an evaporator, in which the source and all of the wafers are on the surface of a sphere cannot be accurately carried out with wafers, since they are flat and planar rather than having a curved surface. This problem gets worse as the wafers get larger relative to the size of the planetary. Consider the geometry shown in the accompanying diagram. A wafer of radius A is located in a planetary of radius r. The wafer is centered over the crucible. The center of the wafer is on the spherical surface of the planetary.

    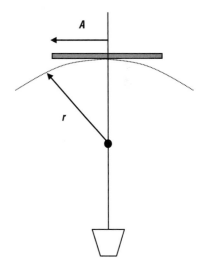

    (a) Calculate an expression for the ratio of the deposition rate at the edge of the wafer to the deposition rate at the edge of the wafer.

    (b) If $r = 20$ cm and $A = 10$ cm, find a numerical value for this ratio.

8. It has been suggested that directing a broad-beam laser could improve the step coverage of evaporated films. Discuss possible mechanisms for this improvement.

9. Suggest why the sputter yield might increase with ion mass for a fixed ion energy. (*Hint*: This may require a discussion of inelastic collisions.)

10. A dc sputtering system is used to deposit aluminum. The system has two large circular parallel plates (see Figure 12.12). If the diameter of the plates is much larger than the plate spacing, near the center of the system, sputtering is essentially a one-dimensional process and transport losses are very low. At 20 mtorr of argon, the maximum sputter deposition rate on the wafer for any power in the plasma is found to be 100 nm/min in this chamber.

    (a) The number density of aluminum is 6.0×10^{22} cm⁻³. Find the flux of the aluminum atoms that arrive at the surface of the wafer.

    (b) Using simple kinetic theory arguments, calculate the approximate flux of neutral argon atoms to the surface of the target (Take the argon temperature to be 400 K.)

    (c) If the maximum deposition rate occurs at the bias that produces the maximum sputter yield, what is the sputter yield for this process?

    (d) If every Al atom ejected from the target eventually arrives at the wafer, and the Al atoms are ejected by argon ions only, what fraction of the argon in the plasma is ionized?

11. Discuss why the sputter yield has a maximum at some energy.

12. Why would the reflectivity of a metal film be a concern?

13. Explain why the stress in a film depends on the temperature of the film when it is measured.

References

1. C. B. Alcock, V. P. Iktin, and M. K. Horrigan, *Can. Metallurg. Q.* **23**:309 (1984).
2. T. Iida, Y. Kita, H. Okano, I. Katayama, and T. Tanaka, "Equation for the Vapor Pressure of Liquid Metals and Calculations of Their Enthalopies of Evaporation," *J. High Temp. Mater. Process* **10**:199 (1992).
3. L. I. Maissel and R. Glang, eds., *Handbook of Thin Film Technology*, McGraw-Hill, New York, 1970.
4. L. Holland, *Vacuum Deposition of Thin Films*, Wiley, New York, 1956.
5. M. Aceves, J. A. Hernández, and R. Murphy, "Applying Statistics to Find the Causes of Variability in Aluminum Evaporation: A Case Study," *IEEE Trans. Semicond. Manuf.* **5**:165 (1992).
6. K. Chopra, *Thin Film Phenomena*, McGraw-Hill, New York, 1969, p. 91.
7. C. Lu and A. W. Czanderna, *Applications of Piezoelectric Quartz Crystal Microbalances*, Elsevier, Amsterdam, 1984.
8. I. A. Blech, B. D. Fraser, and S. E. Haszko, "Optimization of Al Step Coverage Through Computer Simulation and Scanning Electron Microscopy," *J. Vacuum Sci. Technol.* **15**:13 (1978).
9. Y. Homma and S. Tsunekawa, *J. Electrochem. Soc.* **132**:1466 (1985).
10. M. A. Lardon, H. P. Bader, and K. J. Hoefler, "Metallization of High Aspect Ratio Structures with a Multiple Cycle Evaporation/Sputter Etching Process," *Proc. IEEE VLSI Multilevel Intercon. Conf.*, 1986, p. 212.
11. B. T. Jonker, "A Compact Flange-Mounted Electron Beam Source," *J. Vacuum Sci. Technol. A* **8**:3883 (1990).
12. J. Bloch, M. Heiblum, and J. J. O'Sullivan, "Ultra-High Vacuum Compatible Electron Source," *IBM Tech. Disc. Bull.* **27**:6789 (1985).
13. L. C. Hecht, "Use of Successive Dilution for Reproducible Control of Al–Cu Alloy Evaporation," *J. Vacuum Sci. Technol.* **14**:648 (January/February 1977).
14. M. L. Kniffin and C. R. Helms, "The Synthesis and Properties of Low Barrier Ag–Ga Intermetallic Contacts to n-Type GaAs," *J. Appl. Phys.* **68**:1367 (1990).
15. J. Villalobos, R. Glosser, and H. Edelson, "A Quartz Crystal-Controlled Evaporator for the Study of Metal Film Alloy Hydrides," *Meas. Sci. Technol.* **1**:365 (1990).
16. L. Esaki, in *Novel Materials and Techniques in Condensed Matter*, G. W. Crabtree and P. Vashishta, eds., North-Holland, Amsterdam, 1982.
17. W. Sevenhans, J.-P. Locquet, and Y. Bruynseraede, "Mass Spectrometer Controlled Electron Beam Evaporation of Multilayer Materials," *Rev. Sci. Instrum.* **57**:937 (1986).
18. N. Uetake, T. Asano, and K. Suzuki, "Measurement of Vaporized Atom Flux and Velocity in a Vacuum Using a Microbalance," *Rev. Sci. Instrum.* **62**:1942 (1991).
19. W. R. Grove, *Philos. Trans. Faraday Soc.* 87 (1852).
20. I. Langmuir, *General Electric Rev.* **26**:731 (1923).
21. G. K. Wehner, *Adv. Electron. Electron Phys.* **VII**:253 (1955).
22. M. L. Tarng and G. K. Wehner, "Alloy Sputtering Studies with *in-situ* Auger Electron Spectroscopy," *J. Appl. Phys.* **42**:2449 (1971).
23. R. L. Wilson and L. E. Terry, "Application of High-Rate ExB or Magnetron Sputtering in the Metallization of Semiconducting Devices," *J. Vacuum Sci. Technol.* **13**:157 (1976).
24. M. Konuma, *Film Deposition by Plasma Techniques*, Springer-Verlag, Berlin, 1992.
25. L. I. Maissel and R. Glang, eds., *Handbook of Thin Film Technology*, McGraw-Hill, New York, 1970, pp. 3–23.
26. G. K. Wehner and G. S. Anderson, "The Nature of Physical Sputtering," in *Handbook of Thin Film Technology*, L. I. Maissel and R. Glang, eds., McGraw-Hill, New York, 1970.

27. H. H. Anderson and H. L. Bay, *Sputtering by Particle Bombardment I*, R. Behrisch, ed., Springer-Verlag, Berlin, 1981.

28. R. V. Stuart and G. K. Wehner, "Sputtering Yields at Very Low Bombarding Ion Energies," *J. Appl. Phys.* **33**:2345 (1962).

29. G. K. Wehner, "Controlled Sputtering of Metals by Low-Energy Hg Ions," *Phys. Rev.* **102**:690 (1956); "Sputtering Yields for Normally Incident Hg$^+$ Ion Bombardment at Low Ion Energy," *Phys. Rev.* **108**:35 (1957); "Low-Energy Sputtering Yields in Hg," *Phys. Rev.* **112**:1120 (1958).

30. G. K. Wehner and D. Rosenberg, "Hg Ion Beam Sputtering of Metals at Energies 4–15 keV," *J. Appl. Phys.* **32**:1842 (1962).

31. H. H. Anderson and H. L. Bay, in *Sputtering by Particle Ion Bombardment I*, R. Behrisch, ed., Springer-Verlag, Berlin, 1981, p. 202.

32. G. K. Wehner and D. L. Rosenberg, "Angular Distribution of Sputtered Material," *J. Appl. Phys.* **31**:177 (1960).

33. O. Almen and G. Bruce, "Collection and Sputtering Experiments with Noble Gas Ions," *Nucl. Instrum. Methods* **11**:257, 279 (1961).

34. K. Wasa and S. Hayakawa, Jpn. Patent 642,012, assigned to Matsushita Electric Ind. Corp. (1967).

35. F. M. Penning, U.S. Patent 2,146,025 (February 1935).

36. R. S. Rastogi, V. D. Vankar, and K. L. Chopra, "Simple Planar Magnetron Sputtering Source," *Rev. Sci. Instrum.* **58**:1505 (1987).

37. P. J. Clarke, U.S. Patent 3,616,450 (1971).

38. T. Van Vorous, "Planar Magnetron Sputtering: A New Industrial Coating Technique," *Solid State Technol.* 62 (December 1976).

39. D. B. Fraser, "The Sputter and S-Gun Magnetrons," in *Thin Film Processes*, J. L. Vossen and W. Kern, eds., Academic Press, New York, 1978, p. 115.

40. J. A. Thornton and A. S. Penfold, "Cylindrical Magnetron Sputtering," in *Thin Film Processes*, J. L. Vossen and W. Kern, eds., Academic Press, New York, 1978.

41. B. A. Movchan and A. V. Demchishin, "Study of the Structure and Properties of Thick Vacuum Condensates of Nickel, Titanium, Aluminum Oxide, and Zirconium Dioxide," *Phys. Met. Metallogr.* **28**:83 (1969).

42. J. A. Thornton, "Influence of Apparatus Geometry and Deposition Conditions on the Structure and Topology of Thick Sputtered Coatings," *J. Vacuum Sci. Technol.* **11**:666 (1974).

43. R. J. Gnaedinger, "Some Calculations of the Thickness Distribution of Films Deposited from Large Area Sputtering Sources," *J. Vacuum Sci. Technol.* **6**:355 (1969).

44. I. A. Blech and H. A. Vander Plas, "Step Coverage Simulation and Measurements in a dc Planar Magnetron Sputtering Systems," *J. Appl. Phys.* **54**:3489 (1983).

45. W. H. Class, "Deposition and Characterization of Magnetron Sputtered Aluminum and Aluminum Alloy Films," *Solid State Technol.* **22**:61 (June 1979).

46. K. Wasa and S. Hayakawa, *Handbook of Sputter Deposition Technology*, Noyes, Park Ridge, NJ, 1992.

47. P. Burggraaf, "Sputtering's Task: Metallizing Holes," *Semicond. Int.* 28 (December 1990).

48. J. L. Vossen, J. J. O'Neill, K. M. Finlayson, and L. J. Royer, "Backscattering of Materials Emitted from RF-Sputtering Targets," *RCA Rev.* **31**:293 (1970).

49. L. Holland, "The Cleaning of Glass in a Glow Discharge," *Br. J. Appl. Phys.* **9**:410 (1958).

50. J. L. Vossen and E. B. Davidson, "The Interaction of Photoresist with Metals and Oxides During RF Sputter Etching," *J. Electrochem. Soc.* **119**:1708 (1972).

51. G. W. Sachse, W. E. Miller, and C. Gross, "An Investigation of RF-Sputter Etched Silicon Surfaces Using He Ion Backscattering," *Solid-State Electron.* **18**:431 (1975).

52. J. C. Bean, G. E. Becker, P. M. Petroff, and T. E. Seidel, "Dependence of Residual Damage on Temperature During Ar$^+$ Sputter Cleaning of Silicon," *J. Appl. Phys.* **48**:907 (1977).

53. M. J. Whitcomb, "Sputter Etch Profiles of Spheres, Cylinders, and Slab-like Silica Targets," *J. Mater. Sci.* **11**:859 (1976).

54. J. L. Vossen, "Control of Film Properties by RF-Sputtering Techniques," *J. Vacuum Sci. Technol.* **8**:512 (1971).

55. Y. H. Park, F. T. Zold, and J. F. Smith, "Influence of dc Bias on Aluminum Films Prepared with a High Rate Magnetron Sputtering Cathode," *Thin Solid Films* **129**:309 (1985).

56. J. E. Houston and R. D. Bland, "Relationships Between Sputter Cleaning Parameters and Surface Contamination," *J. Appl. Phys.* **44**:2504 (1973).

57. L. D. Hartsough and P. S. McLeod, "High-Rate Sputtering of Enhanced Aluminum Mirrors," *J. Vacuum Sci. Technol.* **14**:123 (1977).

58. T. N. Fogarty, D. B. Fraser, and W. J. Valentine, "MOS Metallization Via Automatic 'S-Gun' Planetary Deposition System," *J. Vacuum Sci. Technol.* **15**:178 (1978).

59. D. S. Herman, M. A. Schuster, and R. M. Gerber, "Hillock Growth on Vacuum Deposited Aluminum Films," *J. Vacuum Sci. Technol.* **9**:515 (1972).

60. S. M. Rossnagel, "Deposition and Redeposition in Magnetrons," *J. Vacuum Sci. Technol. A* **6**:3049 (1988).

61. S. M. Rossnagel, I. Yang, and J. J. Cuomo, "Compositional Changes During Magnetron Sputtering of Alloys," *Thin Solid Films* **199**:59 (1991).

62. F. J. Cadieu and N. Cheneinski, "Selective Thermalization in Sputtering to Produce High T_c Films," *IEEE Trans. Magnet.* **11**:227 (1975).

63. S. Kobayashi, M. Sakata, K. Abe, T. Kamei, O. Kasahara, H. Ohgishi, and K. Nakata, "High Rate Deposition of $MoSi_2$ Films by Selective Co-Sputtering," *Thin Solid Films* **118**:129 (1984).

64. V. Hoffman, "High Rate Magnetron Sputtering for Metallizing Semiconductor Devices," *Solid State Technol.* **19**:57 (December 1976).

65. D. R. Denison, "Sputtering System Design for Optimum Deposited Film Quality," *Microelectron. Manuf. Testing*, July 1985, p. 12.

66. P. S. McLeod and L. D. Hartsough, "High Rate Sputtering of Aluminum for Metallization of Integrated Circuits," *J. Vacuum Sci. Technol.* **14**:263 (1977).

67. K. Kamoshida, T. Makino, and H. Nakamura, "Preparation of Low Reflectivity Al–Si Films Using DC Magnetron Sputtering and Its Application to Multilevel Metallization," *J. Vacuum Sci. Technol. B* **3**:1340 (1985).

68. R. S. Nowicki, "Influence of Residual Gases on the Properties of DC Magnetron Sputtered Al–Si," *J. Vacuum Sci. Technol.* **17**:384 (1980).

69. R. J. Wilson and B. L. Wiess, "The Sputtered Reflectivity of DC Magnetron Sputtered Al–1%-Si Films," *Vacuum* **42**:987 (1991).

70. R. J. Wilson and B. L. Wiess, "The Structure of DC Magnetron Sputtered Al-1%-Si Films," *Thin Solid Films* **203**:147 (1991).

71. J. Visser, *Le Vide* (suppl.), no, 157.

72. J. E. Sundgren, "Structure and Properties of TiN Coatings," *Thin Solid Films* **128**:21 (1985).

73. K. Wasa and S. Hayakawa, *Microelectron. Rev.* **6**:213 (1967).

74. W. Tsai, J. Fair, and D. Hodul, "Ti/TiN Reactive Sputtering: Plasma Emission, X-ray Diffraction and Modeling," *J. Electrochem. Soc.* **139**:2004 (1992).

75. J. M. Molarius and M. Orpana, "Titanium Nitride Process Development," in *Issues in Semiconductor Materials and Processing Technologies*, S. Coffa, F. Priolo, E. Rimini, and J. M. Poate, eds., Kluwer, Dordrecht, 1991.

76. A. Kohlhase, M. Mändl, and W. Pamler, "Performance and Failure Mechanisms of TiN Diffusion Barrier Layers in Submicron Devices," *J. Appl. Phys.* **65**:2464 (1989).

77. I. Suni, M. Mäenpä; M. A. Nicolet, and M. Luomajärvi, "Thermal Stability of Hafnium and Titanium Nitride Diffusion Barriers in Multilayer Contacts to Silicon," *J. Electrochem. Soc.* **130**:1215 (1983).

78. M. A. Nicolet, "Diffusion Barriers in Thin Films," *Thin Solid Films* **52**:415 (1978).

79. T. Hara, A. Yamanoue, H. Ito, K. Inoue, G. Washidzu, and S. Nakamura, "Properties of Titanium Nitride Films for Barrier Metal in Aluminum Ohmic Contact Systems," *Jpn. J. Appl. Phys.* **30**:1447 (1991).

80. S. Berg, N. Eguchi, V. Grajewski, S. W. Kim, and E. Fromm, "Effect of Contamination Reactions on the Composition and Mechanical Properties of Magnetron Sputtered TiN Coatings," *Surf. Coatings Technol.* **49**:127 (1991).

81. P. Jin and S. Maruno, "Bias Effect on the Microstructure and Diffusion Barrier Capability of Sputtered TiN and TiO_xN_y Films," *Jpn. J. Appl. Phys.* **31**:1446 (1992).

82. S. S. Ang, H. M. Le, and W. D. Brown, "Sputtering-etching and Plasma Effects on the Electrical Properties of Titanium Nitride Contacts on n-Type Silicon," *Solid-State Electron.* **33**:1387 (1990).

83. G. J. Kominiak, "Silicon Nitride by Direct RF Sputter Deposition," *J. Electrochem. Soc.* **122**:1271 (1975).

84. R. V. Stuart, *Vacuum Technology, Thin Films, and Sputtering, An Introduction*, Academic Press, New York, 1983.

85. B. Chapman and S. Mangano, in *Handbook of Thin-Film Deposition Processes and Techniques*, K. K. Schuegraf, ed., Noyes, Park Ridge, NJ, 1988.

86. J. L. Vossen and W. Kern, *Thin Film Processes II*, Academic Press, Boston, 1991.

Chapter 13

Chemical Vapor Deposition

The last chapter discussed physically based methods for depositing thin films: evaporation and sputtering. They are called physical processes because the techniques did not involve chemical reactions. Rather, they produce a vapor of the material to be deposited by heating (evaporation) or by energetic ion bombardment (sputtering). Although most metal films for silicon ICs are deposited using these methods, they have major problems associated with step coverage. This is a particular concern with deeply scaled technologies, where very small contacts require the coverage of high aspect ratio features. Furthermore, these techniques are not well suited to the deposition of insulating or semiconducting films. This chapter will discuss methods of thin film deposition based on chemical reactions. By definition, the chemical bonding state of the final product is different than the sources.

Chemical vapor deposition (CVD) has become extremely popular and is the preferred deposition method for a wide range of materials. Thermal CVD also forms the basis for most epitaxial growth in IC manufacturing. Modifications of simple thermal CVD processes provide alternate energy sources such as plasmas or optical excitation to drive the chemical reactions, allowing the deposition to occur at low temperature. For a comprehensive review of CVD for IC fabrication, the reader is referred to Sherman [1] or Sivaram [2]. Unfortunately, CVD does not lend itself to simple analytic explanation. Both the gas flow in the reactor and the chemical reactions require a detailed numerical analysis that is both reactor and process dependent. This chapter will begin by introducing a simple CVD system that will be used in the following two sections to discuss the equations that would have to be solved to understand the chemical reactions and flows in the chamber. Next the chapter will discuss various types of CVD systems and specific gas chemistries used to deposit materials of interest. Finally, the chapter will cover plating, the dominant technique for depositing copper.

13.1 A Simple CVD System for the Deposition of Silicon

To begin to understand CVD processes, consider the simple reactor shown in Figure 13.1. The reactor consists of a tube with a rectangular cross section. The walls of the tube are maintained at a temperature T_w. A single wafer rests on a heated susceptor in the center of the tube. This susceptor is maintained at

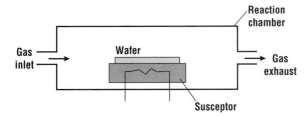

Figure 13.1   A simple prototype thermal CVD reactor.

T_s, where normally $T_s \gg T_w$. To discuss a simple but representative process, we will use the decomposition of silane gas (SiH_4) to form polycrystalline silicon. Assume that the gas flows through the tube from left to right. Since the silane will begin to decompose when it approaches the hot susceptor, the concentration of silane, and therefore the deposition rate, will decrease along the length of the tube. To improve the uniformity of the deposition, the silane can be mixed in an inert carrier gas. A common diluent for silane is molecular hydrogen (H_2). Assume the chamber is fed a mixture of 1% SiH_4 in H_2. Not only is the use of diluents common practice in real systems, it also avoids further complications in the chemistry of the reaction, since at typical deposition conditions very little of the hydrogen can decompose. Finally assume that the temperature of the gas as it enters the tube is the same as the wall temperature. The reaction products and any unreacted silane flow out of the tube at the right. The flows in the chamber will be slow enough (see Section 13.3) that the pressure in the chamber can be considered uniform.

The overall reaction that must occur is

$$SiH_4(g) \rightarrow Si(s) + 2H_2(g) \tag{13.1}$$

where the quantity in parentheses is (g) for gas phase and (s) for solid. The detailed process by which this overall reaction occurs is much more complex. One of the distinctions drawn regarding CVD is the location of the reaction that liberates a solid atom or cluster of atoms from a gaseous source. If that reaction occurs spontaneously in the gas above the wafer, it is called a *homogeneous process*. Such processes are generally undesirable if they produce solids. In deposition from silane, for example, excessive homogeneous reactions will result in large silicon particles in the gas phase that gradually accumulate on the wafer. The result is a deposit with poor surface morphology and inconsistent properties. In real systems, such deposits have a poorly controlled composition and may have significant contamination from residual gases in the chamber. This chapter will therefore emphasize processes that are heterogeneous. That is, processes that operate in a manner that greatly favors the formation of solids at surfaces only. Even for processes that are run this way, homogeneous reactions are still important. For example, in deposition from silane, the homogeneous production of silylene (SiH_2) is a crucial process because it is generally believed that over some ranges of temperature and pressure it is the silylene rather than silane itself that adsorbs on the surface of the wafer and produces most of the solid silicon. The distinction here is that the homogeneous reaction produces a gaseous, not a solid, product. The chapter will focus initially on the simplest type of heterogeneous reaction, which is run in a cold wall chamber (like the one shown in Figure 13.1), where all deposition reactions occur at the surface of the wafer.

The steps that occur during a chemical vapor deposition process include (1) transport of the precursors from the chamber inlet to the proximity of the wafer, (2) reaction of these gases to form a range of daughter molecules, (3) transport of these reactants to the surface of the wafer, (4) surface reactions to release the silicon, (5) desorption of the gaseous by-products, (6) transport of the by-products away from the surface of the wafer, and (7) transport of the by-products from the reactor. Even if the discussion is limited to thermal CVD in this very simple deposition system, understanding each of these steps is a formidable task. To simplify matters, the problem is often divided in half. The next section will focus on the chemical reactions that occur in the reactor, both in the gas phase and at the surface of the wafer. The following section will discuss the flow of gases in the reactor. Choosing to study a system that contains only a small concentration of the reactant gas (1% SiH_4 in H_2) allows this separation to be fairly realistic. The thermal and mechanical properties of the gas are relatively unaffected by any chemical reaction in the small, active component.

13.2 Chemical Equilibrium and the Law of Mass Action°

Focusing for the moment on CVD processes that involve long times and many collisions between molecules, the chemical composition at each point in the reactor approaches equilibrium. To understand chemical equilibrium, consider a unit volume of the gas somewhere in the chamber (Figure 13.2). Assume that the volume is small enough that the temperature and chemical composition in this volume are uniform. One of the reactions that one might think of is

$$SiH_4(g) \rightleftharpoons SiH_2(g) + 2H(g) \tag{13.2}$$

The double arrow indicates that the reaction proceeds in both directions. Chemical equilibrium is reached when the concentration of each species is constant, even if the gases take an arbitrarily long time traversing this unit volume. (Actually this reaction is unlikely, but it is instructive.)

Assume for the moment that this is the only reaction that occurs. The law of mass action says that

$$K_p(T) = \frac{p_{SiH_2}p_H^2}{p_{SiH_4}} \tag{13.3}$$

where p refers to the partial pressure of the subscripted species, and $K_p(T)$ is a reaction equilibrium constant that depends only on the temperature. The atomic hydrogen term is squared because of the 2 in front of the atomic hydrogen term of the species balance equation (Equation 13.2). The equilibrium constant generally follows an Arrhenius function

$$K_p(T) = K_o \, e^{-\Delta G/kT} \tag{13.4}$$

where ΔG is the change in the Gibbs free energy in the reaction. K_p may be greater or less than one and is independent of pressure, including inert gases such as He.

Assume that $K_p(T)$ is known for this process. There are three unknowns (the three partial pressures) and only one equation. Solving for them requires two more equations. The total pressure of the reactor P is a constant whose value is normally known. It is the sum of the partial pressures

$$P = p_{SiH_4} + p_{SiH_2} + p_H + p_{H_2}. \tag{13.5}$$

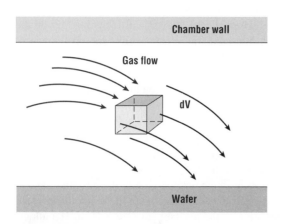

Chamber wall

Gas flow

dV

Wafer

Figure 13.2   A volume element dV at some point in the gas above the surface of the wafer.

Assuming, for example, that the chamber is run at atmospheric pressure, the partial pressure of H_2 is the same as its inlet partial pressure ($0.99P$) since it is assumed to be inert. The final equation comes from the inlet flow. One can use the Si/H ratio as

$$\frac{Si}{H} = \frac{f_{SiH_4}}{4f_{SiH_4} + 2f_{H_2}} = \frac{p_{SiH_4}}{4p_{SiH_4} + 2p_{H_2} + p_H} \tag{13.6}$$

where the f terms are the inlet flows, also assumed to be known.

This expression has not taken any other reactions into account. For example, in actual CVD, silicon is consumed from the gas phase. In that case, the partial pressure of silicon cannot be completely determined by the inlet flows. Instead, one must consider the inlet flow to be a source of silicon-containing molecules and the deposition surface to be a sink. Then, the silicon-containing molecule flux, which depends on the flow field and diffusion, must be calculated.

To begin to develop a more realistic picture, some of the reactions that would have to be included are [3]:

$$SiH_4(g) \rightleftharpoons SiH_2(g) + H_2(g) \tag{13.7}$$

$$SiH_4(g) + SiH_2(g) \rightleftharpoons Si_2H_6(g) \tag{13.8}$$

$$Si_2H_6(g) \rightleftharpoons HSiSiH_3(g) + H_2(g) \tag{13.9}$$

Other reactions are, of course, possible, and a priori one cannot decide which reactions to include. Instead, one must find the equilibrium constant for each possible reaction and ignore only those reactions for which the $K_p(T)$ values are negligibly small. Finding the equilibrium partial pressures, therefore, requires an equilibrium constant for each of the three preceding reactions and the solution of a set of coupled algebraic equations.

Example 13.1

Assume that the gas AB is introduced into a reactor and that the only chemical reaction that occurs in the chamber is

$$AB \rightleftharpoons A + B \tag{13.10}$$

If the process is run at 1 atm (760 torr) and a temperature of 1000 K and the process reaches chemical equilibrium, calculate the partial pressure of each species. The equilibrium constant for this reaction is given by

$$K(T) = 1.8 \times 10^9 \text{ torr } e^{-2.0 \text{ eV}/kT} \tag{13.11}$$

Solution

At 1000 K the equilibrium constant is calculated to be 0.15 torr. Then

$$0.15 = \frac{p_A p_B}{p_{AB}} \tag{13.12}$$

and the total pressure P is the sum of the partial pressures:

$$P = p_A + p_B + p_{AB} \tag{13.13}$$

We have three unknowns but only two equations. Since A and B are both created by the dissociation of the inlet gas, there must be an equal number of each. It would be reasonable to assume that the partial pressures are equal. Then, from Equations 13.12 and 13.13,

$$p_A^2 + 0.3p_A - 0.15 \cdot P = 0 \tag{13.14}$$

Setting $P = 760$ torr and solving gives $p_A = p_B = 10.5$ torr and $p_{AB} = 739$ torr.

The discussion thus far makes a crucial approximation: that all species are in chemical equilibrium. To understand the limitation of this approximation, consider what happens as the pressure of the chamber is reduced. At a low enough pressure, the mean free path of the molecules approaches the width of the chamber (see Section 10.1). If these gas phase collisions do not occur, the species in the gas often do not reach thermal equilibrium and therefore cannot achieve chemical equilibrium. Furthermore, since the gas molecules have a distribution of energies, a large number of collisions must occur in each unit volume for the gases to potentially reach chemical equilibrium. Therefore, a characteristic chamber length such as the distance between the gas injector and the susceptor must be at least a few orders of magnitude larger than the mean free path. Depending on the particular reaction equations involved, some processes may reach equilibrium, others will not. Processes that do not reach chemical equilibrium are called *kinetically controlled processes*. Typically, low pressure CVD is kinetically controlled, while atmospheric CVD may be in equilibrium.

To begin to understand this class of deposition, reconsider Reaction 13.7. This is a prototypical reaction in silicon CVD. When the reaction is kinetically controlled, it is written as

$$\text{SiH}_4(g) \underset{k_r}{\overset{k_f}{\rightleftharpoons}} \text{SiH}_2(g) + \text{H}_2(g) \tag{13.15}$$

where k_f and k_r are the forward and reverse reaction rate coefficients.

By writing a similar expression for each of the chemical reactions, it is possible to construct a differential equation for the time rate of change of the concentration (or partial pressure) of all of the chemical species. For example, if only Reactions 13.7 through 13.9 are considered, the time rate of change of silane is given by

$$\frac{d}{dt} C_{\text{SiH}_4} = -k_{f1}C_{\text{SiH}_4} + k_{r1}C_{\text{SiH}_2}C_{\text{H}_2} - k_{f2}C_{\text{SiH}_4}C_{\text{SiH}_2} + k_{r2}C_{\text{Si}_2\text{H}_6} \tag{13.16}$$

where the subscript 1 on the first two right-hand-side terms refers to the reaction given in Equation 13.7, and the subscript 2 refers to the reaction listed in Equation 13.8. Equation 13.9 does not involve silane and so does not enter into Equation 13.16. One can construct similar equations for the other species. If the $k(T)$s are known, we are left with a set of coupled first-order differential equations that can be solved for the rate of change of each chemical species in the unit volume. If one also takes into account diffusion of chemical species due to concentration gradients, the residence time of the gases in the unit volume is known (through the flow velocities), and the value of the temperature for each of the unit volumes is known, one could begin to solve for a map of the chemical species in the chamber. Although it is possible to solve for the species balance in very simple systems, real CVD systems can involve dozens of species and hundreds of reactions [4]. Furthermore, many of the important rate coefficients, even in the most commonly used chemical systems, are not always known accurately enough to produce meaningful information. For that reason, we will set aside attempts at meaningful quantitative discussion of the CVD reactions and use this introduction as a basis for a more qualitative discussion.

Although the chemical reactions in the gas phase are at least qualitatively understood, the situation at the surface is less clear. Part of the problem is the tools that can be used to investigate the gas sample over some finite volume. When these same techniques are applied to the surface, there is insufficient signal. Methods that have been used to develop a detailed understanding of the surface during high vacuum growth cannot be used in chambers under typical CVD conditions. Although new techniques have emerged in the past few years, the picture that we have is still tentative and qualitative.

The last two sections suggest that for the deposition of silicon from silane, several types of silicon-containing species may strike the surface of the wafer. In the boundary layer model, one can readily calculate the flux of the molecules if a boundary condition is established for the concentration at the surface. It is common practice to define a phenomenological parameter called the *sticking coefficient*, which varies from zero for molecules that are completely reflected from the surface to unity for molecules that irreversibly adsorb. In the case of radicals such as silylene, it is commonly assumed that the sticking coefficient is one. On the other hand, it has been shown that the probability of an incident silane molecule sticking and reacting is given by [5]

$$\gamma_{SiH_4} = 0.054\, e^{-0.81\, \text{eV}/kT} \tag{13.17}$$

At typical deposition conditions, the sticking coefficient is only about 10^{-6}; however, the flux of silane to the surface is much larger than the flux of silylene, so that both molecules may contribute to the deposition rate.

Once the molecule is on the surface, a chemical reaction must occur to remove the silicon atom and free the hydrogens. Taking silylene as an example, the molecule first adsorbs:

$$SiH_2(g) \rightleftharpoons SiH_2(a) \tag{13.18}$$

The overall surface reaction must take the form

$$SiH_2(a) \rightleftharpoons Si(s) + H_2(g) \tag{13.19}$$

where (a) refers to adsorbed species and (s) refers to atoms that have been incorporated into the solid. Because of the high concentration of H_2 in the gas, it is believed that the surface is covered with either physically adsorbed H_2 (at low temperature) or chemically adsorbed H (at higher temperature). These surface species must be desorbed to allow the reaction described in Equation 13.19 to proceed. The desorption process also follows an Arrhenius behavior, so that the surface will have a concentration of vacancies, whose density increases with increasing temperature. The adsorbed silylene can diffuse across the passivated surface (Figure 13.3) until it finds such a vacancy, at which point it will bond and eventually lose its hydrogen atom. This diffusion across the surface plays an important role in CVD processes. When the surface diffusion is large (of order millimeters), the deposition is very uniform. When the surface diffusion length is short, a less uniform deposit will result. As with physical deposition processes, surface diffusion increases exponentially with temperature, and so film uniformity can generally be improved by heating the wafer.

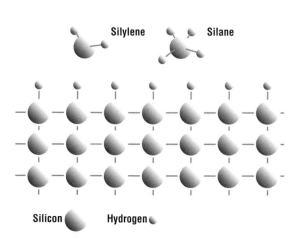

Figure 13.3   A simple model of the surface of the wafer during silane CVD includes adsorbed SiH_4 and SiH_2.

13.3 Gas Flow and Boundary Layers°

The second area that needs to be understood for CVD is gas flow dynamics. The gas flow in the reactor is important because it determines the transport of the various

Figure 13.4   Flow development in a tubular reactor. The gas enters with a simple plug flow on the left and exits with a fully developed parabolic flow.

chemical species in the chamber, and it plays a significant role in the temperature distribution in the gas in many reactors. The temperature distribution will also affect the flow. If the mean free path of the gas is much smaller than the chamber geometries, the gas can be treated as a viscous fluid. Furthermore, if the flow velocities are much less than the speed of sound (low Mach numbers) the gas can be considered incompressible. Nearly all CVD systems operate in pressure and flow regimes that make these approximations valid. Finally, as a starting point assume that the gas velocity is low enough that the gas flows along the contours of the chamber. The flow is said to be laminar, and it can be well described by the mechanical properties of the gas.

If the reactor is a circular tube and all surfaces are at the same temperature, the problem can be simplified considerably. Assume that the gas is introduced with a uniform velocity U_∞ at the left end of the tube (Figure 13.4). One important feature of gas flows is that the gas velocity must be zero at all surfaces. Because of the finite gas viscosity, the flow velocity must vary smoothly from zero at the walls to some maximum value at the center. This change from a uniform or plug flow, to a fully developed tube flow occurs over a distance z_v

$$z_v \approx \frac{a}{25} N_{Re} \tag{13.20}$$

where a is the radius of the tube, and N_{Re} is a dimensionless quantity known as the Reynolds number. The Reynolds number (N_{Re}) is given by

$$N_{Re} = U_\infty \frac{L}{\mu} = U_\infty \frac{L\rho}{\eta} \tag{13.21}$$

where L is a characteristic length of the chamber (such as the radius a), μ is the kinematic viscosity, ρ is the mass density of the gas, and η is the dynamic viscosity of the gas [6]. When N_{Re} is low, the flow in the tube is dominated by the finite viscosity effects and so is parabolic across the chamber. The velocity then is given by

$$v(r) = \frac{1}{4\eta} \frac{dp}{dz} (a^2 - r^2) \tag{13.22}$$

where dp/dz is the pressure gradient across the tube, which is assumed to be small. At very large N_{Re}, the gas cannot support the large velocity gradients required for fully developed laminar flow, and so the flow becomes turbulent. The transition between laminar and turbulent flows depends on the gas. For example, when $N_{Re} > 2300$ in H_2, the flow is turbulent [7].

Example 13.2

An LPCVD tube operating at 10 torr has an inlet gas flow of 1000 standard cubic centimeters per minute (sccm) of nitrogen. At the reactor temperature of 1000 K, the dynamic viscosity of nitrogen is 0.04 g/cm-sec. The reactor is 200 mm in diameter. Estimate the length required for fully developed flow and calculate $v(r)$ after the flow is fully developed.

Solution

According to Equations 13.20 and 13.21,

$$z_v \approx \frac{a}{25} U_\infty \frac{\rho}{\eta}$$

The mass density can be calculated using the ideal gas law

$$\rho = m_{N_2} \frac{P}{kT} = 4.5 \times 10^{-6} \text{ g/cm}^3$$

If U_∞ is taken as the plug flow velocity,

$$U_\infty \approx 1000 \text{ cm}^3/\text{min} \frac{1 \text{ min}}{60 \text{ sec}} \frac{1000 \text{ } K}{273 \text{ } K} \frac{760 \text{ torr}}{10 \text{ torr}} \frac{1}{\pi(10 \text{ cm})^2} = 15 \text{ cm/sec}$$

and, if $L \approx a$,

$$z_v \approx \frac{(10 \text{ cm})^2}{25} \, 15 \text{ cm/sec} \, \frac{4.5 \times 10^{-6} \text{ g/cm}^3}{0.04 \text{ g/cm-sec}} = 0.0068 \text{ cm}$$

Since the total flow must be constant,

$$\int 2\pi r v(r) dr = 15 \text{ cm/sec } \pi 100 \text{ cm}^2$$

one can readily show that

$$v(r) = 30 \text{ cm/sec} \left[1 - \frac{r^2}{a^2} \right]$$

To begin to understand the flow fields of a more complicated chamber, return to Figure 13.1, but for the moment keep the system at a uniform temperature. The wafers will rest on the bottom surface of the chamber. The previous discussion suggests that the gas flow velocity must go to zero at the surface of the wafer. The standard textbook picture has the height of the chamber large enough to have a large N_{Re} and so a broad region of uniform gas velocity approximately equal to U_∞. To further simplify the behavior of flow, it is customary to approximate the parabolic falloff of the gas velocity in the vicinity of the wafer as a boundary layer of width $\delta(z)$ where, for a flat surface whose normal is perpendicular to the flow direction,

$$\delta \approx 5 \sqrt{\frac{\mu z}{U_\infty}} \tag{13.23}$$

In this model, the gas flow in the boundary layer is zero, while outside the boundary layer the flow velocity is U_∞ (Figure 13.5).

If deposition is occurring at the surface of the wafer, the deposition gases must diffuse through the stagnant boundary layer. The boundary layer thickness, therefore, may play a critical role in

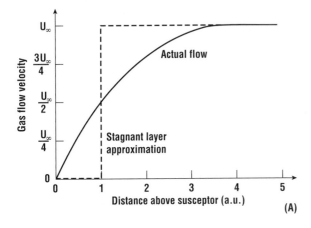

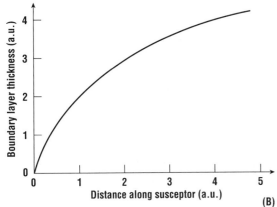

Figure 13.5   (A) Gas flow as predicted by a parabolic flow model and in the stagnant layer approximation. (B) Stagnant layer thickness versus position assuming a plug flow inlet.

determining the deposition rate. Notice that, according to Equation 13.23, for a flat surface the boundary layer thickness increases as $z^{1/2}$. To maintain a uniform boundary layer thickness, CVD systems in which gas transport plays a significant role in the deposition rate often tilt the deposition surface with respect to the flow direction. The wafers then rest on a wedge-shaped susceptor. The tilt angle of the wedge must be optimized to obtain the best uniformity for a particular CVD process.

Gas phase diffusivities are much less temperature sensitive than bulk diffusivities. One common form is given by Hammond [8]

$$D_e \propto T^{3/2} \frac{p_g}{P} \qquad (13.24)$$

where p_g and P are the partial pressure of the diffusing species and the total pressure, respectively, at the edge of the stagnant layer. One of the ways to distinguish CVD processes that are limited by diffusion through the gas is to measure the temperature dependence of the deposition rate. If p_g can be considered nearly independent of temperature, the deposition rate of such a mass transport limited reaction would increase only weakly with increasing temperature.

The gas flow in CVD reactors can be nonlaminar and may involve recirculations and roll cells. One of the most common sources of these flows is natural convection. As the gas flows past hot surfaces, it expands. The expansion is described by an equation of state such as the ideal gas law

$$\rho = \frac{nm}{V} = \frac{Pm}{kT} \qquad (13.25)$$

where m is the molecular mass. As the gas expands, the mass density decreases. (Recall that the pressure is fixed.) The hot gas tends to float or rise in the reactor with respect to cooler gases. This effect, called *natural convection*, must also be considered when calculating the flow of gases in a real chamber under deposition conditions [9]. The effects are most pronounced when heavy molecules are used at near atmospheric pressures. Conversely, natural convection has almost no effect for low pressure H_2 ambients.

Figure 13.6A shows the calculated flow fields for a horizontal reactor with a square cross section. The calculations were done by discretizing the reactor in 3-D and solving for momentum conservation in the form of the Navier–Stokes equation over this grid. Gases are injected at the back (upper left) end of the box and flow toward the front (lower right) end of the box. The lower surface is a hot susceptor. The hot susceptor causes the gas at the center of the reactor to rise, leading to

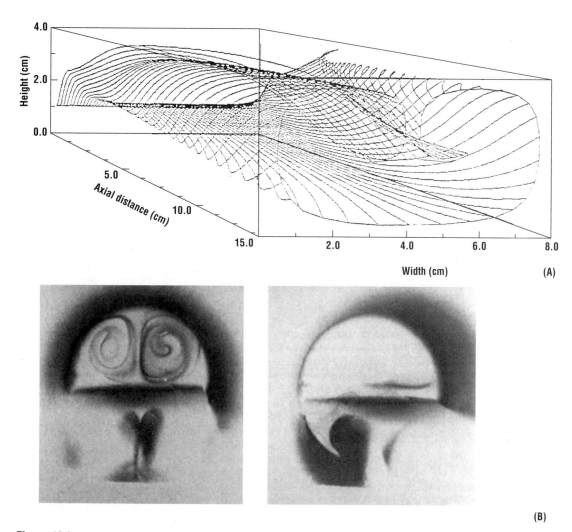

Figure 13.6   Flow field for a square cross section CVD reactor showing roll cells *(after Moffat)*. Roll cells in hemispherical tubes at 760 torr (left) and 160 torr (right) *(after Takahashi et al., reprinted by permission of the publisher, The Electrochemical Society)*.

transverse roll cells that develop along the length of the reactor. Again, a quantitative discussion of the exact flow in any reactor is well beyond the scope of this text, but these principles can be used to develop a qualitative understanding of a variety of chamber geometries and CVD processes. Figure 13.6B shows TiO_2 particles in a carrier gas as they flow through semicircular tubes. These small particles allow one to view the gas flow. The left-hand figure shows the flow at 760 torr and the right-hand picture shows the flow at 160 torr [9]. Obviously the roll cells are greatly diminished at low pressure.

In other reactors, natural convection shows up as recirculation cells in which part of the gas flow has a significant velocity component in the opposite direction of the intended flow (Figure 13.7). Recirculation cells may also arise when the cross-sectional area of the chamber changes abruptly. These cells can trap gas components that would otherwise be swept from the growth system.

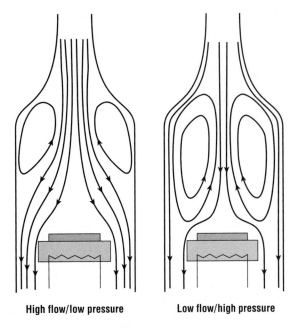

High flow/low pressure   **Low flow/high pressure**

Figure 13.7   Buoyancy-driven recirculation cells.

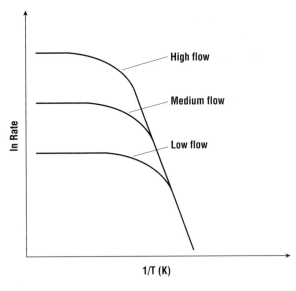

Figure 13.8   Typical deposition rates for CVD as a function of the temperature with flow rate as a parameter.

The effect can lead to an incomplete flushing of the chamber. If the composition of the deposit changes, this change will be much less abrupt than if the reactor and process are designed to avoid recirculation cells.

13.4 Evaluation of the Simple CVD System

Now examine what happens if the deposition rate in our example CVD reactor is measured as a function of temperature. The prediction of the growth rate and growth rate uniformity requires extensive fluid dynamics and chemical concentration calculations. Instead we will work backward by developing a qualitative understanding of experimental results. The deposition rate of the silicon-containing precursor (silane) will be used as a growth parameter. Figure 13.8 shows the results. At low wafer temperature the deposition rate increases exponentially with decreasing inverse temperature. In this regime the limiting step of the process is some reaction rate. This may be in the gas phase or at the surface. The information provided is insufficient to determine whether the gas is in chemical equilibrium or is kinetically controlled. Processes operated in this region are called *reaction rate limited*. CVD systems that operate in this regime must have excellent temperature control and temperature uniformity. These are primarily large batch systems in which many wafers are processed at relatively low rates. Flow dynamics in this chamber are of less concern, except to the extent that they contribute to temperature nonuniformity across the wafer. For that reason, this type of reactor often heats not only the wafer but the walls as well. It is therefore called a *hot wall batch CVD reactor*.

At high wafer temperature, the growth is limited by the arrival rate of the growth species. It is often assumed that the limitation is diffusion across the boundary layer. As already pointed out, the temperature dependence of the gas phase diffusivity is much lower than the activation energy of the chemical reaction. Processes operating in this regime are referred to as mass transport limited. Then the concentration of the deposition gas or gases controls the deposition rate. Remember that the rate of production of these gases from the precursors may also have a temperature dependence. CVD systems that operate in mass transport limited regime must have excellent control of the gas flows, and the chamber geometry must be designed to ensure a uniform transport to all parts of all wafers. As a result, these systems often are single-wafer or small-batch systems.

In addition to deposition rate and uniformity, CVD films must be examined with regard to stress, step coverage, and composition. Just as is the case for sputtered films, CVD layers with large compressive or tensile stress may crack, particularly when they cover steps. The step coverage of CVD films is usually very good. Plasma-enhanced CVD films, however, may be somewhat reentrant. This will be discussed later in the chapter. One of the major concerns with CVD layers is chemical composition. In the silane decomposition process that has been discussed so far, the deposited silicon may have a high concentration of hydrogen. This leads to low density films that may etch more quickly than pure silicon films. Residual gases in the chamber, such as oxygen or water vapor, may also react with the silicon to form SiO_x layers with high resistivities. In many processes, compounds such as SiO_2 or Si_3N_4 are the desired result. In that case, not only are contaminants a possibility, but also the stoichiometry of the deposited films may differ from the ideal. For example, it is common to anneal a CVD oxide in oxygen at high temperature to densify the film. This step adds oxygen and moves the film closer to stoichiometric SiO_2. Plasma-enhanced CVD (PECVD) films are particularly prone to stoichiometry concerns. One common and easy method to check the density of deposited oxides is to measure the wet chemical etch in dilute HF solutions. As deposited, PECVD films frequently have etch rates 10 times those of thermal oxides.

13.5 Atmospheric CVD of Dielectrics

Some of the earliest CVD processes were done at atmosphere pressure (APCVD) because of the large reaction rates and the simplicity of the CVD system, particularly for the deposition of dielectrics. Although the deposition of silicon from silane, as discussed earlier in the chapter, has been run at atmospheric pressure, the uniformity is poor. It is easy to obtain good uniformity at low pressure, and so APCVD is generally reserved for thick dielectrics, where deposition rates in excess of 1000 Å/min make the process very attractive.

Figure 13.9 shows a simple continuous-feed atmospheric CVD reactor. Wafers travel from cassette to cassette on a heated chain track. Wafer temperatures may be anywhere from 240 to 450°C [10]. The gases are injected from a showerhead above the wafers. When the oxygen-to-silane gas flow ratio is at least 3:1, stoichiometric SiO_2 will result. A very volatile mixture, which reacts readily, is produced by SiH_4 and O_2. Without a sufficient flow of a diluent such as N_2, this reaction proceeds in the gas phase, resulting in poor morphology.

As will be discussed in Chapter 15, it is often desirable to deposit silicon dioxide films with 4 to 12% phosphorus. These phosphosilicate glasses (PSG) soften and reflow at moderate temperatures, smoothing wafer topology and gettering many impurities. PSG can be formed in an atmospheric process by adding phosphine (PH_3). Figure 13.10 shows a typical plot of the deposition rate of PSG versus temperature and oxygen to hydride flow rates. For high oxygen concentration ambients (30:1), the deposition rate increases sharply with temperature and is probably reaction rate limited. For low oxygen containing ambients (2.5:1), the growth rate actually decreases slightly with increasing temperature. The phosphorus content of the film can be controlled by changing the phosphine-to-silane ratio. Due to the toxicity of phosphine and silane, APCVD systems designed for PSG

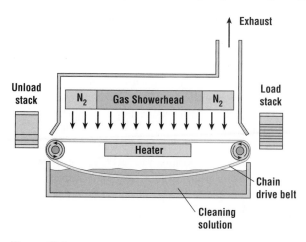

Figure 13.9 Simple continuous-feed atmospheric pressure reactor (APCVD).

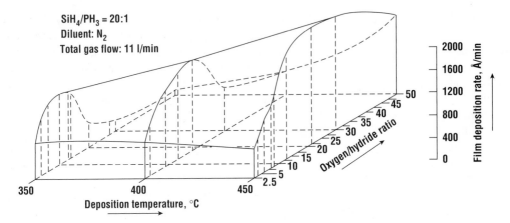

SiH$_4$/PH$_3$ = 20:1
Diluent: N$_2$
Total gas flow: 11 l/min

Figure 13.10 Deposition rate of PSG in an APCVD system *(afte Kern and Rosler,* [®]*1977, AIP)*.

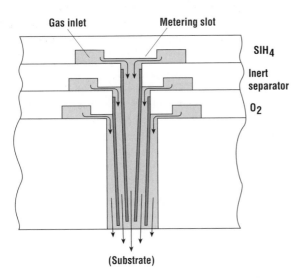

Figure 13.11 Showerhead design used to minimize deposition at the nozzle by maintaining an inert curtain between the reactants.

deposition are generally housed in a vented cabinet. To improve uniformity and step coverage, many PSG and borophosphosilicate glass (BPSG) processes now use organometallic sources such as TEOS [tetraethyloxysilane or Si(OC$_2$H$_5$)$_4$]. TEOS and ozone can also be used to deposit SiO$_2$ at about 400°C [11]. TEOS is supplied as a stable, inert, high vapor pressure liquid that is used in a bubbler. One of the advantages of using TEOS is the elimination of the need for some of the hazardous chemical handling. The lines from the bubbler must be heated to prevent deposition on the walls of the tubing. Various alternate organometallics have also been investigated. Hexamethyldisiloxane [(CH$_3$)$_3$–Si–O–Si–(CH$_3$)$_3$], a linear disiloxane, has also shown excellent characteristics comparable to TEOS [12], but the deposition rate was found to depend on the substrate material. Similar films have also been deposited from hydridospherosiloxanes [13] with wet oxygen at about 500°C.

The major drawback of APCVD is particle formation. Although particle formation in the gas phase can be controlled by adding a sufficient amount of N$_2$ or another inert gas, heterogeneous deposition can also occur at the gas injectors. Even if the growth rate of these particles is low, after a number of wafers the particles will become large enough to flake off and fall on the wafer surface. To avoid this problem, the showerhead may be segmented to keep the reactant gases separated until they are injected into the chamber. Figure 13.11 shows a showerhead arrangement designed to reduce this problem. Due to the high deposition rate, APCVD of BPSG has been used as a premetal dielectric for DRAMs and other cost-sensitive commodity components.

13.6 Low Pressure CVD of Dielectrics and Semiconductors in Hot Wall Systems

A variety of system geometries have been used for low pressure CVD (LPCVD). Figure 13.12 shows a sample of some of the most common. We can divide the reactors into hot and cold wall systems.

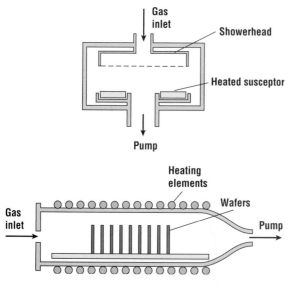

Figure 13.12   Common LPCVD reactor geometries.

Figure 13.13   Standard bank of horizontal LPCVD tubes.

Hot wall systems have the advantages of uniform temperature distributions and reduced convection effects. Cold wall systems are able to reduce deposition on the walls. These deposits can lead to depletion of the deposition species and the formation of particles, which may flake off the walls and fall on the wafers. Deposits on the walls also lead to memory effects: the deposition on the wafer of material previously deposited on the walls. For that reason hot wall reactors must be dedicated to the growth of a particular film.

Virtually all polycrystalline silicon deposition and a considerable amount of dielectric deposition are done in hot wall systems. Instead of using an inclined susceptor, the wafers are close packed, like wafers in thermal oxidation systems. To achieve reasonable deposition uniformity in such a system, the process must be designed so as to keep the reaction strictly controlled by the deposition kinetics [14]. Instead of a diluent gas, the use of low pressures (0.1 to 1.0 torr) reduces gas phase nucleation. This process is commonly called LPCVD.

Figure 13.13 shows a photograph of a horizontal LPCVD system. Like furnaces, they generally are built in banks of four tubes. In this bank, the upper tube has been removed to show the heater coils. The gases are controlled at the back of the tube using mass flow controllers and routed to the front of the furnace. Depending on the process to be run, the gases are either injected through a ring at the front of the tube, or plumbed to a tube that runs the length of the furnace and injects gases uniformly across the load. Due to the amount of deposition on the walls, most production systems have soft landing cantilever loaders to minimize particulate formation and flaking during the load/unload process. After loading the furnace is closed with an O-ring-sealed door. The tube is flushed with an inert gas such as N_2 and pumped to a medium vacuum. The furnace is ramped to the deposition temperature if it is not already idling at that temperature, and the deposition gases are switched on. The deposition is allowed to proceed for a predetermined time, then the furnace is again flushed with N_2, the pressure raised to an atmosphere, and the wafers unloaded.

A recent innovation in the LPCVD area is the introduction of vertical chambers (Figure 13.14). Similar to vertical oxidation/diffusion tubes, these systems have several advantages over standard tubes. Since the wafers are all held by gravity, the wafer-to-wafer spacing in the reactor is more uniform. Convective effects are more uniformly distributed across the wafer. As a result of these advantages, vertical LPCVD systems can routinely achieve uniformities of better than 2% in the deposition of undoped poly and silicon nitride [15]. Vertical CVD systems are more easily integrated into automated factories, since the wafers do not have to be tipped to the vertical allowing easier robotic handling. Perhaps the most important advantage of vertical LPCVD systems is the reduced particle counts. The cost of these systems is considerably higher than conventional LPCVD systems, however.

Another type of hot wall LPCVD system that has widespread application to silicon IC manufacturing is the hot wall cross-flow reactor. In this reactor, the wafers are placed vertically in closely spaced cassettes, arranged so that fresh gas flows past each wafer. This reduces particle formation and improves uniformity. The system requires extensive quartzware maintenance, however.

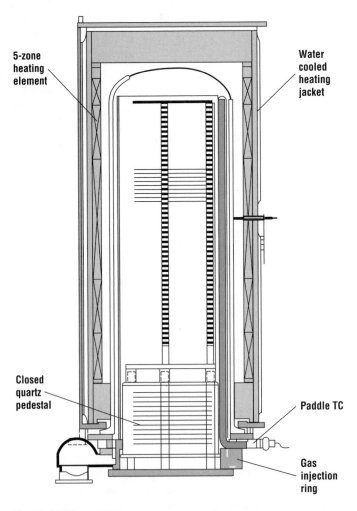

Figure 13.14   LPCVD process section: a low pressure process tube showing the heater, quartz tube, boat, and various gas and mechanical control mechanisms *(ASM Europe)*.

Most LPCVD poly is done with silane in furnaces at temperatures ranging from 575 to 650°C. The activation energy for poly deposition is about 1.7 eV. It is believed that the deposition rate is limited by hydrogen desorption from the silicon surface. Typical polysilicon deposition rates are 100 to 1000 Å/min, so common deposition times are tens of minutes. When the gas is injected at the front of the tube, the furnace may be programmed with a small temperature gradient (25°C) from front to back. This allows the higher reactivity at the back of the tube to compensate for silane depletion. Temperature ramping has the undesirable effect of producing wafers with larger poly grains at the back of the tube than at the front; however, after high temperature annealing the crystal structure of the films is indistinguishable [16]. It is common to obtain thickness uniformities of 5% across a batch of 100 large-diameter wafers [17].

Figure 13.15 shows the deposition rate of poly as function of pressure for a range of temperatures [18]. The morphology of the deposit is a sensitive function of the deposition conditions. When the deposit is performed at temperatures below 600°C, it is usually amorphous [16], but may be polycrystalline if the deposition rate is low enough. These layers can be crystallized into poly layers when annealed at low temperatures [19].

LPCVD polysilicon can be N^+ doped by implantation, solid or $POCl_3$ diffusion, or *in situ* doping through the addition of arsine (AsH_3) or PH_3. Impurity concentrations approaching 10^{21} cm^{-3} can be achieved in poly layers doped in this manner [20]. It is common to achieve resistivities of less than 1 mΩ-cm in both *in situ*-doped and diffused polysilicon. The major difficulty with *in situ* doping is that the addition of PH_3 degrades the deposition uniformity, particularly near the edge of the wafer. Modeling efforts have suggested that this may be the result of an increased silylene concentration [21]. It is also critical to include the effects of temperature variations across the load, particularly when considering the first and last few wafers [22].

Aside from particle control, one of the major limitations of traditional LPCVD furnaces is the difficulty of integrating them into automated, cluster tool environments. The most common example is polysilicon. It would be desirable to grow a gate oxide, nitride the oxide, and deposit the poly in a single cluster tool. Such systems can be made by using a number of simple single-wafer reactors that can be fed by a central robot arm. In the Applied Materials Precision 5000®, for example, wafers can be shuttled among up to five chambers (Figure 13.16). The wafer enters the load-locked area, where it is transferred to a rapid thermal chamber for oxidation and nitridation. It is then transferred to one of two poly deposition chambers.

Various chemistries have been used for hot wall deposition of oxide in LPCVD reactor including silane and oxygen, dichlorosilane ($SiCl_2H_2$ or DCS) and nitrous oxide (N_2O), and the decomposition of TEOS. The silane and oxygen process can be run at substrate temperatures of 300 to 450°C. This means that it can often be used to deposit a layer of oxide over an aluminum layer. Films deposited at these temperatures are found to contain significant quantities of silanol (SiOH), hydride (SiH), and water [23]. As with the corresponding APCVD process, the major limitations are particulate generation and low deposition rates [13]. Furthermore, due to the low substrate temperature, the step coverage of these films is often unacceptable.

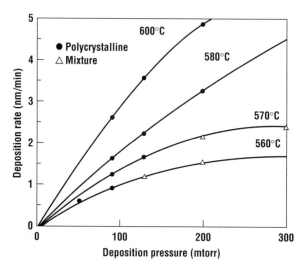

Figure 13.15 Deposition rate of polysilicon as a function of temperature and of silane flow rate *(after Voutsas and Hatalis, reprinted by permission of the publisher, The Electrochemical Society)*.

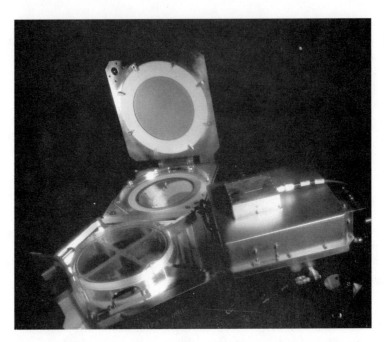

Figure 13.16   A CVD cluster tool showing the central robot and one of the single-wafer processing stations *(photo courtesy Applied Materials)*.

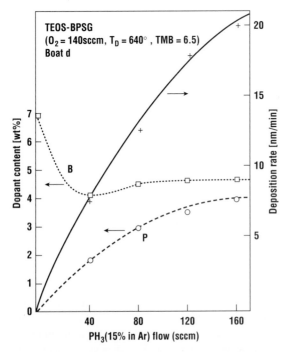

Figure 13.17   BPSG deposition rate and the boron and phosphorus concentrations as a function of phosphine flow in the TEOS/TMB system *(after Becker et al., ©1986, AIP)*.

The DCS and nitrous oxide process [24] must be run at temperatures of about 900°C. Uniformity and step coverage are excellent, and etch rates are close to those of thermal SiO_2. Although these films are virtually free of hydrogen, they do contain measurable amounts of chlorine, which can lead to etching of underlying polysilicon layers. The DCS and nitrous oxide deposition process has a strong nonlinear pressure dependence. Deposition nonuniformities across the tube are common. Special injectors and temperature gradients can be used to attempt to provide a more uniform deposition.

A popular LPCVD oxide process is the thermal decomposition of TEOS. Figure 13.17 shows the deposition rate of oxide as a function of TEOS partial pressure [25]. Typical deposition temperatures of 650 to 750°C are low enough that redistribution of dopants in the substrate is not a concern, but the temperature is still too high to use over aluminum or copper layers. Lower temperature depositions have been reported with alternate organometallics such as diacteoxyditertiarybutoxysilane, which decomposes at temperatures as low as 450°C [26] and provides excellent step coverage, but is not yet used extensively.

TEOS decomposes as

$$Si(OC_2H_5)_4 \rightarrow SiO_2 + 2H_2O + 4C_2H_4$$

where all of the species (except for SiO_2) are gaseous at normal processing temperatures. Thus TEOS can be used by itself (unimolecular decomposition) to deposit SiO_2. However, the temperature required to run the process, particularly without excessive residual carbon in the film, is typically about 700°C. The addition of oxygen allows the decomposition as

$$Si(OC_2H_5)_4 + 12O_2 \rightarrow SiO_2 + 10H_2O + 8CO_2$$

This can proceed at temperatures between 400 and 600°C, where the best films are obtained at higher temperatures. To further reduce the temperature one can use TEOS–ozone [27]. These pure films deposited by CVD are now often called undoped silicate glass (USG). TEOS deposition processes can achieve almost perfect conformality due to the fact that the process is surface reaction controlled [28]. As a result, the process can fill even fairly high aspect ratio gaps, as long as the profile is not reentrant.

The deposited stress in undoped LPCVD oxides are about $1-3 \times 10^9$ dynes/cm². Low temperature depositions are tensile, while high temperature depositions are compressive. The refractive index of thermal SiO_2 is 1.46 [29], while deposited oxides are generally higher. It is often found that the higher refractive index of deposited oxides correlates with a low mass density and high etch rates in buffered HF. As with APCVD, dilute mixtures of phosphine or phosphine and diborane [30] in hydrogen can be added to the TEOS LPCVD process [31] or the SiH_4 and O_2 process [32] to produce doped oxide layers. The addition of phosphine generally increases the deposition rate but degrades the deposition uniformity, while adding diborane to form BPSG sometimes increases the deposition rate and has little effect on the uniformity. This loss of uniformity is due to the change from a surface-controlled process to one that is controlled by gas phase reactions [33]. It also leads to a significant reduction in the ability of doped glasses to fill high aspect ratio gaps. Instead, the deposition rate is slightly higher at the top edges, leading to the formation of a void in the gap. This can be eliminated by a high temperature anneal, which reflows the glass and closes the void, but the temperature required is excessive for very small devices.

Although silicon nitride can be deposited from silane and a nitrogen-containing precursor, LPCVD silicon nitride is most commonly deposited from mixtures of DCS and ammonia (NH_3). Typical deposition temperatures are 700 to 900°C [13]. The activation energy for this process is 1.8 eV. The deposition rate increases with the DCS flow rate. Due to DCS depletion effects, a ramped profile is usually required for this process. Typical deposition rates are 10 to 20 Å/min. The films are highly conformal due to the surface reaction nature of the process. Stoichiometric films have a refractive index between 2.0 and 2.1 and high tensile stress, typically 12 to 18 GPa. Due to this stress, films thicker than about 2000 Å tend to crack.

Two common techniques to check the composition of the films are the measurement of the refractive index with ellipsometry and the measurement of the etch rate in buffered HF. High refractive indices indicate a silicon-rich film. Low refractive indices indicate the presence of oxygen, often due to a vacuum leak, a contaminated gas, or an incomplete pumpdown. The presence of oxygen is also indicated if the etch rate of the film exceeds 1 nm/min in 49% HF. Other common impurities in Si_3N_4 include hydrogen and approximately 0.4% Cl [34]. The authors of this work also found that these films have a 10 to 20-Å layer of SiO_2. Increasing the concentration of DCS in the growth ambient can decrease the stress but results in films with a considerable amount of excess silicon. Due to the high concentration of silicon-containing species (primarily $SiCl_2$), gas phase nucleation is likely during the LPCVD of Si_3N_4, leading to a high concentration of particles formed in the gas.

13.7 Plasma-enhanced CVD of Dielectrics

In many applications, it is necessary to deposit films at very low substrate temperatures. The deposition of SiO_2 over aluminum and the deposition of Si_3N_4 capping layers over GaAs are two common examples. To accommodate these lower substrate temperatures, an alternate energy source must be applied to the gaseous and/or adsorbed molecules. Although photoenhanced deposition has been experimentally demonstrated, and even seen some limited use in production, the primary nonthermal energy source used to drive CVD reactions is the RF plasma. Plasma-enhanced chemical vapor deposition (PECVD) systems have the added advantage of using ion bombardment of the surface to provide energy to the adspecies to allow them to diffuse further along the surface, without a high substrate temperature. As a result, the process is very good at filling small features. Chapter 10 introduced the concept of plasmas. The application of glow discharges to sputtering and etching has already been discussed. This section will discuss the use of plasmas to enhance the deposition rate in CVD processes. Since the deposition of insulating layers is of primary interest, only RF discharges need to be considered.

There are three basic types of PECVD systems as shown in Figure 13.18. In each system, the RF frequency chosen is normally less than 1 MHz, although PECVD oxide can be deposited at 13.56 MHz. The first PECVD systems were cold wall parallel-plate reactors. Either gases are injected at the edge or through an upper electrode showerhead and exhausted through a port at the center, or gas is injected at the center and exhausted around the edges. As the wafer diameter has increased, the low throughput and marginal uniformity of these systems has obviated their use for silicon IC production. Due to the small wafer diameters and the limited number of wafers per batch, however, this style of reactor is often preferred for GaAs technologies.

For silicon IC manufacturing with large-diameter wafers, one of the currently preferred techniques for conventional PECVD is a parallel-plate hot wall system. Similar in appearance to an LPCVD tube, the wafers are mounted vertically on conductive graphite electrodes of alternating polarity. The substrate temperature is controlled as in any furnace, although it is much cooler than it would be for a comparable LPCVD process. Although the throughput of this type of reactor is better than the parallel-plate plasma reactor, it is much less than that of a standard LPCVD system.

Hot wall batch PECVD systems suffer from the same types of gas depletion/uniformity and particle problems as their thermal counterparts. For that reason, there has been a resurgence of interest in cold wall PECVD systems. To increase the throughput, a number of deposition stages may be put into a single vacuum system, or several single-wafer chambers may be run in parallel with a robot arm to feed the

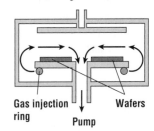

Gas injection ring   Wafers

Pump

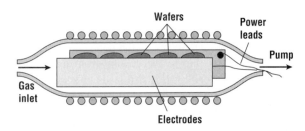

Wafers

Power leads

Pump

Gas inlet

Electrodes

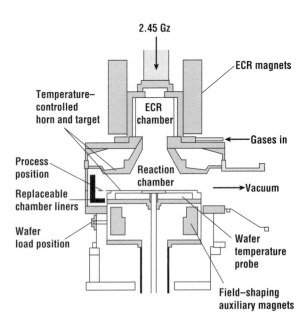

2.45 Gz

ECR magnets

Temperature–controlled horn and target

ECR chamber

Gases in

Process position

Reaction chamber

Vacuum

Replaceable chamber liners

Wafer load position

Wafer temperature probe

Field–shaping auxiliary magnets

Figure 13.18 Basic PECVD geometries: cold wall parallel plate, hot wall parallel plate, and ECR.

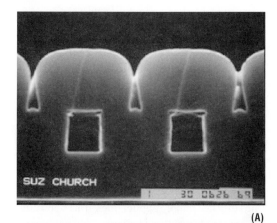

(A)

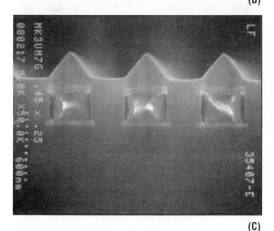

(B)

(C)

Figure 13.19   Profiles of SiO$_2$ deposited by (A) PECVD, (B) thermal CVD, and (C) HDP CVD *(courtesy IBM).*

chambers. One manufacturer does the deposition sequentially in five stages. Not only does this improve throughput, uniformity of close to 1% is seen as well [35].

To deposit high-quality layers at low substrate temperatures, high-density plasmas (HDP) have been introduced. These reactors use a variety of high density plasma configurations including electron cyclotron resonance (ECR) [36] to dissociate or crack one or more of the precursors. These sources were covered in Chapter 10. One application of HDP is to dissociate N$_2$ to form atomic nitrogen that readily reacts with silane to form Si$_x$N$_y$ with virtually no ion bombardment of the substrate. The silane can be introduced outside the plasma [37]. Because of the high reactivity of the atomic species, a large substrate temperature is not necessary to drive the reaction and obtain dense films [38]. Good silicon dioxide films have also been demonstrated at temperatures as low as 120°C [39].

The low pressure of a HDP plasma (about 0.01 torr) results in long mean free paths and, therefore, poor step coverage. If the system is designed to allow a significant amount of ion bombardment of the surface, however, the deposited species will be continuously sputtered, allowing the fill of high aspect ratio features. This has proven to be one of the most attractive applications for the technology, particularly for depositing SiO$_2$. One of the primary limitations of HDP deposition systems is the high concentration of particles produced in the plasma. This has been addressed recently by creating particle traps and/or particle absorbent chamber surfaces. To improve the low deposition rates of ECR, production systems can be fabricated with a large number of the remote plasma injectors in parallel in the same vacuum chamber. Properly designed, however, HDP CVD systems provide exceptionally high quality films deposited at low temperature.

The PECVD of silicon nitride has been used in GaAs implantation for many years. In silicon technologies, this process has also been used as a final passivation or scratch protect layer as one of the final steps of the process [40]. The process is run at 300 to 400°C using mixtures of a diluent gas such as Ar or He, SiH$_4$, and either NH$_3$ or N$_2$. Both hot wall and cold wall reactors have been used (Figure 13.18).

Figure 13.19 demonstrates the film coverage for PECVD (A), thermal subatmospheric (B), and HDP CVD (C) of silicon dioxide. The bread loaf profile of the PECVD film can be adjusted by changing substrate temperature, power, and pressure. The HDP profile is a result of simultaneous deposition and etch, and results in excellent fill properties [41]. If the etch rate is too large, corner features will be removed or "clipped." Often the etch is due to sputtering by argon gas in the reactor.

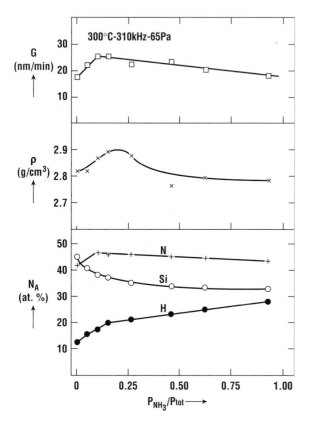

Figure 13.20 Deposition rate, density, and stoichiometry in the PECVD of Si₃N₄ *(after Claasen et al., reprinted by permission of the publisher, The Electrochemical Society).*

Figure 13.20 shows plots of the deposition rate, mass density, and atomic composition as a function of the fraction of ammonia in the gas flow [42]. The deposition rate is relatively insensitive to gas composition. The maximum in the density occurs at a Si/N ratio of 0.75, the correct value for stoichiometric silicon nitride. Increasing ammonia flow, however, increases the hydrogen concentration, which, at about 20%, is typical for PECVD Si_3N_4. Increasing the substrate temperature decreases the hydrogen content in the film. The use of N_2 rather than NH_3 also decreases the hydrogen content [43].

One of the major applications for deposited dielectrics is to form the insulator between metal interconnect levels in an IC. Particularly when aluminum is used for the metal, PECVD provides the necessary low temperature deposition. A problem with PECVD nitrides in this application is the relatively large dielectric constant of Si_3N_4. When used as an insulator between two metal layers, this will result in a large node capacitance and, therefore, a slower circuit speed. To improve circuit performance, the nitride can be replaced with a PECVD oxide. PECVD silicon dioxide processes can be run using silane and an oxidizer. Oxygen gas can be used, but the reaction between silane and O_2 does not need a plasma to drive it. As a result, considerable homogeneous nucleation occurs in the inlet nozzles and in the gas above the wafer resulting in high particle counts and poor morphology. Carbon dioxide can be used, but N_2O is the preferred oxidant to avoid carbon incorporation. It has also been reported that the addition of He as a diluent improves deposition uniformity and reproducibility [44].

Oxides deposited with this technique have high concentrations (1 to 10%) of hydrogen [45]. It is also common to find substantial amounts of water and nitrogen [46]. The exact composition depends critically on the chamber power and the gas flows. As shown in Figure 13.21, increasing the plasma power increases the deposition rate but also reduces the density [47]. As a result of the ease of silicon oxidation reaction, low plasma power densities provide large deposition rates. It has also been shown that if the power density of the plasma is large enough to ensure a sufficient concentration of atomic oxygen in the plasma to completely oxidize the silane, good films with a high dielectric strength are observed [48]. Postdeposition high temperature bakes may be used to reduce the hydrogen content and densify the film. These bakes can also be used to control film stress [49], but often PECVD processes are chosen precisely because such high temperature steps cannot be tolerated.

One of the interesting features of PECVD films is that it is possible to change the composition of the films continuously from oxide to nitride by changing the gas flow. By adding increasing amounts of N_2O to a mixture of SiH_4, NH_3, and He in a 13.56-MHz cold wall PECVD system, films were deposited whose refractive index varies smoothly from that of nitride to that of oxide [50]. This allows the technologically interesting possibility of stacks and graded composition films.

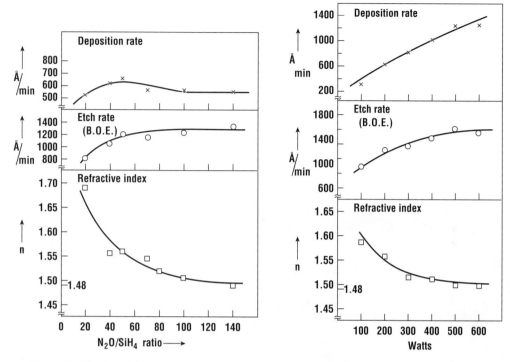

Figure 13.21   Results for the PECVD of SiO$_2$ *(after Van de Ven, ©1981, Solid State Technology).*

PECVD of silicon dioxide may also be done using a TEOS source [51, 52]. One of the motivations for this approach is the danger associated with silane use. Organometallics such as TEOS are not usually used for plasma processes because they tend to incorporate carbon in the film. At sufficiently large O$_2$-to-TEOS inlet flow ratios, the residual carbon contamination in the deposited films can be made extremely small [53], and reasonable values of the index of refraction and dielectric constant [54] can be obtained. Substrate temperatures are well below 400°C. As with thermally deposited TEOS films, the stress of the layer can be controlled over a broad range [55] by a postdeposition bake. Doped layers such as PSG and BPSG are also increasingly looking toward the use of organometallics such as tetrabutylphosphine and trimethylborane to reduce the use of hydrides.

13.8 Metal CVD$^+$

The last sections described the attributes of CVD: excellent step coverage and the potential for low substrate temperature depositions. One of the most serious step coverage concerns is that of metal as it goes into contacts. Particularly for deep submicron devices, the step coverage of sputter deposited films into increasingly high aspect ratio features is becoming extremely difficult. Furthermore, to ensure coverage of the metal over the contact, sidewalls must be carefully tapered during the etch process. To accommodate this widening of the contact, in older technologies, metal lines had caps (Figure 13.22). These caps can significantly reduce wiring density. Finally, the topology created by tapering the contact accumulates so that coincident contacts or "plugs" are not allowed. If, on the other hand, metal CVD is available, vertical contact structures can be used, filling

Top
view

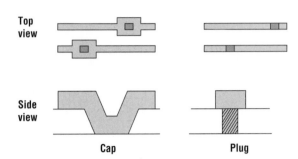

Side
view

Cap   **Plug**

Figure 13.22   The use of caps versus plug-filled contacts.

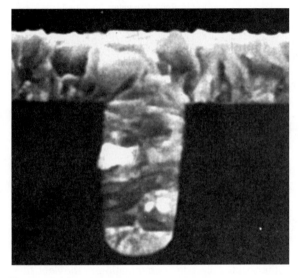

Figure 13.23   A filled contact showing the CVD tungsten morphology *(after Schmitz, reprinted by permission, Noyes).*

the contact and decreasing the surface topology. No cap is required, and step coverage is of much less concern. It is therefore highly desirable to develop processes for metal CVD. Of the various metals that have been attempted, the greatest success has been achieved with tungsten [56].

Much of the early work in tungsten CVD was done in standard horizontal LPCVD tubes [57, 58]. Tungsten was found not to adhere to the tube walls and so particulation was a serious problem. Of even more concern, however, was the fact that thin layers of tungsten could effectively block the IR radiation from the coils that heat the wafers. This effect resulted in poor uniformity and reproducibility. As a result of these problems, most tungsten CVD is now carried out in cold wall reactors [56]. Keeping the chamber walls below about 150°C is critical to the success of the process due to the high reactivity of the precursors. Tungsten sources include WCl_6 [59, 60], $W(CO)_6$ [61, 62], and WF_6 [63]. Of these, only WF_6, which boils at 25°C, is a liquid at room temperature. The others are high vapor pressure solids. Most tungsten CVD is therefore run using WF_6 with a H_2 carrier gas. Deposition temperatures are usually less than 400°C.

The simplest type of tungsten deposition process is the blanket CVD of tungsten. Since blanket CVD W films do not adhere well to the oxide, a thin adhesion layer must first be deposited. One of the most commonly used layers is sputtered or CVD TiN, a barrier metal under aluminum interconnect [64, 65]. It has been shown that the deposition of W over TiN has a substantial initiation time. No film is formed unless SiH_4 is used during the growth initiation phase [66]. The reduction of WF_6 by SiH_4 is found to deposit pure W when the gas mixture is rich in WF_6 and tungsten silicide when it is SiH_4 rich.

Although typical deposition rates are rather slow, the best step coverage for W CVD is obtained using WF_6 and H_2 [67]. This is a major concern, since one of the primary applications for CVD tungsten films is the filling of high aspect ratio contacts. If the films have insufficient coverage, a void will be left in the center of the contact as the top closes. Even a small taper of the contact will ensure complete filling (Figure 13.23). The overall reaction for this process is

$$WF_6 + 3H_2 \rightleftharpoons W + 6HF$$

Below 400°C this process has an activation energy of approximately 70 kJ/mole. It is believed that the rate-limiting step is the desorption of HF from the surface [68]. The tungsten layer deposited in this manner may be left in place and used for the interconnect level, or it may be etched back, leaving only the filled contacts or plugs. In the latter case, a sacrificial nitride layer may be used on top of the oxide to avoid roughening the surface of the oxide during the etchback (Figure 13.24).

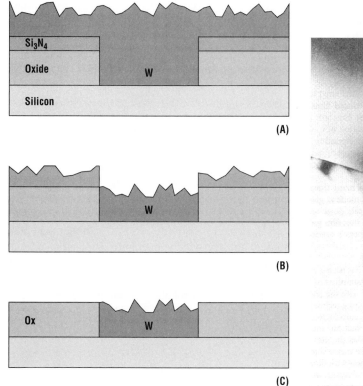

Figure 13.24 The use of a sacrificial nitride layer with tungsten CVD to avoid surface roughening *(after Schmitz, ©1992, Noyes).*

Figure 13.25 Damage to the substrate produced by selective tungsten.

To avoid the etchback altogether, it is also possible to deposit tungsten selectively in the exposed contact windows. Selective tungsten makes use of the fact that W nucleates preferentially on conducting substrates such as Si over insulating substrates such as SiO_2. Selective tungsten is expected to fill any aspect ratio structure, since the deposition occurs from the bottom of the structure up, rather than from the sides. This has the additional advantage of reducing the cost of the deposition, which is considerable. Maintaining selectivity is extremely difficult, however. Deposits tend to form on the dielectric and may also creep up the walls of the contact. A second problem with selective tungsten processes is the formation of silicon voids called worm holes. In many cases tungsten filaments fill these worm holes (Figure 13.25).

Aside from tungsten processes, there are two primary areas of application for metal CVD: barrier metals and copper to replace sputtered AlSi and AlCuSi. Common CVD barrier metals include titanium nitride and tungsten nitride. These metals are deposited under a low resistance interconnect layer to provide adhesion to the substrate and to prevent chemical interactions between layers. Barrier metals may also be deposited on top of the interconnect layer to serve as antireflective layers and to improve the reliability of the metal.

Titanium nitride can be deposited from NH_3 and $TiCl_4$ following the overall reaction

$$6TiCl_4 + 8NH_3 \rightarrow 6TiN + 24HCl + N_2$$

Typically the deposition is run at 700 to 800°C. The film resistivity and chlorine concentration both drop as the deposition temperature increase [69]. Lower deposition temperature can be used in the same chemistry used by PECVD, but corrosion due to residual chlorine remains a concern. Alternatively, one can use organometallic titanium sources such as tetrakisdiethylaminotitanium, $Ti[N(CH_2CH_3)_2]_4$, or TDEAT, or with tetrakisdimethylaminotitanium, $Ti[N(CH_3)_2]_4$, or TDMAT, at temperatures less than 500°C [70]. Either process can produce smooth films with excellent step coverage over severe topographies. The TDEAT/NH$_3$ process is generally preferred, however, since it is less prone to particulate contamination.

As will be discussed in a later chapter, sputtered aliminnum alloys is being replaced by lower resistivity copper. Virtually all CVD copper is done with organometallics. A typical source is copper II hexafluoroacetylacetonate, $(CF_3—CO—CH—CO—CF_3)$ Cu, or $Cu(hfa)_2$ for short. Typically $Cu(hfa)_2$ is reduced in an H$_2$ ambient at temperatures of 250 to 450°C with the high temperatures required at pressures close to an atmosphere or when the H$_2$ carrier is replaced by an inert gas [71]. As with nearly all organometallic sources, carbon can be seen in the deposited films, but purities of 99% can be achieved if H$_2$ is used as the carrier [72]. Copper resistivities only 10% higher than bulk can be achieved [73], and the resistivity decreased linearly with temperature, indicating normal phonon scattering was the dominant scattering mechanism. These low resistivity films require the use of an H$_2$ carrier and deposition temperatures of at least 300°C. Film resistivity also depends strongly on film thickness for films less than 300 nm thick. This probably represents the influence of grain boundaries, and at very thin films, the breakdown of continuity [74]. Due to the success of plating, however, little copper CVD is used in production.

13.9 Atomic Layer Deposition

Atomic layer deposition (ALD) is a variation of chemical vapor deposition. The ALD process has been existence since the 1970's, but interest increased markedly in the last few years [75]. Similar to chemical vapor deposition, the process injects gases containing material to be deposited into a reactor containing the sample to be coated. In CVD, heat or a plasma provides the energy to overcome significant kinetic barriers to one or more chemical processes. This barrier is a necessary part of the process, since excessive spontaneous gas phase reactions can lead to the homogeneous nucleation of solid particles, producing a poor quality film. The ALD process relies on the sequential introduction into the chamber of two gases that have two critical properties: (1) for at least one of the gases (step A), the surface of the sample will saturate once a monolayer coverage has been achieved—no further coating is possible, and (2) the second gas (step B) will react with the first to give the desired material. This reaction typically has very low or no kinetic barrier, but since the first gas has been flushed from the chamber before the introduction of the second gas, there is no possibility of reaction except on the surface. To prevent any possibility of gas phase reaction, the chamber is typically flushed with an inert species after each gas flow. Thus the actual sequence is A/inert/B/inert/A/inert/B/inert/ . . . . The growth rate of ALD process, plotted as a function of temperature, is U shaped, with increases at both low and high temperature. The central region over which the deposition rate is nearly temperature independent is called the ALD process window [76]. ALD's ability to coat high aspect ratio features makes it an attractive candidate for use in several steps in deeply scaled silicon technologies [77, 78].

A common prototype for the ALD process is the deposition of Al$_2$O$_3$ from trimethylaluminum [Al(CH$_3$)$_3$] and H$_2$O [79]. It has been assumed that the exposure of the surface to water vapor leads to the formation of a monolayer of surface hydroxyl (OH) groups and that these groups react with the aluminum source molecules, which lose one of the methyl groups and in the process form a bond to the substrate [Al(CH$_3$)$_2$—O—Si]. At the right temperature, the trimethylaluminum does not bond to itself, nor can it dissociate, so once all of the hydroxyls have reacted, the process self-limits. Upon the next water cycle, the water vapor will react with the remaining methyl ligands, forming methane (CH_4) and Al_2O_3 and leaving more hydroxyls on the surface. Although the actual chemistry is more complex [80] than this simple picture, the process is very effective. Metal nitrides are commonly formed from metal halides like $TiCl_4$ and ammonia (NH_3) [81], or organometallics (OMs) and ammonia [82]. Often the OMs can be used at lower temperature. Hydrogen can be used as a reducing agent for depositing pure metals. In some cases plasmas have been used to improve the process [83].

Ideally this A/B sequence will coat the surface with exactly one monolayer of film. Films of the desired thickness are built up digitally, by switching back and forth between A and B. This sort of process, with essentially digital control over the film thickness, is a significant advantage over conventional CVD. In general, CVD does not use any type of rate-monitoring equipment. Deposition rates are determined by running test wafers, and the deposition time is calculated from the ratio of the desired thickness and the estimated rate. For films such as gate insulators and barrier metals, where very thin (3–6 nm) films must be deposited, and control of the thickness is essential, ALD provides a significant advantage. Equally important is the ability of ALD to provide uniform coating of extreme surface geometries. CVD can provide conformal films over significant topography by ensuring that the process is run well into a surface reaction limited regime, allowing adatoms and admolecules to diffuse long distances before reacting and becoming incorporating into the film. The more extreme the topography, however, the more difficult it is to ensure that surface reactions and not gas transport limit the deposition. Typically this limitation is seen in coating high aspect ratio holes or gaps, where the deposition rate drops near the bottom of the hole. In the ALD process, the chemical reaction and the transport are separated in time. Because of this, ALD can be set up in such a way that almost arbitrarily high aspect ratio features can be coated uniformly.

ALD can be used to deposit a variety of materials and multiple sets of precursors exist for some of these materials. Table 13.1 shows some of the materials that have been demonstrated, categorized by their typical applications.

ALD has several drawbacks, however. One of the most commonly cited is the low deposition rate. Ideally one should get one monolayer (~0.2 nm) for each complete cycle. If a cycle can be run in 5 secs, this is 12 cycles, or 2.4 nm/min. Typical CVD rates are an order of magnitude larger. In reality, ALD rates are typically somewhat less than one monolayer per cycle, and the process may have a significant growth initiation step. As a result, the deposition of a 5-nm film typically takes about 100 cycles or ~3 min/wafer [84]. Batch systems in which multiple wafers are run simultaneously can improve throughput. Practically speaking, therefore, ALD is limited to applications such as the gate insulator and metal diffusion barriers and adhesion layers, where the final film thickness is less than about 10 nm. A second concern is low material consumption and therefore a high cost of operation. Complete monolayer coverage, especially for high aspect ratio features, may require many

Table 13.1   Some of the materials for which ALD processes have been demonstrated

Elemental Metals	**Cu, Ru, Ir, Pt, Pd, Co, Fe, Ni, Mo**
Diffusion barriers	WN, TiN, NgN, MoN, TaN, Al_2O_3
Gate insulators	Al_2O_3, ZrO_2, HfO_2, $HfAlO_2$, Ta_2O_5, La_2O_3, $PrAlO$, TiO_2, $HfSiO_2$, HfSiON, laminates
Sensors and piezoelectrics	SnO_2, Ta_2O_5, ZnO, AlN, ZnS
Optical applications	AlTiO, SnO_2, ZnO, SrS:Cu, ZnS:Mn, ZnS:Tb, SrS:Ce, Al_2O_3, SnO_2, ZnS, Ta_2O_5, Ta_3N_5, TiO_2
Transparent conductors	ZnO:Al, InSnO

seconds of gas flow. The vast majority of this gas is not used due to the nature of the process. The organometallics gases used in some of the processes are quite expensive and so drive up the cost of using the system.

Unwanted and/or incomplete reactions form a third area of concern. For use as contact diffusion barriers, the by-products such as hydrogen released by ALD can substantially degrade the low permittivity materials on which the deposition is run [85]. If porous insulators are chosen to reduce the permittivity of the insulators used for interconnect, the high conformality of ALD presents a problem because metal films will penetrate into the insulator unless a capping layer is used, and that capping layer will coat all exposed porous surfaces including contact/via sidewalls. Finally, ALD can have significant residual contamination in the film. This is especially a concern in gate insulators, where the device is sensitive to ppm levels of charge trapping. Common contaminants are Cl, C, and H. ALD process must be run at low temperature. If the substrate temperature is too large, undesired chemical reactions can occur. Under these conditions, reaction by-products may have insufficient energy to be completely removed from the surface, leading to their incorporation in the film.

13.10 Electroplating Copper

Electroplating is the process of depositing metal on a conductive surface in a solution containing the ions of the metal to be deposited. From a microelectronics perspective, the material of interest for electroplating is interconnect copper, where electroplating has become the standard process. To electroplate copper on a silicon wafer, the wafer is connected to the cathode and the copper piece (source) is connected to the anode as shown in Figure 13.26. The anode and the cathode are immersed in a conductive solution containing cupric ions or some electrolyte. A common solution is a cupric sulfate mixture. More complex arrangements, including multiple shaped electrodes and various rings or shields, may be used to improve uniformity. When an electrical connection is made to the seed layer, there are two reactions that occur near the surface of the wafer. In general the following two steps account for the copper deposition on the wafer [86, 87]:

$$Cu^{2+} + e \Leftrightarrow Cu^+$$
$$Cu^+ + e \Leftrightarrow Cu$$

An important problem with copper interconnect is that it can easily diffuse through both Si and SiO_2. Once in the substrate, Cu can kill devices as it introduces an energy level deep in the silicon bandgap. As a result, the use of a barrier layer to prevent diffusion of Cu is mandatory. There are several materials that can be used for the barrier layer. Some of the more common materials are the refractory metals and their nitrides such as Ta and TaN or W and WN. Barrier layers may or may not stick well to the insulator. If the metal does not adhere, an adhesion layer is used. The adhesion layer is often the pure of a corresponding metal nitride barrier. Examples include Ti/TiN and Ta/TaN. An adhesion layer may also be used on top of the diffusion barrier to ensure that the copper adheres.

For the electroplating process to start, a seed layer needs to be deposited. The most common method to deposit the seed layer is physical vapor deposition (PVD) of copper. Poor step coverage is obtained over the barrier and seed layer using

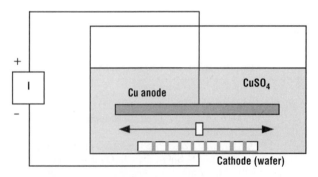

Figure 13.26   Basic electroplating setup.

PVD [88]. To alleviate this problem, advanced PVD techniques have been developed to improve the step coverage. One technique is ionized metal sputtering, a form of ionized physical vapor deposition (IPVD) that has obtained usable step coverage for vias down to 0.1 μm. In one study [89], IPVD and collimated PVD were compared as seed layers. Both were covered using electroplating, and void-free filling in the contact structures resulted. However collimation voids were left near the bottom after the damascene process had been filled. IPVD generated approximately double the coverage on the bottom and the sidewalls compared to the collimation process. It is also possible to deposit the copper seed using an electroless plating method [90]. The texture of the seed layer is quite important, since the seed can serve as a template for the electroplated growth.

When one considers all of the layers needed (Figure 13.27), a complex stack structure is needed for copper. For very small vias (<65 nm) it is challenging to even design a stack structure that preserves a reasonable amount of copper at the core. For that reason, there has been a great deal of research into continuous ultrathin barriers such as the ALD metal nitrides discussed in the last section. It has been shown that one can electroplate directly on a barrier metal [91], but this is not commonly done due to the high resistivity of more barrier metals.

One of the most important characteristics of electroplated copper is the resistivity, since that is the reason for moving from aluminum to copper [92]. In considering resistivity in copper, one of the most important factors is the film roughness. Improperly plated, copper has large grains and an extremely rough texture. The governing factors in copper electroplating roughness are the electrolyte surface reaction and the Cu^{2+} diffusion from the electrolyte solution to the electrode surface [87]. In general, when the Cu^{2+} flux from the solution is slower than the reaction rate at the electrode, polarization occurs. This polarization causes the deposition to slow and results in a smooth surface. Polarization can also occur when the current density is high. Again, the causes the surface reaction to be faster than the Cu_2^+ supply from the solution. Figure 13.28 shows a plot of the effect of plating current on the resistivity. A lower concentration of $CuSO_4$ in the solution also helps to improve the resistivity for the same reason.

Bath additives can be used to improve film properties. Sulfuric acid is commonly used as an additive to cupric sulfate. Increasing the H_2SO_4 concentration decreases the Cu^{2+} diffusion and causes an increase in polarization. Both effects help in forming a smooth surface, as does increasing the

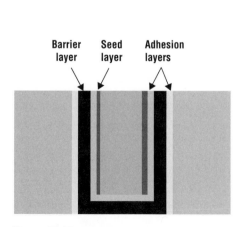

Figure 13.27 Most extreme stack in a contact, showing all of the needed layers. The crosshatched area is electroplated Cu.

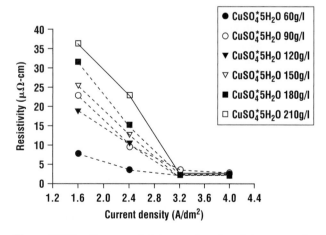

Figure 13.28 Copper resistivity as a function of the current density for various bath concentrations *(After Gau, et al., reprinted with permission from Journal of Vacuum Science and Technology. Copyright 2000 IOP Publishing).*

sulfuric acid concentration in the bath solution. Another additive that has been studied is hydroxy-lamine sulfate [$(NH_2OH)_2H_2S_O4$]. Hydroxylamine sulfate was used because it dissolves into small molecules with a lone pair of electrons around the nitrogen atom. It produces a complete fill with no voids in 0.3-μm-wide high aspect ratio trenches [88]. Acidic gluconate has also been added to baths to increase the polarization in copper electroplating. The study found that the use of gluconate increased the polarization at the cathode and decreased the current density.

There are some disadvantages with copper electroplating. One is the control of the electrolyte solution used for electroplating, but there have been advances in systems monitoring the chemical composition of the solution. One study used high performance chromatography to monitor active components of the inlet flow path in the plating bath. Another disadvantage with copper electroplating is the need for a seed layer and a barrier for copper diffusion. The lowest resistivity is achieved by annealing the copper at low temperatures ($\sim$100°C), with thinner films annealing better than thicker ones. Upon temperature cycling, however, plated contacts can blister. This is caused by the release of hydrogen trapped in the film. The problem is most acute for films plated on electroless Cu seed layers. Swirl defects occur when the wafer does not uniformly wet the solution, but can be eliminated with the correct bath [93]. The other major concern is the fill topology for very small contacts. The desired profile in that of a superfill, filling predominantly from the bottom. Again, this can be achieved with the proper bath [94].

13.11 Summary

This chapter reviewed the basic chemistry and fluid mechanics of chemical vapor deposition. Although an accurate quantitative description of any reaction is highly process and reactor dependent, this introductory material can be used to develop a qualitative understanding of the deposition process. Atmospheric pressure CVD (APCVD) is widely used only for SiO_2 deposition. The major problem with these processes is particle formation either at the injector nozzles or through homogeneous nucleation. Low pressure CVD (LPCVD) of polysilicon and silicon nitride is widely used in silicon technologies. Standard furnace tube arrangements provide good uniformity and high throughput. To improve the uniformity further, cross-flow reactors have been developed. The deposition of oxide can be done with silane and oxygen or nitrous oxide, but the currently preferred method is the thermal decomposition of TEOS. Plasma-enhanced CVD is also popular when low deposition temperatures are required. These films tend to have poor stoichiometry, and they suffer from high etch rates unless subsequent anneals are done. To obtain high density PECVD films, afterglow reactors have been developed. The CVD of metals is a new and important area that allows the fabrication of high density interconnect. Currently, tungsten is the most popular CVD metal system.

Problems

1. Repeat the calculations done in Example 13.1 if the reaction of interest is

$$AB_2 \rightleftharpoons A + 2B$$

   Assume that the equilibrium constant is unchanged.
2. Repeat the calculations done in Example 13.2 assuming that the gas is hydrogen (η = 30 g/cm-sec).
3. Briefly describe the advantages and disadvantages of APCVD.
4. Assume that you wanted to deposit NaCl (table salt) on a wafer. What precursors might you use? What problems could you foresee?

5. A particular process is reaction rate limited at 700°C and the activation energy is 2 eV. At this temperature the deposition rate is 1000 Å/min. What would you guess that it would be at 800°C? If the measured deposition rate at 800°C is much less than this prediction, what might your conclusion be? How would you prove it?

6. The deposition rate of some material in a CVD reactor is measured as a function of both temperature and total gas flow rate. The results, in angstroms per minute, are summarized in the following table.
   (a) Find the expected deposition rate at 100 sccm and 250°C.
   (b) Find the expected deposition rate at 300 sccm and 550°C.
   Be certain to show your work.

	300°C	350°C	400°C	450°C	500°C	550°C
100 sccm	0.012	0.70	22	430	470	520
200 sccm	0.012	0.70	25	490	620	730

7. The accompanying figure shows the deposition rate of silicon from silane at a gas flow of 100 sccm.
   (a) Is the deposition reaction rate limited regime at 650°C? Why or why not?
   (b) What is the activation energy (in eV) in the reaction rate limited regime?
   (c) At the highest temperature, the deposition rate is 0.85 μm/min. What would you expect for a deposition rate at this temperature if the flow rate was reduced to 50 sccm?

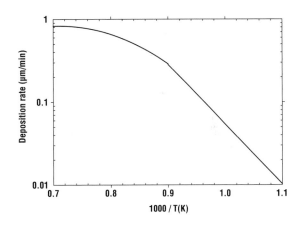

8. A process is run in a standard horizontal LPCVD tube, where the wafers are loaded on edge in a standard slotted boat. What factors might explain a reduction in the deposition rate from the front of the tube to the back? From the edge of each wafer to the center? What would you do to try to improve the uniformity in each case?

9. A postdeposition anneal called a *densification step* is often used to reduce the etch rate of CVD SiO_2. The densification is typically run at 900–1000°C. The step is not normally done for PECVD films, although they would benefit from the anneal. Explain why the process is not done for these films.

References

1. A. Sherman, *Chemical Vapor Deposition for Microelectronics: Principles, Technology, and Applications*, Noyes, Park Ridge, NJ, 1987.
2. S. Sivaram, *Chemical Vapor Deposition*, Van Nostrand Reinhold, New York, 1995.

3.   J. Dzarnoski, S. F. Rickborn, H. E. O'Neal, and M. A. Ring, "Shock Induced Kinetics of the Disilane Decomposition and Silylene Reactions with Trimethylsilane and Butadiene," *Organometallics* **1**:1217 (1982).

4.   H. K. Moffat, *Numerical Modeling of Chemical Vapor Deposition Processes in Horizontal Reactors*, Thesis, University of Minnesota, 1992.

5.   M. E. Coltrin, R. J. Lee, and J. A. Miller, "A Mathematical Model of Silicon Vapor Deposition Further Refinements and the Effects of Thermal Diffusion," *J. Electrochem. Soc.* **133**:1206 (1986).

6.   For the viscosities of common carrier gases, see L. J. Giling, in *Crystal Growth of Electronic Materials*, E. Kaldis, ed., Elsevier Science Publisher, Amsterdam, 1985, p. 71.

7.   G. B. Stringfellow, *Organometallic Vapor-Phase Epitaxy Theory and Practice*, Academic Press, Boston, 1989.

8.   M. L. Hammond, "Introduction to Chemical Vapor Deposition," *Solid State Technol.* **22**:61 (December 1979).

9.   R. Takahashi, Y. Koga, and K. Sugawara, "Gas Flow Patterns and Mass Transfer Analysis in a Horizontal Flow Reactor for Chemical Vapor Deposition," *J. Electrochem. Soc.* **119**:1406 (1972).

10. W. Kern and R. S. Rosler, "Advances in Deposition Processes for Passivation Films," *J.Vacuum Sci. Technol.* **14**:1082 (1977).

11. K. Shankar, "Thermal CVD Equipment: Competition Heats Up," *Solid State Technol.* April 1991, p. 47.

12. K. Fujino, Y. Nishimoto, N. Tokumasu, and K. Maeda, "Low Temperature, Atmospheric Pressure CVD Using Hexamethylsilixane and Ozone," *J. Electrochem. Soc.* **139**:2282 (1992).

13. S. B. Desu, C. H. Peng, T. Shi, and P. A. Agaskar, "Low Temperature CVD of SiO_2 Films Using Novel Precursors," *J. Electrochem. Soc.* **139**:2682 (1992).

14. R. S. Rosler, "Low Pressure CVD Production Processes for Poly, Nitride, and Oxide," *Solid State Technol.* **20**:63 (1977).

15. R. Iscoff, "Hotwall LPCVD Reactors: Considering the Choices," *Semicond. Int.* 60 (June 1991).

16. T. I. Kamins, "Structure and Properties of LPCVD Silicon Films," *J. Electrochem. Soc.* **127**:686 (1980).

17. A. C. Adams, "Dielectric and Polysilicon Film Deposition," in *VLSI Technology*, S. M. Sze, ed., McGraw-Hill, New York, 1988.

18. A. T. Voutsas and M. K. Hatalis, "Structure of As-Deposited LPCVD Silicon Films at Low Deposition Temperatures and Pressures," *J. Electrochem. Soc.* **139**:2659 (1992).

19. M. K. Hatalis and D. W. Greve, "Large Grain Polycrystalline Silicon by Low-Temperature Annealing of Low Pressure Chemical Vapor Deposited Amorphous Silicon Films," *J. Appl. Phys.* **63**:2260 (1988).

20. A. Baudrant and M. Sacilotti, "The LPCVD Polysilicon Phosphorus Doped in situ as an Industrial Process," *J. Electrochem. Soc.* **129**:2620 (1982).

21. P. Duverneuil and J.-P. Couderc, "Two Dimensional Modeling of Low Pressure Chemical Vapor Deposition Hot Wall Tubular Reactors," *J. Electrochem. Soc.* **139**:296 (1992).

22. T. A. Badgwell, T. F. Edgar, I. Tractenberg, and J. K. Elliott, "Experimental Verification of a Fundamental Model for Multiwafer Low-pressure Chemical Vapor Deposition of Polysilicon," *J. Electrochem. Soc.* **139**:524 (1992).

23. W. A. Pliskin, "Comparison of Dielectric Films Deposited by Various Methods," *J. Vacuum Sci. Technol.* **14**:1064 (1977).

24. K. Watanabe, T. Tanigaki, and S. Wakayama, "The Properties of LPCVD SiO_2 Film Deposited by SiH_2Cl_2 and N_2O Mixtures," *J. Electrochem. Soc.* **128**:2630 (1981).

25. F. S. Becker, D. Pawlik, H. Schafer, and G. Standigl, "Process and Film Characterization of Low Pressure Tetraethylorthosilicateborophosphosilicate Glass," *J. Vacuum Sci. Technol. B* **4**:732 (1986).

26. G. Smolinsky, "The Low Pressure Chemical Vapor Deposition of Silicon Oxide Films in the Temperature Range 450°C to 600°C from a New Source: Diacteoxyditertiarybutoxysilane," *Proc. 1986 Symp. VLSI Technol.*, San Diego, May, 1986.

27. E. J. Kim and W. N. Gill, *J. Electrochem. Soc.* **141**:3463 (1994).

28. D. M. Dobkin, S. Mokhtari, M. Schmidt, A. Pant, L. Robinson, and A. Sherman, *J. Electrochem. Soc.* **142**:2332 (1995).

29. W. A. Pliskin, "Comparison of Properties of Dielectric Films Deposited by Various Methods," *J. Vacuum Sci. Technol.* **14**:1064 (1977).

30. C. L. Ramiller and L. Yau, "Borophosphosilicate Glass for Low Temperature Reflow," *Semicond. West Technol. Proc.* **5**:29 (1982).

31. A. C. Adams and C. D. Capio, "The Deposition of Silicon Dioxide Films at Reduced Pressure," *J. Electrochem. Soc.* **126**:1042 (1979).

32. A. J. Learn, "Phosphorus Incorporation Effects in Silicon Dioxide Grown at Low Pressure and Temperature," *J. Electrochem. Soc.* **132**:405 (1985).

33. L.-Q. Xia, E. Yieh, P. Gee, F. Campana, and B. C. Nguyen, *J. Electrochem. Soc.* **144**:3209 (1997).

34. F. H. P. M. Habraken, A. E. T. Kuiper, A. V. Oostrom, and Y. Tamminga, "Characterization of Low Pressure Chemical Vapor Deposited and Thermally Grown Silicon Nitride Films," *J. Appl. Phys.* **53**:404 (1982).

35. R. D. Compton, "PECVD: A Versatile Technology," *Semicond. Int.* 60 (July 1992).

36. S. Matsuo and M. Kiuchi, "Low Temperature Deposition Apparatus Using an Electron Cyclotron Resonance Plasma," *Proc. Symp. VLSI Sci., Technol.*, Electrochem. Society, Pennington, NJ, 1982, p. 83.

37. P. D. Lucovsky, D. V. Richard, S. Tsu, Y. Lin, and R. J. Markunas, "Deposition of Silicon Dioxide and Silicon Nitride by Remote PECVD," *J. Vacuum Sci. Technol. A* **4**:681 (1986).

38. V. Herak and D. J. Thomson, "Effects of Substrate Temperature on the Electrical and Physical Properties of Silicon Dioxide Films Deposited from Electron Cyclotron Resonance Microwave Plasmas," *J. Appl. Phys.* **67**:6347 (1990).

39. F. Plais, B. Agius, F. Abel, J. Siejka, M. Puech, G. Ravel, P. Alnot, and N. Proust, "Low Temperature Deposition of SiO_2 by Distributed Electron Cyclotron Resonance Plasma Enhanced Chemical Vapor Deposition," *J. Electrochem. Soc.* **139**:1489 (1992).

40. R. S. Rosler, W. C. Benzing, and J. A. Baldo, "A Production Reactor for Low Temperature Plasma Enhanced Silicon Nitride Deposition," *Solid State Technol.* **19**:45 (1976).

41. D. R. Cote, S. V. Nguyen, A. K. Stamper, D. S. Armbrust, D. Többen, R. A. Conti, and G. Y. Lee, "Plasma Assisted Chemical Vapor Deposition of Dielectric Thin Films for ULSI Semiconductor Circuits," *IBM J. Res. Dev.* **43**(1/2):5 (1999).

42. W. A. P. Claasen, W. G. J. N. Valkenburg, M. F. C. Willemsen, and W. M. v. d. Wijgert, "Influence of the Deposition Temperature, Gas Pressure, Gas Phase Composition, and RF Frequency on Composition and Mechanical Stress of Plasma Silicon Nitride Layer," *J. Electrochem. Soc.* **132**:893 (1985).

43.   R. Chow, W. A. Lanford, W. Ke-Ming, and R. S. Rosler, "Hydrogen Content of a Variety of Plasma-Deposited Silicon Nitrides," *J. Appl. Phys.* **53**:5630 (1982).

44.   J. Batey and E. Tierney, "Low-Temperature Deposition of High-Quality Silicon Dioxide by Plasma Enhanced Chemical Vapor Deposited," *J. Appl. Phys.* **60**:3136 (1986).

45.   A. C. Adams, F. B. Alexander, C. D. Capio, and T. E. Smith, "Characterization of Plasma Deposited Silicon Dioxide," *J. Electrochem. Soc.* **128**:1545 (1981).

46.   A. Sherman, *Chemical Vapor Deposition for Microelectronics: Principles, Technology, and Applications*, Noyes, Park Ridge, NJ, 1987.

47.   E. P. G. T. van de Ven, "Plasma Deposition of Silicon Dioxide and Silicon Nitride Films," *Solid State Technol.* **24**:167 (April 1981).

48 .   D. L. Smith and A. S. Alimonda, "Chemistry of SiO_2 Plasma Deposition," *J. Electrochem. Soc.* **140**:1496 (1993).

49.   H.-J. Schliwinski, U. Schnakenberg, W. Windbracke, H. Neff, and P. Lange, "Thermal Annealing Effects on the Mechanical Properties of Plasma Enhanced Chemical Vapor Deposited Silicon Oxide Films," *J. Electrochem. Soc.* **139**:1730 (1992).

50.   S. Nguyen, S. Burton, and P. Pan, "The Variation of Physical Properties of Plasma-Deposited Silicon Nitride and Oxynitride with Their Compositions," *J. Electrochem. Soc.* **131**:2348 (1984).

51.   D. R. Secrist and J. D. Mackenzie, "Deposition of Silica Films by the Glow Discharge Technique," *J. Electrochem. Soc.* **113**:914 (1966).

52.   U. Mackens and U. Merkt, "Plasma Enhanced Chemical Vapor Deposition of Metal-Oxide-Semiconductor Structures on InSb," *Thin Solid Films* **97**:53 (1982).

53.   F. Fracassi, R. d'Agostino, and P. Favia, "Plasma-Enhanced Chemical Vapor Deposition of Organosilicon Thin Films from Tetraethoxysilane-Oxygen Feeds," *J. Electrochem. Soc.* **139**:2636 (1992).

54.   W. J. Patrick, G. C. Schwartz, J. D. Chapple-Sokol, R. Carruthers, and K. Olsen, "Plasma Enhanced Chemical Vapor Deposition of Silicon Dioxide Films Using Tetraethoxysilane and Oxygen: Characterization and Properties of Films," *J. Electrochem. Soc.* **139**:2604 (1992).

55.   K. Ramkumar and A. N. Saxena, "Stress in Thermal SiO_2 Films Deposited by Plasma and Ozone Tetraethylorthosilicate Chemical Vapor Deposition Processes," *J. Electrochem. Soc.* **139**:1437 (1992).

56.   John E. J. Schmitz, *Chemical Vapor Deposition of Tungsten and Tungsten Silicides for VLSI/ULSI Applications*, Noyes, Park Ridge, NJ, 1992.

57.   E. K. Broadbent and C. L. Ramiller, "Selective Low Pressure Chemical Vapor Deposition of Tungsten," *J. Electrochem. Soc.* **131**:1427 (1984).

58.   Y. Pauleau and P. Lami, "Kinetics and Mechanism of Selective Tungsten Deposition by LPCVD," *J. Electrochem. Soc.* **132**:2779 (1985).

59.   C. M. Melliar-Smith, A. C. Adams, R.-K. Kaiser, and R. A. Kushner, "Chemical Vapor Deposition of Tungsten for Semiconductor Metallizations," *J. Electrochem. Soc.* **121**:298 (1974).

60.   N. Hashimoto and Y. Koga, "The Si-WSi_2-Si Epitaxial Structure," *J. Electrochem. Soc.* **114**:1189 (1967).

61.   L. Kaplan and F. d'Heurle, "The Deposition of Molybdenum and Tungsten Films from Vapour Decomposition of Carbonyls," *J. Electrochem. Soc.* **117**:693 (1970).

62.   M. Diem, M. Fisk, and J. Goldman, "Properties of Chemically Vapor Deposited Tungsten Thin Films on Silicon Wafers," *Thin Solid Films* **107**:39 (1983).

63.   R. Hogle and K. Aitcheson, *Tungsten Workshop I*, 1985, p. 225.

64. S. R. Kurtz and R. G. Gordon, "Chemical Vapor Deposition of Titanium Nitride at Low Temperatures," *Thin Solid Films* **140**:277 (1986).

65. A. Sherman, "Growth and Properties of LPCVD Titanium Nitride as a Diffusion Barrier for Silicon Device Technology," *J. Electrochem. Soc.* **137**:1892 (1990).

66. M. Iwasaki, H. Itoh, T. Katayama, K. Tsukamoto, and Y. Akasaka, *Tungsten Workshop V*, 1990, p. 187.

67. J. E. J. Schmitz, R. C. Ellwanger, and A. J. M. van Dijk, *Tungsten Workshop III*, 1988, p. 55.

68. S. Sivaram, *Chemical Vapor Deposition*, Van Nostrand-Reinhold, New York, 1995.

69. N. Yokoyama, K. Hinode, and Y. Homma, "LPCVD Titanium Nitride for ULSIs," *J. Electrochem. Soc.* **138**(1):190 (1991).

70. J. Baliga, "Depositing Diffusion Barriers," *Semiconductor Int.* **20**(3):76 (March 1997).

71. T. Kodas and M. Hampden-Smith, *The Chemistry of Metal CVD*, VCH, Weinheim, 1994.

72. A. E. Kaloyeros, A. Feng, J. Garhart, K. C. Brooks, S. K. Ghosh, A. N. Sexena, and F. J. Luehers, *Electronic Mater.* **19**:271 (1990).

73. C. Oehr and H. Suhr, "Thin Copper Films by Plasma CVD Using Copper-Hexafluoro-Acetylacetonate," *Appl. Phys. A* **45**:151 (1998).

74. W. G. Lai, Y. Xie, and G. L. Griffin, "Atmospheric Pressure Chemical Vapor Deposition of Copper Thin Films. I. Horizontal Hot Wall Reactor," *J. Electrochem. Soc.* **138**:3499 (1991).

75. M. Leskela and M. Ritala, *Thin Solid Films* **409**:138 (2002).

76. H. Kim, "Atomic Layer Deposition of Metal and Nitride Thin Films: Current Research Efforts and Applications for Semiconductor Device Processing," *J. Vacuum Sci. Technol. B Microelectron. Nanometer Struct.* **21**(6):2231 (November/December 2003).

77. Adrien R. Lavoie, "ALD as Enabling Technology for the Next Generation of Microprocessors," in *Nanofabrication: Technologies, Devices, and Applications II*, *Proc. SPIE* **6002**:60020J (2005).

78. Bijan Moslehi, "Reviewing Process Technology Challenges," *Micro* **23**(8):82 (October/November 2005).

79. A brief ALD animation can be found at http://www.cambridgenanotech.com/animation/.

80. Riikka L. Puurunen, "Correlation Between the Growth-per-Cycle and the Surface Hydroxyl Group Concentration in the Atomic Layer Deposition of Aluminum Oxide from Trimethylaluminum and Water," *Appl. Surf. Sci.* **245**(1–4):6 (2005).

81. L. Hiltunen, M. M. Leskela, M. Makela, L. Ninisto, E. Nykanen, and P. Soinen, *Thin Solid Films* **166**:154 (1988).

82. *Handbook of Semiconductor Manufacturing Technology*, Y. Nishi and R. Doering, eds., Dekker, New York, 2000.

83. H. S. Sim, Y. T. Kim, and H. Jeon, in *Proceedings of the American Vacuum Society Topical Conference on Atomic Layer Deposition*, Seoul, Korea, 2002.

84. A . P. Paranjpe, B. McDougall, K. Z. Zhang, and W. Vereb, in *Proceedings of the American Vacuum Society Topical Conference on Atomic Layer Deposition*, Seoul, Korea, 2002.

85. G. Beyer, A. Satta, J. Schuhmacher, K. Maex, W. Besling, O. Kilpela, H. Sprey, and G. Tempel, *Microelectron. Eng.* **64**:233 (2002).

86. S. S. Abd El Rehim, S. M. Sayyah, and M. M. El Deeb, "Electroplating of Copper Films on Steel Substrates from Acidic Baths," *Appl. Surf. Sci.* **165**(4):249–254 (2000).

87. V. M. Dubin, C. D. Thomas, N. Baxter, C. Block, V. Chikarmane, P. McGregor, D. Jentz, K. Hong, S. Hearne, C. Zhi, D. Zierath, B. Miner, M. Kuhn, A. Budrevich, H. Simka, and S. Shankar, "Engineering Gap Fill, Microstructure and Film Composition of Electroplated Copper for On-Chip Metallization," *Proc. IEEE, International Interconnect Technology Conference*, 2000.

88. C. H. Ting, and I. Ivanov, *"Advances in Copper Metallization Technology," Proc. IEEE, International Interconnect Technology Conference*, 2001.

89. E. C. Cooney, D. C. Strippe, and J. W. Korejwa, "Effects of Copper Seed Layer Deposition Method for Electroplating," *J. Vacuum Sci. Technol. A Vacuum, Surf. Films* **18**(4) II:1550–1554 (2000).

90. Tohru Hara, and Takumi Takachi, "Deposition of Low Resistivity Copper Interconnection Layers Electroplated on Electroless Plating Copper Seed Layer," Meeting Abstracts, 2004 Joint International Meeting—206th Meeting of the Electrochemical Society/2004 Fall Meeting of the Electrochemical Society of Japan.

91. Rajesh Baskaran, and Thomas Ritzdorf, "Electrodepositing a Copper Seed Layer Directly on Diffusion Barriers for Damascene Interconnects," Advanced Metallization Conference, 2004, pp 517–523.

92. W. C. Gau, T. C. Chang, Y. S. Lin, J. C. Hu, L. J. Chen, C. Y. Chang, and C. L. Cheng, "Copper Electroplating for Future Ultralarge Scale Integration Interconnection," *J. Vacuum Sci. Technol. A Vacuum, Surf. Films* **18**(2):656–660 (2000).

93. Ted Cacouris, Chee Ping Lee, Augustine Teo, Li Chaoyong, He Xin, "Identifying and Eliminating Unique Copper Electroplating Defects," *Micro* **24**(5):49–57 (2006).

94. S. Dasilva, T. Mourier, P.H. Haumesser, M. Cordeau, K. Haxaire, Passemard, Chainet, "Gap Fill Enhancement with Medium Acid Electrolyte for the 45 nm Node and Below," Advanced Metallization Conference 2005, pp. 513–517.

Chapter 14

Epitaxial Growth

Many techniques have been used to deposit single-crystal silicon and GaAs [1]. A partial list would include a number of CVD variations, including plasma enhanced, rapid thermal, metallorganic, ultrahigh vacuum, atomic layer, and laser, optical, and x-ray assisted. Non-CVD methods include molecular beam, ion beam, and clustered ion beam epitaxy, to name just a few. All of these methods have the capability to grow single-crystal layers. Very few, however, have demonstrated the capability to grow material of sufficient quality to economically fabricate high density ICs. This chapter will first concentrate on thermal chemical vapor deposition or vapor phase epitaxy (VPE), the most commonly used approach for silicon epitaxy in IC production. In this process, heat provides the energy necessary to drive the chemical processes. Later, the chapter will discuss a VPE method for growing epitaxial layers of GaAs, as well as advanced forms of VPE, including metallorganic VPE (MOCVD) and rapid thermal VPE (RTCVD). The chapter will conclude with a discussion of molecular beam epitaxy (MBE) and its variants.

A prototype VPE system is shown in Figure 14.1. Much like a low pressure CVD reactor, the wafer is in a vacuum chamber resting on a heated susceptor. This heating is usually done either by radiation from filament lamps or by inductively heating the graphite susceptor. The system is normally operated at reduced pressure, although some silicon epitaxial growth systems have no vacuum capability at all. The gas flows are controlled with mass flow controllers and pneumatic valves.

At traditional VPE temperatures the dopants diffuse significantly, making very thin layers and abrupt transitions impossible. As the temperature is reduced, the diffusivity decreases more rapidly than the growth rate, making lower temperature epitaxy very desirable. A great deal of progress has been made in recent years in reducing the temperature of silicon epitaxy. These improvements have generally involved reducing the impurity concentrations in the growth chamber. Small loading chambers called *load locks* that are attached to, and can be pumped independent of, the main chamber, allow the introduction of new wafers without venting the growth chamber after each run. Also, better pumps and seals are now being used that improve the quality of the vacuum in the chamber. Finally filters are being used to remove chemical contaminants from the gas feed. Some production growth is now done at temperatures below 1000°C. Research has demonstrated high quality growth in

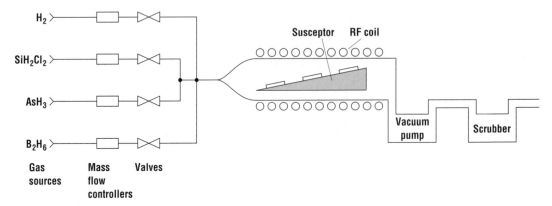

Figure 14.1   A simple VPE system. The susceptor in this chamber is inductively heated using RF power in the external coil.

single-wafer systems at temperatures of about 800°C [2]. Ultrahigh vacuum growth from the vapor has been done at much lower temperatures. Given this progress and the high defect density of the alternative growth techniques, the future of these VPE approaches looks very attractive for the growth of thick layers. Their largest liability for growing very thin structures is their lack of *in situ*, real-time growth diagnostics [3].

Current mainstream silicon manufacturing, however, uses epitaxial growth only for thick layers (1 to 10 μm). Devices may be fabricated in these layers after their growth, but the basic device regions are not formed by the epitaxy. In contrast, virtually all compound semiconductor technologies require the use of advanced epitaxial growth processes as an integral part of the device. These processes have the capability to control the vertical dimension in the device to the atomic scale, and they allow the fabrication of nearly perfect interfaces of dissimilar semiconductors. The use of these heterojunction layers provides a new degree of freedom in transistor design. Critical to their application, however, is the ability of epitaxial techniques to simultaneously provide excellent control of composition, doping, and thickness, as well as near zero defect density.

14.1 Wafer Cleaning and Native Oxide Removal

Before growth, the substrate must be cleaned; that is, the native oxide and any residual impurities and particles must be removed. This is a common requirement for many processes [4]. The ideal goal of this step would be to produce a perfect silicon surface with no impurities or change in atomic position. This is impossible to achieve even if all impurities are removed, since the surface itself has unsatisfied bonds. The atoms on the surface move or reconstruct in order to minimize the density of these bonds. Furthermore, in actual semiconductor fabrication the wafer is often exposed to air as it is moved from the preclean station to the growth chamber. During this exposure the surface may oxidize and often collects carbon contamination from the air. Thus, even a pure reconstructed surface is normally only achieved in ultrahigh vacuum (UHV) equipment.

The types of impurities likely to exist on the surface of the wafer include organics from photoresist residues or polymeric materials deposited during etches, metallic particles deposited in plasma chambers such as implanters and etchers, and thin chemical oxides. The removal of all types of impurities often requires a sequential operation. To isolate the operations, each step normally ends with a rinse in deionized water. The purity of the water is measured by the resistivity and the total organic content (TOC). Typical resistivities are 14 to 18 MΩ-cm, and TOC counts are a few parts per million. One of the most popular wet chemical preclean steps for silicon is the RCA clean [5], commonly used when only bare silicon or silicon and thick oxides are exposed on the surface of the wafer. This process will be used as a standard example. For a detailed explanation of the RCA process, see Kern and Gale [6]. When mixed in the proper proportions and held at the correct temperatures, the RCA clean will not significantly attack the silicon surface. The simplest arrangement for an RCA clean is a series of baths for immersion of the wafers. To ensure a fresh supply of chemicals, the solutions may also be plumbed into a spray-cleaning system. These cleaning systems also use less chemicals and ensure a more reproducible cleaning environment.

Organic residues are most often removed in an oxidation/reduction solution. A typical solution, called SC1, is a 1:1:5 mixture (by volume) of ammonium hydroxide, hydrogen peroxide, and water ($NH_4OH/H_2O_2/H_2O$) [7], although more recently, lower ammonium hydroxide concentrations have been found to be as effective in reducing contamination and produce less surface roughening. To increase the rate of chemical attack, the bath temperature is held at 60 to 80°C. This solution also dissolves elements from the group IB and IIB columns of the periodic table, and other metals including gold, silver, nickel, cadmium, zinc, cobalt, and chromium [8]. Because of the depletion of the reactants, the mixture should be prepared just before using it. Other acids including dilute mixtures of HCl, HNO_3, and CH_3COOH are also effective in reducing the concentration of some types of metallic impurities [9].

Since hydrogen peroxide is a strong oxidant, the wafer surface will oxidize during this clean. To remove this chemical oxide, the wafers are dipped in a dilute HF or buffered HF solution. The time required for HF removal of the chemical oxide depends on the solution strength, but is usually less than 10 sec. Complete removal of a surface oxide can be detected by dewet. This is the tendency of water to rapidly sheet off of the surface of the wafer when bare silicon is exposed. When an oxide remains, the surface stays wet for several seconds as the water slowly flows off of the wafer. On patterned wafers, the bare silicon areas will dewet if they are large enough and if there is a continuous downward path to the edge of the wafer. Typically these are the scribe lines or streets between the die.

Finally, the heavy alkali ions and cations are stripped in a halogen-containing solution. A common mixture is 6:1:1 of H_2O:HCl:H_2O_2 heated to 70 to 80°C. As with the organic step, typical immersion times are 10 to 15 min and end with a I rinse in deionized water. A drying step using compressed N_2 gas and/or a spin dryer completes the process. It is important in designing an RCA clean to keep the ammonium hydroxide solution and the hydrochloric solution well separated. The vapors of these solutions can react to form ammonium chloride, a salt that acts as particle contamination to the bath.

The wet chemical clean described often leaves the wafer with a thin chemical oxide that is free of metallic and organic impurities. This oxide must be removed if good epitaxial growth is to be achieved. Virtually all VPE includes as its first step a thermal preclean. This can be conducted in H_2, an H_2/HCl mixture, or in vacuum. A 1% HCl/H_2 mixture etches silicon nonpreferentially and was preferred when the subsequent growth was to be done using silane or disilane without chlorine [10], since it would also volatilize from and other heavy metals. There is no direct measurement in most VPE that indicates whether precleaning has been done well. The grower must simply standardize on a process that achieves good films.

The prebake must be carried out in a temperature-oxidizing ambient partial pressure regime that makes the native oxide thermodynamically unstable. Figure 14.2 shows a plot of data taken from Ghidini and Smith [11, 12] and extrapolated to lower temperatures. For oxygen, a clear boundary exists between regions that are thermodynamically stable. At high temperature and low oxygen partial pressure, the native oxide on the wafer forms volatile suboxides such as SiO that desorb [13]. The process is believed to proceed nonuniformly, with pinholes opening up first, allowing the semiconductor to evaporate and react with the oxide [14]. This leads to some surface roughening, suggesting that it is desirable to minimize the thickness of the oxide before prebake. In the presence

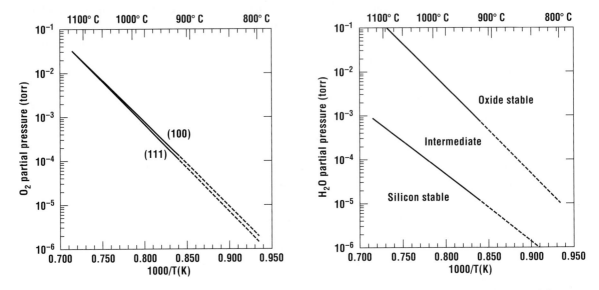

Figure 14.2 Stability diagram for the surface of silicon and silicon dioxide as functions of the temperature and the partial pressure of oxygen and water vapor. The curve for water vapor has an intermediate region where both surfaces can coexist (*redrawn using data from Ghidini and Smith, and Smith and Ghidini*).

Example 14.1

A preclean must be done using H_2 contaminated with 10 ppm of O_2. If the anneal is done at 1.0 atm, what is the minimum anneal temperature? What if it is done at 0.1 atm?

Solution

If the hydrogen prebake is done at 1 atm (760 torr), the H_2 gas contains 10 ppm of O_2 and no other impurities are present, according to Figure 14.2 the wafer must be heated to at least 1070°C to desorb the native oxide. Given a sufficient amount of time, all of the oxide will desorb from the surface if the wafer is held at or above this temperature. If the chamber pressure is reduced to 76 torr using the same gas, the preclean could be done at temperatures as low as 980°C.

---

of water vapor, there is an intermediate region where both Si and SiO_2 may exist on the surface of the wafer. If the wafer is held in this region, significant surface roughening can occur. The use of H_2 as a prebake gas reduces the temperature required, since it can help reduce SiO_2.

The limit to gas purity in Example 14.1 is that the chamber contributes impurities as well, particularly when the wafer is at temperature. As a result, both gas purity and chamber base vacuum are important concerns in epitaxial growth. This careful attention to gaseous and chamber impurities and the ability to heat the wafers to prebake temperatures for relatively short times (of order minutes) are two of the primary differences between epitaxial growth and other CVD equipment.

To allow lower temperature precleans, the chamber may be pumped to a high base vacuum before the preclean. The hydrogen source can also be scrubbed to remove any residual gases. Hydrogen is usually purified through palladium diffusion (Figure 14.3). In this method, a stream of hydrogen gas is

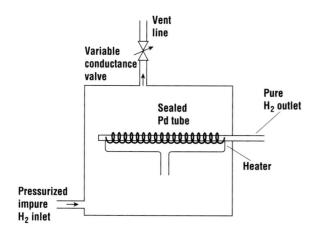

Vent line

Variable conductance valve

Pure H₂ outlet

Sealed Pd tube

Heater

Pressurized impure H₂ inlet

Figure 14.3 Schematic diagram of a palladium diffuser used for purifying hydrogen. Hydrogen diffuses through the high temperature palladium barrier, while the impurities are swept out the vent line.

delivered to a canister containing a heated (400°C) palladium tube that has been sealed on one end. Due to the high diffusivity of hydrogen in palladium, the hydrogen diffuses into the tube and from there is delivered to the growth system. Gaseous impurities that cannot diffuse through the tubing wall are swept out the vent line by the hydrogen flow. Palladium diffusers can lower the residual water and oxygen partial pressures to 1 ppb. For other gases, chemically active resins must be used. Often combined with particle filtration, these units contain a readily oxidizable species that getters oxidants from the gas stream.

There is considerable interest in improving silicon precleaning processes [15]. This research involves variations in both the wet chemical cleans and the *in situ* part of the preclean. Motivation for this work includes a reduction in the production of waste chemicals, a reduction in particle contamination, and a reduction in the prebake time and temperature to minimize dopant diffusion in the semiconductor. For applications in which even moderate temperatures cannot be tolerated, the surface can be cleaned *in situ* by sputtering in an argon [16] or hydrogen [17] ambient. A preclean chamber designed for this cleaning can be incorporated into the load lock itself, or it may be in line between the chamber and the load lock. Alternately, an Ar$^+$ beam can be set up to impinge off the surface of the wafer while it is in position for growth. The process is carried out at room temperature at low power densities (1 mA/cm^2). A major problem with the technique is residual sputter damage that is very difficult to remove by thermal annealing. This damage can be reduced by using very low energy ions such as those formed in high density plasmas [18]. Sputter damage and metallic contamination cannot be completely eliminated, however.

Another method that receives considerable attention is the surface passivation of silicon through the use of an HF dip [19]. This may be done after a standard RCA clean by removing the final chemical oxide in HF. This process is believed to leave a monolayer of silicon monohydride (SiH) on the (100) silicon surface [20] that leaves the surface resistant to oxidation in air [21]. The Si–F bond is believed to be highly polar due to the large electronegativity difference between the Si and F atoms. This polarization in turn induces a polarization in the Si–Si back bond when a fluorine atom attaches to the surface. The Si–Si polarization allows the bond to be attacked by HF [22], which results in the release of SiF$_x$ and the replacement of the surface Si–Si bond with a stable Si–H bond. Exposure of the wafer to UV light before immersion in HF helps to produce very low levels of residual carbon on the surface. It has also been shown that a brief deionized water rinse can be tolerated if the dissolved oxygen concentration in the water is kept low [23] and the wafers are lightly doped. This is an important consideration since rinses of only a few minutes are sufficient to remove most of the residual F left on the surface after an HF etch [24]. The hydrogen termination lasts from a few hours to a few days depending on the lab atmosphere.

Once the wafer has been hydrogen terminated, it can be transferred into the growth chamber, and growth may be commenced immediately without a high temperature preclean step. The hydride layer desorbs at about 500°C. This type of process then is preferred primarily for low temperature processes such as MBE. The problem of H desorption during the ramp up to deposition temperatures make this method less attractive for standard thermal VPE. This process is also not preferred for cleaning the surface before thin oxide growth because of a slight surface roughening that reduces the breakdown characteristics of the MOS capacitor [25].

14.2 The Thermodynamics of Vapor Phase Growth

The simplest model that can be used to describe the VPE process is that of Deal (see Ref. 26). It is essentially the same model as described in Chapter 3 for oxidation. The flux of the deposition species across a gaseous boundary layer is set equal to the flux of reactant consumed by the growing surface

$$F = h_g(C_g - C_s) = k_s C_s \qquad (14.1)$$

where h_g is a mass transport coefficient that depends on the flow in the chamber, k_s is a surface reaction rate, and C_g and C_s are the concentrations of the growth species in the gas and on the surface of the wafer, respectively. Just as was done for oxidation, one can solve for the growth rate R

$$R = \frac{k_s h_g}{k_s + h_g} \frac{C_g}{N} \qquad (14.2)$$

where N is the number density of silicon (5×10^{22} cm^{-3}) divided by the number of silicon atoms in the growth molecule. The mass transport limited regime occurs when $k_s \gg h_g$, and the reaction rate-limited regime when $h_g \gg k_s$.

Example 14.2

Assume in Equation 14.2 that h_g and C_g are independent of temperature and

$$k_s = k_o e^{-E_A/kT}$$

If $h_g = 0.1$ cm/sec, $C_g = 10^{15}$ cm^{-3}, $E_A = 2.0$ eV, $k_o = 10^8$ cm/sec, and $N = 5 \times 10^{22}$ cm^{-3}, plot log R versus $1/T$.

Solution (at right)

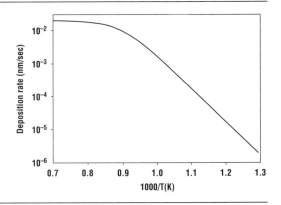

Although the Deal model provides a semiquantitative model for the growth, it vastly oversimplifies the growth process. Unlike oxidation, the growth species is not the same as the feed gas. Instead, many chemical reactions may occur in the gas phase and on the surface of the wafer. To further complicate matters, a number of processes may be going on simultaneously that assist or compete with the growth process. In the Si–H–Cl system, for example, the silicon-containing molecules on the surface of the wafer may be SiCl$_2$, SiCl$_4$, SiH$_2$, or other Si-containing species. At low chamber pressure or low inlet gas flows, the production of these reactive species limits the growth rate. Silicon atoms that are adsorbed on the surface and the substrate atoms themselves may be etched by reaction with Cl. The Deal model has no mechanism for taking this complicated chemistry into consideration. It should therefore be considered a simple parameterization of a very complicated process.

Just as for CVD in the previous chapter, a more detailed model for the growth of a layer in a VPE process can begin by dividing the process into a sequence of steps (Figure 14.4), each of which can potentially determine the growth rate. The precursor gas enters the chamber, where it partially decomposes to several more reactive daughter species. These growth species must move through the chamber until they reach the vicinity of the wafer. Near the wafer the growth species diffuse through the stagnant or boundary layer until they reach the surface. Here they adsorb, diffuse across the

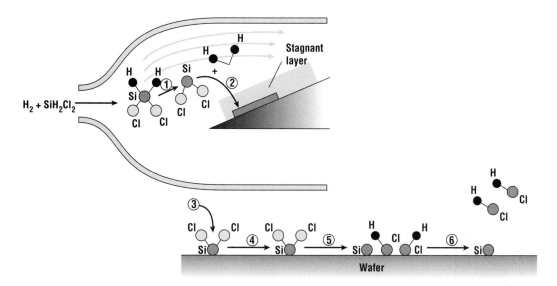

Figure 14.4   VPE steps include (1) gas phase decomposition and (2) transport to the surface of the wafer. At the surface the growth species must (3) adsorb, (4) diffuse, and (5) decompose; and (6) the reaction by-products must desorb.

surface, and decompose further to atomic silicon and volatile reaction products, which must then desorb and be pumped away. The first part of this section will describe the gas phase reactions of some of the common homogeneous processes and popular silicon epitaxy chemistries. The following section will discuss the reactions that occur on the wafer surface.

The simplest chemical system for the growth of silicon is the pyrolytic decomposition of silane. The overall process is

$$SiH_4(g) \rightarrow Si(s) + 2H_2(g) \tag{14.3}$$

This reaction is essentially irreversible and can be driven at temperatures as low as 600°C. As discussed in the last chapter, this process is often used for polysilicon deposition. Although silane can be used to epitaxially grow silicon at low temperature (600 to 800°C), its tendency to decompose to form particles in the gas (homogeneous nucleation) means that it is difficult to obtain high quality growth except at very low pressures. Homogeneous nucleation occurs continuously in the gas phase for all pressures; however, the nucleation rate increases sharply with the partial pressure of silane in the vapor. If the size of the particle created in the vapor is small, its surface energy is large enough to make the particle energetically unfavorable. The particle must therefore achieve a minimum size, or it will shrink and ultimately disappear. This critical size is given by [27]

$$r^* = \frac{2UV}{kT \ln(\sigma_o)} \tag{14.4}$$

where U is the surface interfacial free energy, V is the atomic volume, and σ_o is the ratio of the pressure of the growth species to the equilibrium pressure of the growth species. This ratio, called the *degree of saturation*, will be discussed later in the chapter. The production of these particles sets a limit on the maximum pressure of SiH_4 that is allowable for a given temperature. Most epitaxial

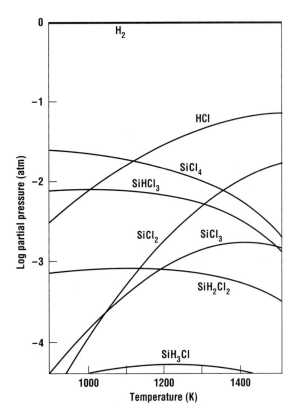

Figure 14.5 Equilibrium partial pressures in the Si–Cl–H system at 1 atm and a Cl: H ratio of 0.06 *(after Bloem and Claasen, reprinted by permission, Philips).*

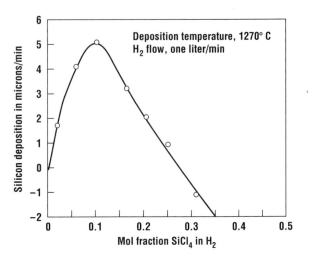

Figure 14.6 Growth rate as a function of the $SiCl_4$ flow. At high concentrations, the chlorine in the chamber leads to etching *(after Theuerer, reprinted by permission of the publisher, The Electrochemical Society).*

processes dilute the growth species with H_2 to form 1 to 5% growth gas mixtures. Even with this constraint, epitaxial silicon is not commonly grown from silane. Silicon will deposit on all surfaces in the reactor. These films may peel off and particulate during movement of the fixtures or during pumpdown and heating. Some of these particles deposit on the wafer, leading to high defect densities and potentially poor growth. Of course, epitaxial processes are far more sensitive to these defects than standard CVD because the films will be made into active device instead of simple insulators or conductors.

Nearly all silicon epitaxy for IC fabrication is done by the reduction of chlorosilanes (SiH_xCl_{4-x}, where $x = 0$, 1, 2, or 3), heavily diluted in hydrogen. The smaller the number of chlorine atoms in the precursor molecule, the lower the temperature required for the same growth rate. The first commonly used precursor was silicon tetrachloride ($SiCl_4$). To obtain useful growth rates with $SiCl_4$, however, the substrate must be held at temperatures in excess of 1150°C. Because of extreme dopant redistribution at this temperature, growth rates from $SiHCl_3$, SiH_2Cl_2, and SiH_3Cl have all been developed. Dichlorosilane (SiH_2Cl_2 or DCS) is currently the most commonly employed source [28].

The chlorosilanes all follow similar reaction paths. Figure 14.5 shows the equilibrium partial pressures for various species for $SiCl_4$ at 1 atm and 1500 K [29, 30]. As discussed in the last chapter, these calculations come from a minimization of the total free energy of the system. In real epitaxy systems, the gases may not actually reach equilibrium [31], but the tendencies remain the same. For low atomic percentages of feed gas, the primary components at the growth temperature are H_2, HCl, and dichlorosilylene ($SiCl_2$). The latter is believed to be the primary growth species in most chlorosilane VPE processes. As shown in Figure 14.6, in the reaction rate limited regime, the growth rate is reasonably well fit to [32, 33]

$$R = c_1 p_{SiCl_2} - c_2 p_{HCl}^2$$

(14.5)

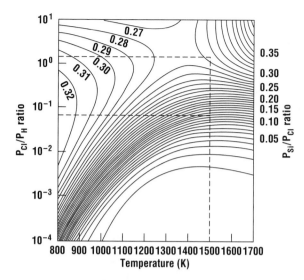

Figure 14.7   The equilibrium ratio of silicon to chlorine at 1 atm *(after Arizumi, reprinted by permission, Elsevier Science).*

where c_1 and c_2 show Arrhenius behavior. The negative term on the right is due to chlorine etching of the substrate. The etch mechanism depends on the square of the HCl partial pressure because the etching reaction requires two HCl molecules

$$Si(s) + 2HCl(g) \rightleftharpoons SiCl_2(g) + H_2(g) \qquad (14.6)$$

where $Si(s)$ refers to silicon atoms already incorporated into the crystal. Adding gaseous HCl slows the net deposition rate by increasing the etch rate.

One can define σ as the degree of supersaturation in the growth ambient given by

$$\sigma = \left[\frac{p_{Si}}{p_{Cl}}\right]_{feed} - \left[\frac{p_{Si}}{p_{Cl}}\right]_{eq} \qquad (14.7)$$

where p_{Si} and p_{Cl} are the partial pressures of the silicon and chlorine, respectively. The supersaturation is just the saturation (σ_o) minus one. The first term in Equation 14.7 refers to the ratio of the feed gases, and the second is the equilibrium ratio given as a function of temperature in Figure 14.7 [34]. To find the equilibrium ratio, first determine the ratio of chlorine to hydrogen in the chamber. Next, read across Figure 14.7 to the chamber temperature to obtain the equilibrium ratio of silicon to chlorine. Finally, subtract this ratio from the feed gas ratio to obtain the degree of supersaturation. If σ is positive, the vapor is said to be supersaturated, and under equilibrium conditions, growth will result. If σ is negative, on the other hand, the system is undersaturated and etching will result.

A problem arises if σ is very large. Unlike standard CVD, the films grown in VPE are single crystal. The ability of the incident atoms to be correctly incorporated into lattice sites is essential to obtaining high quality growth. The picture of the surface often used is the terrace-kink model. A quantitative description of the model will be covered later in the chapter. For now, assume that the model predicts the existence of two planes (terraces). Molecules landing on the lower plane diffuse to the edge of the terrace, where they dissociate and incorporate into the crystal. This type of growth is called *two dimensional* (2-D), since it proceeds primarily across the surface of the wafer. If the arrival rate is too large (the maximum growth rate for epitaxial silicon is about 0.3 μm/min at 1000°C and about 20 μm/min at 1100°C) [35], the admolecules will agglomerate in the center of the plane, forming a growth island that never reaches the terrace edge. This type of growth is called *islanding, three-dimensional* (3-D) *growth*, or Stranski–Krastanov growth. It produces a lower quality film than 2-D growth. The degree of saturation then is an important first-order description of the growth process.

Example 14.3

For the data shown in Figure 14.6, calculate the supersaturation as a function of SiCl₄ concentration. Compare this value to the growth rate. Does the model predict the peak in the growth rate? Does it predict the transition to etching?

Solution

Since the growths use $SiCl_4$, the ratio of Si/Cl in the feed gas is 0.25 for all growths. The equilibrium ratio of Si/Cl can be found using Figure 14.7 by first calculating the Cl/H ratio. For a 5% mixture, the chlorine-to-hydrogen ratio is $0.05 \times 4/(0.95 \times 2) = 0.11$. According to Figure 14.7, at the growth temperature (1270°C = 1543 K), this produces an equilibrium Si/Cl ratio of 0.14. For increasing $SiCl_4$ concentrations, see Table 14.1. The model does indeed predict a change from growth to etching between 20 and 30% $SiCl_4$ due to the large amount of Cl released. It does not predict the maximum shown at 10% because the small growth rate at low concentrations is due to mass transport limitations. The model does not take this into account.

Table 14.1

$SiCl_4/H_2$	$[Si/Cl]_{feed}$	$[Cl/H]_{feed}$	$[Si/Cl]_{eq}$	$\sigma_{\chi\chi\chi}$	Growth Rate (μm/min)
0.05	0.25	0.11	0.14	+0.11	+3.7
0.10	0.25	0.22	0.21	+0.04	+5.0
0.20	0.25	0.50	0.24	+0.01	+2.1
0.30	0.25	0.86	0.28	−0.03	−0.6

14.3 Surface Reactions

An atomistic picture of what is happening at the surface of the wafer during VPE is not very clear. The fact that the epitaxial growth of silicon from all of the precursors in the Si–H–Cl system has the same activation energy in the reaction rate limited regime (see Figure 14.8) suggests that the same rate-limiting process may be involved for each reaction [36]. This is generally believed to be the desorption of hydrogen from the surface of the wafer. In this model, a large part of the surface is hydrogen terminated. These hydrogen atoms must be desorbed before silicon atoms can be incorporated. The model has been supported by the observations of Claassen and Bloem [32] that the growth rate of silicon in dichlorosilane saturates with increasing DCS flow in a hydrogen ambient, but does not saturate in a nitrogen ambient. Coon et al. [37] have used temperature-programmed desorption to measure the surface coverage of the various species. Their results suggest that HCl, and not H, desorption is the rate-limiting step at low temperature.

A second important clue about the surface chemistry was obtained by Aoyama and coworkers [38]. Their calculations indicated that the primary growth species on the surface of the wafer is $SiCl_2$. Adsorbed molecules such as $SiCl_2$ are not chemically bonded, but instead are only physically adsorbed (physisorbed) to this passivated surface in VPE. As such there is a

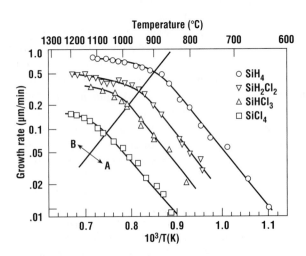

Figure 14.8   Arrhenius behavior of a variety of silicon-containing growth species *(after Eversteyn, reprinted by permission, Philips).*

Table 14.2 The major surface components for a (111) silicon surface for Cl/H = 0.01 at 1500 K

Species	% Coverage
H	63
Cl	20
$SiCl_2$	16
Vacancies	1.5
H_2	0.01

After Chernov [39].

constant flux of $SiCl_2$ to and from the surface. Chernov modeled the (111) surface of silicon in a typical epitaxial growth environment and found that under equilibrium conditions, the major surface components are H, Cl, and $SiCl_2$ (Table 14.2) [39].

To develop a model of the surface during VPE, one generally assumes that the gas phase reactions only produce growth species that are then adsorbed on the surface of the wafer, where they react and become incorporated into the growing film. It is also assumed that a limited number of sites exist on the surface of the wafer, where these molecules can adsorb. θ_x is defined as the fraction of these sites occupied by the admolecule x. Finally, assume that these fractions are proportional to their partial pressure in the gas phase, which in turn is proportional to the mass flow rate of the parent precursor

$$\theta_x \propto p_x \propto flow_x \tag{14.8}$$

For small fractions of dichlorosilane in a hydrogen carrier the growth rate can then be approximated by [32]

$$R = k_1 \frac{p_{H_2}p_{SiCl_2}}{1 + k_2 p_{H_2}}\theta - k_3 \frac{p_{HCl}^2}{p_{H_2}} \tag{14.9}$$

where θ is the fraction of free sites given by

$$\theta = \frac{1}{1 + k_4 p_{SiCl_2} + k_5 \dfrac{p_{HCl}^2}{p_{H_2}^2} + k_6 p_{H_2}^2} \tag{14.10}$$

where the k terms are constants. As with Equation 14.5, the negative term in Equation 14.9 corresponds to etching of the substrate. For most growth conditions, the gas is primarily H_2. Then, the partial pressure of hydrogen is approximately the same as the chamber pressure. The $SiCl_2$ partial pressure is proportional to the dichlorosilane inlet flow rate. The HCl partial pressure, on the other hand, is proportional to the square of the dichlorosilane inlet flow rate, since each dichlorosilane molecule contributes two chlorine atoms.

14.4 Dopant Incorporation

Both intentional and unintentional dopants may be incorporated during epitaxial growth. The primary unintentional sources are solid state diffusion from the substrate and gas phase autodoping. Solid state diffusion dominates the profile near the epi/substrate interface. It follows a complementary error function dependence if the growth rate obeys the relation

$$R > 2\sqrt{\frac{D}{t}}$$

where D is the diffusivity of the impurity and t is the growth time. The diffusion coefficients used in this calculation may be somewhat different from those calculated in Chapter 3 due to the effects of the growth process on the local concentration of self interstitials and vacancies.

Gas phase autodoping occurs when impurities desorb from the wafer travel through the gas, and readsorb elsewhere on the wafer. It generally follows a doping profile of

$$C(x) = fN_{os}e^{-x/x_m} \tag{14.11}$$

where f is the trapping density, N_{os} is the number of trapping sites on the surface, and x_m is a transition width, which depends on growth rate and the desorption coefficient.

For silicon epitaxy, the most common gaseous dopant sources are diborane (B_2H_6), arsine (AsH_3), and phosphine (PH_3). The doping concentration in the epitaxial layer is controlled by metering the flow of these gases into the reaction chamber. To achieve low partial pressures for lightly doped layers, the dopant gas is often diluted with hydrogen. One can define a segregation coefficient K as

$$K_{eff} = \frac{P_{Si}}{P_{dopant}}\frac{C_{dopant}}{5 \times 10^{22}~\text{cm}^{-3}} \tag{14.12}$$

where P_{Si} and P_{dopant} are the partial pressures of the silicon growth species and dopant, respectively, and C_{dopant} is the concentration of the dopant in the growing film due to intentional doping. When K_{eff} is less than 1, the growing film rejects the dopant. For dilute flows, the pressure ratios can be approximated by the ratios of the inlet flows multiplied by the dilution of the dopant gas. Then, the doping concentration in the film is just linearly proportional to the dopant gas flow rate.

At high dopant gas concentrations, the incorporation rate usually falls below a simple linear dependence on the dopant gas flow. In the case of PH_3, this occurs due to the parasitic gas phase reaction:

$$2PH_3 \rightarrow P_2 + 3H_2 \tag{14.13}$$

Thus, at low concentrations the phosphorus incorporation is proportional to the partial pressure, but at high concentrations it is proportional to the square root of the partial pressure. Boron doping shows similar effects [40]. The concentration at which these competing gas phase reactions become important is a sensitive function of temperature. Lower growth temperatures generally allow the growth of more heavily doped films. High concentration doping has also been shown to affect the growth rate. Arsenic doping at high concentration, in particular, is known to dramatically reduce the growth rate by poisoning the surface. That is, arsenic occupies a high concentration of surface sites, preventing silicon incorporation in the normal 2-D growth process, but not becoming incorporated itself.

14.5 Defects in Epitaxial Growth

Epitaxial growth can introduce defects and it can propagate defects. If these defects are in the active region of the wafer where the transistors are fabricated, they will often lead to device failures. These failures can be caused directly by electronic states associated with the defects, which lead to excessive leakage. Failures may also be less direct. During processing, the defects may trap other impurities in the wafer that contribute to these electronic states. The defects may also lead to excessive impurity diffusion during processing, which changes the physical device structure.

The number and density of defects are sensitive functions of the growth conditions, including substrate temperature, chamber pressure, growth precursor, and wafer precleaning procedure. The most common types of defect in epitaxial silicon layers are dislocations and stacking faults. As discussed in Section 2.3, dislocations are extra or missing lines of atoms; stacking faults are an extra plane of atoms inserted into the crystal or a missing plane of atoms in the crystal. Stacking faults in silicon normally occur in the <111> directions. Figure 14.9 shows an electron micrograph of a stacking fault in a (111) wafer, along with four scans of the surface morphology of the defect [41]. At one end of the defect is a pit, while at the other side, the stacking fault extends above the surface of the wafer. With (100) wafers the stacking fault appears as a line along the <110> directions. In either case, an obvious crystal orientation is seen.

Spikes are protrusions from the epitaxial layer that show little or no alignment with the crystal directions. Spikes may be related to the onset of 3-D growth [42]. Stacking faults and spikes often originate from a defect on the original wafer surface [43]. These original defects include oxygen, metallic impurities, oxidation-induced stacking faults in the wafer, and particles deposited on the surface of the wafer. Improvements in cleaning procedures have dramatically reduced stacking fault densities in production silicon VPE.

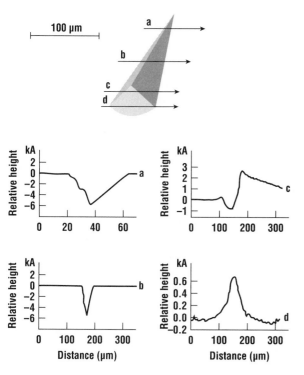

Figure 14.9 An electron micrograph of a stacking fault in an epitaxial layer grown on a (111) wafer. The lower plot shows mechanical scans of the surface topology at four sites on the fault *(reprinted from Liaw and Rose by permission, Academic Press).*

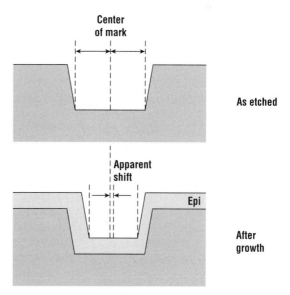

Figure 14.10 A schematic diagram of pattern shift. Alignment marks are etched into the wafer, but nonuniform growth rates cause the mark position to shift after growth.

Dislocations are the 2-D analog of the stacking fault. Dislocations are much less obvious on the surface of the wafer but are still a serious yield concern. Dislocations may simply propagate from substrate dislocations. Dislocations may also form in the epitaxial layer due to plastic deformation caused by thermal nonuniformity and excessive growth rates. Subtle defects such as dislocations are often measured by selective etching. As discussed in Chapter 11, these methods are destructive and employ stains that etch more rapidly at defect sites.

Often the reason for growing the epitaxial layer is to reduce a parasitic resistance. This is commonly done by growing a lightly doped epi layer on top of a heavily doped wafer or localized buried region. The transistor is formed in the epi layer, and the heavily doped region is essentially a buried contact to the bottom of the transistor. For bipolar transistors it is necessary to be able to align the upper device layers with this buried layer. To do this alignment marks may be etched into the substrate before epitaxial growth. Pattern shift is the tendency for position of the alignment marks to appear to move after the growth (Figure 14.10). The cause of pattern shift is the dependence of the growth rate on the exposed crystal orientation. Pattern shift is much less pronounced in (100) wafers than in (111). Generally, the lower chlorine content precursors (SiH_2Cl_2 and $SiHCl_3$) also show less pattern shift as does growth at higher temperature.

For relatively thick epitaxial layers, the preferred method of measuring the film thickness is Fourier transform infrared (FTIR) spectroscopy. As shown in Figure 14.11, an infrared source is sent through a beam splitter to the surface of the wafer and to a movable mirror. The reflected radiation from both surfaces is added and sent to a detector. The distance of the mirrored path is swept, and the intensity of the reflected beam as a function of the position of the mirror is monitored. The separation between these peaks is proportional to the thickness of the epitaxial layer.

The other popular methods for determining the epitaxial thickness are electrical. All of these techniques depend on measuring the differences in the doping levels of the substrate and the epitaxial layer. The quickest measurement is a simple four-point probe of the resistivity of the epitaxial layer. More detailed information can be obtained by measuring the doping profile. This can be done using capacitance–voltage analysis, spreading resistance, or staining techniques, as discussed in Chapter 3. The accuracy of these techniques, however, is limited by autodoping.

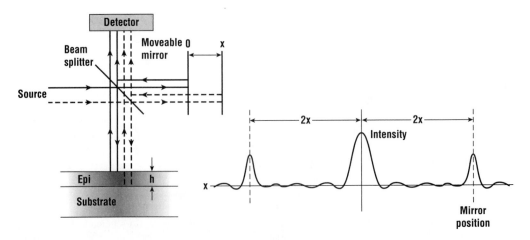

Figure 14.11   A Fourier transform infrared spectroscopy system. Infrared light is used, since silicon is nearly transparent in this region.

14.6 Selective Growth[+]

It is possible to grow selectively from the vapor: that is, to grow the epitaxial layers in some regions of the wafer and not to deposit any material in other regions. The process, known as selective epitaxial growth (SEG), is not in widespread production use, although it is extremely attractive in some applications, including device isolation, planarization of contacts, trench isolation filling, and the formation of elevated source/drain structures for MOS transistors [44]. An overview of the history of the process has been presented by Borland [45].

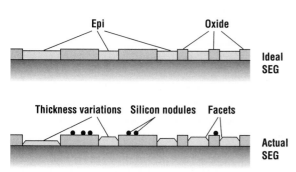

Figure 14.12   A cross-sectional view of selective epitaxial growth.

SEG capitalizes on the fact that the nucleation rate for silicon growth increases with substrate type in the order $SiO_2 < Si_3N_4 < Si$. For example, if an attempt is made to deposit polysilicon over thermal oxide using dichlorosilane, the deposition rate will be very slow until a layer of silicon has been able to accumulate over the surface of the oxide. If a thermal oxide is first grown over the surface of the wafer and holes are etched to the substrate (Figure 14.12), silicon can be made to grow selectively in these holes. To perform selective epitaxial growth, nucleation in the gas phase and on the surface of the masking layer must also be actively suppressed by adding enough chlorine to the growth chemistry. By adjusting the Si/Cl ratio, the process goes from nonselective growth to selective growth to etching. It is believed that higher concentrations of HCl in the gas also improve the selectivity by etching any small nodules that begin to grow on the oxide surface [46]. The best SEG films of silicon are grown at low pressure in SiH_2Cl_2 and HCl mixtures [47].

Problems with selective growth include faceting (a nonplanarity of the surface due to lower growth rates along the various crystal planes) [48], the formation of defects along the semiconductor/oxide interface, and the dependence of the growth rate on the size and distribution of windows. The interface defects are particularly important if a doping junction is bounded by the oxide/semiconductor interface. They can be dramatically reduced if the windows are oriented in the <100> directions [49].

One interesting application of SEG is extended lateral overgrowth (ELO) [50]. If the epitaxy is allowed to continue past the height of the hole, it will begin to spread across the surface of the oxide. If several holes are placed close together, the approaching growth fronts coalesce, and a region of single-crystal silicon is produced over the oxide. If the seed windows are along the <100> directions (i.e., the same as the wafer orientation), the growth rates in the vertical and horizontal directions are identical. Devices fabricated in these overgrowth regions have lower parasitic capacitance and can be easily isolated from one another. The problem with the method is defects. In addition to defects along the interface with the oxide, twins will occur where the growth front changes direction [51]. Defects will also be present at the line where the two growth fronts meet. New applications for ELO include structures that do not require the growth fronts to meet. Instead, the lateral overgrowth regions are used to form low capacitance contacts to the active device formed in the growth window.

14.7 Halide Transport GaAs Vapor Phase Epitaxy

When VPE is used to grow GaAs, the process is somewhat different from the VPE growth of silicon. The hydride AsH_3 is a stable, though very toxic, gaseous arsenic source. Gallium, however, does not form stable chlorohydrides. All of the major gallium chlorides are solids at room temperature. It is

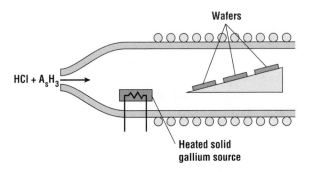

Wafers

HCl + A$_s$H$_3$ →

Heated solid
gallium source

Figure 14.13 A halide transport system for growing
GaAs epitaxially. Gaseous gallium chlorides are created in
the chamber and are transported to the wafer surface.

known that GaCl and GaCl$_3$ can be made gaseous at
high temperature. This fact was the basis of one of the
early and very successful GaAs growth techniques,
halide transport [52, 53] (Figure 14.13). In this
method, a hot (700 to 750°C) solid serves as the gallium source. A dilute mixture of HCl in H$_2$ flows over
the hot Ga or GaAs to form GaCl. The carrier gas
picks up the gaseous GaCl and transports it to the
wafer surface, where it forms GaAs heterogeneously
on the surface. Some practitioners also react solid As
with H$_2$ to form AsH$_3$ *in situ* so that large tanks of AsH$_3$
are not required. In that case, the solid arsenic is maintained at about 800 to 850°C. Substrate temperatures
for halide transport growth tend to be in the 650 to
800°C range.

There are two primary liabilities of the halide transport growth of GaAs. The first is that the
substrate temperatures required for device-quality growth are so high that abrupt junctions are difficult to achieve. The second problem is that it is difficult to control the growth thickness for thin structures due to the nature of the process. Halide transport is still used for the large-scale fabrication of
light-emitting diodes. For other applications, however, these two factors, combined with the emergence
of organometallic gallium precursors, have decreased the popularity of the process.

14.8 Incommensurate and Strained Layer Heteroepitaxy

The first half of this chapter discussed processes that are sometimes collectively called *homoepitaxy*,
the growth of the same material as the substrate. The rest of the chapter will discuss heteroepi-

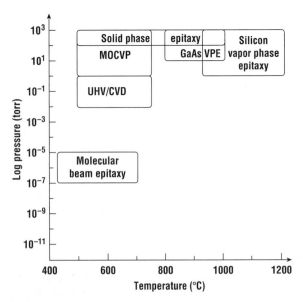

Figure 14.14 Temperature and pressure regimes of various
epitaxial growth techniques.

taxy, the growth of dissimilar materials. The various
approaches to epitaxial growth can be classified on a
pressure–temperature chart (Figure 14.14). The older
methods (VPE and LPE) are done at high temperature
at or near 1 atm. These methods will be briefly discussed here in connection with the growth of thick
layers of silicon on insulating substrates. The remainder of the chapter will deal with techniques used to
grow thin layers for the active regions of the device. In
that application, it is very desirable to be able to grow
layers at low temperature to limit diffusion. Molecular
beam epitaxy (MBE) has demonstrated the ability to
grow device-quality layers at very low temperature,
but requires ultrahigh vacuum (UHV). The extreme
vacuum requirements of MBE make its use in large-
scale IC manufacturing difficult. Because of the diagnostics available in a UHV environment, however,
MBE is probably the best understood epitaxy process.
Other device growth techniques, such as metallorganic
CVD (MOCVD), rapid thermal CVD (RTCVD), and
ultrahigh vacuum CVD (UHVCVD) can be viewed as
extensions of the VPE processes that have already

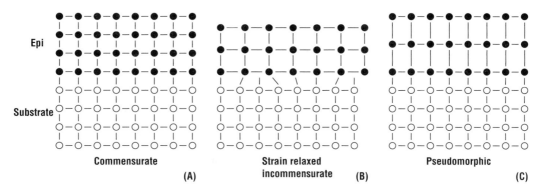

Figure 14.15   Epitaxial growth processes can be divided into (A) commensurate, (B) strain relaxed incommensurate, and (C) incommensurate but pseudomorphic.

been described. The remainder of the chapter will therefore begin with these advanced VPE approaches and then proceed to MBE.

Advanced growth processes would not be of great technological interest if not for their capability to fabricate high quality heterostructures, that is, the growth of epitaxial layers of one material on top of another. There are three types of heteroepitaxial growth (Figure 14.15). It is common to say that the growth is commensurate if the substrate has the same crystal structure and lattice constant as the epi layer. All homoepitaxial growth is by definition commensurate. Incommensurate growth is the production of thick layers that are not lattice matched to the substrate. In this type of growth, the misfit between the two crystals must be accommodated by defects at or near the interface. Many of these defects may propagate into the epi layer, making the successful manufacture of even simple circuits extremely difficult. A third possibility is pseudomorphic growth. This occurs when one is growing thin layers that are not lattice matched. Instead of forming defects at the interface, the epi layer strains to match the substrate. The critical thickness or maximum thickness that can be grown in this manner before the strain is released in the form of dislocations depends on the lattice mismatch and the mechanical properties of the layers. For a lattice mismatch of a few percent, it is typically of order hundreds of angstroms. Strictly speaking, under this definition all heteroepitaxial growth must be pseudomorphic, since one cannot have perfect lattice matching with dissimilar materials. A more practical description suggests that commensurate growth occurs when the adatom/adatom interaction energy is small compared to the adatom/substrate energy. Incommensurate growth then occurs when the adatom/substrate energy is small compared to the adatom/adatom energy. Pseudomorphic growth occurs when the energies are comparable.

Most of the earliest heteroepitaxial growth was incommensurate. A popular example of incommensurate growth is silicon on sapphire (SOS). Experiments by Cadoff and Bicknell [54] and later work by Manasevit and Simpson [55] demonstrated that silicon, which has cubic symmetry with a lattice parameter of 5.43 Å, can be grown successfully on sapphire (Al_2O_3), which is rhombohedral with lattice parameters of 4.75 Å (*a*-axis) and 12.97 Å (*c*-axis). Sapphire is insulating, and so circuits built in SOS substrates have high speed and excellent device isolation. Device isolation can be done simply by removing the conducting silicon between the transistors, etching down to the insulating substrate. A common application of SOS is radiation-hardened circuits.

Most SOS is done using VPE from SiH_4 diluted in H_2. Although chlorosilanes may be used, Cl attacks Al and produces heavily doped p-type films. Traditionally, the deposition temperatures are around 1000°C, although more recently some success has been achieved at lower temperatures. Of course, supersaturation and homogeneous nucleation are significant problems for SOS growth. The most basic

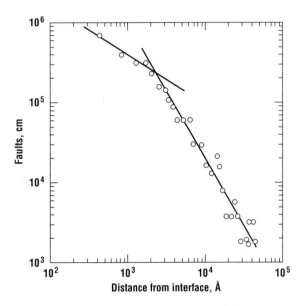

Figure 14.16 Defect density as a function of the epitaxial layer thickness for the growth of Si on sapphire (Al₂O₃) *(after Vasudev).*

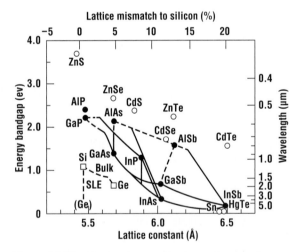

Figure 14.17 The bandgap and lattice parameter of a variety of semiconductor compounds and alloys.

problem for SOS, however, is the presence of defects in the epitaxial layers caused by the sapphire substrate. These defects arise both because of the lattice mismatch and because the thermal expansion coefficient of Al_2O_3 is more than twice that of silicon. The mismatch at the interface creates defects that will propagate up into the growing film. One way to reduce this problem is by growing thick films. Defects may propagate upward, but they do not necessarily penetrate the entire layer. Figure 14.16 shows a plot of defect density per unit length as measured by cross-sectional transmission electron microscopy (TEM) [56]. The defects fall off sharply with increasing thickness. Thick films are therefore usually required for incommensurate growth. It has also been found that high dose Si implantation can be used to amorphize part of the epi layer. This can be regrown by solid phase annealing. By repeating this process for different depths, the material quality can be further improved. SOS quality is currently suitable for LSI and some high cost VLSI circuits, but is strongly challenged by the recent advances in SIMOX technologies (Section 5.7) and bonded SOI (Section 15.4).

The other primary example of incommensurate epitaxy is GaAs on silicon [57, 58]. Silicon and GaAs share the same type of crystal structure, but the GaAs lattice constant is about 4% larger. As with SOS, thick epitaxial layers are required in order to reduce the defect densities [59] enough to allow IC fabrication. It has also been reported that a large number of very thin layers with alternating composition, called a *strained layer superlattice*, can be used to distribute the strain more gradually [60]. The GaAs-on-silicon technology is attractive for three reasons: The silicon substrates are inexpensive and large. It is therefore possible to fabricate GaAs ICs on 8-in. rather than 4-in. substrates. The silicon substrate has a higher thermal conductivity than GaAs. This is an important consideration in high power ICs and devices where the performance is often limited by the maximum junction temperature. Finally, it is possible that GaAs devices could be integrated with silicon on the same chip. In this application a silicon technology such as CMOS would be fabricated first. The GaAs wafers would be grown selectively in oxide windows etched down to the silicon substrate. At present, GaAs can be used to form lasers for on-chip optical interconnect. This type of integration has been demonstrated on simple systems but has not seen appreciable commercial application.

In virtually all incommensurate growth, the interface must be far from the active device region since it contains so many defects. From a device design standpoint, it would be much more desirable to grow layers that can be used as part of the device structure itself. Figure 14.17 shows a plot of the

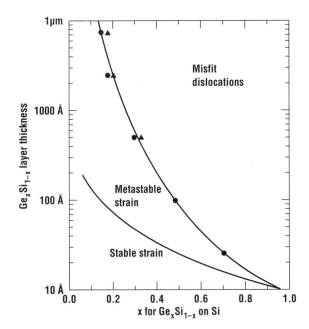

Figure 14.18   For pseudomorphic GeSi on Si there are three regions: mechanically stable, metastable, and unstable *(after People and Bean)*.

bandgap of various ternaries as a function of the lattice parameter. By drawing a line vertically, one can predict which usable heterojunctions might be feasible. For example, GaAs, is lattice matched to the proper mole fraction of $Ga_xIn_{1-x}P$. The most heavily studied commensurate growth system is GaAs on $Al_xGa_{1-x}As$. Although a very small lattice mismatch exists, these two compounds can generally be regarded as perfectly lattice matched. Although gross defects such as dislocations are relatively rare in these lattice-matched systems, a number of interface states exist. These states may serve as recombination sites that degrade the performance of the heterostructure.

An example of pseudomorphic growth is Ge_xSi_{1-x} layers grown epitaxially on silicon. This system is interesting because it suggests the possibility of combining the high speed of heterostructures with the density of silicon ULSI. Figure 14.18 shows a plot of the critical thickness of GeSi as a function of the germanium mole fraction [61]. As long as the film thickness lies to the left and below this line, the GeSi layer will accept the strain and be thermodynamically stable. The resultant heterostructure is very useful for making high performance bipolar devices. By using a smaller bandgap material and grading the concentration of germanium in the base, it is possible to dramatically improve the performance of the device. This type of epitaxy has also been shown to be useful for MOS devices. While early MOSFETs used a structure in which the channel was formed in pseudomorphic GeSi, it proved difficult to form an oxide with an acceptable interface to the GeSi. More recently it has been shown that one can grow a strain-relaxed layer of GeSi on top of the silicon wafer, and then grow a thin strained layer of Si on top of that. The structure allows one to change the mobility of the electrons and to a lesser extent holes in the silicon. This will be discussed in Chapter 16. Naturally the strain-relaxed layer will have some dislocations, and these dislocations may propagate into the active silicon layer. This problem has prompted investigations of other ways of straining the silicon channel.

14.9 Metal Organic Chemical Vapor Deposition (MOCVD)

Metal organic chemical vapor deposition (MOCVD or OMVPE) began in the late 1960s with the research of Manasevit and Simpson [62]. MOCVD is sometimes regarded as a competitor to MBE. Both are capable of growing high quality III–V devices with atomically abrupt or nearly atomically abrupt interfaces. The process is not generally used for the epitaxial growth of silicon since suitable inorganic sources are readily available.

Classic MOCVD processes were developed using group III organic compounds and group V hydrides. Historically this occurred because the organic molecules had been developed for other applications and were therefore available, although not in the purity that is now required. Figure 14.19 shows examples of some of the relevant precursors. Although a wide variety of organic compounds have been used, this chapter will restrict its attention to the most commonly used sources, methyl (M) and ethyl (E) compounds. For example, TMG is one of the most common gallium sources, trimethylgallium $[Ga(CH_3)_3]$.

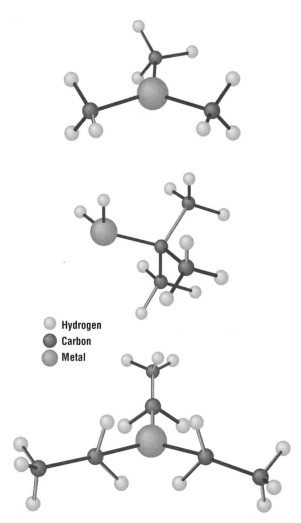

Hydrogen
Carbon
Metal

Figure 14.19 Examples of common organometallics used in MOCVD include (from top to bottom): trimethylgallium, tetrabutylarsine, and triethylgallium.

To understand the bonding arrangements in the organometallics, start from the valence arrangement of the metal atom [63]. Gallium has three valence electrons. In the ground state, gallium has two s electrons and one p electron. In TMG, the carbon atom from each of the methyl groups bonds covalently with the gallium, forming a hybridizing sp^2 bond. The three methyl groups form in a plane with the Ga–C bonds 120° from each other. This leaves an empty p orbital perpendicular to the plane of the methyl groups. This unfilled orbital easily accepts electrons. The group V hydrides have two s electrons and three p electrons in the valence. The three As–H bonds in arsine form covalently with bond angles of about 109.5°. A single pair of unbonded electrons sits on top of the molecule away from the hydrogen atoms. One of these electrons is easily lost or donated. The ability of the molecules to easily accept and lose electrons plays an important role in the chemical reactions that occur during growth.

A typical group V hydride source is AsH_3. Unfortunately these hydrides are extremely toxic. Exposure to arsine leads to red blood cell destruction, renal failure, and ultimately death. Since III–V growth uses these chemicals not as dopants but as the growth species themselves, they must be stored in large quantities. The concentration at which arsine is immediately dangerous to life and health (IDLH) is 6 ppm. A cylinder of AsH_3 at 1500 psi will contaminate about 10^9 L of air to the IDLH concentration. This corresponds to roughly one square block. Of course, in the event of an accident, personnel close to the release are in much greater danger than those further removed. The cost and discipline of the proper gas handling for such toxic precursors have led to the investigation of less toxic alternatives, typically group V organometallics [64]. Potential group V sources must have a sufficiently high vapor pressure. They must not have parasitic reaction paths with each other or the group III precursor. They must be stable. They must be extremely pure. On reaction they must not introduce impurities into the epitaxial films. It is generally found that replacing H bonds with C bonds decreases the toxicity but increases the amount of unintentional doping in the epitaxial layers. One of the most promising OM arsenic sources is tertiary butylarsine [65] (TBA), which is nearly as toxic as AsH_3. Since it is stored as a liquid rather than a high pressure gas however, it is generally regarded as safer than AsH_3.

An MOCVD system must be capable of the controlled growth of thin, atomically abrupt, extremely perfect epitaxial layers. The growth system can be divided into three subcomponents: gas delivery, reaction chamber, and gas exhaust. The gas delivery system is the easiest to specify. It is desired that the growth chamber be provided a well-mixed, well-controlled flow of a number of

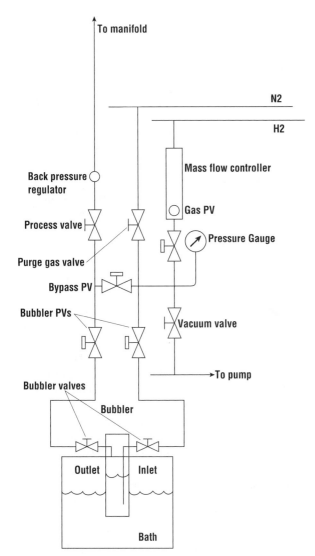

Figure 14.20   A typical OM source for MOCVD includes valves that isolate and bypass the bubbler during changeout, allowing cycle purging of the process line with nitrogen.

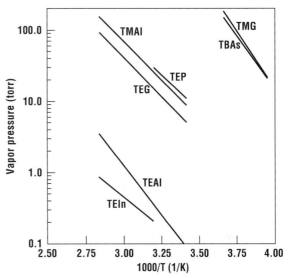

Figure 14.21   Vapor pressure curves for some common organometallics *(after Stringfellow).*

gases. For growing GaAs, this is typically H_2, TMG or TEG, AsH_3, and a dopant, typically DMZ for p-type doping or SiH_4 for n-type doping [66].

The OM sources consist of a bubbler immersed in a controlled temperature bath of glycerin and water (Figure 14.20). The OM vapor is obtained by heating the liquid in a constant temperature bath and bubbling H_2 through the bath. The temperature of the bubbler must be carefully controlled, since the vapor pressure depends exponentially on the temperature of the bath. The back-pressure regulator is used along with the temperature of the bath to establish the mole fraction of the OM in the gas that is sent to the reactor. Figure 14.21 shows the vapor pressure of several commonly used OM sources. The OM source must also include valves so that the source can be isolated from the line and flushed with N_2 when it is to be changed.

The details of the gas delivery system depend on the particular organometallic being used. Some OM compounds such as triethylaluminum (TEAl) require substantial heating in order to obtain a sufficient vapor pressure for deposition. The vapor of these compounds can recondense in the tubing before it reaches the chamber. This results in poorly controlled deposition conditions unless the tubing and valves also are heated to a temperature greater than the bubbler temperature. Other compounds such as dimethylzinc (DMZ), which is commonly used as a p-type dopant in GaAs MOCVD, has a very high vapor pressure at room temperature and may actually have to be chilled to obtain the desired vapor pressure. DMZ pyrolyzes at only slightly above room temperature, however. To prevent a reaction in the supply tubes, particularly the part of the lines close to the reactor, they may have to be cooled. Other compounds, such as TEIn, which can be used to grow InGaAs and InP, are unstable and so will begin to decompose in the liquid cylinder.

The hydride sources are delivered as high pressure gases just as in standard CVD systems. The gas pressure is reduced to a few atmospheres, and the gas is plumbed to a valve and mass flow controller. Due to the toxicity of the hydrides, these lines are typically double contained, leak checked, and continually sniffed for leaks. Furthermore, only the most reliable gas seals are typically allowed on such a system. The design of the actual gas introduction system is one of the most critical parts of the chamber. The simplest method is to simply merge the various lines at a manifold near the entrance to the growth chamber. This arrangement is not satisfactory for growing thin, abrupt layers, since the growth during a switching transient is very poorly controlled. The solution is to include valves as an integral part of the manifold. These valves are designed to have the minimum dead volume, the volume between the chamber and the valves. The second concern with gas switching transients is that a flow burst or bolus occurs when a valve is switched on. To avoid this problem, gas manifolds for MOCVD may be designed in a run/vent configuration (Figure 14.22). The manifold of such a system uses three-way valves that supply gas either to the growth chamber or to a bypass. This redirection provides a much more stable flow than can be achieved by turning the gas on and off. Of course, the amount of gas consumed by this process may be considerably larger, but this is considered an acceptable cost for many applications.

The gas exhaust must allow the system to operate at a controlled pressure. Most of the early MOCVD systems were run at 1 atm. It has since been found that growth at reduced pressure (10 to 100 torr) can improve uniformity and layer abruptness. Most MOCVD systems now include vacuum components to maintain the chamber pressure. The exhaust gases must also be scrubbed prior to release. This is done either in a high temperature furnace in which the hydrides are oxidized or in wet chemical scrubbers, where the effluent reacts with caustic solutions to form water-soluble salts that can be diluted and flushed.

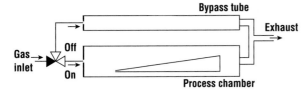

Figure 14.22 Run/vent gas switching used to avoid large gas transients.

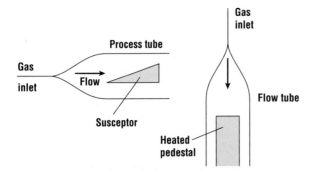

Figure 14.23 Common MOCVD chambers include the horizontal flow and the stagnation point flow reactors.

A great deal of research has gone into the design of the reactor itself. Figure 14.23 shows cross sections of some typical chambers. The two most common are the wedge-shaped horizontal reactor, similar to those described earlier, and the vertical impinging jet (also called the *stagnation point flow*) reactor. For low flow velocities, it is usually assumed that the gas enters the chamber in a fully developed flow. The gas velocity is zero at the walls, and it increases parabolically toward the center of the inlet tube. It is of critical importance to the success of the MOCVD process that the flow through the chamber be laminar, with no turbulence or recirculation. Nonlaminar and recirculative flows trap gases in the chamber. This concern was also considered during the discussion of CVD, but is much more important in MOCVD, since typical growth runs involve multiple thin layers with different compositions and dopings. Turbulence usually arises at high gas velocities due to shear forces. Recirculation cells arise from two sources. The first is the sudden change in the cross-sectional area of the chamber. The second source of recirculation cells, mentioned previously,

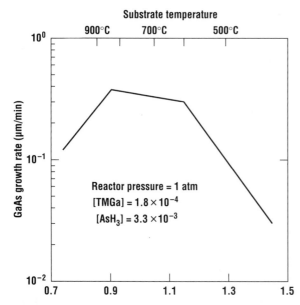

Figure 14.24   Growth rate of GaAs during MOCVD as a function of temperature *(after Tischler [75], ©1990 IBM).*

is buoyancy forces caused by nonuniform temperatures. The effect of recirculation cells can be reduced by reducing the mass density of the gas. This is a common reason for using H_2 as the carrier gas. By also reducing the pressure one to two orders of magnitude below 1 atm, the effect can be virtually eliminated. Another method for overcoming the effects of convective flow is to rotate the substrate at several thousand rpm. This transfers momentum to the molecules near the surface of the susceptor that opposes the recirculation flow.

A plot of the growth rate of GaAs in MOCVD as a function of the substrate temperature [67] is shown in Figure 14.24. At low temperature the growth rate increases exponentially with temperature. The activation energy for this process has been measured to be between 13 and 22 kcal/mole (0.55 to 0.95 eV). At high temperatures the growth rate begins to drop. One reason for this reduction is the emergence of a parasitic reaction pathway that scavenges the TMG. It is believed that homogeneous pyrolysis of TMG produces CH_3 radicals. These radicals in turn attack both the AsH_3 and H_2. In the case of H_2, the remaining atomic hydrogen attacks the TMG, freeing another methyl group:

$$Ga(CH_3)_3 \rightleftharpoons Ga(CH_3)_2 + CH_3$$

$$CH_3 + H_2 \rightleftharpoons CH_4 + H \qquad \textbf{(14.14)}$$

$$Ga(CH_3)_3 + H \rightleftharpoons GaH(CH_3)_2 + CH_3$$

This reaction path causes gallium to deposit on the walls of the chamber and deplete the TMG source. It is also possible however, that the reduction in the reaction rate is related to desorption of AsH_x admolecules from the surface of the wafer.

Most GaAs growth is done in the intermediate temperature regime, where the growth rate is determined by the arrival rate of the TMG. The reaction process for TMG and AsH_3 in this regime was first postulated by Schlyer and Ring [68] and later supported by others [69]. Although it has not yet been proven, it is the most widely accepted. In this model, at temperatures less than 650°C very little homogeneous reaction occurs at any pressure. As a result, both TMG and AsH_3 independently adsorb on the surface of the wafer. Once on the surface, TMG appears to react spontaneously to form monomethylgallium ($GaCH_3$) and two methyl groups. Then AsH_3 reacts with a CH_3 group to form AsH_2 and methane (CH_4). The latter is quickly desorbed and pumped away:

$$AsH_3 + CH_3 \rightleftharpoons AsH_2 + CH_4 \qquad \textbf{(14.15)}$$

The AsH_2 and the monomethylgallium diffuse independently on the surface until they come into contact. The methyl group reacts with the hydride, yielding GaAs, atomic hydrogen, and methane, which also desorb from the surface. It now appears that the atomic hydrogen produced by AsH_2 dissociation plays an important role in removing the methyl groups from the surface of the wafer.

One of the first major problems of MOCVD was carbon contamination. This results in unintentional p-type doping. Early MOCVD films [70] had hole concentrations of 10^{19} cm^{-3}. Although they can be counterdoped, the films have very low mobilities. It was found, however, that by increasing the group V/group III gas flow ratio, the unintentional doping could be dramatically decreased. High group V partial pressures are needed because AsH$_3$ dissociation is much less efficient than dissociation of the organometallics at normal growth temperatures. By the late 1970s, material quality was comparable to MBE, with heterojunction undoped liquid nitrogen temperature mobilities routinely reported to be in excess of 100,000 cm^2/V-sec. It has also been shown that the use of TEG rather than TMG also produces layers with little carbon incorporation [71] because the ethyl organometallic pyrolyzes to form ethylene (C$_2$H$_4$), which, unlike CH$_3$, is nonreactive.

Within the last decade it has been found that this early liability of carbon contamination can be used to considerable advantage. When heavily doped p-type layers are required for a particular device, the films can be made by choosing the V/III ratio and the temperature to incorporate carbon [72] or by adding an appropriate carbon-containing gas [73]. Carbon doping is preferred to Be and Zn for many applications, since carbon has a lower diffusion coefficient [74], and the source is not toxic.

It is possible to grow Al$_x$Ga$_{1-x}$As by adding an aluminum-containing organic such as trimethylaluminum (TMAl) to the gallium-containing organic. The concentration of aluminum in the film is roughly proportional to the ratio of the injected gases [75] and is nearly independent of the substrate temperature. Additional care must be taken in growing AlGaAs, because of the extreme reactivity of aluminum, particularly for oxygen. To obtain high-quality films in MOCVD systems, the substrate must be heated to about 750°C [76]. It is believed that in this temperature range, the aluminum oxide more readily desorbs from the growing film, and so the oxygen contamination drops sharply. It is also critically important in growing these layers to obtain a high base vacuum, to palladium-diffuse the H$_2$ source, and to use any other method possible to getter out residual O$_2$ and H$_2$O. Liquid metal bubblers have been found to be effective for gettering oxygen from hydride source gases [77].

14.10 Advanced Silicon Vapor Phase Epitaxial Growth Techniques

The first part of the chapter discussed the epitaxial growth of silicon from the vapor. The application for these processes is the growth of thick layers above a substrate with a different dopant type and/or concentration. The process must have a low defect density, there must be good control over thickness and uniformity, and it must be done so as to minimize the dopant redistribution during the growth. Recall that one can define a characteristic diffusion length as $\sqrt{Dt}$, where D is the diffusivity and t is the time at temperature. There are two approaches to solving this problem. The first uses rapid thermal processing to extend standard VPE using chlorosilanes to very short and well-controlled growth times. The second uses ultrahigh vacuum systems with a silane chemistry to grow at very low temperature, limiting the diffusivity.

Figure 14.25 shows a typical time/temperature cycle for growing 1 μm of epitaxial silicon in a VPE reactor. The wafer and susceptor must be heated to the

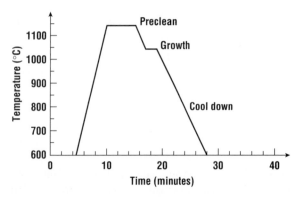

Figure 14.25   A typical time/temperature cycle for growing 1 μm of epitaxial silicon in a VPE reactor.

preclean temperature, then held at that temperature for some period to allow the native oxide to desorb. Next, the wafer is ramped to the epitaxial growth temperature, the dichlorosilane turned on, and the growth proceeds. After the desired growth time, the growth gas is turned off, the chamber is purged with an inert gas, and the wafer is ramped to room temperature. If one wants to grow a 1000-Å layer instead of 1 μm in order to be able to directly fabricate devices, the thickness of the layer is reduced by a factor of 10, but because of the overhead associated with precleaning the wafer and ramping the temperature, the overall time at temperature may be reduced by only a factor of 2. As a result, the layer doping will be dominated by diffusion from the lower layers and will be very poorly controlled.

Gibbons et al. [78] were the first to combine rapid thermal processing and epitaxial growth in a technique called limited-reaction processing (LRP). Other groups have called similar techniques rapid thermal vapor phase epitaxy (RTVPE) [79] or rapid thermal chemical vapor deposition (RTCVD) [80]. The chamber for doing this work is like the quartz walled RTP systems described in Chapter 6, although a load lock may be used to reduce residual gas impurities. A cross section of a system is shown in Figure 14.26A. The wafer rests on three quartz pins in a quartz flow tube. Gases are injected at one side, flow across the wafer, and exit at the other end. The wafer is heated by lamps situated above and below the flow tube. Because of the rapid thermal response of the wafer in such a system, the actual growth time for thin layers can now be the majority of the overall time at temperature. Figure 14.26B shows a commercial system.

A wide variety of structures can be grown in an RTCVD system [81], and it is an excellent, low cost method of developing advanced device structures. From a manufacturing standpoint, however, the technique has several important liabilities. As with all VPE, there is no direct control over the thickness. Instead, the growth rate must be assumed from some previous calibration run and the growth time calculated. Furthermore, in standard VPE the growth temperature is measured using a thermocouple embedded in the susceptor. In an RTCVD system the growth temperature is most often inferred using pyrometry, a method with significant reproducibility problems. This is particularly a concern during growth when deposited layers may coat the quartz surfaces, changing the optical path to the lamps and the sensor. It is also well known that the edges of a wafer are cooler than the center of the wafer in simple RTP systems at growth temperatures [82]. Additional power can be focused on the edge of the wafer to help compensate, but unless dynamic control is used, the temperature uniformity can be optimized only for a single process condition.

To help overcome the temperature uniformity and control problems of RTP, one can run the epitaxial growth process in a mass transport limited regime. Once again this presents difficulties. Recall that in a standard horizontal CVD reactor the susceptor is tilted to provide a uniform stagnant layer thickness. This is not possible in RTCVD, since there is no susceptor. An impinging jet system such as those used for MOCVD can be employed, but it is difficult to design such a system that is radiatively heated and can reach the temperatures necessary for silicon VPE. As a result, commercial RTP epitaxy systems sometimes use thin susceptors to hold the wafer, providing some compromise between rapid response and good uniformity.

The first part of this chapter emphasized the importance of reducing the residual water and oxygen concentrations in the growth chamber and the relationship to these partial pressures and the minimum allowable growth temperature. To obtain growth at the lowest possible temperatures, it is necessary to obtain a growth ambient with extremely small partial pressures of oxygen and water. One silicon epitaxial growth process that has been developed to meet this objective is ultrahigh vacuum chemical vapor deposition (UHVCVD) [83]. As shown in Figure 14.27, the system is similar to a load-locked LPCVD tube. Up to 35 wafers can be loaded into the chamber at a time in the current generation research reactors. The base vacuum for the system is typically at least 10^{-9} torr.

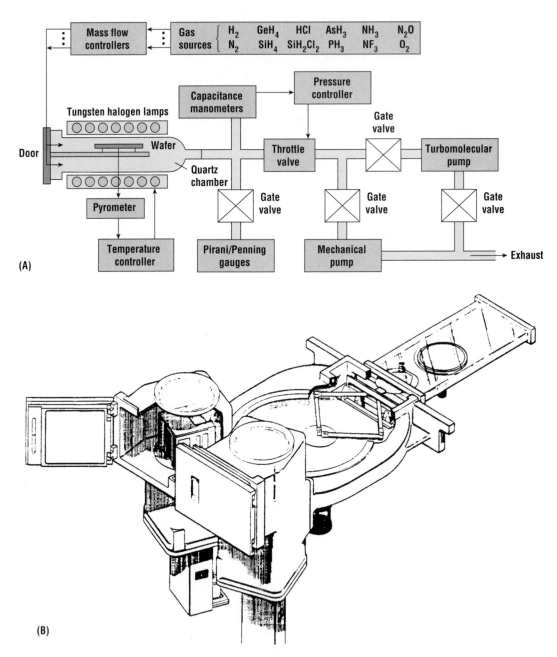

Figure 14.26   (A) Schematic of a rapid thermal epitaxial growth system *(after Hsieh et al., reprinted by permission of the publisher, The Electrochemical Society).* (B) Commercial rapid thermal epitaxial growth system showing load and unload stations, robot arm, and rapid thermal flow sleeve *(courtesy of ASM Corp.).*

Growth is carried out at temperatures below 800°C in about 1 mtorr of silane. Under these conditions, it is believed that little decomposition of silane occurs in the gas phase, since this reaction generally requires a collision with another gas molecule [84]. The growth rate is only a few angstroms per minute for undoped or p-type doped layers [85], but the low rate is tolerable due to the batch

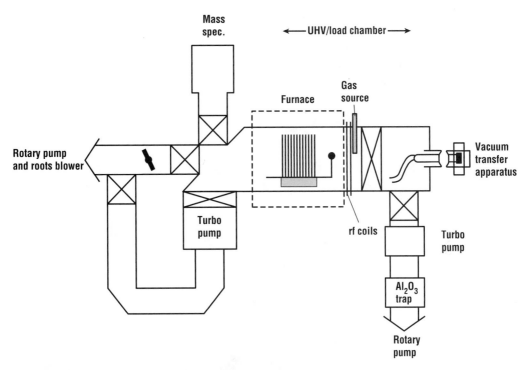

Figure 14.27   Schematic of an ultrahigh vacuum CVD epitaxial growth system *(after Mayerson, ©1986 AIP).*

loading. Because of the very low temperature, heavily doped n-type layers cannot be practically grown using the technique. The process is run well into the reaction rate limited regime. Since the temperature uniformity of a hot wall system is very good, the deposition rate uniformity is often better than 2%.

One of the major concerns with UHVCVD is wafer precleaning. Because of the low pressure in the chamber, most of the heating is radiative. Since the wafers are loaded by the batch, heat must flow into or out of the boatload of wafers, primarily through radiation exchange between the wafer edges and the tube walls. Temperatures must therefore be slowly ramped to avoid excessive thermoplastic stress in the wafers. This makes the use of a high temperature oxide desorption step rather difficult. The use of hydrogen-terminated surfaces, however, has made UHVCVD a promising technology.

14.11 Molecular Beam Epitaxy Technology

One of the premier growth techniques for GaAs heterostructures is molecular beam epitaxy (MBE). The technique has the capability of growing device-quality layers of semiconductors with atomic resolution of the growth thickness. Furthermore, typical growth temperatures virtually preclude dopant diffusion. MBE has also been applied to silicon epitaxy. Figure 14.28 shows a diagram of a typical MBE system [86]. The basic requirements for such a system are ultrahigh vacuum (approximately 10^{-10} torr base vacuum), *in situ* sample heating and cleaning, and independently controlled thermal and/or electron beam sources for all materials and dopants. Most systems also include several *in situ* analysis capabilities, including electron diffraction and Auger spectroscopy. A separate chamber

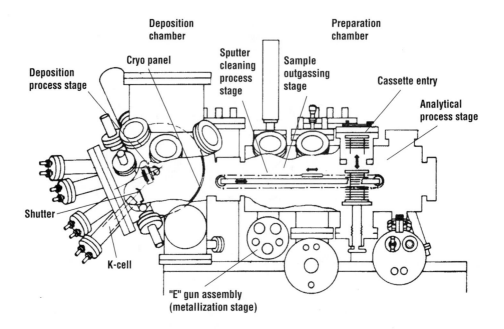

Deposition
chamber

Preparation
chamber

Cryo panel

Sputter
cleaning
process
stage

Sample
outgassing
stage

Cassette entry

Deposition
process stage

Analytical
process stage

Shutter

K-cell

"E" gun assembly
(metallization stage)

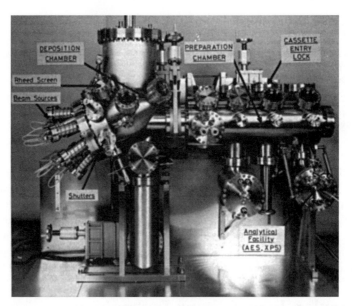

Figure 14.28   Schematic and photograph of a molecular beam epitaxy growth system *(after Davies and Williams).*

connected through UHV load locks is often provided to evaporate various metallizations without breaking vacuum after the growth is complete.

To achieve these high base vacuums, the MBE chamber (including gaskets) must be suitable for baking at 150 to 250°C. Baking at these elevated temperatures increases the vapor pressure of gases, such as water adsorbed on the inside walls and surfaces of the system. When the system is pumped hot and then cooled, the pressure in the chamber may drop by an order of magnitude or more. Heating jackets are often provided with the system for this purpose. Particular care must be taken during high temperature bakeouts for the durability of valves and fused silica viewports. These components often will not tolerate repeated high temperature cycling. Critical components such as the evaporation sources may be held at somewhat higher temperatures during bakeout to ensure an even higher level of cleanliness in these areas. Bakeout times vary with the level of exposure. If the entire system must be opened to air such as must be done to refill a source, it is common to require extended pumping with bakeout before device-quality growth can begin.

A variety of pumps can be used to maintain the high vacuum state. Early systems used diffusion pumps, but extreme care must be used to prevent backdiffusion of the hot diffusion pump oil into the growth chamber. Some small systems use liquid nitrogen–cooled adsorption pumps. These pumps operate by physical adsorption of gases at the surface of molecular sieves or other adsorption material such as activated Al_2O_3. This material is typically a large surface area solid. The pump is cooled to 77 K, then opened to the area where pumping is desired. After pumping is complete, the pump is isolated from the chamber, heated, and evacuated. A number of adsorption pumps may be used to provide an adequate base vacuum. The most popular MBE high vacuum pump is the cryopump. As with adsorption pumps, the cryopump must be isolated from the chamber occasionally and heated to remove the pumped gases. Newer MBE systems often use turbomolecular pumps for producing high vacuum. This option is particularly attractive if toxic sources are being used, since material from these sources tends to accumulate in the cryopumps between temperature cyclings.

To achieve routine ultrahigh vacuum, wafers must be introduced into the MBE system using a load lock. The load lock can be evacuated to high vacuum in a short time. Typical pumpdown characteristics for an MBE load lock are 10^{-6} torr in 15 min. Some high volume MBE systems allow a cassette or platen of wafers to be kept under UHV. This increases the wafer throughput, since the total pumpdown time per wafer decreases. Many GaAs MBE systems also make use of cryopanels in the growth area. The walls of the chamber have liquid nitrogen jackets, making them large pumps themselves. Although capable of maintaining very high vacuums, these systems also consume so much liquid nitrogen (particularly if the substrate must be heated to a high temperature) that it must be plumbed directly to the chamber through double-walled, vacuum-jacketed pipes.

Substrate heating is typically provided through the use of chucks or susceptors that can be either radiantly or resistively heated. Resistive heaters can be either refractory metal wire wound on a form or graphite films that are etched into a heating pattern and covered with SiC. Wafers are often attached to the chuck through clips, clamps, or indium-based solders, although the latter is becoming less common. The temperature of the wafer can be monitored by measuring the temperature of the chuck with a thermocouple. It can also be measured by siting the wafer with a pyrometer. Pyrometers can measure the temperature of the wafer directly; however, to be accurate it must be calibrated in the growth chamber. Furthermore, one must be careful to avoid measuring radiation of the heater directly. Radiation from the heater may come from reflections off of the walls of the chamber or it may come directly through the wafer, particularly in the infrared below the bandgap of the semiconductor.

There are three types of sources most commonly used in MBE: Knudsen cells, electron beam sources, and gas sources. Figure 14.29 shows the cross section of a Knudsen cell. A crucible loaded with a source charge is radiantly heated by a foil heater. Thermocouples pressed against the outside of the crucible can be used to control the temperature and therefore the material flux. To avoid contamination, the heater foil is typically made of tantalum and is baffled to prevent a line of sight to the wafer. For III–V growth, most crucibles are made from pyrolytic boron nitride (pBN). Quartz can also be used for silicon MBE. The empty crucible must be annealed at high temperature before use to ensure that the beam, once produced, will be pure. To achieve reasonable fluxes in silicon MBE source,

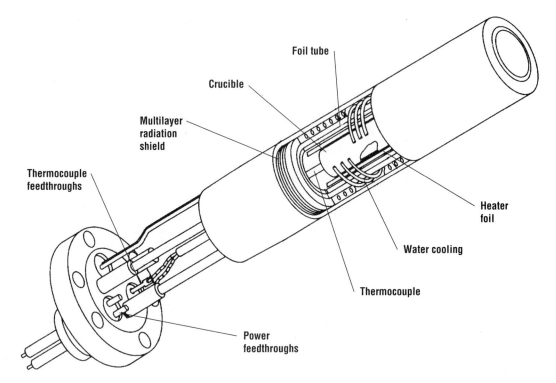

Figure 14.29   Cutaway of a simple Knudsen cell for producing thermal beam for MBE *(after Davies and Williams).*

a temperature of about 1700°C is required. At this temperature, silicon is molten and highly reactive. As a result, the films are contaminated with impurities from the crucible. To avoid this problem, most silicon MBE systems use electromagnetically focused electron beam sources. This type of source has been discussed as e-beam evaporation (see Section 12.4).

For both e-beam and thermal cells, the source flux is determined by the temperature, which in turn establishes a vapor pressure. Since the charge has a finite thermal mass, this flux cannot be turned off instantaneously. To provide this on/off control, most MBE systems use shutters in front of the cells. One difficulty with these shutters is material creep. Since the source is still at temperature, deposition will continue onto the back of the shutter. This material may surface diffuse around the shutter and eventually reevaporate. Shutters may be cooled to reduce this problem, but they are being constantly heated by the exposure to the hot cell. To avoid this problem, as well as to improve uniformity, eliminate chamber venting to replenish the charge, and improve material quality, MBE has adopted gaseous sources (GSMBE). Figure 14.30 shows a typical gas source MBE.

One of the most popular techniques in research applications for real-time growth monitoring is reflective high energy electron diffraction (RHEED). In this method, high energy electrons are diffracted off of the growing surface and imaged on a screen on the opposite side of the chamber. Since the films should grow a layer at a time, the spot intensity should grow, then diminish, then grow again as each monolayer of coverage is achieved (Figure 14.31). For production systems where uniformity is of paramount importance, the substrate must be rotated during growth, making RHEED impractical.

A major source of defects in MBE is particulation from other surfaces that are coated during the deposition process. Particles that fall from the walls are actually pushed by the growth beam

Figure 14.30   GSMBE growth chamber. The solid effusion and gas sources are aimed at the growth position *(courtesy of RIBER S.A.)*.

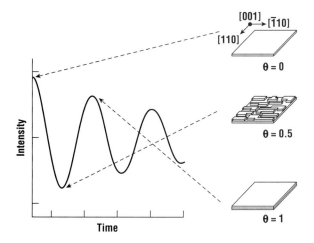

Figure 14.31   Electron diffraction oscillations during MBE growth. The peaks correspond to nearly complete layers.

toward the wafer surface [87]. This is a particularly serious concern in silicon MBE. It has been found that replacing liquid nitrogen cooling of the chamber walls with simple water cooling can dramatically reduce the defect density. Also, by using high voltage plates near the evaporation source, dust can be kept from entering the beam, and defect levels as low as 30 cm^{-2} can be achieved [88]. Metal contamination is another serious concern in MBE. In most VPE, chlorinated species are present that react with residual metals to clean the surface. These impurities may reduce the minority carrier lifetime for MBE films unless great care is taken in designing the growth system and preclean cycles [89].

One of the manufacturing concerns in GaAs MBE is the oval defect: a region of excess gallium concentration [90]. The oval defect density is commonly reported [91] to be in the range 10^3 to 10^4 cm^{-2}. By carefully cleaning the system [92], the oval defect densities can be made as low as 300 cm^{-2}. In fact, oval defects can be completely eliminated by holding the gallium source in a boron nitride crucible that has previously been used for aluminum [93]. The reason for this is not known but may be related to oxygen gettering. Another common cure for oval defects is to install a heater at the top of the gallium source to reevaporate droplets that formed after the initial evaporation.

The growth rate of an MBE system is simply determined by the flux of atoms leaving the source and the fraction of those atoms that strike the surface of the wafer and stick. This flux of atoms from the source for a thermal source is given by Equation 10.10

$$I_n = AJ_n = A\sqrt{\frac{n^2 kT}{2\pi m}} = A\sqrt{\frac{P_e^2}{2\pi kTm}} \qquad (14.16)$$

where A is the cross-sectional area of the source, P_e is the equilibrium vapor pressure that depends on the source temperature, and m is the atomic mass. Knudsen cells are designed so that the cross-sectional area remains nearly constant as the source is consumed.

Example 14.4

Assume that aluminum is being evaporated at 1150 K in a 25-cm^2 cell. What is the atomic flux at a distance of 0.5 m if the wafer is directly above the source? What would the growth rate be?

Solution

Referring to Chapter 10, the flux at a distance r is

$$J = A\sqrt{\frac{P_e^2}{2\pi kTm}}\frac{\cos\theta\cos\phi}{\pi r^2}$$

From Chapter 12, the vapor pressure of Al at 1150 K is about 10^{-6} torr. Inserting and solving with $m = 27 \times 1.67 \times 10^{-27}$ kg, the flux is about 4.8×10^{14} cm^{-2} sec^{-1}. The growth rate can be found using

$$R = \frac{J}{N}$$

where N is the number density of aluminum (6×10^{22} cm^{-3}). Then $R \approx 48$ Å/min.

For compound semiconductor systems, such as GaAs, both gallium and arsenic cells must be used. The solid arsenic source emits As$_4$, which can be converted to As$_2$ if the cell includes a high temperature stage [94]. To grow GaAs, the solid As source is kept at a sufficiently high temperature that the growth is limited by the arrival of the gallium atoms, typically of order 10^{19} atoms/m^2-sec. The arsenic flux, in the form of As$_2$ and As$_4$, is 5 to 10 times higher. Practically all of the gallium atoms stick to the surface at normal growth temperatures, but only one arsenic atom adheres for each gallium atom.

14.12 BCF Theory[+]

Unlike VPE, a reasonably accurate model of the surface during MBE can be developed. This model starts by examining the opposite of MBE: evaporation. Consider the surface of the wafer to consist of two levels or terraces. The upper terrace is one atomic layer higher than the lower terrace. Kink sites occur where the terrace edge forms an angle with itself. Here, few bond-breaking events are required

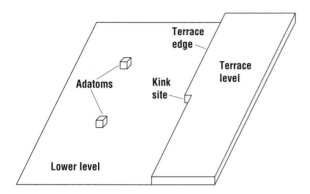

Figure 14.32   A microscopic view of a semiconductor surface during MBE growth or evaporation.

for an atom to liberate itself from the upper level. A lone atom away from the terrace edge, however, is the most loosely bound. The model for low temperature sublimation, then, is that atoms break loose from kink sites, diffuse across the surface of the wafer, and eventually desorb (Figure 14.32). Assume that the activation energy for liberating an atom from the kink site is W_s. Then the equilibrium density of adsorbed atoms is

$$n_{seq} = N_s e^{-W_s/kT} \tag{14.17}$$

where N_s is the density of surface positions. Once freed from the kink sites, the adsorbed atoms have only a small energy barrier U_s to move from site to site. If the distance between sites is a, and the frequency of lattice vibrations is f, the surface diffusivity is

$$D_s = a^2 f e^{-U_s/kT} \tag{14.18}$$

Assume that the adsorbed atom desorbs after some mean time τ. If the energy barrier for this desorption is E_d,

$$\frac{1}{\tau} = f' e^{-E_d/kT} \tag{14.19}$$

where f' is a preexponential factor related to the lattice vibrational frequency f. If the mean migration length before desorption is given by

$$\lambda_s = \sqrt{D_s \tau} \approx a e^{(E_d - U_s)/2kT} \tag{14.20}$$

the desorbing flux is given by

$$F_o = \frac{n_{seq}}{\tau} = N_s f' e^{-(E_d + W_s)/kT} \tag{14.21}$$

Generally W_s and E_d are comparable and much larger than U_s. Due to their exponential temperature dependence, n_{seq}, P_e, and F_o increase strongly with increasing temperature. The surface residence time and the mean migration distance both decrease rapidly with increasing temperature. For typical growth conditions, the surface diffusion length is of order tens or hundreds of micrometers.

The terraces in this model may be caused by dislocations, previous growth, or by small deviations from perfect crystal orientation. On <111> surfaces, this intentional misorientation is required. If the crystal misorientation is $\Delta\theta$, the terrace separation is given by

$$L_o = \frac{h}{\sin(\Delta\theta)} \tag{14.22}$$

where h is the step height, typically one atomic layer thickness. Since h is about 1 Å, the mean terrace separation is usually of order microns or less. This separation must be much less than the mean diffusion distance to ensure 2-D growth.

The best known theory regarding 2-D growth was developed by Burton, Cabrera, and Franks [95]. BCF theory assumes that most adatoms reach the surface between the surface steps and diffuse to the edge of the step, where they are captured. The adatoms may then diffuse along the step edge until they reach a kink site, where they are incorporated into the crystal. Terrace edge movement occurs by lateral movement of the kink sites. Crystal growth then occurs as the terrace edges sweep across the surface of the wafer.

If the terrace edge is considered to be a perfect sink for excess adatoms, the adatom concentration near the terrace edge drops to the equilibrium value, n_{seq}. That is, the terrace is not a perfect sink for adatoms but will emit a small number at nonzero temperatures. The net flux of adatoms is given by the difference between the incident flux and the evaporating flux,

$$j_v = j_{inc} - \frac{n_s}{\tau} \tag{14.23}$$

The surface current density of adatoms is described by the diffusion equation

$$j_s = -D_s \nabla n_s \tag{14.24}$$

For steady state growth, the divergence of the surface current density is the net growth flux

$$\nabla \cdot j_s + j_v = 0 \tag{14.25}$$

One can define a supersaturation parameter σ, where

$$\sigma = \frac{j_{inc}}{n_{sec}/\tau} - 1 \tag{14.26}$$

Just as in VPE, when $\sigma > 0$ epitaxial layers will grow. When $\sigma < 0$, the wafer material evaporates. Combining these equations and assuming that the surface diffusivity is independent of position and direction, one can derive the differential equation

$$\lambda_s^2 \nabla^2 n_s + n_s = n_{seq}(\sigma + 1) \tag{14.27}$$

In reality the diffusivity is strongly directional due to surface reconstruction [96], but we will ignore this refinement of the basic BCF theory.

Example 14.5

Assume that MBE is used to grow a material at a substrate temperature of 500°C. Assume that the material has the following properties: $W_s = 2.0$ eV, $E_d = 1.8$ eV, $U_s = 0.2$ eV, $N_s = 10^{15}$ cm^{-2}, $f = f' = 10^{14}$ sec^{-1}, $a = 2$ Å, $h = 0.5$ Å, and $\Delta\theta = 1°$. Calculate the mean surface diffusion length. Compare this with the mean terrace separation. If the flux on the surface of the wafer is the same as that calculated in Example 14.4, calculate the supersaturation.

Solution

Inserting the given parameters into Equations 14.18–14.20 produces $D_s = 2 \times 10^{-3}$ cm^2/ sec, $T = 5.7$ msec, and $n_{seq} = 86$ cm^{-2}. Inserting the surface diffusivity and mean desorption time, $\lambda_s = 34$ μm. In comparison, the average terrace separation (given by Equation 14.22) is 28 Å. This separation is much less than the characteristic surface diffusion length. The desorbing flux F_o can be found directly as the ratio of n_{seq} and τ. It is 1.5×10^4 cm^{-2}-sec^{-1}. Inserting into Equation 14.26, the supersaturation is approximately 3×10^{10}. These large values of supersaturation are quite typical for MBE. They indicate that the system is run far from equilibrium. As a point of comparison, typical values for growth in a VPE system are less than 2.

Equation 14.27 was solved by BCF, subject to the boundary condition that n_s is equal to n_{seq} at the step edge position. (Step movement can be neglected when the average distance traveled by an atom is much larger than the distance traveled by a step.) Then, assuming that at the step ($y = 0$) $n_s = n_{seq}$ and as $y \rightarrow \infty$, $n = (1 + \sigma)n_{seq}$,

$$n_s = n_{seq}[1 + \sigma(1 - e^{-y/\lambda_s})] \tag{14.28}$$

A surface supersaturation can be defined as

$$\sigma_s = \frac{n_s}{n_{seq}} - 1 \tag{14.29}$$

For a series of steps separated by L_o,

$$\sigma_s = \sigma \left[1 - \frac{\cosh(y/\lambda_s)}{\cosh(L_o/2\lambda_s)} \right] \tag{14.30}$$

If one knows the step distribution, then the saturation and the mean diffusion length, the local surface saturation, the step velocity, and ultimately the growth rate can be calculated. Finally it should be noted that BCF theory applies only when the degree of saturation is not extremely large. At high surface admolecule concentrations, 3-D growth dominates the process, and this terrace–kink model is no longer valid.

We are now in a position to appreciate the need for a clean environment. Recall that there is a background pressure of impurity gases in the chamber. When one of these impurity atoms strikes the surface of the wafer, it may adsorb or react chemically so as to remain on the surface. Water, for example, is likely to react with the surface of the semiconductor to form an immobile oxide. As the step approaches this oxide, it must grow around it and coalesce on the opposite side. Subsequent layers may not align themselves on this imperfect substrate, resulting in defects. It is also possible that this surface defect will be a getter site for admolecules, leading to island formation. A partial pressure of 10^{-8} torr of water results in an arrival flux of water molecules of 10^{13} atoms/cm^2-sec. Assume a sticking coefficient of one. A typical growth rate for MBE is 10000 Å/hr. Then, the arrival rate of growth atoms must be about 10^{15} atoms/cm^2-sec. That is, nearly one in 100 atoms is an impurity. Fortunately, most of the residual gas in the chamber is not water vapor. The semiconductor atoms in the beam flux rapidly react with the residual water and getter it from the chamber (Figure 14.33).

It is also not necessary for defects to be caused by impurities. Anything other than a terrace edge on which admolecules can condense can be a defect site. In VPE, there is a constant flux of chemical reactions, removing and adding the growth species from the surface of the wafer. The degree of supersaturation, therefore, is quite modest. In MBE, the primary mechanism for removal of the adatoms once they stick to the surface is evaporation. (In GaAs MBE, the As adatoms may actually form As$_4$, which has a small but not negligible vapor pressure at typical growth temperatures.) Since the substrate temperature during MBE is low, P_e is small and the supersaturation is extremely high. This can also lead to higher defect densities in MBE growth due to self-agglomeration. Once nucleated, these islands will grow, producing dislocations that propagate through the epitaxial layer. This effect limits the maximum growth rate that can be achieved in MBE.

For the epitaxial layer to be useful in device structures, the structures must be doped. In a good MBE system, unintentional doping can be as low as 10^{13} to 10^{15} cm^{-3}. For silicon, n-type dopants include arsenic, phosphorus, and antimony. For heavily doped layers, P and As are preferred due to their high solid solubility and low activation energy. Antimony has reasonable vapor pressures at typical source temperatures, so that standard Knudsen cells can readily be used for doping. On the other

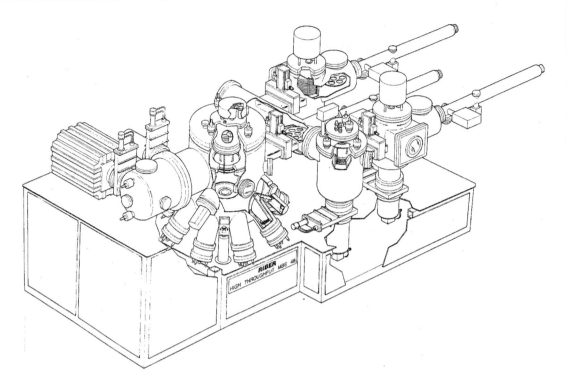

Figure 14.33 A high throughput automated MBE system includes load and unload locks, preparation, and growth chambers, all connected under UHV. Three 4-in. wafers are held on circular platens *(courtesy of RIBER S.A.).*

hand, As and P, have very high vapor pressures at room temperature. This results in memory effects. For p-type layers in silicon, boron and aluminum have been used.

One of the problems with silicon MBE is that all of the common dopants have large segregation coefficients. The dopant concentration in the film is much less than the concentration on the surface of the wafer [97]. This excess dopant concentration on the surface continues to act like a dopant source even after the actual dopant evaporation source has been shut off. Several techniques have been developed to solve this problem. The most common method is to flash off these surface impurities by raising the temperature of the wafer high enough to evaporate the dopant atoms. The high temperature required, however, is not desirable due to dopant redistribution. In p-type doping, molecular boron sources can also be used. In particular, HBO_3 produces good control and only modest oxygen incorporation at substrate temperatures as low as 650°C [98].

Producing heavily doped layers in either material system is difficult, particularly if high quality epi layers with abrupt junctions are also desired. This problem arises because the sticking coefficients of common dopants decrease as the substrate temperature increases. Moderate or high substrate temperatures are necessary to reduce the defect density. The third method of introducing dopants in MBE is the use of low energy implantation during the growth [99] (Figure 14.34). This may be done using an actual ion gun or by ionizing the dopant atoms and biasing the substrate to imbed the dopant atom into the growing film. Energies for these processes are kept below a few kiloelectron-volts to maintain well-controlled profiles and greater than 100 eV to minimize sputtering. Alternatively, the growth atoms themselves may be ionized and accelerated toward the growing substrate. Upon reaching the substrate, some of them will impact impurities on the surface, driving them into the substrate. This is

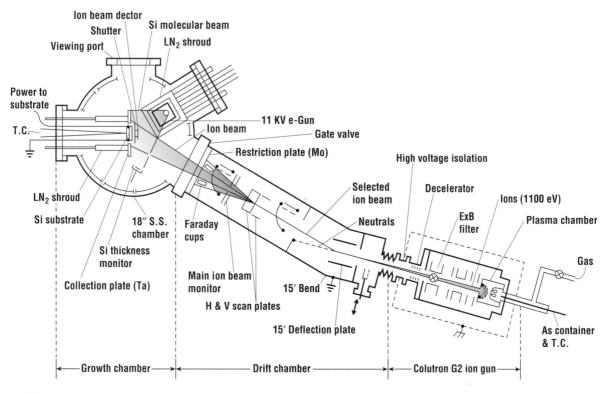

Figure 14.34   An ion implantation system for introducing impurities during silicon MBE *(after Fujiwara et al. [90].*

referred to as knock-on implantation. Substrate temperatures for implanted silicon MBE growth must be kept above 650°C to minimize radiation damage [100].

Group IV elements such as Si and Sn are typical dopants for producing n-type layers in GaAs. Silicon is the most common source, since silicon doped layers have the highest mobilities for a given doping level. Optimization of the growth process must be done to minimize the amount of doping compensation for group IV dopants. Zinc and beryllium are commonly used p-type dopants for GaAs. Recently carbon doping has been investigated as an alternate p-type doping source for MBE [101]. Carbon can easily be introduced by using a serpentine graphite filament in the chamber.

14.13 Gas Source MBE and Chemical Beam Epitaxy$^+$

The distinction between vapor phase epitaxy and physical techniques begins to blur as VPE crystal growth techniques are being extended to ultrahigh vacuum and MBE crystal growth techniques begin to use gaseous sources. Depending on the details of the procedure, they may be called chemical beam epitaxy [102] (CBE), metal organic molecular beam epitaxy (MOMBE), or gas source molecular beam epitaxy (GSMBE). In each case, gases are injected into an ultrahigh vacuum environment at very low flow rates. Table 14.3 differentiates the processes [103].

In early use of GSMBE, it was found that gallium droplets formed at the surface when substrate temperatures were held below 550°C [104]. To solve this problem, most UHV growth systems that use gaseous group V sources have an oven or cracking furnace to decompose the source [105]. In the case of AsH_3, the cracking product is As_2. One of the interesting features of using organometallic sources in

Table 14.3   Ultrahigh vacuum growth techniques that use gaseous sources

Technique	Group III Source	Group V Source
MBE	Elemental	Elemental
CBE	Gaseous	Gaseous
MOMBE	Gaseous	Elemental
GSMBE	Elemental	Gaseous

See Tsang [103].

UHV is that the molecular sticking coefficients depend strongly on the surface coverage. In growing GaAs, for example, it has been shown that the sticking coefficient for trimethylgallium is almost zero when the surface is gallium terminated but is much higher when it is arsine terminated [106]. This observation suggests that the growth of GaAs can be accomplished in a digital manner by alternate exposures to TMG and an arsine source. These processes are called atomic layer epitaxy (ALE).

14.14 Summary

The epitaxial growth of silicon from the vapor is a well-established process for fabricating advanced semiconductor technologies. An understanding of the process requires a detailed knowledge of both the gas phase chemistry and the events at the surface. Cleanliness, both of the chamber and of the growth gases, is critical to the success of the process. Chlorosilanes are the preferred precursors for silicon epitaxy, which can be done uniformly across the wafer and selectively in windows etched in dielectrics. The growth of GaAs from the vapor, however, is complicated by the lack of suitable inorganic gallium source gases.

The second half of the chapter presented growth techniques that have been developed for controllably producing thin epitaxial layers, often in the form of heterostructures. Some of these processes allow atomic level control of the growth thickness. These methods have been divided into physical (MBE) and chemical (MOCVD, RTCVD, and UHV/CVD) techniques. This distinction is becoming more blurred as UHV processes are developed that use chemical reactions occurring on the surface of the wafer rather than in the gas phase.

Problems

1. A silicon epitaxial growth process is attempted with a chamber temperature of 1050°C, and gas flows of 200 sccm $SiCl_4$ and 100 sccm of Si_2H_6. Assume that the mixture attains chemical equilibrium. What is the supersaturation in the chamber? Will these conditions grow an epitaxial layer or etch? If the temperature is increased to 1300°C, by what percentage would you expect the growth (or etch) rate to increase or decrease?
2. Starting from Equation 14.1, derive Equation 14.2.
3. Compare Equation 14.2 to Figure 14.8. For each compound, identify the approximate temperature at which the growth becomes mass transport limited. If

$$k_s = k_o e^{-E_a/kT}$$

h_g is independent of temperature and C_g is 10^{15} cm^{-3} at all temperatures and for all gases, find h_g and k_o.

4. An epitaxial growth chamber has a background O_2 pressure of 2×10^{-5} torr. What would the minimum annealing temperature be to ensure a Si-stable surface if only the background gases are present? Alternatively, one can flow H_2, which contains 5 ppb of O_2. If this gas flushes the chamber of its background contaminants, what would the minimum annealing temperature be at 760 torr of H_2?
5. Explain why the growth curve shown in Figure 14.6 becomes negative.
6. Explain why low temperature growth is desirable. To assist you in this explanation, use Figure 14.8 to determine the amount of time necessary to grow 1 μm of epitaxial silicon at 800, 900, 1000, and 1100°C. Also calculate the intrinsic diffusivity of As at these temperatures (see Chapter 3), and from this find the characteristic diffusion length $\sqrt{Dt}$.
7. A 5% mixture of $SiCl_4$ in H_2 is used to grow epitaxial silicon at 1100°C. Calculate the supersaturation (σ) for this condition. Will growth or etching result?
8. Why isn't halide transport used for silicon epitaxial growth?
9. Briefly contrast commensurate, incommensurate, and pseudomorphic growth techniques. Describe the advantages and limitations of each.
10. The growth of GaAs on silicon wafers is able to provide large wafers for IC fabrication. If you wanted to extend this process to InP, HgCdTe, or other compound systems, what would you need to find out to determine whether or not it would be possible?
11. Let us assume that an MOCVD reactor is able to grow 1.0 μm/hr and that the utilization efficiency of the TMG is about 2% (common for run/vent systems). If the V/III ratio is 50:1 and the system grows 2 μm/day, how much arsine (in moles) is required per day? What volume does this correspond to at standard temperature and pressure (*Note*: 1 mole = 22.8 L at STP). How many runs can be done with a 25-L cylinder containing 10% AsH_3 at 2000 psi? The deposition area is 200 cm^2.
12. What are the advantages and disadvantages of using AsH_3 in MOCVD of GaAs?
13. Compare the advantages of both RTCVD and UHV/CVD.
14. Repeat Example 14.4 if $E_d = 1.5$ eV, $U_s = 0.1$, and $W_s = 1.7$ eV, and all other parameters are unchanged. Estimate n_{seq}, λ_s, and σ.
15. What are some of the driving forces for using gas sources in MBE?

References

1. B. J. Baliga, ed., *Epitaxial Silicon Technology*, Academic Press, Orlando, FL, 1986.
2. S. A. Campbell, J. D. Leighton, G. H. Case, and K. Knutson, "Very Thin Silicon Epitaxial Layers Grown Using Rapid Thermal Vapor Phase Epitaxy," *J. Vacuum Sci. Technol. B* **7**: 1080 (1989).
3. C. Norris, "Monitoring Growth with X-ray Diffraction," *Philos. Trans. R. Soc. London Series A: Phys. Sci. Eng.* **344**:1673 (1993).
4. For a comprehensive review of wafer-cleaning techniques the reader is referred to W. Kern, ed., *Handbook of Semiconductor Wafer Cleaning Technology: Science, Technology, and Applications*, Noyes, Park Ridge, NJ, 1993.
5. W. Kern and D. A. Puotinen, "Cleaning Solution Based on Hydrogen Peroxide for Use in Semiconductor Technology," *RCA Rev.* 187, (June 1970).
6. F. W. Kern and G. W. Gale, "Surface Preparation," in *Handbook of Semiconductor Manufacturing Technology*, Y. Nishi and R. Doering, eds., Marcel Dekker, New York (2000).
7. W. Kern, "Purifying Si and SiO_2 Surfaces with Hydrogen Peroxide," *Semicond. Int.*, April 1984, p. 94.

8.   W. Kern, "The Evolution of Silicon Wafer Cleaning Technology," *J. Electrochem. Soc.* **137**:1887 (1990).

9.   O. J. Antilla and M. V. Tilli, "Metal Contamination Removal on Silicon Wafers Using Dilute Acidic Solutions," *J. Electrochem. Soc.* **139**:1751 (1992).

10. H. R. Chang, in *Defects in Silicon*, W. M. Bullis and L. C. Kimberling, eds., Electrochemical Society, Pennington, NJ, 1983.

11. G. Ghidini and F. W. Smith, "Interaction of H_2O with Si(111) and (100)," *J. Electrochem. Soc.* **131**:2924 (1984).

12. F. W. Smith and G. Ghidini, "Reaction of Oxygen with Si(111) and (100): Critical Conditions for the Growth of SiO_2," *J. Electrochem. Soc.* **129**:1300 (1982).

13. G. W. Rubloff, "Defect Microchemistry in SiO_2/Si Structures," *J. Vacuum Sci. Technol. A* **8**:1857 (1990).

14. Y.-K. Sun, D. J. Bonser, and T. Engel, "Spatial Inhomogeneity and Void-Growth Kinetics in the Decomposition of Ultrathin Oxide Overlayers on Si (100)," *Phys. Rev. B* **3**:14,309 (1991).

15. R. Iscoff, "Wafer Cleaning: Wet Methods Still Lead the Pack," *Semicond. Int.*, July 1993, p. 58.

16. J. H. Comfort, L. M. Garverick, and R. Reif, "Silicon Surface Cleaning by Low Dose Argon-Ion Bombardment for Low Temperature (750°C) Epitaxial Silicon Deposition," *J. Appl. Phys.* **62**:3388 (1988).

17. M. Ishii, K. Nakashima, I. Tajima, and M. Yamamoto, "Properties of Silicon Surface Cleaned by Hydrogen Plasma," *Appl. Phys. Lett.* **58**:1378 (1991).

18. S. Salimian and M. Delfino, "Removal of Native Silicon Oxide with Low-Energy Argon Ions," *J. Appl. Phys.* **70**:3970 (1991).

19. Y. J. Chabal, G. H. Higashi, K. Raghavachari, and V. A. Burrows, "Infrared Spectroscopy of Si(111) and Si(100) Surfaces after HF Treatment: Hydrogen Termination and Surface Morphology," *J. Vacuum Sci. Technol. A* **7**:2104 (1989).

20. G. S. Higashi, Y. J. Chabal, G. W. Trucks, and K. Raggavachari, "Ideal Hydrogen Termination of the Si(111) Surface," *Appl. Phys. Lett.* **56**:656 (1990).

21. P. Dumas, Y. J. Chabal, and G. S. Higashi, "Coupling of an Adsorbate Vibration to a Substrate Surface," *Phys. Rev. Lett.* **65**:1124 (1990).

22. H. Ubara, T. Imura, and A. Hiraki, "Formation of Si–H Bonds on the Surface of Microcrystalline Silicon Covered with SiO_x by HF Treatment," *Solid State Commun.* **50**:673 (1984).

23. M. Morita, T. Ohmi, E. Hasegawa, M. Kawakami, and K. Suma, "Control Factor of Native Oxide Growth on Silicon in Air or in Ultrapure Water," *Appl. Phys. Lett.* **55**:562 (1989).

24. D. Gräf, M. Grundner, and R. Schultz, "Reaction of Water with Hydrofluoric Acid Treated Silicon (111) and (100) Surfaces," *J. Vacuum Sci. Technol. A* **7**:808 (1989).

25. M. Offenberg, M. Liehr, and G. W. Rubloff, "Ultraclean Integrated Processing of Thermal Oxide Structures," *Appl. Phys. Lett.* **57**:1254 (1990).

26. A. S. Grove, *Physics and Technology of Semiconductor Devices*, Wiley, New York, 1967.

27. J. P. Hirth and G. H. Pond, "Condensation and Evaporation: Nucleation and Growth," *Prog. Mater. Sci.* **11**:1 (1963).

28. R. Pagliaro Jr., J. F. Corboy, L. Jastrzebski, and R. Soydan, "Uniformly Thick Selective Epitaxial Silicon," *J. Electrochem. Soc.* **134**:1235 (1987).

29. P. van der Putte, L. J. Giling, and J. Bloem, "Growth and Etching of Silicon in Chemical Vapor Deposition Systems: The Influence of Thermal Diffusion and Temperature Gradients," *J. Cryst. Growth* **31**:299 (1975).

30. J. Bloem and W. A. P. Claassen, "Nucleation and Growth of Silicon Films by Chemical Vapour Deposition," *Philips Technol. Rev.* **41**:60 (1983).

31. J. W. Medernach and P. Ho, *Mater. Res. Soc. Conf. Proc.*, Honolulu, 1987, p. 101.

32. W. A. P. Claassen and J. Bloem, "Rate-Determining Reactions and Surface Species in CVD of Silicon II. The SiH_2Cl_2-H_2-N_2-HCl System," *J. Cryst. Growth* **50**:807 (1980).

33. H. C. Theuerer, "Epitaxial Silicon Films by the Reduction of $SiCl_4$," *J. Electrochem. Soc.* **108**:649 (1961).

34. T. Arizumi, *Curr. Topics Mater Sci.* **1**:343 (1975).

35. J. Bloem, *J. Cryst. Growth* **18**:70 (1973).

36. F. C. Eversteyn, "Chemical Reaction Engineering in the Semiconductor Industry," *Philips Res. Rep.* **19**:45 (1974).

37. P. A. Coon, M. L. Wise, and S. M. George, "Modelling Silicon Epitaxial Growth with SiH_2Cl_2," *J. Cryst. Growth* **130**:162 (1993).

38. T. Aoyama, Y. Inoue, and T. Suzuki, "Gas Phase Reactions and Transport in Silicon Epitaxy," *J. Electrochem. Soc.* **130**:204 (1983).

39. A. A. Chernov, "Growth Kinetics and Capture of Impurities During Gas Phase Crystallization," *J. Cryst. Growth* **42**:55 (1977).

40. P. Rai-Choudhury and E. Salkovitz, "Doping of Epitaxial Silicon: The Effect of Dopant Partial Pressure," *J. Cryst. Growth* **7**:361 (1970).

41. H. M. Liaw and J. W. Rose, "Silicon Vapor-Phase Epitaxy," in *Epitaxial Silicon Technology*, B. J. Baliga, ed., Academic Press, Orlando, FL, 1986.

42. C. H. J. van den Breckel, "Characterization of Chemical Vapor Deposition Processing," *Philips Res. Rep.* **32**:118 (1977).

43. M. J. Stowell, in *Epitaxial Growth, Part B*, J. W. Matthews, ed., Academic Press, New York, 1975, p. 437.

44. B. J. Ginsberg, J. Burghartz, G. B. Bronner, and S. R. Mader, "Selective Epitaxial Growth of Silicon and Some Potential Applications," *IBM J. Res. Dev.* **34**:816 (1990).

45. J. O. Borland, in *Proc. 10th Int. Conf. Chemical Vapor Deposition*, Electrochemical Society, Pennington, NJ, 1987, p. 307.

46. D. M. Jackson, "Advanced Epitaxial Processes for Monolithic Integrated Circuit Applications," *Trans. Metall. Soc. AIME* **233**:596 (1965).

47. K. Tanno, N. Endo, H. Kitajima, Y. Kurogi, and H. Tsuya, "Selective Silicon Epitaxy Using Reduced Pressure Techniques," *Jpn. J. Appl. Phys.* **21**:L564 (1982).

48. C. I. Drowley, G. A. Reid, and R. Hull, "Model for Facet and Sidewall Defect Formation During Selective Epitaxial Growth of (100) Silicon," *Appl. Phys. Lett.* **52**:546 (1988).

49. J. T. McGinn, L. Jastrzebski, and J. F. Corboy, "Defect Characterization in Monocrystalline Silicon Grown over SiO_2," *J. Electrochem. Soc.* **136**:398 (1984).

50. L. Jastrzebski, J. F. Corboy, and R. Soydan, "Issues and Problems Involved in Selective Epitaxial Growth of Silicon for SOI Fabrication," *J. Electrochem. Soc.* **136**:3506 (1989).

51. R. Pagliaro, F. Corboy, L. Jastrzebski, and R. Soydan, "Uniformly Thick Selective Epitaxial Silicon," *J. Electrochem. Soc.* **134**:1235 (1987).

52. B. P. Jain and R. K. Purohit, "Physics and Technology of Vapour Phase Epitaxial Growth of GaAs—A Review," *Prog. Cryst. Growth, Charact.* **9**:51 (1984).

53. J. L. Gentner, "Vapour Phase Growth of GaAs by the Chloride Process Under Reduced Pressure," *Philips J. Res.* **38**:37 (1983).

54. A. Cadoff and J. Bicknell, "The Epitaxy of Silicon on Alumina—Structural Effects," *Philos. Mag.* **14**:31 (1966).

55. H. Manasevit and R. Simpson, "A Survey of the Heteroepitaxial Growth of Semiconductor Films on Insulating Substrates," *J. Cryst. Growth* **22**:125 (1974).

56. P. K. Vasudev, "Silicon-on-Sapphire Heteroepitaxy," in *Epitaxial Silicon Technology*, B. J. Baliga, ed., Academic Press, Orlando, FL, 1986.

57. W. I. Wang, "Molecular Beam Epitaxial Growth and Materials Properties of GaAs and AlGaAs on Si (100)," *Appl. Phys. Lett.* **44**:1149 (1984).

58. W. T. Masselink, T. Henderson, J. Klem, R. Fischer, P. Pearah, H. Morkóc, M. Hafich, P. D. Wang, and G. Y. Robinson, "Optical Properties of GaAs on (100) Si Using Molecular Beam Epitaxy," *Appl. Phys. Lett.* **45**:1309 (1984).

59. T. Soga and S. Hattori, "Epitaxial Growth and Material Properties of GaAs on Si Grown by MOCVD," *J. Cryst. Growth* **77**:498 (1986).

60. T. Soga, S. Hattori, S. Sakai, M. Takeyasu, and M. Umeno, "MOCVD Growth of GaAs on Si Substrates with AlGaP and Strained Layer Superlattice Layers," *J. Appl. Phys.* **57**:4578 (1985).

61. R. People and J. C. Bean, "Calculation of Critical Layer Thickness Versus Lattice Mismatch for $Ge_xSi_{1-x}As$: Strained-Layer Heterointerfaces," *Appl. Phys. Lett.* **47**:322 (1985); **49**:229 (1986).

62. H. M. Manasevit and W. I. Simpson, *J. Appl. Phys. Lett.* **116**:1725 (1969).

63. G. B. Stringfellow, *Organometallic Vapor-Phase Epitaxy*, Academic Press, Boston, 1989.

64. R. M. Lum, J. K. Klingert, and M. G. Lamont, "Comparison of Alternate As-sources to Arsine in the MOCVD Growth of GaAs," in *Fourth Int. Conf. MOVPE*, 1988, p. P1–3.

65. C. H. Chen, C. A. Larsen, G. B. Stringfellow, D. W. Brown, and A. J. Robertson, "MOVPE Growth of InP Using Isobutylphosphine and *tert*-Butylphosphine," *J. Cryst. Growth* **77**:11 (1986).

66. R. J. Field and S. K. Ghandhi, "Doping of GaAs in a Low Pressure Organometallic CVD System," *J. Cryst. Growth* **74**:543 (1986).

67. T. F. Kuech, "Metal-Organic Vapor Phase Epitaxy of Compound Semiconductors," *Mater. Sci. Rep.* **2**:1 (1987).

68. D. J. Schlyer and M. A. Ring, "An Examination of the Product-Catalyzed Reaction of Trimethylgallium with Arsine," *J. Organometall. Chem.* **114**:9 (1976).

69. D. H. Reep and S. K. Ghandhi, "Deposition of GaAs Epitaxial Layers by Organometallic CVD," *J. Electrochem. Soc.* **130**:675 (1983).

70. P. Rai-Chaudhury, "Epitaxial Gallium Arsenide from Trimethylgallium and Arsine," *J. Electrochem. Soc.* **116**:1745 (1969).

71. Y. Seki, K. Tanno, K. Iida, and E. Ichiki, "Properties of Epitaxial GaAs Layers from a Triethylgallium and Arsine System," *J. Electrochem. Soc.* **122**:1108 (1975).

72. T. F. Kuech, M. A. Tischler, P.-J. Wang, G. Scilla, R. Potemski, and F. Cardone, "Controlled Carbon Doping of GaAs by Metallorganic Vapor Phase Epitaxy," *Appl. Phys. Lett.* **53**:1317 (1988).

73. B. T. Cunningham, M. A. Haase, M. J. McCollum, J. E. Baker, and G. E. Stillman, "Heavy Carbon Doping of Metallorganic Chemical Vapor Deposition Grown GaAs Using Carbon Tetrachloride," *Appl. Phys. Lett.* **54**:1905 (1989).

74. B. T. Cunningham, L. J. Guido, J. E. Baker, J. S. Major, Jr., N. Holonyak, Jr., and G. E. Stillman, "Carbon Diffusion in Undoped *n*-type and *p*-type GaAs," *Appl. Phys. Lett.* **55**:687 (1989).

75. M. A. Tischler, "Advances in Metallorganic Vapor-Phase Epitaxy," *IBM J. Res. Dev* **34**:828 (1990).

76. T. F. Kuech, E. Veuhoff, D. J. Wolford, and J. A. Bradley, "Low Temperature Growth of $Al_xGa_{1-x}As$ by MOCVD," *GaAs, Related Compounds, 11th Int. Symp.*, 1985, p. 181.

77. J. R. Shealey and J. M. Woodall, "A New Technique for Gettering Oxygen and Moisture from Gases in Semiconductor Processing," *Appl. Phys. Lett.* **68**:157 (1984).

78. J. F. Gibbons, C. M. Gronet, and K. E. Williams, "Limited Reaction Processing: Silicon Epitaxy," *Appl. Phys. Lett.* **47**:721 (1985).

79. S. A. Campbell, J. D. Leighton, G. H. Case, and K. Knutson, "Very Thin Silicon Epitaxial Layers Grown Using Rapid Thermal Vapor Phase Epitaxy," *J. Vacuum Sci. Technol. B* **7**:1080 (1989).

80. M. L. Green, D. Brasen, H. Luftman, and V. C. Kannan, "High Quality Homoepitaxial Silicon Films Deposited by Rapid Thermal Chemical Vapor Deposition," *J. Appl. Phys.* **65**:2558 (1989).

81. T. Y. Hsieh, K. H. Jung, and D. L. Kwong, "Silicon Homoepitaxy by Rapid Thermal Processing Chemical Vapor Deposition," *J. Electrochem. Soc.* **138**:1188 (1991).

82. K. L. Knutson, S. A. Campbell, and F. Dunn, "Three Dimensional Temperature Uniformity Modeling of a Rapid Thermal Processing Chamber," *IEEE Trans. Semicond. Manuf.* 7(1) pp. 68–72 (1994).

83. B. S. Mayerson, "Low-Temperature Silicon Epitaxy by Ultrahigh Vacuum/Chemical Vapor Deposition," *Appl. Phys. Lett.* **48**:797 (1986).

84. B. S. Mayerson, E. Ganin, D. A. Smith, and T. N. Nguyen, "Low Temperature Silicon Epitaxy by Hot Wall Ultrahigh Vacuum/Chemical Vapor Deposition Techniques: Surface Optimization," *J. Electrochem. Soc.* **133**:1232 (1986).

85. B. S. Meyerson, "Low Temperature Si and Ge:Si Epitaxy by Ultrahigh Vacuum/Chemical Vapor Deposition: Process Fundamentals," *IBM J. Res. Dev.* **34**:806 (1990).

86. J. Davies and D. Williams, "III–V MBE Growth System," in *The Technology and Physics of Molecular Beam Epitaxy*, E. H. C. Parker, ed., Plenum, New York (1985).

87. D. Bellevance, "Industrial Application: Perspective and Requirements," in *Silicon Molecular Beam Epitaxy* **2**, E. Kasper and J. C. Bean, eds., CRC, Boca Raton, FL, 1985, p. 153.

88. T. Tatsumi, H. Hirayama, and N. Aizaki, "Si Particle Density Reduction in Si Molecular Beam Epitaxy Using a Deflection Electrode," *Appl. Phys. Lett.* **54**:629 (1989).

89. A. von Gorkum, "Performance and Processing Line Integration of a Silicon Molecular Beam Epitaxy System," *Proc. 3rd Int. Symp. Si MBE, Thin Solid Films* **184**:207 (1990).

90. K. Fujiwara, K. Kanamoto, Y. N. Ohta, Y. Tokuda, and T. Nakayama, "Classification and Origins of GaAs Oval Defects Grown by Molecular Beam Epitaxy," *J. Cryst. Growth* **80**:104 (1987).

91. A. Y. Cho, "Advances in Molecular Beam Epitaxy (MBE)," *J. Cryst. Growth* **111**:1 (1991).

92. J. Saito, K. Nanbu, T. Ishikawa, and K. Kondo, "In situ Cleaning of GaAs Substrates with HCl Gas and Hydrogen Mixture Prior to MBE Growth," *J. Cryst. Growth* **95**:322 (1989).

93. N. Chand, "A Simple Method for Elimination of Gallium-Source Related Oval Defects in Molecular Beam Epitaxy of GaAs," *Appl. Phys. Lett.* **56**:466 (1990).

94. J. H. Neave, P. Blood, and B. A. Joyce, "A Correlation Between Electron Traps and Growth Processes in n-GaAs Prepared by Molecular Beam Epitaxy," *Appl. Phys. Lett.* **36**:311 (1980).

95. W. K. Burton, N. Cabrera, and F. C. Franks, *Philos. Trans. R. Soc. London,* Ser. *A* **243**:299 (1951).

96. M. G. Legally, "Atoms in Motion on Surfaces," *Phys. Today* **46**:24 (1993).

97. J. C. Bean, "Silicon Molecular Beam Epitaxy: Highlights of Recent Work," *J. Electron. Mater.* **19**:1055 (1990).

98. T. L. Lin, R. W. Fathauer, and P. J. Grunthaner, "Heavily Boron-doped Si Layers Grown Below 700°C by Molecular Beam Epitaxy Using a HBO$_2$ Source," *Appl. Phys. Lett.* **55**:795 (1989).

99. P. Fons, N. Hirashita, L. C. Markert, Y.-W. Kim, J. E. Green, W.-X. Ni, J. Knall, G. V. Hansson, and J.-E. Sundgren, "Electrical Properties of Si(100) Films Doped with Low-Energy (<150 eV) Sb Ions During Growth by Molecular Beam Epitaxy," *Appl. Phys. Lett.* **53**:1732 (1988).

100. J.-P. Noel, J. E. Greene, N. L. Rowell, S. Kechang, and D. C. Houghton, "Photoluminescence Studies of Si(100) Doped with Low Energy (<1000 eV) As$^+$ Ions During Molecular Beam Epitaxy," *Appl. Phys. Lett.* **55**:1525 (1989).

101. R. J. Malik, R. N. Nottenberg, E. F. Schubert, J. F. Walker, and R. W. Ryan, "Carbon Doping in Molecular Beam Epitaxy of GaAs from a Heated Graphite Filament," *Appl. Phys. Lett.* **53**:2661 (1988).

102. W. T. Tsang, "From Chemical Vapor Epitaxy to Chemical Beam Epitaxy," *J. Cryst. Growth* **95**:121 (1989).

103. W. T. Tsang, "Current Status Review and Future Prospects of CBE, MOMBE, and GSMBE," *J. Cryst. Growth* **107**:960 (1991).

104. E. Veuhoff, W. Pletschen, P. Balk, and H. Lüth, "Metallorganic CVD of GaAs in a Molecular Beam Epitaxy System," *J. Cryst. Growth* **55**:30 (1981).

105. M. B. Panish, "Molecular Beam Epitaxy of GaAs and InP with Gas Sources for As and P," *J. Electrochem. Soc.* **127**:2729 (1980).

106. M. Uneta, Y. Watanabe, and Y. Ohmachi, "Desorption of Triethylgallium During Metallorganic Molecular Beam Epitaxial Growth of GaAs," *Appl. Phys. Lett.* **54**:2327 (1989).

Part V

Process Integration

If an elderly but distinguished scientist says that something is possible, he is almost certainly right; but if he says that it is impossible, he is very probably wrong.[1]

Any IC technology can be divided into three basic components, whether it is FET or bipolar, Si or GaAs. These components are device fabrication, interconnection, and isolation. When deciding which processes are to be used, the technologist must ensure that all three of these tasks are done. The structure of this section will be first discuss groups of processes, called *modules*, for isolation and interconnection (including ohmic contacts) for both silicon and GaAs-based technologies. This is done both to avoid repetition and so that the student can compare these modules across technologies.

In deciding on the order of the process flow, there are two important considerations. The first is the function of the particular process. As an example, interconnect could be done first by using diffused resistors in the substrate. The problem with such an approach is that the interconnect would have a large resistance and a large capacitance. This type of interconnect, while functional, would dramatically reduce the circuit speed. The second consideration is the effect that each process might have on later processes or on the structures that have been fabricated before them. One obvious example of this consideration is thermal budget. Since most technologies involve the use of selectively doped regions of semiconductors, diffusion is a serious concern if high temperature processes are required. It is therefore desirable to do the highest temperature processes first. If aluminum is to be used for interconnection, after aluminum deposition the wafer temperature certainly must not exceed 660°C, the melting temperature of aluminum. These considerations mean that it may not be possible for the three modules to be done sequentially but that in some cases they may have to be intertwined.

After this introductory chapter, some of the most important individual technologies will be presented, which might wonder how the technologies were selected. As a useful reference, refer to the figure, which shows the fraction of the IC market for variety of technologies. Clearly, MOS, which includes part of the analog and discrete devices

---

[1]Arthur C. Clarke.

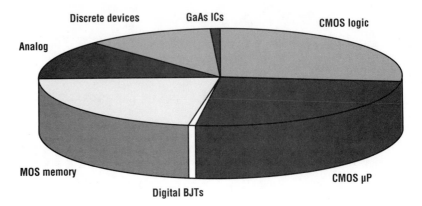

Estimated market share in 2005 *(McClean Report, 2006 Edition).*

section, is dominant. Chapter 16 will therefore begin this section of the book by covering CMOS. GaAs, has a small but slowly increasing market share. Chapter 17 will review other transistor technologies. It discusses GaAs field effect transistors (FETs) both because they are representative of GaAs technologies in general and because they present some interesting contrasts with silicon CMOS. The bipolar market share is shrinking, particularly in the areas of low power/moderate performance digital. A large fraction of analog and some of the high end digital, however, continue to be done in bipolar, and so they are included in Chapter 17. Chapter 17 will conclude with a brief overview of thin film transistors. Still a niche area, this field is poised for significant expansion. Chapter 18 will cover optical devices: light-emitting diodes and lasers. Chapter 19 will introduce a newer area of application for microelectronic fabrication, micromechanical structures (MEMS). Although still small commercially, this new area is expected to open many new applications for conventional process technology. Finally, some material common to all technologies will be presented: yield, statistical process control, and design of experiments.

Each of the technology chapters begins with a brief review of the ideal device equations. The emphasis here is not a complete description of the operation of the device. It is assumed that the student has taken a basic introductory semiconductor device course. This device understanding is necessary to be able to point out how the process must be designed to optimize device performance. Next, the series of unit processes used to fabricate a simple version of the technology will be presented. The extension of these simple technologies to higher speed applications is then presented, along with a description of the process modules necessary to obtain the desired performance. Finally, the concerns and process directions for each technology will be discussed.

Chapter 15

Device Isolation, Contacts, and Metallization

This chapter will cover two of the most basic functions of an IC. The first half will discuss device isolation, that is, the ability of the technology to allow each device to operate independently of the state of the other devices. Unless the technology is limited to building discrete devices, this is an essential function. The second half of the chapter will discuss device interconnection, including the metal-to-semiconductor contact. Again, only a brief review of the device physics will be provided to point out the implications for the technology.

15.1 Junction and Oxide Isolation

To fabricate ICs, some sort of isolation module must be developed [1]. The metrics of such a module are density, process complexity, yield, planarity, and parasitic effects. Trade-offs exist among these parameters. The first ICs were bipolar. To understand the necessity for isolation, consider Figure 15.1. From the discussion on unit processes, it is easy to understand how a modest bipolar technology can be constructed. To do this, simply diffuse a deep p-layer and a shallower N^+-layer into an n-type substrate. The substrate acts as the common collector. Assume that an insulating layer is then deposited on the substrate, contacts are patterned and etched, and the interconnect is applied and patterned. The first question is, How close can the transistors be placed? That is, what is the packing density?

The emitters of two adjacent transistors are automatically isolated from each other by the fact that each emitter is totally enclosed in the base diffusion. For the base layers to be isolated from each other, there must be a large energy barrier between the holes in the two base regions. As the height of this barrier decreases, the leakage between the bases increases exponentially. A simple and convenient measure of

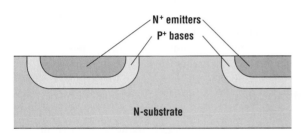

Figure 15.1 Simple junction isolation in a bipolar transistor technology with a common collector.

isolation is that the depletion layers associated with the two base collector junctions do not touch. To get an estimate, assume a simple one-sided step junction. Then the depletion layer thickness is

$$W_D = \sqrt{\frac{2k_s\varepsilon_o}{qN_D}(V_{bi} + V_{CB})} \tag{15.1}$$

where k_s is the relative permittivity of silicon, V_{bi} is the built-in voltage of the junction (typically ≈ 0.7 V), which for nondegenerate doping is given by

$$V_{bi} = \frac{kT}{q} \ln \frac{N_A N_D}{n_i^2} \tag{15.2}$$

and n_i is the intrinsic carrier concentration (about 10^{10} cm^{-3} at room temperature). For a typical collector concentration of 10^{16} cm^{-3}, this reduces to

$$W_D \approx 0.36 \ \mu m \sqrt{V_{bi} + V_{CB}} \tag{15.3}$$

The isolation distance then is just $2W_D$, or about 1.8 μm for a maximum V_{BC} of 5 V, or about 2.4 μm, to ensure isolation for a 10-V V_{BC}. Of course, one must also consider lateral diffusion. The base regions must be at least this far apart at the end of the process, not just on the photomasks. Note that this depleted region introduces a capacitance that is inversely proportional to W_D. Thus, a thinner W_D allows closer transistor spacing, but a larger parasitic capacitance per unit area of device.

To extend the approach to completely isolate the transistors, simply add a third diffusion to form the collector and use p-type substrates. To minimize the capacitance associated with the collector and to ensure that the collector doping is independent of variations in the substrate doping, a lightly doped substrate must be employed. Assume that the substrate is now 10^{15} cm^{-3} p-type. Then

$$W_D \approx 1.14 \ \mu m \sqrt{V_{bi} + V_{CS}} \tag{15.4}$$

For a 10-V bias ($2\times$ a supply voltage of 5 V), this would require about a 4-μm spacing around each diffusion or a total of 8 μm between devices after processing is complete. Taking into account lateral diffusion, a 12-μm collector-to-collector spacing might be required. In these processes, it is essential to ensure that the junctions remain reverse biased. This idea of using reverse-biased junctions to isolate the devices, first patented in 1959 [2], was the earliest practical method of device isolation.

To get an estimate for the isolation length required for a technology, consider the densest portion of the circuit. Then referring to Figure 15.2, read off the maximum permissible isolation distance. For example, if the densest portion of the circuit has 10^5 transistors/cm^2 and 50% of the area is active in that region, the device separation must be 10 μm or less. In many designs, the circuit density is limited not by the isolation distance, but by the metal density. The

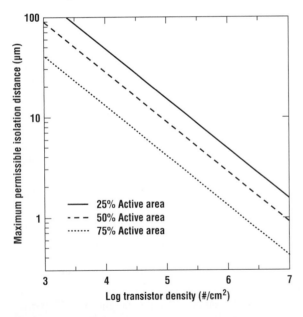

Figure 15.2 Simple calculation of average isolation distance required between transistors as a function of the device density.

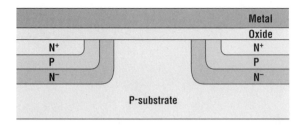

Figure 15.3 Cross section of simple bipolar technology with a metal line crossing the junction isolation region, forming a parasitic MOSFET.

density goal of the isolation module, then, is that the separation necessary to make the interconnect, and not the device isolation, is the limiting factor for the circuit density. Generally, for a random logic circuit, the isolation distance should be less than twice the first metal pitch (pitch is minimum linewidth + minimum line space). For memories and other highly structured logic, the isolation distance should be no more than the first metal pitch.

The other important point to make about Figure 15.2 is that for isolation-limited circuits, the circuit density is a sensitive function of the isolation distance. Reducing the isolation distance by a factor of 3 allows an order-of-magnitude increase in the circuit density. Consequently, companies that deal in isolation-driven circuits, such as memories, have developed elaborate isolation techniques.

Density, however, is not the only consideration in designing the isolation module. Go back to Figure 15.1, but add an insulating layer and a metal line that happens to pass over, but does not contact, both transistors (Figure 15.3). A parasitic MOS transistor now exists. The collectors of the two transistors act as the source and drain of the MOSFET, and the metal line acts as the gate. If a large enough positive bias exists on the line, the surface of the area under the line may invert, turning on the parasitic MOSFET. This effect will short out the two collectors, even if they are sufficiently far apart that the depletion regions do not touch.

Neglecting oxide charge, the MOSFET threshold for NMOS transistors on p-type substrates is given by (see Section 16.1)

$$V_T = \phi_{ms} + 2\phi_f + \frac{k_s t_{ox}}{k_{ox}} \sqrt{\frac{4qN_A}{k_s \varepsilon_o} \phi_f} \tag{15.5}$$

where

$$\phi_f = \frac{kT}{q} \ln\left[\frac{N_A}{n_i}\right] \tag{15.6}$$

and ϕ_{ms} is the metal semiconductor work function that depends on the type of metal used in the interconnect and the doping concentration in the substrate. For aluminum on p-type substrates, ϕ_{ms} varies from about -0.8 to -1.0 V. Since the intrinsic carrier concentration increases with temperature, the parasitic threshold voltage may shift by several volts when operating at high temperatures. Furthermore, the source-to-drain current of an MOS device increases exponentially with gate voltage in the subthreshold regime. Normal practice is to make the threshold of these parasitic transistors at least 2× (and preferably 3×) the supply voltage. This ensures that they will not turn on, even in the presence of an excess supply voltage and/or voltage spikes on the supply line, nor will the leakage currents be excessive.

Figure 15.4 shows a plot of threshold voltage versus the substrate concentration with oxide thickness as a parameter. The oxide thickness was varied from 0.2 to 1.0 μm in 0.2-μm increments. Also included is the threshold voltage if a total fixed charge of 10^{11} cm^{-2} is present. To achieve a suitably large parasitic threshold, one must select a thick field oxide and/or a large substrate concentration. The large substrate concentration degrades the performance due to junction capacitance. The thicker oxide improves both performance and parasitic turn-on; but obviously, due to the effect of oxide charge, some substrate doping is required even for very thick field oxides.

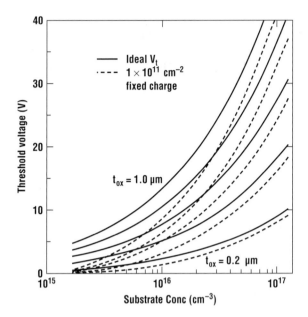

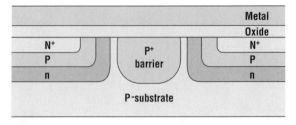

Figure 15.5   Guard ring isolation for the bipolar technology from Figure 15.3 .

Figure 15.4   Plot of V_T as a function of the N_A assuming $\phi_{ms} = 0$. Solid lines are a perfect interface, dashed lines are for $N_{it} = 10^{11}$ cm^{-2}.

The device separation can be reduced and the parasitic threshold increased by adding a p$^+$ diffused barrier ring around each device (Figure 15.5). Guard rings require an additional mask that must be aligned to the transistor. Furthermore, the guard ring must be deep (at least 2 μm), or the depletion layer will simply extend beneath the guard ring, shorting out the devices. The large thermal cycle needed to produce such a deep junction must be done early in the process. Assuming a 2-μm resolution lithography such as might be used in such a simple technology and recognizing that lateral diffusion that would occur, we see that the final width of the guard ring is over 5 μm. This requirement ultimately limits improvements in packing density.

We can now analyze this result in terms of the metrics just described. Obviously, junction isolation is simple and produces a planar isolation. Because of its simplicity, it would have a high yield. Density, however, is not large and must be traded off against the parasitic capacitance of the collector substrate junction. Increasing the substrate concentration increases the density but also increases the capacitance. Guard rings can improve the situation, particularly insofar as they prevent the turn on of parasitic MOS devices.

15.2 LOCOS Methods

The most straightforward way to produce a thick field oxide is by growing one before device fabrication, then etching holes in the oxide and fabricating the devices in these holes. This approach has two serious shortcomings. The first is the topology that is created. The step coverage for subsequent depositions will be poor and the photolithography will suffer. This is extremely serious if small features are to be printed. The second drawback is less obvious. On lightly doped substrates, a guard ring must be implanted to increase the parasitic threshold voltage. Unless very high energies are used, the implant must be done before the oxidation. Diffusion during oxidation may also be enhanced by point defects released during the oxidation process. Combined with alignment tolerance requirements, this will significantly reduce the density of the IC.

The isolation approach that has become the standard of silicon IC fabrication is local oxidation of silicon or LOCOS [3]. Local oxidation is essentially an outgrowth of junction isolation and addresses both the isolation and the parasitic device formation concerns. A thin oxide is first grown and a layer of Si$_3$N$_4$ deposited on the wafer, usually by LPCVD. After the nitride is patterned, a field

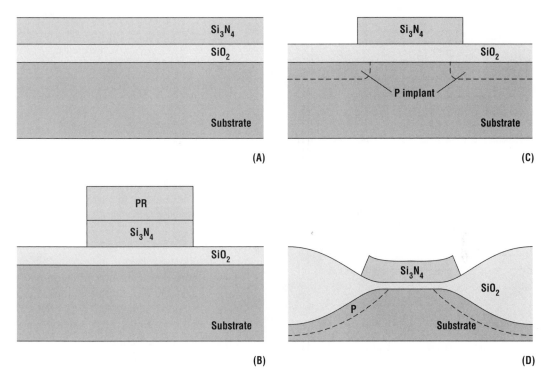

Figure 15.6   Cross-sectional views of a standard local oxidation of silicon (LOCOS) process.

implant may be done to increase the threshold voltage of the parasitic MOSFET. Then the photoresist is stripped and the wafer is oxidized (Figure 15.6). The nitride acts as a barrier to the diffusion of the oxidant, preventing oxidation in selected regions of the silicon. A thin oxide will also be grown on top of the nitride. This is important because it limits the minimum nitride thickness to about 1000 Å and because the oxide must be removed before the nitride can be stripped after the field oxidation.

Since oxidation consumes 44% as much silicon as it grows, the resultant oxide is partially recessed and has a gradual step onto the field that is easy for photoresist and subsequent layers to cover. If the silicon is etched before the field implant, the field oxide can be made fully recessed, resulting in a nearly planar surface. Figure 15.7 shows the growth of a local oxide along with a cross-sectional view of a completed LOCOS structure. The process leaves a characteristic bump on the surface, followed by a gradually narrowing oxide tail into the active area. The structure is called a *bird's beak* for obvious reasons. The bump, or bird's head, is particularly pronounced in recessed structures.

The purpose of the thin pad oxide layer under the nitride is to reduce the stress that occurs in the silicon substrate during oxidation. This stress is due to the mismatch of the thermal expansion coefficients of the substrate and the nitride and due to the volumetric increase of the growing oxide. At high temperature, viscous flow of the oxide greatly reduces the stress. A great deal of work has gone into optimizing the thicknesses of the oxide and nitride layers. If the stress exceeds the yield strength of silicon, it will generate dislocations in the substrate. A thicker pad oxide will lower the stress in the substrate. The minimum pad oxide thickness that can be tolerated without dislocation formation is about one-third the thickness of the nitride [4]. This defect protection must be traded off

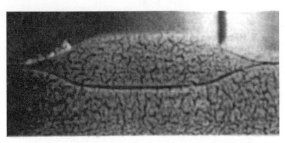

(A) Standard LOCOS—800Å Nitride/300Å Oxide

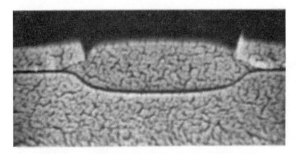

(B) LOPOS—2000Å Nitride/400Å Poly/300Å Oxide

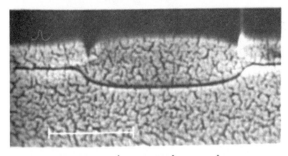

(C) LOPOS—2000Å Nitride/400Å Poly/300Å Oxide

Figure 15.7   Cross-sectional scanning electron micrograph of a typical LOCOS isolation *(after Ghezzo et al. [8], reprinted by permission, The Electrochemical Society).*

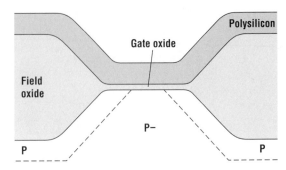

Figure 15.8   Cross-sectional schematic of an MOS transistor cut along its width, illustrating the origin of narrow channel effects.

against increased lateral encroachment of the oxide, which occurs due to lateral diffusion of the oxidizing species through the pad oxide. A nitride to thermal pad oxide thickness ratio of 2.5:1 produces a lateral encroachment or bird's beak, approximately equal to the thickness of the field oxide.

One concern of the LOCOS process is the white ribbon or Kooi nitride effect [5]. In this situation, a thermal oxynitride forms at the surface of the silicon under the edges of the nitride pad. White ribbon is caused by the reaction of Si_3N_4 with the high temperature wet ambient to form NH_3, which diffuses to the silicon/ silicon dioxide interface where it dissociates. When the effect is severe, the surface texture caused by these nitrides can be seen as a white ribbon around the edges of the active area. This defect leads to a reduced breakdown voltage in subsequent thermal oxides (such as gate oxides) in the active region.

The existence of the bird's beak has two important consequences from a device standpoint. Often the active region defines the edge of the device in at least one direction. Then encroachment reduces the active width of the device, reducing the amount of current that a transistor will drive. A more subtle effect is due to the field doping. The field oxidation causes the field implant to diffuse into the edge of the active region. Figure 15.8 shows a schematic of an MOS transistor along its width, immediately under the gate. If the transistor is narrow enough, the additional dopant from the field diffusion will increase the threshold voltage of the device, reducing its drive current. Known as the narrow channel effect, it is important in extremely dense technologies such as memories.

Various methods have been proposed to modify the LOCOS process to reduce the bird's beak length. The simplest is the use of materials other than a thermal oxide for the pad layer. A sandwich of thermal oxide and polysilicon is extremely effective in reducing lateral encroachment [6] and has become quite popular. The polysilicon absorbs the excessive stress resulting from the use of thin pad oxides and very thick nitrides. Typically about 500 Å of polysilicon is deposited on top of 150 Å of thermal oxide, followed by 1500 Å of nitride. This structure produces a bird's beak that is less than half the thickness of the field oxide [7]. Experiments describing the optimization of the poly-buffered process can be found in Ghezzo et al. [8] and Guldi et al. [9]. Although the length of the bird's beak is reduced, the poly-buffered process does not solve the problem of lateral diffusion of the field dopant. To gain the full benefit of such a process, the field oxidation should be done at high pressure in a wet ambient to minimize the lateral diffusion. White ribbon effects in poly-buffered LOCOS is a concern, however. Care must be taken when etching back the nitride/poly/oxide layers [10] to avoid it.

15.3 Trench Isolation

It became obvious during the 1980s that neither LOCOS nor any of its variations would be acceptable for ICs with transistor densities much greater than 10^7 cm^{-2}. This can be seen easily if one considers transistor size. Typical internal logic devices have a width-to-length ratio of about 4:1. For a 1-μm gate length, this corresponds to a width of 4 μm. In that case, a lateral encroachment of 0.3 μm per side might be undesirable, since it reduces the width of the transistor by about 15%, but it is probably acceptable. On the other hand, when the gate length is 0.18 m, this same encroachment is almost 85%, a clearly unacceptable, amount. It is now believed that the absolute minimum isolation distance for advanced LOCOS processes is about 0.8 μm from the edge of one N$^+$/p junction to another [11]. The ultimate limitation is not surface inversion or simple punchthrough, but a reachthrough effect known as drain-induced barrier lowering.

Many new isolation approaches have been developed around the idea of etching away part of the substrate and refilling it with an insulator. These can be divided into two classes. Shallow etches with small aspect ratios were considered first [12]. These techniques can be thought of as similar to recessed LOCOS, but a deposited rather than a thermal oxide is used to fill the field regions. After deposition, planarization must be done to remove the unwanted oxide from over the active regions (Figure 15.9). To be effective, an additional photomask step was initially, required for oxide removal. Due to problems of misalignment, this was difficult to achieve reliably. Furthermore it is difficult to prevent inversion along the etched sidewalls [13], since the field implant is normally done at close to normal incidence. In a CMOS process this inversion may cause excess leakage between source and drain.

Deep trench isolation modules (Figure 15.10) use trenches of fixed width. Typical dimensions are 65 nm to 0.5 μm, in width and 2 to 5 μm in depth, although trenches with depths of 10 μm have been demonstrated [14] (see Figure 15.11). The smaller trench widths are particularly attractive for memory applications. The process is fabricated by starting from a standard LOCOS structure. After nitride patterning the trenches are etched. Trench isolation puts extreme demands on the etch process. It must have smooth walls at no more than 85° with respect to the plane

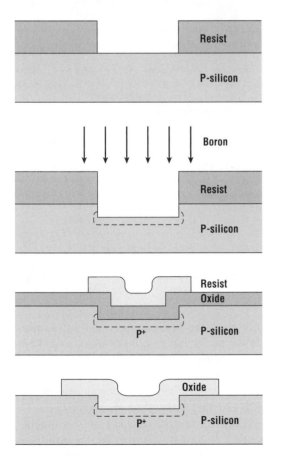

Figure 15.9 Simple shallow trench isolation process.

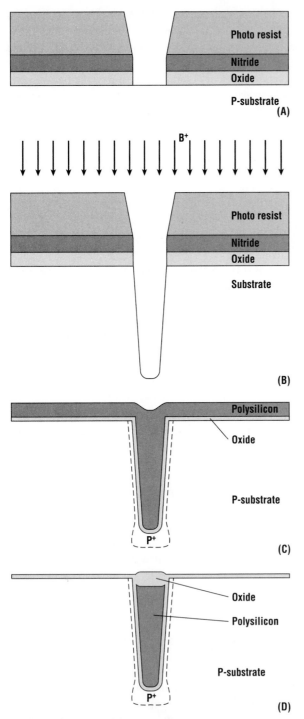

Figure 15.10 Deep trench isolation process schematic.

Figure 15.11 Cross-sectional scanning electron micrograph of an extreme aspect ratio deep isolation process *(after Rajeevakamur et al., ©1991 IEEE).*

of the wafer. More taper is highly desirable [15]. The trench etch is typically done by simultaneously depositing SiO_2 while etching silicon anisotropically. This creates a small cusp of SiO_2 at the top of the trench. The thickness of this cusp increases with time, producing the desired taper. The walls cannot undercut the mask, and should end in a rounded bottom. Sharp corners at the bottom of the trench will result in excess stress during oxidation, and ultimately defects in the oxide [16]. Next a field implant is done. Preventing sidewall inversion becomes even more difficult as the aspect ratio of the structure increases. An important feature of the implant, therefore, is that the beam be perpendicular to the surface of the wafer. As discussed in Chapter 5, new implanters have been developed to accomplish this [17]. Another, albeit much less popular, approach to solving this problem is the use of rapid thermal processing combined with planar dopant sources [18] or a doped CVD glass [19].

The implant is followed by a thin local oxidation. In applications in which the trench fill is to be used as a charge storage capacitor, it is common practice to use thin oxides to increase the capacitance. Finally, a layer of polysilicon or an oxide is deposited and etched back. If the layer is thick enough, it will fill the groove. Etching back to the substrate will leave this fill in place. A second thermal oxidation can be used to complete the process by oxidizing the upper part of the polysilicon in the groove. Often this is done as part of a LOCOS process. This minimizes leakage, it allows the fabrication of arbitrary isolation lengths (see discussion of Figure 15.12), and it offsets the junctions from the wall of the trench. To minimize lateral diffusion, the channel stop implant associated with the LOCOS may be done after oxidation [20].

It is important to emphasize that the trench must be of a fixed width for a deep trench process to work. This can be made to work with the minimum N^+/P^+ separation, but if the design also calls for larger separations in some areas, an additional isolation (such as a standard LOCOS) technique must also be used. The result is a complex process that is difficult to control. If the deposition process is not done properly, a void may form in the center of the trench, trapping some material. This may also occur if too steep an angle is etched into the substrate during the silicon etch. The void may pose a reliability problem. Deep trench isolation technology has demonstrated N^+-to-P^+ spacings less than 2 μm and N^+-to-N^+ spacings less than 0.5 μm. Referring to Figure 15.2, such an isolation technique will be adequate for device densities well in excess of 10^7 cm^{-2}.

Deep trench isolation has proven difficult to manufacture and it is difficult to integrate with random logic when arbitrary device spacings must be accommodated. The development of chemical mechanical polishing (CMP) has made previously rejected shallow trench isolation (STI) a viable process, since it can remove the excess deposited oxide without a lithography step. As shown in Figure 15.12, the process begins with a pad oxide of 100 to 150 Å, followed by a layer 1500 to 2000 Å of LPCVD Si_3N_4. Next a field is patterned and the nitride, oxide, and silicon are etched. Typical etch depths are about 0.5 μm. Trench sidewalls are etched at 75 to 80°. If desired, a field implant can then be done now to prevent inversion under the trench, or it can be delayed until after the polishing step is sufficiently high energy is used. Next a thin (150 to 200 Å) layer of SiO_2 is grown thermally to reduce

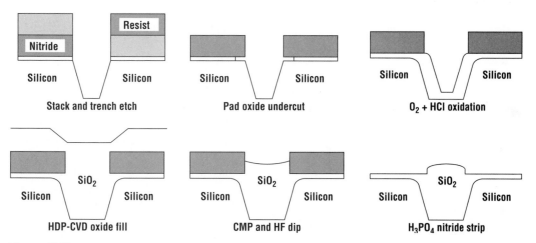

Figure 15.12 Schematic of a shallow trench isolation module (*after Chaterjee et al., used with permission, APS, 1997*).

the etch damage on the sidewalls and round off some of the corners. A 0.9- to 1.1-μm layer of SiO$_2$ is then deposited, usually by high density PECVD, and CMP is used to remove the excess oxide. The nitride serves as a polish stop for this step. Finally the nitride is removed and the pad oxide stripped in HF [21].

The integration of STI presents a number of challenging problems. Some of these problems are associated with the upper corner. Typically in an MOS transistor, the gate polysilicon stripe extends onto the field oxide to ensure a separation between the source and drain. If the corner of the STI is too sharp, the trench sidewall will invert (due to field concentration), leading to excess subthreshold leakage. This is especially a problem if the CMP is overdone, that is, if the top of the planarized oxide is below the top of the silicon [22]. To avoid this problem, the trench walls must be properly tapered and the top corner must be rounded. To achieve the desired rounding, the pad oxide is selectively removed by undercutting the nitride layer, then oxidized using high oxidation temperatures (~1100°C) and/or an ambient containing HCl. CMP dishing leads to a thinning of the field oxide [23] and may lead to a design rule for the maximum isolation distance and/or the use of dummy active areas [24]. The trench may also act as a gate insulator for a lateral parasitic device, shorting isolated junctions [25].

15.4 Silicon-on-Insulator Isolation Techniques

The ideal method of device isolation would be to completely encase each device in an insulating material. Several techniques exist for doing this in silicon. Generically they are called silicon on insulator (SOI). All of these methods have traxitionally suffered from problems related to defect density. For that reason SOI technologies in silicon have traditionally been relegated to small markets such as radiation-hardened devices, where extreme isolation is required. In some measure, this has changed in the last few years. Several techniques have improved dramatically, and ULSI applications have been demonstrated. Furthermore, the use of a buried insulator can reduce device parasitic capacitance, increasing circuit speed. The most promising method, SIMOX, has been presented in Chapter 5 and so will not be repeated here. Instead we will concentrate on several other processes.

One of the first SOI methods developed is called dielectric isolation (DI). The DI process sequence, shown in Figure 15.13 [26], was developed to build high voltage telecommunication ICs that required electrically isolated bidirectional switches. It has since gained popularity for other high voltage and radiation-hard digital applications. Deep grooves are first etched in the surface of the wafer. The wafer is oxidized and a very thick (>200-μm) layer of polysilicon is deposited. If substrate contact is needed in some islands, windows can be opened in the oxide before deposition. The deposition can be done by conventional CVD or by using a molten silicon spray deposition (MSSD) process [27]. The wafer is turned over and mechanically ground until the grooves penetrate through the wafer. Finally, the wafer is chemically polished and devices are fabricated in the isolated islands.

Dielectric isolation has several severe drawbacks. The wafers are not as planar as normal starting material. This not only impacts further processes such as lithography, but also results in varying silicon island thicknesses across the wafer. Wafers made using the DI process are expensive. Typical costs are well over $100

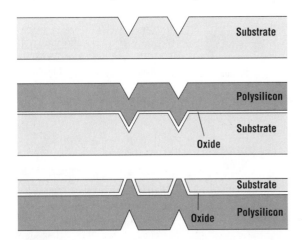

Figure 15.13   Dielecric isolation (DI) process for forming silicon on insulator.

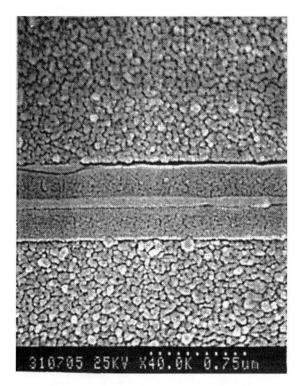

Figure 15.14 Wafer bonding method for achieving silicon on insulator *(after Quenzer and Benecke, reprinted with permission, Elsevier Sequoia S.A.).*

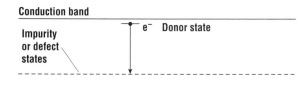

Figure 15.15 Deep levels necessary for forming a semi-insulating substrate.

per 100-mm wafer, before any device fabrication. Finally, if KOH is used to etch the v-grooves, the isolation density is not very large.

Wafer bonding is an alternate approach to producing SOI wafers (Figure 15.14). In this process, two wafers are pressed together at high temperature until they fuse [28]. Alternatively, the wafers can be fused at low temperature by anodic bonding [29, 30]. If the wafers are oxidized before bonding, a layer of oxide remains at the center of the fused wafer. The wafer can be ground back down to thicknesses of 2 to 3 μm using standard grinding and polishing techniques. If thinner layers are required, additional processing can be done to produce submicrometer semiconducting films on top of the oxide. Devices can be isolated with a simple etch process that produces single-crystal islands on top of the insulating oxide. Also, as with DI isolation, wafer bonding suffers from high cost, but can produce high density, well-isolated structures in CZ-grade silicon.

15.5 Semi-insulating Substrates

There are far fewer papers considering device isolation in GaAs technologies than there are for silicon isolation. Due to the availability of semi-insulating substrates, one can easily achieve high density, planar isolation. The only remaining question is how to separate the conducting islands from each other. The fact that a semi-insulating substrate is used in most GaAs ICs means that the devices are inherently radiation hard. A significant market for GaAs ICs involves defense and space-based applications that demand such hardening.

Most of the early semi-insulating GaAs material was made by chromium doping. Both chromium and oxygen have an energy level very close to the center of the GaAs band. If chromium is added in high concentrations (about 10^{17} cm^{-3} in older material, now more commonly in the mid-10^{16} range) to the boule as it is grown, semi-insulating wafers will result. The chromium adds discrete states near the center of the band (Figure 15.15). Consider what happens if a single donor atom is added to a semi-insulating wafer with no other dopants. Normally, the dopant atom will ionize and the electron will go into the conduction band, even though the conduction band of the crystal is a higher energy state. The probability of ionization is nearly one at room temperature for most dopants because of the large number of empty states near the conduction band edge as opposed to the single state at the donor atom. For chromium-doped GaAs, much lower energy states exist at the chromium sites. The sites will therefore ionize, accepting the donor atom's electron. Since the states are discrete and widely spaced, the electron cannot move. The crystal is therefore semi-insulating and the carrier concentration is approximately the intrinsic carrier concentration. Resistivities of 10^8 Ω-cm have been reported. Not until the dopant concentration approaches that of the chromium will there be significant conduction in the crystal. Conduction will also occur when the injected carrier density exceeds the trap concentration. This may occur under illumination, or in GaAs metal semiconductor field effect transistors (MESFETs) in a process known as *sidegating*. The effect is seen as a reduction in the current of a device if an adjacent device is turned on. Sidegating is a problem for high density GaAs ICs.

Many GaAs technologies, therefore, begin by producing a thin conducting layer on top of the semi-insulating substrate. Device isolation can be achieved by simply etching through the conducting layer, leaving islands of semiconducting GaAs. This simple method is known as *mesa isolation*. It has the disadvantage that the surface is no longer planar. To avoid this problem, many GaAs technologies use ion implantation. This method implants the field regions of the wafer with hydrogen or another ion. A 10^{13}-cm^{-2} dose and 100-keV energy are typical implant parameters. This implant is sufficient to severely damage the lattice, producing resistivities as high as 10^7 Ω-cm. The surface is also quite planar. A drawback of proton implantation is that the crystal will be regrown if any further processing is done at temperatures above 350°C. Proton implantation must therefore be done near the end of the process.

Semi-insulating substrates are not produced in silicon. Due to the smaller bandgap of silicon, it has a maximum resistivity of less than 10^6 Ω-cm at room temperature. Many of the noble and transition metals have states near the center of the silicon gap, but their diffusivities are very large (10^4 μm^2/hr at 800°C and 10^6 μm^2/hr at 1100°C for gold). As a result, it is very difficult to keep the impurity in the substrate from diffusing into the active region during processing. If the impurity reaches device junctions, enormous increases in leakage will result, since the states that render the wafer semi-insulating are also very efficient recombination centers. Semi-insulating silicon has been grown epitaxially using oxygen diluted in argon along with silane [31].

In many GaAs applications, chromium doping has been replaced by nonstoichiometric liquid-encapsulated Czochralski (LEC: see Section 2.4) wafers. Figure 15.16 shows the resistivity of the GaAs produced this way as a function of the arsenic mole fraction in the melt [32]. For compositions close to stoichiometric, the wafers are semi-insulating with bulk resistivities of 10^7 to 10^8 Ω-cm. The semi-insulating property is believed to be due to the presence of mid-10^{16}-cm^{-3} deep donor levels called *EL2 sites* [33]. The atomic structure of these sites is believed to be related to arsenic atoms on gallium sites (As$_{Ga}$). Although these defects introduce a large etch pit density in undoped semi-insulating wafers, they do not seem to affect IC yield. The B$_2$O$_3$ plug used in LEC growth also tends to getter silicon impurities that would otherwise dope the GaAs substrate n-type. The EL2 center has the interesting property that when illuminated by visible or near-IR light while at low temperature, the defect state is transformed into a higher energy metastable state [34]. This second state has very different properties than the original defect. High temperature hydrogen anneals may be used to remove EL2 centers near the surface of the wafer.

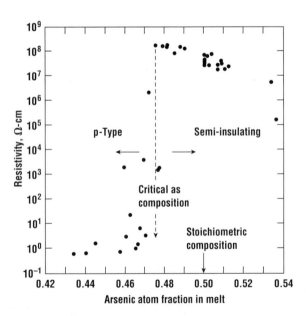

Figure 15.16   The resistivity of GaAs substrates as a function of the melt stoichiometry *(after Ferry)*.

One interesting comparison between GaAs and silicon isolation methods is that GaAs technologies do not generally use low dielectric constant insulators, whereas they are emphasized in most silicon technologies to minimize parasitic capacitance. Many GaAs technologies are designed for microwave applications. Then, it is not the absolute capacitance that is critical but the characteristic impedance of the interconnect. This must be impedance matched to the devices to ensure the maximum power transfer. To achieve this characteristic impedance, the GaAs wafers are often thinned after processing, and a ground plane is deposited on the back of the wafer. Through-wafer vias are used to contact the ground plane.

15.6 Schottky Contacts

Later sections of the book will deal with the fabrication of specific transistor structures. To efficiently connect semiconductor devices to the outside world, contacts must be made between the semiconductor and metal lines. Two types of contacts are often used: ohmic contacts and rectifying or Schottky contacts. In ideal ohmic contacts, the current varies linearly with the applied voltage. It is also implicitly assumed that ohmic contacts have low resistance. To transfer as much current as possible from the device to charge the various capacitances of the circuit, the contact resistance must also be a small fraction of the device resistance. Schottky contacts, however, should act as perfect diodes. They should have a very low resistance in the forward direction and an infinite resistance in the reverse direction. They should have a well-defined and perfectly reproducible "on voltage." The reader should be aware that although semiconductor technologists often use these descriptions for metal-to-semiconductor contacts, real contacts are neither perfectly ohmic nor perfectly rectifying. If due care is taken in the fabrication, however, good approximations of the ideal can be achieved. This section will deal with the fabrication of Schottky contacts. The next section will discuss process modules for making ohmic contacts.

Schottky barrier contacts have a variety of applications related to voltage clamping and controlled diode drops. It has the advantage of turning on (and off) more quickly than a p–n diode, and the turn-on voltage can be selected during the fabrication sequence. Figure 15.17 shows a typical application of a Schottky diode in a bipolar silicon technology. The base and collector of an NPN bipolar transistor are clamped through the Schottky diode. The turn-on voltage of the diode is set to a few tenths of a volt less than the turn-on of the base collector diode. If the base is suddenly driven higher than the collector, the base voltage will be limited to the Schottky diode on voltage. This prevents the transistor from being driven into a strong saturation condition from which it would take a long time to recover.

Figure 15.18 shows a band diagram for a Schottky diode on a p-type substrate [35]. ϕ_m is the metal work function, the voltage necessary to remove an electron from the surface of the metal; χ is the electron affinity, the voltage necessary to remove an electron from the conduction band minimum of the semiconductor; and ϕ_s is the voltage difference between the Fermi energy of the semiconductor and the vacuum level. When the metal and the semiconductor are brought into contact, charge will flow from one to the other until an electric field builds up that is sufficient to prevent further current flow. This occurs when the barrier at the interface reaches ϕ_{ms}. If a Schottky contact is needed, the metal must be chosen to create such a barrier. If, for example, this metal were chosen for a Schottky contact to a lightly doped n-type semiconductor, electrons in the metal, which has a higher Fermi energy than the semiconductor, flow readily into the semiconductor. This leads to an electric field that will assist the movement of

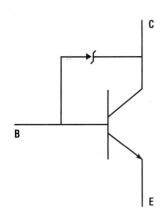

Figure 15.17 Schottky shunted bipolar transistor used for nonsaturating bipolar logic.

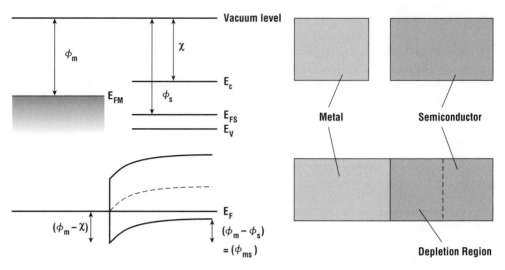

Figure 15.18 Band diagram for an ideal Schottky contact: before contact (right) and after contact (left).

carriers across the interface. Then a Schottky diode will not be formed. Ideally, if $\phi_m > \phi_s$ and the semiconductor is n-type, or if $\phi_m < \phi_s$ and the semiconductor is p-type, a Schottky diode will result.

The current across a Schottky diode is determined by thermionic emission. That is, it is determined by the fraction of carriers that has sufficient energy to surmount the barrier. It is given by the equation

$$I = I_o\left(\exp\frac{qV}{nkT} - 1\right)$$

(15.7)

where

$$I_o = RT^2A\,\exp\left[-\frac{\phi_{ms}}{kT}\right]$$

(15.8)

In this equation A is the area and n is the ideality factor of the diode; R is Richardson's constant, which has values of 110 and 32 A cm^{-2} K^{-2} for n- and p-type silicon, respectively, and 8 and 74 A cm^{-2} K^{-2} for n- and p-type gallium arsenide. It is clear from this equation that the current density depends strongly on the barrier height.

Figure 15.19 shows the measured barrier heights for both silicon and gallium arsenide as a function of the metal work function. The experimental points should fall on a line with unity slope. The offset would be due to the different electron affinities and (potentially) Fermi levels for the two semiconductors. Instead of a unity slope, a slope of about 0.25 is seen for silicon and about 0.1 is seen for GaAs. From an empirical point of view, we need to modify Equation 15.8 to read

$$I_o = RT^2A\,\exp\left[-\frac{\phi_b}{kT}\right]$$

(15.9)

where ϕ_b is the effective barrier height.

To understand this discrepancy, one needs to go back to the basic assumptions in the model. In presenting these equations a perfect interface was implicitly assumed. In most processes that make

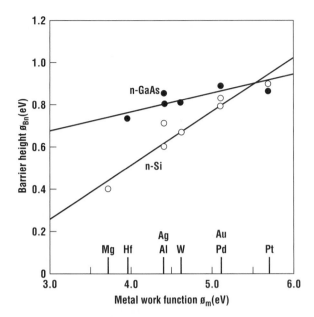

Figure 15.19 Experimentally measured Schottky barrier heights for silicon and GaAs *(from Sze, reprinted by permission, Wiley).*

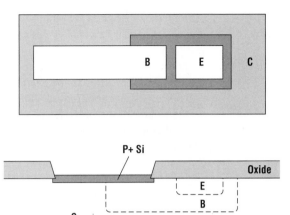

Figure 15.20 A PtSi Schottky used to shunt the base and collector of a bipolar transistor.

Schottky diodes, interface is not formed under ultra-high vacuum conditions. Rather, an active region of the semiconductor is first exposed, for example, by etching a hole in an insulating layer. Then the wafer is placed in a vacuum system to deposit the metal layer. During this time the surface of the wafer will oxidize and receive a coating of carbon and other contaminants from the residual pump oil vapor in the chamber and from the air in the room. To improve the process, many systems can presputter the surface of the wafer before deposition. This, however, does not ensure that a perfect surface is achieved. The sputtered species are not volatile; many will redeposit on the surface of the wafer. Furthermore, material will be sputtered wherever the plasma touches a surface. Thus, it would not be surprising to find Fe, Ni, and Co on the surface of the wafer. The plasma also damages the surface of the wafer. Both contamination and damage have an impact on the Schottky diode. Even without these defects, the semiconductor surface itself represents an important "imperfection" from the standpoint of the electronic structure. Unless the metal can be epitaxially grown on the surface of the semiconductor, in such a way that there is no charge accumulation at the interface, the barrier height and the metal semiconductor work function may be considerably different.

PtSi is commonly used as a Schottky contact for lightly doped n-type silicon. The cross section of a Schottky clamped bipolar transistor is shown in Figure 15.20. The diode is formed by first etching holes down to bare silicon. A dip in dilute HF is normally done just before deposition. A platinum layer is then sputtered onto the surface of the wafer. Typical thicknesses are 300 to 600 Å. It is critically important in obtaining a good surface morphology that the silicon surface be clean before deposition [36], and that little oxygen and water vapor be present during the deposition step. Recall from kinetic theory that the flux of residual gases is given by

$$J = \frac{p}{\sqrt{3kTm}} \qquad \text{(15.10)}$$

For nitrogen at 10^{-6} torr, the bombardment rate is 0.17 atom Å^{-2} $\sec^{-1}$. Thus, if one assumes that all of the atoms that strike the surface of the wafer are going to stick, a monolayer of atoms will deposit within 1 sec. It is therefore impossible to ensure even a reasonably clean surface unless ultrahigh vacuum methods are used. Platinum will reduce a thin oxide, however, breaking it up in the anneal. As a result, a standard presputter is sufficient to ensure that a uniform reaction will proceed. It is also critically important to ensure that no oxygen leak exists into the sputtering system. The deposition rate is normally kept low to achieve good process control and uniformity. A small air leak or a significant water partial pressure in the chamber can produce highly oxygen-contaminated films. This leads to high resistivity and poor reproducibility in the diode characteristics.

After deposition, the wafer is annealed at about 550°C in a furnace to allow the film to react with the silicon. Toward the end of the process, the furnace is switched to an oxidizing ambient and a thin layer of SiO_2 is grown over the platinum silicide. The platinum itself does not oxidize. After reaction, the wafer is immersed in a dilute solution of aqua regia at 85°C to selectively remove the unreacted platinum. The oxide over the silicide protects it during this etch. Finally, the wafer is dipped in HF to remove the oxide. After silicide formation, the wafer temperature must not be allowed to exceed 800°C, otherwise the morphology of the PtSi layer quickly degrades. Since the silicon surface is consumed during the reaction process, a clean interface results between the semiconductor and the wafer. Typical PtSi to lightly doped n-type Si barrier heights are 0.85 ± 0.05 V.

Schottky diodes are more widely used in GaAs technologies than in silicon. The fundamental reason is that GaAs does not have the exceptional oxide that silicon has. Thus, the basic GaAs field effect device is not the MOSFET but the MESFET. For the MESFET to work well, the gate electrode must make a Schottky contact with the channel. The pinchoff voltage of the FET depends directly on the barrier voltage of the Schottky diode. Furthermore, all of the metals used in GaAs must be fully deposited; they are not reacted with the substrate. Figure 15.21 shows a typical cross section of a GaAs MESFET. Commonly used gate electrodes WSi and WSi/W, although aluminum, chromium, titanium, and molybdenum have also been used. Also, CoAl alloys can be produced that have good thermal stability and Schottky barrier heights as large as 0.9 V [37]. Deposition of WSi_x can be achieved by using CVD [38], by sputtering from a composite target, or more commonly by cosputtering W and Si [39]. The optimal composition appears to be $WSi_{0.4}$ [40]. Gallium does not diffuse rapidly in WSi unlike many metals, preventing degradation of the interface; WSi is also attractive for this application for the same reason polysilicon is attractive for MOSFETs in silicon. A refractory metal, WSi will tolerate subsequent high temperature processing [41]. This means that the gate can be used as an implant mask for the source and drain, self-aligning these diffusions to the gate (see Chapter 17). An 0.8% atomic alloy of Al in W is also an attractive gate material for GaAs MESFETs. It is thermally stable, has a larger barrier than W [42], and has lower resistance than WSi [43].

As shown in Figure 15.19, the barrier for a Schottky diode in GaAs is about 0.8 eV and is nearly independent of the metal. Many compound semiconductors have similar behavior. This represents a serious limitation for MESFET circuits. For example, InP-based materials, have several electronic advantages over GaAs, but there is no suitable material that will provide a Schottky barrier larger than 0.4 eV [44], virtually precluding their use in MESFETs. Even in GaAs, the gate cannot be forward biased much beyond 1 V or significant gate leakage will result. This leakage increases power consumption and reduces performance. This greatly limits the potential for complementary circuits in GaAs. The Schottky barrier also depends strongly on the interface quality. A poor interface results in a larger ideality factor and a reduced barrier height. It has been shown that a sputter cleaning of the interface can greatly reduce the nonuniformity of the pinchoff voltage of GaAs MESFETs [45].

Several methods have been attempted to increase the Schottky barrier height for GaAs. One technique is to use a narrow band of doping called a *charge sheet*

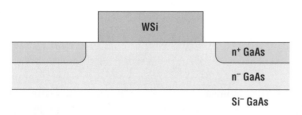

Figure 15.21   Typical GaAs MESFET structure.

or *delta doping* at the surface of the semiconductor [46]. The carriers will be depleted, but the remaining ionic charges are sufficient to shift the barrier height by as much as a few tenths of an electron-volt [47]. Another method is to use a monolayer of alternate materials [48] or heterostructures with a wide bandgap material (AlGaAs) between the GaAs and the WSi. This latter technique produces a barrier height of 1.0 to 1.2 eV [49]. Either method requires the use of MBE or MOCVD growth.

15.7 Implanted Ohmic Contacts

Often one is interested in making low resistance ohmic contacts to the semiconductor. The specific contact resistance can be defined as

$$R_c = \left[\frac{\partial J}{\partial V}\right]^{-1}_{V=0} \qquad (15.11)$$

From Equation 15.9, for thermionic emission this becomes

$$R_c = \frac{k}{qRT} \exp\left[\frac{\phi_b}{kT}\right] \qquad (15.12)$$

Since the barrier height depends logarithmically on the dopant concentration in the substrate, R_c should decrease linearly as the doping concentration increases. This is true up to doping concentrations of 10^{18} to 10^{19} cm^{-3}. For more heavily doped substrates, R_c depends strongly on the doping concentration, falling by orders of magnitude. The reason for this behavior is that thermionic emission over the energy barrier is no longer the dominant transport mechanism.

The width of the depletion region at zero applied bias can be calculated using

$$W_D = \sqrt{\frac{2k_s \varepsilon_o \phi_b}{qN_A}} \qquad (15.13)$$

For heavily doped substrates this width becomes sufficiently small that carriers can tunnel through the barrier (Figure 15.22) rather than be limited by thermionic emission over the barrier. The specific contact resistance in this region can be approximated by

$$R_c \approx A_o \exp\left[\frac{C_2 \phi_b}{\sqrt{N_D}}\right] \qquad (15.14)$$

where

$$C_2 = \frac{4\pi}{h}\sqrt{m_n^* k_{si} \varepsilon_o} \qquad (15.15)$$

h is Planck's constant, and m_n^* is the effective mass of the electron in the semiconductor. In contacts to p-type substrates m_n^* is replaced by m_p^*, the effective mass of the hole.

From Equations 15.14 and 15.15, it is apparent that one should make the substrate doping as large as possible to reduce the contact resistance. Alternately, it is possible to make the barrier height as small as possible. This approach

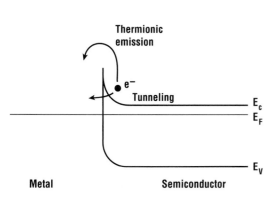

Figure 15.22   Two carrier transport mechanisms typically found in metal semiconductor contacts.

has as its limitation the fact that both polarities are contacted in many technologies. One or the other will have a large barrier height. To form ohmic contacts then, very heavily doped junctions must be made in one or both types of semiconductor. Generally, the higher the concentration, the lower the specific contact resistance. Values approaching 10^{-8} Ω-cm^2 are often obtained in semiconductor fabrication.

Most of the work in silicon ohmic contacts has been done with aluminum. Aluminum has the property that it will react readily with SiO_2 to form a thin layer of Al_2O_3, which promotes adhesion between the silicon dioxide and the aluminum. This reaction also can assist in the formation of an ohmic contact.

To form a low resistance contact with silicon, most aluminum metallization processes include as the last step a low temperature anneal or sinter. Typically done at 450°C, this sinter has several effects. As the aluminum reduces the native oxide, it breaks it up, diffusing the oxygen atoms back into the bulk of the aluminum and allowing fresh Al to diffuse to the metal semiconductor interface. Figure 15.23 shows the diffusion rate of aluminum through aluminum oxide. For a 30-min sinter at 450°C, the aluminum will penetrate about 10 Å into the aluminum oxide. Just as with Pt deposition, it is important to ensure that the oxide at the contact interface is as thin as possible. Contacts are often dipped in a very dilute HF solution immediately before the wafers are loaded into the deposition chamber. Once in the chamber, a sputter clean is often carried out immediately before deposition.

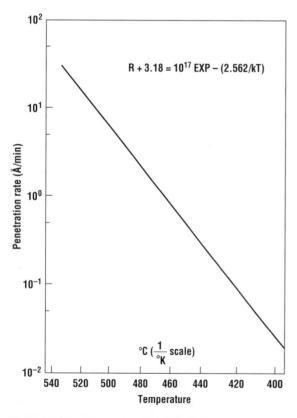

Figure 15.23   Aluminum penetration rate through aluminum oxide as a function of temperature *(after Wolf [1], reprinted by permission, Lattice Press).*

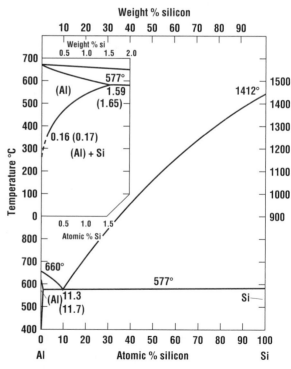

Figure 15.24   Phase diagram of Al/Si. Inset shows the low concentration region.

While the sinter step is essential to obtaining low resistance contacts, it has undesirable side effects. Figure 15.24 shows a phase diagram for the Al/Si system. As the inset shows, if pure aluminum is heated to 450°C and a source of silicon is present, silicon, will begin to dissolve in the aluminum until it reaches a concentration of about 0.5%. If the sample is heated to 525°C, silicon will dissolve to about 1%. The source for this silicon, of course, is the wafer. Although this does not seem to be a large amount of silicon, the metal line is an enormous sink. Once dissolved in the aluminum, silicon will diffuse rapidly along the grain boundaries, moving silicon away from the contact. The aluminum in turn moves into the holes to fill the voids left by the silicon. Spikes of aluminum can penetrate into the wafer as deep as 1 μm. If the spike penetrates an electrical junction, the result is a short circuit. Figure 15.25 shows a sequence of diagrams that depicts junction spiking.

Several methods have been used to reduce spiking liability in semiconductors. The simplest is to ensure that deep junctions are employed. This is not always desirable from a device standpoint. The next solution is the use of dilute alloys of aluminum/silicon instead of pure aluminum. If the concentration of silicon in the aluminum exceeds the solid solubility at the anneal temperature, little spiking will occur. For 500°C anneals typical Si alloy concentrations range from 1 to 2%. The use of Al/Si alloys also presents problems. The first is silicon condensation. The fact that the dissolved silicon concentration exceeds the solid solubility at low temperature means that there will be a driving force to condense silicon nodules, which are typically 0.5–1.5 μm in diameter. These nodules form between the aluminum grain boundaries and at the metal/semiconductor interface. Since aluminum is an acceptor in silicon, these nodules are heavily doped p-type. If the contact is between n-type silicon and metal, this p layer can significantly increase the contact resistance, particularly if the contact size is comparable to the nodule. Furthermore, the existence of silicon nodules between the grain boundaries of the metal line represents a serious reliability concern. For narrow lines, the nodules approach the full cross-sectional area of the wire. When large currents are passed through the line, significant local heating can occur, eventually leading to a failure of the line.

For very shallow junctions (less than about 0.2 μm), silicon-doped aluminum is no longer effective. This can occur due to silicon condensation that lowers the silicon concentration in the metal. It can also be related to electromigration-induced contact failures. In this phenomenon, electrons moving under the influence of an imposed electric field accelerate to high energies and then collide with aluminum atoms. In some cases, the collision transfers enough momentum to the aluminum atom to drive it into the substrate. The effect is known to increase as the junction area decreases. To make reliable ohmic contacts to very shallow junctions, a barrier metallization must be employed. A thin layer of the barrier metal is deposited beneath the aluminum. This metal

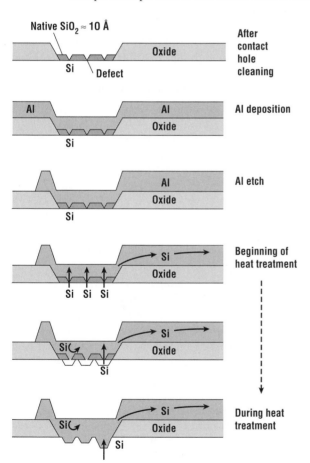

Figure 15.25   Cross-sectional diagrams of Al on Si contact formation process *(after Wolf [1], reprinted by permission, Lattice Press).*

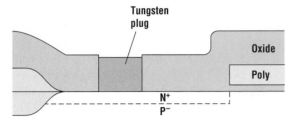

Figure 15.26   Tungsten plug structure.

must have the property that the diffusivity of silicon and the interconnect metal are both low at typical sintering temperatures. It must also be highly conductive and have good adhesion between the semiconductor and the metal. One of the first widely used barrier metal is an alloy of titanium and tungsten. Typically 1000 to 2000 Å of 10 wt% Ti/90 wt% W is first sputter deposited. The aluminum is then deposited without breaking vacuum. Several common improvements to this basic process include the use of PtSi or $TiSi_2$ under the TiW to improve the contact resistance. Whereas PtSi is formed as discussed previously, $TiSi_2$ can be formed in a furnace [50] or in a multistep process in a rapid thermal processor [51]. One disadvantage of TiW processes is that the material flakes off the walls of the sputtering system [52]. These flakes can reduce the IC yield. Refractory metal nitrides are now more widely used as barrier metals. The most widely used, is TiN, but TaN and WN are also used as barriers. These films can be deposited through reactive sputtering or through rapid thermal annealing of a deposited TiN layer in a nitridizing ambient. In this process, the Ti or TiW is sputtered in the presence of nitrogen. The nitrogen is incorporated in the grain boundaries and further reduces the diffusion rate. It is believed that this reduction in diffusivity is due to a passivation effect that dramatically reduces the grain boundary diffusion coefficient. Contacts made this way are stable up to temperatures of 500°C. Alternatively one can use amorphous metal nitrides to avoid grain boundary effects.

All sputtered barrier metals suffer from a common shortcoming for deep submicron devices. In these devices, the contact holes have large aspect ratios. In such a topology, sputtering will not produce a uniform deposit at the bottom of the contact. The barrier metal will be very thin near the edges. Furthermore, the barrier metal does little to improve the planarity of the contact. As a result, the deposited metal will also have poor step coverage going into and out of the contact. One contact method that has received considerable use is chemical vapor deposited (CVD) tungsten (Figure 15.26), as discussed in Section 13.8.

15.8 Alloyed Contacts

Unlike silicon, which uses implanted contacts almost exclusively, most ohmic contacts in GaAs and other compound semiconductors are alloyed [53]. To form an ohmic contact to n-type GaAs, the wafers must first be cleaned. A typical process is to rinse the wafer in organic solvents, followed by a deionized water rinse. Various metallizations have been used for ohmic contacts to n-type GaAs, including GeMoW [54] and $GeWSi_2Au$ [55]. The most common contact, however, is NiAuGe. In this process, the wafer is placed in an evaporator where 1000 to 1500 Å of a eutectic of 88 wt% Au/12 wt% Ge is deposited, followed by 100 to 500 Å of Ni. Layers of a refractory barrier metal and gold may be added to the top of the Ni to reduce the resistance of the metal lines. The ohmic contact is formed by annealing the wafer for 30 min at 450°C in an H_2/N_2 mixture. Contact resistivities of 10^{-6} Ω-cm^2 have been reported using this technique.

During alloyed contact formation, Au reacts with substrate Ga to form various alloys, leaving behind a large concentration of Ga vacancies. Germanium diffuses into the GaAs, occupying the Ga sites and doping the GaAs heavily n-type [56]. The specific contact resistance of ohmic contacts formed in this way depends inversely on the doping concentration in the lightly doped substrate [57] instead of depending on the doping concentration at the contact, which can be as large as 5×10^{19} cm^{-3}. It has been found that the germanium does not penetrate the surface of the contact uniformly. Instead, contact is made in small pockets of hemispherical radius r

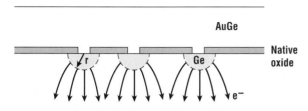

Figure 15.27 Schematic of ohmic contact to n-type GaAs showing the formation of small hemispherical contact regions and subsequent current crowding.

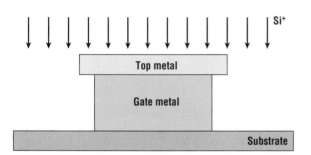

Figure 15.28 Cross-sectional schematic of a low resistance MESFET using ion implantation to reduce the contact resistance.

(Figure 15.27) that have been observed electrically [58] and by using TEM [59]. The specific contact resistance will have two components, one due to the tunneling contact resistance that depends on the Ge doping level in the GaAs (N_D) and one due to the spreading resistance of the Ge inclusions. The total specific contact resistance then is

$$R_c \approx A_o \exp\left[\frac{C_2\phi_b}{\sqrt{N_D}}\right] + D^2 \frac{\rho}{\pi r} \tag{15.16}$$

where D is the mean distance between the pockets and ρ is the resistivity of the substrate. This suggests that it is desirable to have a small separation between these inclusions. In practice, however, that is difficult to achieve.

Alternatively, the specific contact resistance can be lowered by increasing the dopant concentration near the contact (as opposed to at the contact). The most common method is ion implantation. Implanted alloyed contacts are now commonly used in GaAs technologies. One problem that arises in such technologies is simultaneously maintaining both Schottky and ohmic contacts. Figure 15.28 shows the cross section of a MESFET with ion-implanted contacts. It is important that the implantation be offset from the gate. If it extends under the gate, gate leakage will increase substantially. Although a deposited dielectric can be used, followed by an implantation directly into the contact, this approach leaves a high series resistance between the FET and the contact. As will be described in a later section, this seriously reduces the device performance. To achieve this self-aligned implantation for the contact formation, several schemes have been developed. These will be covered in Chapter 17.

Alloyed ohmic contacts to p-type GaAs have been studied in less detail, in part due to the low hole mobility of GaAs. The same general procedure is used, but the usual metallization employed is Au/Zn [60] with 5 to 15 wt% zinc. The metal structure may be evaporated or sputtered [61]. Since Zn is a group II material, it will dope the GaAs p-type when it resides on a Ga site. An ohmic contact may be formed in a furnace or in a rapid thermal processor [62]. As with n-type GaAs, specific resistances less than 10^{-6} Ω-cm^2 have been achieved. The thermal stability and the reliability of these contacts are somewhat suspect, however, due to the large vapor pressure of Zn.

An alternate approach to forming low resistance ohmic contacts to GaAs is to use indium in the deposited layers, usually close to the GaAs surface. The indium does not dope the GaAs. Rather, it forms Ga$_x$In$_{1-x}$As compounds, where x varies smoothly from 1 to 0 through the interface. These compounds have lower barriers and so improved ohmic behavior [63]. This type of contact is also more thermally stable than traditionally formed contacts [64].

15.9 Multilevel Metallization

This section will deal with the problem of making the connection between the contact and the bonding pad. Silicon and GaAs technologies have taken very different approaches in this area. In large part, this is because the tasks for which these technologies have been designed are so different. Most silicon technologies have been designed to achieve high levels of integration. Many GaAs technologies have been optimized for high speed analog operation, with only a secondary emphasis on density. These technologies often simply place a single layer of gold on top of the Ni to reduce the interconnect resistance. This layer is deposited directly on top of the wafer after the transistors have been fabricated. The most important criterion is that the interconnect have a controlled characteristic impedance that is matched to the device input and output impedance. As the number of transistors on digital GaAs circuits has increased however, digital GaAs-based technologies have had to use multiple layers of interconnect. In doing so, they have sometimes grafted the same metallization approaches that have been used in silicon technologies for many years onto the basic MESFET technology. For that reason, we will approach the metallization process primarily as technology independent.

A few comments about the metrics for the interconnect process module can be made. The most critical for digital circuits is capacitance. Digital switching speed is proportional to the capacitance on each node as

$$\tau \approx \frac{V_{swing}C_{node}}{<I_{drive}>} \tag{15.17}$$

Alternatively, for very long runs, the speed may be controlled by the RC time constant of the wire. In either case, controlling the node capacitance is essential.

As technologies have improved, an increasingly large part of this node capacitance is due to the wire. This can be capacitance between the wire and the substrate or the capacitance between the wires. As the wiring density increases, the wires get narrower. In principle, the wire-to-substrate capacitance should be proportional to the wire width. In practice, edge effects make the capacitance larger than what is predicted from a simple parallel-plate model. This occurs when the linewidth approaches the oxide thickness. Due to fringing capacitance, the reduction of metal linewidth does not reduce the line-to-substrate capacitance as much as one would expect. Furthermore, the wire-to-wire capacitance increases as the metal spacing decreases. This line-to-line capacitance is inversely proportional to the line-to-line spacing. The result of these considerations is that there is a minimum in capacitance for a particular density. There are two ways to accommodate the amount of interconnect required: decreasing the pitch or increasing the number of interconnect layers. Due to crosstalk noise, metal pitch cannot be decreased too severely [65]. As a result, the number of levels of interconnect on modern IC processing is constantly increasing. The state of the art for silicon is eight to ten layers of metal [66].

As lithography and etching have continued to improve, it has become possible to form lines whose spacings are much less than their thickness. In this situation, one could substantially lower the wire-to-wire capacitance by using thinner metal. As described in Chapter 13, this has driven leading-edge technologies to abandon Al in favor of Cu. A 0.40-μm-thick layer of pure Cu has the same sheet resistance as 0.65 μm of Al/Cu. Naturally, one must take severe precautions to ensure that Cu does not get into the silicon. Fabricators will dedicate not only process equipment and wafer-handling supplies, but often entire bays to copper work to minimize any chance of cross-contamination.

Example 15.1

A 0.25-μm metal line is 500 μm long. It is on top of 0.5 μm of SiO_2, and there are two more identical lines, one on each side. The line-to-line spacing is 0.25 μm. This space is also filled with SiO_2. Neglecting fringing effects, calculate wire-to-wire and wire-to-substrate capacitances for 0.40-μm-thick Cu and 0.65-μm-thick Al/Cu.

Solution

In either case, the wire-to-substrate capacitance is

$$C_{w-s} = \frac{(5.0 \times 10^{-2} \text{ cm}) \times (2.5 \times 10^{-5} \text{ cm}) \times 3.9 \times 8.84 \times 10^{-14} \text{ F/cm}}{5 \times 10^{-5} \text{ cm}}$$

$$= 8.6 \text{ fF}$$

For the aluminum wire,

$$C_{w-w} = 2 \times \frac{(5.0 \times 10^{-2} \text{ cm}) \times (6.5 \times 10^{-5} \text{ cm}) \times 3.9 \times 8.84 \times 10^{-14}}{2.5 \times 10^{-5} \text{ cm}}$$

$$= 90 \text{ fF}$$

For the copper wire,

$$C_{w-w} = 55 \text{ fF}$$

In microwave circuits, the important parameter is the wire impedance. To achieve efficient power transfer, the characteristic impedance of the wire must match the input impedance of the device. Two primary methods have been employed to achieve a controlled impedance and a stable ground connection. One common approach is called a *microstrip line*. To do this, the wafer is first thinned to achieve the desired characteristic impedance for the design wire width. Next, a layer of metal is deposited on the backside of the wafer. Holes are etched through the wafer to make contact to this solid ground plane. These through holes are patterned and etched from the backside of the wafer using an infrared aligner. A second method for fabricating controlled impedance lines uses a coplanar waveguide. Here two ground wires are run on either side of the signal wire. This does not require any backside processing, but it decreases the density of the IC and does not provide as stable a ground plane.

Most millimeter microwave ICs (MMICs) often have only a small number of transistors. Many discrete microwave components are also manufactured. In these technologies, wire density is not typically an issue. One layer of metallization is used to interconnect the devices. The Schottky gate metal is also often .used as a first layer for parallel-plate capacitors and, where necessary, the Ni/AuGe serves as the second layer of metallization.

In modern silicon technologies a distinction is drawn between global or "true" interconnect and local interconnect. Although the resistivity of aluminum is low enough that long runs do not degrade performance in most circumstances, the same cannot be said of polysilicon. Its resistivity is typically 10^{-4} Ω-cm. Silicides, which can be run directly on top of the polysilicon to shunt the poly resistance, still have resistivities that are much larger than aluminum (Table 15.1). A rough estimate of the delay can be made using a lumped capacitance model. In this model, the line is thought of as one side of a parallel-plate capacitor. Then

$$C = \frac{LWk_{ox}\varepsilon_o}{t_{ox}} \qquad \text{(15.18)}$$

Table 15.1   Properties of commonly used interconnected materials in both Si and GaAs technologies

Material	Bulk Resistivity ($\mu\Omega$-cm)	Melting Point (°C)
Au	2.2	1064
Al	2.7	660
Cu	1.7	1083
W	5.7	3410
PtSi	30	1229
TaSi$_2$	40	2200
TiSi$_2$	15	1540
WSi$_2$	40	2165

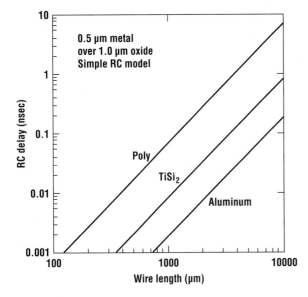

Figure 15.29   Simple lumped parameter model for the time constant of various interconnect materials.

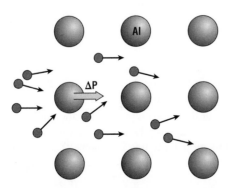

Figure 15.30   Momentum transfer between electrons and matrix atoms in an interconnect material.

and

$$R = \rho_{met}\frac{L}{W t_{met}} \qquad (15.19)$$

Therefore,

$$RC \approx \rho_{met}\frac{\varepsilon_{ox} L^2}{t_{met} t_{ox}} \qquad (15.20)$$

Figure 15.29 shows the lumped delay as a function of length for several metals running over 1 μm of oxide. The important feature to understand here is that the delay depends on the square of the length of the interconnect. Designers can therefore use moderate resistance materials to join adjacent transistors or even adjacent cells if they are close together. These materials, often called *local interconnects*, have several advantages over aluminum or copper that is used on the upper layers of interconnect. They can tolerate high temperature processing. Local polysilicon interconnect is often available for no additional processing cost, since it may already be used for the gates of MOSFETs or the emitter of a bipolar structure. Similarly modern devices often use silicides, as will be discussed in later chapters. These materials can also be used for local interconnect with little additional process complexity.

Aluminum-based alloys were the metallization of choice for silicon IC technologies for many years. One of their limitations, junction spiking, has already been described. A second limitation for Al and AlSi alloys is electromigration. Electromigration is the movement of conducting atoms as a result of momentum transfer from current-carrying electrons (Figure 15.30). This movement gives rise to a net flux of metal atoms.

If a nonzero divergence in the atomic flux exists anywhere along the line, metal atoms will be depleted or accumulated. Upon depletion, open circuits form. Hillocks are formed by accumulation. If these hillocks get large enough, adjacent lines or even overlying lines can be shorted together. For aluminum metallization systems in particular, this represents a serious reliability problem. A circuit that works well initially will wear out and eventually fail in the field.

A phenomonological description of the electromigration process is given by Black's equation [67]:

$$MTF = AJ^{-n} \exp\left[\frac{-E_A}{kT}\right]$$ **(15.21)**

where MTF is the median time to failure, J is the current density, n is a fitting parameter, typically about 2 [68], E_A is the activation energy, and A is a constant. The value of E_A depends on the diffusivity of the metal atom. At high temperatures (above 350°C), the activation energy for aluminum closely matches the self-diffusivity of aluminum. At lower temperatures where ICs operate, however, the activation energy for electromigration is smaller. At these temperatures grain boundary diffusion along the facets of the aluminum crystals dominates.

A common metal atom flux divergence mechanism is shown in Figure 15.31. At some point in the line, three grains of aluminum come together. One crystal face leads into this intersection and two lead out. As a result, atoms experiencing grain boundary diffusion will tend to diffuse more easily out of this point in the line than into it. Over time, a void will form. A second source of flux divergence is the presence of a temperature gradient. Because of the temperature dependence of the diffusivity, there will be a net movement of material away from the hotter portion of the wire. This will ultimately produce a void. Electromigration failures of this sort are often seen near bonding pads or contacts where a thermal sink exists. Electromigration in contacts is further aggravated by metal thinning as it goes over the topology.

Accelerated testing for electromigration can be done by holding the wafer at high temperatures and/or high current densities, measuring the MTF, and extrapolating back to use conditions. The result is a rule for the maximum dc current that can be allowed in a line to ensure a certain lifetime. A rather conservative number often used for Al/Si is 10^5 A/cm^2 for a 20-yr life. For pulsed dc stress, Black's equation can be modified by multiplying the current density by the duty factor [69]. Various models have been used to explain ac electromigration results. In the commonly used average current model, the same formula that is used for pulsed dc testing [70] is found appropriate for both unidirectional and bidirectional stressing.

An interesting question in electromigration is what happens when the linewidth approaches the grain size. Typical grains in aluminum metallization after the final sinter are of order 1 μm, so many advanced technologies have linewidths less than the grain size. Early reports suggested a sharp rise in lifetime due to this so-called bamboo effect [71]. On closer examination, however, it was found that while the median time to failure increased, there was still a sizable fraction of wires that failed in much shorter times. These failures were attributed to the statistical distribution of grain sizes in the wire and ultimately to the existence of an intersection between three crystals.

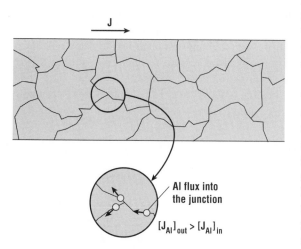

Figure 15.31 Typical point of flux divergence in an interconnect material.

To increase current-carrying capability, a different metallization system must be used. One commonly employed method is the addition of 1 to 4% copper to the aluminum. The copper atoms reduce grain boundary diffusion effects in aluminum and so greatly increase the activation energy for electromigration. Current densities as high as 10^6 A/cm^2 are considered reliable with many Al/Cu metallurgies. A difficulty with the addition of Cu is plasma etching. Normally, aluminum interconnect is etched in a chlorine plasma. Silicon also etches well in these systems due to the high vapor pressure of the various chlorosilanes. Copper chlorides, on the other hand, do not tend to form gases readily at temperatures below 175°C. Since most photoresists will not tolerate such high temperatures, the wafers cannot be heated sufficiently during the etch to remove all of the copper. As a result, most of the copper in the metallization is left behind after a plasma etch. Chlorinated plasma etch processes tend to corrode the remaining metal due to the presence of AlCl$_2$ etch residue and water vapor in the air. When an Al/Cu alloy is used, AlCu$_2$ is formed in the metal. This compound has a galvanic response with Al that drives the corrosion process. As a result, care must be taken to ensure that all of the chlorine has been removed from AlCu wafers that have been plasma etched. This is sometimes done with an inert plasma after the chlorine plasma is completed.

Another reliability concern for fine pitch metallization systems is stress-induced voiding [72]. This effect can open narrow slits [73] or wider features called wedges [74]. The effect arises due to the difference in the thermal expansion coefficients of the metal and the encapsulating dielectric layers and is aggravated at very small metal linewidths. As with electromigration, stress-induced voiding is significantly reduced by the introduction of small amounts of Cu in Al [75] or by using refractory metals instead of aluminum.

Several other metal systems have been proposed for use in silicon ICs. Layers of Al and a refractory metal such as Ti or TiN [76] have been suggested to reduce electromigration. The processing is more difficult, however. The most promising material at this point is tungsten. As mentioned in a previous chapter, the increasing contact aspect ratio is making the attainment of an adequate sputtering process extremely difficult. One possible solution is the use of contact plugs made from CVD tungsten [77, 78]. As discussed in Section 13.8, CVD tungsten is typically deposited in a vacuum system from WF$_6$ or WF$_6$/SiH$_4$ mixtures. It is possible to deposit the tungsten either selectively [79] or nonselectively and etch back the film as with trench isolation (Figure 15.26). It is found that tungsten deposition consumes some silicon; however, this can be reduced by the addition of SiH$_4$ during deposition. The process can be incorporated into a single multichamber system [80]. In this application, an underlayer of TiW or TiN is first sputtered to seal the source/drain surfaces from attack during the tungsten deposition [81] and to serve as an adhesion promoter between the metallization and the oxide [82]. A second possibility is to keep the tungsten layer and to use it as a layer of interconnect. This is very attractive, particularly for local interconnect. Since tungsten has a very low self-diffusivity at operating temperatures, electromigration is not a concern. It appears likely that the use of tungsten as the first layer of metallization will become increasingly more common. Selective nickel silicidation of polysilicon can also be used to form these contact plug structures [83].

15.10 Planarization and Advanced Interconnect

Planarized interconnect process modules have become very popular as the convergence of several factors has driven digital technologies into the nanoscale range. The most obvious factor is photolithography. As the imaging lenses have tended to higher numerical apertures, the depth of focus has decreased. Furthermore, not only does nonplanarized interconnect typically have the highest steps in the process, but these steps are often nearly vertical. One way of coping with the problem was to impose large-design rules for the minimum pitch (linewidth plus line space) that resulted in increasing the number of required interconnect layers. If the lowest level of interconnect had a pitch of x, the

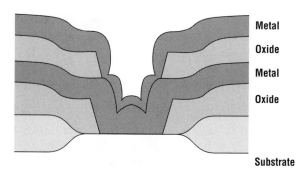

Figure 15.32 Nonplanarized plug structure consisting of a coincident contact and via.

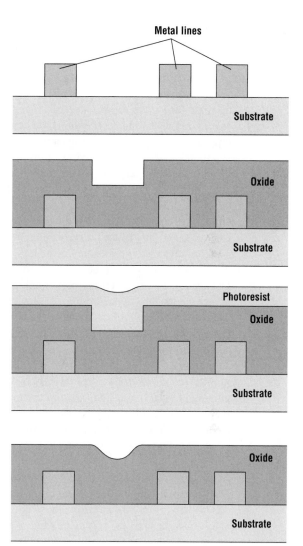

Figure 15.33 Etch-back planarization process.

second level might have a pitch of 1.5*x*, the third level might have a pitch of 3*x*, and so on. If finer pitch can be achieved with planarized interconnect modules, however, their additional process complexity may be more than offset by their ability to reduce the number of interconnect layers that are required.

At the same time, the step height often cannot be reduced. To do so would increase parasitic capacitance and/or resistance. The result is higher aspect ratio structures that are difficult to cover. Not only must the metal have sufficient step coverage into and out of contacts, it must also have good step coverage as it goes over metal lines on lower levels. There can be particularly severe topology as the second level metal exits a via down to the first layer of metal. In an extreme example called a *plug* (Figure 15.32), the via may be coincident with a contact from the first layer metal down to the substrate.

An early, popular planarization approach was the sacrificial oxide etchback method [84]. As shown in Figure 15.33, a thick oxide is first deposited over a patterned metal layer. Next, a layer of photoresist is spun onto the wafer. The wafer is then put into a plasma etch system. The etch ambient is a mixture of O_2 and CF_4 or another fluorinated species. The mixture is set to provide nearly identical etch rates of resist and oxide. Since the resist is thinner over the first metal lines, the oxide will be exposed here first. The etch is done until most of the resist has been removed. A second layer of oxide may then be deposited. Another etchback cycle can also be run if required.

Since these processes follow the resist profile, they are only locally planarizing. That is, they smooth out steps but do not produce a truly planar surface. For feature sizes greater than about 50 μm, the planarizing effect is lost [85], although temperature baking of the resist tends to improve this somewhat. The thickness of the oxide will also vary across the device. This means that a subsequent contact etch process must have a high selectivity and must not significantly etch laterally. This second criterion can be difficult to achieve if the etch process must also slope the sidewalls of the oxide. These local planarization techniques have been replaced by chemical mechanical polishing, as discussed in Chapter 11.

From the previous discussion it is apparent that for many integrated circuits, the interconnect capacitance is the dominant term in the delay equation. Decreasing the interconnect capacitance decreases the node capacitance and therefore increases the circuit speed. The most straightforward way to do this is to lower the dielectric constant (permittivity) of the insulator used between the metal layers, often called the interlayer dielectric (ILD). When an external electric field is applied to a simple dielectric, the valence electron cloud is displaced from the nuclei and core electrons. This creates a dipole that changes the electric field. The permittivity is a measure of the strength of the field caused by this dipole. The more tightly bound the valence electrons, the lower the permittivity. Of course, the lowest relative permittivity possible, one, is that of empty space. There are two approaches then to decreasing the dielectric constant of a film: (1) use a material with tightly bound electrons, and/or (2) use an open film structure with a large amount of void space. Both have been tried for advanced interconnect insulators.

The simplest approach is to modify CVD SiO_2, lowering the permittivity by increasing the localization of the electrons participating in the silicon bond. Replacing O with F does this. The higher the fluorine concentration, the lower the permittivity. Fluorine can be added readily in a CVD process by using NF_3 or some other fluorine-containing species in the reactor. As shown in Table 15.2, "pure" CVD SiO_2 has a permittivity of 4.1 to 4.2. This can be reduced to almost 3.2 by adding enough fluorine, but the etch rate becomes very large. The addition of F weakens the glass, noticeably reducing the hardness and changing the elastic modulus for F concentrations above 6%. At high fluorine concentrations films also absorb water, which can lead to metal corrosion [86]. This ultimately makes it difficult to stop the metal etch and to CMP the insulator controllably. Very high F-concentration films may be used with upper and/or lower "hard" layers consisting of SiO_2 without any fluorine; otherwise the lowest usable permittivity of fluorinated oxide is about 3.4.

The next step in reducing the permittivity of the ILD is to abandon SiO_2-based materials altogether. The next lowest permittivity material is diamond-like carbon (DLC). DLC can be deposited from CH_4 or other carbon-containing species in a PECVD reactor (Applied Materials's trade name for DLC is Black Diamond™). DLC has a permittivity ranging from 2.7 to 3.3 depending on deposition conditions. Diamond-like carbon does not adhere well to most layers, however, so a stack of adhesion layers must be used [87]. Typically this includes thin ($\sim$100-Å) layers of SiO_2 and SiO_x where $x < 2$. To further reduce the permittivity of DLC, one can again add fluorine to the film. Fluorinated carbon films are amorphous and can reduce the dielectric constant to a value as low as 2.0. Insulators with

Table 15.2   Effect of adding fluorine using C_2F_6 in a TEOS/O_2 PECVD process

		C_2F_6 **Flow Rate (slm)**				
		0.8	**2**	**3**	**4**	**5**
Si–F (atom%)	0	2.4	4.2	5.8	7.5	8.9
ε	4.2	3.92	3.67	3.45	3.28	3.19
Stress (MPa)	−176	−135	−68	−36	15	35
Etch rate (nm/min in 10:1 BHF)	110	310	492	915	1180	1350

The permittivity decreases linearly with F concentration.

permittivities in this range can also be formed using spin-on organic insulators such as some types of polyimide [88]. Fluorinated polyimide can be used for extremely low permittivity films [89]. Generically these very low permittivity ILDs have problems: low mechanical strength, moisture absorption, poor dimensional stability, low breakdown strength, increased leakage current, poor thermal stability, increased thermal expanison coefficients, low thermal conductivity, and outgassing leading to an effect known as via poisoning [90]. Polyimides also have problems with CVD tungsten, since the fluorine that is formed as a decomposition by-product can easily attack the insulator, leading to void formation and/or low breakdown strength.

The alternate approach to low permittivity insulators is the use of low density materials. In that case the dielectric constant is well described by a simple effective media approximation

$$\varepsilon = x \times \varepsilon_{\text{oxide}} + (1 - x) \times 1$$

where x is the filling factor. Thus an SiO_2 film with a filling factor of 0.3 has an effective permittivity of about 1.9. A simple way to form this type of film is to apply a slurry of SiO_2 particles in a liquid carrier. After spin-on application, the film is heated to form a gelatinous material (Figure 15.34) [91]. Next, higher temperature steps are used to drive off most of the solvent, and the film is annealed to set up the desired electrical properties. The result is sometimes called a silica aerogel (more properly, that is, a xenogel. Aerogels are dried supercritically at high pressure and show less shrinkage.) This type of film can be produced with up to 95% porosity ($x = 0.05$) resulting in a permittivity of 1.1 [92]. The structural stability of these films, however, is extremely questionable.

As suggested earlier, the interconnect capacitance may be lowered by using thinner layers of low resistivity interconnect. Silver has the lowest room temperature resistance, but it does not adhere well to SiO_2, Instead, it diffuses rapidly through it. Furthermore it has poor electromigration resistance. The low resistivity and improved electromigration resistance of copper [93] (about 10× the maximum current density of AlCu) have driven its use as an interconnect material. One of the major problems with copper is the lack of an adequate copper etch process. This has been solved through the use of the damascene process. The interlayer dielectric is first etched anistropically. Next a thin barrier metal such as Ti, Ta, TaN, or TiN is deposited [94]. Copper is then seeded and deposited to a thickness greater than the oxide recess, filling the trenches. Finally the excess copper is removed by CMP [95].

The simple damascene process has been extended to a dual damascene process by using the inlaid copper to form both interconnect lines and via fills. There are two ways to run a dual damascene process. As shown in Figure 15.35, the ILD is first etched down to the previous metal layer, the via is filled, and the excess removed by CMP. Next the interconnect recess is patterned and etched, and a second copper fill and CMP are done. Alternatively, both the via and line recess can be done before copper deposition. Often a thin Si_3N_4 etch stop is inserted in the ILD to mark the top of the via. This process has the disadvantage that the smallest feature, the via,

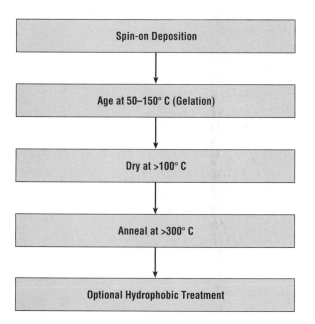

Figure 15.34   Process sequence for a typical aerogel film *(after Ramos et al.).*

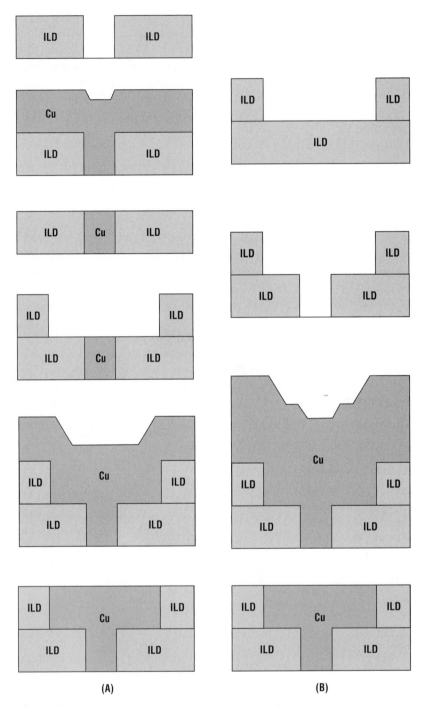

Figure 15.35   Two options for performing a dual damascene process: (A) single damascene; (B) dual damascene *(after Price et al., Used with permission. Thin Solid Films, 1997).*

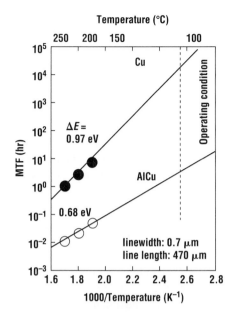

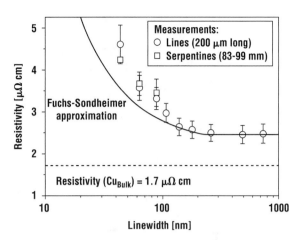

Figure 15.37 Example of the effect of linewidth on the measured resistivity of copper lines (*from Steinhogl et. al., reprinted with permission from Journal of Applied Physics. Copyright 2005, American Institute of Physics*).

Figure 15.36 Median time to failure of Cu and AlCu lines (*from Sun, reprinted with permission IEEE*).

must be patterned in the bottom of the interconnect recess. For small vias this can be extremely challenging.

Electromigration in copper is a somewhat less serious issue than it is in aluminum. Due to the higher melting temperature (and correspondingly lower self-diffusivity), the median time to failure (MTF) for copper is typically more than 100 times longer (Figure 15.36) [96]. This allows higher current densities, a feature sorely needed by the power requirements of modern integrated circuits. As with Al, one area where electromigration can be seen is the contact between copper and the refractory tungsten studs or the metal nitride liners. Diffusion in copper is primarily a surface effect unless the lines are clad top, bottom, and sides. As a result, the electromigration can depend on the linewidth, with smaller lines being more prone to electromigration failures than wide ones.

One of the current outstanding unsolved interconnect problems the effect of scaling on copper resistivity. The copper formed by electroplating is polycrystalline. The lowest resistivity is typically obtained after a low temperature post deposition anneal. When the line size is of order the grain size or less, the grain boundary stretches all the way across the line. Under these circumstances, the resistivity of the copper increases substantially (Figure 15.37) [97]. This is well into the regime of current capabilities and so represents a significant limitation for the technology. This is true not only on flat surfaces, but in contacts as well.

15.11 Summary

In this chapter, the process modules of device isolation, contact formation, and interconnection were presented. The simplest isolation techniques involve junction isolation. Various LOCOS-based methods have been widely used, but they suffer from lateral encroachment and incomplete isolation at small junction separations. Trench-based methods have become popular for submicron technologies. For GaAs technologies, nearly all device isolation is accomplished with semi-insulating substrates. Conducting islands can be created via proton implantation or mesa etching.

Contacts are divided into rectifying and ohmic. For rectifying contacts, the barrier height is a sensitive function of the nature of the metal/semiconductor interface. For ohmic contacts, achieving a low contact resistivity requires a large doping concentration at the metal/semiconductor interface. Most silicon technologies achieve heavy doping by implantation. Self-aligned silicides (salicides) have been developed to reduce the series resistance of shallow junctions in silicon. Many GaAs technologies use alloyed contacts, sometimes with an implantation, to achieve acceptably low specific contact resistivities.

High performance interconnect requires the use of low resistivity metal on top of low capacitance dielectrics. Aluminum alloys are the most widely used for silicon-based technologies, although copper is now beginning to replace it. Gold is generally used for GaAs technologies. CVD SiO_2 is the most commonly used dielectric, although lower permittivity films such as polyimide are being developed for future applications.

Problems

1. In some microwave applications, the collector of a bipolar transistor is heavily doped to allow the device to be run at large dc-bias currents. Assume that for the simple IC shown in Figure 15.1, the n-type collector concentration is 2×10^{17} cm^{-3}. If the maximum reverse bias applied to the base collector junction is ever 5 V, what is the minimum base-to-base spacing required to isolate the transistors?

2. If the field oxide is 5000 Å thick, what will the threshold voltage of the parasitic device be? Assume no interface states or metal semiconductor workfunction and a substrate concentration of 2×10^{15} cm^{-3}.

3. A LOCOS process produces a field doping concentration of 4×10^{16} cm^{-3}. It is used to separate two diffused regions. The maximum reverse bias that either region would ever experience is 5.0 V.
   (a) Apply this bias to both diffusions and find the minimum distance that these diffusions can be spaced.
   (b) If the field oxide is 0.6 micron thick and the oxide is ideal (no work function, no charges, no interface states), what is the parasitic threshold voltage?

4. A LOCOS structure is used to isolate two N$^+$ regions. A metal line runs over the 0.5-μm-thick LOCOS oxide forming a parasitic MOSFET. Assume that both N$^+$ junctions and the line are all at the same voltage. The substrate concentration (N_A) is 2×10^{16} cm^{-3} (p-type), the oxide is ideal, and the junctions are 3.0 μm apart. Calculate and show the parasitic threshold voltage and the voltage necessary to punch through between the two N$^+$ regions. Which one will occur at the lower voltage? What substrate concentration would you need to ensure that neither occurred below 10 V.

5. Consider the deep trenches shown in Figure 15.12. If the substrate is uniformly doped at 1×10^{16} cm^{-3} and the trenches are 5 μm deep, how large a reverse bias is required to deplete the trench all the way to the bottom?

6. Calculate the specific contact resistance at room temperature for a contact to 10^{17} cms^{-3} n-type GaAs if the metallization has a barrier height of 0.8 eV. (*Hint*: Differentiate Equation 15.7 and evaluate at $V = 0$.)

7. An implanted ohmic contact is formed to n-type silicon. When the doping concentration at the metal/semiconductor interface is 1×10^{20} cm^{-3}, the specific contact resistance 5×10^{-6} Ω-cm^2. A new source/drain implant is being implemented to increase the doping concentration at the metal/semiconductor interface to 2×10^{20} cm^{-3}. Assume that A_o is fixed, the barrier height is

0.6 eV, and $m* = 1.18 \times m_o$. Calculate the specific contact resistance that you would expect for the new technology. Calculate the resistance of a vertical-flow 0.5×0.5-μm contact for each of the two technologies.

8. A triple-diffused bipolar technology is fabricated using 10^{14} p-type substrates. The maximum bias that either collector would ever see is 10 V. Assume simple-step junctions and calculate the required separation between devices. (Calculate the distance between the collectors after fabrication. The design of the collectors must be further apart than this to allow lateral diffusion during the collector drive.) To improve matters, a p^+ guard ring will be added to the process. The guard ring must be at least 3 μm deep to be effective. Assume that the guard ring pattern is 2 μm wide, that the final guard ring diffusion must never be allowed to contact the collector, and that the alignment tolerance between the collector and guard ring is 1 μm. What is the new minimum separation between the collector diffusions?

9. Memory makers found that replacing long word lines made from polysilicon with a salicide could decrease the access time of the device. Assume that the line is 1 cm long, the oxide is 1 μm thick, the poly and silicide are each 0.5 μm thick, and the resistivities of the two films are 10^{-3} and 10^{-4} Ω-cm, respectively. Use the simple lumped RC model to determine the delay associated with each of the two lines. (The dielectric constant of SiO_2 is $3.9 \times 8.84E - 14$ F/cm.)

10. In a university lab simple poly gate MOSFETs were made. These devices were n-channel devices (N^+, source and drain in a p-type substrate). The doping concentration at the surface of the contact was 8×10^{19} cm^{-3}. The contact was 3 μm long and used Al. Assume that the width of the transistor and the contact was 50 μm.

   (a) If the A_o for this technology is 10^{-11} Ω-cm^2, find the specific contact resistance for these devices. (*Hint*: the case of silicon NMOS transistors with Al contacts will be solved in Example 16.1—you can recycle some of these results–assume the same barrier height.)

   (b) the contact is 10 μm from the edge of the gate, no self-aligned silicide process is used, and the source/drain sheet resistance is 100 $\Omega/\square$, find the total resistance from the metal contact to the edge of the channel.

11. An ohmic contact is to be made to n-type GaAs using an alloyed Ni/Au/Ge contact on an ion-implanted channel layer. The resistivity of the conducting GaAs layer is 0.01 Ω-cm. In the region of the alloyed contact, the surface concentration is 1×10^{19} cm^{-3}. The mass of an electron in GaAs is $0.067 \times m_o$. The barrier height of the metal is 0.75 V, and the constant A for GaAs in the exponential expression for the ohmic contact resistance is 10^{-8} Ω-cm^2. The contact is found to form pits at the metal/semiconductor interface. These pits have radii of 300 Å with a pit density of 10^9 cm^{-2}.

   (a) What is the specific contact resistance in ohm-cm^2?

   (b) What will the specific contact resistance be if the diffused impurity concentration under the contact is raised to 3×10^{19} cm^{-3}?

12. An advanced metallization process is proposed for high density silicon-based ICs. This process will use several new materials. Identify one advantage and one disadvantage for each new material: (a) CVD tungsten, (b) electroplated copper, and (c) spin-on polyimide.

References

1. For additional discussion on silicon process integration the reader is referred to S. Wolf, *Silicon Processing for the VLSI Era Vol. 2, Process Integration*, Lattice Press, Sunset Beach, CA. 1990.
2. U.S. Patent 3,029,366, K. Lehovec (1959).
3. E. Kooi and J. A. Appels, in *Semiconductor Silicon 1973*, H. R. Huff and R. Burgess, eds., *The Electrochemical Symposium Series*, Pennington, NJ, 1973.

4. A. Bogh and A. K. Gaind, "Influence of Film Stress and Thermal Oxidation on the Generation of Dislocations in Silicon," *Appl. Phys. Lett.* **33**:895 (1978).

5. E. Kooi, J. G. van Lierop, and J. A. Appels, "Formation of Silicon Nitride at an Si/SiO$_2$ Interface During the Local Oxidation of Silicon in NH$_3$ Gas," *J. Electrochem. Soc.* **123**: 1117 (1976).

6. U.S. Patent 4,541,167, R. H. Havemann and G. P. Pollack (1986).

7. Y. Han and B. Ma, "Poly Buffered Layer for Scaled MOS," *VLSI Sci., Technology*, Electrochemical Society, Pennington, NJ, 1984, p. 334.

8. M. Ghezzo, E. Kaminski, Y. Nissan-Cohen, P. Frank, and R. Saia, "LOPOS: Advanced Device Isolation for a 0.8 μm CMOS/Bulk Process Technology," *J. Electrochem. Soc.* **136**:1992 (1989).

9. R. L. Guldi, B. McKee, G. M. Damminga, C. Y. Young, and M. A. Beals, "Characterization of Poly-Buffered LOCOS in a Manufacturing Environment," *J. Electrochem. Soc.* **136**:3815 (1989).

10. T.-H. Lin, N.-S. Tsia, and C.-S. Yoo, "Twin White Ribbon Effect and Pit Formation Mechanism in PBLOCOS," *J. Electrochem. Soc.* **138**:2415 (1991).

11. J. W. Lutze and J. P. Krusius, "Electrical Limitations of Advanced LOCOS Isolation for Deep Submicrometer CMOS," *IEEE Trans. Electron Dev.* **38**:242 (1991).

12. M. Mikoshiba, T. Homma, and K. Hamano, "A New Trench Isolation Technology as a Replacement of LOCOS," *IEDM Tech. Dig.*, 1984, p. 578.

13. T. Iizuka, K. Y. Chiu, and J. L. Moll, "Double Threshold MOSFETs in Bird's-Beak Free Isolation," *IEDM Tech. Dig.*, 1981, p. 380.

14. T. V. Rajeevakumar, T. Lii, Z. A. Wienberg, G. B. Bronner, P. MacFarland, P. Coane, K. Kwietniak, A. Megdanis, K. J. Stein, and S. Cohen, "Trench Storage Capacitors for High Density DRAMs," *IEDM Tech. Dig.*, 1991, p. 835.

15. K. Shibahara, Y. Fujimoto, M. Hamada, S. Iwao, K. Tokashiki, and T. Kunio, "Trench Isolation with ∇ Shaped Buried Oxide for 256 Mega-bit DRAMs," *IEDM Tech. Dig.*, 1992, p. 275.

16. C. W. Teng, C. Slawinski, and W. R. Hunter, "Defect Generation in Trench Isolation," *IEDM Tech. Dig.*, 1984, p. 586.

17. R. Kakoshke, R. E. Kaim, P. F. H. M. van der Meulen, and J. F. M. Westendorp, "Trench Sidewall Implantation with a Scanned Ion Beam," *IEEE Trans. Electron Dev.* **37**:1052 (1990).

18. W. Zagodon-Wosik, J. C. Wolfe, and C. W. Teng, "Doping of Trench Capacitors by Rapid Thermal Diffusion," *IEEE Electron Dev. Lett.* **12**:264 (1991).

19. F. S. Becker, H. Treichel, and S. Röhl, "Low Pressure Deposition of Doped SiO$_2$ by Pyrolysis of Tetraethylorthosilicate (TEOS)," *J. Electrochem. Soc.* **136**:3033 (1989).

20. K. Sunouchi, F. Horiguchi, A. Nitayama, K. Hieda, H. Takato, N. Okabe, T. Yamada, T. Ozaki, K. Hashimoto, S. Takedai, A. Yagishita, A. Kumagae, Y. Takahashi, and F. Masuoka, "Process Integration for 64M DRAM Using an Asymmetrical Stacked Trench Capacitor (AST) Cell," *IEDM Tech. Dig.*, 1990, p. 647.

21. S. Nag and A. Chatterjee, "Shallow Trench Isolation for Sub-0.25 μm IC Technologies," *Solid State Technol*, September 1997, p. 129.

22. M. Nandakumar, "Shallow Trench Isolation for Advanced ULSI CMOS Technologies," *Proc. IEDM*, 1998, p. 133.

23. J. P. Benedict, "Shallow Trench Isolation with Oxide-Nitride/Oxynitride Liner," U.S. Patent 5,763,315, IBM.

24. A. Chatterjee, I. Ali, K. Joyner, D. Mercer, J. Kuehne, M. Mason, A. Esquivel, D. Rogers S. O'Brien, P. Mei, S. Murtaza, S. P. Kwok, K. Taylor, S. Nag, G. Harnes, M. Hanratty,

H. Marchman, S. Ashburn, and I.-C. Chen, "Integration of Unit Processes in a Shallow Trench Isolation Module in a 0.25 mm Complementary Metal-Oxide Semiconductor Technology," *J. Vacuum Sci. Technol. B* **15**(6):1936 (1997).

25. S. Wolf, *Silicon Processing for the ULSI Era*, Vol. 4, Lattice Press, Sunset Beach, CA, 2000, p. 441.

26. Y. Sugawara, T. Kamei, Y. Hosokawa, and M. Okamura, "Practical Size Limits of High Voltage IC's," *IEDM Tech. Dig.*, 1983, p. 412.

27. T. Aso, H. Mizuide, T. Usui, K. Akahane, N. Ishikawa, I. Hide, and Y. Maeda, "An Application of MSSD to Dielectrically Isolated Intelligent Power ICs," *1991 IEEE Int. Symp. Power Semiconductor Devices, and ICs*, M. A. Shibib and B. J. Baliga, eds., IEEE, Piscataway, NJ, 1991.

28. J. B. Lasky, S. R. Stiffler, F. R. White, and J. R. Abernathey, "Silicon on Insulator (SOI) by Bonding and Etch Back," *IEDM Tech. Dig.*, 1985, p. 684.

29. H. J. Quenzer and W. Benecke, "Low-Temperature Silicon Wafer Bonding," *Sensors, Actuators A* **32**:340 (1990).

30. J. G. Fleming, E. Roherty-Osmun, and N. A. Godshall, "Low Temperature, High Strength, Wafer-to-Wafer Bonding," *J. Electrochem. Soc.* **139**:3300 (1992).

31. P. V. Schwartz, C. W. Liu, and J. C. Sturm, "Semi-Insulating Crystalline Silicon Formed by Oxygen Doping During Low-Temperature Chemical Vapor Deposition," *Appl. Phys. Lett.* **62**:1102 (1993).

32. D. K. Ferry, *Gallium Arsenide Technology*, Harold W. Sams, Indianapolis, 1985.

33. G. M. Marin, J. P. Farges, G. Jacob, J. P. Hallais, and G. Poublaud, "Compensation Mechanisms in GaAs," *J. Appl. Phys.* **51**:2840 (1980).

34. H. J. von Bardeleben and B. Pajot, eds., "Recent Developments in the Study of the EL2 Defect in GaAs," *Rev. Phys. Appl.* **23**:727 (1988).

35. S. M. Sze, *Physics of Semiconductor Devices*, Wiley, New York, 1981.

36. C. A. Crider et al., "Platinum Silicide Formation Under Ultra High Vacuum and Controlled Impurity Ambients," *J. Appl. Phys.* **52**:2860 (1981).

37. H. C. Cheng, C. Y. Wu, and J. J. Shy, "Excellent Thermal Stability of Cobalt-Aluminum Alloy Schottky Contacts on GaAs Substrates," *Solid-State Electron.* **33**:863 (1990).

38. T. Hara, A. Suga, and R. Ichikawa, "Properties of CVD WSi_x Films and CVD WSi_x/GaAs Schottky Barrier," *Phys. Stat. Sol. A* **113**:459 (1989).

39. J. Willer, M. Heinzle, N. Arnold, and D. Ristow, "WSi_x Refractory Gate Metal Process for GaAs MESFET's," *Solid-State Electron.* **33**:571 (1990).

40. M. Kanamori, K. Nagai, and T. Nozaki, "Low Resistivity W/Wsi_x Bilayer Gates for Self-Aligned GaAs Metal-Semiconductor Field Effect Transistor Large Scale Integrated Circuits," *J. Vacuum Sci. Technol. B* **6**:1317 (1987).

41. J. Willer, M. Heinzle, L. Schleicher, and D. Ristow, "Characterization of WSi_x Gate Metal Process for GaAs MESFET's," *Appl. Surf. Sci.* **38**:548 (1989).

42. H. Nakamura, Y. Sano, T. Nonaka, T. Ishida, and K. Kaminishi, "A Self-Aligned GaAs MESFET with W-Al Gate," *IEEE GaAs IC Symp. Tech. Dig.*, IEEE, Piscataway, NJ, 1983, p. 134.

43. T. Ohnishi, N. Yokoyama, H. Onodera, S. Suzuki, and A. Shibatomi, "Characterization of Wsi_x/GaAs Schottky Contacts," *Appl. Phys. Lett.* **43**:600 (1983).

44. C. A. Mead and W. G. Spitzer, *Phys. Rev.* **134**:A173 (1964).

45. Y. Sekino, T. Kimura, K. Inokuchi, Y. Sano, and M. Sakuta, "Effect of Bias Sputtering on W and W–Al Schottky Contact Formation and Its Application to GaAs MESFETs," *Jpn. J. Appl. Phys.* **27**:L2183 (1988).

46.   T.-H. Shen, M. Elliott, R. H. Williams, D. A. Woolf, D. I. Westwood, and A. C. Ford, "Control of Semiconductor Interface Barriers by Delta-Doping," *Appl. Surf. Sci. B* (**56–58**) 749 (1992).

47.   R. A. Kiehl, S. L. Wright, J. H. Margelin, and D. J. Frank, *Proc. IEEE Cornell Conf.*, 1987, p. 28.

48.   S. T. Ali and D. N. Bose, "Improved Au/n-GaAs Schottky Barriers Due to Ru Surface Modification," *Mater. Lett.* **12**:388 (1991).

49.   N. C. Cirillo, J. K. Abrokwah, and M. S. Shur, "Self-Aligned Modulation-Doped (Al,Ga)As/GaAs Field-Effect Transistors," *IEEE Electron Dev. Lett.* **EDL-5**:129 (1984).

50.   C. Y. Ting, S. S. Iyer, C. M. Osborn, G. J. Hu, and A. M. Schweigart, *Electrochem. Soc. Ext. Abstr.* **82-2**:254 (1982).

51.   C. Mallardeau, Y. Morand, and E. Abonneau, "Characterization of TiSi$_2$ Ohmic and Schottky Contacts Formed by Rapid Thermal Annealing Technology," *J. Electrochem. Soc.* **136**:238 (1989).

52.   E. Waterman, J. Dunlop, and T. Brat, "Tungsten-Titanium Sputtering Target Processing Effects on Particle Generation and Thin Film Properties for VLSI Technology," *IEEE VLSI Multilevel Interconnect Conf.*, 1991, p. 329.

53.   T. Sands, "Compound Semiconductor Contact Metallurgy," *Mater. Sci., Eng. B* **1**:289 (1989).

54.   C. Dubon-Chevallier, P. Blanconnier, and C. Bescombes, "GeMoW Refractory Ohmic Contact for GaAs/GaAlAs Self-Aligned Heterojunction Bipolar Transistors," *J. Electrochem. Soc.* **137**:1514 (1990).

55.   R. P. Gupta, W. S. Khokle, J. Wuerfl, and H. L. Hartnagel, "Design and Characterization of a Thermally Stable Ohmic Contact Metallization on n-GaAs," *J. Electrochem. Soc.* **137**:631 (1990).

56.   J. H. Pugh and R. S. Williams, "Boundary Driven Loss of Gas-Phase Group V Sources for Gold/III-V Compound Semiconductor Systems," *J. Mater. Res.* **1**:343 (1986).

57.   J. B. B. Oliveira, C. A. Olivieri, and J. C. Galzerani, "Characterization of NiAuGe Ohmic Contacts on n-GaAs Using Electrical Measurements, AUGER Electron Spectroscopy and X-ray Diffractometry," *Vacuum* **41**:807 (1990).

58.   M. Kamada, T. Suzuki, T. Taira, and M. Arai, "Electrical Inhomogeneity in Alloyed AuGe-Ni Contact Formed on GaAs," *Solid-State Electron.* **33**:999 (1990).

59.   T. S. Kuan, P. E. Baston, T. N. Jackson, H. Rupprecht, and E. L. Wilke, "Electron Microscope Studies of an Alloyed Au/Ni/Au-Ge Ohmic Contacts to GaAs," *J. Appl. Phys.* **54**:6952 (1983).

60.   T. Sanada and O. Wada, "Ohmic Contacts to p-GaAs with Au/Zn/Au Structures," *Jpn. J. Appl. Phys.* **19**:L491 (1980).

61.   H. J. Gopen and A. Y. C. Yu, "Ohmic Contacts to Epitaxial p-GaAs," *Solid-State Electron.* **14**:515 (1971).

62.   Y. Lu, T. S. Kalkur, and C. A. Paz de Araujo, "Reduced Thermal Alloyed Ohmic Contacts to p-Type GaAs," *J. Electrochem. Soc.* **136**:3123 (1989).

63.   A. A. Lakhani, "The Role of Compound Formation and Heteroepitaxy in Indium-Based Contacts to GaAs," *J. Appl. Phys.* **56**:1888 (1984).

64.   T. S. Kalker, "Preliminary Studies on Mo-In-Mn Based Ohmic Contacts to p-GaAs," *J. Electrochem. Soc.* **136**:3549 (1989).

65.   D. C. Thomas and S. S. Wong, "A Planar Interconnect Technology Utilizing the Selective Deposition of Tungsten—Multilevel Implementation," *IEEE Trans. Electron Dev.* **39**:901 (1992).

66. S. R. Wilson, J. L. Freeman Jr., and C. J. Tracey, "A Four-Metal Layer, High Performance Interconnect System for Bipolar and BiCMOS Circuits," *Solid State Technol.*, November 1991, p. 67.

67. J. R. Black, "Electromigration: A Brief Survey and Some Recent Results," *IEEE Trans. Electron Dev.* **ED-16**:338 (1969).

68. P. B. Ghate, "Electromigration Induced Failures in VLSI Interconnects," *Proc. IEEE 20th Int. Rel. Phys. Symp.*, 1982, p. 292.

69. J. M. Towner and E. P. van de Ven, "Aluminum Electromigration Under Pulsed D.C. Conditions," *21st Annu. Proc. Rel. Phys. Symp.*, 1983, p. 36.

70. J. A. Maiz, "Characterization of Electromigration Under Bidirectional (BDC) and Pulsed Unidirectional Currents," *Proc. 27th Int. Rel. Phys. Symp.*, 1989, p. 220.

71. S. Vaidya, T. T. Sheng, and A. K. Sinha, "Line Width Dependence of Electromigration in Evaporated Al-0.5%Cu," *Appl. Phys. Lett.* **36**:464 (1980).

72. T. Turner and K. Wendel, "The Influence of Stress on Aluminum Conductor Life," *Proc. IEEE Int. Rel. Phys. Symp.*, 1985, p. 142.

73. H. Kaneko, M. Hasanuma, A. Sawabe, T. Kawanoue, Y. Kohanawa, S. Komatsu, and M. Miyauchi, "A Newly Developed Model for Stress Induced Slit-like Voiding," *Proc. IEEE Int. Rel. Phys. Symp.*, 1990, p. 194.

74. K. Hinode, N. Owada, T. Nishida, and K. Mukai, "Stress-Induced Grain Boundary Fractures in Al-Si Interconnects," *J. Vacuum Sci. Technol. B* **5**:518 (1987).

75. S. Mayumi, T. Umemoto, M. Shishino, H. Nanatsue, S. Ueda, and M. Inoue, "The Effect of Cu Addition to Al-Si Interconnects on Stress-Induced Open-Circuit Failures," *Proc. IEEE Int. Rel. Phys. Symp.*, 1987, p. 15.

76. P. Singer, "Double Aluminum Interconnects," *Semicond. Int.* **16**:34 (1993).

77. R. A. Levy and M. L. Green, "Low Pressure Chemical Vapor Deposition of Tungsten and Aluminum for VLSI Applications," *J. Electrochem. Soc.* **134**:37C (1987).

78. R. J. Saia, B. Gorowitz, D. Woodruff, and D. M. Brown, "Plasma Etching Methods for the Formation of Planarized Tungsten Plugs Used in Multilevel VLSI Metallization," *J. Electrochem. Soc.* **135**:936 (1988).

79. D. C. Thomas, A. Behfar-Rad, G. L. Comeau, M. J. Skvarla, and S. S. Wong, "A Planar Interconnect Technology Utilizing the Selective Deposition of Tungsten—Process Characterization," *IEEE Trans. Electron Dev.* **39**:893 (1992).

80. T. E. Clark, P. E. Riley, M. Chang, S. G. Ghanayem, C. Leung, and A. Mak, "Integrated Deposition and Etchback in a Multi-Chamber Single-Wafer System," *IEEE VLSI Multilevel Interconnect Conf.*, 1990, p. 478.

81. K. Suguro, Y. Nakasaki, S. Shima, T. Yoshii, T. Moriya, and H. Tango, "High Aspect Ratio Hole Filling by Tungsten Chemical Vapor Deposition Combined with a Silicon Sidewall and Barrier Metal for Multilevel Metallization," *J. Appl. Phys.* **62**:1265 (1987).

82. P. E. Riley, T. E. Clark, E. F. Gleason, and M. M. Garver, "Implementation of Tungsten Metallization in Multilevel Interconnect Technologies," *IEEE Trans. Semicond. Manuf.* **3**:150 (1990).

83. T. Iijima, A. Nishiyama, Y. Ushiku, T. Ohguro, I. Kunishima, K. Suguro, and H. Iwai, "A New Contact Plug Technique for Deep-Submicrometer ULSIs Employing Selective Nickel Silicidation of Polysilicon with a Titanium Nitride Stopper," *IEEE Trans. Electron Dev.* **40**:371 (1993).

84. A. C. Adams and D. D. Capio, "Planarization of p-Doped Silicon Dioxide," *J. Electrochem. Soc.* **128**:423 (1981).

85. C. Jang et al., *IEEE VLSI Multilevel Interconnect Conf.*, 1984, p. 357.

86. W. T. Tseng, Y. T. Hsieh, and C. F. Lin, "CMP of Fluorinated Silicon Dioxide: Is It Necessary and Feasible?" *Solid State Technol.* **40**(2):4 (February 1997).

87. Y. Matsubara et al., "Low-K Fluorinated Amorphous Carbon Interlayer Technology for Quarter Micron Devices," *Proc. IEDM*, 1996, p. 369.

88. R. J. Gutmann, J. M. Steigerward, L. You, and D. T. Price, "Chemical-Mechanical Polishing of Copper with Oxide and Polymer Interlevel Dielectrics," *Thin Solid Films* **270**:596 (1995).

89. N. Hendricks, "Low Dielectric Constant Materials for IC Intermetal Dielectric Applications: A Status Report on the Leading Contenders," *Mater Res. Soc. Symp. Proc.* **443**:91 (1997).

90. K. J. Taylor, S. P. Jeng, M. Eissa, J. Gaynor, and H. Nguyen, "Polymers for High-Performance Interconnects," *Microelectron Eng.* **37–38**:255 (1997).

91. T. Ramos, K. Roderick, A. Maskara, and D. M. Smith, "Nanoporous Silica for Low *k* Dielectrics," *Mater. Res. Soc. Symp. Proc.* **443**:91 (1997).

92. Nanoglass, *Solid State Technol.* February 1997, p. 36.

93. S. P. Murkara, R. J. Gutmann, A. E. Kaloyeros, and W.A. Lanford, "Advanced Multilayer Metallization Schemes with Copper as Interconnection Metal," *Thin Soild Films* **236**:257 (1993).

94. Y. Gotkis, D. Schey, S. Alamgir, and J. Yang, "Cu CMP with Orbital Technology: Summary of the Experience," *IEEE/SEMI Adv. Semicond. Manuf. Conf.*, 1998, p. 364.

95. D. T. Price, R. J. Gutman, and J. Yang, "Demascene Copper Interconnects with Polymer ILD," *Thin Solid Films* **308–309**:523 (1997).

96. C. S. Sun, *Proc. Int. Electronic Devices Meeting*, 1997, p. 765.

97. W. Steinhogl et al., *Phys. Rev. B.* **66**:15414 (2002).

Chapter 16

CMOS Technologies

T
he last chapter discussed the process modules of device isolation and interconnection. This chapter will apply these modules to the first of the technologies to be considered, CMOS. The discussion begins with a 3-μm CMOS technology. This sort of process was state of the art during the mid-1970s. Although it may seem odd to begin with such old techniques, modern CMOS processing is primarily a set of extensions to this basic technology. Device scaling and short-channel effects will then be discussed. In highly scaled devices, parasitics are an important element, so methods of reducing series resistance and source/drain capacitance are covered next. Device reliability and drain engineering and latchup will then be covered, and the chapter will finish with a discussion of deeply scaled CMOS technologies. For a more extensive review of CMOS devices and technology the reader is referred to Wolf [1].

16.1 Basic Long-Channel Device Behavior

This section will review the properties of the ideal long-channel MOSFET. It is not intended to serve as a complete description of MOS device physics; that occupies books in itself. (For detailed descriptions of MOS device operation see Sze [2], Nicollian and Brews [3], or Tsividis [4].) Instead, it will review the material necessary to develop an appreciation for the requirements of MOS technologies. To do this we will first review the concept of threshold voltage and then discuss the three operating regimes of the MOS transistor.

The basic MOS transistor has four terminals: source, drain, gate, and substrate (Figure 16.1). One of the most important parameters in determining the electrical characteristics of the MOSFET is the "turn-on" or threshold voltage, V_t. The threshold voltage is normally defined to be the gate voltage at which the surface of the semiconductor is at the onset of strong inversion. That is, the carrier concentration of the inversion layer at the surface is equal to, but of the opposite type as, the bulk carrier concentration. At this point, the surface potential is $2\phi_f$, where ϕ_f, the Fermi level, is defined as

$$\phi_f = \frac{kT}{q} \ln \frac{N_a}{n_i}$$
(16.1)

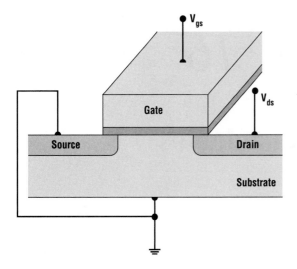

Figure 16.1 Cross-sectional schematic of a four-terminal MOSFET. In this transistor the source and substrate are both grounded.

for an NMOS transistor and

for a p-type semiconductor and

$$\phi_f = -\frac{kT}{q} \ln \frac{N_d}{n_i} \tag{16.2}$$

for an n-type semiconductor. In these equations, N_a and N_d are the acceptor and donor concentrations in the substrate, and n_i is the intrinsic carrier concentration.

For a p-type semiconductor, the gate voltage can be expressed in terms of the surface potential (ϕ_s) as

$$V_g = \phi_s + \frac{k_s t_{ox}}{k_{ox}} \sqrt{\frac{2qN_a}{k_s \varepsilon_o} \phi_s} \tag{16.3}$$

where k_s and k_{ox} are the relative dielectric constants of silicon and silicon dioxide (11.8 and 3.9, respectively) and t_{ox} is the gate oxide thickness. Inserting for the surface potential at threshold ($\phi_s = 2\phi_f$), one can obtain for the threshold voltage

$$V_t = 2\phi_f + \frac{k_s t_{ox}}{k_o} \sqrt{\frac{4qN_a}{k_s \varepsilon_o} \phi_f} \tag{16.4}$$

$$V_t = 2\phi_f - \frac{k_s t_{ox}}{k_o} \sqrt{\frac{4qN_d}{k_s \varepsilon_o} (-\phi_f)} \tag{16.5}$$

for a PMOS transistor. If one also takes into account the metal semiconductor work function and the presence of oxide and interface charges, an additional term of

$$\Delta V_t = \phi_{ms} - \frac{Q_f + Q_{it}(0) + Q_{MI}\gamma_{MI}}{C_{ox}} \tag{16.6}$$

must be added to the right-hand sides of Equations 16.4 and 16.5. In Equation 16.6, Q_f and Q_{MI} represent the area density of fixed and mobile ion charges, γ_{MI} is a constant that depends on the distribution of mobile charge in the oxide, which varies between zero and one, and $Q_{it}(0)$ represents the interface state charge; C_{ox} is the oxide capacitance per unit area.

From a process standpoint, it is clear that a wide variety of well-controlled threshold voltages are available if one can control the oxide thickness, the charges in the oxide, the metal semiconductor work function, and the doping concentration in the channel. Of these, the doping concentration is most easily manipulated, since reasonable adjustments of this parameter can be accomplished by simply changing an implantation dose. Furthermore, changing the substrate concentration does not have the first-order effect on performance and reliability that changing the oxide thickness has, (multiple gate oxide thicknesses are sometimes used, however, if the chip has multiple supply voltages.)

Detailed equations have been developed for MOS operations that are valid for any set of biasing conditions. The derivation of these equations is beyond the scope of this book. They can be simplified by dividing the operation of the MOSFET (Figure 16.2) into three regions: subthreshold ($V_{gs} < V_t$), linear ($V_t < V_{gs}$ and $0 < V_{ds} < V_{gs} - V_t$), and pinchoff ($V_t < V_{gs}$ and $V_{gs} - V_t < V_{ds}$). Near the transition points, these equations will be only approximate, but they will serve our purpose.

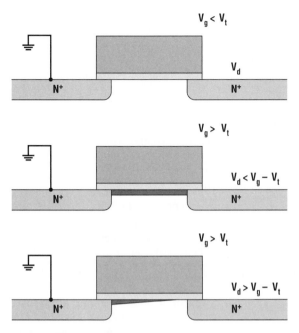

Figure 16.2 Three regions of n-channel MOSFET operation are subthreshold, linear, and saturation.

The linear region is the easiest to derive. In this region the channel forms a thin, nearly uniform layer under the gate. As with any resistor, the conductance is proportional to the width-to-length ratio, the mobility, and the number of carriers. The number of carriers is proportional to the difference between the gate voltage and the threshold voltage. Collecting these terms, in the simplest (square law) theory, one can write the drain current as

$$I_{ds} \approx \frac{W}{L} \mu_{\text{eff}} C_{\text{ox}} (V_{gs} - V_t) V_{ds} \tag{16.7}$$

where μ_{eff} is the effective mobility and C_{ox} is the oxide capacitance per unit area. When $V_{ds} > 0$, the carrier concentration shrinks as one approaches the drain edge. Taking this into account,

$$I_{ds} = \frac{W}{L} \mu_{\text{eff}} C_{\text{ox}} \left[(V_{gs} - V_t) V_{ds} - \frac{V_{ds}^2}{2} \right]$$

$$V_{gs} - V_t > V_{ds} \quad V_{gs} > V_t \tag{16.8}$$

These expressions use an effective mobility to represent the fact that the carriers in the channel may scatter off each other and the semiconductor/oxide interface in addition to the scattering mechanisms seen in the bulk. The mobility that must be used is substantially less than the bulk value and at high values of vertical electric field, depends on the gate bias.

In the pinchoff region, the channel does not extend fully from source to drain. The current is limited by the movement of carriers from the pinched off channel edge to the drain. As a result, the current in this region is nearly independent of the drain voltage. Since this occurs at the onset of the pinchoff region, we can find the value of the linear current at $V_{ds} = V_{gs} - V_t$ from Equation 16.8. This gives us the drain to source current in pinchoff

$$I_{ds} = \frac{W}{L} \mu_{\text{eff}} C_{\text{ox}} \frac{(V_{gs} - V_t)^2}{2} \quad 0 < V_{gs} - V_t < V_{ds} \tag{16.9}$$

The analysis of the subthreshold region is more complicated than either the linear or saturation regions. The drain-to-source current depends strongly on the surface potential, since the minority carrier concentration increases exponentially with surface potential. The exact equation for the drain-to-source current is also beyond the range of this text, but for an ideal device,

$$I_{ds} \propto \exp \left[K_{\text{sub}} \frac{q V_{gs}}{kT} \right] \quad V_{gs} < V_t \tag{16.10}$$

where K_{sub} is a constant that depends on the gate oxide thickness and the substrate doping.

16.2 Early MOS Technologies

The first MOS technologies were PMOS. By the early 1970s, they were replaced by NMOS, due to the higher mobility of the electron. Resistor load and enhancement depletion (E/D) NMOS

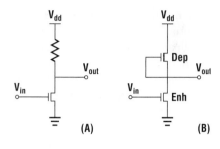

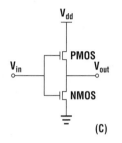

Figure 16.3   MOS inverters: (A) resistor load, (B) enhancement/depletion or E/D, and (C) CMOS.

technologies, shown in Figure 16.3A, were widely fabricated through the 1970s and into the early 1980s. In comparison to bipolar, the technology was viewed as cheap (it required only five masks for resistor load and seven masks for E/D) and capable of high density (in the range of 1000 to 20,000 gates per chip) but was fairly slow, with digital clock cycles of a few megahertz. As the density increased from 1000 to 10,000 gates, the chip power required grew from hundreds of milliwatts to several watts.

The idea of complementary logic (Figure 16.3B) was first proposed by Wanlass and Sah in 1963 [5]. Their devices showed gate delays of over 100 ns per stage. Early work was led by RCA, with the first demonstration of an IC in 1966. These technologies were not very dense and were slow. They were also prone to latchup, a self-destructive condition to be reviewed later in this chapter. Since LOCOS had not yet been developed, guard rings were used, which had a large area penalty and a large capacitance. During the early 1970s, CMOS was much derided and used primarily in toys, watches, and calculators where such slow speeds could be tolerated, and in a few military applications, due to its small power dissipation and large noise margins. The invention of the LOCOS process module, the reduction of mobile ionic charge, the introduction of ion implantation, which could provide precise control of the well concentration and the threshold voltage, and improvements in lithography have dramatically improved the acceptance of CMOS.

CMOS has the basic advantage that for an inverter structure, only one of the transistors is on when the input is either high or low. Ideally appreciable power is dissipated only during switching transients. Part of this power is capacitive charging current, which can be represented as

$$P \approx \bar{n} \times \frac{\text{freq}}{2} \times \overline{C_{\text{node}}} \times V_{\text{swing}}^2 \qquad (16.11)$$

where $\bar{n}$ is the average number of gates switching at any given time, C_{node} is the average node capacitance, and V_{swing} is the voltage difference between a logical high and a logical low. An additional current arises from the fact that during the switch transient, both transistors are on. Because only a small fraction are likely to switch at any given clock cycle, CMOS power dissipation is much lower than NMOS. (For deeply scaled CMOS, junction leakage, gate leakage, and off-state leakage are all significant, so standby power is considerable.) The tremendous increase in density completely offset the cost of the additional process complexity required for CMOS. As a result, it is the dominant semiconductor technology today.

16.3   The Basic 3-μm Technology

In designing a CMOS technology, attention must be paid to the concept of thermal budget. Therefore, those processes that involve the formation of deeply diffused wells (also called *tubs*) must be done first. The choices for well type are n-well, p-well, and twin well [6]. The substrate material for CMOS is normally lightly doped. This means that the depletion regions of the source and drain junctions for the MOSFET fabricated in the substrate will be large and the parasitic capacitance of these junctions will be small. The well, on the other hand, is typically doped several orders of magnitude higher than the substrate so that any uncertainty in the substrate concentration will not affect the well concentration. As a result of this higher doping, the device placed in the well is inherently slower than the same device would be if placed in the substrate. If the NMOS devices are placed in a p-well, their degraded performance will more closely match the inherently lower mobility holes in PMOS transistors [7]. The most commonly used tub strategy for submicron technologies is the twin tub. That is, both n-type and p-type "wells" are formed in a lightly doped substrate. The doping concentration required to obtain acceptable device behavior increases as the gate length shrinks. As a result, neither device really operates in a lightly doped substrate. Although it adds process complexity, the twin-tub approach allows the doping profile for each device to be set independently to optimize the performance of both device types [8].

The first step in a simple n-well technology is the growth of a 1000-Å initial oxide. This serves two functions. It prevents photoresist from contacting the silicon, and it will later provide an alignment mark to locate the position of the well. After implantation, the initial oxide is etched, the photoresist is stripped, and the well is driven in a high temperature process (Figure 16.4).

The well dose and drive cycle are set by two basic considerations: surface concentration and depth. The surface concentration is determined by the device designer. This is usually not what is required for the threshold voltage, but rather the concentration required to prevent source-to-drain punchthrough. Thus, the sum of the source and drain depletion widths must be less than the channel length when V_{ds} is at least as large as the supply voltage. To provide a margin of safety, the minimum acceptable punchthrough voltage is often established by biasing the drain at two or three times the supply voltage with the source shorted to the substrate. For this calculation, one must recognize that the channel length is not identical to the drawn gate length. During the gate etch process the gate length may be reduced. The source and drain will also diffuse laterally under the gate. Assume, therefore, that the shortest permissible effective channel length devices in this 3-μm technology is 2 μm. The channel doping must be set high enough to ensure that the sum of the depletion regions of the source and drain cannot exceed 2 μm. For a single-sided step junction, a substrate concentration of 5×10^{15} cm^{-3} would provide a total depletion width of 1.8 μm for an unbiased source and a drain biased at 10 V (2 times a supply voltage of 5 V). Common well surface concentrations then are slightly less than 10^{16} cm^{-3}.

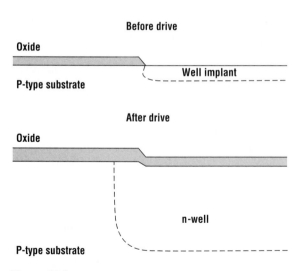

Figure 16.4 Cross-sectional views of an inverter structure before and after well drive.

The well depth is determined by ensuring that the source and drain do not completely deplete the well vertically. The full supply bias must be applied between the well and the substrate to ensure a proper reverse bias on all of the junctions. For the n-well technology discussed, just designed to run at 5 V, the depletion layer of the source (at 0 V) in the well is about 1.2 μm. The depletion layer of the well/substrate junction will be primarily in the substrate, owing to its light doping. However, approximately 0.2 μm will be in the well. Thus, the well depth must be at least the source/drain junction depth (which we will take to be 0.5 μm), plus the source/drain depletion width (1.2 μm), plus the width of the well/substrate depletion that lies inside the well (Figure 16.5). The well then must be at least 2.0 μm deep. In reality, the well profile is not uniform. Instead, the carrier concentration decreases near the well/substrate junction. Because of this, the depletion width of the substrate junction is considerably larger

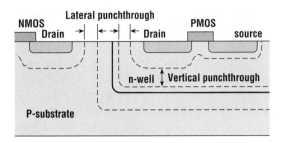

Figure 16.5 The minimum well depth must be chosen to prevent punchthrough from the reverse-biased source/well and source/substrate junctions.

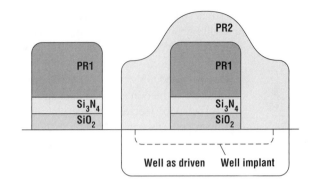

Figure 16.6 The field implant mask is the inverse of the well, but the features are bloated to account for lateral diffusion.

than 0.2 μm. For a 3-μm process, a typical final well depth, then, is about 4 μm. To achieve this deep well, phosphorus is normally chosen as the well dopant due to its high diffusivity.

The well drive must be done at least partially in an oxidizing ambient. This oxide consumes silicon in the exposed well window. When the oxide is later removed, this creates a depression that can be used to align the next level. As an alternative, a separate masking level to etch deep pits in the wafer, which will serve as alignment mark, can be done before well photolithography. Many submicron technologies are done this way to improve registration accuracy. It is important to understand that the well diffuses laterally as well as vertically. The diffusion is not isotropic in this case, however, because the oxidation process creates vacancies and interstitials that enhance diffusion. Since the oxidation is proceeding more rapidly in the exposed region, the well diffuses deeper than laterally. A typical ratio of width-to-depth diffusion is about 0.7.

After the well has been driven, the oxide is stripped, a pad oxide is grown, and the nitride layer is deposited for the LOCOS process. The nitride is patterned, and if a recessed LOCOS is desired, so are the oxide and the substrate. Field implants are normally done before oxidation. During the field oxidation phosphorus is rejected and boron is absorbed by the growing oxide. For n-well processes, this means that the threshold voltage of the parasitic PMOS transistors formed in the n-well is large and gets larger due to this dopant redistribution. The lightly doped p-type substrate, however, becomes even more lightly doped due to dopant redistribution. Furthermore, the fixed oxide charge also decreases the threshold voltage of the parasitic NMOS device and increases the threshold voltage of the parasitic PMOS device.

To avoid the formation of parasitic channels in the substrate in an n-well technology, boron implants are often used. (Twin-tub approaches have sufficient doping.) The n-well regions are often protected against this implant with an additional photomask. The active regions of the substrate must also be protected from the implant. This can either be done by relying on the thickness of the nitride layer to prevent implantation in the active region or by using a double-coat resist process (Figure 16.6). In the former case, BF_2 rather than B is often used as the field implant species so that the implant can be stopped by the thin nitride layer. In a double-coat process, the resist used to pattern the nitride is left on the wafer and hardbaked. A second layer of resist is applied and patterned for the implant mask. The difficult part of the process is to ensure that the first resist is stable upon the application of the second layer. This is primarily a matter of selecting the proper resists and bake cycles. Once the field implant is completed, the resist is stripped and the field oxide is grown (Figure 16.7). Following the oxidation, the residual nitride is stripped. The wafers are first dipped in HF to remove any SiO_2 that

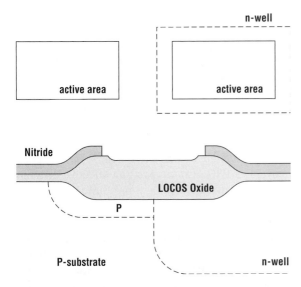

Figure 16.7 Top and cross-sectional views of a CMOS inverter after field oxidation.

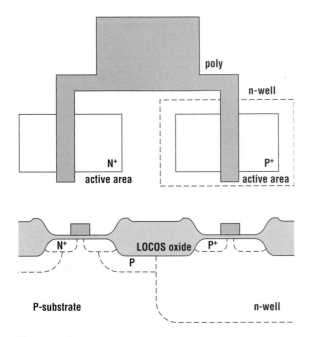

Figure 16.8 Top and cross-sectional views of a CMOS inverter after source/drain implants.

has grown on top of the nitride. They are then placed in a hot phosphoric bath to remove the nitride and thereby complete the device isolation.

The pad oxide is removed in HF and a thin oxide is regrown. This layer is used as a screen oxide for the threshold adjust implants. The implant masks are patterned in photoresist, and the implants are done. In the case of n-well technology, multiple implants are often done into the substrate, a deep one to prevent source to drain punchthrough of the NMOS transistor, and a shallower implant to set the threshold voltage. Boron is also typically implanted into the channel region of the PMOS transistor in an n-well. This occurs because the heavily n-doped polysilicon has a work function of about -0.4 V for 10^{16} cm^{-3} n-type layers (PMOS transistors) and about -1.1 V for similarly doped p-type layers (NMOS transistors). Because of this difference, the PMOS device has a large negative threshold voltage that must be raised toward zero. Typical implant doses are about 10^{12} cm^{-2} of boron. Often, part of the shallow NMOS implant is done without a mask, allowing it to implant the PMOS device as well.

Following the channel implants, the sacrificial oxide is stripped and the gate oxide is grown. Next the polysilicon is deposited and patterned. The polysilicon is usually heavily doped with phosphorus. This may be done by the addition of phosphine to silane in the low pressure CVD tube, or it may be done by doping the wafer with solid or liquid sources after undoped polysilicon has been deposited. Of course, the fidelity of the pattern transfer during this step is critical to the performance of the circuit. If the final poly gate length is too short, the source and drain may punch through. If the gate length is too long, I_{ds} will be decreased and the IC will operate slowly or not at all.

After polysilicon patterning, the channel masks are reapplied in sequence and the NMOS and PMOS source/drain regions are implanted (Figure 16.8). The polysilicon gate acts as an implant stop, preventing dopant species from reaching the channel. The field oxide defines the other three sides of the source/drain regions. The source/drain implant mask simply protects some of the active regions from receiving the implant.

The p-type impurity that is used almost exclusively is boron. The n-type implant species is normally arsenic, since it has a lower diffusion coefficient than phosphorus or antimony. Typical implant doses for the source/drain range from 1 to 5×10^{15} cm^{-2} for older technologies with relatively deep junctions. A great deal of work has been done to optimize the anneal cycles used to activate these high concentration implants and repair the implant damage. Typical processes for the 3-μm technology are 1000°C for 30 to 60 min. This may also include a thermal cycle to grow a thin oxide on top of the polysilicon.

After activation, a contact oxide is deposited. This usually contains phosphorus (PSG) and frequently boron as well (BPSG). Concentrations of phosphorus range from 4 to 12%, although high concentrations are generally avoided because of their ability to form phosphoric acid in the presence of moisture. Following deposition, the oxide is densified through a high temperature annealing process. Densification lowers the oxide etch rate making uniform, reproducible etching much more feasible. Again, temperatures of 1000°C are typical, this time in an oxidizing ambient. As mentioned in the last chapter, this oxide has the capability to reflow at high temperature to reduce the sharpness of vertical steps and so improve the metal step coverage. PSG also serves as an effective gettering agent. Ionic and metallic impurities become trapped in the oxide where they have little effect on the electrical operation of the device.

Next, the contact holes are patterned and etched into the wafer. The contact glass may be reflowed after etch by a final high temperature step. This has the effect of rounding the top edge of the contact, further improving the metal step coverage. The first level of metallization is applied, patterned, and etched. If desired, a second set of insulator and metal is applied and patterned. Finally, a sinter is done in H_2 or H_2/N_2 mixtures to form the ohmic contact. This step also reduces the interface state density of the FET devices. A thick layer of insulator, either SiO_2 or Si_3N_4, is then deposited for scratch protection and final passivation. Large vias are opened up so that contact can be made to the bonding pads. Figure 16.9 shows top and cross-sectional views of a completed CMOS inverter.

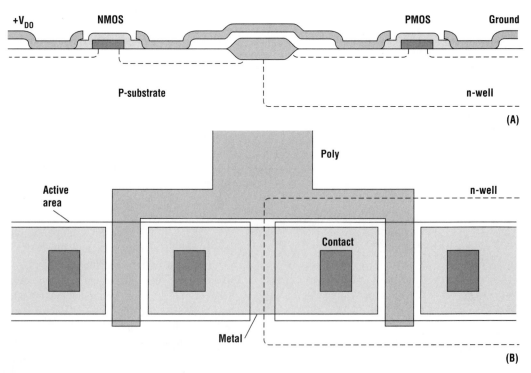

Figure 16.9   Top (A) and cross-sectional (B) views of a completed CMOS inverter fabricated with a simple 3-μm process.

16.4 Device Scaling

The performance of the baseline 3-μm process can be improved dramatically through device scaling. In this method, an improvement in the speed and packing density can be achieved through a reduction in the various physical lengths in the device. Several schemes can be constructed from the scaling rules shown in Table 16.1. The supply voltage is reduced by a factor of $1/k$. The transistor width, length, oxide thickness, and junction depth are all reduced by the factor $1/\lambda$. The doping in the channel is increased by a factor of λ^2/k. One of the first scaling methodologies that was proposed was constant-field scaling [9]. The intent of this approach is to keep the electric fields in the transistor unchanged from their long channel values. To do this, k is set equal to λ. Then, to first order the drive current of the MOSFET, as given by Equations 16.8 and 16.9, does not change if W is held fixed. In this method the increase in speed comes from two sources: the reduction in the voltage swing and the reduction in the capacitance that must be driven due to the smaller device sizes.

Although intellectually appealing, true constant-field scaling was not been widely applied originally. The approach that was used in scaling to 1.0 μm is closer to constant voltage scaling [10] in which $k = 1$. In this type of scaling, the voltage swing stays the same, but the drive current increases due to the increase in C_{ox}. Since the drive current increases roughly as the square of the supply voltage, constant-voltage scaling produces more speed improvement than constant-field scaling. Furthermore, since the supply voltage is held constant there is no need to reduce the threshold voltage. This avoids the increased "off-state" leakage that always occurs when V_t is reduced. The constant-voltage scaling approach is ultimately subject to limitations relating to the reliable operation of the devices at very high electric fields. As a result, commercial practice has used constant-voltage scaling until adequately reliable operation can no longer be maintained. Commonly, this is considered to be a 10- or 20-year life under some sets of worst-case operating conditions. The current standard has moved from 5.0 to 3.3 to 2.0 V with some devices now operating at 1.5 V. It is expected to make additional steps to 1.2 V and ultimately 0.8 V.

If the gate length of a MOSFET is reduced but these scaling rules are not followed, the MOSFET will operate in the short-channel regime. Short-channel MOSFETs have several liabilities for IC applications. As shown in Figure 16.10, the output impedance of a short-channel MOSFET is much smaller. The subthreshold slope also depends on the drain voltage in a short-channel MOSFET. Finally, the threshold voltage of the device drops sharply with the gate length. This effect further increases subthreshold leakage. The result is very poor control of the device characteristics.

Table 16.1 Generalized scaling theory for MOS transistors

Parameters	Variables	Scaling Factor
Dimensions	W, L, t_{ox}, x_j	$1/\lambda$
Potentials	V_{ds}, V_{gs}	$1/k$
Doping concentration	N_a, N_d	λ^2/k
Electric field	$\mathscr{E}$	λ/k
Current	I_{ds}	λ/k^2
Gate delay	T	k/λ^2

$1/\lambda$ is the factor by which the dimensions are scaled and $1/k$ is the factor by which the voltages are scaled.

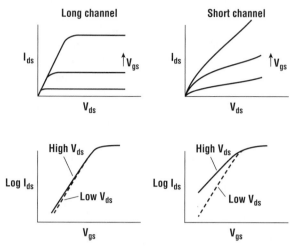

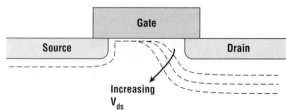

Figure 16.11 Schematic of device cross section showing enlarged depletion regions that give rise to short-channel behavior.

Figure 16.10 Current–voltage characteristics of MOSFETs comparing long- and short-channel behavior; typical threshold voltage dependence on gate length.

Figure 16.11 illustrates the cause of short-channel effects. Ideally, the charge under the gate should be controlled only by the gate electrode. For short-channel devices, the depletion layer associated with the drain removes a significant fraction of the charge under the gate. Brews et al. [11] have developed an empirical equation that has been used as a guide for the limits of short-channel behavior. It has been verified over the range of 1 to 10 μm and has also been applied successfully to deep submicron devices. Brews's rule is given as

$$L_{min} \approx 0.4[x_j t_{ox}(W_d + W_s)^2]^{1/3} \tag{16.12}$$

where x_j is the junction depth in microns, t_{ox} is the oxide thickness in angstroms, and W_s and W_d are the source and drain depletion widths, respectively, in microns. The depletion widths shrink due to increases in the channel doping upon device scaling. The oxide thickness can be reduced by changing the oxidation process. In so doing, care must be exercised to minimize the oxide fixed charge and interface state density (Sections 4.7 and 4.8). The reduction of the junction depth, however, is a more difficult problem, since it usually necessitates increasing the series resistance of the device.

To understand the effect of the parasitic series resistance on MOSFET performance, it is convenient to first introduce the idea of a mobility reduction factor θ [12]. As the vertical electric field increases, the carrier density in the channel increases. This in turn leads to an increase in both the electron–electron and electron–surface scattering rates. To model this effect, the basic equations that describe MOSFET operation are modified by the addition of an empirical mobility reduction parameter. For example, in the linear regime of operation for small V_{ds}, the drain-to-source current is given by

$$I_{ds} \approx C_{ox} \frac{W}{L} \cdot \frac{\mu_o}{1 + \theta(V_{gs} - V_t)} (V_{gs} - V_t)V_{ds} \tag{16.13}$$

Now consider the equivalent circuit shown in Figure 16.12. Inserting the effects of the parasitic resistance and rearranging terms,

$$I_{ds} \approx C_{ox} \frac{W}{L} \frac{\mu_o}{1 + \theta(V_{gs} - R_s I_{ds} - V_T)} [(V_{gs} - R_s I_{ds}) - V_t][V_{ds} - (R_d + R_s)I_{ds}]$$

$$\approx C_{ox} \frac{W}{L} \mu_o \frac{(V_{gs} - V_t)V_{ds}}{1 + [\theta + C_{ox}(W/L)\mu_o(R_s + R_d)](V_{gs} - V_t)} \tag{16.14}$$

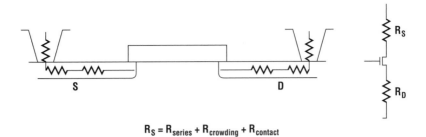

$$R_S = R_{series} + R_{crowding} + R_{contact}$$

Figure 16.12   Equivalent circuit for MOSFET with source/drain series resistances displayed.

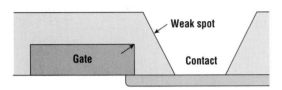

Figure 16.13   Gate to source/drain shorts caused by misalignment and/or contact blowout during etch.

Thus, if the parasitic resistance associated with the source/drain diffusions is large, it has the effect of substantially increasing the apparent mobility reduction factor and so reducing the transconductance of the device [13]. The effect is particularly acute at small gate lengths due to the increase in the intrinsic device performance. It has been shown that traditional source/drain structures can reduce I_{ds} by a factor of 2 for gate lengths of 0.2 μm [14], even if the source and drain resistance can be held constant during device scaling.

To achieve shallower junctions, the technologist may decrease the characteristic diffusion length of the activation step, reduce the implant dose, and for boron implantation, preamorphize and reduce the effective implant energy through the use of molecular implants. Of course, the reduction of the implant dose is undesirable since it would increase the parasitic resistance. Furthermore, if the surface concentration decreases as a result of the reduced dose, the contact resistance will increase sharply as discussed in Chapter 15. This resistance has exactly the same effect as the series resistance of the source/drain diffusion.

As discussed in Chapters 3 and 6, a great deal of effort has gone into the development of processes that reduce the thermal cycle during the activation process. Of course, this is subject to the limitation that the implant damage must also be sufficiently annealed. This in turn implies high temperature annealing. Dislocation loops and other extended defects resist annealing unless the temperature is about 1000°C. To allow these high temperature anneals with minimal diffusion, rapid thermal processing techniques have proliferated (Chapter 6).

One way to reduce the series resistance is to minimize the separation between the contact and the gate. Figure 16.13 shows the concern of having a metal contact too close to the gate of the MOSFET. The top edge of the gate is typically a 90° angle that produces a large electric field. If the contact is sufficiently close, the oxide between the gate and the contact can easily be ruptured. The contact may come too close due to misalignment or overetching of the contact. The latter is particularly likely if the contact edges must be tapered, and an oxide planarization process that results in varying oxide thicknesses has been used. Since this breakdown is likely to occur while the device is in use, device designers have been reluctant to push this rule very far in spite of the density and performance consequences.

The second effect that further reduces the performance of the MOSFET for submicron devices is the spreading resistance of the contact. For bipolar transistors and other devices with a vertical current flow, the contact resistance is proportional to the specific contact resistance as

$$R = \frac{R_c}{\text{area}} \tag{16.15}$$

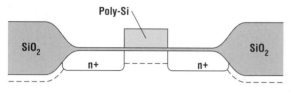

(A) Form standard device up to diffusion

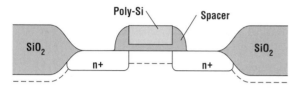

(B) Form sidewall oxide spacers

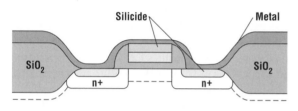

(C) Deposit metal, react to form silicide

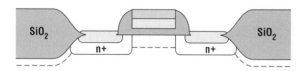

(D) Selectively remove unreacted metal

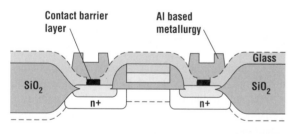

(E) Final structure after glass passivation, reflow, contact opening and metallization

Figure 16.14   Common salicide process sequence (*after Ting, © 1984 IEEE*).

assuming that the contact size is not very small. In the case of an MOS device, however, if one assumes that the contact extends the entire width of the device, the effective contact resistance is given by

$$R = \frac{\sqrt{R_c \rho_D}}{W} \coth\left[L \sqrt{\frac{\rho_D}{R_c}} \right] \qquad (16.16)$$

where ρ_D is the sheet resistance (in $\Omega/\square$) of the diffusion, L is the length of the contact, and W is the width of the transistor and contact. This equation is the 1-D transmission line model. It has two limiting forms that can be used for estimates with about 10% accuracy:

$$R \approx \frac{R_c}{WL} \qquad L < 0.6 \sqrt{\frac{R_c}{\rho_D}}$$

and

$$R \approx \frac{\sqrt{R_c \rho_D}}{W} \qquad L > 1.5 \sqrt{\frac{R_c}{\rho_D}} \qquad (16.17)$$

Small contacts that are required to increase the circuit density will therefore increase the parasitic resistance. Several process solutions have been developed for this problem. The most widespread is the use of the self-aligned silicidation (salicide) of the contact area. As shown in Figure 16.14, in most salicide processes the source/drain diffusions are formed first. Then, a thin metal film is deposited and allowed to react with the exposed active regions of the device. The result is a substantial increase in the metal/semiconductor contact area. The success of this approach depends critically on diffusion and silicon consumption that occur during the silicidation process. For example, one of the most popular silicides, $TiSi_2$, was originally restricted to junctions greater than 0.35 μm [15] to avoid excessive leakage. It was found that to use $TiSi_2$ on more shallow junctions, the Ti thickness must be no more than about one-fifth of the original junction depth [16]. Silicides formed over polysilicon may also invert at high temperature. That is, the polysilicon diffuses through the silicide layer to the surface, and the silicide sinks until parts of it come into contact with the gate oxide. Due to the work function difference, these parts of the transistor will have a different threshold voltage, and the device will have very poorly controlled behavior. This problem can be minimized by ensuring a large grain structure in the poly before silicide formation [17].

When this problem is avoided, high temperature steps after $TiSi_2$ formation must still be limited to 900°C because of agglomeration of the silicide layer [18, 19]. Agglomeration effects are aggravated if the thin native silicon oxide, and therefore the silicide reaction, is nonuniform. To solve this problem, the silicide can be formed through a 50-Å thermal oxide. Silicides formed this way are stable against agglomeration up to 1100°C [20]. It is also possible to form the silicide first, then implant the source and drain regions into the silicide [21]. This technique can be used to form very shallow silicided junctions of $CoSi_2$ [22]. Another liability for self-aligned silicides is oxygen incorporation during the thermal silicidation, leading to higher than expected resistivity and poor morphology. Specially designed furnaces can be used to minimize oxygen [23]. The more popular technique, however, is to react the silicide in a rapid thermal processor. The reduced time at temperature reduces and compact chamber size reduce the oxygen incorporation [24].

The use of a salicide process can significantly reduce the parasitic series resistance of the MOSFET. For a salicide, the total resistance between the metal contact and the edge of the reachthrough is given by [25]

$$R = \frac{L_{ms}}{W} \frac{\rho_D \rho_S}{\rho_D + \rho_S} + \frac{2\rho_S \rho_D + (\rho_S^2 + \rho_D^2) \cosh \beta L_{ms}}{W\beta(\rho_S + \rho_D) \sinh \beta L_{ms}} \tag{16.18}$$

where W is the width of the transistor, L_{ms} is the distance between the metal contact and the edge of the silicide, ρ_S and ρ_D are the sheet resistivities of the silicide and diffusion (in $\Omega/\square$), and β is given by

$$\beta = \sqrt{\frac{\rho_D + \rho_S}{R_c}} \tag{16.19}$$

where R_c is the specific contact resistance (in $\Omega\text{-cm}^2$) between the silicide and the substrate.

The choice of which silicide to use for a salicide process hinges on the resistivity required, the number of contacts expected to n- and p-type substrates, and the subsequent processing that must be done (Table 16.2). In the case of a predominantly NMOS application for example, $MoSi_2$ may be a better choice than $TiSi_2$, since its lower barrier height provides a lower R_C that more than compensates for the higher series resistance. A second consideration is the dominant diffuser. Platinum will readily form rather good contacts, since the platinum diffuses to the interface, providing a low contamination contact. Silicides in which Si is the dominant diffuser have a problem with the salicidation process called *wicking*. Wicking occurs during the silicide formation anneal when the diffusing silicon results in a thread of silicide that short circuits the gate to the source or drain. Furthermore, the oxide spacer itself, particularly if the oxide is heavily damaged during the sidewall spacer formation, can be a source of silicon for silicide formation.

Table 16.2   Properties of commonly used materials for self-aligned silicides

Silicide	Resistivity ($\mu\Omega$-cm)	Anneal Temperature (°C)	Dominant Diffuser	n-Type Barrier (eV)
$TiSi_2$	13–16	900	Si	0.6
$TaSi_2$	35–45	1000	Si	0.59
$MoSi_2$	90–100	1100	Si	0.55
WSi_2	70	1000	Si	0.65
$CoSi_2$	16–20	900	Co	0.65
PtSi	28–35	600–800	Pt	0.86

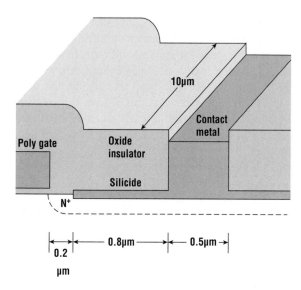

Figure 16.15   Cross section and dimensions of MOSFET example.

For very deeply scaled transistors, however, $TiSi_2$ poses a significant problem. The process for forming a self-aligned silicide is to anneal the Ti at about 600°C to form the C49 phase of $TiSi_2$. The unreacted Ti is then stripped off in aqua regia. Finally the wafer is annealed at 800 to 900°C to form the low resistance (C54) phase [26]. This phase transformation begins at triple grain boundaries. Since the C49 phase typically has a grain size of 0.17 to 0.22 μm, however, very small devices may not have a triple grain boundary [27]. Under these conditions, the film will be highly resistant to phase transformation, and some transistors will have unacceptable high series resistance. Numerous attempts to solve this problem have been proposed, including pre-amorphization of the silicon to force smaller C49 grains [28] and low dose molybdenum implants before Ti deposition to attempt to form epitaxial $TiSi_2$ [29].

The semiconductor industry has replaced the $TiSi_2$ process with $CoSi_2$ for devices well below 0.25 μm in gate length. Unfortunately Co oxidizes readily in air at room temperature. Thus one cannot simply sputter Co on top of the wafer and then react it with the silicon in a separate system. To solve this problem, the cobalt film may be capped with TiN in the deposition system [30]. Silicide formation is typically done at low temperature, followed by a

---

Example 16.1

An NMOS transistor has the dimensions shown in Figure 16.16. The source/drain diffusion has a sheet resistance of 100 $\Omega/\square$, and the surface concentration is 10^{20} cm^{-3} n-type. A 1-μm-wide contact had been used on both the source and drain of an MOS transistor. This was replaced by a self-aligned 10-$\Omega/\square$ silicide that extends to 0.2 μm from the gate edge. Calculate the resistances before and after the implementation of the silicided process. The contact metal and the silicide both have barriers of 0.72 V to the silicon. The mass of an electron in silicon is 1.1 m_o, and the constant A_o in Equation 15.14 is 10^{-12} Ω-cm^2.

Solution

The constant C_2 in Equation 15.15 is given by

$$C_2 = \frac{4\pi}{6.6 \times 10^{-34} \text{ J-s}} \sqrt{1.1 \times 9.1 \times 10^{-31} \text{ kg} \times 11.9 \times 8.84 \times 10^{-12} \text{ F-m}}$$

$$= 1.95 \times 10^{14} \text{ m}^{-3/2} \text{ V}^{-1} \tag{16.20}$$

Then

$$R_c = 10^{-12} \text{ }\Omega\text{-cm}^2 \exp\left[\frac{1.95 \times 10^{14} \text{ m}^{-3/2} \text{ V}^{-1} \times 0.72 \text{ V}}{\sqrt{10^{26} \text{ m}^{-3}}}\right]$$

$$= 1.24 \times 10^{-6} \text{ }\Omega\text{-cm}^2 \tag{16.21}$$

For aluminum the contact resistance is given by

$$R = \frac{\sqrt{100\ \Omega/\square}\ 1.24 \times 10^{-6}\ \Omega\text{-cm}^2}{1 \times 10^{-3}\ \text{cm}}\ \coth\left[5 \times 10^{-5}\ \text{cm}\ \sqrt{\frac{100\ \Omega/\square}{1.24 \times 10^{-6}\ \Omega\text{-cm}^2}}\right]$$

$$= 26.5\ \Omega \tag{16.22}$$

The total resistance to the edge of the gate is

$$R = 26.5\ \Omega + 100\ \Omega/\square\ \frac{10^{-4}\ \text{cm}}{10^{-3}\ \text{cm}} = 36.5\ \Omega \tag{16.23}$$

For silicided contacts

$$\beta = \sqrt{\frac{100\ \Omega/\square + 10\ \Omega/\square}{1.2 \times 10^{-6}\ \Omega\text{-cm}^2}} = 9.6 \times 10^3\ \text{cm}^{-1} \tag{16.24}$$

The total resistance for this type of contact is given by

$$R = 100\ \Omega/\square\ \frac{0.2\ \mu\text{m}}{10\ \mu\text{m}} + \frac{0.8\ \mu\text{m}}{10\ \mu\text{m}}\ \frac{100\ \Omega/\square \times 10\ \Omega/\square}{100\ \Omega/\square + 10\ \Omega/\square}$$

$$+ \frac{\begin{array}{c}2 \times 100\ \Omega/\square \times 10\ \Omega/\square + [(100\ \Omega/\square)^2 \\ + (10\ \Omega/\square)^2]\ \cosh(9.42 \times 10^3\ \text{cm}^{-1} \times 8 \times 10^{-5}\ \text{cm})\end{array}}{\begin{array}{c}1 \times 10^{-3}\ \text{cm} \times 9.42 \times 10^3\ \text{cm}^{-1} \times (100\ \Omega/\square \\ + 10\ \Omega/\square) \times \sinh(9.42 \times 10^3\ \text{cm}^{-1} \times 8 \times 10^{-5}\ \text{cm})\end{array}} \tag{16.25}$$

$$R = 2.0\ \Omega + 0.7\ \Omega + 17.6\ \Omega = 20.3\ \Omega \tag{16.26}$$

The dimensions in this example are compatible with a 0.5-μm device. If one assumes an L_{eff} of 0.4 μm, a gate oxide thickness of 100 Å, and an electron mobility of 400 cm^2/V-sec, one can show that even this modest reduction in resistance can have a significant effect on the device performance. Furthermore, the resistance of the salicide is independent of the size of the metal to silicide contact and so can be scaled to very small feature sizes.

---

750 to 850°C anneal to complete the reaction [31]. Alternatively, the Co can be deposited at an elevated substrate temperature to form the silicide *in situ*, which may then be followed by an *ex situ* high temperature anneal [32]. In either case, the anneal cycles have to be optimized to avoid the problem of leakage current due to nonuniform silicide formation. This leads to spikes of CoSi$_x$ that can penetrate the N$^+$/p junction. Spikes are found to grow dramatically for films annealed between 400 and 450°C, but are sharply suppressed by anneals between 800 and 850°C [31].

As devices scaled into the deep submicron, a serious concern was raised about the operation of the p-channel device. If N$^+$ poly continues to be used for the gate electrode, the well concentration necessary to prevent punchthrough produces a threshold voltage of −1.5 to −2.0 V. To obtain the desired value of about −0.5, boron (a p-type dopant), must be implanted into the surface of the channel. The concentration of boron required normally forms a metallurgical junction with the n-type dopant of the well, although at zero gate bias, this region is depleted. As the gate voltage is moved toward V_t, this doping profile causes the channel to form not at the surface, but in the bulk. The PMOS device, therefore, has a buried channel for gate voltages close to threshold. Although a buried channel has some minor advantages of improved mobility and some types of reliability, the threshold voltage of the device is much more difficult to control, the subthreshold leakage is larger, and the devices are much more susceptible to punchthrough.

Although it is possible to achieve some reduction in buried channel behavior by minimizing the range of the boron threshold voltage adjust implant or adding an n-type implant [33], the most effective way to eliminate the problem is to ensure that both NMOS and PMOS are surface channel devices. This can be achieved with a symmetric gate; that is, the gate electrode material is chosen to have a work function that puts its Fermi energy near the center of the silicon bandgap. Three materials that have been investigated are $MoSi_2$, Mo [34], and W [35]. Although MOS fabrication is possible using these materials, stress in the deposited layer tends to reduce the oxide quality. Furthermore, at very short gate lengths, these materials allow buried channels to form in both NMOS and PMOS devices.

As a result, the industry has adopted the use of dual gate materials. A simple example of this approach is the use of N^+ poly for the NMOS gate and P^+ poly for the PMOS gate [36]. This can be done by depositing undoped polysilicon and doping it during the source/drain ion implantation step. This approach presents several difficulties. Of course, the P^+ poly has a higher resistivity for the same dopant concentration, since the hole mobility in silicon is lower than the electron mobility. A problem also occurs in polysilicon interconnect. If the N^+ poly and the P^+ poly contact each other directly, as is commonly done for inverters and most other basic logic functions, a tunnel junction will be formed. Also, interdiffusion of the dopants can lead to undesirable work function variations. This is particularly a problem when a self-aligned silicide is used, as the diffusivity of dopants is extremely rapid in most silicides [37]. The polysilicon on the field oxide must therefore remain undoped and thermal processing minimized after dopant introduction. Of course, the polysilicon layer must then be silicided to shunt the large resistivity of the undoped regions. The most serious problem with this technique, however, is boron penetration [38]. Grain boundary diffusion in the poly allows the boron to diffuse rapidly to the poly/SiO_2 interface. The thermal cycles, poly thickness, oxide thickness, poly grain size, implant dose, and implant energy must all be optimized to prevent boron punchthrough of the oxide. Fluorine incorporation in the oxide is known to exacerbate this problem [39]; however, nitrided oxides appear to be effective in reducing boron penetration [40]. This dual gate doping approach is now widely used in CMOS manufacture.

16.5 Hot Carrier Effects and Drain Engineering

To maximize performance, device scaling from 3 μm to about 1 μm primarily followed a constant voltage rather than a constant-field approach. As a result, electric fields in the devices rose sharply. The elastic mean free path for excited electrons in silicon at room temperature is about 100 Å. When the electric field is high enough that the energy gained between collisions exceeds the bandgap of the semiconductor by a sufficient amount ($\sim$30%), impact ionization events can occur (Figure 16.16). Some of the holes so created can recombine with other high energy electrons, creating a population of extremely high energy electrons.

$$e^* \rightarrow e + e + h$$
$$e^* + e + h \rightarrow e^{**}$$

(16.27)

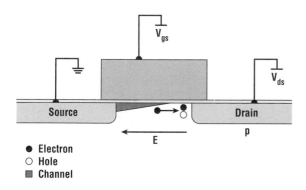

● Electron
○ Hole
■ Channel

Figure 16.16   Impact ionization process in a MOSFET operated in pinchoff.

It is also possible for a small number of electrons to gain enough energy directly by traveling a number of mean free paths before suffering an inelastic collision. In either case, these electrons have sufficient energy to change the physical and therefore electrical structure of the device. This is known as hot carrier effects. The remainder of the holes escape into the substrate and are collected in the form of substrate current.

A few of these high energy electrons can surmount the oxide/semiconductor barrier. Once in the oxide, they can cause damage directly, or they can create positive charge either by impact ionization in the oxide or by positive hydrogen ion creation at the gate electrode. This charge may become trapped in the oxide, leading to shifts in the threshold voltage and in the transconductance. It is currently believed that at least one important degradation mechanism involves the displacement of hydrogen atoms at the oxide/semiconductor interface [41]. These atoms passivate the interface, tying up unsatisfied interface bonds. Only about 0.3 eV is required to break such a bond. Once released, the bonds leave interface states or other defects behind that may trap charge. The effect is seen primarily at the drain edge [42] and occurs most quickly when the device is held in saturation with a large drain bias and the gate biased near $V_{ds}/2$. The damage is particularly evident when the transistor is used in both forward and reverse configurations and when dealing with small devices at large electric fields. Modeling work has suggested that if constant-field scaling is employed, these effects will become less important for gate lengths less than 0.2 μm [43], due to the small supply voltages. However, hot electron effects have been observed in devices as small as 0.07 μm [44]. Due to incomplete scaling of the voltage, this effect, along with power consumption, is expected to limit the maximum allowable supply voltage for deep submicron devices [45]. In recognition of this limit, scaling methods have been developed that keep the maximum electric field, rather than the average electric field, constant [46].

One of the most effective ways to minimize hot carrier effects is to develop techniques that reduce the maximum electric field in the device. Since the total voltage that must be dropped from drain to source is not to be changed, the distance over which that voltage is dropped must be increased. In particular, at the drain edge when the device is in saturation, the drain/substrate junction operates like a reverse-biased p–n diode. The maximum electric field occurs near the metallurgical junction. For a simple n-channel MOSFET with a gate oxide thickness less than 150 Å and gate lengths less than 0.5 μm, one can estimate the peak field as [47, 48]

$$\mathscr{E}_{max} \approx \frac{V_{ds} - V_{sat}}{I} \approx \frac{V_{ds} - \dfrac{\varepsilon_{sat}L[V_{gs} - V_t]}{V_{gs} - V_t + \varepsilon_{sat}L}}{1.7 \times 10^{-2} t_{ox}^{1/8} x_j^{1/3} L^{1/5}} \tag{16.28}$$

where ε_{sat} is the critical field for velocity saturation (~40 kV/cm), L is the gate length, and x_j is the junction depth.

Example 16.2

A transistor has L = 0.5 μm, t_{ox} = 125 Å, x_j = 0.2 μm, and V_t = 0.7 V. Calculate the maximum electric field at several gate voltages for V_{ds} of 5 V and 3.3 V. Assume $V_{gs} = V_{ds}/2$.

Solution

Inserting the values given into Equation 16.28, the maximum electric field will be

$$\mathscr{E}_{max} = 8.6 \times 10^4 \, cm^{-1} \left[V_{ds} - \frac{2[V_{gs} - V_t]}{V_{gs} - V_t + 2} \right] \qquad (16.29)$$

Then the peak fields are approximately 3.6×10^5 V/cm for $V_{ds} = 5$ V and 2.3×10^5 V/cm for $V_{ds} = 3.3$ V.

At large electric fields, a hot electron need travel only about four mean free paths in order to achieve sufficient energy for impact ionization to occur. (Due to momentum conservation the electron must gain more energy than the bandgap to suffer impact ionization. This is usually taken to be about an additional 30%.) In more realistic 2-D simulations, fields approaching 10^6 V/cm have been predicted for this type of structure. The class of process modules used to reduce this peak electric field near the drain/substrate metallurgical junction is referred to as *drain engineering*, although the process modules are usually applied to both sides of the device. These techniques are usually applied only to NMOS devices, since they are far more prone to hot carrier damage than PMOS devices.

The process module most commonly used to control hot carrier effects is the lightly doped drain (LDD) technique [49]. Since the concept of sidewall spacers was already introduced for salicide formation, it is simple to add a moderate dose implant step before sidewall formation and to reserve the heavier source/drain implant for after the sidewall formation (Figure 16.17). The moderate implant step is called the *reachthrough*. The effect of this process sequence is to put, between the contact drain and the channel, a region whose length and doping level can be easily controlled to reduce the peak electric field and with it hot electron degradation. The LDD structure, however, adds additional parasitic resistance. An optimization process for the reachthrough doping, depth, and length must be carried out in which the transistor performance must be balanced against the device reliability. Furthermore, if the doping in the reachthrough is too low, the hot electrons that are trapped in the oxide above the reachthrough can deplete the reachthrough region, substantially increasing the resistance and bringing about a premature failure. To provide the best hot electron resistance, the n-region should be completely under the gate. Notice that this is not the case for the simple LDD process presented here.

Several variations of the basic LDD process, including moderately doped LDD (MLDD) [50], profiled LDD [51], and the gate–drain overlap (GOLD) structure [52] have been proposed to improve the performance of the basic LDD. The latter is designed to provide a large overlap of the gate electrode and the reachthrough. This improvement can also be obtained in a conventional LDD device through the use of high dielectric constant spacers. The larger capacitance causes the fringing fields to modulate the field in the reachthrough as it would if the reachthrough were under the gate in a conventional spacer approach [53]. The most commonly used material is Si_3N_4. Since the interface of nitride and silicon often contains a high density of interface states, a thin thermal oxide is often grown before nitride deposition. One interesting proposed device is the buried LDD. In this process, a heavily doped reachthrough is formed with a high energy implant. This forces the current to flow down away from the $Si–SiO_2$ interface. Although hot carriers are still produced, they lose sufficient energy before reaching the interface so that few can penetrate the oxide. An even larger improvement can be made if the buried LDD is also graded. This is done by performing both high energy arsenic and high energy phosphorus implants [54].

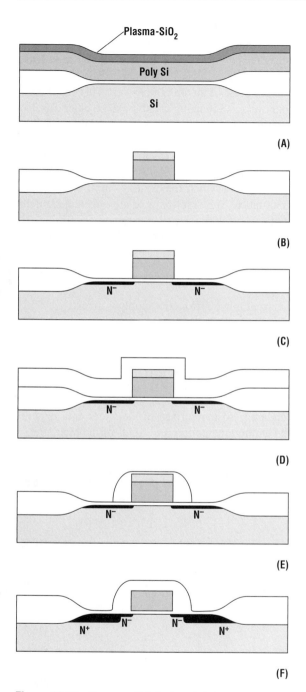

Figure 16.17 Common lightly doped drain (LDD) process sequence.

The LDD process has three drawbacks. The first is the added process complexity, although when a self-aligned silicide is formed, this is minimal. The second drawback is that the spacer adds a parasitic resistance between the metal contact and the channel. As a result, LDD devices are slower than non-LDD devices operating at the same voltage. The third problem with the LDD process is shadowing. Recall that most implantation is done off axis to avoid the problem of channeling. In the case of the LDD structure, the polysilicon gate acts as an effective shadow that blocks implantation of one side of the device. The result is that the reachthrough may not extend all the way under the gate. This effect results in unusual MOSFET electrical characteristics [55]. When the shadowed side is used as the drain, little performance degradation is seen. When it is used as the source, the drive current in saturation may be reduced by 30 to 40%. This effect is avoided by either implanting on axis or by ensuring that the final poly gate profile has at least a 7° slope to it. Shadowing may also be caused by a polysilicon oxidation process (called a *reox*) that is done after the gate level patterning.

16.6 Latchup

One of the early limitations of CMOS technology was latchup. Although now well understood, latchup continues to be a concern, and scaled CMOS technologies must be constructed to avoid this problem. Latchup is seen as a sudden massive increase in the current draw of an IC. The current may be high enough to rupture metal interconnect lines. As a result, latchup is often destructive. If the supply current is limited, the latchup condition can be eliminated by powering down the circuit, then reapplying power.

The most common form of latchup occurs as a result of parasitic elements in a CMOS IC [56]. Figure 16.18A shows a cross section of an n-well CMOS wafer with the various parasitic elements drawn in. The transistors are connected as an inverter. Explicitly added to this picture are the substrate contacts for both the well and the p-type wafer. A lateral NPN bipolar transistor ($Q_{\parallel}$) is formed from the drain of the NMOS transistor, the p-substrate, and the n-well. A vertical PNP bipolar transistor ($Q_{\perp}$) is formed from the p$^+$ source of the PMOS transistor, the n-well, and the p-substrate. The collector of the lateral NPN is connected to the base of the PNP through the n-well. The collector of the vertical PNP is connected to the base of the NPN through the

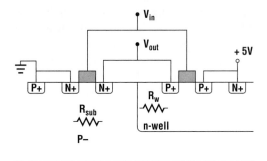

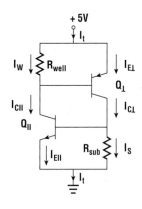

(A)

(B)

Figure 16.18 (A) Cross section of an n-well CMOS inverter with parasitic elements superimposed. (B) Equivalent circuit.

p-substrate. The equivalent circuit for the parasitics in this inverter is shown in Figure 16.18B [57].

To understand how this circuit latches, consider that at some time, a sufficiently large current is allowed to flow through the collector of the NPN transistor. This could occur by a voltage spike, an ionizing event, or some other transient [58]. The current through the well resistance biases the base of the PNP. If this bias is large enough, transistor action forces current to flow in the collector of the PNP transistor. This current flowing through the substrate resistance can bias the base of the NPN, which forces more current to flow into the well resistance. The result is a positive feedback circuit that will not turn off until the power is removed [59].

To analyze the circuit, assume forward active operation of the transistors. Then neglecting leakage,

$$I_c = \alpha I_e \tag{16.30}$$

The total current flowing from the supply is given by

$$
\begin{aligned}
I_t &= I_{c\parallel} + I_{c\perp} \\
&= \alpha_{\parallel} I_{e\parallel} + \alpha_{\perp} I_{e\perp} \\
&= \alpha_{\parallel}(I_t - I_s) + \alpha_{\perp}(I_t - I_w)
\end{aligned} \tag{16.31}
$$

where the $\parallel$ and $\perp$ subscripts refer to the transistors parallel and perpendicular to the surface of the wafer, and α is the ratio of the emitter and collector currents of the bipolar transistors. Collecting terms,

$$I_t \left[\alpha_{\perp} + \alpha_{\parallel} - 1 - \alpha_{\perp} \frac{I_w}{I_t} - \alpha_{\parallel} \frac{I_s}{I_t} \right] = 0 \tag{16.32}$$

For a nontrivial solution,

$$\alpha_{\perp} + \alpha_{\parallel} - 1 - \alpha_{\perp} \frac{I_w}{I_t} - \alpha_{\parallel} \frac{I_s}{I_t} = 0 \tag{16.33}$$

If we express this equation in terms of β, the ratio of the base and collector currents of the transistors, we can show that

$$\beta_{\perp} \beta_{\parallel} = 1 + \frac{I_w}{I_t} \beta_{\perp}[\beta_{\parallel} + 1] + \frac{I_s}{I_t} \beta_{\parallel}[\beta_{\perp} + 1] \tag{16.34}$$

where $\beta = \alpha/(1 - \alpha)$.

Equation 16.34 points out several important effects. Since the right-hand side of Equation 16.33 must be greater than 1, the product $\beta_{\perp} \beta_{\parallel}$ must be greater than 1 for latchup to occur. If this product can be made less than unity, latchup cannot occur. For a uniformly doped ideal bipolar transistor, the gain is given by [60]

$$\beta = \frac{D_B N_E L_E}{D_E N_B W} \tag{16.35}$$

where D_B and D_E are the minority carrier diffusivities in the base and emitter, N_B and N_E are the doping concentrations in the base and emitter, L_E is the characteristic diffusion length of minority carriers in the emitter, and W is the width of the base. More generally this can be rewritten as

$$\beta = \frac{D_B N_E L_E}{D_E \int_0^W N(x)\, dx} \qquad (16.36)$$

The lateral transistor typically has a gain less than unity and is sometimes as low as 0.01, but the vertical device often has a gain much larger than one.

　　The most direct ways to avoid latchup, therefore, are to increase the base width, which is the separation between the NMOS and PMOS transistor and the well depth, to increase the base doping so that the gain product of the two transistors is less than unity, or to use guard rings that can absorb the injected charge, thereby preventing the bipolar action. Any of these approaches incurs a penalty in density. There are a number of ways to minimize this penalty. These techniques either tend to reduce the parasitic resistance, and therefore increase I_{sub} and/or I_{well}, which allows the technology to avoid latchup with large $\beta_\perp \beta_\parallel$ products, or they tend to reduce the gain of the parasitic transistors. The two goals sometimes overlap. Increasing the doping concentration in the substrate and the well decreases both the gain and the resistance. Of course, the well concentration at the surface must satisfy the device constraints needed by the MOSFET. This has led to the development of retrograde wells in which the doping concentration rises deep in the well. This is done using high energy implantation of phosphorus for the well shunt and a lower energy implant for the surface profile. Although this technique is capable of making some improvement in latchup susceptibility, it does not eliminate it.

　　Two techniques have come into widespread use that dramatically reduce latchup susceptibility. In the simplest, the devices are fabricated in lightly doped epitaxial layers grown on heavily doped substrates [61]. This technique provides two avenues of relief. The low resistance substrate effectively shunts the substrate resistance, substantially increasing I_{sub} and thereby increasing the maximum allowable gain product. The substrate also serves as an effective sink for the minority carriers in the base. Since the substrate is heavily doped, minority carriers that reach the substrate quickly recombine. This prevents the bipolar action of the device and effectively reduces latchup susceptibility. The effectiveness of the epi increases as the thickness of the epi decreases. Typical CMOS epi thicknesses start at about 5 μm. After the well drive, updiffusion may shrink this to about 3 μm. Figure 16.19 shows the result of epitaxial thickness on latchup susceptibility. Problems with the epi approach include the cost of the starting material and defects in the epi that reduce the IC production yield. These considerations preclude the use of epi in very cost-sensitive products such as DRAMs.

　　Trench isolation has also proven effective in improving latchup susceptibility, particularly at small N-to-P spacings [63]. If the trench depth is considerably greater than the well depth, the carriers are forced to travel around the trench. This increases the effective base length and forces the current to flow in a 2-D manner that reduces the transistor gain [64]. Trench depth is found to have a much stronger impact

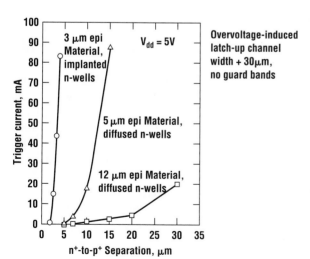

Figure 16.19　Effects of epitaxial layer thickness on latchup (from ref. 62).

than trench width. If a P^+ ring is implanted into the bottom of the trench, latchup is further suppressed. Proper design can force the voltage necessary to maintain the latched condition to well above the supply voltage used at deep submicron, making the devices latchup free, even when fabricated on bulk substrates without the implanted plug [65].

16.7 Shallow Source/Drains and Tailored Channel Doping

As CMOS devices continue to scale into the nanoscale, there is a continuing need to reduce the depth of the source/drain junction [66]. It has been shown that [67] drain-induced barrier lowering (DIBL) decreases the threshold voltage of the device as

$$\Delta V \propto \exp[-kL x_j^{-1/3} t_{ox}^{-1}]$$

where L is the gate length and k is a constant that is technology dependent. Thus one must reduce the junction depth as the transistor is scaled. Some increase in the sheet resistance of the source/drain diffusion before silicidation is inevitable. Most deeply scaled technologies have N^+ sheet resistance of 500 to 1000 $\Omega/\square$. The surface concentration in particular is extremely significant since the specific contact resistance is a sensitive function of this parameter. A second need for advanced doping technology is to adjust the doping profile, both in depth and along the length of the channel, to optimize the device performance or to control short-channel effects. This becomes increasingly important at short gate lengths. In some cases the doping profile can be used to minimize short-channel behavior.

Shallow source/drain formation requires two steps: the introduction of impurities close to the surface of the wafer, and the minimization of the redistribution of those impurities. The ideal result is a box profile, one with a large, constant concentration up to the junction at which point the concentration drops discontinuously to zero. An additional goal is often attaining active dopant concentrations that exceed the solid solubility. This requires annealing the wafer for such short times that the impurities atoms cannot cluster and precipitate. Generally, the further from equilibrium (i.e., the solid solubility), the shorter the allowable time at temperature. Extreme nonequilibrium results require anneal times much less than 1 sec, and dopant loss can occur in subsequent thermal cycles.

The formation of shallow profiles was discussed in Chapter 5. To briefly review, one can use molecular implants to reduce the effective energy of the impurity atom, or one can use a newer type of implanter capable of implant energies in the one kiloelectron-volt range. The challenge of making low sheet resistance extremely shallow junctions is particularly acute for forming P^+ profiles, since the projected range and the diffusivity of boron are substantially larger than that of arsenic at the same energy. The dose of extremely shallow implants is often reduced from the low to mid 10^{15} cm^{-2} to the mid to high 10^{14} cm^{-2} to avoid the formation of silicon dopant atom compounds such as SiB_4, which will make poor ohmic contacts [68].

The goal of the annealing step is to activate a high percentage of the impurities while minimizing residual defects and redistribution. As discussed in Chapter 6, this has motivated much of the work on rapid thermal annealing. A single ion can cause up to 400 interstitial-vacancy pairs [69]. These excess defects lead directly or indirectly to transient-enhanced diffusion, particularly for large doses. After the initial low temperature ($\sim$550°C) annealing of nearby point defects, approximately one additional point defect exists for every implanted ion [70]. One approach to minimizing this transient-enhanced diffusion relies on coimplantation of carbon or some other impurity that will form defect clusters outside the active device region that will scavenge point defects such as self-interstitials [71].

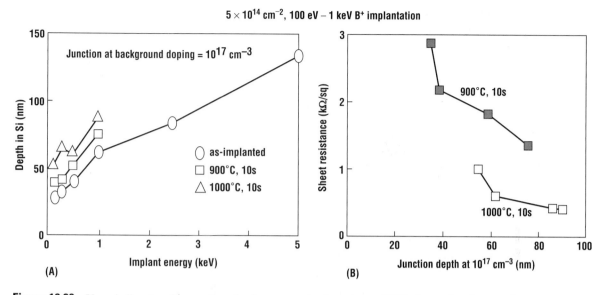

Figure 16.20   Very shallow junction results using low energy implantation and RTA. Increasing the anneal temperature increases the junction depth (A), but produces lower sheet resistance (B).

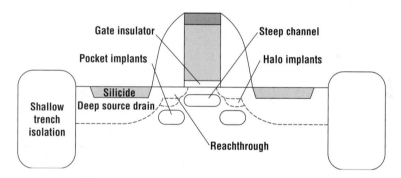

Figure 16.21   Cross section of deep submicron MOSFET showing optional channel engineering implants.

Figure 16.20 shows the effect of implant energy and anneal cycle on junction depth and sheet resistance for a 5×10^{14} cm^{-2} implant of boron. Lowering the implant energy is, to first order, the most important process variation for reducing the junction depth [72]. Higher temperature anneals will produce lower sheet resistance diffusions, but anneal times must be kept as short as possible. Some technologists believe that millisecond anneals at temperatures as high as 1100°C will be needed for very small devices. Several new approaches to dopant introduction are also under intensive investigation, including pulsed gas immersion laser doping (P-GILD), rapid vapor doping (RVD), plasma immersion ion implantation (PIII), and very high molecular weight implants such as decaborane ($B_{10}H_{14}$). The reader is referred to Jones and Ishida [66] for a review of each technique. Alternatively one may have to use some form of raised source/drain technique in which additional silicon is added to the FET structure. A number of alternative strategies exist for this approach [73].

A variety of schemes can be used to improve the operation of the device by tailoring the doping profile in and around the channel (Figure 16.21). All of these implants have peak concentrations and depths well below those of the deep source/drain. The electrical channel width is less than 100 Å. Retrograde channel doping processes are used to reduce the dopant concentration in this region to minimize ion scattering [74, 75]. The basic idea is that electrical carriers (electrons and holes) will diffuse away from areas of high concentration to areas of low concentration until that diffusion is balanced by an electric field caused by the resulting charge dipole. If the ions can be kept away from the interface where the inversion layer forms, one might expect to see higher carrier mobilities, particularly at the large channel dopant concentrations used in deep submicron devices. Often indium rather than boron is used as the dopant. Although indium has a low solid solubility (5×10^{17} cm^{-3} at 1000°C), it has a much lower diffusivity that boron, particularly at low temperatures. This is essential in forming an effective retrograde profile.

For NMOS devices, halo implants, which are deeper than the reachthrough but not as deep as the contact/source drain, are used on LDD structures to reduce short-channel effects by shielding the channel from the deep source/drain diffusion [76]. Halo implants are done at a large angle to the vertical while the wafer is rotated. This produces a pocket of impurity that prevents the spread of the depletion region associated with the deeper contact source/drain, reducing the DIBL effect. This can be seen as a reduced dependence of threshold voltage on channel length, and a reduced dependence of subthreshold current on drain bias.

16.8 The Universal Curve and Advanced CMOS

As CMOS has scaled deeply into the nanometer regime, there have been a variety of efforts at improving the performance of the basic device beyond scaling. As a starting point for this discussion, we need to introduce the universal curve. As shown in Figure 16.22, it has been found that the mobility of an ideal silicon inversion layer follows two simple curves, one for n-channel and one for p-channel, and that the mobility is simply a function of the vertical electric field [77–79] as shown in Figure 16.23 [80]. At low fields the mobility is limited by Coulomb scattering. At moderate fields it is limited by phonon scattering. At high fields the mobility is limited by surface scattering.

It is generally believed that one route to higher performance circuits is through higher mobility inversion layers. Even when the gate length is so short that electrons are capability of traveling ballistically from source to drain, the low field mobility plays an important role in the switching transient. This has been called the tyranny of the universal curve. One way to get better devices then is to be able to operate at lower vertical electric fields. One of the first devices to do this was the double-gate FET [81]. Although generally successful, the process for making the devices was difficult until the introduction of the FinFET [82]. The FinFET uses a narrow fin of silicon etched into a substrate, often using SIMOX, as the active part of the device. A single polysilicon layer is deposited over a silicon fin and patterned to form perfectly aligned front and back gates straddling the fin structure and providing gate electrodes on two sides of the conducting channel. This allows one to reduce the doping concentration required in the channel and, more important, operate the transistor at very low vertical electric fields (Figure 16.23) and therefore with high mobility [83]. FinFETs have been scaled successfully to 10-nm gate

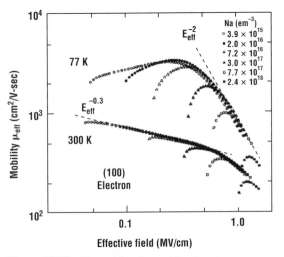

Figure 16.22   The universal curve for electrons at room temperature and at 77 K for electrons for various doping concentrations *(Takagi et al., reprinted by permission of IEEE).*

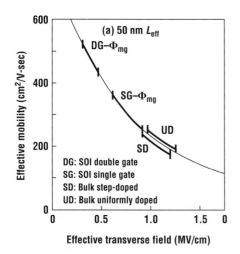

Figure 16.23 Effective transverse electric field ranges one can expect from bulk devices either uniformly doped or with a step profile, SOI single gate, and double gate devices (*from Lochtefeld and Antoniadis, reprinted with by permission from IBM J. Res. Dev, Copyright 2000, IBM*).

lengths [84], and a variety of circuits have been successfully prototyped in the technology. Although nonplanar technology has proven successful in demonstrating increased performance, manufacture of the devices is challenging.

There has also been a great deal of interest in the use of silicon nanowires for making surround gate devices. This technology is at a research stage. Nanowires can be grown from silicon in place, or harvested and deposited. These have a variety of problems, including controlling the position, length, direction, and parasitic effects in the nanowire.

Another technology very recently developed is the use of cubic single-crystal semiconductor nanoparticles to replace the fin in a FinFET. The particles are created in a high density silane plasma [85] and are deposited on a substrate at predetermined locations (to within a few nanometers) using electrostatics [86]. The transistor is then built using this nanoparticle [87]. The advantage of this approach is that any substrate can be used, including fully fabricated CMOS wafers. Thus one can add layers of transistors on top of already fabricated transistors to make a 3-D chip. This work, however, is still in the early research stage with only discrete device demonstrations.

As discussed in Chapter 14, it is possible to use strain to change the band structure of a semiconductor and therefore its mobility. Initially this was demonstrated through the use of Si/GeSi heteroepitaxy. The first work used GeSi grown pseudomorphically on Si. Later it was shown that a better gate oxide could be formed by growing strained Si on strain-relaxed GeSi. To do this, one grows a layer of GeSi sufficiently thick that the film strain relaxes (creating defects at the Si/GeSi interface), and then grows strained silicon on top of that. It was found [88] that the inversion layer mobility depended strongly on the degree of strain, however, and that the threshold voltage in NMOS devices shifted. This type of strain is called biaxial, as the lattice strains in two directions to accommodate the lattice constant mismatch.

The mobility in strained n-channel FETs was found to increase significantly with biaxial strain. A factor of 2 in mobility is achieved for Si grown on relaxed $Ge_{0.28}Si_{0.72}$, and this increase is roughly independent of the transverse field [89]. For p-channel FETs, however, the situation is not as good. For small amounts of strain, the mobility actually decreases, increasing over the control sample only at high levels of strain. Furthermore, at high transverse fields, the strain enhancement of the mobility decreases significantly at all levels of strain [89]. Finally, there has been concern with this type of structure that the dislocations in the substrate due to the strain-relaxed GeSi layer may propagate into the silicon layer and lead to defects in the active region, particularly at the high strain levels (and therefore high Ge content films) needed for high performance p-channel devices.

More recently it has been shown that one-dimensional (uniaxial) strain has many advantages over biaxial strain. The first is that uniaxial strain can be induced in the channel by layers added onto the device. As such n-channel and p-channel devices can be independently tuned to give the optimal performance. Unlike biaxial strain, p-channel devices under uniaxial strain show increased mobility even for small amounts of strain [90]. Furthermore, the p-channel mobility enhancement is not degraded at high field [91]. Figure 16.24 shows an example of a uniaxial strain technology where the strain is caused by silicon nitride [92]. It is easy to control the strain in this film by controlling the composition of the nitride. This type of approach is sometimes called a process strain since the strain is introduced by the process technology as opposed to being introduced by the wafer or epitaxial layer

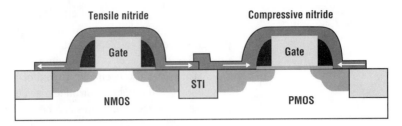

Figure 16.24   Example of uniaxial strained technology using nitrides with different stoichiometries to induce stress in the channel (*from Thompson et al., reprinted by permission at IEEE*).

growth. It should be noted that this is not the only way to create process strain. Many authors have demonstrated the use of GeSi raised source/drain structures and the effects of shallow trench isolation [92]. Finally, for this type of strain, the strain level depends on the channel length. As a result, for technologies of these types, the drive current does not scale simply with gate length. Instead, the drive current rises faster than $1/L$. Due to the greatly enhanced performance that strain provides, it is being widely applied at and below the 90-nm node. The only really question is how best to apply the strain.

16.9 Summary

This chapter reviewed the basic long-channel device equations for MOSFET operation and qualitatively described the effects of short-channel behavior. The description of the technology began with the basic 3-μm flow and presented scaling options for shrinking the technology. Parasitic effects, in particular series resistance and contact resistance, were discussed and their impact on device behavior reviewed. Salicide process modules were presented as a way to reduce these effects. Large electric fields due to nearly constant voltage scaling have impacted MOS device reliability. Hot electron effects were described semiquantitatively. Double-diffused drain and lightly doped drain process modules were presented as ways to improve the device reliability. Finally, CMOS latchup was reviewed, along with options for designing the technology to avoid this problem.

CMOS and BiCMOS are expected to continue to dominate the IC market. Table 16.3 lists the future of CMOS as projected by the Semiconductor Industry Association.

Table 16.3   Projected characteristics of CMOS technology

	2007	2010	2013	2016	2020
DRAM half-Pitch (nm)	65	45	32	22	14
DRAM Intro generation (Gbit)	16	32	64	128	256
DRAM chip area at intro (mm²)	757	563	560	557	440
μP Physical gate length (nm)	25	18	13	9	6
μP Extra/chip at intro ($\times 10^6$)	386	773	1546	3092	12368
μP Mask levels	33	35	37	37	39
μP Supply voltage (V_{dd})	1.1	1.0	0.9	0.8	0.7
μP High performance power (W)	189	198	198	198	198

Semiconductor Industry Association, 2005.

Problems

1. For the nonsilicided transistor described at the end of Example 16.1, calculate the current that would flow, ignoring the parasitic resistance effects at $V_{gs} = V_{ds} = 3.3$ V. Assume $V_t = 0.6$ V. From this, calculate the voltage drops across the parasitic resistances and recalculate the current flow in the transistor. Repeat the process until the solution converges. Repeat the problem for the silicided transistor.

2. A particular metallization has a specific contact resistance of 10^{-6} Ω-cm². Assume that the contact is 0.5 µm long and 10 µm wide, and that ρ_D is 80 $\Omega/\square$.
   (a) If the current flow is vertical, what is the contact resistance?
   (b) Repeat part (a) if the current flow is horizontal. Assume that R_s is 80 $\Omega/\square$. How much resistance does current crowding add?
   (c) Assume that the contact described in part (b) is used for an MOS transistor. If the contact edge is 1 µm from the edge of the gate, how large is the series resistance between the gate and the contact?

3. To improve the performance of the transistor described in Problem 2(c), it is decided to form a self-aligned silicide on the source and drain. Assume that the silicide resistivity is 10 $\Omega/\square$ and that the silicide extends to 0.2 µm from the gate (0.8 µm from the contact edge). Calculate the series resistance of this contact.

4. For the following two process sequences, explain the function of the indicated step. For each of the two sequences, sketch the wafer cross section after the sequence is done. Be sure to show all relevant features.
   > N-well implantation
   > N-well drive
   > Deep trench mask
   > Reactive ion etch of silicon
   >> Field implant
   >> (a) Thermal oxidation
   >> Chemical vapor deposition of polysilicon
   >> (b) Plasma or reactive ion polysilicon etch

   > Gate mask
   >> Polysilicon etch
   >> PMOS source/drain implant mask
   >> PMOS source/drain implant
   >> (c) Reachthrough implant
   >> Oxide deposition
   >> (d) Oxide reactive ion etch
   >>> NMOS source/drain implant mask
   >>> NMOS source/drain implant
   >>> Source/drain activation anneal
   >>> Tantalum deposition
   >>> (e) Tantalum reaction/anneal
   >>> Aqua regia strip

5. The following process sequence is to be used to fabricate a simple LOCOS isolated poly gate CMOS technology. Find and explain four (there are five) fatal errors that are embedded in the

sequence. Be sure to be specific about why the step is wrong. (It may be helpful to draw some cross sections of the transistors, but this is not required.)

> Starting wafer: p-type, concentration $< 10^{15}$ cm^{-3}
> Initial oxidation (1000-Å thermal oxide)
> N-well mask
> Implant n-well (P, 1×10^{13} cm^{-2}, 60 keV)
> Strip resist
> Etch oxide to dewet (10 : 1 HF)
> N-well drive (1100°C for 3 hr in N$_2$, 1 hr in O$_2$)
> Strip all oxide (10 : 1 HF)
> Pad oxidation (400-Å thermal oxide)
> Nitride deposition (9000-Å LPCVD nitride)
> Field mask lithography
> Etch nitride
> Strip resist
> Field oxidation (5000-Å wet thermal oxide)
> Strip nitride
> PMOS select mask
> Implant PMOS devices for threshold voltage control
> NMOS select mask
> Implant NMOS devices for threshold voltage and punchthrough control
> Strip resist
> Strip pad ooxide
> Gate poly deposition

6. The following lists part of a 0.25-μm CMOS process flow. This sequence includes a lightly doped drain and a self-aligned silicide. The "select" mask produces a photoresist image that opens up that device for an implant or etch and protects everything else on the chip. Embedded in this sequence are five **major** mistakes that will prevent the device from working. Find four of them and explain why they are incorrect.

> Complete device through well formation, isolation, and channel implants
> Strip screen oxide to silicon
> Gate oxidation (dry O$_2$, 800°C, 10 min)
> Gate mask
> Reactive ion etch gate electrode to gate oxide
> Resist strip
> PMOS transistor select mask
> Implant (P, 10 keV, 1×10^{15} cm^{-2})
> Resist strip
> NMOS transistor select mask
> Implant (As, 10 keV, 1×10^{14} cm^{-2})
> Resist strip
> Deposit aluminum for sidewall spacer (0.2 μm)
> Etch reactive ion etch to silicon
> NMOS transistor select mask
> Implant (As, 20 keV, 1×10^{15} cm^{-2})

Anneal (N_2, 1000°C, 5 sec)
Etch gate oxide down to silicon
Sputter Ti (5 μm thick)
Anneal (N_2, 550°C, 60 sec)
Etch in aqua regia
Anneal (N_2, 750°C, 60 sec)
Finish wafer with interconnect

7. A state-of-the-art CMOS device has a source/drain diffusion sheet resistance of 800 $\Omega/\square$, and a $CoSi_2$ sheet resistance of 20 $\Omega/\square$. The metal contact is 90 nm wide, and the distance from the edge of this contact to the edge of the reachthrough is 120 nm. If the barrier height to n-type silicon is 0.6 V in both cases, the effective mass is 1.1 m_o, the surface doping concentration is 2×10^{20} cm^{-3}, the device (and contact) are 1 μm wide, and A_o is 2×10^{-11} Ω-cm^2, find the change in the resistance from the edge of the contact to the edge of the reachthrough for an unsilicided device and for a silicided device

8. An LDD with a self-aligned titanium silicide n-channel MOSFET is made with the dimensions shown on the figure. The doping concentration under the silicide is 1×10^{20} cm^{-3}. The width of the device is 5 μm. The sheet resistance of the reachthrough region, source/drain diffusion, and silicide are 250, 80, and 5 Ω, respectively. The barrier height for the silicide-to-semiconductor contact is 0.6 V, and the specific contact coefficient A_o is 10^{-11} Ω/cm^2. Find the total resistance from the edge of the metal contact to the edge of the channel.

9. To reduce hot electron effects, an LDD is implemented in the transistor described in Problem 2. The LDD is doped 10^{17} cm^{-3} n-type. Ignoring depletion regions, calculate the resistance that this LDD adds. Assume that the reach-through is 0.1 μm deep and 0.2 μm long.

10. Hot electron effects can also be reduced if the power supply is reduced. Discuss the trade-offs involved in such a decision.

11. The first CMOS devices were not as prone to latchup as some of their scaled counterparts. Discuss the reasons.

References

1. S. Wolf, *Silicon Processing for the VLSI Era*, Vol. 3, *The Submicron MOSFET*, Lattice Press, Sunset Beach, CA, 1995.
2. S. M. Sze, *Physics of Semiconductor Devices*, Wiley, New York, 1981.
3. E. H. Nicollian and J. R. Brews, *Metal Oxide Semiconductor Physics and Technology*, Wiley, New York, 1982.
4. Y. P. Tsividis, *Operation and Modeling of the MOS Transistor*, McGraw-Hill, New York, 1987.
5. F. M. Wanlass and C. T. Sah, "Nanowatt Logic Using Field-Effect Metal-Oxide-Semiconductor Triodes," *IEEE Int. Solid-State Circuits Conf.*, February 1963.
6. J. Y. Chen, *CMOS Devices and Technology for VLSI*, Prentice-Hall, Englewood Cliffs, NJ, 1989.
7. R. Chwang and K. Yu, "CHMOS—An n-Well Bulk CMOS Technology for VLSI," *VLSI Design*, p. 42 (Fourth Quarter 1981).

8. L. C. Parrillo, L. K. Wang, R. D. Swenumson, R. L. Field, R. C. Melin, and R. A. Levy, "Twin-Tub CMOS II," *IEDM Tech. Dig.* 1982, p. 706.

9. R. H. Dennard, F. H. Gaensslen, H. N. Yu, V. L. Rideout, E. Barsous, and A. R. LeBlanc, "Design of Ion-Implanted MOSFETs with Very Small Physical Dimensions," *IEEE J. Solid-State Circuits* **SC-9**:256 (1974).

10. Y. El-Maney, "MOS Device and Technology Constraints in VLSI," *IEEE Trans. Electron Dev.* **ED-29**:567 (1982).

11. J. R. Brews, W. Fichtner, E. H. Nicollian, and S. M. Sze, "Generalized Guide for MOSFET Miniaturization," *IEEE Electron Devices Lett.* **EDL-1**:2 (1980).

12. M. H. White, F. Van de Wiele, and J. P. Lambot, "High-Accuracy Models for Computer-Aided Design," *IEEE Trans. Electron Dev.* **ED-27**:899 (1980).

13. P. L. Suciu and R. I. Johnston, "Experimental Derivation of the Source and Drain Resistance of MOS Transistors," *IEEE Trans. Electron Dev.* **ED-27**:1846 (1980).

14. M.-C. Jeng, J. E. Chung, P.-K. Ko, and C. Hu, "The Effects of Source/Drain Resistance on Deep Submicrometer Device Performance," *IEEE Trans. Electron Dev.* **37**:2408 (1990).

15. C.-Y. Lu, J. M. J. Sung, R. Liu, N.-S. Tsai, R. Singh, S. J. Hillenius, and H. C. Kirsch, "Process Limitation and Device Design Tradeoffs of Self-Aligned $TiSi_2$ Junction Formation in Submicrometer CMOS Devices," *IEEE Trans. Electron Dev.* **38**:246 (1991).

16. B. Davari, W.-H. Chang, K. E. Petrillo, C. Y. Wong, D. Moy, Y. Taur, M. W. Wordeman, J.Y.-C. Sun, C. C.-H. Hsu, and M. R. Polcari, "A High Performance 0.25-μm CMOS Technology: II—Technology," *IEEE Trans. Electron Dev.* **39**:967 (1992).

17. S. Nygren and F. d'Heurle, "Morphological Instabilities in Bilayers Incorporating Polycrystalline Silicon," *Solid State Phenom.* **23&24**:81 (1992).

18. A. Ohsaki, J. Komori, T. Katayama, M. Shimizu, T. Okamoto, H. Kotani, and S. Nagao, "Thermally Stable $TiSi_2$ Thin Films by Modification in Interface and Surface Structures," *Ext. Absstr. 21st SSDM*, 1989, p. 13.

19. C. Y. Ting, F. M. d'Heurle, S. S. Iyer, and P. M. Fryer, "High Temperature Process Limitations on $TiSi_2$," *J. Electrochem. Soc.* **133**:2621 (1986).

20. H. Sumi, T. Nishihara, Y. Sugano, H. Masuya, and M. Takasu, "New Silicidation Technology by SITOX (Silicidation Through Oxide) and Its Impact on Sub-Half-Micron MOS Devices," *Proc. IEDM*, 1990, p. 249.

21. F. C. Shone, K. C. Saraswat, and J. D. Plummer, "Formation of a 0.1 μm n^+/p and p^+/n Junction by Doped Silicide Technology," *IEDM Tech. Dig.*, 1985, p. 407.

22. R. Liu, D. S. Williams, and W. T. Lynch, "A Study of the Leakage Mechanisms of Silicided n^+/p Junctions," *J. Appl. Phys.* **63**:1990 (1988).

23. M. A. Alperin, T. C. Holloway, R. A. Haken, C. D. Gosmeyer, R. V. Karnaugh, and W. D. Parmantie, "Development of the Self-Aligned Titanium Silicide Process for VLSI Applications," *IEEE J. Solid-State Circuits* **SC-20**:61 (1985).

24. R. Pantel, D. Levy, D. Nicholas, and J. P. Ponpon, "Oxygen Behavior During Titanium Silicide Formation by Rapid Thermal Annealing," *J. Appl. Phys.* **62**:4319 (1987).

25. D. B. Scott, W. R. Hunter, and H. Shichijo, "A Transmission Line Model for Silicided Diffusions: Impact on the Performance of VLSI Circuits," *IEEE Trans. Electron Dev.* **ED-29**:651 (1982).

26. P. Liu, T. C. Hsiao, and J. C. S. Woo, "A Low Thermal Budget Self-Aligned Ti Silicide Technology Using Germanium Implantation for Thin-Film SOI MOSFETs," *IEEE Trans. Electron. Dev.* **45**(6):1280 (1998).

27. J. A. Kittl and Q. Z. Hong, "Self-aligned Ti and Co Silicides for High Performance sub-0.18 μm CMOS Technologies," *Thin Solid Films* **320**:110 (1998).

28. Q. Xu and C. Hu, "New Ti-Salicide Process Using Sb and Ge Premorphization for sub 0.2-μm CMOS Technology," *IEEE Trans. Electron Dev.* **45**(9):2002 (1998).

29. J. A. Kittl, Q. Z. Hong, M. Rodder, and T. Breedijk, "Novel Self-Aligned Ti Silicide Process for Scaled CMOS Technologies with Low Sheet Resistance at 0.06 μm Gate Lengths," *IEEE Electron. Dev. Lett.* **19**(5):151 (1998).

30. Q. F. Wang, K. Max, S. Kubivek, R. Jonckeere, B. Kerwijk, R. Verbeeck, S. Biesemans, and K. De Meyer, "New CoSi$_2$ Salicide Technology for 0.1 Micron Processes and Below," *IEEE Symp. VLSI Technol.* p. 17 (1995).

31. K. Goto, A. Fushida, J. Wantanabe, T. Sukegawa, Y. Tada, T. Nakamura, T. Yamazaki, and T. Sugii, "New Leakage Mechanism of Co Salicide and Optimized Process Conditions," *IEEE Trans. Electron. Dev.* **46**(1):117 (1999).

32. K. Inoue, K. Mikagi, H. Abiko, S. Chikaki, and T. Kikkawa, "A New Cobalt Salicide Technology for 0.15-μm CMOS Devices," *IEEE Trans. Electron Dev.* **45**(11):2312 (1998).

33. K. M. Cham and S. Y. Chiang, "Device Design for the Submicrometer p-channel FET with N+ Polysilicon Gate," *IEEE Trans. Electron Dev.* **ED-31**:964 (1984).

34. R. F. Kwasnick, E. B. Kaminsky, P. A. Frank, G. A. Franz, K. J. Polasko, R. J. Saia, and T. B. Gorczya, "An Investigation of Molybdenum Gate for Submicrometer CMOS," *IEEE Trans. Electron Dev.* **ED-35**:1432 (1988).

35. S. Iwata, N. Yamamoto, N. Kobayashi, T. Terada, and T. Mizutani, "A New Tungsten Gate Process for VLSI Applications," *IEEE Trans. Electron Dev.* **ED-31**:1174 (1984).

36. C. Y. Wong, J. Y.-C. Sun, Y. Tsuar, C. S. Oh, R. Angelucci, and B. Davari, "Doping of N$^+$ and P$^+$ Polysilicon in a Dual-Gate CMOS Process," *IEDM Tech. Dig.*, 1988, p. 238.

37. C. L. Chu, C. Saraswat, and S. S. Wong, "Characterization of Lateral Dopant Diffusion in Silicides," *Proc. IEDM*, 1990, p. 245.

38. J. Y. Sun et al., "Study of Boron Penetration Through Thin Oxide with P$^+$ Polysilicon Gate," *Proc. 1989 Symp. VLSI Technol.*, 1989, p. 17.

39. J. M. Sung, C. Y. Lu, M. L. Chen, S. J. Hillenius, W. S. Lindenberger, L. Manchanda, T. E. Smith, and S. J. Wang, "Fluorine Effect on Boron Diffusion of P+ Gate Devices," *IEDM Tech. Dig.* 1989, p. 447.

40. Y. Sato, K. Ehara, and K. Saito, "Enhanced Boron Diffusion Through Thin Silicon Dioxide in a Wet Oxygen Atmosphere," *J. Electrochem. Soc.* **136**:1777 (1989).

41. C. Hu, S. C. Tam, F.-C. Hsu, P. K. Ko, T.-Y. Chan, and K. W. Terrill, "Hot-electron-induced MOSFET Degradation—Model, Monitor, and Improvement," *IEEE Trans. Electron Dev.* **ED-32**:375 (1985).

42. S. Baba, A. Kita, and J. Ueda, "Mechanism of Hot Carrier Induced Degradation in MOSFETs," *IEDM Tech. Dig.*, 1986, p. 734.

43. L. Hendrickson, Z. Peng, J. Frey, and N. Goldsman, "Enhanced Reliability of Si MOSFETs with Channel Lengths Under 0.2 Micron," *Solid-State Electron.* **33**:1275 (1990).

44. H. Kurino, H. Hashimoto, Y. Hiruma, T. Fujimori, and M. Koyanagi, "Photon Emission from 70 nm Gate Length MOSFETs," *IEDM Tech. Dig.*, 1992, p. 1015.

45. H. Hazama, M. Iwase, and S. Takagi, "Hot Carrier Reliability in Deep Submicrometer MOSFETs," *Proc. IEDM,* 1990, p. 569.

46. K. Taniguchi, K. Sonoda, and C. Hamaguchi, "Physical Limitations of Ultrasmall MOSFETs: Constant Energy Scaling and Analytical Device Model," *Proc. 22nd Int. Conf. Solid State Devices, Mater. (SSDM)*, 1990, p. 825.

47. C. Sodini, P. K. Ko, and J. L. Moll, "The Effect of High Fields on MOS Devices and Circuit Performance," *IEEE Trans. Electron Dev.* **ED-31**:1386 (1986).

48. J. Chung, M.-C. Jeng, G. May, P. K. Ko, and C. Hu, "New Insight into Hot-Electron Currents in Deep-Submicrometer MOSFETs," *IEDM Tech. Dig.*, 1988, p. 200.

49. S. Ogura, P. J. Tsang, W. W. Walker, P. L. Critchlow, and J. F. Shepard, "Design and Characteristics of the Lightly Doped Drain-source (LDD) Insulated Gate Field-effect Transistor," *IEEE Trans. Electron Dev.* **ED-27**:1359 (1980).

50. M. Kinugawa, M. Kakuma, S. Yokogama, and K. Hashimoto, "Submicron MLDD NMOSFET's for 5 V Operation," *VLSI Symp. Tech. Dig.*, 1985, p. 116.

51. Y. Toyoshima, N. Nihira, and K. Kanzaki, "Profiled Lightly Doped Drain (PLDD) Structure for High Reliable NMOS-FETs," *VLSI Symp. Tech. Dig.*, 1985, p. 118.

52. R. Izawa, T. Kure, S. Iijima, and E. Takeda, "The Impact of Gate-Drain Overlapped LDD (GOLD) for Deep Submicron VLSI's," *IEDM Tech. Dig.*, 1987, p. 38.

53. T. Mizuno, T. Kobori, Y. Saitoh, S. Sawada, and T. Tanaka, "Gate-Fringing Field Effects on High Performance in High Dielectric LDD Spacer MOSFETs," *IEEE Trans. Electron Dev.* **39**:982 (1992).

54. C. Wei, J. M. Pimbley, and Y. Nissan-Cohen, "Buried and Graded/Buried LDD Structures for Improved Hot-Electron Reliability," *IEEE Electron Dev. Lett.* **EDL-7**:380 (1986).

55. J. Pfiester and F. K. Baker, "Assymmetrical High Field Effects in Submicron MOSFETs," *IEDM Tech. Dig.*, 1987, p. 51.

56. G. H. Hu, "A Better Understanding of CMOS Latchup," *IEEE Trans. Electron Dev.* **ED-31**:62 (1984).

57. J. E. Hall, J. A. Seitchik, L. A. Arledge, P. Yang, and P. K. Fung, "Analysis of Latchup Susceptibility in CMOS Circuits," *IEDM Tech. Dig.*, 1984, p. 292.

58. T. Ohzone and H. Iwata, "Transient Latchup Characteristics in n-well CMOS," *IEEE Trans. Electron Dev.* **39**:1870 (1992).

59. A. Herlet and K. Raithel, "Forward Characteristics of Thyristors in the Fired State," *Solid-State Electron.* **9**:1089 (1966).

60. For example, G. W. Neudeck, *The Bipolar Junction Transistor, Modular Series on Solid State Devices*, Addison-Wesley, Reading, MA, 1989.

61. R. R. Troutman, *Latchup in CMOS Technology*, Kluwer, Norwell, MA, 1986.

62. R. A. Martin, A. G. Lewis, T. Y. Huang, J. Y. and Chen," A New Process for One Micron and Finer CMOS," *IEDM* **31**:403–406 (1985).

63. R. D. Rung, "Trench Isolation Prospects for Application in CMOS VLSI," *IEDM Tech. Dig.*, 1984, p. 574.

64. S. Bhattacharya, S. Banerjee, J. Lee, A. Tasch, and A. Chatterjee, "Design Issues for Achieving Latchup-Free Deep Trench-Isolated, Bulk, Non-Epitaxial, Submicron CMOS," *IEDM Tech. Dig.*, 1990, p. 185.

65. S. Bhattacharya, S. Banerjee, J. C. Lee, A. F. Tasch, and A. Chatterjee, "Parametric Study of Latchup Immunity of Deep Trench-Isolated, Bulk, Non-Epitaxial CMOS," *IEEE Trans. Electron Dev.* **39**:921 (1992).

66. For a good review of this area, consult E. C. Jones and E. Ishida, "Shallow Junction Doping Technologies for ULSI," *Mater. Sci. Eng.* **R24**:1 (October 1998).

67. Z. H. Liu, C. Hu, J.-H. Huang, and T.-Y. Chan, *IEEE Trans. Elec. Dev.* **40**:86 (1993).

68. A. Agarwal, H. J. Grossman, D. J. Eaglesham, D. C. Robinson, T. E. Haynes, J. Jackson, Y. E. Erokin, and J. M. Poate, *Proc. Symp. Meas., Char., Modeling Ultra-Shallow Doping Profiles in Semiconductors*, 1997, p. 39.1.

69. M. J. Caturla, T. Diaz de la Rubia, L. A. Marques, and G. H. Gilmer, *J. Appl. Phys.* **80**(11):6160 (1996).

70. E. Chason, S. T. Picraux, J. M. Poate, J. O. Borland, M. I. Current, T. Diaz de la Rubia, D. J. Eaglesham, O. W. Holland, M. E. Law, C. W. Magee, J. M. Mayer, J. Meingailis, and A. Tasch, "Ion Beams in Silicon Processing and Characterization," *J. Appl. Phys.* **81**:6513 (1997).

71. A. Cacciato, J. G. E. Klappe, N. E. B. Cowern, W. Vandervost, L. P. Biro, J. S. Custer, and F. W. Saris, "Dislocation Formation and B Transient Diffusion in C Coimplanted Si," *J. Appl. Phys.* **79**:2314 (1996).

72. E. J. H. Collart, K. Weemers, D. J. Gravesteijn, and J. G. M. van Berkum, *Proc. Symp. Meas., Char., Modeling Ultra-Shallow Doping Profiles in Semiconductors*, 1997, p. 6.1.

73. Y. Nakahara, K. Takeuchi, T. Tatsumi, Y. Ochiai, S. Manako, S. Samukawa, and A. Furukawa, "Ultra-shallow In-situ-doped Raised Source/Drain Structure for Sub-tenth micron CMOS," in *Digest of Technical Papers—Symposium on VLSI Technology*, IEEE, Piscataway, NJ, 1996, p. 174.

74. H. Tian, R. B. Hulfachor, J. J. Ellis-Monaghan, K. W. Kim, M. A. Littlejohn, J. R. Hauser, and N. A. Masnari, "Evalution of Super-Steep-Retrograde Channel Doping for Deep-Submicron MOSFET Applications," *IEEE Trans. Electron Dev.* **41**:1880 (1994).

75. Y. Mii, S. Rishton, Y. Taur, D. Kern, T. Lii, K. Lee, K. A. Jenkins, D. Quinlan, T. Brown, Jr., D. Danner, F. Sewell, and M. Polcari, "Experimental High Performance Sub-0.1 μm Channel nMOSFET's," *IEEE Electron Dev. Lett.* **15**:28 (1994).

76. H. Hwang, D.-H. Lee, and J. M. Hwang, *Proc. IEDM,* 1996, p. 567.

77. A. G. Sabnis and J. Clemens, "Characterization of the Electron Mobility in the Inverted Less than 100 Greater than Si Surface," *Proc. IEEE Int. Electron Device Meet.* 1979, p. 18.

78. J. Watt and J. D. Plummer, "Universal Mobility-Field Curves for Electrons and Holes in MOS Inversion Layers," *Proc. Symp. VSLI Technology.* 1987, p. 81.

79. S. C. Sun and J. D. Plummer, "Electron Mobility in Inversion and Accumulation Layers on Thermally Oxidized Silicon Surfaces," *IEEE Trans. Electron. Devices* **ED-27**: 1497–1508 (1980).

80. S. Takagi, A. Toriumi, and H. Tango, "On the Universality of Inversion Layer Mobility in Si MOSFETs: Part I. Effects of Substrate Impurity Concentration," *IEEE Trans. Electron Dev.* **41**(12):2357 (1994).

81. H.-S. P. Wong, D. J. Frank, and P. M. Solomon, "Device Design Considerations for Double-Gate, Ground-plane, and Single-Gated Ultra-thin SOI MOSFETs at the 25 nm Channel Length Generation," *IEDM Tech. Dig.*, 1998, pp. 407–408.

82. D. Hisamoto, W.-C. Lee, J. Kedzierski, E. Anderson, H. Takeuchi, K. Asano, T.-J. King, J. Bokor, and C. Hu, "A Folded-Channel MOSFET for Deep-subtenth Micron Era," *IEDM Tech. Dig.*, 1998, pp. 1032–1034.

83. Djomehri Lochtefeld, and Samudra Antoniadis "New Insights into Carrier Transport in n-MOSFETs," *IBM J. Res. Dev.* **46**(2/3):347–357 (2002).

84. Bin Yu, Leland Chang, Shibly Ahmed, Haihong Wang, Scott Bell, Chih-Yuh Yang, Cyrus Tabery, Chau Ho, Qi Xiang, Tsu-Jae King, Jeffrey Bokor, Chenming Hu, Ming-Ren Lin, and David Kyser, "FinFET Scaling to 10 nm Gate Length", *IEDM Tech. Dig.* 2002, pp. 251–282.

85. Ameya Bapat, Christopher R. Perrey, Stephen A. Campbell, C. Barry Carter, and Uwe Kortshagen, "Synthesis of Highly Oriented, Single-Crystal Silicon Nanoparticles in a Low-Pressure Inductively Coupled plasma," *J. Appl. Phys.* **94**(3):1969–1974 (2003).

86. Chad R. Barry, Steven Campbell, and O. Heiko Jacobs, "Nanoxerography: Electrostatic Force Directed Printing of Nanomaterials," *Digital Fabrication 2005, Final Program and Proceedings*, 2005, p. 25.

87. Yongping Ding, Ying Dong, Ameya Bapat, Julia Deneen, C. Barry Carter, Uwe R. Kortshagen, and Stephen A. Campbell, "A Single Nanoparticle Silicon Transistor," *IEEE Trans. Electron. Dev.* **53**(10):2525–2531 (2006).

88. For detailed information on Si heterojunction devices in general, the reader is referred to John D. Cressler, *Silicon Heterostructure Handbook*, CRC Taylor & Francis, Boca Raton, FL, 2006.

89. K. Rim, J. Chu, H. Chen, K. A. Jenkins, T. Kanarsky, K. Lee, A. Mocuta, H. Zhu, R. Roy, J. Newbury, J. Ott, K. Petrarca, P. Mooney, D. Lacey, S. Koester, K. Chan, D. Boyd, M Ieong, and H.-S. Wong, *Tech. Dig. Symp. VLSI Technol.* 2002, pp. 98–99.

90. S. E. Thompson, M. Armstrong, C. Auth, S. Cea, R. Chau, G. Glass, T. Hoffman, J. Klaus, Zhiyong Ma, B. McIntyre, A. Murthy, B. Obradovic, L. Shifren, S. Sivakumar, S. Tyagi, T. Ghani, K. Mistry, M. Bohr, and Y. El-Mansey, "A Logic Technology Featuring Strained Silicon," *IEEE Electron. Devices Lett.* **25**:191–193 (2004).

91. S. E. Thompson, M. Armstrong, C. Auth, M. Alavi, M. Buehler, R. Chau, S. Cea, T. Ghani, G. Glass, T. Hoffman, C. H. Jan, C. Kenyon, J. Klaus, K. Kuhn, Z. Ma, B. McIntyre, K. Mistry, A. Murthy, B. Obradovic, R. Nagisetty, P. Nguyen, R. Shaheed, L. Shifren, S. Sivakumar, B. Tuffs, S. Tyagi, M. Bohr, and Y. El-Mansey, "A 90 nm Logic Technology Featuring Strained Silicon," *IEEE Trans. Electron. Dev. Lett.* **51**:1790–1797 (2004).

92. Scott E. Thompson, Guangyu Sun, Youn Sung Choi, et al., "Uniaxial-Process-Induced Strained-Si:Extending the CMOS Roadmap," *IEEE Trans. Electron Devices* **53**(5):1010 (2006).

Chapter 17

Other Transistor Technologies

The last chapter discussed the dominant silicon technology, CMOS. This chapter will review common GaAs and other transistor technologies. Each of these areas currently has a small part of the overall market and can be viewed as being important in certain niches. This is not intended to be an exhaustive list. Gallium arsenide can be made in the form of a very high resistivity semi-insulating substrate. This gives it unique advantages for high speed analog applications such as amplifiers and receivers for communications and radar. This also makes GaAs very useful for building digital integrated circuits that might be exposed to radiation such as is found on satellites. The low field electron mobility is large, and the material is amenable to the growth of heterostructures. Both favor the fabrication of very high speed n-channel transistors, although defect densities and power dissipation limit the density compared with CMOS. Amorphous and organic transistors have low mobility and so relatively poor performance. They do not require high temperature processing and so are capable of being built on a variety of substrates. This has important implications for displays and flexible electronics. Finally, we will summarize the construction and operation of a very different kind of transistor: the bipolar device. This device has several advantages over CMOS, particularly for analog applications, but it consumes more power than CMOS and because of the continuously improving performance of CMOS, its market share has shrunk considerably over the last 10 years.

17.1 Basic MESFET Operation

Figure 17.1 shows the cross section of a typical mesa isolated GaAs MESFET [1]; n-channel devices are used almost exclusively because of the small mobility of holes in GaAs. The threshold voltage of MESFETs is simpler than that of MOSFET transistors, since there is no oxide to support a voltage drop. In a simple depletion mode (d-mode) MESFET, the threshold voltage is given by [2]

$$V_T = V_{bi} - V_{po} \qquad (17.1)$$

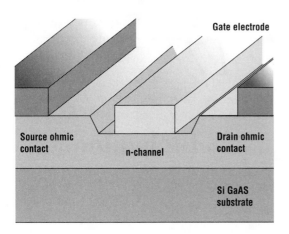

Figure 17.1 Cross section of a simple mesa-isolated MESFET.

The pinchoff voltage V_{po} is

$$V_{po} = \frac{qN_d d^2}{2k_{GaAs}\varepsilon_o} \tag{17.2}$$

where d is the thickness of the channel and N_d is the channel doping concentration.

The current–voltage characteristics of MESFETs are very similar to those of MOSFETs. Assuming for the moment that the contact and series resistance are small (not generally a very good assumption in MESFETs), in the linear region the drain-to-source current is given by

$$I_{ds} = \frac{q\mu_o N_d Wd}{L}\left[V_{ds} - \frac{(V_{ds} + V_{bi} - V_{gs})^{3/2} - (V_{bi} - V_{gs})^{3/2}}{V_{po}^{1/2}}\right] \tag{17.3}$$

In the saturation region the drain-to-source current is given by

$$I_{ds} = \frac{q\mu_o N_d Wd}{L}\left[\frac{V_{po}}{3} + \frac{2}{3}\frac{(V_{bi} - V_{gs})^{3/2}}{V_{po}^{1/2}} + V_{gs} - V_{bi}\right] \tag{17.4}$$

17.2 Basic MESFET Technology

A wide variety of GaAs MESFET technologies have been developed. This section will introduce a basic d-mode technology [3]. The process flow is shown in Figure 17.2. Either three or five masking levels are required, depending on whether one or two levels of metal are needed. A semi-insulating substrate is first coated with a thin layer of Si_3N_4 and then implanted with silicon to form the active layer. The silicon implant must be heavy enough to swamp the Cr or EL2 density that makes the substrate semi-insulating. Typically, this requires a dopant concentration of 1 to 6×10^{17} cm^{-3}. Next, the implant is activated in a high temperature anneal. The activation can be done in a furnace or in a rapid thermal processor [4]. A typical furnace activation process is 850° C for 20 min. The nitride layer protects the wafer from contamination during the implant and serves as a barrier to the outdiffusion of arsenic. The state of the semi-insulating substrate, particularly with regard to its boron and carbon impurity concentration and its residual defect density, is critically important in the uniformity of the resistivity of the activated channel [5]. Following the anneal, the nitride is removed. Alternatively, the conducting channel can be formed by epitaxial growth. This can be done by either MOCVD or MBE. Diffused ohmic contacts are then formed by evaporating an Ni/AuGe sandwich (see Section 15.7) using a liftoff (see Section 11.9) process, then sintering the contacts at about 450° C. The source-to-drain contact separation is usually 3 to 4 μm. This is the last high temperature step of the technology.

After ohmic contact formation, the mesas are isolated by wet chemically etching the field regions through the active layer to the semi-insulating substrate. At this point, the pinchoff voltage of the transistor is measured using a mercury probe, or the drain-to-source ungated current–voltage characteristic is measured directly. If the pinchoff voltage must be adjusted, it is done by wet

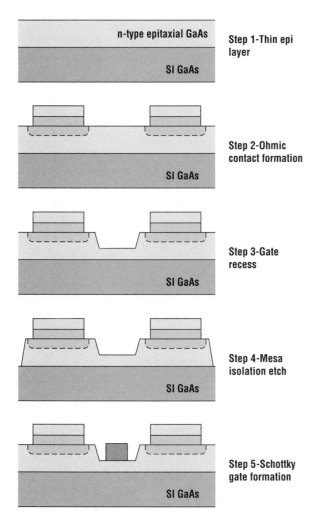

Step 1-Thin epi layer

n-type epitaxial GaAs

SI GaAs

Step 2-Ohmic contact formation

SI GaAs

Step 3-Gate recess

SI GaAs

Step 4-Mesa isolation etch

SI GaAs

Step 5-Schottky gate formation

SI GaAs

Figure 17.2 Process flow for a simple mesa-isolated MESFET. The first step can be replaced with an Si implant and anneal as discussed.

chemically recessing the channel to the desired value. As described in Chapter 11, common wet etchants for GaAs include various proportions of H_2SO_4, H_2O_2, and H_2O [6, 7]. For maximum repeatability, a slow etch is required. Since the ohmic contacts have already been formed, this measurement can be done during the wet etch and the current can be used as an end point detection method.

The Schottky gate electrode is then deposited. As discussed in Chapter 15, almost any metal will form a Schottky barrier to moderately doped GaAs. Because of the high diffusivity of Ga in most metals, however, the metal layer is chosen for its stability with the GaAs channel [8]. Furthermore, the metal must have excellent adhesion to GaAs. Two of the most popular gate metal systems for simple MESFETs are Ti/Pt/Au and Ti/Pd/Au. For these two structures, the Schottky barrier metal (Ti) is very thin, usually 500 to 1000 Å. The barrier metal (Pt or Pd) is also about 500 Å thick. The gold serves as the low resistance interconnect and is typically 2000 to 5000 Å thick. This metal stack may also be applied to the top of the ohmic contacts to further lower their series resistance, if a separate gold deposit was not done on top of the Ni/Au/Ge layers.

The gate electrode is not necessarily positioned midway between the source and drain electrodes as it is in most MOSFETs. It is essential to minimize the gate-to-drain capacitance, particularly in microwave technologies, as this parasitic property is multiplied by the Miller effect, dramatically degrading performance. It is often a good trade-off in these designs to opt for the increased series resistance of placing the gate further from the drain than the source. In 1-μm gate length microwave devices, gate-to-source spacings may be 1 μm and gate-to-drain spacings may be 2 μm.

After gate formation, it is only necessary to add a second layer of interconnect if desired. An insulating layer of 0.5 to 1.0 μm of polyimide, SiO_2, or Si_3N_4 is deposited. SiO_2 and Si_3N_4 must be deposited by plasma-enhanced CVD (Section 13.7), since thermal CVD in these materials must be carried out at high temperatures. Finally, contacts are opened up and a second layer of metal, typically Ti/Au, is deposited. The second level of metal is often ion milled rather than lifted off, so that thick layers (about 1 μm) of metal can be used to lower the resistivity.

17.3 Digital Technologies

This section will review the three design approaches commonly used in GaAs circuits. It will then cover enhancements to the basic FET technology currently being used to fabricate these circuits. The basic process technology that was described in the previous section may be used to build two

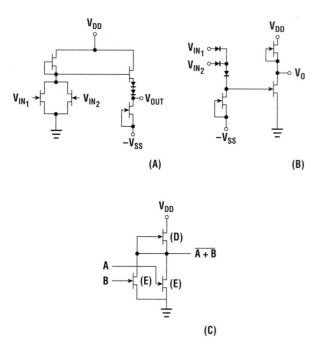

Figure 17.3   (A) Buffered FET logic (BFL). (B) Schottky diode FET logic (SDFL). (C) Direct coupled FET logic (DCFL).

important classes of GaAs ICs. The fastest, but highest power approach is buffered FET [9] logic (BFL), shown in Figure 17.3A. The two diodes near the output terminal are required to shift the output voltage back to the input level. Due to its large power dissipation (typically more than 5 mW per gate) and the large area required to lay out a gate, BFL is usually restricted to small and medium levels of integration.

A typical gate designed with Schottky diode FET [10] logic (SDFL) is shown in Figure 17.3B. The level shifting necessitated by the use of (d-mode) FETs is done by using diodes at the input stage to provide the logical OR input function. Power dissipation of SDFL is typically 1 mW per stage, but the gate delay is about twice as large as BFL. Due to the reduced power consumption, ICs with as many as 10,000 gates can be built in SDFL.

Direct-coupled FET logic [11] (DCFL), shown in Figure 17.3C, has the least power consumption of the three approaches, less than 0.5 mW per gate. The name comes from the fact that the output of the basic inverter structure does not involve any diode drops. Because of this, no level shifting is required. Combined with its compact layout, the low power consumption allows ICs with over 65,000 transistors per chip to be built [12], although complexities of current commercial devices are about half that. DCFL requires the fabrication of both enhancement (e-mode) and depletion mode transistors on the same chip, however. Noise margins are small, and the gate delay is somewhat larger than SDFL.

One of the most popular design approaches for VLSI GaAs, DCFL, requires the simultaneous fabrication of enhancement and depletion mode transistors. In Equations 17.1 and 17.2, it was shown that the threshold voltage depends on the built-in voltage, the dopant concentration in the channel, and the channel thickness. In principle, any of these three parameters could be varied in order to achieve an E/D technology. Varying the channel thickness was initially the most popular approach. As shown in Figure 17.4, both devices are recessed to the thickness desired for the d-mode transistors. The d-mode transistors are then masked off with photoresist and the e-mode devices are further etched to obtain the desired V_{po}. The difficulty of this approach is that the pinchoff voltage varies as the square of the thickness. DCFL designs have small noise margins and therefore require very careful control of the pinchoff voltage. Simple wet chemical etching, the most common way to achieve this recess, is notoriously nonuniform and has only moderate reproducibility. Plasma etch recesses have improved uniformity but tend to leave residual damage that shifts the barrier voltage.

An alternate approach to selective recessing is selective implantation. In this approach the initial implant step of the basic FET process is replaced by a lower concentration implant to set the e-mode threshold. The e-mode device areas are masked off, and the d-mode device areas given an additional implant to achieve the desired threshold. Due to its improved uniformity, the selective implantation process is now the most widely used for E/D technologies.

As discussed for MOSFETs, the transistor performance can be degraded by the presence of a series resistance between the source and drain metal and the channel. This can be particularly severe in GaAs technologies, since the traditional MESFET process is not self-aligned. Furthermore, the

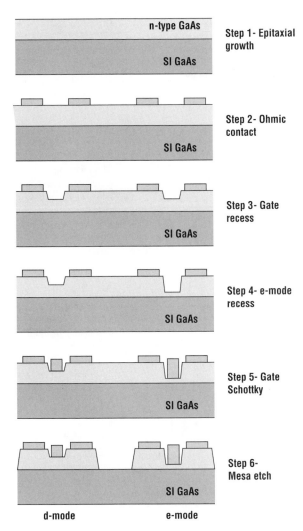

Figure 17.4   Process flow for an etched-channel E/D technology.

alloyed contact resistance is a strong function of the doping concentration in the channel layer. The primary difficulty in forming self-aligned source/drain structures is the lack of an insulating layer between the gate and the channel. If the heavily doped region formed by the alloyed ohmic contact comes into contact with the gate, it will form an ohmic contact to the gate as well. In recent years a great deal of effort has gone into forming advanced gate structures in GaAs. Collectively these devices are sometimes called self-aligned gate FETs (SAGFETs).

Two methods of producing self-aligned source/ drain contacts to the channel are shown in Figure 17.5. In the self-aligned implantation N^+ technology (SAINT) [13], the surface is Si_3N_4 passivated, and a trilayer resist sandwich is used to define the N^+ implant. The bottom resist is then undercut between 0.1 and 0.2 μm per side, and a dielectric is deposited at low temperature. The dielectric is lifted off to define the gate region. At this point, a high temperature anneal can be done to activate the N^+ implant. Finally, a Schottky metal is deposited for the gate electrode. Since the dielectric isolates the N^+ from the gate, excess leakage is avoided. This process flow does not expose the gate electrode to a high temperature step.

In the T-gate approach, a procedure more similar to traditional silicon technologies is used [14]. The Schottky metal is applied first. The second layer above the metal may be another metal, or it may be the photoresist itself. As with the SAINT process, the lower level is now intentionally undercut 0.1 to 0.2 μm per side and the N^+ implantation done. The upper masking layer is then removed and a thin Si_3N_4 anneal capping layer is deposited. The implant is then activated, and the nitride is patterned to allow the ohmic contact formation. Finally, the ohmic metal is deposited and annealed. The difficulty of the T-gate process is in finding a suitable Schottky gate material that will tolerate the high temperature implant activation cycle. Most of the materials used for this purpose are refractory metals or their silicides; WSi_2 and PtSi are two of the most popular choices.

A T-gate variation widely used in state-of-the-art GaAs ICs is a sidewall spacer offset of the source/drain implants [15]. This process is similar to the silicon salicide LDD. The major advantage of these processes is that the ohmic contact is formed self-aligned to the gate electrode, minimizing the parasitic series resistance. When very low resistance gate electrodes are desired, a substitutional gate process can be used. In these processes, the gate electrode spacing is reserved by a photoresist image of the gate [16]. This image is used to lift off an insulating layer, leaving a window for the eventual deposition of the gate electrode. The process has also been combined with the LDD method to produce a self-aligned substitutional gate [17].

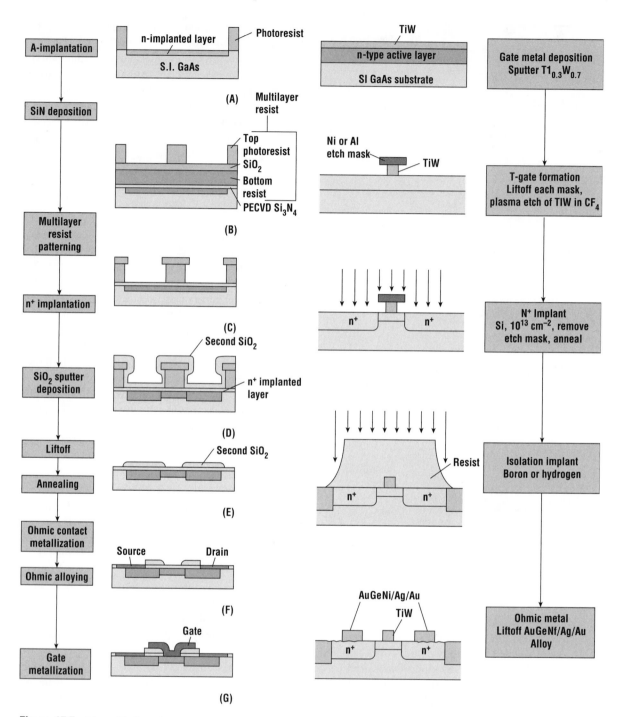

Figure 17.5 The self-aligned gate process implantation N⁺ technology (SAINT) *(after Yamaski et al., ©1982 IEEE)* and the T-gate process. The order of the last two steps of the T-gate may be reversed to reduce any isolation degradation.

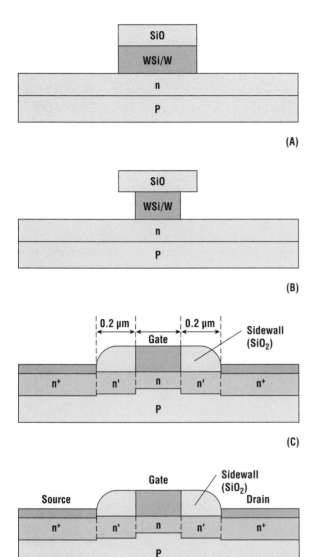

Figure 17.6 Process flow that employs both a buried p-layer and an LDD *(after Shimura et al., ©1992 IEEE).*

At deep submicron gate lengths (<0.5 μm), MESFET characteristics are severely degraded by the same short-channel effects as those that silicon MOSFET scaling is intended to prevent. Recall that in MOS devices, short-channel behavior arose because the depletion region of the reverse-biased drain junction modulated the effective charge under the gate electrode. In the case of MESFETs, there is no junction. The ohmic contacts are the same polarity as the channel. The charge under the gate can still be affected by the source/drain diffusions, however. This effect arises because the carriers are spread over a characteristic distance known as the Debye length. Thus, even if the ohmic regions are atomically abrupt, the carriers they contribute will spill over into the region under the Schottky gate. At sufficiently short source-to-drain separations, this effect gives rise to short-channel behavior. Since the Debye length is much less than the depletion width, short-channel behavior is not usually seen in MESFETs unless the gate length is less than 0.5 μm.

In self-aligned gate processes, the minimum alloyed contact-to-gate spacing results in more pronounced short-channel and gate-leakage effects. For the smallest devices, the source/drain implant must be done in stages. The LDD like SAGFETs are particularly well suited for this. To further prevent short-channel behavior, a buried p-layer can be used under the channel [18]. This layer is commonly formed by high energy implantation of Mg. Figure 17.6 shows a process that combines the buried p-layer and the LDD [19].

17.4 MMIC Technologies

Monolithic microwave IC (MMIC) technologies span a broad range of circuits, from power amplifiers to mixers to transmit/receive modules. Applications include communication such as cellular telephone, direct-broadcast satellite, data links, CATV, radar transmission and detection, and even automobile collision avoidance systems [20]. The technology must often be tuned to the application. This section will therefore describe a "generic" MMIC technology that approximately fits a number of applications. Figure 17.7 shows a cross section of a typical MMIC showing some common analog components [21].

The MMIC technology starts with the basic FET technology described in Section 17.2. As previously mentioned, the gate electrode may be noncentered, although self-aligned technologies are now quite common in MMICs. For power applications, a comb structure may be used for the gate electrode, with alternating sources and drains. Of critical importance is the ability to define very short

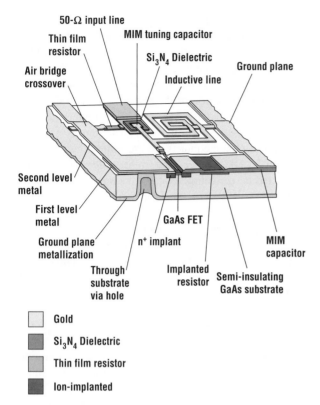

Figure labels:
50-Ω input line
Thin film resistor
Air bridge crossover
MIM tuning capacitor
Si₃N₄ Dielectric
Inductive line
Ground plane
Second level metal
First level metal
Ground plane metallization
Through substrate via hole
n⁺ implant
GaAs FET
Implanted resistor
Semi-insulating GaAs substrate
MIM capacitor

Legend:
☐ Gold
■ Si₃N₄ Dielectric
☐ Thin film resistor
■ Ion-implanted

Figure 17.7 Cross-sectional view of a typical MMIC *(after Decker).*

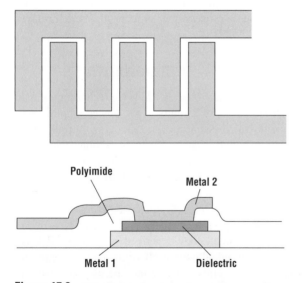

Polyimide
Metal 2
Metal 1
Dielectric

Figure 17.8 Interdigitated and simple overlay capacitor.

gate lengths, since the unity gain frequency f_T is proportional to the inverse of the effective channel length. Shorter gate lengths have the advantage of lower noise figures as well. As a result, the current generation of MMICs has gate lengths of 0.1 μm. Since the transistor count is low, the gate level of many MMICs is often fabricated by electron beam lithography. In addition to the basic transistor structure, several lumped and distributed passive elements must typically be developed.

Many analog circuits require the use of capacitors and inductors. They may be used to adjust the signal phase, to impedance match a source and load, or to filter a signal. Capacitors may be formed in two ways. Interdigitated capacitors can be formed on a single layer of metal, but typically have capacitances of less than 1 pF. Furthermore, the value of the capacitance is determined by the lithographically defined spacing. This dimension is difficult to control. When a larger or more precisely controlled capacitance is required, an overlay capacitor (Figure 17.8) can be used. The dielectric in overlay capacitors is most commonly Si_3N_4, although SiO_2, Al_2O_3, and polyimide have also been used.

There are three methods for making inductors in MMICs. Metal thickness in all three types is typically several microns to reduce the resistivity and minimize skin losses. Straight line inductors are used at the highest frequencies, but typically have too low an inductance (<1 nH) to be useful in most applications. Single loop "Ω" inductors are also easy to form but are limited in inductance to a few nanohenries. The spiral inductor can be made with inductances of up to 50 nH but requires two levels of metal with an underpass. This underpass also represents an unwanted capacitance that must be minimized.

An air bridge process module is often used in forming spiral inductors and elsewhere in the MMIC technology to minimize parasitic capacitance. The air bridge process module is shown in Figure 17.9. A thick polymer such as photoresist is patterned, opening up holes to the substrate. The openings must be large, and the polymer sidewalls must be well tapered. A thick metal layer is then deposited and the polymer dissolved. If the step coverage is adequate, the metal will not lift during the dissolution, but rather will be left behind. If long runs of metal are required, a series of these support posts may be used.

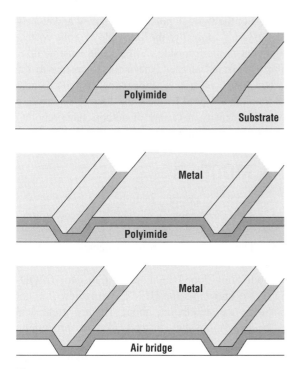

Figure 17.9 Air bridge process flow.

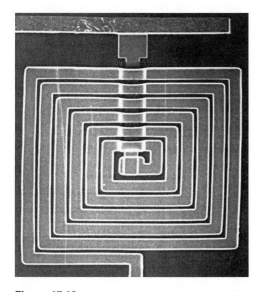

Figure 17.10 Completed spiral inductor using an air bridge crossover *(after Sciater, reprinted by permission, TAB Books. Inc.)*.

Air bridges have the lowest possible dielectric coefficient. They have also proven to be rugged and reliable. No final passivation is allowed on MMICs that employ air bridges, however, since the CVD process is conformal and would refill the gaps. Figure 17.10 shows a completed spiral inductor using an air bridge crossover [22].

Gold air bridges are often formed for small MMICs because of gold's low resistivity. Due to gold's malleability, these structures require a high density of support posts when the technology is extended to LSI levels of integration. The maximum distance between pillars is typically 50 μm for gold lines. Bridges of RhAuRh have been demonstrated to allow pillar spacings of up to 300 μm [23], due to the larger Young's modulus of rhodium. It has been shown that U-shaped wires can be used for very long stretches. Even gold can be run up to 3 mm with no loss in yield using this technology [24].

At high frequency it is essential that the metal interconnects have a controlled, reproducible impedance. Lines must also be well shielded from each other to avoid crosstalk. Furthermore, the line loss at frequency must be minimized. Finally, it is essential that a stable ground voltage be provided. At high source inductances, the FETs themselves become less stable. There are two choices for fabricating these interconnects: coplanar and microstrip waveguides (Figure 17.11). Coplanar waveguides terminate the field lines associated with the waveguide with parallel ground lines. These lines should be made as wide as possible and close to the signal line. For low density MMICs, they may be wire bonded directly to a large ground plane surrounding the chip.

Microstrip waveguides use the back of the wafer as the ground plane. The wafer must first be thinned from over 500 μm to about 100 μm. This is commonly done by lapping the wafer in a slurry of water and abrasive materials such as alumina and silicon carbide. After lapping, the backside is commonly polished using a fine abrasive and/or wet chemicals. Wafer thinning also has the advantage of reducing the thermal resistance of the wafer. This is an important consideration, since high f_T values require large dc currents, and GaAs has poor thermal conductivity. A trade-off must be made between the depth of the through holes and the mechanical strength of the wafer. Control of the final thickness uniformity is critical to obtain a consistent impedance.

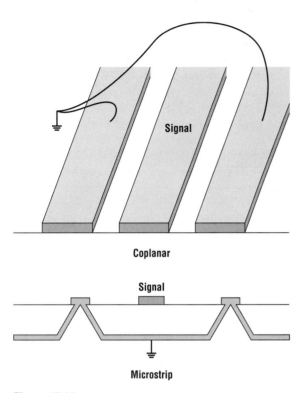

Figure 17.11   Coplanar and microstrip waveguides for MMIC interconnect.

After the through holes are patterned and etched, usually wet chemically, the backside is gold coated. Alignment of the frontside and backside must be done using an infrared aligner. The metal is opaque to the IR, while the substrate is transparent. An IR camera and monitor are used to view the image. To increase the device density, a variety of authors have reported anisotropic dry etching of through holes [25, 26].

17.5 MODFETs

An extremely important feature of several of the III–V materials systems is the ability to grow epitaxial heterostructures (see Chapter 14). Although many devices have sprung from heterostructure technology, this section will very briefly review one such device, the modulation-doped field effect transistor (MODFET), also known as the HEMT. We review this not from a device standpoint, about which a great deal has been written, but instead from a processing perspective.

A cross section of a simple MODFET is shown in Figure 17.12. The basic concept of the device is to heavily dope the wide bandgap material (typically $Al_{.25}Ga_{.75}As$) to provide the carriers. Due to the proximity of the narrower bandgap GaAs, the free carriers from the AlGaAs will become trapped by the band discontinuity at the GaAs/AlGaAs interface

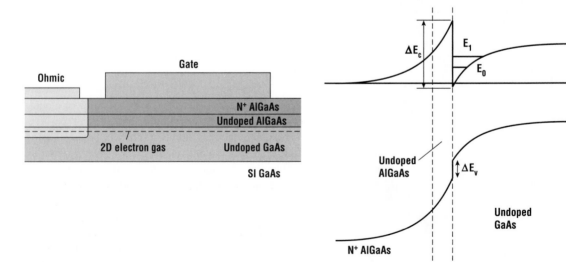

Figure 17.12   Structure and band diagram for a simple MODFET.

(Figure 17.12), creating a 2-D electron gas. This has the effect of separating the mobile carriers from their ionic impurity atoms, reducing Coulomb scattering and improving the device performance, particularly at low temperature, where the scattering contribution from phonons is no longer dominant. To further improve performance, thin (50 Å) undoped AlGaAs spacer layers are frequently grown between the GaAs and AlGaAs layers to further isolate the carriers from the charge centers. It is now common to observe electron mobilities in excess of 100,000 cm^2/V-sec at 77 K for the 2-D gas. The ability of the heterointerface to trap carriers close to the gate, along with the enhanced mobility, results in a large transconductance.

MODFETs are not without problems. The defect density of MBE material is a serious issue for manufacturing GaAs ICs based on MODFETs. Deep level traps associated with silicon doping in the AlGaAs can also shift the threshold voltage of the device by several tenths of a volt when they are charged. The effect is most evident at low temperature when the devices are exposed to light. The absorbed optical energy is enough to deplete the traps and increase the carrier concentration. Indeed, V_t shifts of this order of magnitude are particularly troublesome when operating voltages are limited to less than 2 V, as they are for many GaAs digital design approaches. Finally, MODFETs require several process changes from the basic MESFET technology.

The cross-sectional view of the MODFET (Figure 17.12) shows that the structure is quite comparable to that of a MESFET. The major portion of the effort in developing a MODFET technology is in the growth and in the device design. If a good self-aligned MESFET technology is available, only minor modifications are required to form a MODFET technology. Instead of ion implanting the surface or growing a simple conducting epitaxial layer, the technology starts with the epitaxy of the heterostructure on a semi-insulating substrate. The normal MESFET process sequence follows. Device isolation is quite straightforward, since the epitaxial layer thickness is typically 2500 Å or less.

To extract the optimal performance of the device, low resistance, self-aligned source and drain contacts are essential [27] to MODFETs. These may be accomplished using a process such as the T-gate. Special care must be used during the implant activation cycle. Typically done at 800 or 850°C, dopant diffusion from the AlGaAs to the GaAs can completely eliminate the desired mobility enhancement. To avoid this problem, rapid thermal annealing (Section 6.8) processes have been developed. The use of undoped spacer layers also allows some minor diffusion without dramatic performance degradation. Compositional gradients to provide a built-in electric field that opposes dopant diffusion have also been tried. These compositional gradients can also be used to lower the effective barrier height for ohmic contact formation.

17.6   Review of Bipolar Devices: Ideal and Quasi-ideal Behavior

The bipolar junction transistor (BJT) consists of two pn junctions in close proximity. The three regions of the device are referred to as emitter, base, and collector (Figure 17.13). Devices can be NPN or PNP. Transistor action refers to the process by which the current flow across one of the pn junctions is modulated by the bias of the other junction. Often a small change in the bias of the first junction can result in a large change in the current through the second junction.

In the forward active region [28], the emitter–base junction is forward biased and the base collector junction is reverse biased. In this case, the majority carriers in the emitter diffuse across the emitter–base depletion region, where they enter the neutral base, becoming minority carriers. These carriers diffuse across the base until they reach the base–collector junction, where the electric field sweeps them into the collector. Since the emitter acts as a minority carrier source and the collector acts as a minority carrier sink, a large concentration gradient exists across the base, which leads to the large collector current. For the ideal device, the base current is primarily due to base majority

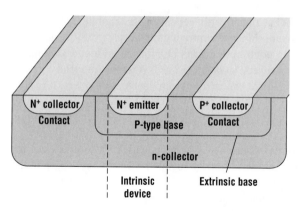

Figure 17.13   Cross-sectional view of a simple bipolar transistor showing the active or intrinsic device region and the parasitic or extrinsic regions.

carriers that are back injected into the emitter. By Kirchhoff's law, the collector current can be found by

$$I_c = I_e - I_b \qquad (17.5)$$

Two important figures of merit for the transistor are the common base current gain α, and common emitter current gain β:

$$\alpha = \frac{I_C}{I_E} \quad \text{and} \quad \beta = \frac{I_C}{I_B} \qquad (17.6)$$

The two gains are related by

$$\beta = \frac{\alpha}{1 - \alpha} \qquad (17.7)$$

Consider a simple bipolar transistor, which has uniformly doped emitter, base, and collector regions. Ignoring generation/recombination, high current effects, and various 2-D effects, the transistor can be considered ideal. Then

$$\beta = \frac{N_E D_B L_E}{N_B D_E W} \qquad (17.8)$$

where D is the diffusivity of the minority carrier in the subscript region, N is the doping concentration in the subscript region, L_E is the characteristic diffusion length of minority carriers in the emitter, and W is the nondepleted base width. Since it is generally desirable to have large gains in bipolar transistors, N_E is typically made much larger than N_B and W is kept as small as practical, as long as the base resistance is not too large and the base is not fully depleted during normal operation.

If the concentration in the base is nonuniform, Equation 17.8 is replaced by

$$\beta = \frac{N_E D_B L_E}{D_E G_B} \qquad (17.9)$$

where G_B is the base Gummel number given by

$$G_B = \int_{\text{base}} N(x) \, dx \qquad (17.10)$$

The next level of approximation is the quasi-ideal device. In this model, a small amount of electron–hole recombination is allowed in the base. The recombination rate is represented in this model by the minority carrier diffusion in the base, L_B. The quasi-ideal device is a good approximation of the intrinsic behavior of real devices under moderate current conditions. In this approximation

$$\beta \approx \frac{D_B L_E N_E}{G_B \left[D_E + \dfrac{D_B W_B^3 L_E N_E}{2 G_B L_B^2} \right]} \qquad (17.11)$$

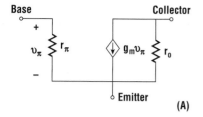

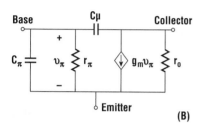

Figure 17.14 Small-signal models of the bipolar transistor at (A) low frequency and (B) high frequency.

If L_B goes to infinity (recombination in the base is not allowed), this form reverts to the ideal BJT.

17.7 Performance of BJTs

Figure 17.14A shows a low frequency, small-signal hybrid-pi model of a high gain BJT. The input resistance is

$$r_\pi = \frac{\beta}{g_m} \qquad (17.12)$$

where g_m is the transconductance, given by

$$g_m = \frac{qI_C}{kT} \qquad (17.13)$$

The output impedance is shown as r_o. To improve the analog performance of the device, one would like a high input impedance and a large transconductance. This requires a large value of I_C and a very large value for β. This implies that for high performance ac applications, the device should be biased to conduct as large a dc collector as possible with the transistor still behaving well (i.e., I_C controlled by I_B, not V_{EB}).

If high frequency effects and the parasitic resistances are added to the simple model, the equivalent circuit is as shown in Figure 17.14B. Of particular importance in this model is the parasitic capacitance, C_μ. This capacitance can be split into components in parallel with the input resistance and the output resistance using a Miller decomposition. Due to the multiplication effect, the input capacitance component will be very large at typical bias conditions. Neglecting the parasitic resistances, the device will have a pole at a frequency of

$$\omega_p \approx \frac{1}{r_\pi(C_\pi + g_m r_o C_\mu)} \qquad (17.14)$$

In practice, both capacitive elements are important. Since the emitter is much more heavily doped than the collector, the emitter–base junction capacitance per unit area is much larger than that of the base–collector junction. Minimizing both of these capacitances substantially improves the ac performance of the transistor. One commonly used measure of the device performance can be obtained by shorting out the emitter and collector. The frequency at which the gain drops to unity in this configuration is given by

$$f_T \approx \frac{1}{2\pi r_\pi(C_\pi + C_\mu)} \qquad (17.15)$$

In a more exact treatment, the transit time of carriers across the base must be added to the RC delay

$$f_T \approx \frac{1}{2\pi} \frac{1}{r_\pi(C_\pi + C_\mu) + (W_B^2/\eta D_B)} \qquad (17.16)$$

where η is a field enhancement factor, usually between 2 and 20.

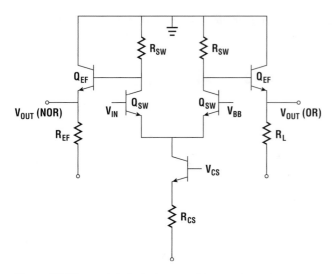

Figure 17.15   Basic ECL logic gate with inverting and noninverting output.

Thus, for high performance of bipolar junction transistors in ac applications, the following transistor characteristics are very desirable:

1. High gain without punchthrough
2. Capacitances as small as possible
3. Low parasitic resistances
4. Short base widths
5. The ability to operate at high currents

The requirements for digital applications are somewhat different. Most modern digital BJT applications use ECL and/or CML (Figure 17.15) circuit design approaches. A CML gate is an ECL gate without the emitter followers (Q_{EF}). What is of primary importance here is not just speed, but the speed–power product. In these design approaches, additional speed can often be gained for a given technology if the user is willing to pay the penalty of increased power consumption (and heat dissipation). Roughly speaking, the gate delay of an ECL circuit is given by

$$t_{pd} \approx R_L C_C + k_1 \tau_B + R_B C_B \tag{17.17}$$

where

$$C_C = k_2 C_{BC} + k_3 C_{CS} + k_4 C_{wire} + k_5 C_{EF}$$

and

$$C_B = k_6 C_{diff} + k_7 C_{EB} + k_8 C_{BC} \tag{17.18}$$

In these equations the k_x variables are the various weighting factors (Table 17.1), R_L is the pulldown resistor, τ_B is the base transit time, R_B is the base resistance, C_{BC} is the base collector capacitance, C_{CS} is the collector substrate capacitance, C_{EF} is the input capacitance of the emitter follower, and C_{diff} is the diffusion capacitance associated with the stored minority charge in the base.

A general plot of gate delay versus switching current is shown in Figure 17.16. At very high levels of current injection, the diffusion capacitance of the device increases and the transistor gain

Table 17.1   Typical weighting factors for digital bipolar gate delay contributions

Collector base capacitance (k_2)	$2.50 + 0.41 \times FO$
Collector substrate capacitance (k_3)	1.10
Wire capacitance (k_4)	0.23
Emitter–follower input capacitance (k_5)	FO
Diffusion capacitance (k_6)	2.6
Emitter–base capacitance (k_7)	$0.24 \times FO$
Base–collector capacitance (k_8)	1.27

FO is the fanout: the number of gates being driven.

decreases. For increased current, the diffusion capacitance associated with minority charge storage actually increases the gate delay with increasing current. Most circuits operate on the left-hand side of plot where the gate delay is proportional to the $R_L C_C$ product. Furthermore, most digital circuits are power limited. A particular switching current is budgeted, based on the gate density required for the chip. To improve speed, one must therefore reduce C_C. This in turn means that one must reduce the capacitance associated with the collector–base (CB) junction, with the collector–substrate (CS) junction, the input of the emitter follower (a combination of the EB and CB capacitances), and the wire capacitance. As these capacitances are reduced, the other components, namely the base transit time and the $R_B C_B$ terms, become increasingly important.

Scaling the bipolar transistor is considerably more complicated than scaling the MOS transistor. In bipolar ECL, the voltage swing is essentially fixed, therefore one must scale the capacitances. Table 17.2 gives a generalized bipolar scaling methodology [29]. In this methodology, the vertical and horizontal dimensions must both be scaled in order to improve the gate delay performance.

There are several limits that one must be aware of in applying these scaling rules. The first has to do with increasing the base doping. As the base becomes more heavily doped, the mobility in the base decreases due to ion scattering, limiting the reduction of the base resistance. The second effect is bandgap narrowing. Once the doping density is high enough to allow a significant overlap of the dopant atoms, the semiconductor bandgap will begin to shrink. This begins at about 10^{17} cm^{-3} and rises to about 120 meV at 10^{20} cm^{-3}. Bandgap narrowing in the emitter reduces the gain of the transistor by a factor of $\exp[\Delta E_g/2kT]$.

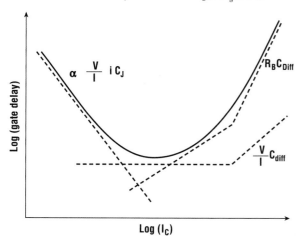

Figure 17.16 Log/log plot of ECL gate delay versus the collector current.

17.8 Early Bipolar Processes

One of the earliest bipolar technologies that was used widely in IC manufacturing was the triple-diffused (3-D) method. This is still used in some applications due to its low cost, ease of manufacture, and high yields. A process flow for a simple 3-D technology is shown in Figure 17.17. Only seven masks are required. An initial oxide is grown on a p$^-$ substrate and patterned for the guard ring diffusion. The guard rings are driven, the oxide is stripped, and a second oxide is grown. This oxide is patterned for the collector implant. The collector is implanted and driven, the oxide stripped, and a third oxide is grown. This oxide is patterned for the base implant.

Table 17.2 Bipolar scaling methodology

Emitter stripe width	a
Base doping	$a^{-1.6}$
Collector doping	a^{-2}
Base width	$a^{0.8}$
Collector current density	a^{-2}
Circuit delay	a

After Solomon and Tang [29].

The base is implanted, the oxide stripped, the base is activated, and a fourth oxide is grown. This oxide is then patterned for the emitter implant. The emitter is implanted, the oxide is stripped, and a final oxide is grown. This oxide is patterned for the base contact (also known as the extrinsic base). A heavy p^+ implant is done, and the base contact and the emitter are activated in a final thermal step. A contact glass is deposited, contacts are made, and metallization layers are applied and patterned. As a final note, the extrinsic base implant mask can be eliminated by implanting the extrinsic base at the same time as the intrinsic base. If this implant is kept shallower and at a lower concentration than the emitter, the emitter diffusion will swamp the extrinsic base in the active region of the device, but will not affect it in the extrinsic device region. The penalties that must be paid are a larger emitter–base capacitance, a lower break-down voltage, and a higher base contact resistance.

One of the first improvements in the basic 3-D technology is the addition of a buried collector. This is a heavily doped diffusion under the collector that shorts out the otherwise large collector series resistance. The use of buried collectors implies that the collector must be grown epitaxially on the substrate. This technology has been called *standard buried collector* (SBC). Oxide-isolated SBC was the mainstay of the bipolar IC industry through the mid-1970s. A typical SBC bipolar transistor doping profile is shown in Figure 17.18 [30].

As shown in Figure 17.19, the process flow for SBC starts with growing an oxide on a lightly doped p-type substrate [31].

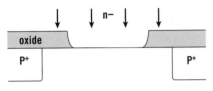

(1) Initial oxidation and guard ring implantation

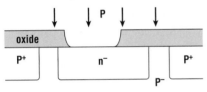

(2) Guard ring drive, collector oxidation, collector implant

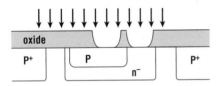

(3) Collector drive, base oxidation, base implant

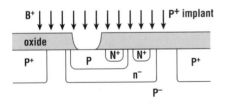

(4) Base drive, emitter oxidation, emitter implant

(5) Base contact oxidation, base contact implant

Figure 17.17   Process flow for a simple junction-isolated, triple-diffused bipolar technology.

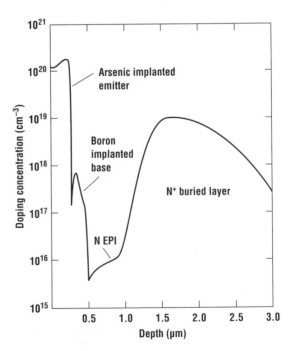

Figure 17.18   Doping profile of a typical buried collector BJT *(after Ko et al., ©1983 IEEE).*

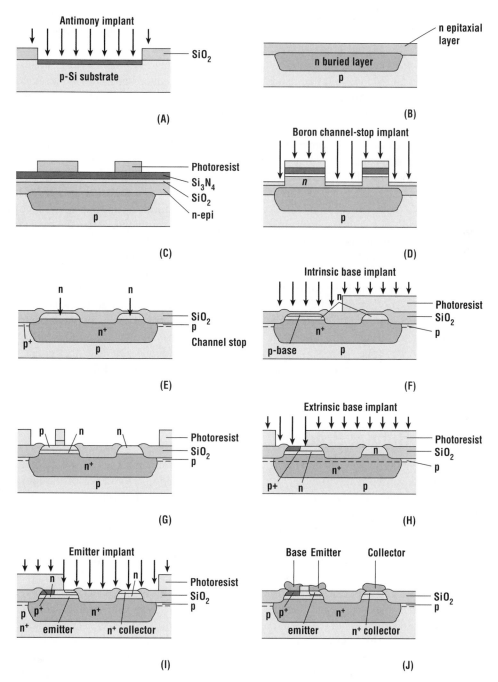

Figure 17.19 Process flow for an oxide-isolated, triple-diffused bipolar technology without sinkers: (A) buried layer formation, (B) epitaxial growth, (C) LOCOS patterning, (D) silicon recessing and channel stop implants, (E) local oxidation, (F) intrinsic base implant, (G) contact mask, (H) extrinsic base implant, (I) emitter and collector contact implant, and (J) metallization *(after Sze, reprinted by permission of John Wiley & Sons).*

The first mask opens up windows in this oxide for implantation of the buried layer. Implant doses of 2 to 10×10^{15} cm^{-2} are typical. Antimony is typically used for this implant for NPN transistors, since it tends to diffuse into the epitaxial layer less than arsenic does. Because of the large size mismatch between antimony and silicon, arsenic-doped buried layers tend to produce fewer defects in the epi than antimony buried layers. It has also been shown that at reduced pressure, the arsenic updiffusion is greatly diminished. Arsenic buried layers may therefore see a revival. The oxide is then stripped, and an anneal step is done to remove the implant damage and activate the impurities. Another oxide is typically grown in this step. The oxide grows more rapidly over the exposed heavily doped regions. The differential oxidation rate provides a step that can later be used to register the upper layers of the device to the buried collector.

After annealing, the oxide is stripped and the epitaxial layer grown. As discussed in Chapter 14, this is typically done in a radiantly heated reactor with a H$_2$/SiH$_2$Cl$_2$ ambient. The epitaxial growth step is quite critical in an SBC technology. The collector concentration is set during this growth. The collector width is determined by the epitaxial layer thickness less the base diffusion and the updiffusion from the buried collector. Frequently this involves the subtraction of several numbers that are of the same order of magnitude. Control of the collector thickness, therefore, requires tight control of the epi growth process and subsequent thermal processes.

Following the epitaxial growth, the LOCOS isolation module is begun by growing a pad oxide and depositing Si$_3$N$_4$. A thick LOCOS is grown, often down to the buried layer. Usually the silicon is etched along with the nitride to produce a recessed field oxide. Boron is implanted into the bottom of the recess to prevent parasitic MOS formation, and the field oxide is grown. Next, a sinker implant may be done (not shown) to connect the buried layer and the collector contact. This involves a high dose n-type implant or diffusion of phosphorus, followed by a high temperature drive-in diffusion. The processing then continues up through the metallization layers as for the 3-D process.

17.9 Advanced Bipolar Processes

The preceding discussion suggests that due to the trade-offs inherent in the BJT, only modest improvements can be made to either the digital or the analog performance of the intrinsic device. Reducing the width of the emitter stripe helps somewhat, since it effectively lowers the base resistance, but for most digital circuits, R_B is not a factor of the dominant term in the performance equation. This section will discuss process architectures that have been used to improve the performance of bipolar devices. Many of these improvements are not in the intrinsic transistor, but instead are aimed at reducing the parasitic resistances and capacitances that degrade the performance of the transistor.

One clear improvement in the intrinsic device came with the introduction of poly emitter contacts [32]. When a layer of heavily doped ($>10^{20}$) polysilicon, at least 500 Å thick, is placed between the single-crystal emitter and the metal contact, the gain of the device increases. This effect has been studied in considerable detail and can be attributed to several effects. Any minority carriers in the emitter will immediately recombine at the metal semiconductor contact. Thus, if the emitter thickness is less than L_E, the presence of the contact will increase the minority carrier concentration gradient and thereby increase the base current and so reduce the gain. The addition of the polysilicon moves the metal contact further away from the E–B junction and reduces this effect. Since the minority carrier diffusion length in the emitter is about 0.2 to 0.4 μm [30], increasing the effective emitter width reduces the base current [33]. Even for wide emitter transistors, however, a poly emitter increases the gain. Furthermore, the use of a polysilicon emitter contact allows for the use of very shallow emitter–base junctions. Typical E–B junctions are now in the range of 200 to 500 Å.

In some cases an interfacial dielectric exists between the polysilicon and the substrate. TEM studies have shown that a discontinuous interfacial oxide that is a few atomic layers thick exists near the poly/single-crystal interface [34]. This explains the nonexponential collector and base currents often seen in these devices [35]. The oxide is often due to native oxidation before poly deposition, although a thermal oxide can also be intentionally grown [36]. The gain may be due to the differential tunneling of electrons and holes through this thin oxide if the oxide is thicker than about 10 Å. Such oxides are typically formed during the wet cleaning process or during the temperature ramp-up in the poly deposition reactor. For much thicker oxides, however, the emitter resistance becomes excessive [37]. The use of rapid thermal processing during the drive steps has been shown to retain the high gain characteristics while minimizing the emitter resistance [38]. This is believed to be due to oxide thinning for low temperature RTP and agglomeration of the oxide at high temperature RTP [39]. It has also been shown that thermal nitrides can be used as an interfacial layer [40]. Thermal nitrides are self-limiting and therefore quite uniform, resulting in good control over the current gain and low emitter resistances if phosphorus-doped emitters are used.

A gain enhancement can be observed even if the surface has been carefully prepared to minimize the growth of any interfacial oxide. At high concentrations, arsenic segregates to the grain boundaries in polysilicon films. In poly emitter structures, the poly/single-crystal interface is a large grain boundary that will be very heavily doped [41]. Even in the absence of an interfacial oxide, this dopant pileup can be observed in TEM, SIMS, and STEM [42, 43]. This excess n-type dopant leads to a barrier to hole injection but does not affect electron ejection. This effect has been reinforced by recent observations that phosphorus-doped emitters have a larger gain enhancement than arsenic-doped emitters. This has been observed in both *in situ* doped [44] and implanted poly emitters [45]. It has also been speculated that the difference in carrier mobility between the substrate and the poly may also provide this barrier [46, 47].

In addition to the gain enhancement, polysilicon emitters have considerable technological advantages that alone would justify their application. The emitter contact to first metal can be made on the field oxide, minimizing the necessary poly stripe width. Furthermore, as will be described in the next section, the topology of the stripe can be used to self-align the transistor. Finally, the emitter and sometimes the base can be implanted into the polysilicon and diffused into the single-crystal substrate. This avoids the problem of implant damage that may serve as recombination sites and degrade device performance.

A variety of process architectures have been developed that exploit the poly emitter structure to further improve bipolar performance. The earliest ones were simple extensions of SBC technologies to include polysilicon-contacted emitters, often called *poly emitters*. A common variant is shown in Figure 17.20. A layer of undoped

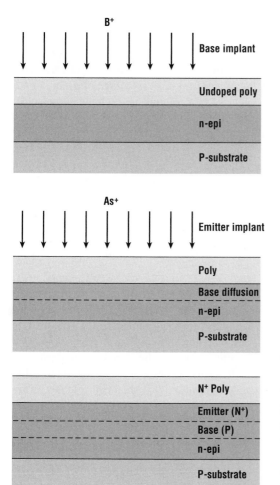

Figure 17.20   In the simple poly emitter technology, the poly is used to diffuse the emitter and intrinsic base layers.

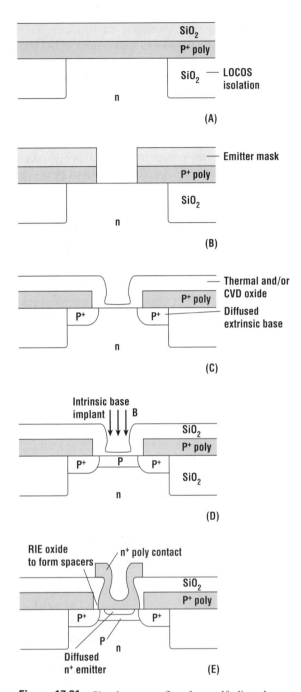

(A)

(B)

(C)

(D)

(E)

Figure 17.21   Simple process flow for a self-aligned double-poly bipolar technology.

polysilicon is deposited before the base implant. The base and emitter implants are then done into the poly and diffused into the single-crystal substrate. If arsenic is used for the N^+ implant, the base must be diffused before the introduction of the high dose arsenic, since boron diffusion is severely restricted in the presence of high concentrations of arsenic, owing to the large electric field that the arsenic produces. The high temperature step that drives the impurities is quite critical. If it is too high, the benefits of the poly contact are lost. If it is not high enough, the emitter contact resistance will be too large. Of course, the final diffused profile depends on this temperature as well.

After the emitter and base have been driven, the poly must be etched back down to the substrate. This is a difficult process to accomplish, since a selective plasma (or RIE) is required for polysilicon on silicon. Of course, both are heavily doped n-type silicon and so must etch in the same chemistries. A selective etch process must therefore take advantage of grain boundary erosion. This generally means etching at low power in a chlorine ambient. Selectivities as large as 7:1 can be achieved in this system. Amorphous Si can also be used for poly 2. It can be later crystallized in a high temperature anneal [48].

To obtain further improvement in the performance of bipolar technologies, self-aligning must be used. That is, the base and emitter contacts must align themselves automatically due to the surface topology, without the need for a critical alignment step. The most successful of these technologies use double-poly methods. A typical one, referred to as self-aligned (SA) bipolar [49], is shown in Figure 17.21. It became the standard digital bipolar technology from most of the late 1980s and is still being refined today. After buried collector and isolation formation, a P^+ layer of polysilicon is deposited. This may be doped *in situ* during the deposition or it may receive a blanket implant after deposition. A layer of oxide is then deposited or grown on the p^+ poly.

The first poly is then patterned to open a window in an active region. In the SA technology, this alignment is critical. If the overlap of the poly to the active area is large enough to ensure that the poly makes contact with the substrate even with a poor lithographic alignment, the device can suffer from an overly large base–collector capacitance. If the overlap is too small, however, an alignment error will produce a fatal defect, since after lateral encroachment of the field oxide, the first poly will terminate on the field oxide and there will be no base contact. After first poly patterning, the extrinsic base is diffused out of the first poly. A second oxide is then deposited or grown and reactive ion etched back down to the substrate, leaving sidewall spacers on the base contacts. The intrinsic base is then implanted through the hole, and the second poly is deposited, and doped n-type; then the emitter is driven. Finally, some of the base and emitter areas may be silicided to reduce the series resistance.

There are several advantages of the SA structure. The first is that the base contact to metal is now done through the p^+ poly over the field oxide. This reduces the base–collector capacitance dramatically. The extrinsic base collector junction area of a double-contacted base BJT is typically 10 to 12 μm times the length of the transistor. In the SA process, it may be less than a half-micron times the length of the transistor. The second advantage is that due to the sidewalls, the emitter stripe width is actually less than the minimum lithographic feature size. Typical SA emitter widths are about 0.3 μm for 0.5 μm lithographies.

Aggressive scaling of the SA structure to deep submicron stripe widths along with the use of polysilicon-filled deep trench isolation to reduce parasitic capacitances [50] have resulted in ECL gate delays below 20 psec [51] and speed–power products of 92 fJ [52]. To achieve very high current densities along with high speed, the selectively doped collector can be used [53]. In this process, a high energy n-type implant is done after the first poly to increase the collector concentration only directly under the intrinsic base (Figure 17.22). This allows the extrinsic base–collector capacitance to remain low while obtaining a large intrinsic collector concentration. To reduce the collector--substrate capacitance, it is possible to grow a thick, undoped epitaxial layer under the buried collector. This has been grown unpatterned on p^+ substrates, using a deep trench that penetrates to a heavily doped substrate to provide excellent isolation and high speed [52]. Another concern in scaling the double-poly structure is the increased resistance of the extrinsic base. Techniques used to minimize the impact of any reduction in the first poly thickness include using larger grain polysilicon [54] and siliciding as much of the extrinsic base as possible. Finally many advanced double-poly processes are using shallow trench isolation to allow very small overlaps of the first poly and the active area.

Most of the problems with SA technologies relate to making the link between the intrinsic and extrinsic base regions. Just as in CMOS, the reliance on surface topology along with minimal dopant redistribution during the activation means that implant shadowing may present problems [55]. Thus there may be an incomplete linkup between the intrinsic and extrinsic base. If enough shadowing occurs, the linkup region may be completely depleted, resulting in an extremely large base resistance. The simplest method of improving the linkup is to increase the diffusion of the extrinsic base region to extend almost completely under the sidewalls. The problem with this method is that if the heavily doped extrinsic base comes into contact with the heavily doped emitter, not only will the capacitance increase, the base–emitter breakdown voltage will decrease. Furthermore, the injection

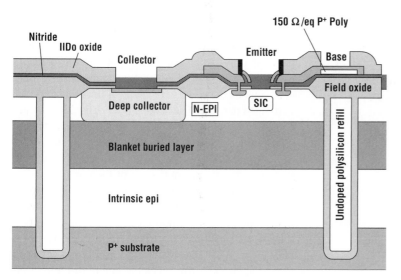

Figure 17.22   A cross section of a self-aligned bipolar technology called Mosaic *(after de la Torre et al., ©1991 IEEE).*

efficiency of the majority carriers in the base into the emitter depends on the doping concentration in the base at the emitter–base junction. If the extrinsic base comes into contact with the emitter, the device can be thought of as two bipolar transistors in parallel. The parasitic device has a very low gain due to the back–injection of carriers into the emitter.

The linkup problem can be solved with several approaches. The first is to implant the intrinsic base on axis directly into the substrate before the formation of the sidewall spacers. This is even less desirable than in CMOS LDD formation, however, in that the channel tail in BJTs will dramatically change the performance of the device. Since the collector concentration is typically about two orders of magnitude less than the base, a small change in the base profile can significantly increase the base width. The second method of improving linkup is to implant the link separately before the formation of the sidewall spacers, then diffuse the intrinsic base from the second poly [56]. In this process, the second poly must be deposited undoped, the intrinsic base implanted and diffused, and then the emitter implanted and diffused. Alternatively, a boron-doped glass [57] or P$^+$-doped polysilicon with a thin oxide [58] can be used to form the sidewall spacers and the boron diffused from the spacers themselves.

In addition to the linkup problem, care must be taken in the isolation region of an SA technology. When the BJT stripe is terminated by the isolation, as is commonly done for MOSFETs, the structure is called a *walled transistor*. Walled transistors can have better performance than their unwalled counterparts due to the reduction in the base–collector capacitance. They are also more compact, increasing the device density. In simple 3-D structures, the major concern is leakage in junctions terminated along the wall due to boron depletion during oxidation. This is relatively easy to overcome by increasing the doping concentration as evidenced by the density of CMOS currently in manufacture. In SA technologies, however, one of the goals is the minimization of the base collector capacitance. This is done by reducing the length of the P$^+$ poly base length that is in contact with the collector. A high concentration of boron under the field oxide, as shown in Figure 17.23, however, can reduce the emitter–base breakdown voltage [59]. A trade-off therefore must be made when selecting the length of the P$^+$-poly contact.

To further improve the performance of BJTs, a series of devices has been proposed that reduce C_{BC} without a critical alignment between the base and isolation levels. The first, shown in Figure 17.24, was originally called the Antipov emitter [60], but is now more commonly known as the super-self-aligned transistor [61] (SST). The technology is essentially an extension of the SA transistor. A thin (1000-Å) oxide is grown before the deposition of the first poly. After patterning poly 1, the oxide is undercut by a controlled amount using an isotropic etch such as dilute wet HF. Typical undercuts are 0.1 to 0.3 μm per side. The second poly deposition occurs conformally so that it fills the undercut. The second poly is then oxidized to form the sidewall spacers as in the typical SA process. Care must be taken during this oxidation to ensure that all of the second poly on the planar areas is oxidized, but that the poly plug formed in the undercut region is not. Next, the technology proceeds as a normal SA with the exception that the extrinsic base is driven through the undoped poly plug. This reduces the parasitic C_{BC} junction area to about 0.25 μm times the length of the transistor. The process also removes the necessity of selectively etching the polysilicon down to the substrate.

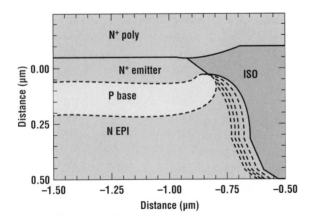

Figure 17.23   Walled transistors may have emitter–collector leakage along the isolation edge *(after Ratnam et al., © 1992 IEEE).*

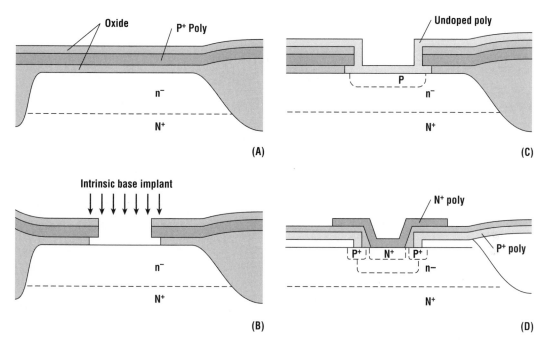

Figure 17.24   The super-self-aligned or Antipov emitter bipolar transistor.

Advances in the SA process such as shallow trench isolation and in the overlay accuracy of lithographic tools have reduced the advantages of the SST. As a result, it has not been accepted into widespread manufacture.

Several very high speed bipolar structures have also been fabricated through the use of selective epitaxy. There are two primary advantages of this approach. The first is that all of the layers are grown at the same time. Unlike most BJT technologies, where the collector width is determined by the epitaxial collector thickness less the sum of the diffused base–collector–junction depth and the buried collector updiffusion, the collector width is directly determined by the thickness grown. Similarly, the base width is determined directly by the growth rate. As a result, very thin base transistors can be fabricated. Furthermore, the device parasitics can be reduced sharply by the nature of the structure. Devices with f_T values of over 50 GHz have been fabricated using epitaxial growth.

One heavily studied type of epitaxial BJT, known as the sidewall base contact (SICOS) structure [62], is shown in Figure 17.25. After the formation of the buried collector and LOCOS isolation, a sandwich structure of oxide, P^+ polysilicon, and oxide is formed. This structure is then etched back down to the substrate inside of an active area window, and an n-collector is selectively grown epitaxially in the window. In this structure, the base contacts are diffused laterally from the P^+ poly. As with the other self-aligned structures, one problem is the linkup of the intrinsic and extrinsic base regions [63]. An additional difficulty is the epitaxy itself. Growth inside a window tends to produce a high density of defects along the sidewalls [64] (at the epi/oxide interface), facets at the upper surface, and nonuniform growth rates [65]. Growth may also be nucleated at the P^+ poly. The orientation of this growth is unlikely to match the growth from the substrate. Unless the boundary between the epi and the poly is contained in the heavily doped region, the device will suffer from high leakage and low gain. Furthermore, it is found that the selective epitaxy process often depends on the amount of exposed silicon in the reactor and so is difficult to control. Although many reports of SICOS transistors have

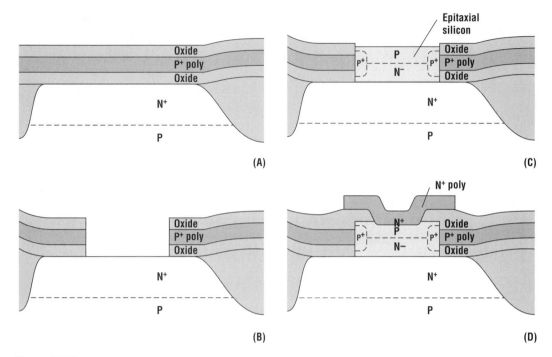

Figure 17.25   The sidewall contacted (SICOS) structure provides the least parasitic base–collector capacitance but is difficult to manufacture.

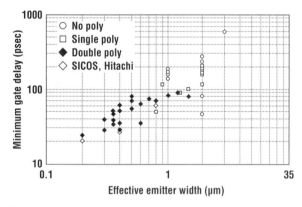

Figure 17.26   Ring oscillator gate delay of various bipolar technologies as a function of the emitter stripe width *(After Zbedel, used by permission, IEEE).*

appeared in the literature, it is yet to be used in any mainstream manufacturing process. Figure 17.26 shows the gate delay of various technologies as a function of the gate length [66]. Thinner bases provided higher speeds, and III–V devices are faster than silicon BJTs at the same base width, since they are generally heterojunction devices.

Perhaps one of the most promising trends for further improvements in silicon bipolar transistors is the use of silicon-on-insulator (SOI) substrates to further reduce the collector-to-substrate capacitance. The most commonly discussed substrates are SIMOX and bonded SOI. Although still at the demonstration phase [67], as the prices of the material decrease and the quality improves, these technologies are likely to move into the mainstream. One of the disadvantages of this approach is the concern with thermal effects. Silicon dioxide, which is often used as the buried insulator in these structures, has a much lower thermal conductivity than silicon. As a result, BJTs fabricated in SOI cannot be operated at the same power density as devices built in bulk silicon, assuming that both devices have the same maximum junction temperature [68].

Alternatively, one can use epitaxial growth processes to make high speed heterojunction bipolar transistors (HBTs). The key to this approach is to use a base material whose bandgap is less than

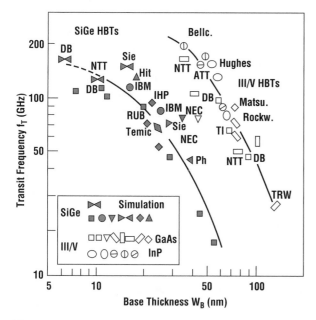

Figure 17.27 Transit frequency of GeSi, and III–V HBTs *(after König, used by permission, IEEE, 1998).*

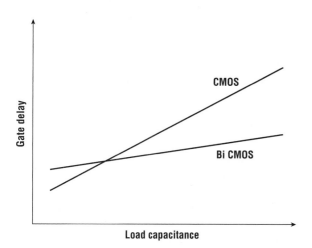

Figure 17.28 Relative gate delays for equal-area CMOS and BiCMOS technologies.

the bandgap of the emitter material. This discontinuity increases the gain by a factor $e^{\Delta E_g/kT}$. This occurs because the band discontinuity inhibits the back-injection of base carriers into the emitter. Often this increase in gain is used to allow increased base doping to decrease base resistance, or decreased emitter doping to decrease C_{EB}, or other changes in the device structure.

The HBT can be built in AlGaAs/GaAs, but low hole mobility is a concern. Alternatively it can be built by growing alloys of Ge and Si on top of Si. These alloys have bandgaps less than those of Si and so can serve as the base of a HBT. Pure Si is grown on top of the GeSi as an emitter. Figure 17.27 shows the performance of GeSi and III–V HBTs. Although the III–V devices have better performance at the same base width, they suffer from more noise [69].

17.10 BiCMOS

BiCMOS is a combination of both bipolar and CMOS that allows the designer to use both devices on a single chip [70]. This has three primary applications: memory applications such as SRAMs, where the bipolar devices are used in sense amps to detect small voltage changes on the lines; high speed digital, where the bipolar devices are used to drive large capacitive loads; and merged digital/analog. Bipolar transistors have a very large transconductance and show much less performance degradation upon capacitive loading. In digital circuits, this is particularly useful to drive output pads and long interconnect lines between the circuit elements on a chip that may be separated by as much as several millimeters. Figure 17.28 shows a typical plot of gate delay versus load capacitance for equal-area CMOS and BiCMOS circuits. For very lightly loaded gates, CMOS is faster. For large capacitive loads, however, BiCMOS has a significant advantage. The crossover point is technology dependent.

As CMOS devices have improved, the crossover point has shifted further to the right, reducing the need for bipolar devices. The additional process complexity and the additional power consumption, however, must be weighed against the improved performance of many ICs. Often these technologies start from a CMOS technology, then graft on bipolar devices. In analog ICs, bipolar is used for its close matching and low noise. Additionally, these circuits frequently require PNPs, precision resistors, and precision capacitors. As a result, BiCMOS technologies for analog can become quite complex and are often developed from bipolar technologies. The section will focus on digital BiCMOS technologies, as these are much more common than analog BiCMOS technologies.

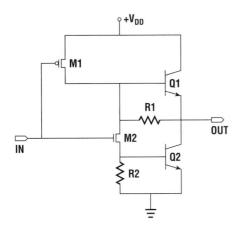

Figure 17.29   A common example of a merged gate BiCMOS structure.

BiCMOS technologies are extremely malleable to a wide variety of design methodologies. It is generally observed that they provide speed and performance somewhere between ECL bipolar and CMOS. At one extreme, it is common to design the chip completely in CMOS, with only a small number of drivers designed in ECL/CML, or using a merged gate approach. The driver, for example, might be used to interconnect blocks of logic, which are internally CMOS. If the transmitted signal levels are ECL, the reduced voltage swing further minimizes the effects of the large capacitance and can even result in a lower dynamic power dissipation than CMOS [71]. Furthermore, the small device sizes that the bipolar devices require relative to the large CMOS drivers actually reduce the node capacitance. The use of ECL signals requires some level shifting, but is straightforward. This approach has become popular with traditonal CMOS designers as it also allows them to place BiCMOS elements at critical points of the logic path. This can be done after the design is nearly complete, since the BiCMOS gate will often fit into the space required by a moderate to large sized CMOS driver. It is common to have a very small percentage of bipolar devices (<1%) in a digital BiCMOS design, yet these devices make a significant impact on the overall chip performance.

It is also possible to embed the BJTs in a basic CMOS inverter structure and use these gates as an integral part of the design. Figure 17.29 shows a common example of this type of merged gate structure. This is a push-pull or totem pole structure in that BJTs are used to both charge and discharge the output node. The two resistors are used to drain the charge from bases after the MOSFETs have been turned off. The CMOS front end ensures a high input impedance and a low quiescent power dissipation. This type of design provides a voltage swing on the output from V_{SS} to V_{DD} (full rail). Mixtures of CMOS and BiCMOS gates have been used in gate array or sea of gate logic arrays [72], where the capacitance on each node is unknown.

There are a variety of ways to fabricate BJTs starting from a CMOS technology. An attempt is usually made to use as much of the existing MOS processes for the bipolar transistor as possible; but the higher the performance required, the greater the additional process complexity. The simplest type of BJT, shown in Figure 17.30A, uses an n-well as the collector, the PMOS source/drain as the base, and the NMOS source/drain as the emitter. It is fortuitous that typical n-well concentrations (1×10^{17} cm^{-3}) are about the same as required for digital BJT collectors. Furthermore, the shallow, very heavily doped NMOS source/drain makes a good lowresistance emitter. The difficulty with this device is that it has very low gain since the base is heavily doped. If, as shown in Figure 17.30, an intrinsic base implant mask is added, the PMOS source/drain can be used as the extrinsic base and the intrinsic base profile can be controlled independently of the CMOS transistors.

As with the early BJT 3-D technology, the basic limitation of this device is the collector resistance. To improve these simple devices, a buried layer mask must also be added. This implies that a p-layer must be grown on the wafers. The n-well must drive deep enough to merge with the buried layer (Figure 17.30C). This can have the additional advantage of reducing CMOS latchup susceptibility. Thus, with only two additional masking steps one can add a simple, moderate performance NPN device to a CMOS process flow.

High performance BiCMOS technologies have attempted to produce uncompromised BJTs on-chip with CMOS [73]. A trade-off must always be made between the amount of acceptable process complexity and the device performance required. As there is no single way to accomplish this, we

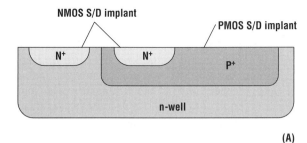

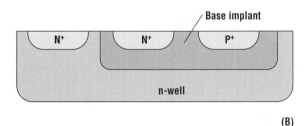

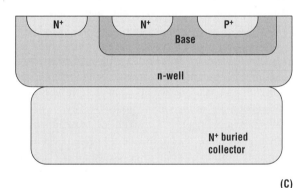

Figure 17.30 Diffused bipolar transistors that can be fabricated in a CMOS technology: (A) the emitter and base are formed by the N+ and P+ source/drain diffusions, respectively, while the n-well serves as a collector, (B) a separate intrinsic base implant is done to set the BJT gain, and (C) a buried layer is added beneath the n-well.

will describe a representative BiCMOS technology [74]. The process begins with the high energy implantation of an N+ buried layer in a p− substrate. To minimize outdiffusion from the buried layer, arsenic may be chosen as the dopant. To further compensate for the outdiffusion, a shallow boron implant may also be done. This blanket implant also forms a moderate P+ buried layer under the epi that reduces latchup susceptibility. Even with this counterdoping, the final depth of the buried layer is only about 0.5 μm.

The n-well/collector is formed next. The concentration of the well must be carefully optimized for obtaining acceptable PMOS device characteristics as well as high performance NPNs. Concentrations of about 10^{17} cm^{-3} allow reasonably high BJT drive currents, and so f_T values, while maintaining a suitably small base–collector capacitance and providing adequate punchthrough protection from the PMOS transistors at 0.35-μm gate lengths. After well formation, a conventional LOCOS or trench and LOCOS process is used to isolate the devices. If LOCOS is used, it is recessed about 0.3 μm and is about 0.5 μm thick. To minimize diffusion, this oxidation may be done at high pressure. A field or channel stop implant must be done as with normal CMOS, to prevent inversion of the p− epi. If deep trenches are used, the trenches should extend into, but not through, the buried layers.

To avoid excessive collector resistance, a series of high-dose n-type implants is done through a section of the n-well to form a sinker that contacts the buried layer. A trench can also be etched through the epi down to the buried layer. An oxide layer is deposited and reactive ion etched to isolate the sidewalls. Next, an N+ polysilicon layer is deposited and etched back to the substrate. Although somewhat more complex, the trench and refill process offers a lower parasitic collector resistance.

A standard CMOS process then proceeds. A thin screen oxide is grown and a series of channel implants is done to adjust the threshold voltage and punchthrough characteristics of the FETs. Next, the gate oxide is grown and the N+ poly gate electrode material is deposited and patterned. If LDD structures are required, the n-reachthrough is implanted, and the dielectric is deposited and etched back to form the sidewall spacers. Finally, the NMOS and PMOS source/drains are implanted. To avoid the need for an extra mask, the intrinsic base implant can be done at this point if the energy of the base implant is kept low enough that it does not penetrate the gate electrode. Since the intrinsic base concentration is about two orders of magnitude lower than the source/drain regions of the FETs, counterdoping is not significant. Alternatively, the base implant can be done later in the process into the emitter poly layer, followed by an intrinsic base diffusion. This requires more careful attention to the previously discussed linkup problems.

Finally, the source/drain regions can be salicided, completing the formation of the MOS devices. The silicide must be selected so as to withstand the high temperature step of the emitter drive. To reduce this problem, the emitter may be driven with a rapid thermal process.

The base poly is now deposited and a heavy P$^+$ implant is done to dope the extrinsic base region. The PMOS source/drain implant can also be used to dope the base poly. A thermal and/or deposited oxide is formed on top of the base poly, which is then patterned anisotropically to open the emitter window. Another oxide is grown or deposited, and the oxide is anisotropically etched to form the sidewall spacers. Finally, the emitter poly is deposited, implanted, and patterned, and the emitter is driven. If desired, the emitter and base poly layers can be silicided at this point. Interconnect then completes the processing. For three layers of metal, this requires at least 20 masks (compared to 15 masks for the equivalent CMOS) and produces an SA transistor along with high performance CMOS. Additional metal layers and device refinements often push this to almost 30 masks.

17.11 Thin Film Transistors

The transistors that have been discussed thus far use a single-crystal wafer as the starting point. Doing so provides high mobility and low leakage p-n junctions, but also presents some significant limitations, particularly when one wants to integrate multiple active materials on the same substrate, particularly if that substrate must be larger than a standard wafer or if it must be flexible. The classic example is the fabrication of flat panel displays using electro-optical devices such as light-emitting diodes (LEDs) or active matrix liquid crystals (AMLCDs) as pixels. One needs to have transistors at each pixel to drive the light emitting (or transmitting) device. For a typical display one needs of order 10^6 transistors over a sheet that might be 50 to 100 cm on a side. Unlike digital logic, high performance and high density may not be necessary, although power considerations suggest that one would want to have very low on-resistances. Thin film transistors (TFTs) have been developed to meet these goals [75].

TFT work began in 1962 with the work of Weimer at RCA [76]. Although the materials have changed, his basic device structure is not dramatically different than those being produced today. Figure 17.31 shows a simple top-gate structure. Table 17.3 shows the process sequences for a staggered TFT. Such a structure would be called a coplanar device (Table 17.4). By now the reader should be familiar with the basic elements: an active semiconductor layer, source and drain contacts, a gate insulator, and a gate electrode. Both n-channel and p-channel devices can be made; however, the vast majority of the work has centered on n-channel devices. The physics of operation is similar to the devices already presented. Compared to CMOS, this is a very simple structure. There is no need for a complicated isolation structure since simple mesa isolation works very well. Furthermore,

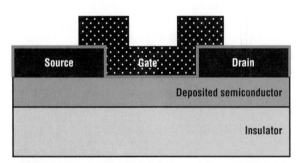

Figure 17.31   Schematic of a simple top-gate structure. Weimer's device had the source/drain contacts under the semiconductor in what is now called a staggered structure.

because the thermal cycle for this type of device is severely limited, low resistance implanted ohmic contacts are typically not used. Instead one must choose a metallization with an acceptably low barrier height, that is, a suitable work function, and live with a large contact resistance. If we are dealing with long channel lengths (~10 μm is typical for display applications) and poor channel mobility, however, one does not need very high performance contacts. An additional consideration is that for display applications, the device may have to be transparent. In that case thin layers and transparent metals such as indium tin oxide are used. This simple figure points out the three materials-related technology hurdles that one must address to

Table 17.3   Typical process flow for a typical back-channel-etched staggered TFT

1. Base coat deposition (optional diffusion barrier)
2. Cr deposition
3. Mask 1: gate pattern
4. SiN deposition
5. a-Si:H deposition
6. n + a-Si:H deposition
7. Postdeposition anneal (optional)
8. Mask 2: transistor definition (removes all of the a-Si:H between transistors)
9. Cr deposition for source drain
10. Mask 3: etch Cr and back channel recess (removes n + a-Si:H from the channel)
11. Final anneal (optional)

Table 17.4   Layout of the four basic TFT topologies indicating the position of the electrodes with respect to the semiconducting layer

	Gate	Source/Drain
Staggered	Top	Bottom
Inverted staggered	Bottom	Top
Coplanar	Top	Top
Inverted coplanar	Bottom	Bottom

obtain adequate performance from TFTs: (1) the development of a reliable semiconductor material that can be deposited at low temperature, but still has a sufficiently large mobility; (2) the development of a low interface state density, low charge density, deposited insulator for this semiconductor; and (3) the development of reliable low resistance contacts.

For many years the dominant semiconductor for TFTs has been hydrogenated amorphous silicon (a-Si:H) deposited by high frequency (~60 MHz) PECVD from silane and hydrogen. Oxygen is known to severely degrade performance, so the process must be set up to be oxygen-free. Due to the large number of unsatisfied bonds, amorphous silicon has a large trap density in the silicon bandgap. The use of hydrogen passivates most of these defect states reducing their concentration from $\sim 10^{20}$ cm^{-3} to $\sim 10^{16}$ cm^{-3}. Films are typically deposited at $\sim 300^{\circ}$C. Exposure to temperatures above 400°C leads to dehydrogenation, and so poor electrical performance. To get slightly improved performance one can create nanocrystalline silicon under these conditions by heavily diluting with hydrogen. Of course, if one deposits at much higher temperatures (~600°C) or postdeposition anneals at high temperature, one can get polysilicon, which can have substantially better mobility, but the temperature required prohibits many substrates of interest. It has been shown that laser annealing is also quite effective, but cost is a concern.

The insulator of choice for most TFTs is silicon nitride deposited by PECVD using silane and hydrogen [77]. This allows one to use the same tool that is used for the a-Si:H deposition. An advantage of nitride is that it serves as a diffusion barrier for sodium, oxygen, water, and other impurities. One must take care in this process to avoid the formation of particles in the plasma that can lead to shorts in the film. Some manufacturers have used a bilayer gate to avoid pinholes and other defects that may propagate through a single layer. Under typical process conditions the hydrogen content in the PECVD SiN is 20 to 40 atomic percent. The hydrogen in the film saturates the traps normally present in stoichiometric nitride. Interface state densities as low as 2×10^{11} eV^{-1}-cm^{-2} have been reported with better interfaces generally requiring higher deposition temperatures and/or RF power [78].

Currently the contacts to a-Si:H provide a specific contact resistance of 1.0 to 0.1 Ω-cm^2. This is about seven orders of magnitude higher than state-of-the-art CMOS. The limitation largely reflects the inability to heavily dope a-Si:H. Kanicki [79] showed that for a-Si:H layers, when the resisitivity can be reduced from 100 Ω-cm to 10 Ω-cm, the specific contact resistance dropped from 30 Ω-cm^2 to 0.1 Ω-cm^2. The latter corresponds to effective doping concentrations in the 10^{17} to 10^{18} cm^{-3} range. The doped layers can be deposited between the active channel layer and the metal contact and removed in the channel region by etching. To avoid etching into the channel layer, one sometimes deposits a thin etch stop between the doped and updoped films. This must be patterned before the deposition of the second a-Si:H layer [80]. This can be done in a semi-self-aligned manner for inverted staggered structures by illuminating from the bottom and using the gate electrode itself as the mask. Then the resist in areas not above the gate are exposed, and one produces an image of the gate in this resist semi-self aligned with the gate. This is basically proximity printing with all of the limitations that represents. To overcome the large contact resistance, many TFTs have very large contacts, and often operate with supply voltages of tens of volts. This provides the required dc current to drive the pixel, but adds capacitance that must be switched in operation and therefore increases power consumption. One can also go to a fully self-aligned implant salacide process similar to modern CMOS [81]. The a-SiH layer degrades at temperatures above 300°C, so low temperatures for implant activation and for silicidation are required. Most of the interest has been in Cr and Ni silicides. In addition to reducing the series resistance, the source and drain are now self aligned to the gate and so the overlap capacitance is reduced.

Alternatively one can use organic materials for making TFTs. These devices can be made at very low temperatures and have the potential for a fully wet process. That is, one can make the complete transistors and even full circuits using a roll-to-roll printing technology without the need for any lithography or vacuum processing [82]. This has the potential for extremely low cost circuits and, when combined with organic light emitting devices, which will be covered in the next chapter, for fully integrated circuits and optoelectronic systems. Although printing resolution and overall accuracy would limit the transistor gate lengths to ~20 μm, this is in many cases an acceptable trade-off. This is a new and rapidly evolving area. While many materials have been demonstrated, a popular one is pentacene. This consists of five aromatic rings condensed into a linear molecule. Pentacene and other fused-ring molecules tend to pack into parallel layers, allowing high carrier mobilities (up to 1.5 cm^2/V-sec has been reported). As with other TFTs, common concerns are mobility, stability, interface states, and contact resistance.

17.12 Summary

Basic MESFET processes have been presented for both digital and MMIC technologies. To obtain the maximum in performance, self-aligned structures are required. To avoid excessive gate leakage, careful attention must be paid with regard to the spacing between the gate edge and the source/drain implant. Several process flows were presented to accomplish this. MMICs also require the use of such unique structures as air bridges, through-hole vias, and coil inductors. MODFETs make use of advanced growth processes to improve device performance, particularly at low temperature. In these devices, careful control of dopant diffusion and device parasitics is essential.

The performance of bipolar transistors has been limited to a significant degree by device parasitics. Primary among these is the junction capacitance associated with the contact or extrinsic regions. Simple bipolar technologies, such as the triple-diffused or standard buried collector, have very large extrinsic capacitances. More advanced structures use self-aligned polysilicon structures to contact the emitter, the base, or both. The metal-to-polysilicon contact in these structures is formed on top of the thick field oxide, allowing the junction areas to be shrunk considerably. Furthermore, the use of polysilicon to contact the emitter produces intrinsically larger device gains. Recently the combination of bipolar and CMOS has been used for a variety of applications, but the technology is more expensive than conventional CMOS due to the increased process complexity.

Finally, new approachs to transistor fabrication are presented for thin film devices. These approaches have the potential to make moderate performance devices on large-area substrates at very low temperature. One of the most interesting variations of this process allows the fabrication of fully printed technologies with no need for clean rooms, vacuum systems, or lithography. Compared to mainstream CMOS, density and performance are extremely poor, but the low cost and the ability to use the technology in new kinds of applications are extremely interesting.

Problems

1. Why isn't gold used as the Schottky gate? What would you expect to happen if the Pt barrier metal in a Au/Pt/Ti sandwich is too thin?

2. A spiral inductor is to be made using a metal layer that is 1.5 μm thick. Assume that the minimum metal linewidth and space is 1.0 μm, and the inductor is 50 μm in diameter. Estimate the resultant inductance. (*Hint*: Consider the inductor to be made up of a series combination of inductors with different radii. Look up a formula for the inductance of a loop of wire.) Compare this to the figure given in Section 17.4 (50 nH). What could be done to increase the inductance if for example, a 1-μH inductor were needed?

3. The following is a digital GaAs enhancement/depletion FET technology. Explain the reason for the steps marked with an asterisk in one or two sentences.

   Starting material: semi-insulating undoped GaAs
   Implant: silicon dose $= 1 \times 10^{13}$ cm^{-2}, energy $= 60$ keV
   Deposit: silicon nitride thickness $= 1000$ Å
   *Anneal: N$_2$ temperature $= 800°$C
   Etch: Silicon nitride (wet chemical)
   Pattern mask no. 1
   Clean: wet chemical clean of GaAs surface
   *Evaporate: AuGe/Ni thickness $= 300$ Å/1200 Å
   Liftoff
   Anneal: 90% N$_2$/10% H$_2$ temperature $= 450°$C
   Electrical test and etch GaAs (wet chemical)
   *Pattern mask no. 2
   Electrical test and etch GaAs (wet chemical)
   Stripresist
   Pattern mask no. 3
   *Evaporate: Ti/Pt/Au thickness $= 750$ Å/500 Å/4000 Å
   Liftoff
   Pattern mask no. 4
   *Implant: hydrogen dose $= 10^{13}$ cm^{-2}, energy $= 100$ keV
   Stripresist

4. The contacts formed in a GaAs MESFET process were found to have an excessively large resistance. Using special test structures, it was determined that $R_c = 4 \times 10^{-5}$ Ω-cm^2. Further experiments have shown that the specific contact resistance is independent of the Ge doping of the GaAs. The channel doping concentration is 2×10^{16} cm^{-3} (corresponds to a resistivity of 0.08 Ω-cm) and an SEM showed that the pit density was 10^8 cm^{-2}.

(a)   What is the average pit radius?

(b)   Without adding any implants, describe at least two process steps that could be optimized in an attempt to lower the specific contact resistance and what you might try to improve them.

(c)   If an additional implant is used, indicate what additional processes steps should be used and where they would go in the sequence listed in the last question.

5.   Refer back to Chapter 3. If a MODFET with 50-Å spacers is doped with sulfur (S) and annealed at 800°C, how long can it be annealed before the characteristic diffusion length ($\sqrt{Dt}$) is equal to the spacer length? Assume intrinsic diffusivity only.

6.   Air bridges have very low capacitance, but are rarely applied to silicon technologies. Why not?

7.   In the following NPN triple-diffused (3-D) bipolar technology, identify the purpose of the steps marked with an asterisk:

> Initial oxidation
> Pattern and etch oxide
> *Boron implant
> Strip and regrow oxide
> Pattern and etch oxide
> *Implant phosphorus
> (and so on)

8.   In the following NPN standard buried collector bipolar technology, identify the purpose of the steps marked with an asterisk:

> Initial oxidation
> Pattern and etch oxide
> *Antimony implant
> Oxidation and anneal
> Strip oxide
> Epitaxial growth
> Oxidation
> Pattern and etch oxide
> Boron implant
> Strip and regrow oxide
> Pattern and etch oxide
> *Medium and high energy phosphorus implant
> Anneal
> (and so on)

9.   In the following NPN single–poly bipolar technology, identify the purpose of the steps marked with an asterisk:

> (Skip many steps . . . start in the middle of the process)
> Undoped polysilicon deposition
> *Boron implantation
> Anneal
> Arsenic implantation
> Polysilicon etch
> *Anneal
> (and so on)

10.   (a)   The delay time of an ECL circuit is found to depend primarily on the base–collector capacitance. Which of the following parameters would you also expect to be very important to the delay equation: base resistance, load resistance, emitter resistance, collector resistance?

(b) If this ECL circuit had been implemented in a standard buried collector (SBC) technology, what type of technology improvement could be done to increase the speed of the circuit?

11. The base concentration of a 3-D technology was 1×10^{18} cm^{-3}, but when the technology was changed to add a poly emitter, the base doping was increased to 5×10^{18} cm^{-3}. What benefits are gained by this increase, and why does the poly emitter allow it?

12. Explain why the base transit time is more important in a SICOS transistor than it is in an SBC transistor. The base doping in very high performance bipolars may be graded to form an internal field that reduces the base transit time.

13. Design a process flow that would produce an oxide-isolated complementary bipolar technology (both NPN and PNP), where both transistors are double-poly self-aligned. Try to minimize the number of masks required.

14. In the following double-poly self-aligned (SA) NPN bipolar process, the wafers have already gone through a buried layer formation, epi growth, and recessed LOCOS process. The following process sequence begins.

(a) Identify the purpose of the steps marked with an asterisk in one or two sentences:
(skip many steps)
Undoped polysilicion deposition (1500 Å)
Boron (p-type) implant, 30 keV, 5×10^{15} cm^{-2}
Oxide deposition (1500 Å)
Emitter mask I (minimum feature size $= 0.5$ μm)
Emitter stripe reactive ion etch of oxide and poly
Oxide deposition (1000 Å)
Oxide reactive ion etch (removes 1000 Å)
Undoped polysilicon deposition (1500 Å)
* Boron (p-type) implant (60 keV, 5×10^{13} cm^{-2})
* Thermal diffusion (900°C, 60 min)
Arsenic (n-type) implant 30 keV, 5×10^{15} cm^{-2}
* Rapid thermal diffusion/activation (1000°C, 60 sec)
Emitter mask II
Reactive ion etch of polysilicon (1500 Å)
Reactive ion etch of oxide (1000 Å)
Oxide deposition (3000 Å)
Oxide reactive ion etch (removes 3000 Å)
* Titanium deposition (300 Å)
Thermal reaction (750°C, 120 sec)
(many additional steps)

(b) Approximately what do you expect the minimum emitter width to be? Draw a picture of the emitter region to justify your answer.

References

1. For a thorough review of GaAs device physics, see M. Shur, *GaAs Devices and Circuits*, Plenum, New York, 1987.

2. C. A. Mead, "Schottky Barrier Gate Field Effect Transistors," *Proc. IEEE* **54**:307 (1966).

3. J. Mun, "GaAs Digital Integrated Circuits," in *GaAs for Devices and Integrated Circuits*, H. Thomas, D. V. Morgan, J. E. Aubrey, and G. B. Morgan, eds., IEE—Peregrinus, London, 1986.

4. M. Kuzuhara, H. Kohzu, and Y. Takayama, "Infrared Rapid Thermal Annealing of Si-Implanted GaAs," *Appl. Phys. Lett.* **41**:755 (1982).

5. F. Orito, K. Watanabe, Y. Yamada, O. Yamamoto, and F. Yajima, "Effects of Semi-Insulating GaAs Substrate Properties on Silicon Implanted Layer," *Proc. GaAs IC Symp.*, 1990, p. 321.

6. S. Adachi and K. Oe, "Chemical Etching Characteristics of (001) GaAs," *J. Electrochem. Soc.* **130**:2427 (1983).

7. D. N. McFadyen, "On the Preferential Etching of GaAs by H_2SO_4-H_2O_2-H_2O," *J. Electrochem. Soc.* **130**:1934 (1983).

8. R. E. Williams, *Gallium Arsenide Processing Techniques*, Artech, Dedham, MA, 1984.

9. R. V. Tuyl and C. A. Liechti, "High Speed Integrated Logic with GaAs MESFET," *IEEE J. Solid State Circuits* **SSC-9**:269 (1974).

10. R. C. Eden, B. M. Welch, and R. Zucca, "Planar GaAs IC Technology: Applications for Digital IC's," *IEEE J. Solid-State Circuits* **SSC-13**:419 (1977).

11. T. Mizutani, N. Kato, M. Ida, and M. Ohmori, "High Speed Enhancement Mode GaAs MESFET Logic," *IEEE Trans. Microwave Theory, Tech.* **MTT-28**:479 (1980).

12. R. B. Brown, P. Barker, A. Chandna, T. R. Huff, A. I. Kayssi, R. J. Lomax, T. N. Mudge, D. Nagle, K. A. Sakallah, P. J. Sherhart, R. Uhlig, and M. Upton, "GaAs RISC Processors," *GaAs IC Symp.*, 1992, p. 81.

13. K. Yamasaki, K. Asai, and K. Kurumada, "MESFET's with a Self Aligned Implantation for n$^+$ Layer Technology (SAINT)," *IEEE Trans. Electron Dev.* **ED-29**:1772 (1982).

14. M. Abe, T. Mura, N. Y. Ama, and H. Ishikawa, "New Technology Towards GaAs LSI/VLSI for Computer Applications," *IEEE Trans Electron Dev.* **ED-29**:1088 (1982).

15. S. Asai, N. Goto, M. Kanamori, M. Tanaka, and T. Furutsuka, "A High Performance LDD GaAs MESFET with a Refractory Gate Metal," *Proc. 18th Annu. Conf. Solid State Dev., Mater. (SSDM)*, 1986, p. 383.

16. C. F. Wan, H. Schichijo, R. D. Hudgens, D. L. Plumton, and L. T. Tran, "A Comparison Study of GaAs E/D MESFETs Fabricated with Self-Aligned and Non-Self-Aligned Processes," *Proc. GaAs IC Symp.*, 1987, p. 133.

17. S. Shikata, S. Sawada, J. Tsuchimoto, and H. Hayashi, "A Novel Self-Aligned Gate Process for GaAs LSI Using ECR-CVD," *Proc. GaAs IC Symp.*, 1990, p. 257.

18. M. Noda, K. Hosogi, K. Sumitani, H. Nakano, K. Nishitani, M. Otsubo, H. Makino, and A. Tada, "A GaAs MESFET with a Partially Depleted p Layer for SRAM Applications," *IEEE Trans. Electron Dev.* **38**:2590 (1991).

19. T. Shimura, K. Hosogi, Y. Khono, M. Sakai, T. Kuragaki, M. Shimada, T. Kitano, N. Nishitani, M. Otsubo, and S. Mitsui, "High Performance and Highly Uniform Sub-Quarter Micron BPLDD SAGFET with Reduced Source to Gate Spacing," *Proc. GaAs IC Symp.*, 1992, p. 165.

20. T. Noguchi, "Commercial Applications of GaAs ICs in Japan," *Proc. GaAs IC Symp.*, 1990, p. 263.

21. K. Sciater, *Gallium Arsenide IC Technology*, TAB Books, Blue Ridge Summit, PA, 1988.

22. D. R. Decker, in *VLSI Electronics: Microstructure Science* **11**, N. G. Einspruch, ed., Academic Press, New York, 1985.

23. T. Inoue, K. Tomita, Y. Kitaura, T. Terada, and N. Uchitomi, "A Rh/Au/Rh Rigid Air Bridge Interconnection Technique for Ultra High Speed GaAs LSIs," *Proc. GaAs IC Symp.*, 1990, p. 253.

24. M. Hirano, I. Toyada, M. Tokumitsu, and K. Asai, "Folded U-Shaped Micro-Wire Technology for GaAs IC Interconnections," *Proc. GaAs IC Symp.*, 1992, p. 177.

25. L. G. Hipwood and P. N. Wood, "Dry Etching of Through Substrate Via Holes for GaAs MMICs," *J. Vacuum. Sci. Technol. B* **3**:395 (1985).

26. K. Sumitano, M. Komaru, M. Kobiki, Y. Higaki, Y. Mitsui, H. Takano, and K. Nishitani, "A High Aspect Ratio Via Hole Dry Etching Technology for High Power GaAs MESFET," *Proc. GaAs IC Symp.*, 1989, p. 207.

27. R. Dingle, M. D. Feuer, and C. W. Tu, "The Selectivity Doped Heterostructure Transistor: Materials, Devices, and Circuits," in *VLSI Electronics: Microstructure Science* **11**, N. G. Einspruch, ed., Academic Press, New York, 1985.

28. More complete descriptions of bipolar devices at an undergraduate level can be found in G. W. Neudeck, *Modular Series on Solid State Devices*. Vol. III: *The Bipolar Junction Transistor*, Prentice-Hall, Englewood Cliffs, NJ, 1989; and B. Streetman, *Solid State Electronic Devices*, Prentice-Hall, Englewood Cliffs, NJ, 1990. Advanced level bipolar texts include D. J. Roulston, *Bipolar Semiconductor Devices*, McGraw-Hill, New York, 1990; and R. M. Warner, Jr. and B. L. Grung, *Transistors: Fundamentals for the Integrated-Circuit Engineer*, Wiley-Interscience, New York, 1983.

29. P. M. Solomon and D. D. Tang, "Bipolar Circuit Scaling," *Int. Solid State Circuits Conf. Tech. Dig.*, 1979, p. 96.

30. W. C. Ko, T. C. Gwo, P. H. Yeung, and S. J. Radigan, "A Simplified Fully Implanted Bipolar VLSI Technology," *IEEE Trans. Electron Dev.* **ED-30**:236 (1983).

31. S. M. Sze, *Physics of Semiconductor Devices*, Wiley, New york, 1981.

32. J. Graul, A. Glasl, and H. Murrmann, "High Performance Transistors with Arsenic Implanted Polysil Emitters," *IEEE J. Solid-State Circuits* **SSC-11**:291 (1976).

33. P. Ashburn, "Polysilicon Emitter Technology," *Proc. IEEE Bipolar Circuits, Technol. Mtg.*, 1989, p. 90.

34. H. C. deGraaf and J. G. de Groot, "The SIS Tunnel Emitter: A Theory for Emitters with Thin Interface Layers," *IEEE Trans. Electron Dev.* **ED-26**:1771 (1979).

35. H. Schaber and T. F. Meister, "Technology and Physics of Polysilicon Emitters," *Proc. IEEE Bipolar Circuits, Technol. Mtg.*, 1989, p. 75.

36. G. R. Wolstenholma, N. Jorgensen, P. Ashburn, and G. R. Booker, "An Investigation of the Thermal Stability of the Interfacial Oxide in Polycrystalline Silicon Emitter Bipolar Transistors by Comparing Device Results with High Resolution Transmission Electron Microscopy Observations," *J. Appl. Phys.* **61**:225 (1986).

37. H. Schaber, B. Benna, L. Treitinger, and A. Weider, "Conduction Mechanisms of Polysilicon Emitters with Thin Interfacial Oxide Layers," *IEDM Tech. Dig.*, 1984, p. 738.

38. J. E. Turner, D. Coen, G. Burton, A. Kapoor, and S. J. Rosner, "Interface Control in Double Diffused Polysilicon Bipolar Transistors," *Proc. IEEE Bipolar Circuits, Technol. Mtg.*, 1989, p. 33.

39. T. M. Liu, Y. O. Kim, K. F. Lee, D. Y. Jeon, and A. Ourmazd, "The Control of Polysilicon/Silicon Interface Processed by Rapid Thermal Processing," *Proc. IEEE Bipolar Circuits, Technol. Mtg.*, 1991, p. 263.

40. F. Nouri and B. Scharf, "Polysilicon-Emitter Bipolar Transistors with Interfacial Nitride," *Proc. IEEE Bipolar Circuits Technol. Mtg.*, 1992, p. 88.

41. C. C. Ng and E. S. Yang, "A Thermionic Diffusion Model of Polysilicon Emitter," *IEDM Tech. Dig.*, 1986, p. 32.

42. C. Y. Wong, C. R. M. Grovenor, P. E. Batson, and D. A. Smith, "Effects of Arsenic Segregation on the Electrical Properties of Grain Boundaries in Polycrystalline Silicon," *J. Appl. Phys.* **57**:438 (1985).

43. H. Schaber, R. V. Criegern, and J. Weitzel, "Analysis of Polycrystalline Silicon Diffusion Sources by Secondary Ion Mass Spectrometry," *J. Appl. Phys.* **58**:4036 (1985).

44. M. Nanba, T. Kobayashi, T. Uchino, T. Nakamura, M. Kondo, Y. Tamaki, S. Iijima, T. Kure, and M. Tanabe, "A 64 GHz Si Bipolar Transistor Using in-situ Phosphorus Doped Polysilicon Emitters," *IEDM Tech. Dig.* 1991, p. 443.

45. G. Streutker, A. Pruijmboom, D. B. M. Klaasen, and J. W. Slotboom, "Thermionic Emission Limited Recombination in Phosphorus-Implanted Polysilicon Emitters," *Proc. IEEE Bipolar Circuits, Technol. Mtg.*, 1992, p. 50.

46. A. A. Eltoukhy and D. J. Roulston, "Minority Carrier Injection into Polycrystalline Emitters," *IEEE Trans. Electron Dev.* **ED-29**:961 (1982).

47. Z. Yu, B. Ricco, and R. W. Dutton, "A Comprehensive Analytical and Numerical Model of Polysilicon Emitter Contacts in Bipolar Transistors," *IEEE Trans. Electron Dev.* **ED-31**:773 (1984).

48. T. Hashimoto, T. Kumachi, T. Jinbo, K. Watanabe, E. Yoshida, T. Miura, T. Shiba, and Y. Tamaki, "Interface Controlled IDP Process Technology for 0.3 μm High Speed Bipolar and BiCMOS LSIS," *Bipolar Circuits, Technol. Mtg.*, 1996, p. 181.

49. D. D. Tang, P. M. Solomon, T. H. Ning, R. D. Isaac, and R. E. Burger, "1.25 μm Deep-Groove-Isolated Self-Aligned ECL Circuits," *Proc. Int. Solid State Circuits Conf.*, 1982, p. 242.

50. T. M. Liu, G. M. Chin, D. Y. Jeon, M. D. Morris, V. D. Archer, H. H. Kim, M. Cerullo, K. F. Lee, J. M. Sung, K. Lau, T. Y. Chiu, A. M. Voshchenkov, and R. G. Swartz, "A Half-Micron Super Self-Aligned BiCMOS Technology for High Speed Applications," *IEEE IEDM Tech. Dig.*, 1992, p. 23.

51. J. D. Warnock, "Silicon Bipolar Device Structures for Digital Applications: Technology Trends and Future Directions," *IEEE Trans. Electron Dev.* **42**:377 (1995).

52. V. de la Torre, J. Foerstner, B. Lojek, K. Sakamoto, S. L. Sundaram, N. Tracht, B. Vasquez, and P. Zdebel, "MOSAIC V—A Very High Performance Bipolar Technology," *IEEE Bipolar Circuits, Technol. Mtg.*, 1991, p. 21.

53. A. Felder, R. Stengl, J. Hauenschild, H. M. Rein, and T. F. Meister, "25 to 40 Gb/s Si IC in Selective Bipolar Technology," *Int. Solid State Circuits Conf. Tech. Dig.*, 1993, p. 156.

54. T. Shiba, Y. Tamaki, T. Kure, T. Kobayashi, and T. Nakamure, "A 0.5 μm Very-High-Speed Silicon Devices Technology U-Groove-Isolated SICOS," *IEEE Trans. Electron Dev.* **38**:2505 (1991).

55. C.-T. Chuang, G. P. Li, and T. H. Ning, "Effect of Off-axis Implant on the Characteristics of Advanced Self-aligned Bipolar Transistors," *IEEE Electron Dev. Lett.* **EDL-8**:321 (1987).

56. T. Yuzuhira, T. Yamaguchi, and J. Lee, "Submicron Bipolar-CMOS Technology for 16 GHz f_T Double Poly-Si Bipolar Devices," *IEDM Tech. Dig.*, 1988, p. 748.

57. M. Sugiyama, H. Takemura, C. Ogawa, T. Tashiro, T. Morikawa, and M. Nakamae, "A 40 GHz f_T Silicon Bipolar Transistor LSI Technology," *IEDM Tech. Dig.*, 1989, p. 221.

58. J. D. Hayden, J. D. Burnett, J. R. Pfiester, and M. P. Woo, "An Ultra-Shallow Link Base for a Double Polysilicon Bipolar Transistor," *Proc. IEEE Bipolar Circuits, Technol. Mtg.*, 1992, p. 96.

59. P. Ratnam, M. Grubisich, B. Mehrotra, A. Iranmanesch, C. Blair, and M. Biswal, "The Effect of Isolation Edge Profile on the Leakage and Breakdown Characteristics of Advanced Bipolar Transistors," *Proc. IEEE Bipolar Circuits, Technol. Mtg.*, 1992, p. 117.

60. I. Antipov, "Bipolar Transistor with Minimized Collector-to-Base Junction Area," *IEEE Trans. Electron Dev.* **ED-30**:723 (1983).

61. Y. Tamaki, T. Shiba, I. Ogiwara, T. Kure, K. Ohyu, and T. Nakamura, "Advanced Device Process Technology for 0.3 μm Self-Aligned Bipolar LSIs," *IEEE Bipolar Cicuits, Technol, Mtg.*, 1990, p. 166.

62. T. Nakamura, T. Miyazaki, S. Takahashi, T. Kure, T. Okabe, and M. Nagata, "Self-Aligned Transistor with Sidewall Base Electrodes," *IEEE Trans. Electron Dev.* **ED-29**:596 (1982).

63. J. Van der Veldedn, R. Dekker, R. van Es, S. Jansen, M. Koolen, H. Maas, and A. Pruijmbuch, "Basic: An Advanced High Performance Bipolar Process," *IEDM Tech. Dig.*, 1989, p. 233.

64. J. O. Borland and C. I. Drowley, "Advanced Dielectric Isolation Through Selective Epitaxial Growth Techniques," *Solid State Technol.* **28**:141 (1985).

65. A. Ishitani, N. Endo, and H. Tsuya, "Local Loading Effects in Selective Silicon Epitaxy," *Jpn. J. Appl. Phys.* **23**:L391 (1984).

66. P. J. Zbedel, "Current Status of High Performance Silicon Bipolar Technology," *Proc. GaAs IC Symp.*, 1992, p. 15.

67. E. Bertagnolli, H. Klose, R. Mahnkopf, A. Felder, M. Kerber, M. Stolz, G. Schutte, H.-M. Rein, and R. Kopl, "An SOI-Based High Performance Self-Aligned Bipolar Technology Featuring 20 psec Gate Delay and a 8.6 fJ Power-Delay Product," *1993 Symp. VLSI Tech. Dig.*, 1993, p. 63.

68. P. R. Ganci, J.-J. J. Hajjar, T. Clark, P. Humphries, J. Iapham, and D. Buss, "Self-Heating in High-Performance Bipolar Transistors Fabricated on SOI Substrates," *IEDM Tech. Dig.*, 1992, p. 417.

69. U. König, "SiGe and GaAs as Competitive Technologies for RF Applications," *IEEE Bipolar Circuits Technol Conf.*, 1998, p. 87.

70. A. R. Alvarez, *BiCMOS Technology and Applications*, Kluwer, Norwell, MA, 1989.

71. T. Oguri and T. Kimura, "A New 0.8V Logic Swing, 1.6V Operational High Speed BiCMOS Circut," *IEEE Bipolar Circuits, Technol. Mtg.*, 1992, p. 187.

72. "BiCMOS, Is It the Next Technology Driver?" *Electronics*, February. 4, 1988.

73. G. Shahidi, J. Warnock, B. Davari, B. Wu, Y. Taur, C. Wong, C. Chen, M. Rodriquez, D. Tang, K. Jenkins, P. McFarland, R. Schulz, D. Zicherman, P. Coane, D. Klaus, J. Sun, M. Polcari, and T. Ning, "A High Performance BiCMOS Technology Using 0.25 μm CMOS and Double Poly 47 GHz Bipolar," *Proc. Symp. VLSI Technol.* 1992, p. 28.

74. C. K. Lau, C.-H. Lin, and D. L. Packwood, "Sub-Micron BiCMOS Process Design for Manufacturing," *IEEE Bipolar Circuits, Technol. Mtg.*, 1992, p. 76.

75. For a review of TFT technology the reader is referred to *Thin Film Transistors*, C. R. Kagan and P. Andry, eds., Dekker, New York, 2003.

76. P. K. Weimer, "The TFT—A New Thin Film Transistor," *Proc. IEEE* **50**:1462 (1962).

77. J. Mort and F. Jansen, *Plasma-Deposited Thin Films*, CRC Press, Boca Raton, FL, 1986, MP. 29–33. (1986).

78. R. Ishihara, H. Kanoh, Y. Uchida, O. Suguira, and M. Matsumura, "Low Temperature Chemical Vapor Deposition of Silicon Nitride from Tetrasilane and Hydrogen Azide," *Mater. Res. Soc. Symp. Proc.* **284**:3–8 (1992).

79. J. Kanicki, "Contact Resistance to Undoped and Phosphorus-doped Hydrogenated Amorphous Silicon Films," *Appl. Phys. Lett.* **53**:1943–1945 (1988).

80. A. Ban, Y. Nishioka, T. Shimada, M. Okamoto, and M. Katayama, "A Simplified Process for SVGA TFT-LCDs with Single-Layered ITO Source Buss-Lines," *SID 96 Digest*, 1996, pp. 93–96.

81.   N. Hirano, N. Ikeda, H. Yamaguchi, S. Nishida, Y. Hirai, and S. Kaneko, "A 33 cm-diagonal High-resolution Multi-color TFT-LCD with Fully Self-aligned a-Si:H TFTs," *Proc. Int. Display Res. Conf*, Monterey, CA, 1994, pp. 369–372.

82.   G. Horowitz, X. Peng, D. Dischou, and F. Garnier, "Role of Semiconductor Insulator Interface in the Characteristics of π-conjugated Oligomer Based Thin Film Transistors," *Synth. Met.* **51**:419–424 (1992).

Chapter 18

Optoelectronic Technologies

The last two chapters technologies centered on forming transistors, typically for use in integrated circuits. Another important area is the fabrication of various optoelectronic (OE) devices: light emitters, modulators, detectors, and so on. This chapter will review a few of the most common light emitter technologies. Although there has been considerable research into integrated optical systems and some systems are commercially available, most current commercial OE devices are discrete. A large part of the discrete components of the chart shown on page 436 represent OE devices. As a result, there is not much need to be concerned about isolation or interconnect (aside from the actual contacts) for most OE technologies, and the process to build the devices is fairly straightforward once the materials issues related to the active part of the device and any embedded mirrors have been resolved. The devices with the best performance are made from single-crystal GaAs or some of the other III–V or II–VI inorganic semiconductors that are direct gap. For this type of material, electron–hole recombination is very likely to give off a photon. Therefore, GaAs is a popular material for making various light-emitting structures. Recently there has been considerable interest in combining transistors with OE devices. Application areas include very high speed on-chip and off-chip interconnect. Due to the requirement of epitaxial growth, monolithically integrating these semiconductor inorganic devices with a silicon wafer is extremely challenging. Another important and rapidly growing application of integrated OE systems is large-area displays and lighting. Here high speed is not the principal requirement, but one must make systems that are more than a meter across and contain millions of devices. These systems are far more suited to organic and polymer-based OE devices.

18.1 Optoelectronic Devices Overview

The physics of device operation will not be reviewed here. Instead, the reader is referred standard texts such as Schubert [1], Mullen and Scherf [2], Sands [3], and Toshiaki [4]. It was mentioned earlier that most GaAs production is that of light-emitting diodes (LEDs) and lasers. LEDs are becoming popular for a wide variety of applications due to their extremely long life and high efficiency in comparison to filament bulbs. If this technology can be applied to general illumination (i.e., solid state lighting [5]), the world could realize tremendous energy savings. Other LED applications include displays ranging from small disposable plastic screens to computer monitors to televisions to very large format (>10 m) billboards. Solid state lasers, on the other hand, allow light to be focused to a small spot size and are much more amenable to efficient injection into optical fibers and wave-guides. They are widely used in various types of information storage systems such as compact discs. Both devices present extremely interesting challenges with regard to material growth. Defects typically introduce energy levels into the bandgap of the material. This creates a nonradiative recombination pathway called Shockley–Read recombination that competes with the radiative pathway. Each process has a characteristic lifetime. Trap lifetimes are often long, particularly if they are not too close to the middle of the gap, so if the trap density can be made sufficiently low, most of the recombination is radiative.

These nonradiative defect states may be in the bulk of the material, at internal semiconductor material interfaces, as discussed shortly, or at the surface of the semiconductor. As discussed in Chapter 16, surfaces and interfaces invariably lead to interface states in the semiconductor bandgap. Passivation of these dangling bonds is critical to the success of many devices. Since GaAs in particu-lar has a very high surface recombination velocity (about 10^5 that of Si), it is susceptible to this problem. Surface recombination can only occur, of course, if both carrier types are present. One way to control recombination, then, is to design the device in such a way that one carrier type never approaches the surface.

The basic concept of light-emitting devices is to force a high concentration of electrons and holes in close proximity. In direct-gap semiconductors, as long as the thermal energy (kT) is much less than the bandgap of the semiconductor, a significant part of this recombination will result in the emission of a photon whose wavelength is approximately

$$\lambda \cong h \times c/E_g = 1.24 \text{ eV} - \mu\text{m}/E_g$$

where h is Planck's constant, c is the speed of light, and E_g is the semiconductor bandgap. In bulk semiconductors and in conventional transistors, the minority carrier concentration is so low that negligible light is emitted. (One exception is direct-gap bipolar devices. Under certain bias regimes, one can see some light emission from the base; since, however, the device is designed to minimize electron–hole recombination, emission is normally very weak.) If one applies forward bias to a simple pn diode, the two regions are flooded with high concentrations of minority carriers. If the semiconductor is direct, it should emit light. For GaAs (E_g = 1.42 eV), the wavelength of the light emitted is 0.88 μm, which is in the near-infrared part of the spectrum.

A figure of merit for the material is the internal quantum efficiency η, which is the ratio of photons out divided by the number of electron–hole pairs created. A simple way to make this measurement is to illuminate the material with a light source of shorter wavelength than that given by the preceding equation. Theoretically, it should be given by

$$\eta_{\text{int}} = \frac{1/\tau_r}{1/\tau_r + 1/\tau_{nr}}$$

where τ_r and τ_{nr} are the net radiative and nonradiative lifetimes, respectively. For single-crystal, direct-gap inorganic semiconductors like GaAs, τ_r is typically a few nanoseconds, while for polymers and organic films, which may be polycrystalline, τ_r is typically of order tens or hundreds of microseconds. Not all of the light created can be collected, and the design of the device will play a role in how many carriers are delivered to the recombination region, so a second figure of merit is the power efficiency:

$$\eta_{power} = \frac{P_{emitted}}{IV}$$

Both efficiency figures of merit are dimensionless. To obtain high efficiencies, it is important to balance electron and hole fluences. If one of the two is much less than the other, the excess fluence represents wasted power.

Finally, for visible light devices, one sometimes sees the luminous efficiency quoted. This is the visible light (luminous flux) as perceived by the human eye delivered by the LED per unit supplied power. The best LEDs exhibit more than 100 lumens per watt, which is slightly better than the best compact fluorescent bulbs and 5× better than incandescent bulbs. Many LEDs are in the range of 1 to 50 lumens/watt, but this is an area where tremendous progress has been made in recent years.

18.2 Direct-Gap Inorganic LEDs

Although light emission was observed from SiC as early as 1891 [1], the earliest practical LEDs were made from $GaAs_xP_{1-x}$. For the correct value of x, the material is still direct and it has a bandgap of 2.03 eV, which produces a photon with wavelength 0.611 μm, in the visible red part of the spectrum. Thus, early LEDs were all red; 0.65-μm red is now also obtained with $Al_xGa_{1-x}As$. To get other colors such as blue and green, one can use direct semiconductors with still larger bandgaps, such as SiC (0.48 μm blue) and $Al_xGa_yIn_{1-x-y}As$ (0.57 to 0.60 μm, yellow to amber). The latter, although difficult to grow, is extremely efficient and can produce very bright LEDs. Another rapidly emerging material for blue LEDs is GaN (0.45 μm). Alternatively, one can introduce defect states into materials to control the emission wavelength. A prominent example is the introduction of nitrogen in GaP producing a material commonly called GaP:N (0.57 μm) [6].

The latter is an interesting example of an indirect-gap light emitter. Nitrogen produces a deep level in GaP that emits in the yellow. Since the material is indirect, it needs a phonon (lattice vibration) to simultaneously balance energy and momentum and so allow the process to proceed. The phonon density increases with temperature, and so heating increases the ability of the device to emit radiatively. Quantum efficiencies as high as several percent can be achieved. For these indirect transitions, the energy of the photon is slightly different

$$h\nu = \frac{E_g + h^2k^2}{8\pi^2 m_r^*}$$

where k is the phonon wavenumber and m_r^* is a reduced effective mass.

The basic fabrication process for building the diode is extremely simple. Consider for example, a top-emitting device (Figure 18.1) Taking the GaP:N as the first example, one starts with a GaP wafer, then epitaxially grows a thick layer of heavily doped N^+ GaP:N. One then grows a p^+-type layer of GaP:N. Finally, the front and back of the wafer are metallized and the front is patterned. The front pattern must be thick enough and have a low enough resistivity to distribute the current, but must not block too much of the light. One can use a high conductivity semiconductor layer called a current-spreading layer to distribute the current away from the vicinity of the contact. To be most effective, the material selected should have a larger bandgap than the emitting material so that a thick layer can be used but will not absorb a significant amount of light. For example, GaP may be used on top of the GaAs LED. A low resistance ohmic contact is needed to get efficient carrier injection.

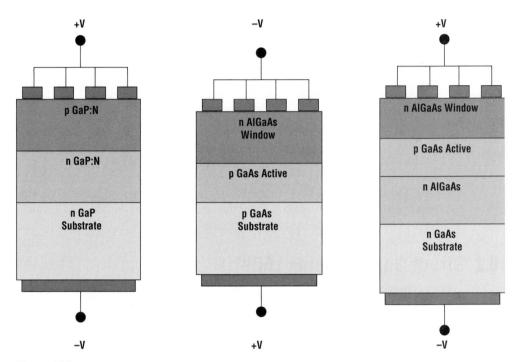

Figure 18.1   Typical LED structures: (A) homojunction, (B) single barrier, and (C) double barrier. If backside contact is not desired, the stack can be etched down to the substrate and a top-side substrate contact added.

For GaAs/AlGaAs devices (red emitters), one can use the Ni/AuGe and Ni/AuZn alloyed contacts discussed in the last chapter. For ZnSe (blue emitters), graded semiconductors such as $ZnTe_xSe_{1-x}$ can be used to reduce the band discontinuity.

The simple structure just described has a serious limitation: if the top layer is too thin, the carriers injected from the junction will recombine at the top electrode and will not produce photons. If the photons are produced too deep in the structure, they will be reabsorbed before they are emitted. This is particularly a problem in direct semiconductors, where the absorption lengths are short. This limitation can be overcome by using window layers or confining layers of a wider bandgap material.

In Figure 18.1B and C a wide bandgap material, AlGaAs, creates a quantum well in the active GaAs layer. Because of the wide bandgap of the AlGaAs, electrons in the GaAs are confined in an active region of the device. Photons emitted by the GaAs, however, are not absorbed by the AlGaAs, producing highly efficient diodes. If the AlGaAs is heavily doped, it can also serve as a current-spreading layer. These double-heterojunction devices have optical conversion efficiencies better than those of filament lamps. In some cases, the lower confining layer is grown so thick (~100 μm) that the substrate has very little effect. Much of the light of LEDs, and particularly those with thick lower layers, is emitted out the sides of the structure. The stack of layers needed to make an advanced LED is usually done by MBE or MOCVD (see Chapter 14 for descriptions of these processes). Normally the layer stack is grown before LED processing is begun.

For a pure, direct-gap semiconductor, the thermally broadened peak has a full width at half-maximum (FWHM) of about $1.8kT$. For alloys, statistical variations of the composition lead to a width of 3 to $8kT$. One can also use a resonant cavity structure to narrow the linewidth of the emission. This

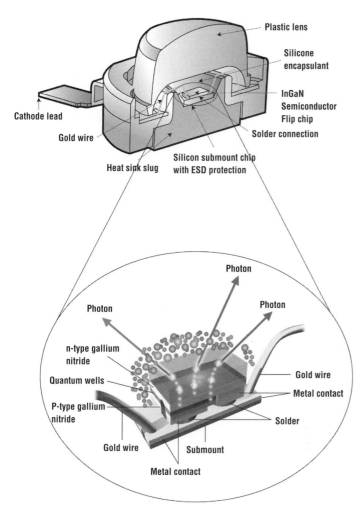

Figure 18.2 Cross section view of a LUXEON power LED. The active device is mounted upside down in the package. Considerable care is taken in heat sinking and encapsulation to allow high power density *(used by permission, Philips Lumileds Lighting Company).*

consists of a top and bottom reflectors designed to be resonant somewhere in the range of wavelengths emitted by the LED. The effect can reduce the FWHM by 5 to 10 × for a well-designed cavity. Resonant cavities will be discussed in more detail shortly.

Finally, we need to say a bit about the problems associated with extracting light from an LED. The semiconductor has an optical index much larger than 1. As a result, Snell's law says that a critical angle exists θ_c, given by

$$\theta_c = \frac{1}{n_{\text{semi}}}$$

where n_{semi} is the real index of refraction of the semiconductor. For GaAs, $n_{\text{semi}} = 3.3$ (Appendix II), so θ_c is $1/3.3 = 0.3$ radian, or 17°. Any light emitted at an angle more than 17° away from the surface normal will therefore be internally reflected back into the semiconductor. There are various ways of dealing with this problem, including the use of resonant cavities that tend to emit preferentially along the vertical axis, and texturing or shaping the surface to increase the probability of emission. The latter can be done by cutting the die to form a widened pedestal at the top, or simply by using a hemispherical drop of an appropriate high index epoxy on top of the device to aid in the emission. This epoxy may also serve as a sealant/encapsulant.

The final result for a state-of-the-art high performance LED is shown in Figure 18.2. A series of quantum wells is grown on n-type GaN (or InGaN), followed by the growth of a p-type GaN layer. On part of the wafer, these layers are etched to expose the substrate. Low resistance contacts are formed to both polarities, and solder bumps are formed on top of the ohmic contacts. The wafers are cut into die and mounted upside down (flip chip). Considerable care is taken to ensure that the die is well heat sunk and that suitable encapsulation and light extraction is done. These LEDs can emit more than 100 lumens per single source.

18.3 Polymer/Organic Light-Emitting Diodes

The direct bandgap inorganic materials discussed earlier can be used to make excellent optoelectronic devices, including all of the solid state lasers currently being produced. Inorganic devices are capable of high power densities, allowing intense point sources. Furthermore, these materials have short carrier lifetimes, allowing the light intensity to be modulated at high speed. As a result, inorganic light sources can be used for high speed communications. These devices are, however, expensive to manufacture. When a large array is desired, they must be sorted and assembled onto a suitable backplane. For applications like displays, this is not economically viable. In the last 20 years technologies for producing light-emitting diodes from polymers (PLEDs) and small-molecule organics (OLEDs) have been developed, and incredible progress has been made. In 2006 these devices were still not as as efficient, nor was the lifetime as great as the best inorganic devices; however their performance is suitable for many array applications.

Some of the lowest cost devices have been made using polymers (see Chapter 8 for a brief overview of polymers). Since the primary market is displays, a great deal of effort has gone into red, green, and blue emitters [7, 8]. The active materials used for these devices are semiconductor polymers. Typically one uses π-conjugated polymers such as PVK or PFE, where the aromatic rings contribute delocalized electrons that can move along the chains in response to an electric field. Rather than speaking about conduction and valence bands, one speaks about the lowest unoccupied molecular orbital (LUMO) and highest occupied molecular orbital (HOMO) states or for continuous bands, the π^* and π bands, respectively. Dyes may be used to tune the emission wavelength.

To maximize performance, one needs to minimize contact barriers. It is very difficult to use heavy doping or graded bands or alloyed contacts, all of which are common in organic devices, to get low resistance. PLED and OLED technologies rely on low barrier height contacts. Thus, a low work function metal must be used as the cathode contact and a high work function metal must be used as the anode contact indium tin oxide. ITO, polyaniline, polypyrrole, and PEDOT, a complex polyethylene polystycene, are the most commonly used anode materials. Although the work function is not as high as one might like, all are transparent in the visible. Common cathode materials are Ca, Ba, and Mg, owing to their small work function. Unfortunately, all of the low work function materials are highly reactive. To obtain reliable devices, one must provide a robust hermetic seal [9]. Otherwise, lifetimes can be measured in minutes or even seconds, especially when the device is energized and the dissipated power heats the LED. Device test stands may be contained in vacuum or nitrogen glove boxes to allow sorting.

If one chooses the electrodes properly, the polymer will emit when biased. The current in the device is typically limited by space chage effects (the field caused by the carriers traveling through the LED). On-voltages are roughly comparable to the difference in the work functions of the two contact electrodes. Due to interference effects from electrode reflections, there is an optimal thickness of the polymer layer where the best external quantum efficiency is achieved, typically around 100 nm.

One of the problems with this basic design is that the efficiency is usually not very high. It is difficult to find electrode materials that are well matched to the polymer and are chemically and thermally stable, particularly for wide bandgap semiconductors. To assist in the recombination, one can use electron and/or hole transport layers (ETL and HTL) (see Figure 18.3). These materials also act as hole and electron blocking layers. Electrons emitted at the cathode easily transport through the ETL and reach the active polymer layer. They are impeded in entering the HTL due to the band discontinuity. For that reason the HTL also acts as an electron blocking layer. Similar processes go on with the holes. The design of the device increases the probability that electrons and holes will recombine. The HTL and ETL must have an energy gap

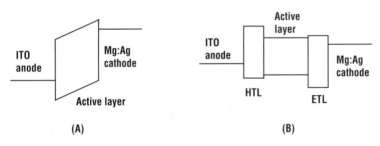

Figure 18.3 Band diagrams of (A) simple OLEDs and (B) OLEDs with electron transport layers (ETL) and hole transport layers (HTL).

larger than the energy of the emitting photon to prevent absorption; if thin layers are used, however, this is not absolutely essential.

In processing PLEDs, one can use conventional optical lithography and etching. Geometries are often quite large, with typical features as large as a few mm. There is a great deal of interest in using ink-jet or other printing processes to make these devices, since the process will be much less expensive and is easily compatible with flexible plastic substrtates.

It is also common to build OLEDs using small-molecule organics that can be thermally evaporated. This can be done in a manner very similar to those described in Chapter 12 (although at much lower temperatures), or one can mount the material to be deposited in a bubbler and use a carrier gas in a method similar to MOCVD. A variety of materials have been demonstrated. Such a system allows one much better control over the thickness of the layers, and so is much more amenable to building complex LED structures with HTLs/ETLs, windows, and other layers. The process of course requires vacuum processing and so more expensive than printing, but the highest performance devices are typically built this way.

18.4 Lasers

Lasers consist of a light-emitting element and a resonator cavity, usually consisting of a conventional mirror and weak or half-silvered mirror. The light is trapped in this cavity until it reaches a critical intensity, at which point it is emitted through the half-silvered mirror. Due to the long radiative lifetimes of organic devices, it is very difficult to get sufficient photon flux to create a laser. As a result, nearly all solid state lasers use inorganic direct bandgap semiconductors.

The first solid state lasers were edge emitting (Figure 17.14A). The layer stack is etched or cleaved into a mesa structure, and light is emitted out the sides of the mesa. Since the length of the resonator is many times the wavelength of the light, the emission is typically multimode [10]. Furthermore, the light is highly astigmatic, with beam divergences up to 50°. The beam is elliptical and is difficult to match to an optical fiber. If it is desired to turn the light beam perpendicular to the surface of the wafer, metal can be used to form mirrors on etched surfaces outside the active mesa.

To improve the device performance, some edge-emitting lasers (e.g., Figure 18.4A) are gain-guided. That its, some type of structure is etched into the top of the device in a way that creates a waveguide that traverses the laser. Typical widths are about 7 μm. This can be done using a layer of deposited SiO_2 (index guiding) or using doped semiconductor blocking layers (gain-guiding) [11].

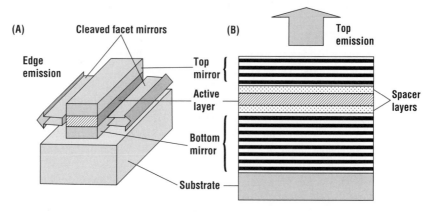

Figure 18.4   (A) Edge-emitting and (B) VCSEL laser structures.

Many flavors of this type of device exist. The interested reader to referred to Suematsu and Adams for a detailed taxonomy [12].

Soda et al. were the first group to experiment with vertical cavity surface emitting lasers (VCSELs) (e.g., Figure 18.4B) [13]. The key to this technology is the ability to grow a mirror under the semiconductor stack. The most common mirror in a VCSEL is a distributed Bragg reflector (DBR). The DBR is formed by epitaxially growing many repetitions of the basic period. This period consists of a quarter-wavelength-thick layer of a high index of refraction material followed by a quarter-wavelength-thick layer of a low index of refraction material. The simplest example of this is GaAs/AlGaAs; however, the ratio of the indices is only about 1.2 for this pair, so many repeats (~20) are required. The half-silvered mirror is obtained by growing fewer repeats of the structure on top of the active light-emitting layers. For GaAs emission, λ is 0.88 μm, so each layer must be 0.22 μm. A concern with these reflectors is the voltage drop across these layers when the device is forward biased, particularly since the heterojunctions will have associated band discontinuities that impede current flow. As with LEDs, high power lasers must be packaged in such a way as to dissipate the high power density in the device.

18.5 Summary

A wide variety of optoelectronic devices is available. These devices are often very challenging from a materials perspective, but much simpler than integrated circuits with regard to fabrication. When the highest performance and/or high speed modulation is required, single-crystal, direct-gap inorganic materials such as GaAs are used. When large areas, lower costs, or integration with other devices and active materials are required, polymer-based and organic materials are used. Thus far, electrically pumped lasers have not been demonstrated in these materials.

References

1. E. Fred Schubert, *Light-Emitting Diodes*, Cambridge University Press, Cambridge, 2006.
2. K. Mullen and U. Scherf, eds., *Organic Light Emitting Devices*, Wiley-VCH, New York, 2006.
3. D. Sands, *Diode Lasers*, Institute of Physics, Bristol, 2005.
4. S. Toshiaki, *Semiconductor Laser Fundamentals*, Dekker, New York, 2004.
5. For example, see www.netl.doe.gov/ssl/. To find many other references, simply search on the term "solid state lighting".
6. K. Warner, "Higher Visibility for LEDs," *IEEE Spectrum*, July 1994, p. 30.
7. U. Scherf and E. J. W. List, *Adv. Mater.* **14**:477 (2002).
8. S. Setayesh, D. Marsitzky, and K. Mullen, *Macromolecules* **33**:2016 (1999).
9. X. Gong, D. Moses, and A. J. Heeger, "Polymer-Based Light-Emitting Diodes (PLEDS) and Displays Fabricated from Arrays of PLEDs," in *Organic Light Emitting Devices*, K. Mullen and U. Scherf, eds. Wiley-VCH, New York, 2006.
10. W. W. Chow, K. D. Choquette, M. H. Crawford, K. L. Lear, and G. R. Hadley, "Design, Fabrication, and Performance of Infrared and Visible Vertical-Cavity Surface-Emitting Lasers," *IEEE J. Quantum Electron.* **33**:1810 (1997).
11. E. Kapon, *Semiconductor Lasers II, Materials and Structures*, Academic Press, London, 1999.
12. Y. Suematsu and A. R. Adams, *Semiconductor Lasers and Photonic Integrated Circuits*, 320, Chapman & Hall, London, 1994, p. 320.
13. H. Soda, K. Iga, C. Kitahara, and Y. Suematsu, "GaInAs/InP Surface Emitting Injection Lasers," *Jpn. J. Appl. Phys.* **18**:2329 (1979).

Chapter 19

MEMS

G. Cibuzar

In the early 1960s, researchers realized that the fabrication techniques developed for standard silicon integrated circuit (IC) processing could be extended to fabricate nontraditional silicon devices. Unlike ICs, which rely on the electrical properties of silicon, these devices utilized silicon's mechanical properties to form flexible membranes capable of moving in response to pressure changes. By detecting this motion, and converting the motion to an electrically measurable signal, a pressure sensor was created. Soon after these early sensors came the development of actuators, which are miniature electromechanical devices that move in response to an electrical input. These were the early years of the field now known as micromachining or MEMS (microelectromechanical systems) [1–3].

Traditional MEMS devices are characterized as either sensors or actuators. Sensors generate an electrical signal from physical stimuli such as pressure, acceleration, heat, and radiation. Actuators convert electrical energy to some form of controlled motion. Examples of MEMS sensors include acceleration sensors used for automobile air bag deployment control, pressure sensors mounted on the tip of catheters for use in intracardiac (within the heart) monitoring of blood pressure, and chemical sensors that quantitatively detect gaseous compounds. Examples of MEMS actuators include video display systems using digital mirror devices consisting of over one million individually controlled micromirrors, ink-dispensing nozzles used in ink-jet printers, and valves and pumps used in miniature fluidic systems (fluid volumes in the microliter to nanoliter range). Traditional MEMS devices rely on materials typically used in silicon IC fabrication, such as single-crystal silicon, polysilicon, silicon dioxide, and silicon nitride. Due to the mechanical nature of MEMS devices, intrinsic material properties such as Young's modulus, temperature coefficient of expansion, and yield strength are important in their design. Since MEMS structures often have unsupported (or freestanding) elements, thin film stress and stress gradients within the film need to be tightly controlled or the unsupported elements will break or curl, rendering the structure useless. Thin film deposition conditions directly impact the level of stress in thin films, and must be tightly controlled to yield functioning sensors and actuators. Even though silicon IC fabricators and traditional MEMS researchers use the same processing tools, each has unique processing issues. IC device processing issues, such as gate oxide integrity, device isolation, and submicron gate formation, are not of much concern for MEMS. Similarly, MEMS processing concerns, such as controlled etching for membrane formation, mechanical friction of microscopic parts, and surface tension effects, do not normally trouble the IC process engineer. Many processing issues do affect both standard IC and MEMS fabrication, including thin film stress, planarization, and selective wet and dry etching. A thorough knowledge of the lithographic and thin film fabrication techniques common to both technologies is important for the MEMS device design and process engineer. As MEMS technology matures, more integration between MEMS devices and signal processing and control circuitry will force designers and process engineers to appreciate the subtleties of both MEMS and IC fabrication.

This chapter is an introduction to the traditional MEMS fabrication processes of bulk and surface micromachining, including examples of common MEMS sensors and actuators. Given the diverse applicability of MEMS devices, one chapter can only touch the surface of the many ways these devices have been used. Similarly, the large number of processing variations cannot possibly be related in one chapter. A grounding in the basic processes and implementations will, however, provide the proper foundation for further study of more advanced MEMS topics.

19.1 Fundamentals of Mechanics

To understand the design and operation of MEMS devices, a basic knowledge of the mechanics of materials is necessary. A brief discussion of basic concepts will provide sufficient background for understanding MEMS devices. These concepts include stress, strain, Hooke's law, Poisson's ratio, and film stress. A complete discussion of this field is beyond the scope of this chapter, but many well-known reference books can be consulted for more complete study [4–6].

When a solid body of a homogeneous material is subjected to a force, the body will respond to the force by changing shape. For nonsolid materials such as liquids, this change is dramatic. For solids, the changes are usually quite small, often too small to notice with the eye, except in cases of large forces that may cause irreversible damage. Consider a solid rod of initial length L_0 and diameter D subjected to a tensile force F uniformly applied to the ends of the rod as shown in Figure 19.1. This tensile force will cause the rod to lengthen by an amount ΔL. This lengthening of the rod is described by the axial strain ε_a, where ε_a is defined by

$$\varepsilon_a = (\Delta L)/L_0 \tag{19.1}$$

Strain is a dimensionless quantity and is usually expressed in units of 10^{-6}, or microstrains, because for most materials strain values are quite small. Stress, denoted by σ, is defined in terms of the force F and the area over which F is applied. For the uniformly applied force F in Figure 19.1,

$$\sigma = \frac{F}{\text{area}} = \frac{F}{\pi(D/2)^2} \tag{19.2}$$

The units of stress σ are commonly newtons per square meter, or pascals (Pa). The standard sign convention for stress is for tensile stresses to be negative and compressive stresses to be positive. Note that the force F in Figure 19.1 is acting perpendicularly to the end of the rod. This type of force is termed an axial force, and it results in an axial strain ε_a and axial stress σ_a. Shear forces, which act parallel to the surface of a body, generate shear stress and strain.

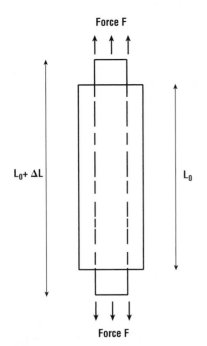

Figure 19.1 Rod elongation due to a tensile force F.

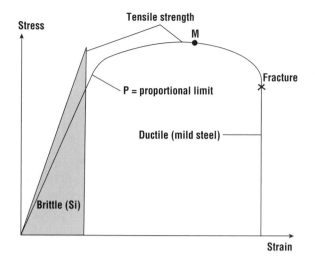

Figure 19.2 Stress–strain curve for a typical metal as well as for a brittle material like silicon (high Young's modulus and no plastic deformation region) *(reprinted with premission from Madou [7]. Copyright CRC Press).*

One important way of characterizing materials is to measure their stress–strain curve. Using a tensile test machine, a bar of the material under test is subjected to a uniform axial loading force. Sensitive measuring devices determine changes in the length as a controlled tensile force is applied to the bar. By measuring the change in length as a function of tensile force, a stress–strain curve is generated for the material. Figure 19.2 shows a typical stress–strain curve for a ductile material such as mild steel, and a brittle material such as silicon or glass [7]. Both materials show a region of low strain values where stress and strain are linearly related. This linear region ends at stresses corresponding to the proportional limit, defined as the highest point on the stress–strain curve, where if the force is removed, the bar will return to its original length. Stresses beyond the proportional limit result in permanent deformation of the bar. For ductile materials, beyond the proportional limit are regions in which the bar will elongate due to plastic deformation and reach a maximum stress value (called the ultimate tensile strength) before eventual fracture (breaking) of the bar. By definition, the yield strength is the stress value that results in a permanent strain of 0.002 when the force is removed. Brittle materials exhibit essentially no plastic deformation, and fracture soon after reaching the proportional limit. Table 19.1 shows material constants for some common materials, including silicon.

Hooke's law describes the linear relationship between the stress σ and ε strain at low stress values,

$$\sigma = \varepsilon E \qquad\qquad (19.3)$$

where E is the slope of the linear region of the stress–strain curve and is commonly called Young's modulus. MEMS devices are generally designed to operate at stresses in the linear region, so E is an

Table 19.1 Properties of materials

	Yield Strength (10^9 Pa)	Young's Modulus (10^9 Pa)	Density (g/cm^3)	Thermal Conductivity (W/cm-°C)	Thermal Coefficient of Expansion (10^{-6}/°C)
Diamond (single crystal)	53.0	1035.0	3.5	20.0	1.0
SiC (single crystal)	21.0	700.0	3.2	3.5	3.3
Si (single crystal)	7.0	190.0	2.3	1.6	2.3
Al$_2$O$_3$	15.4	530.0	4.0	0.5	5.4
Si$_3$N$_4$ (single crystal)	14.0	385.0	3.1	0.2	0.8
Gold	—	80.0	19.4	3.2	14.3
Nickel	—	210.0	9.0	0.9	12.8
Steel	4.2	210.0	7.9	1.0	12.0
Aluminum	0.2	70.0	2.7	2.4	25.0

From Petersen [3].

important material parameter in MEMS design. For single-crystal silicon, E is approximately 190 gigapascals (190 GPa).

Another important mechanical effect is the change in lateral dimensions due to an axial force. The rod in Figure 19.1 is shown as decreasing in diameter as a result of the axial tensile force F. This change in width is characterized by a lateral strain ε_l, which is related to the axial strain ε_a by Poisson's ratio ν as follows:

$$\nu = -\frac{\varepsilon_l}{\varepsilon_a} \tag{19.4}$$

By definition, Poisson's ratio is positive. Since ε_l is a negative value when ε_a is positive, the negative sign in Equation 19.4 is necessary. For silicon, Poisson's ratio is generally taken to be 0.28. For most materials, ν ranges from 0.2 to 0.5, with many metals near 0.3.

19.2 Stress in Thin Films

Control of strain levels in deposited thin films is critical for many MEMS devices. Membrane structures made from polysilicon, a common MEMS mechanical material, require tensile film strains less than -0.001, or the membrane will break. Strain in thin films is difficult to measure directly, so stress levels are generally used to characterize thin films. Film stress can be measured by quantifying the amount of bow or warpage that the deposited thin film causes in a flat substrate. Tensile stress in a thin film will cause the substrate edges to be higher than the center as viewed from the film side, and a compressive stress will cause the center to be higher. By measuring the center deflection δ defined as the difference in height between the center and edge, and assuming the bow is uniform across the substrate, the radius of curvature R of the bow can be calculated based upon the size of the substrate (see Figure 12.30). The stress level can then be calculated using the Stoney equation [8]

$$\sigma = \frac{1}{R} \frac{E}{6(1 - \nu)} \frac{T^2}{t} \tag{19.5}$$

where R is the calculated radius of curvature, E is Young's modulus, ν is Poisson's ratio, T is the substrate thickness, and t is the deposited film thickness. Conditions required for the usage of the Stoney equation are (1) uniform thickness of the thin film, (2) isotropic elastic properties, and (3) thin film thicknesses much less than the substrate thickness ($t \ll T$). The most common thin film stress measurement technique is to measure the center deflection δ using a laser-based interferometer system. A limitation of this technique is that an average stress value for the whole wafer is given, with no indication of local stress variations. Other techniques for measuring the thin film stress involve the fabrication of MEMS structures such as cantilever beams and ring and beam structures. Careful measurement of the change in length of a cantilever beam before and after release from the substrate can be used to quantify stress [9]. By using cantilever beams supported on both ends, compressive stress can be determined by identifying the largest unbroken beam in a series of beams of varying dimensions [10]. Depending on the compressive stress level and the Young's modulus of the beam material, beams of certain sizes will buckle (show displacement out of the plane of the beam). This information can then be used to determine the compressive stress level. Tensile stress levels can be determined from structures that convert the tensile stress to a compressive stress and then use the same beam-buckling technique [11]. Stress measurements using these techniques, although not as easy as the laser-based systems, are considerably less expensive and can be used to determine stress in localized regions.

Stress gradients, or variations in stress level as a function of film thickness, are a serious concern in MEMS device fabrication. Stress gradients lead to curvature in cantilever beams and other structures, even if the overall stress level is sufficiently low. Stress gradients usually result from nonuniformities in the thin film deposition process, which cause atomic structure variations that create uneven strain through the film. Techniques for determining stress gradients include the measurement of the radius of curvature of a cantilever beams and cantilever spiral beams [12]. Normally these structures are used only to qualitatively determine stress gradients in process development.

Factors that lead to thin film stress can be categorized as either intrinsic or extrinsic stresses. Intrinsic stresses usually result from the nonequilibrium nature of thin film deposition processes. During deposition, after reaching the surface, atoms often do not have the necessary kinetic energy or sufficient time to migrate to the desired lowest energy state before more atoms arrive, preventing further migration. The resulting nonequilibrium atomic arrangement is "frozen in," resulting in lattice strain (and stress). Postdeposition annealing at elevated temperatures is commonly used to relieve intrinsic stress. This anneal provides sufficient kinetic energy for atomic rearrangement and reduction of the as-deposited strain. Annealing has been shown to allow a controlled modification of stress level, and even change compressive stress to tensile stress for LPCVD polysilicon [13]. Extrinsic stress is caused by factors external to the film structure. The most common source of extrinsic stress is a difference in the thermal coefficient of expansion (TCE) between the deposited thin film and the substrate. When the deposition temperature is considerably higher than the operating temperature (as is often the case for MEMS films), after completing the deposition, cooling of the substrate/film will cause stress due to the TCE mismatch.

19.3 Mechanical-to-Electrical Transduction

The measurement of physical quantities such as pressure, acceleration, and mass change is based on sensing mechanisms that convert (or transduce) changes in these quantities to electrically measurable parameters such as resistance, capacitance, and changes in characteristic frequencies. MEMS sensors have been fabricated that use many different methods to convert physical changes to an electrical signal. MEMS sensors generally have a structural element that moves in response to the physical quantity being measured, such as a membrane diaphragm moving in response to a change in applied pressure. The most commonly used method to detect this motion involves piezoresistivity, and most of this section will be devoted to its description. Other common transduction methods for mechanical deflections include the piezoelectric effect, capacitance changes, and magnetic effects such as the Hall effect [14–18]. Each of these transduction methods could be also used to sense phenomena other than mechanical deformation. Before focusing on the piezoresistivity effect, these other methods will be briefly described.

The piezoelectric effect [19, 20] is the phenomenon whereby a force applied to certain crystalline materials causes an electrical charge to be generated on the surface of the crystal, with the amount of the charge directly related to the applied force. This happens because the force creates stress in the crystal, leading to strain. On an atomic scale, strain implies a slight change in the positions of the atoms in the crystal. In piezoelectric materials, the unit cell of the material contains positively and negatively charged ions that are not symmetrically oriented with respect to each other within the unit cell. The effective location of the center of charge for the positive ions is not at the same location as that of the negative ions. The application of a stress causes deformation (strain), and electric dipoles are created. These electric dipoles induce surface charges on the crystal that in turn create an electric field. This surface charge induced electric field cancels the stress-induced electric

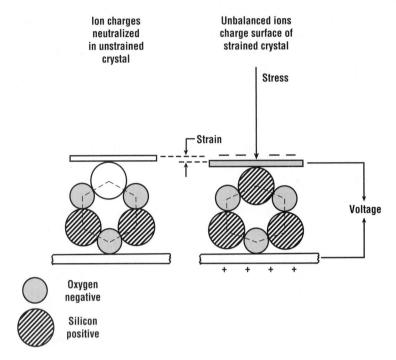

Figure 19.3 Ion position in a piezoelectric crystal lattice such as quartz with and without applied stress *(reprinted with permission from Madou [7]. Copyright CRC Press).*

field from the dipoles. Figure 19.3 is a simplified illustration of this situation for quartz. Piezoelectric crystals also show the reverse effect, where an applied electric field causes a mechanical deformation. This effect is important for MEMS actuators. Piezoelectric materials commonly used in MEMS devices include quartz, lithium niobate, PZT (lead zirconate titanate), and zinc oxide. These materials can be used both in bulk form (macroscopic discs attached after completion of the MEMS device) or as a thin film deposited during the MEMS fabrication process. Compared with other transduction methods, the main advantage of piezoelectric materials is the strong signal transduction for small displacements [21]. The most glaring disadvantage of thin film piezoelectrics is the complication they bring to the device processing. Issues include potential contamination of IC circuitry, nonstandard thin film deposition techniques, and susceptibility to commonly used wet etches.

Capacitance measurements are another common transduction technique for MEMS devices. Consider a pressure sensor made with a flexible membrane that moves in response to changes in the applied pressure [22]. By putting one electrode on the membrane and a second electrode on a fixed surface in close proximity to the membrane electrode, a capacitor is created with capacitance as a function of applied pressure (Figure 19.4). As the pressure increases/decreases, the membrane moves closer to/farther from the fixed electrode, and the measured capacitance increases/decreases. The advantages of capacitance transduction include limited temperature dependence of the measured capacitance, and simplicity of the fabrication process. The major disadvantage is the nonlinearity of the dependence of the capacitance on the pressure. This nonlinearity is a result of the membrane deformation mechanics under varying pressure. Accelerometers have also been made with this sensing technique [23].

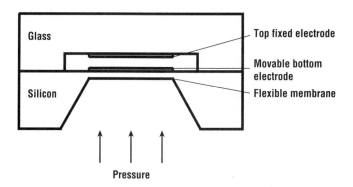

Figure 19.4 Schematic drawing of a pressure sensor designed to correlate measured changes in capacitance with pressure changes.

Signal transduction has used magnetic effects such as the Hall effect for many years [24–26]. Sensors based on the Hall effect generate a voltage proportional to the applied magnetic field. These sensors have been used for detection of electric current, proximity detection, and position of rotating shafts such as crankshafts in engines. Despite their low Hall coefficients relative to some other materials, silicon or gallium arsenide is used for most Hall effect MEMS devices because of the developed processing infrastructure for these materials.

Piezoresistivity is a property of materials that describes the change in electrical resistance as a function of mechanical stress applied to the material. Many materials exhibit piezoresistance, but some semiconductors, including silicon, show a large effect. The discovery of piezoresistance in silicon [27] was instrumental in the development of silicon pressure sensors. The theoretical explanation [28] for piezoresistivity is based on a quantum mechanical description of crystal lattice strain effects on the conductivity of electrons and holes as a function of lattice direction. From a practical standpoint, piezoresistance in single-crystal silicon is primarily dependent on several parameters:

1. The silicon doping type (n- or p-type) and concentration.
2. Temperature.
3. The direction of the current flow relative to the orientation of the crystal lattice.
4. The direction and type of force (tensile or compressive) relative to the orientation of the crystal lattice.

The most important parameter for MEMS devices made with (100) silicon wafers is the doping type. Detailed descriptions of piezoresistivity can be found in the scientific literature [7, 15, 28].

As a simple example, Figure 19.5 shows a bar of single-crystal silicon subjected to a tensile force F, with a current flowing in a direction perpendicular to F. The resistance of the silicon bar would be symbolized by R. The force F generates a stress σ in the bar, and a resistance change due to the piezoresistive effect. The resistance change ΔR is

$$\Delta R/R = \pi_t \sigma \qquad (19.6)$$

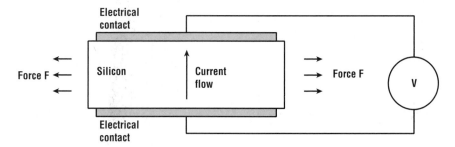

Figure 19.5 Current flowing in a silicon bar with a tensile force applied to the ends of the bar, perpendicular to the current flow.

where π_t is defined as the transverse piezoresistance coefficient. For situations in which the current flow is parallel to the stress, the longitudinal piezoresistance coefficient π_l is used. In the general case in which both transverse and longitudinal stresses are present, we have

$$\Delta R/R = \pi_t \, \sigma_t + \pi_l \, \sigma_l \qquad (19.7)$$

where σ_t and σ_l are the transverse and longitudinal components of the stress σ, respectively.

To use Equation 19.7 to calculate a resistance change, the piezoresistance coefficients need to be calculated. As previously mentioned, the values of π_t and π_l depend on several properties of the crystal, and on the orientation of the stress σ relative to the crystal lattice orientation of the single crystal silicon. For a (100)-oriented silicon wafer, planes parallel to the surface have a [100] orientation. Within this plane, the piezoresistive coefficients for silicon with low doping concentration vary as a function of direction [29]. Figure 19.6 illustrates the directional variation of the piezoresistive coefficients for p-type silicon, with maximum values of approximately -70 and $+70 \times 10^{-11}$ Pa^{-1}, respectively for π_t and π_l along the <110> directions. For n-type silicon the piezoresistive coefficients are largest along the <100> and <010> directions and smaller along the <110> directions. Later we will see that anisotropic silicon etchants create etch features aligned with particular crystal directions. In the case of (100) silicon wafers, these etched features have edges along the <110> directions. For piezoresistors oriented along the <110> directions, p-type resistors will have higher sensitivity than n-type resistors, since piezoresistance values are largest in that direction for p-type silicon and smallest for n-type silicon. Inserting the p-type silicon piezoresistance coefficients for the <110> directions into Equation 19.7, we find the following for a resistor oriented parallel to the transverse stress:

$$\Delta R/R_{\text{transverse}} = \pi_{\text{p-type}} \, (\sigma_l - \sigma_t)$$
$$= 70 \times 10^{-11} \text{ Pa}^{-1} \, (\sigma_l - \sigma_t) \qquad (19.8)$$

For a p-type resistor oriented perpendicular to the transverse stress we find

$$\Delta R/R_{\text{longitudinal}} = -\pi_{\text{p-type}} \, (\sigma_l - \sigma_t)$$
$$= -70 \times 10^{-11} \text{ Pa}^{-1} \, (\sigma_l - \sigma_t) \qquad (19.9)$$

Each resistor changes by the same amount, but with different sign.

The most common implementation of piezoresistors for pressure sensors is to arrange several resistors in a Wheatstone bridge configuration (Figure 19.7) [15]. Wheatstone bridge circuits are popular because voltage measurements are easier than resistance measurements, and the resistors are arranged such that variations in the supply voltage V_b do not affect the result. Also, the Wheatstone bridge effectively cancels the effect of temperature on the piezoresistance if the resistors match each other (have the same resistance value). Figure 19.7A shows a top view of an arrangement of four piezoresistors aligned with the edges of a square membrane oriented along the <110> silicon lattice direction. Application of positive pressure from

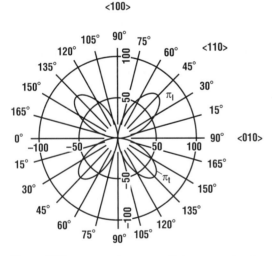

Figure 19.6 Piezoresistance coefficients π_l and π_t for the [100] plane of p-type single crystal silicon. π_l values are shown on the upper half and π_t values on the lower half. Units are 10^{-11} Pa^{-1}. Note that the maximums are in the <110> directions (*after Kanda [29], © 1982 IEEE Press*).

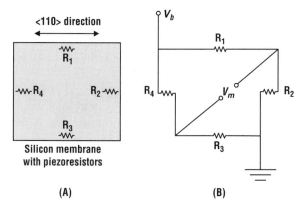

<110> direction

R_1

R_4

R_2

R_3

Silicon membrane
with piezoresistors

(A)

V_b

R_1

R_4

V_m

R_2

R_3

(B)

Figure 19.7 (A) Schematic drawing of the position and orientation of four piezoresistance elements on a silicon membrane with sides defined by anisotropic silicon etching of a (100) silicon wafer; (B) Wheatstone bridge configuration of the four piezoresistive elements.

above will deflect the membrane downward, induce stress in the membrane, and change the resistance of the four resistors. Due to their orientation, R1 and R3 will decrease in resistance, and R2 and R4 will increase. Since for p-type resistors,

$$\Delta R/R_{\text{longitudinal}} = -\Delta R/R_{\text{transverse}} \qquad (19.10)$$

the relative increases in R2 and R4 are equal to the relative decreases in R1 and R3. If all resistors have the same unstressed resistance value, then a simple circuit analysis of the Wheatstone bridge configuration shows that measurement of the two voltages V_b and V_m determines $\Delta R/R$ by the relation

$$\Delta R/R = V_m/V_b \qquad (19.11)$$

The two major advantages of piezoresistive MEMS devices are the simplicity of processing and the maturity of the measurement process. Silicon is by its nature piezoresistive, and by adding dopants, resistors can easily be created either using diffusion or ion implantation. Measurement of resistance changes through the use of Wheatstone bridge configurations to convert the resistance change to a voltage is well established. The major disadvantage of piezoresistive sensors is the $1/T$ temperature dependence of the piezoresistance coefficients for single-crystal silicon. This significant temperature dependence can require temperature compensation circuitry, which complicates the implementation of piezoresistive-based MEMS devices.

19.4 Mechanics of Common MEMS Devices

Many MEMS devices utilize either thin film membranes or cantilever beam structures. The mechanics of thin film membranes is based on the mechanics of thin plates (sometimes called shells), whereas cantilever beam structures are based upon the mechanics of beams. Thin film membranes in MEMS devices are usually square or rectangular in shape, with side lengths ranging from hundreds of microns to more than several millimeters. Less common are circular membranes, primarily due to the difficulty of etching circular structures. The membrane material is usually silicon (either single crystal or polysilicon) or silicon nitride. Membrane thicknesses depend on the area of the membrane and the application, and can range from a few thousand angstroms to 50 μm or more. MEMS cantilever beams are commonly hundreds of microns long, tens to hundreds of microns wide, and less than five microns in thickness. Note that for both membranes and cantilever beam MEMS structures, the thickness is much less than the lateral dimensions of the structure.

The important mechanical properties of membranes and cantilever beams can be derived from a stress/strain analysis using thin plate deflection theory or beam deflection theory [4–6]. Although a complete discussion of these analyses is beyond the scope of this book, the significant results for MEMS devices can be summarized. For membranes, the relevant assumptions of thin plate deflection theory are that (1) the maximum membrane displacement is less than 20% of the membrane thickness, (2) membrane thickness does not exceed 10% of the plate length, and (3) there is no initial stress in the membrane. The first assumption puts a limit on the force that can be applied to the membrane. The second assumption is easily met for most MEMS structures, where the membrane thickness is generally less than 50 µm, which means the membrane size must be at least 0.5 mm on a side. If either of the first two assumptions is not met, thick plate theory must be used. If the third assumption is not met, then the results are qualitatively correct, but a more complete analysis is required to obtain quantitatively accurate values. Although real MEMS films usually have some intrinsic stress, for simplicity we assume no intrinsic stress.

For a square membrane of side a, thickness t, Young's modulus E, density ρ, and Poisson's ratio v, subjected to a uniform pressure P, the maximum membrane deflection W_{max}, maximum longitudinal and transverse stress σ_l and σ_t, and fundamental mode resonant frequency of vibration F_o are defined as

$$\text{Max deflection } W_{max} = 0.001265Pa^4/D \tag{19.12}$$

$$\text{Max longitudinal stress } \sigma_l = 0.3081P(a/t)^2 \tag{19.13}$$

$$\text{Max transverse stress } \sigma_t = v\sigma_l \tag{19.14}$$

$$\text{Resonant frequency } F_o = \frac{1.654t}{a^2}\left[\frac{E}{\rho(1-v^2)}\right]^{1/2} \tag{19.15}$$

where the plate flexural rigidity D, a measure of stiffness of the plate to bending, is defined as

$$D = \frac{Et^3}{12(1-v^2)} \tag{19.16}$$

The maximum longitudinal stress σ_l is at the center of each side at the edge of the membrane, and perpendicular to the membrane edge. The maximum transverse stress σ_t is also at the center of the side along the edge, but is oriented parallel to the edge. The location of the maximum stress points is important for piezoresistive and piezoelectric devices, since maximum stress leads to maximum effect when using these transduction methods.

Example 19.1

Calculate the maximum deflection and maximum stresses for a square silicon membrane of thickness 10 µm and side length 2 mm for an applied pressure of 1000 Pa.

Solution

For silicon, $E = 190$ GPa and $v = 0.28$. Using Equation 19.16 with $t = 10$ µm gives

$$D = 173 \times 10^{11} \, Pa \, (\mu m)^3$$

From Equation 19.12,

$$W_{max} = \frac{0.001265 \, Pa^4}{D} = 1.17 \, \mu m$$

From Equations 19.13 and 19.14,

$$\sigma_l = 0.3081P(a/t)^2 = 12.3 \text{ MPa}$$

$$\sigma_t = v\sigma_l = 3.45 \text{ MPa}$$

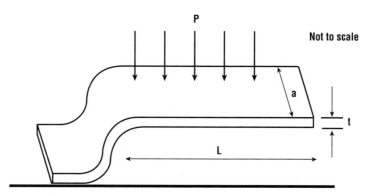

Figure 19.8   Cantilever beam fixed at one end, subjected to downward force per unit length P.

For cantilever beam theory, the major assumption is that the deflection at the end of the beam is small compared with the beam length. For a cantilever fixed at one end, length L, width a, thickness t, Young's modulus E, density ρ, and uniform distributed load H (H is a force per width) as shown in Figure 19.8,

$$\text{Deflection } W(H, x) = \frac{Hx^2}{24EI}(6L^2 - 4Lx + x^2) \qquad (19.17)$$

$$\text{Max stress } \sigma = \frac{HL^2t}{4I} \qquad (19.18)$$

where I is the bending moment of inertia defined as

$$I = \frac{at^3}{12} \qquad (19.19)$$

Note that for a uniform force F applied to the beam surface, the load H is equal to the force F scaled by the beam width a, so

$$H = F/a \qquad (19.20)$$

For a point load Q (Q is a force) at the end of the beam,

$$\text{Deflection } W(Q, x) = \frac{Qx^2}{6EI}(3L - x) \qquad (19.21)$$

$$\text{Max stress } \sigma = QLt/2I \qquad (19.22)$$

The fundamental mode resonant frequency of the beam is

$$F_o = 0.161 \frac{t}{L^2}\left(\frac{E}{\rho}\right)^{1/2} \qquad (19.23)$$

Note that F_o does not depend on the beam width. In terms of the cantilever beam mass M,

$$F_o = 0.161 \frac{t}{L}\left(\frac{Eta}{ML}\right)^{1/2} \qquad (19.24)$$

For the cantilever beam, the maximum stress is at the base of the beam and is longitudinal. The transverse stress for the beam is zero because unlike the membrane where all sides are fixed, the sides of the beam are free to move up and down, so there is no force transverse to the beam for simple up and down motion. More complex cantilever motions involving torsional motions require more complicated analysis.

Example 19.2

A silicon cantilever beam with a piezoresistor located at the point of maximum stress is subjected to a point load Q at the end of the beam. The length of the beam is 1000 μm, the beam thickness is 3 μm, and Q is 10 μN. Calculate the beam width that results in a 3% resistance change for the piezoresistor due to the load Q. Assume the beam lies perpendicular to the silicon <110> lattice direction.

Solution

A p-type piezoresistor will have higher resistance change for this lattice direction, so from Equation 19.8 we have

$$\Delta R/R_{transverse} = 70 \times 10^{-11}\ Pa^{-1}\ (\sigma_l - \sigma_t)$$

$$= 70 \times 10^{-11}\ Pa^{-1}\ \sigma_l$$

since the transverse stress σ_t is zero for the cantilever beam. Inserting the desired resistance change of 3%, gives for the maximum stress

$$\sigma_{max} = \frac{0.03}{70 \times 10^{-11}\ Pa^{-1}} = 4.3 \times 10^7\ Pa$$

From Equations 19.19 and 19.22 we have

$$\sigma_{max} = \frac{QLt}{2a(t^3)/12} = 4.3 \times 10^7\ Pa$$

Plugging in the values for L, Q, and t, converting to common units (MKS system), and solving for side length a yields

$$a = 1.5 \times 10^{-4}\ m = 150\ \mu m$$

The cantilever beam should have a p-type piezoresistor at the base of the beam, and the beam width should be 150 μm.

Changes in the resonant frequency of membranes and cantilever beams are often used to measure mass loading changes. For example, a cantilever beam may be coated with a polymer that absorbs water in proportion to the humidity in the air. Since the resonant frequency of the beam depends on the mass of the beam (through the density term), this increase in beam mass due to the polymer absorbing water leads to a change in resonant frequency. Measurement of this frequency shift can then be correlated with humidity.

Example 19.3

What is the resonant frequency F_o for a silicon cantilever beam 1000 μm long, 100 μm wide, and 3 μm thick? The density of silicon is 2.3 g/cm³.

Solution

Convert all variables to MKS units and use Equation 19.23. Then

$$E = 190 \, \text{GPa} = 190 \times 10^9 \, \text{N/m}^2$$
$$\rho = 2.3 \times 10^3 \, \text{kg/m}^3$$
$$t = 3 \times 10^{-6} \, \text{m} \qquad L = 10^{-3} \, \text{m} \qquad a = 10^{-4} \, \text{m}.$$
$$F_o = 0.161 t/(L^2)(E/\rho)^{1/2}$$

substituting gives

$$F_o = 4.39 \, \text{kHz}$$

19.5 Bulk Micromachining Etching Techniques

Bulk micromachining refers to MEMS fabrication processes that involve removal of significant amounts of the silicon substrate in order to form the desired structure. Etching is the cornerstone of bulk micromachining [7, 14, 30]. Historically, wet etching with both isotropic and anisotropic etchants has dominated MEMS devices, but more recently other techniques, such as isotropic vapor phase etching and high density, plasma-based processes, have been used. This section will focus on bulk micromachining processing using wet etching on standard (100)-oriented silicon wafers.

As discussed in Chapter 11 on etching, isotropic wet etchants for silicon, such as mixtures of hydrofluoric acid–nitric acid–acetic acid, show etch profiles that are difficult to control. The undercutting of features defined by the masking material leads to rounded profiles for holes and trenches. Figure 19.9 shows etch profiles for isotropic etchants with and without agitation of the etchant. With agitation, the profile generally exhibits the shape expected for etch rates that are nearly the same for all crystal orientations. Without agitation, the profile shows a flatter bottom, which is the result of reduced etching in the vertical direction. This reduction in etch rate is caused by a depletion of the etching species near the etch surface (the etch is diffusion rate limited). The agitation assists the diffusion of fresh etchant to the surface, resulting in the expected isotropic etch profile. For bulk micromachined structures, etch depths often approach the full

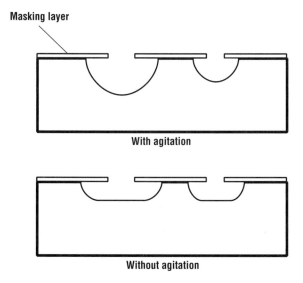

Masking layer

With agitation

Without agitation

Figure 19.9   Isotropic etch cross sections showing undercutting of the etch mask and the effect of agitation on the etch profile.

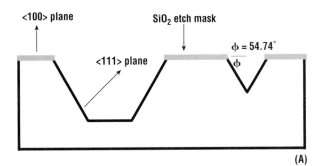

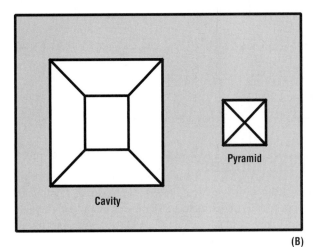

Figure 19.10 Cross section (A) and top view (B) of pyramidal holes and cavities formed in a (100) silicon wafer with an anisotropic etchant.

wafer thickness. In the isotropic etching of these deep structures, the considerable undercutting of the etch mask means the features must be separated by at least the depth of the etch. The two main factors limiting the use of isotropic etchants in MEMS processing are the sensitivity of the etch profile to agitation level and the significant amount of undercutting for masked features.

In addition to wet isotropic etchants, a dry isotropic etchant has been developed: XeF_2 vapor has been shown to etch silicon without any excitation such as heat or plasma [31]. Advantages of XeF_2 vapor phase etching include a high etch selectivity to aluminum, silicon dioxide, silicon nitride, and photoresist, and a relatively simple etch system. These are significant factors when integration with standard IC processing is desired [32]. A significant disadvantage is the very rough silicon surface resulting from the etch.

Anisotropic etching permits the formation of features defined by the crystal planes in the silicon wafer. Potassium hydroxide (KOH) is probably the most common anisotropic etchant for bulk silicon micromachining. Others include ethylenediamine pyrocatechol (EDP) mixed with water, hydrazine (N_2H_4), and water, and tetramethylammonium hydroxide (TMAH). Typical etch profiles for anisotropic etchants are shown in Figure 19.10. These etches are called anisotropic because the etching rate is high in the <100> direction and low in the <111> direction. Etch ratios for these two directions can be as high as 600 to 1. Various factors have been put forth to explain this directionally dependent etch rate, but none completely explains the phenomenon. These include (1) the fact that the <111> crystal plane is the most dense, with three of the four covalent bonds below the plane, and thus less accessible to the etchant, and (2) surfaces with the highest bond density etch fastest. Several chemical models have been proposed [33, 34].

In the structure of the silicon crystal lattice, the [111] planes are oriented at 54.74° relative to the [100] plane. A square mask opening will yield an etched feature in the shape of an inverted pyramid, with the point on the pyramid at a depth determined by the intersection of the [111] planes. If the etch is terminated before the etch stops due to reaching the intersection of the [111] planes, then a truncated pyramidal etch cavity is formed. The edges of the etched structure run in the <110> directions, since these edges are intersections of the [100] and [111] planes (this is an important point for bulk micromachined MEMS devices with piezoresistive elements). If the sides of the square mask opening are long enough, the wafer will be etched through completely, with the size of the square hole on the opposite side determined by the wafer thickness and the initial mask opening size. An important processing point is that the initial mask opening must be aligned with the <111> directions, or else the etched feature will be larger and rotated from the original design (Figure 19.11). Another processing issue is etching of structures with features containing convex corners (corner features

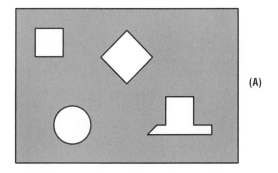

(A)

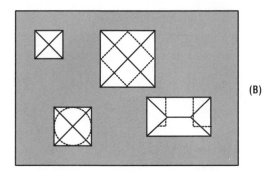

(B)

Figure 19.11 Effect of mask opening orientation on the etch profile. (A) Top view of mask openings as oriented to the <110> direction. (B) Etched structures resulting for an anisotropic etchant on (100) silicon.

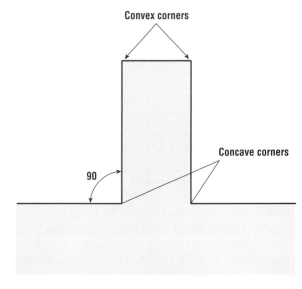

Figure 19.12 Top view of a cantilever beam structure showing the locations of the convex and concave corners.

with interior angles greater than 180°). An example of a structure with both convex and concave corners is a cantilever beam (Figure 19.12). At the end of the beam, the angle of the material to be removed is 270° (convex), whereas at the base of the beam the angle is 90° (concave). Unlike the concave corners, convex corners are not stable because higher order planes are exposed and etch rapidly, leading to rounding and eventually complete elimination of the convex corner. Corner compensation techniques have been developed to eliminate this effect [35]. These techniques typically involve adding masking features to the corners.

At temperatures above 60°C, KOH is commonly used at concentrations of 20 to 50% by weight in water. Figures 19.13 and 19.14 show KOH etching rates as well as selectivity to SiO_2, the most common masking material. When optimized, KOH etched surfaces are nearly as smooth and uniform as that of the starting wafer. At temperatures above 80°C, etching nonuniformities become more prevalent, and at concentrations below 20% small pyramidal defects called hillocks form, increasing surface roughness. KOH etches are quite vigorous, with hydrogen gas as an etch product. Advantages of KOH over other anisotropic etchants are that it is relatively safe and simple. One disadvantage is the relatively high silicon dioxide etching rate (ranges from near 0 to over 150 Å/min, depending on etch conditions). This means thick silicon dioxide films are needed to act as a mask for deep etches. Another serious disadvantage is that KOH, due to the potassium, is not compatible with conventional IC processing unless the circuit area can be adequately protected. Table 19.2 lists important characteristics of KOH. Although KOH is the most common etchant formed from an alkali metal hydroxide, sodium, cesium, and rubidium hydroxides have also been used.

Each of the other common anisotropic etchants, EDP, N_2H_4/water, and TMAH, has advantages and disadvantages [7, 14]. The EDP etch does not contain sodium or potassium, and the SiO_2 etch rate is much lower than KOH. However, EDP is dangerous from a health standpoint, requiring considerable care in mixing and using. The N_2H_4/water mixture is potentially explosive (N_2H_4 has been used as a rocket fuel), and N_2H_4 is a suspected carcinogen. Hydrazine does have some advantages in that SiO_2 etches very slowly, and most metals are not etched. TMAH is a more recently developed etchant that has some nice advantages over

Example 19.4

Find the size of the mask opening that after anisotropic etching will yield a flat rectangular area of size 100 μm by 200 μm, 80 μm below the silicon (100) surface.

Solution

From the side view we find the length X to be

$$X = 100 \ \mu m + 2Z$$

where Z is defined by the relation

$$\tan \phi = \tan 54.74° = \frac{80 \ \mu m}{Z} = 1.41$$

Solving for X gives

$$X = 100 \ \mu m + 2 \frac{80 \ \mu m}{\tan 54.74°} = 213.2 \ \mu m$$

Similarly, solving for Y yields

$$Y = 200 \ \mu m + 2 \frac{80 \ \mu m}{\tan 54.74°} = 313.2 \ \mu m$$

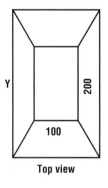

Top view

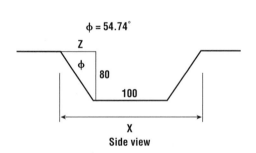

Side view

KOH. TMAH is relatively safe (nontoxic and nonexplosive) and easy to handle, with excellent selectivity to silicon dioxide and silicon nitride. It does not etch Al, a big advantage in terms of compatibility with IC processing. The major disadvantage is that the etch ratio between the [100] and [111] planes is not as high as for the other anisotropic etchants. Despite this, TMAH is growing in popularity and usage.

An important aspect of MEMS bulk micromachining is the capability of forming structures with reproducible mechanical properties such as resonant frequency and deflection. From a processing standpoint, this requires tight control on the dimensions of MEMS elements. Since the lateral dimensions are generally determined by photolithographic tolerances, which are quite accurate relative

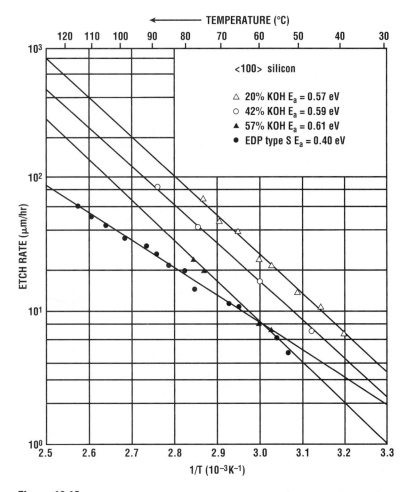

Figure 19.13   Arrhenius plot of the vertical etching rate for (100) silicon wafers for EDP and KOH solutions *(after Seidel [33], reproduced by permission of The Electrochemical Society, Inc.).*

to the size of MEMS devices, thickness is the dimension most difficult to control. Most bulk micromachined structures have their thickness determined by wet etching, so control of etch depth is the key to device yield and performance. Methods for controlling etch depth in bulk micromachining are called etch stop techniques. Four common techniques exist:

1. Timed etches
2. Anisotropic etching of v grooves
3. P^{++} doping
4. Electrochemical etch stop

Timing the etch is the simplest, though least accurate technique. By knowing the etch rate, and the desired etch depth, the etching time can be calculated. One problem with timed etches is the loading effect, where the etch rate changes as silicon is etched, due to dilution of the etch species. This leads to slower etch rates as a function of time. Different feature sizes often etch at different rates due to diffusion effects. Also, variations in initial wafer thickness across the wafer are directly translated

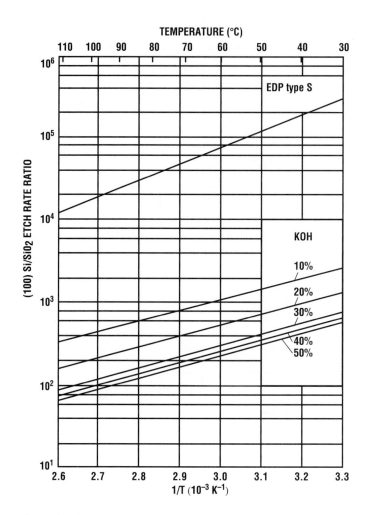

Figure 19.14   Etch rate ratio of (100) silicon to SiO₂ for EDP and KOH etch solutions *(after Seidel [33], reproduced by permission of The Electrochemical Society, Inc.).*

to thickness differences of the etched features. For example, a starting wafer with a total thickness variation (TTV) of ±2 µm, if used to form membranes 10 µm thick, will show membranes with ±2 µm thickness variation even if the etch is perfectly uniform, solely due to the initial TTV. Although the percentage variation of the full wafer thickness due to this thickness nonuniformity is small (less than 1% for a 500-µm-thick wafer), for a 10-µm membrane the variation is 20%. Wafers with low TTV are critical for timed etch processing, or yield will be low on wafers that contain membranes distributed across the wafer.

Anisotropic etching of v-grooves is another etch stop technique that is simple from a processing standpoint [36]. Using simple geometry, the depth of a rectangular pattern can be calculated for a KOH etch. The dimensions of the rectangle are chosen such that the flat bottom of the trench disappears at the desired etch depth, yielding a v-groove. While etching, close monitoring of the rectangular area indicates when the etch is complete. Disadvantages of this technique include the necessity of stopping the etch at the moment indicated by the rectangular v-groove, and the need for wafers of low TTV.

Example 19.5

A 30-μm-thick membrane is needed for a pressure sensor application. Calculate the size of the mask opening W needed for the v-groove if the full wafer thickness is 600 μm.

Solution

From geometry we find

$$\tan 54.74° = \frac{600 \ \mu m - 30 \ \mu m}{W/2}$$

Solving for W yields

$$W = \frac{570 \ \mu m}{(\tan 54.74°)/2} = 808 \ \mu m$$

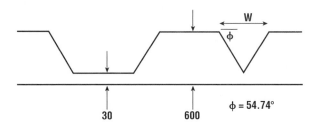

$\phi = 54.74°$

30   600

The P^{++} doping etch stop technique is based on the fact that all anisotropic etchants show a drastic decrease in etch rate when the silicon is heavily doped with boron [7, 14]. Etch rates start to drop at boron concentration in the range of 10^{19}/cm^3, with the etch rate dropping by a factor ranging from 20 to 50 depending on the etchant (see Table 19.2). One proposed mechanism for this effect is that the heavy boron doping reduces the thickness of the space charge region near the surface, and electrons can more easily tunnel through this region and recombine with holes within the silicon [33]. This limits the availability of the electrons necessary for the etching reaction, and slows the rate. Heavily doped boron regions with thicknesses up to 20 μm can be formed by diffusion and thinner regions with ion implantation. The high boron doping creates a tensile stress in the film resulting from the smaller size of the boron atom. The boron doping must be uniform in the layer, or stress gradients will develop, leading to curling of cantilever beam structures. Advantages of the P^{++} etch stop technique include lack of sensitivity to the wafer TTV and no need to remove the wafer at the moment the P^{++} layer is reached. Disadvantages are that (1) process steps are added compared to the previous two techniques, (2) the piezoresistance coefficients decrease significantly at these doping levels, and (3) the high boron concentration is not compatible with IC technologies, so integration with on-chip electronics is difficult.

Electrochemical etch stop techniques are based on the fact that electrically biasing n-type silicon positively with respect to the etchant solution prevents the n-type silicon from etching [37]. The electric field induced in the n region interferes with the chemical reaction involving the hydroxide ions that oxidize the silicon surface. A common usage of this effect is to create an n-type region on a p-type substrate using epitaxial growth or ion implantation. The thickness of the n region corresponds to the desired membrane or cantilever beam thickness. Using an electrochemical etching

Table 19.2 Principal characteristics of four different common anisotropic etchants[a]

Etchant/Diluent/ Additives/ Temperature	Etch Stop	Etch Rate (100) (mm/min)	Etch Rate Ratio (100)/(111)	Remarks	Mask (Etch Rate)
KOH/water, isopropyl alcohol additive, 85°C	Is $>10^{20}$ cm^{-3} reduces etch rate by 20	1.4	400 and 600 for (110)/ (111)	IC incompatible; avoid eye contact; etches oxide fast; lots of H$_2$ bubbles	Photoresist (shallow etch at room temperature); Si$_3$N$_4$ (not attacked); SiO$_2$ (28 Å/min)
Ethylenediamine pyrocatechol (water), pyrazine additive, 115°C	$\geq 5 \times 10^{10}$ cm^{-3} reduces the etch rate by 50	1.25	35	Toxic; ages fast; O$_2$ must be excluded; few H$_2$ bubbles; silicates may precipitate	SiO$_2$ (2–5 Å/min); Si$_3$N$_4$ (1 Å/min); Ta, Au, Cr, Ag, Cu
Tetramethylammonium (TMAH) (water), 90°C	$>4 \times 10^{20}$ cm^{-3} reduces etch rate by 40	1	From 12.5 to 50	IC compatible; easy to handle; smooth surface finish; few studies	SiO$_2$ etch rate is 4 orders of magnitude lower than (100) Si LPCVD Si$_3$N$_4$
N$_2$H$_4$/(water), isopropyl alcohol, 115°C	$>1.5 \times 10^{20}$ cm^{-3} practically stops the etch	3.0	10	Toxic and explosive; okay at 50% water	SiO$_2$ (<2 Å/min) and most metallic films; does not attack Al according to some authors

[a] Given the many possible variables, the data in the table are only typical examples.

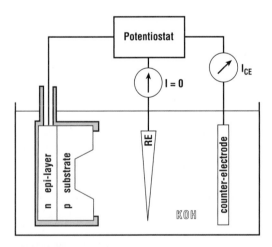

Figure 19.15 Generic electrochemical etch stop setup.

setup (Figure 19.15), proper biasing of the silicon will cause the anisotropic KOH etch to stop when the p-type silicon is gone from the etch hole. For a piezoresistive pressure sensor with low doped p-type piezoresistive elements on the surface of the n-type region, this technique is preferable to the P^{++} doping etch stop, since the piezoresistance coefficients for the low doped p region are higher than for a P^{++} region. Also, the silicon membrane quality is higher for the electro-chemically etched structure since there is no membrane strain induced by the high P^{++} doping.

In the late 1990s a high density plasma etching process was developed that allows etched structures to be formed that have vertical sidewalls, regardless of wafer orientation [38]. As important, these structures can be etched quite deep, even completely through a wafer. Aspect ratios for holes as high as 30:1 have been demonstrated, with etch rates over 5 μm/min. Masking materials include photoresist and silicon dioxide, with etch ratios compared to silicon as high as 1:100 and 1:200, respectively. This technique is known by several names, including deep trench etching, deep silicon reactive ion etching, advanced silicon etching, and Bosch deep silicon etch. The etching process works by alternating a sidewall passivation step and an etching step, through many rapid cycles [39]. The etch step uses

Figure 19.16   Example of structures formed using deep silicon etching with a high density plasma source *(after Picraux and McWharter [40], © 1998 IEEE).*

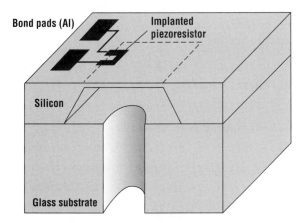

Figure 19.17   Simplified drawing of a typical piezoresistive bulk micromachined silicon pressure sensor *(after Kovacs [14], reprinted with permission, The McGraw-Hill Companies).*

standard fluorine etch gas chemistries. Figure 19.16 shows an array of posts formed using this etching technique, each 3 μm in diameter and approximately 30 μm tall [40]. The advantages of this etching technique over conventional wet etching are (1) lack of crystal orientation effects, (2) vertical features leading to closer spacing of structures, and, most importantly, (3) high aspect ratios, allowing the formation of structures impossible using wet etching. The major disadvantage is the capital cost of the equipment, which is over $500,000 for a research level system [41]. Production systems are considerably more expensive.

19.6 Bulk Micromachining Process Flow

As an example of a bulk micromachined process, consider a piezoresistive pressure sensor, such as may be used to measure air pressure in an automobile air intake system. Figure 19.17 shows a schematic drawing of such a sensor. A membrane is formed using anisotropic wet etching, with piezoresistive elements (only one is shown in Figure 19.17) to detect the membrane deflection. To maximize their resistance change, the piezoresistive elements, or piezoresistors, are positioned at the locations of highest stress. In Section 19.4 the locations of highest stress were shown to be along the edges of the membrane at the center of each side. By locating the piezoresistors at these points, the resistance change is maximized, leading to optimal sensitivity of the sensor.

The design of the piezoresistive membrane pressure sensor is influenced by the membrane material and the process conditions. For a process using (100) silicon wafers, with membranes defined using KOH etching, we know that the membrane sides will be along the <110> direction. As discussed in Section 19.3, the piezoresistance coefficients of p- and n-type silicon are dependent on orientation, and for the <110> direction, p-type piezoresistance coefficients are larger. Therefore our sensor design includes p-type piezoresistive elements located at the center of each of the membrane edges.

The sensor fabrication process will define boron piezoresistors implanted into an n-type silicon (100) substrate, and the membrane will be formed using a timed etch technique. The process involves five photolithography steps, plus dielectric etching and metallization steps. Figure 19.18 outlines the fabrication process. Step 2, the first lithography step, defines the regions where the boron implantation in Step 3 will done. The boron-doped piezoresistive elements form the sensing elements for the pressure sensor. The lithography alignment must be done precisely relative to the crystal orientation of the wafer such that the piezoresistors are along the <110> directions. After the ion implantation and resist removal in Step 3, the implant is activated and 5000 Å of silicon nitride is deposited in Step 4. Silicon nitride is used instead of silicon dioxide because the etch rate of nitride in KOH at 60°C is nearly zero compared with the SiO_2 rate. If SiO_2 is used, a film of thickness of more than 2 μm would be needed to withstand etching of 400 μm of silicon. In preparation for making the metal contacts to

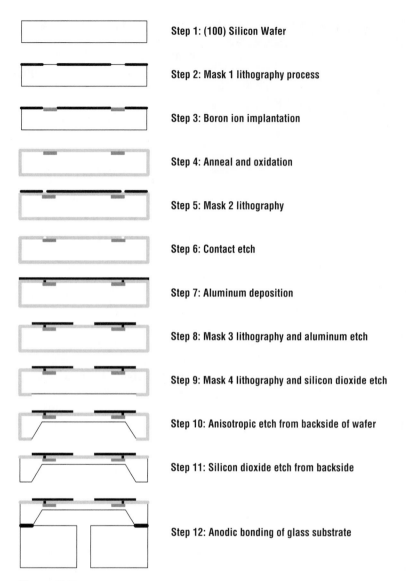

Figure 19.18   Bulk micromachining process for a piezoresistive pressure sensor.

the piezoresistors, resist openings are created in the Step 5 lithography process. Step 6 involves etching the exposed silicon nitride, followed by resist removal. Step 7 is the sputter deposition of aluminum over the entire wafer surface, contacting the silicon piezoresistors only in the contact areas created in Step 6. The third lithography process is performed in Step 8, where the aluminum is patterned and removed in the unwanted areas. Processing now moves to the backside of the wafer for membrane formation. Step 9 is the fourth lithography step, where the silicon nitride on the wafer backside is removed in the appropriate area. The mask alignment for this process is somewhat different from standard lithographic processing. The etch for membrane formation starts from the backside of the wafer, so the opening in the silicon nitride must be oriented properly relative to the piezoresistors already defined on the front side of the wafer. If not, the piezoresistors will not be positioned at the locations of maximum stress and may not even be on the membrane. This alignment is called a backside alignment step, since the pattern on the backside of the wafer must be aligned with the pattern on the front side (top of the wafer). At Step 10 the membrane is formed using a KOH anisotropic etch. Since the etching process will be stopped at a time determined by the etch rate, the wafer thickness, the desired membrane thickness, the etch rate, and the wafer thickness must be known precisely. The final step is removal of the silicon nitride from the backside of the wafer.

At this point the silicon processing is finished, but the wafer, containing many sensor elements, is quite fragile due to the thin membranes. A common packaging method for pressure sensor devices is to bond the finished silicon wafer to a glass substrate that contains an array of holes aligned with the membrane. The holes allow the air or other media to reach the membrane. After bonding, a standard dicing saw process can be used to separate the sensor elements for mounting in a commercial package. The pressure input from the package is attached to the hole in glass. Since the pressure medium is conducted to the silicon membrane through the hole in the glass, the bond between the silicon and the glass must be leakproof (hermetic). This requirement is quite restrictive, considering that the lifetime of many pressure sensor applications is measured in years. One particular technique is used for this type of bond, almost to exclusion of all others. Anodic, or electrostatic bonding makes use of the temperature-dependent conductivity of certain glasses [42]. The most commonly used glass, Corning 7740, contains sodium ions that become mobile at elevated temperatures of 400°C. Placing a wafer of 7740 glass in contact with a silicon wafer, heating to 400°C, then applying a high dc voltage anywhere from hundreds to a thousand volts as shown in Figure 19.19, will form a strong bond between the silicon and glass wafer. The most common explanation of the mechanism of the bonding process centers on the sodium ions. When the voltage is applied at 400°C, the mobile sodium ions are attracted to the cathode contact at the glass surface, leading to the creation of a space charge region at the silicon/glass interface. The resistance of this space charge region is high, so most of the voltage is dropped across this region, leading to a strong electric field. This field generates a large force, drawing the glass and silicon into close contact. This close contact, combined with the elevated temperature, leads to the formation of covalent bonds between the silicon and glass. The resulting anodic bond has high strength and is hermetic. Progress of the bonding process can be monitored during bond formation by measuring the current flowing in the circuit. A successful bond requires the silicon and glass to be within approximately 1 μm of each other, meaning the surfaces must be clean and flat. Any warpage or particles will inhibit a good bond. Anodic bonding can be done with a simple power supply setup and a hot plate. Bond progress can be determined by looking through the glass substrate at the glass/silicon interface, since there is a noticeable darkening of the interface region where the bond has occurred. Once the bond is completed, the electric field can be removed, and the bonded wafers allowed to cool. If the temperature coefficients of expansion (TCE) of the silicon and the glass are significantly different, then a large extrinsic stress will develop, bending and perhaps breaking the silicon. Glasses such as Corning 7740 have been developed specifically to match the TCE of silicon, thus avoiding this problem.

Silicon fusion bonding is another bonding technique often used in MEMS processing [43, 44]. There is no bonding or glue layer used in silicon fusion bonding. As shown earlier in Figure 15.16, two silicon surfaces, each smooth and flat, are modified to be hydrophilic using wet chemistry, such as HF etching or boiling in nitric acid, then are brought into contact. The surfaces of the wafers bond together via hydrogen bonds, and then a high temperature anneal is performed to create a permanent bond that has essentially the same strength as the silicon bulk material [45]. An example of a

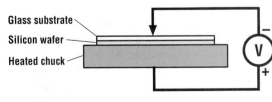

Figure 19.19　Typical electrostatic (anodic) bonding setup.

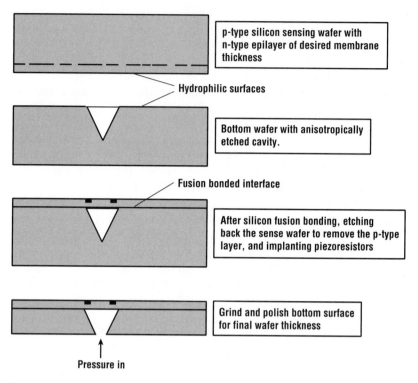

Figure 19.20   Fabrication process for silicon fusion bonded, low pressure sensor *(after Petersen [3], © 1982 IEEE).*

bulk micromachined pressure sensor fabrication process using silicon fusion bonding is shown in Figure 19.20 [46]. The pressure sensors fabricated by this process had square diaphragm areas of 650 μm on a side, with membrane thicknesses of 8 μm. The membrane thickness was determined by the thickness of an n-type epi layer deposited on the p-type sensing wafer. The bottom wafer was anisotropically etched to form a square, pyramidal etch hole 650 μm on a side. Next, both surfaces to be bonded were made hydrophilic, brought into contact to form the initial bond, then annealed at high temperature to complete the fusion bond. The p-type sense wafer is then removed in a selective etching process, leaving the 8-μm n-type epi layer as the membrane. P-type piezoresistors were then formed on the membrane at the high stress locations, and finally the bottom wafer was ground and polished to the final thickness, which allowed the membrane to deflect due to pressure from the bottom side. The advantage to the process is that the area required for the sensor is considerably less than a bulk micromachined pressure sensor of comparable size formed using anisotropic etching only. This can result in an increase in the number of sensors per wafer of 50% or more, which for a manufacturer is significant.

Cantilever beams can also be formed using bulk micromachining. Figure 19.21 is a scanning electron microscopy photograph of two cantilever beams that are to be used as sensors to detect airborne chemicals [47]. Along the length of each beam, resistors have been formed using p^+ diffusion. Periodically heating the resistors used as heaters causes the beams to vibrate due to the different thermal coefficients of expansion of the materials that make up the beam. Piezoresistors are then used to monitor the vibration frequency. The typical resonant frequency for the beams in Figure 19.18 is 140 kHz. To sense airborne chemicals such as volatile organic compounds (VOCs), the beams are coated with thin films of polymers that have been specifically designed to react with VOCs. These

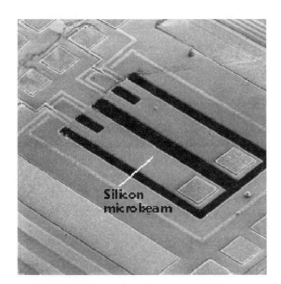

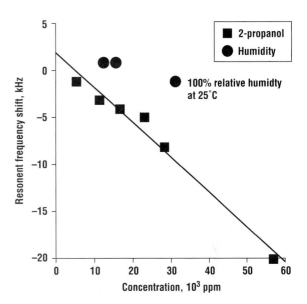

Figure 19.21 A pair of silicon cantilever beams used to detect airborne contaminants prior to deposition of the polymer layer *(after Baltes et al. [47], © 1998 IEEE).*

Figure 19.22 Change in cantilever beam resonant frequency as a function of concentration for isopropanol and water *(after Baltes et al. [47], © 1998 IEEE).*

polymers absorb the VOC, slightly increasing the polymer mass. This mass increase can then be detected as a decrease in the resonant frequency of the beam. Importantly, the polymers can be heated to drive off the VOCs, allowing the polymer film to be used again. The polymer also experiences a change in dielectric constant, and this can be measured with another sensing structure such as an interdigitated capacitor using the polymer as capacitor dielectric. These two pieces of information, the mass as detected by the resonant frequency shift of the cantilever and the polymer capacitance change, determine the airborne chemical identity and concentration. Figure 19.22 shows the change in resonant frequency of the cantilever beam as a function of the concentration of isopropanol. The polymer used for this experiment was specifically designed to absorb isopropanol. Since the number of possible applications for detection of airborne contaminants is large, the sensor technologies being developed for this purpose is extensive. One interesting application is the development of an electronic nose that rivals the sensitivity of biological noses. Considerable work remains to reach that goal [48].

19.7 Surface Micromachining Basics

As illustrated in Section 19.6, bulk micromachining generally uses wet etchants to remove large amounts of silicon in order to fabricate MEMS devices. Surface micromachining is so named because the process takes place on the surface of the wafer, where films used for structural elements are deposited using techniques such as low pressure chemical vapor deposition (LPCVD). Silicon surface micromachining was developed in the 1980s to address several shortcomings of bulk micro-machining. First, the surface area required for bulk micromachined pressure sensors, due to the sloped sidewalls of recesses made using anisotropic etchants, was much larger than the actual membrane area. This means the yield per wafer was reduced over what could be obtained if membrane area were the limiting feature size. Second, with surface micromachining, structures can be made with several deposited layers, and parts can be "released" to allow them to move laterally as well as vertically. With this development, MEMS actuators (MEMS devices with moving parts) were possible.

Third, polysilicon, the most common surface micromachining structural material, was well character-ized from its use in IC processing, and through careful deposition processes, could be deposited with well-controlled, repeatable film stress levels. Polysilicon is also isotropic, which is an advantage for some structures. Lastly, bulk micromachining is not easily integrated with IC processing. Surface micromachining can more easily be integrated with CMOS processing, allowing signal processing circuitry and MEMS devices to exist on the same chip.

There are two key process steps in surface micromachining. The first is deposition of low stress thin films that can be used for structural elements. The second is the use of a sacrificial layer to allow the structural layer to be detached from the substrate, thus allowing motion of the structural layer. The deposition of controlled stress films is an essential process step for surface micromachining [7, 49]. Polysilicon deposited by LPCVD is by far the most common MEMS structural material. Generally the polysilicon is deposited using silane gas at temperatures of 580 to 620°C. Deposition rates are around 100 Å/min, leading to deposition times of 100 to 200 min for the 1- to 2-μm-thick films commonly used in MEMS devices. At temperatures below 600°C, the deposited silicon is gen-erally amorphous in nature. Above 600°C, the deposited polysilicon films have small grain sizes (hundreds of angstroms), a compressive stress, and regions of amorphous silicon as well. Postdeposition annealing at elevated temperatures leads to a slight contraction of the film as a result of the crystal-lization of the amorphous regions [13]. This contraction causes the stress to change from compressive to tensile. Figure 19.23 shows data on how the polysilicon strain changes as a function of time and

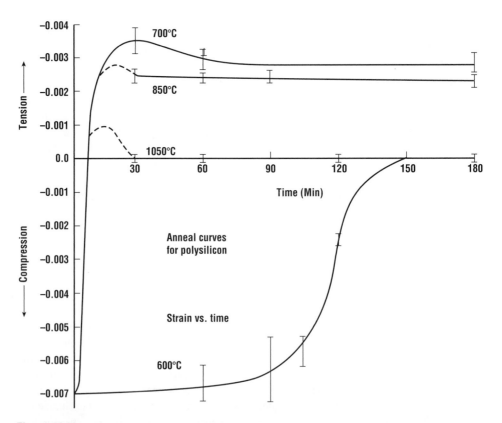

Figure 19.23   Change in strain level for low-stress polysilicon as a function of anneal temperature and time *(after Guckel et al. [13], © 1988 IEEE).*

temperature. Even annealing temperatures as low 600°C eventually lead to a stress-free film. Normally annealing cycles for polysilicon stress reduction are performed at over 1000°C. Polysilicon layers can also be deposited with dopants such as phosphorus and boron, using an *in situ* process (during the LPCVD deposition), or doped afterward using ion implantation or diffusion. The *in situ* doped polysilicon process can eliminate a subsequent doping step for those structures that require conducting polysilicon. The deposition rates for *in situ* doped polysilicon and undoped polysilicon are different, and are dependent on the dopant species (boron doping increases the rate, phosphorus decreases the rate). As with undoped polysilicon, the stress level can be controlled by postdeposition annealing [7].

Table 19.3 is a comparison of the material properties of single-crystal silicon and polysilicon [50–52]. The more disordered nature of the polysilicon leads to a reduction in thermal conductivity, fracture strength, and Young's modulus. The piezoresistance coefficients for polysilicon, although less than for single-crystal silicon, are still large enough to make useful sensors.

In addition to polysilicon, silicon nitride deposited by LPCVD can also be used as a low stress material [53]. Standard LPCVD silicon nitride is deposited at temperatures over 800°C using dichlorosilane (DCS) and ammonia in a flow ratio of approximately 1:5 ($DCS:NH_4$). This process yields a stoichiometric Si_3N_4 film with high tensile stress. By reversing the flow ratio to 5:1, a silicon-rich nitride film is deposited and the tensile stress level is considerably lower, less than 50 MPa for the proper deposition conditions. By proper adjustment of the deposition parameters of temperature, pressure, and flow ratio, compressive films are also possible. Postdeposition annealing of low stress nitride films has little effect on the stress. Silicon nitride has some advantages over polysilicon for MEMS devices, such as a higher hardness and higher Young's modulus (280 GPa for low stress silicon nitride). Increased hardness has advantages in rotating structures, where material wearing due to friction is a leading reliability issue. Higher Young's modulus means that for similar sized structures, those made of silicon nitride will be stiffer than those of polysilicon. Disadvantages

Table 19.3 Materials properties of single-crystal silicon and crystalline polysilicon

Material Property	Single-Crystal Si	Poly-Si
Thermal conductivity (W/cm K)	1.57	0.34
Thermal expansion (10^{-6}/K)	2.33	2–2.8
Specific heat (cal/g K)	0.169	0.169
Piezoresistive coefficients	n-Si ($\pi_{11} = -102.2$); p-Si ($\pi_{44} = +138.1$); e.g., gauge factor of 90	Gauge factor of 30 (>50 with laser recrystallization)
Density (cm^3)	2.32	2.32
Fracture strength (GPa)	6	0.8 to 2.84 (undoped poly-Si)
Dielectric constant	11.9	Sharp maxima of 4.2 and 3.4 eV at 295 and 365 nm, respectively
Residual stress	None	Varies
Temperature resistivity coefficient (TCR) (K^{-1})	0.0017 (p-type)	0.0012 nonlinear, + or − through selective doping, increases with decreasing doping level, can be made 0!
Poisson ratio	0.262 max for (111)	0.23
Young's modulus (10^{11} N/m²)	1.90 (111)	1.61
Resistivity at room temperature (Ω-cm)	Depends on doping	7.5×10^{-4} (always higher than for single-crystal silicon)

Based on Lin [50], Adams [51], and Huerberger [52].

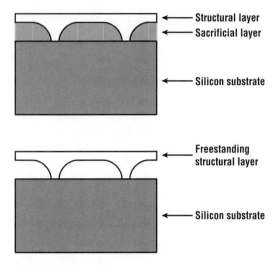

Structural layer
Sacrificial layer

Silicon substrate

Freestanding
structural layer

Silicon substrate

Figure 19.24 Basic surface micromachining sacrificial etch process *(after Howe [49], reprinted with permission, AIP)*.

for silicon nitride are primarily related to the increased processing difficulty, including the need to modify the LPCVD DCS:NH_4 flow ratio from the standard silicon nitride process, which is undesirable from an equipment standpoint if both processes need to be run in the same tube. Also, the low stress LPCVD deposition process generates a large amount of particulates during the deposition, which can cause premature mechanical pump failure.

The second key process step for surface micromachining is the use of sacrificial layers to release structures and allow motion [7, 49]. Figure 19.24 shows a silicon substrate with a patterned sacrificial layer separating the structural layer from the substrate. The removal of the sacrificial layer in the sacrificial etching process step yields a structure with freestanding features capable of moving. Many materials have been used as sacrificial layers, including photoresists and metals such as aluminum. The key requirement for a sacrificial layer is the existence of an etchant that will remove the sacrificial layer without etching the structural layer. When polysilicon is used as the structural layer, silicon dioxide is commonly used for the sacrificial layer. The SiO_2 layers are preferred because unlike photoresist and aluminum, SiO_2 can withstand the 600°C LPCVD deposition temperature of polysilicon. Also, SiO_2 can be etched in hydrofluoric acid (HF) solutions at a rate much higher than polysilicon. LPCVD is commonly used to deposit the SiO_2 with phosphorus doping to form phosphosilicate glass (PSG). PSG layers are desirable since PSG etches in HF solutions 8 to 10 times faster than undoped SiO_2 [54]. Deposition rate limitations as well as film stress issues generally limit the LPCVD PSG layer thickness to 2 μm or less. The relative thinness of this sacrificial layer with respect to the lateral dimensions of MEMS devices (hundreds to thousands of microns) means that the sacrificial etch time for some structures can be long (hours). Appropriate design of the structure to minimize the length of lateral etching is essential. Often holes are designed into the structure specifically to allow faster etching of the sacrificial layer. Common sacrificial etching times are less than 20 min.

In the performance of a sacrificial etch, surface tension of the liquid etchant can cause serious problems [55]. As the etchant removes the sacrificial layer and fills the space between the structural layer and the substrate, the liquid in this space has a high ratio of surface area to volume, and surface tension is the dominant force. The surface tension of most fluids on silicon and polysilicon surfaces causes hydrophilic wetting, meaning that the fluid readily "wets" the surface. During sacrificial etching, as the fluid volume decreases, the surface tension acts to pull the polysilicon and underlying silicon surfaces together. After complete removal of the sacrificial layer, the etchant can be difficult to remove completely due to this surface tension. Water rinses to dilute and remove the HF solution are followed by drying steps to eliminate all the liquid. A phenomena called stiction is often observed at this point, where the structure is not actually released from the substrate, but instead is stuck at one or more points to substrate in the region where the sacrificial layer was removed (Figure 19.25). As the liquid in the space between the structural layer and the substrate is removed by drying, the surface tension forces pull down the structural layer until, when the liquid layer is gone, the structural layer is in contact with the substrate. More often than not the structural layer remains in contact with the substrate and cannot be released, resulting in a nonworking device. Hydrogen bonding between the surfaces is a possible explanation. Stiction is usually the number-one yield-limiting step in surface

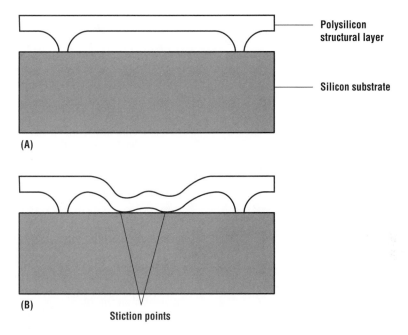

Figure 19.25   Side view of a released structure (A) without and (B) with stiction at two points.

micromachining. Many solutions to the stiction problem have been developed with varying degrees of success. Making changes to the design of the MEMS device can help prevent stiction. For example, by adding bumps to the bottom of a cantilever beam, the number of contact points to the silicon substrate is greatly reduced, and so is the stiction [56]. The disadvantage of this technique is that additional processing steps are required. Freeze-drying techniques, where the liquid is frozen and then removed in a vacuum-drying process, have also been successfully used [57].

19.8 Surface Micromachining Process Flow

A generic surface micromachining process for the fabrication of a cantilever beam with a piezoresistive element is shown in Figure 19.26. Beginning with a silicon wafer, a layer of phosphosilicate glass (PSG) is deposited to a thickness of 2 μm in Step 2. Lithography with mask 1 is performed in Step 3 to define the areas where the beam will be attached to the substrate. Step 4 is the etching of the PSG, either in a wet process using an HF solution or in a dry etch process. If dry etching is used, a high temperature anneal step may be needed to taper the sidewall of the hole in the PSG. This taper is significant because the resulting sharp corner in the deposited beam would have a stress concentration point there, leading to mechanical failure. The polysilicon thickness deposited in Step 5 is determined by the mechanical design of the structure, and is usually between 0.5 and 2 μm. In Step 6, the piezoresistive elements are formed by selective area ion implantation followed by the implant activation anneal. This anneal can also be used for the polysilicon stress relief anneal, which modifies the atomic structure of the deposited polysilicon and reduces the film stress to near zero. In Step 7 the polysilicon is patterned with mask 3 to define the shape of the cantilever beam structure, followed by etching using a dry etch process. Using sputtering, a blanket film of

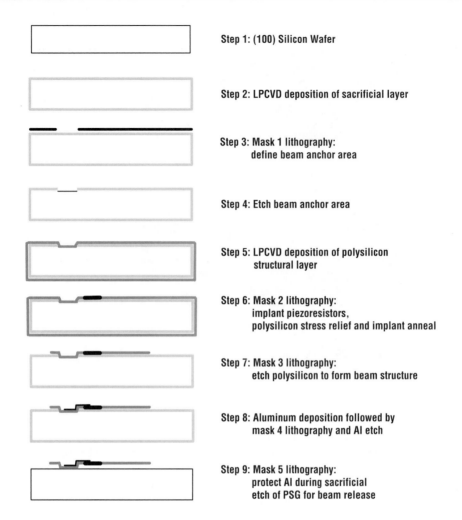

Step 1: (100) Silicon Wafer

Step 2: LPCVD deposition of sacrificial layer

Step 3: Mask 1 lithography:
        define beam anchor area

Step 4: Etch beam anchor area

Step 5: LPCVD deposition of polysilicon
        structural layer

Step 6: Mask 2 lithography:
        implant piezoresistors,
        polysilicon stress relief and implant anneal

Step 7: Mask 3 lithography:
        etch polysilicon to form beam structure

Step 8: Aluminum deposition followed by
        mask 4 lithography and Al etch

Step 9: Mask 5 lithography:
        protect Al during sacrificial
        etch of PSG for beam release

Top view of completed cantilever beam structure

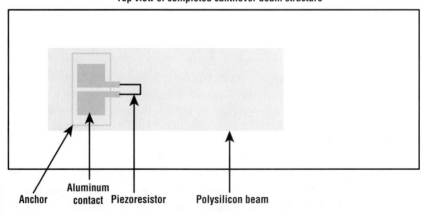

Anchor   Aluminum   Piezoresistor   Polysilicon beam
          contact

Figure 19.26   Surface micromachining process flow for a cantilever beam structure with a piezoresistor.

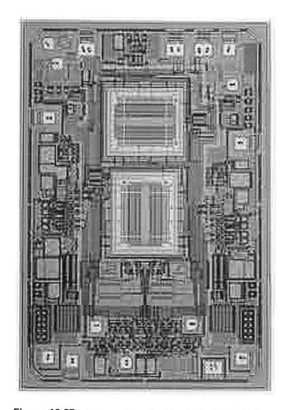

Figure 19.27 Photo of Analog Device's ADXL250 two-axis accelerometer chip. Acceleration measurement range of ± 50 g, with resolution of 10 milli-g. The two MEMS devices are the large structures oriented 90° to each other (*photo courtesy of Analog Devices, Inc.*).

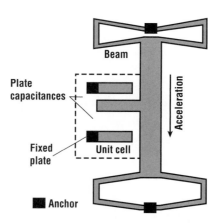

Figure 19.28 Simplified view of the ADXL250 accelerometer MEMS device operating principle (*courtesy of Analog Devices, Inc.*).

aluminum is deposited for electrical contact to the piezoresistive elements, followed by patterning of the aluminum and contact alloying (Step 8). Next the aluminum is protected with a patterned resist layer (mask 5) and the sacrificial PSG layer is removed using an HF etch to release the cantilever beam.

A classic example of a commercially available surface micromachined sensor is the accelerometer used for air bag deployment in automobiles (Figure 19.27). The accelerometer design principle is to use an interdigitated arrangement of fingers, one set of which can move in response to an acceleration (Figure 19.28). As the movable fingers shift position relative to the fixed fingers, the change in spacing between the fingers can be measured as a capacitance change, and correlated with the acceleration. The signal processing to detect this capacitance change is also fabricated on the same chip as the accelerometer, indicating that the surface micromachining process has been integrated into the IC fabrication process. The chip also has self-test capability: by electrically biasing the sets of fingers, motion simulating that of acceleration can be created and detected, thus ensuring the sensor is operating properly [58].

Another example of surface micromachining is in the area of optics. Miniaturized optical systems are important for controlling and modifying light beams from fiber optic cables used in communications systems. An optical system with all the necessary components on the surface of a chip (an optical bench on a chip), fabricated using MEMS techniques, would allow significant improvements over existing systems in terms of performance, size, and cost. To construct MEMS-based optical components suitable for manipulating light, a MEMS fabrication process must be developed to construct mirrors and other optical components that allow the light to travel parallel to the surface of the chip. To accomplish this, MEMS structures using polysilicon hinges [59] have been developed to allow released flat surfaces to be rotated out of the plane of the substrate. Figure 19.29 schematically shows how a plate and hinge structure allows the plate to be moved out of the plane. This structure

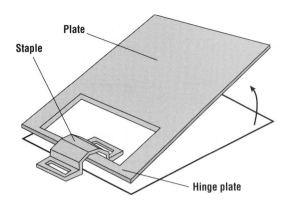

Figure 19.29   Basic hinge structure allowing out-of-plane motion of the plate *(after Pister [59])*.

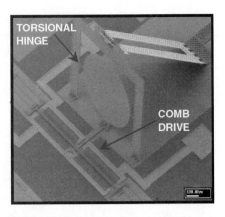

Figure 19.30   SEM photograph of a fast mirror used in a raster scanner. The mirror scans the beam in a direction parallel to the substrate *(after Hagelin [60], reprinted with permission, SPIE).*

can be formed using surface micromachining with two polysilicon structural layers. Figure 19.30 shows an oval-shaped mirror that has been rotated out of the surface plane [60]. The mirror is supported by torsion bars at the sides and is rotated by an electrostatic comb drive actuator, shown below and to the front of the mirror. These torsion bars allow the mirror to rotate in response to the motion of the comb drive actuator. Behind the mirror, attached to the top of the mirror support, the mechanism that provides the force to rotate the mirror out of the substrate plane can be partially seen. Using these techniques, other optical MEMS components can be made using surface micromachining with multiple structural layers [61].

19.9 MEMS Actuators

In addition to sensing applications, many of the mechanical-to-electrical transduction techniques described in Section 19.3 can be used to create motion [14]. Devices capable of motion greatly broaden the potential applications for MEMS. Desirable characteristics of MEMS actuators include

1. Force generation in the millinewton range.
2. Displacements of 10 μm or more.
3. Linear response to input signals.
4. Fabrication compatible with standard surface micromachining.
5. Reliable, with long lifetime.

Actuation methods can be broadly classified by the physical stimulus that underlies the actuation. The most common physical stimuli are electric fields, magnetic fields, and thermal effects. Actuation methods induced by electric fields include electrostatic and piezoelectric. The common magnetic field induced actuation methods are magnetostatic and magnetostrictive. For thermally driven actuation, methods include differences in thermal coefficients of expansion between two materials, shape memory materials, and liquid-to-vapor phase change. Table 19.4 summarizes the common actuation methods and their important properties.

Electrostatic actuation derives from the creation of an electric field by a voltage applied to adjacent conductors. A simple example is a cantilever beam used as an optical switch (Figure 19.31). The conducting polysilicon cantilever beam has an aluminum mirror on the tip that is used to deflect a laser beam. By applying a voltage between the beam and the conducting pad on the substrate under-neath the beam, the beam is deflected downward, shifting the mirror and rotating the reflected beam. The forces generated by electrostatic actuation are generally small, in the nano- to micronewton range. This leads to small displacements as well.

Example 19.6

A polysilicon cantilever fabricated using surface micromachining is 500 μm long, 75 μm wide, and 3 μm thick. The sacrificial layer thickness for the process was 2.0 μm. The electrically grounded beam is electrostatically actuated using a positive voltage V applied to a conducting bottom electrode (length 20 μm, width 75 μm) under the end of the beam. Neglecting fringing effects, estimate the voltage V required to deflect the beam by 0.2 μm.

Solution

First we need to find the amount of force Q needed to deflect the cantilever beam by 0.2 μm, then use electrostatic theory to calculate the voltage V required to generate that force for

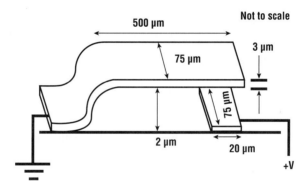

a parallel-plate capacitor arrangement. First, if we approximate the electrostatic force on the cantilever beam as a point force acting on the beam tip, we can use Equation 19.21 to calculate the force Q required to deflect the tip by 0.2 μm:

$$W(Q, x) = \frac{Qx^2}{6EI}(3L-x)$$

$$W(Q, L) = \frac{QL^3}{3EI} = 0.2 \ \mu m$$

Solving for Q gives

$$Q = 0.2 \ \mu m \ \frac{3EI}{L^3}$$

For silicon, E = 190 GPa = 190 × 10⁹ N/m², and the moment of inertia $I = at^3/12$

$$I = \frac{(75 \times 10^{-6} \ m)(3 \times 10^{-6} \ m)^3}{12} = 1.69 \times 10^{-22} \ m^4$$

Solving for Q,

$$Q = 2 \times 10^{-7} \text{ m} \frac{3(190 \times 10^9 \text{ N/m}^2)(1.69 \times 10^{-22} \text{ m}^4)}{(500 \times 10^{-6} \text{ m})^3}$$

$$Q = 154 \times 10^{-9} \text{ N} = 154 \text{ nN}$$

From electrostatic theory, the force Q between two parallel plates of area A, separation d, and applied voltage V is

$$Q = 0.5 \, \varepsilon_0 \varepsilon_r \frac{AV^2}{d^2}$$

where ε_0 is the permittivity of free space [8.85×10^{-12} C^2/(Nm^2)], ε_r is the relative dielectric constant for the capacitor dielectric, which for this case is air, so $\varepsilon_r = 1$. Note that fringing field effects are neglected. Solving for V yields

$$V = \frac{(2Qd^2)^{1/2}}{(\varepsilon_0 \varepsilon_r A)^{1/2}}$$

The bottom electrode area A is

$$A = (75 \times 10^{-6} \text{ m})(20 \times 10^{-6} \text{ m}) = 1.50 \times 10^{-9} \text{ m}^2$$

The separation $d = 2 \times 10^{-6}$ m. Solving for V,

$$V = \frac{[2(154 \times 10^{-9} \text{ N})(2 \times 10^{-6} \text{ m})^2]^{1/2}}{[(8.85 \times 10^{-12} \text{ C}^2/\text{Nm}^2)(1.50 \times 10^{-9} \text{ m}^2)]^{1/2}} = 9.6 \text{ V}$$

Approximately 10 V must be applied to the bottom electrode to deflect the end of cantilever beam 0.2 μm.

Electrostatic actuators generally can be easily integrated into MEMS fabrication processes. Another example of electrostatic actuation is the Digital Micromirror Device™ (DMD) [63] developed by Texas Instruments, shown in Figure 19.32. This device consists of an electrostatically operated mirror (16 μm square) that rotates ±10° from the home position. The spacing between the mirrors is 1 μm. The structural elements of this device are aluminum, and the sacrificial layers are made of photoresist. CMOS control circuitry, fabricated prior to the MEMS device, is located underneath the mirror assembly. These individual DMD elements have been fabricated in arrays of more than a million elements, each individually controlled, to create a video display system.

In piezoelectric actuation [21], a voltage applied to a piezoelectric element generates a strain approximately proportional to the electric field generated by the voltage. The force generated by these actuators is large, millinewtons and higher. Generally the maximum elastic strain of piezoelectric materials is 0.1%, which translates to small displacements. The linearity between the input and output signals is good. The major problem with piezoelectric materials is integration with standard MEMS fabrication processes. Issues include material contamination and complicated deposition techniques. Micromachined microfluid control valves [64] and scanning tunneling microscope tips [65] have been made using piezoelectric actuation.

Magnetostrictive materials respond to a magnetic field by changing shape. This effect is caused by the rearrangement of the magnetic domains in response to the magnetic field. Nickel is a common

Table 19.4 Summary of MEMS actuation methods

Mechanism	Work Density (J/cm³)	Force/ Area (N/cm²)	Displacement	Efficiency	Response Time	Input/ Output	MEMS Fabrication Compatibility
Electrostatic	0.1	100	Microns	High	Tens of micro-seconds	Nonlinear	High
Magnetostatic	1	100	Tens of microns	Moderate	Hundreds of micro-seconds	Nonlinear	Moderate
Thermal	0.1	10^4	Microns	Low	Tens of milli-seconds	Nonlinear	High
Piezoelectric	0.1	10^4	Microns	High	Tens of micro-seconds	Linear	Moderate
Liquid-to-vapor	10	10	10–100 μm	Low	Tens of milli-seconds	Nonlinear	Moderate
Shape memory	10^3	10^6	100 μm	Low	Tens of milli-seconds	Nonlinear	High
Magneto-strictive	0.1	10^4	1–10 μm	Low	Milliseconds	Nonlinear	Low–moderate

From Robbins [62].

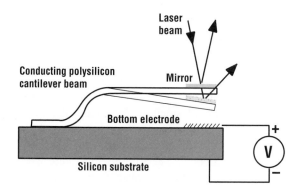

Figure 19.31 Illustration of electrostatic actuation of a cantilever beam micromirror. The insulating SiO_2 layer underneath the bottom electrode is not shown (*after Muller [61], © 1998 IEEE*).

magnetostrictive material, with a magnetic field induced strain as high as -30×10^{-6} (the minus sign indicates that a nickel rod would decrease in length in response to the magnetic field). Strain values in the range of 10^{-3} are possible with some rare earth metal alloy films [66]. Cantilever beams [67] and fluid pumps [68] using magnetostrictive actuation have been developed. The magnetic fields required for actuation can be low, in the range of 30 mT for the cantilever beam [69]. Magnetostrictive actuation can generate forces in the tens of micronewtons with displacements in the micron range. The response time is relatively slow, and compatibility with standard MEMS and IC fabrication processes can be an issue.

Magnetostatic actuation involves using an electric current to generate a magnetic field, which then induces motion in a magnetic material. Magnetic motors with dimensions as small as 2 mm² have been developed, with rotational speeds of 25,000 rpm [70]. Magnetic relays for current switching have also been developed. Magnetostatic actuation is capable of generating relatively large forces, with displacements in the tens of microns. Disadvantages include the inefficiencies associated with planar microfabricated magnetic coils, and difficulties in integrating the processing of magnetic materials into standard MEMS and IC fabrication processes.

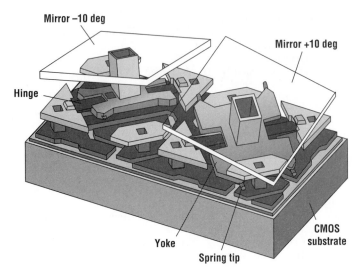

Figure 19.32   Two DMD pixels (mirrors shown as transparent) *(after Van Kessel [63], © 1998 IEEE).*

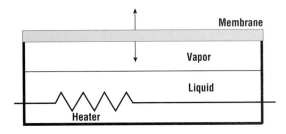

Figure 19.33   Liquid-to-vapor actuated membrane.

Another actuation method is based on the difference in thermal coefficient of expansion (TCE) of two materials. Consider a cantilever beam made of two different materials, with a resistive heater sandwiched between them [71]. This is called a thermal bimorph, and when current is applied to the heater, the temperature increase causes the two materials to expand at different rates and the cantilever moves (curls, in this case). Thermal expansion actuation can generate moderate forces in the tens of millinewtons, but the displacements are relatively small (a few microns). The TCE of most materials is nearly linear over the temperature range of use for most actuators, and this leads to a nearly linear response to the input signal. The major disadvantages of these actuators are the small displacements, the power required to create the heat, and the slow response time.

Shape memory alloy (SMA) materials are another class of thermally activated actuators. These materials possess the ability to repeatably return to a shape "learned" at a high temperature when deformed at a low temperature. The most common shape memory alloys are made from titanium and nickel. SMA actuators can generate forces of millinewtons and larger, and can have large displacements [72]. Integration with standard MEMS fabrication processes is not difficult. Major disadvantages include power required for heating and slow response time.

Liquid-to-vapor actuation can be used to move membranes as shown in Figure 19.33. By heating the liquid with the electric heater, the vapor pressure increases in the space between the top of the liquid and the membrane, forcing the membrane to move upward. This type of actuation is common in microfluidic applications [73, 74]. Microfluidic systems consist of fluid flow channels, valves, pumps, and chemical reaction chambers that are used for chemical and biochemical analyses. These systems are referred to as "micro" because of the small quantities (microliters to nanoliters) of fluids processed. MEMS techniques are often used to fabricate microfluidic systems. Liquid-to-vapor actuation of membranes has been used for sealing valves and for operating micropumps. The force generated with this actuation method can be high, and large displacements are possible. Disadvantages include slow response and fabrication complexities due to the need to make sealed chambers to contain the fluid/vapor.

19.10 High Aspect Ratio Microsystems Technology (HARMST)

MEMS devices fabricated by bulk and surface micromachining processes generally are restricted to thin film technology, which limits the thickness of the resulting structures. For actuators in particular, this reduces the amount of force that can be generated. High aspect ratio microsystems technology (HARMST) refers to methods that have been developed to produce structures with both small, tightly

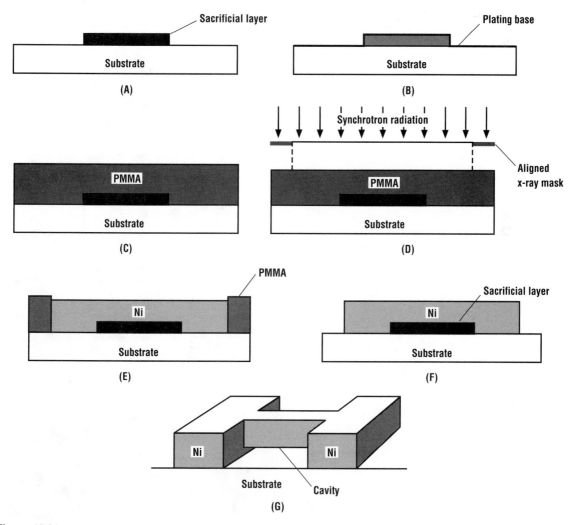

Figure 19.34   Cross-sectional view of LIGA with sacrificial layers. (A) Pattern sacrificial layer; (B) sputter plating base; (C) deposit PMMA; (D) align x-ray mask and expose PMMA; (E) develop PMMA and electroplate Ni; (F) remove PMMA and plating base to clear access to the sacrificial layer; (G) etch sacrificial layer, thereby undercutting and freeing the Ni structure *(from Guckel [75], © 1998 IEEE)*.

controlled lateral size, and heights of hundreds of microns and more [7]. The aspect ratio of these structures, defined as the ratio of the height to the lateral dimension, can be as high as 100:1. Unlike bulk and surface micromachining, HARMST does not have its roots in conventional silicon processing. Although some processing steps, most notably lithography, are shared by HARMST and conventional silicon processing, HARMST has evolved from technologies more common in the macroscopic world, such as metal plating and plastic injection molding.

The standard HARMST process was developed in Germany, where it is known as LIGA, which is the acronym for the German words associated with lithography and galvo, or electroplating. The key step in LIGA processing is the patterning or lithography process. As discussed in Chapter 9 on nonoptical lithographic techniques, x-ray lithography is capable of defining small features due to the short wavelength of the electrons. In addition, the high energy of the electrons translates to large penetration depths in low atomic mass materials such as polymers. PMMA has been found to be a very good resist for electron beam and x-ray lithographic processing. By applying thick layers of PMMA (hundreds of microns) and exposing with x-rays, high aspect ratio structures can be formed in PMMA. The softness of PMMA as a structural material precludes it from acting as a robust actuator. However, the PMMA can act as a mold to define more suitable materials, such as metals, that could be deposited by techniques such as electroplating. The most commonly plated metal in LIGA technology is nickel. After the nickel has been plated up in the PMMA mold, the PMMA is removed. This process can also be combined with sacrificial layer technology to form released structures. Figure 19.34 shows a LIGA process with a sacrificial layer release [75]. Figure 19.35 shows

(A)   (B)

Figure 19.35   (A) Assembled Ni structures for friction and magnetic properties testing. (B) Assembled shaft–gear combination made from Ni with a height of 150 μm. The bushing tolerance is 0.25 μm (*from Guckel [75], © 1998 IEEE*).

LIGA-fabricated nickel gears used for materials characterization. These gears have lateral features controlled to less than a micron tolerance, with heights over 100 μm. LIGA processing has been used to make a variety of devices, including turbines, valves, pumps, optical elements such as lens and prisms, electrostatic actuators, and electromagnetic micromotors.

LIGA processing has some obvious limitations, primarily related to the lithography. The standard x-ray exposure system uses x-rays from synchrotron sources. These sources are large and expensive, and only a few exist in the United States. Also, fabrication of structures that have varying cross-sectional shapes in the vertical direction is difficult.

Other HARMST techniques besides LIGA have been developed. These techniques involve the use of photosensitive polymer films that can be deposited to large thicknesses (several hundred microns) and then patterned using standard IC processing exposure tools [76]. Although not possessing the high precision and perfectly vertical features of LIGA, these materials are easier to use and for many applications are a cheaper and faster way to form high aspect ratio structures.

19.11 Summary

MEMS fabrication processes utilize many of the same process steps as conventional silicon IC processing. Differences arise as a result of the fundamentally different nature of MEMS sensors and actuators from electronic circuits. MEMS devices usually contain a structural component capable of motion, which implies some sort of freestanding or unsupported element. The mechanics of materials, including stress/strain relationships and the motions of membranes and cantilever beams, was presented. The two traditional methods for forming MEMS devices, bulk and surface micromachining, were discussed. Bulk micromachining is a relatively simple processing technology suitable for devices such as piezoresistive-based pressure sensors. Bulk micromachined devices cannot be easily integrated with standard IC processing for on-chip signal processing circuitry for the MEMS device. Surface micromachining is more suited to integration with on-chip electronics, but this integration requires considerable care. Surface micromachining using multiple layers of structural material and sacrificial material enables the fabrication of complex structures not attainable with bulk micromachining, such as small motors, resonators, and optical elements out of the plane of the surface of the wafer. MEMS actuators, which convert an electrical signal to motion using many different transduction methods, are finding application in many fields for small controlled motion. High aspect ratio microsystems (HARMST) processing combines x-ray lithography with electroplating to form structures that can be hundreds of microns high. These high aspect ratio plated structures can then be used as molds for precise formation of plastic parts using molding processes, as well as for mechanical structures such as gears and micromotors.

Problems

1. A wafer of diameter D has a deposited film with tensile stress. The center deflection is measured to be δ. Show that the wafer curvature due to the tensile stress film corresponds to a circle of radius $R \approx (D/2)^2/(2\delta)$. Assume $\delta \ll D/2$.

2. A silicon nitride cantilever beam is defined to be 1000 μm in length by the photomask. After the beam is released (separated from the substrate), the beam changes length due to the tensile stress. Calculate the length change if the tensile stress σ is -20 MPa. Assume $E = 280$ GPa for the silicon nitride beam.

3. Imagine a one-dimensional silicon of length L that has a layer of silicon nitride deposited onto it at 835°C. Calculate the extrinsic stress developed in the film when the bar is cooled to room

temperature. Assume the silicon nitride film thickness is much less than the silicon bar thickness. Use the TCE values for bulk silicon and silicon nitride. Qualitatively describe what happens to the extrinsic stress level in the film as the film thickness increases, approaching the silicon bar thickness.

4. A bar of square cross-sectional area A, length L, and resistivity ρ has an electrical resistance R defined by the relation $R = \rho L/A$. Starting with this relation, derive the relation relating strain to resistance change $dR/R \approx \varepsilon(1 + v) + d\rho/\rho$, where ε is the strain and v is Poisson's ratio. Describe the physical basis for each term.

5. Calculate $\Delta R/R = \pi_t \sigma_t + \pi_l \sigma_l$ for a p-type piezoresistor oriented along the $<110>$ direction for a transverse stress of 10 MPa and long stress of 50 MPa.

6. For the Wheatstone bridge configuration shown in Figure 19.7, assume all four piezoresistors have the same unstressed resistance. Derive the relation $\Delta R/R = V_m/V_b$ for the change in resistance due to a deflection of the membrane.

7. A square membrane of single-crystal silicon is used as a pressure sensor to detect pressure changes from 0 to 5000 Pa. Calculate the length L of the membrane side if the maximum deflection is 15% of the membrane thickness of 25 μm. What is the resonant frequency of this membrane?

8. For a circular membrane of radius a, thickness t, Young's modulus E, density ρ, and Poisson's ratio v, the symmetry allows simpler calculations, yielding the following general solution for the deflection W:

$$W(P, r) = P(a^2 - r^2)^2/(64D)$$

Find the maximum deflection for a single-crystal silicon membrane, 10 μm thick, radius of 1 mm, and applied pressure of 1000 Pa. How does this maximum deflection compare with that of a square membrane of similar thickness and side length 2 mm?

9. A single-crystal silicon cantilever beam of thickness 2 μm, length 750 μm, and width 100 μm has a p-type piezoresistor at the point of maximum stress on the beam. Calculate $\Delta R/R$ for the piezoresistor if there is a load of 10 μN distributed along the length of the beam.

10. For a single-crystal silicon cantilever beam of thickness 2 μm, length 750 μm, width 100 μm, and mass M, calculate the fundamental mode resonant frequency F. How much mass loading ΔM is needed to change F by 1%? What is $\Delta M/M$ for this frequency change?

11. For a (100) silicon wafer of thickness T, an array of square holes through the wafer is to be formed using KOH etching from the topside of the wafer. The holes are equally spaced from each other. Find the minimum separation possible between the holes on the bottom of the wafer as a function of T.

12. A process engineer is specifying a silicon dioxide thickness to use as an etch mask for a 42% KOH etch on (100) silicon wafers. The etching temperature is 80°C, and the desired etching depth is 400 μm. Calculate the required silicon dioxide thickness for these etch conditions. If the engineer wants to have 2000 Å of silicon dioxide remaining after the etch, what oxidation time should be used for a 1100°C wet oxidation? Assume a 25 Å native oxide thickness.

13. Develop a bulk micromachining fabrication process to make a cantilever beam structure using the P^{++} doping etch stop technique to define the cantilever beam.

14. A pressure sensor with a square membrane 500 μm on a side and 25 μm thick is to be fabricated on a 150-μm-thick (100) silicon wafer. If the design calls for a 75-μm border of unetched silicon around the etched hole for each die, calculate how large a die is required for membranes formed (a) using standard anisotropic etching using a timed etch stop process and

(b) using a silicon fusion bonding process. Use the ratio of the areas to make a statement on why the silicon fusion bonding technique is useful.

15. Explain why a freeze-drying technique for removing liquids following sacrificial etching would avoid stiction problems.

16. Develop a surface micromachining fabrication process for a cantilever beam resonator as shown in Example 19.6. Assume the bottom electrode material is doped polysilicon.

17. Develop a surface micromachining process to form a hinged structure as shown in Figure 19.29.

18. A square surface micromachined membrane of polysilicon ($E = 160$ GPa) is electrostatically deflected by a square electrode 1.5 μm below the membrane. The areas of the membrane and bottom electrode are 500 μm by 500 μm. Calculate the voltage required to deflect the center of the membrane downward by 0.3 μm, assuming the membrane thickness is 2 μm.

References

1. W. S. Trimmer, ed., *Micromechanics and MEMS Classic and Seminal Papers to 1990*, IEEE Press, New York, 1997.

2. J. Bryzek, K. Petersen, and W. McCulley, "Micromachines on the March," *IEEE Spectrum* **31**(5):20 (1994).

3. K. Petersen, "Silicon as a Mechanical Material," *Proc. IEEE* **70**:420 (1982).

4. S. Timoshenko and S. Woinowsky-Krieger, *Theory of Plates and Shells*, McGraw-Hill, New York, 1959.

5. J. Gere and S. Timoshenko, *Mechanics of Materials*, PWS Publishing, Boston, 1997.

6. E. P. Popov, *Introduction to the Mechanics of Solids*, Prentice-Hall, Englewood Cliffs, NJ, 1968.

7. M. Madou, *Fundamentals of Microfabrication*, CRC Press, Boca Raton, FL, 1997.

8. P. A. Flinn, "Principles and Applications of Wafer Curvature Techniques for Stress Measurements in Thin Films," *Proc. Materials Research Society Symp.: Thin Films: Stresses and Mechanical Properties*, Boston, November 28–30, 1988. *MRS* **130**:41 (1989).

9. L. Ristic, F. A. Shemansky, M. L. Kniffin, and H. Hughes, "Surface Micromachined Technology," in *Sensors and Actuators*, L. Ristic, ed., Artech House, Boston, 1994, pp. 95–155.

10. H. Guckel, T. Randazzo, and D. Burns, "A Simple Technique for the Determination of Mechanical Strain in Thin Films with Applications to Polysilicon," *J. Appl. Phys.* **57**:1671 (1985).

11. H. Guckel and D. Burns, "Polysilicon Thin Film Process," U.S. Patent 4,897,360, January 30, 1990.

12. L. S. Fan, R. S. Muller, W. Yun, R. T. Howe, and J. Huang, "Spiral Microstructures for the Measurement of Average Strain Gradients in Thin Films," *Proc. 1990 IEEE Conf. Micro-electromechanical Systems,* Napa Valley, CA, 1990, pp. 177–181.

13. H. Guckel, D. W. Burns, C. C. G. Visser, H. A. C. Tilmans, and D. DeRoo, "Fine Grained Polysilicon Films with Built-in Tensile Strain," *IEEE Trans. Electron Dev.* **ED-35**:800 (1988).

14. G. Kovacs, Micromachined Transducers Sourcebook, WCB McGraw-Hill, Boston, 1998.

15. S. Sze, ed., *Semiconductor Sensors*, Wiley-Interscience, New York, 1994.

16. J. W. Gardner, *Microsensors—Principles and Applications*, Wiley, Chichester, England, 1994.

17. L. Ristic, ed., *Sensor Technology and Devices*, Artech House, London, 1994.

18. S. Middlehoek and S. A. Audet, *Silicon Sensors*, Academic Press, Boston, 1989.

19. B. A. Auld, *Acoustic Fields and Waves in Solids*, Wiley-Interscience, New York, 1973.

20. W. G. Cady, *Piezoelectricity*, McGraw-Hill, New York, 1964.

21. D. L. Polla and L. F. Francis, "Ferroelectric Thin Films in Micro-electromechanical Systems Applications," *MRS Bull.*, **7/96**:59–65 (1996).

22. Y. S. Lee and K. D. Wise, "A Batch-fabricated Silicon Capacitive Pressure Transducer with Low Temperature Sensitivity," *IEEE Trans. Electron Dev.* **ED-29**:42 (1982).

23. K. E. Petersen, A. Shartel, and N. F. Raley, "Micromechanical Accelerometer Integrated with MOS Detection Circuitry," *IEEE Trans. Electron Dev.* **ED-29**:23 (1982).

24. S. Kordic, "Integrated Silicon Magnetic-Field Sensors," *Sensors Actuators* **10**:347 (1986).

25. R. S. Popovic, "Hall Effect Devices," *Sensors Actuators* **17**:39 (1989).

26. H. Baltes and R. S. Popovic, "Integrated Semiconductor Magnetic Field Sensors," *Proc. IEEE* **74**:1107 (1986).

27. C. S. Smith, "Piezoresistance Effect in Germanium and Silicon," *Phys. Rev.* **94**:42 (1954).

28. W. G. Pfann and R. N. Thurston, "Semiconducting Stress Transducers Utilizing the Transverse and Shear Piezoresistance Effects," *J. Appl. Phys.* **32**:2008 (1961).

29. Y. Kanda, "A Graphical Representation of the Piezoresistance Coefficients in Silicon," *IEEE Trans. Electron Dev.* **ED-29**:64 (1982).

30. D. L. Kendall, C. B. Fleddermann and K. J. Malloy, "Critical Technologies for the Micromachining of Silicon," in *Semiconductors and Semimetals,* Vol. 17, Academic Press, New York, 1992.

31. H. F. Winters and J. W. Coburn, "The Etching of Silicon with XeF_2 Vapor," *Appl. Phys. Lett.* **34**:70 (1979).

32. E. Hoffman, B. Warneke, E. Kruglick, J. Weigold, and K. S. J. Pister, "3D Structures with Piezoresistive Sensors in Standard CMOS," *Proc. IEEE MEMS Conf.* Amsterdam, Netherlands, 1995, pp. 288–293.

33. H. Seidel, L. Csepregi, A. Huerberger, and H. Baungartel, "Anisotropic Etching of Crystalline Silicon in Alkaline Solutions. I: Orientation Dependence and Behavior of Passivation Layers," *J. Electrochem. Soc.* **137**:3612 (1990).

34. M. Elwenspoek, "On the Mechanism of Anisotropic Etching of Silicon," *J. Electrochem. Soc.* **140**:2075, (1993).

35. B. Puers and W. Sansen, "Compensation Structures for Convex Corner Micromachining in Silicon," *Sensors Actuators* **A23**:1036 (1990).

36. S. Samaun, K. D. Wise, and J. B. Angell, "An IC Piezoresistive Pressure Sensor for Biomedical Instrumentation," *IEEE Trans. Biomed. Eng.* **20**:101 (1973).

37. E. D. Palik, J. W. Faust, H. F. Gray, and R. F. Green, "Study of the Etch-Stop Mechanism in Silicon," *J. Electrochem. Soc.* **129**:2051 (1982).

38. J. Bhardwaj, H Ashraf, and A McQuarrie, "Dry Silicon Etching for MEMS," *Proc. Electrochem. Soc.* **97**(5):118 (1997).

39. F. Larmer and P. Schilp, "Method of Anisotropically Etching Silicon," German Patent DE 4,241,045, issued 1994.

40. S. T. Picraux and P. J. McWhorter, "The Broad Sweep of Integrated Microsystems," *IEEE Spectrum* **35**(12):24 (1998).

41. For further information, contact Unaxis, Inc., St. Petersburg, FL, (www.unaxis.com), STS, Ltd. Gwent, U.K., (www.stsystems.com), or Alcatel Comptech, San Jose, CA.

42. W. H. Ko, J. T. Suminto, and G. J. Yeh, "Bonding Techniques for Microsensors," in *Micromachining and Micropackaging of Transducers*, C. D. Fung, P. W. Cheung, W. H. Ko, and D. G. Fleming, eds., Elsevier, Amsterdam, 1985.

43. L. Tenerz and B. Hok, "Silicon Microcavities Fabricated with a New Technique," *Electron Lett.* **22**:615 (1986).

44. M. Shimbo, K. Furakawa, K. Fukuda, and L. Tanzawa, "Silicon-to-Silicon Direct Bonding Method," *J. Appl. Phys.* **60**:2987 (1986).

45. P. Barth, "Silicon Fusion Bonding for Fabrication of Sensors, Actuators and Microstructures," *Proc. 5th Int. Conf. Solid State Sensors and Actuators*, 1990, pp. 919–926.

46. K. Petersen, P. Barth, J. Poydock, J. Brown, J. Mallon, Jr., and J. Bryzek, "Silicon Fusion Bonding for Pressure Sensors," *Tech. Dig., IEEE Solid-State Sensor and Actuator Workshop,* Hilton Head, SC, 1988, pp. 144–147.

47. H. Baltes, D. Lange, and A. Koll, "The Electronic Nose in Lilliput," *IEEE Spectrum* **35**(9):35 (1998).

48. H. T. Nagle, R. Gutierrez-Osuna, and S. Schiffman, "The How and Why of Electronic Noses," *IEEE Spectrum* **35**(9):22 (1998).

49. R. T. Howe, "Surface Micromachining for Microsensors and Microactuators," *J. Vacuum Sci. Technol. B* **6**:1809 (1988).

50. L. Lin, *Selective Encapsulation of MEMS: Micro Channels, Needles, Resonators and Electromechanical Filters*, Ph.D. Thesis, University of California, Berkeley, 1993.

51. A. C. Adams, "Dielectric and Polysilicon Film Deposition," in *VLSI Technology*, S. Sze, ed., McGraw-Hill, New York, 1988, pp. 233–271.

52. A. Huerberger, *Mikromechanik*, Springer-Verlag, Heidelberg, 1989.

53. M. Sakimoto, H. Yoshihara, and T. Ohkubo, "Silicon Nitride Single-layer X-ray Mask," *J. Vacuum Sci. Technol.* **21**:1017 (1982).

54. R. T. Howe, in *Micromachining and Micropackaging of Transducers*, C. D. Fung, P. W. Cheung, W. H. Ko, and D. G. Fleming, eds., Elsevier, Amsterdam, 1985, p. 169.

55. R. Legtenberg, J. Elders, and M. Elwenspoek, "Stiction of Surface Microstructures After Rinsing and Drying: Model and Investigation of Adhesion Mechanisms," *Proc. 7th Int. Conf. Solid State Sensors and Actuators*, 1993, pp. 198–201.

56. W. C.-K. Tang, *Electrostatic Comb Drive for Resonant Sensor and Actuator Applications*, Ph.D. Thesis, University of California, Berkeley, 1990.

57. G. T. Mulhern, D. S. Soane, and R. T. Howe, "Supercritical Carbon Dioxide Drying of Microstructures," *Proc. 7th Int. Conf. Solid State Sensors and Actuators,* 1993, pp. 296–299.

58. Analog devices specifications sheets for the ADXL150/ADXL250 accelerometers (1998).

59. K. S. J. Pister, M. W. Judy, S. R. Burgett, and R. S. Fearing, "Microfabricated Hinges," *Sensors Actuators* **A33**:249 (1992).

60. P. M. Hagelin, U. Krishnamoorthy, R. Conant, R. Muller, K. Lau, and O. Solgaard, "Integrated Micromachined Scanning Display Systems," *Proc. SPIE* **3749**:472 (1999).

61. R. S. Muller and K. Y. Lau, "Surface-Micromachined Microoptical Elements and Systems," *Proc. IEEE* **86**:1705 (1998).

62. W. P. Robbins, course notes for EE5690, Fundamentals of Microelectromechanical Systems, University of Minnesota, 1997.

63. P. F. Van Kessel, L. J. Hornbeck, R. E. Meier, and M. R. Douglass, "A MEMS-Based Projection Display," *Proc. IEEE* **86**:1687 (1998).

64. S. Shoji, B. van der Schoot, N. de Rooij, and M. Esashi, "Smallest Dead Volume Microvalves for Integrated Chemical Analyzing Systems," *1991 Int. Conf. Sensors and Actuators,* San Francisco, 1991, pp. 1052–1055.

65. S. Akimine, T. R. Albrecht, M. J. Zdeblick, and C. F. Quate, "A Planar Process for Microfabrication of a Scanning Tunneling Microscope," *Sensors Actuators* **A23**:964 (1990).

66. M. V. Ghandi and B. S. Thompson, *Smart Materials and Structures*, Chapman & Hall, London, 1992.

67. T. Honda, K. I. Arai, and M. Yamaguchi, "Fabrication of Actuators Using Magnetostrictive Thin Films," *Proc. 1994 IEEE Workshop on MEMS*, 1994, p. 51.

68. E. Quandt, A. Ludwig, and K. Seemann, "Giant Magnetostrictive Multilayers for Thin Film Actuators," *Proc. 9th Int. Conf. Solid State Sensors and Actuators*, 1997, p. 1089.

69. E. Quandt and K. Seemann, "Fabrication of Giant Magnetostrictive Thin Film Actuators," *Proc. IEEE MEMS-95*, 1995, pp. 273–277.

70. Y. Watanabe, M. Edo, H. Nakazawa, and E. Yonezawa, "A New Fabrication Process of a Planar Coil Using Photosensitive Polyimide and Electroplating," *Proc. 8th Int. Conf. Solid State Sensors and Actuators*, 1995, Vol. 2, pp. 268–271.

71. W. Benecke, "Silicon Microactuators: Activation Mechanisms and Scaling Problems," *6th Int. Conf. Solid State Sensors and Actuators*, 1991, p. 46.

72. W. I. Benard, H. Kahn, A. H. Heuer, and M. A. Huff, "A Titanium-Nickel Shape-Memory Alloy Actuated Micropump," *Proc. 9th Int. Conf. Solid State Sensors and Actuators*, 1997, p. 361.

73. W. K. Schromburg, R. Ahrens, W. Bacher, S. Engemann, P. Krehl, and J. Martin, "Long-Term Performance Analysis of Thermo-Pneumatic Micropump Actuators," *Proc. 9th Int. Conf. Solid State Sensors and Actuators*, 1997, p. 365.

74. A. K. Henning, J. Finch, D. Hopkins, L. Lilly, R. Feath, E. Falskin, and M. Zdeblick, "A Thermopneumatically Actuated Microvalve for Liquid Expansion and Proportional Control," *9th Int. Conf. Solid State Sensors and Actuators*, 1997, p. 825.

75. H. Guckel, "High-Aspect Ratio Micromachining Via Deep X-Ray Lithography," *Proc. IEEE* **86**:1586 (1998).

76. H. Lorentz, M. Despont, M. Fahrnl, H. Biebuyck, and N. Labianca, "SU-8: A Low Cost Negative Resist for MEMS," *J. Microelectron, Microeng.* **7**:121 (1997).

Chapter 20

Integrated Circuit Manufacturing

revious chapters have reviewed the processing steps that make up several popular technologies. The ultimate goal of these technologies is to be able to manufacture functional components at high volume and low cost. As discussed in Chapter 1, the amount of work required to complete a batch of wafers does not depend on the amount of information that the photomasks contain. Clearly, one way to improve the manufacture of ICs is to increase the number of functional die per wafer. To do this, there is a great deal of emphasis on the design of an IC to make it as small as possible without violating the design rules. IC manufacturing has also pursued increasing the number of die per wafer by increasing the wafer size. Thus, silicon wafer diameters have steadily increased from 100 mm, which was common in the mid-1980s, to 200- and 300-mm wafers commonly used today. Due to edge losses and scribe line effects, the usable die count increases slightly faster than the area of the wafer. For example, when a manufacturer increases the wafer diameter from 125 to 200 mm, the number of 7×7-mm die per wafer increases by a factor of 2.8 while the area increases by a factor of 2.6. Gallium arsenide wafers have also increased in diameter, although more slowly. Currently, the most common wafer diameter is 100 mm, although new factories are using 150 mm. The increase in wafer diameter sometimes reduces the number of wafers that a unit process can run simultaneously. Some trade-off must therefore be made to optimize the line throughput.

Perhaps the most fruitful method of increasing functional die count, however, is the focus of this chapter: increasing the die yield, the percentage of possible sites that are successfully fabricated, packaged, and tested. This is a very attractive avenue since, unlike increasing wafer size, considerable improvement can often be made without making large capital equipment investments. The die yield is the product of several factors, including the wafer yield (the fraction of wafers that completes processing), the process yield (the fraction of die on the wafers that completes processing and is considered to be functional), the assembly yield (the fraction of good die that is successfully packaged), and burn-in yield (the fraction of good packaged parts that is still functional after an initial stress test):

$$Y_d = Y_f \times Y_p \times Y_a \times Y_{bi} \tag{20.1}$$

This book does not consider packaging and other back-end processes. For most technologies, however, these yield terms are so close to unity that to a first approximation only the process yield, which will now be called simply the yield, must be considered.

The improvement of process yield has serious implications for both processing and design. The area of design for manufacturability is an active and fruitful one; however, this chapter will concentrate on processing for manufacturability. That is, what considerations must be taken into account in the application of the technologies that have been described so that they can be manufactured with a high probability of success? By now the reader should be familiar with the survey nature of this text. Nowhere is this more true than in this chapter. Each of the sections (yield modeling, particle control, statistical process control, and design of experiments) introduces materials that are the subjects of books themselves. The reader should be aware that only the briefest introductions are given to material that is far more rich than can be presented here. For additional information the reader is referred to de Gyvez and Pradhan [1] and Nishi and Doering [2].

20.1 Yield Prediction and Yield Tracking

A typical chronological yield plot, called a learning curve, is shown in Figure 20.1. When a new technology is first introduced into manufacturing, it is not uncommon for the IC yields to be as low as 20%. With 80% of the product being discarded, the ability to quickly understand and correct the origin of yield losses is absolutely critical to a plant's economic viability. Not only does this understanding allow the manufacturer to concentrate resources at the source of the problem, it also allows an accurate projection of the supply of good die, so that the proper number of wafers can be kept in the fabrication line. As a simple example of the economics of yield, consider a facility that produces 1000 wafers per week. Each wafer contains 100 die, which, when packaged, sell for $50 each. When the facility is yielding 30%, it is producing a gross income of $75,000,000 per year. If the yield can be increased to 50%, the plant's income increases to $125,000,000 per year. This net gain of $50,000,000 is almost pure profit, since very little additional work is involved in packaging and selling the good die. (Remember, the same number of die is produced whether or not they are good.) Furthermore, since it is impossible to completely test every possible mode of operation of many complicated ICs, it is generally believed that the quality of the product improves with increasing yield.

Integrated circuit yield is normally limited by defects. Figure 20.2 shows the defect density trend with generation of the dynamic random access memory or DRAM [3]. A number of defect classification schemes exist. The next two sections will discuss the nature of two common types of defects, but this is not critical to the discussion of yield and yield models. Perhaps the most fundamental distinction is the one between benign and killing defects, although the

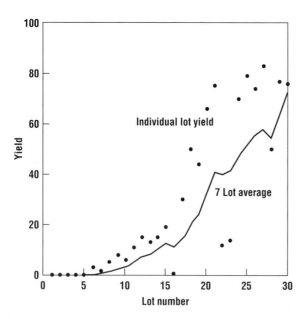

Figure 20.1 A plot of yield versus lot number for a particular technology. Also shown is the running average for the last seven lots.

Defect density trends

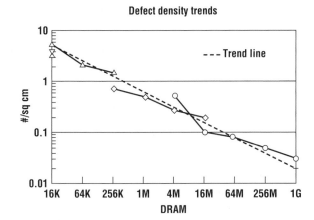

Figure 20.2   Defect–density trend with generation of DRAMs *(after Dance and Gildersleeve)*.

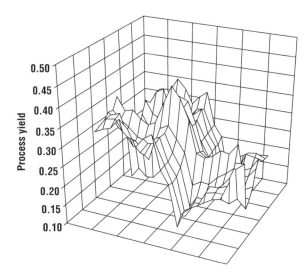

Figure 20.3   Yield plot as a function of die position on the wafer. Yield is often higher near the center of the wafer, but the right side of the wafer shows a strong yield falloff that bears investigation.

difference between the two is sometimes difficult to detect. A large-diameter spot of polysilicon left in the field region far from any other lines may be benign (assuming that the topology it causes for subsequent layers is not too severe). A much smaller defect of exactly the same type is killing if it is present in a region of minimum size lines and spaces. Since killing defects are by definition the only ones contributing to yield loss, the term *defect* will be used as shorthand for *killing defect*.

Defects can also be classified as large-area and point defects. Large-area defects such as scratches are often caused by poor wafer handling. Other large-area defects can be the result of nonuniform processes, such as an incomplete etch that leaves some undesired material after patterning. Nonuniform deposition rates in the preceding process can contribute to this type of large-area defect. Figure 20.3 shows a yield pattern that might be caused by nonuniformities in a single-wafer process. In some processes, all of the wafers are aligned to the flat. Nonuniformities in these processes will result in a distinct yield loss region. If the wafers are randomly oriented in the process in question, the yield loss averaged over a number of wafers would be more difficult to observe, perhaps showing up as an unusually large loss with radial distance. It is more obvious however, in the yield of individual wafers. To look for this effect, engineers use stackmaps. A stackmap is a drawing of a wafer showing all the die locations. Each die is colored according to its yield over some number of wafers run [4].

Point defects are often assumed to be randomly distributed with a well-defined size distribution. This is not necessarily true. Substrate defects like stacking faults, slip, and dislocations are often related to poorly designed thermal processes, incomplete annealing, or low quality epitaxial growth steps. They will lead to highly nonuniform defect distributions in the substrate. A good IC technology will be designed to avoid most of these problems. Point defects associated with deposited layers may be visible and often are major yield limitations. These defects can be categorized as voids (also called pinholes or intrusions), protrusions (also called bridges), and random spot defects. Their size distribution and density can be probed with test structures consisting of meander lines and interdigitated combs with varying sizes, spaces, and areas (Figure 20.4) [5]. These are collectively called process control modules [6].

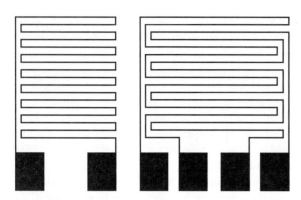

Figure 20.4   Meander lines and interdigitated comb structures used to determine a layer defect density.

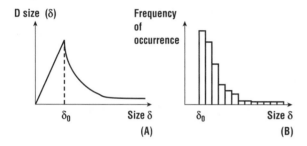

Figure 20.5   Theoretical and measured particle density as a function of particle diameter *(after deGyvez).*

Point defect occurrence is generally found to increase sharply with decreasing size as shown in Figure 20.5 [7]. Ferris-Prabhu [8] has proposed that the distribution of defect sizes can be modeled as

$$D_{size} = c\,\frac{\delta^q}{\delta_o^{q+1}} \quad 0 \le \delta \le \delta_o$$

$$D_{size} = c\,\frac{\delta_o^{p-1}}{\delta^p} \quad \delta_0 \le \delta \le \delta_{max} \qquad \textbf{(20.2)}$$

$$D_{size} = 0 \qquad \delta > \delta_{max}$$

where D_{size} is the defect density at a particular defect size (often quoted in defects/cm²), δ_{max} is the largest observed defect size, δ^o is the defect size with the maximum density, c is a normalization constant, and p and q are fitting parameters. In reality, the part of the curve describing very small defects is often very difficult or even impossible to measure.

The simplest type of yield model assumes Poisson statistics and independent, randomly distributed killer defects. This model uses two parameters to define the yield: the defect density (D) and gross fail area (G), where G is the fraction of the wafer in which all of the circuits fail. Then the yield is given by

$$Y = (1 - G)e^{-AD} \qquad \textbf{(20.3)}$$

where A is the area of the chip. This simple model is often used to project the yield of a new IC that will be fabricated in the same facility as a previous chip. The implicit assumption is that the defect density probably will not change significantly from one chip to the next as long as the facility and the technology are both fixed. Of course, this approximation is valid only if the chips are designed to be equally sensitive to both the types and sizes of defects found in that facility. For example, a low density IC with only a very small fraction of the features at the design rule limit will tend to yield better than the model predicts if the defect density is extracted from a chip that more aggressively pursues minimum feature size geometries.

Example 20.1   Simple Yield Prediction

A particular facility is producing several sizes of the same static random access memory (SRAM) chip. The chips are identical except for the size of the array and therefore the die size. Calculate the defect density and gross fail area.

Solution

Array Size	Die Size	Yield
256 K	0.4 cm × 0.4 cm	75%
1 M	1.4 cm × 0.4 cm	55%
4 M	1.4 cm × 1.4 cm	20%

According to Equation 20.3, plotting the natural log of the yield versus the area produces the defect density (from the slope) and the gross fail area (from the intercept). For this data $D = 0.75 \text{ cm}^{-2}$ and $G = 0.15$.

The predictive capability of this model is limited since it treats all defects as independent and equal. One can do a simple thought experiment to show the problem [9]. Let us assume the numbers quoted before: wafers hold 100 die, and 50% of the die are functional. For purposes of understanding yield, one could construct 50 "superdie" each consisting of two adjacent die. The effective area is double, and one can find the yield of this "chip." One can also divide up the wafer into 25 "colossal die," each one being made up of four adjacent regular die, and so forth [10]. One can then plot the yield as a function of the die size. The result rarely fits Equation 20.3 very well. Instead, one finds that the yield does not fall off as rapidly as predicted. This is because real defects are not random. Instead they tend to cluster. As a simple example, consider particulate defects on the edge of a wafer. This can arise due to improper handling. These defects will be heavily concentrated (or clustered) near the edge of the wafer. To model this behavior yield is more commonly modeled by

$$Y = \frac{1}{\left[1 + \dfrac{DA}{C}\right]^C} \tag{20.4}$$

where D is the defect density, A is the critical area of the chip, and C is the clustering factor [11, 12]. One can show that as C becomes large, the defects are once again independent, and the negative binomial model predictions approach those of the Poisson model.

Thus far, it has been assumed that all defects are equally likely to be killing defects. It is more reasonable to assume that a 1-µm boulder due to some errant CVD process is much more likely to result in a killing defect than is a 5-nm notch in a 90-nm metal line. It also must be pointed out that the 5-nm defect may be far more likely to occur. To take this difference into account, a quantity ϕ is defined to be the averaged probability that the mean size defect will be fatal. In this model

$$Y = (1 - G)e^{-D\phi A} \tag{20.5}$$

The averaged probability is given by

$$\phi = \int_0^\infty D_{\text{size}}(x)K(x)dx \tag{20.6}$$

where $D_{\text{size}}(x)$ is the defect size distribution and $K(x)$, the fault probability kernel, is the fraction of the total die area for which a particular size defect would be fatal. The general shape of a fault kernel curve for defects is shown in Figure 20.6. These curves have three regions of interest. For very large defects, the fault kernel approaches unity. For very small defects, the kernel may or may not go to zero depending on the exact nature of the defect. As an example, gate dielectric defects as small as half the gate oxide

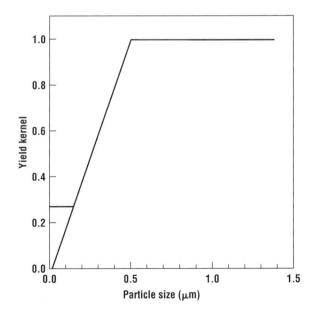

Figure 20.6   Two typical fault kernel curves.

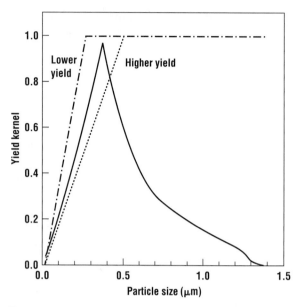

Figure 20.7   An overlay of the fault kernel (broken lines) and the defect distribution (solid lines). Small shifts in the fault kernel curve can lead to dramatic changes in the yield.

thickness may prevent complete oxidation and therefore lead to a short between the gate and the substrate. We refer to this zone as zero size or point defects. Between the two extremes, when the defect size is of order of the feature size, the kernel increases in some manner that is often approximated as linear.

Defect size distributions depend to a great extent on the process itself. Various distribution shapes have been theorized, but as already discussed, it is generally believed that the defect density has a maximum at some size and decreases for both larger and smaller defects. Figure 20.7 shows an overlay of $K(x)$ and a reasonable curve for $D_{size}(x)$. Clearly, the integral of the product depends sensitively on the position of the peak of $D_{size}(x)$ relative to the $K(x)$ curve. As the linewidth is decreased, the $K(x)$ curve shifts to the left, while the $D_{size}(x)$ curve remains relatively constant. Smaller linewidths generally, therefore, imply smaller yields, unless the defect size distribution is also shrunk correspondingly. Since many defects are particle related, a great deal of attention has been paid to reducing the particle counts in the facility and in the processing equipment. It is also of great interest to be able to reliably detect smaller and smaller particles as the linewidths shrink.

The yield equation is sometimes broken down to the product of the individual process level yields. Then, for random defects,

$$Y = \prod_{i=1}^{levels} Y_i = \prod (1 - G_i)e^{-D_i \phi_i A}$$

$$= \left[\prod (1 - G_i) \right] e^{-A \sum \phi_i D_i} \tag{20.7}$$

Some levels, such as implant layers, have very low values of ϕ, since they involve only gross features. It would be a good approximation, for example, in a simple laboratory single-level metal CMOS process, to assume that defects are present primarily in the polysilicon gate, metal, and contact levels, since these are done at or near the photolithographic resolution limit. As such they tend to dominate the yield equation. A model might also include the active area layer, since relatively small defects in the active mask could lead to killing defects. These four are then called *critical layers*. If only a small number of critical layers is present, a special yield test chip can be made up to vary the size of a critical layer slightly, thereby shifting its $K(x)$ curve, to determine the relative sensitivity of Equation 20.7 to the various critical levels. Once it is known, for example, that the IC yield has a strong correlation to a small shift in the metal linewidth, work can be done on reducing the defects at this level. This may involve improving the control of the metal deposition or etching processes, or changing the photolithography process at this level, improving the training of the technicians, or modifying a variety of other variables. The facility and the process tools generally determine the defect number and size distribution in that these components generate the particles that become defects. The process and the design determine the yield kernel, since they determine the sensitivity of the process to a defect. If a design uses a great deal of minimum feature size components, it will have a lower yield than one that is less aggressive. If the process has little latitude, smaller defects will prevent circuit operation. An extensive bibliography of yield models is given in Cheek and O'Donoghue [13].

20.2 Particle Control

Some defects are created by particles. The particle may remain on the wafer and be visible as a short circuit between lines or a boulder that the upper layers cannot cover, or it may be removed from the wafer by a clean after blocking an implant or locally disrupting pattern development during a photolithography step. One estimate suggests that particles are responsible for 75% of the yield loss [14] in volume manufactured VLSI ICs. Because of the importance of defects on IC yield, facilities pay strict attention to particle detection, control, and reduction. The facility itself must be designed and constantly monitored to maintain minimum particle levels in the air. Clean rooms have gone from class 100 to class 10 and are now class 1 in an effort to reduce contamination (Table 20.1). Maintenance of this level of particle control requires a strict discipline of the clean room occupants. Because of the success in controlling airborne particles, the emphasis has now largely shifted to reducing particle counts in the various processes and process equipment. Not only must particle generation be minimized at the time of purchasing new equipment or developing new processes and when process equipment is taken offline for maintenance, but the equipment must be particle qualified before it is returned to service. Periodic particle surveys are also commonly done to ensure that low defect levels are maintained.

There are two common types of particle detection: on the wafer and in the ambient. Both methods use optical scattering of a laser beam by a particle. On-wafer detection is done by sending a blank monitor (witness) wafer through the system to be checked. Blank wafers are usually used in current generation systems to avoid scattering of light by the pattern, which can obscure particle counts. The witness wafer is measured before and after processing, using a surface particle counter to determine the number of particles added by the process. Typical surface particle systems provide particle maps by measuring the scattered light as a function of position by scanning a laser beam across the surface of a wafer [15]. In some cases, the witness wafer may be run through the process a number of times to improve the signal-to-noise ratio. Then the total number of particles added is divided by the number of runs to obtain the particles added per pass [16]. The location of the particles on the wafer and the position of the wafer in the processing equipment (in batch processes) can in some cases be used to help diagnose the source of the contamination. Although the minimum particle size that can be accurately detected depends on the type of surface scanner and the shape and refractive index of the particles, it is

Table 20.1 Maximum numbers of particles per cubic foot of air greater than or equal to various sizes in four classes of clean rooms

	0.1 μm	0.2 μm	0.3 μm	0.5 μm	5.0 μm
Class 1	35	7.5	3	1	NA
Class 10	350	75	30	10	NA
Class 100	NA	750	300	100	NA
Class 1000	NA	NA	NA	1000	7

Federal Standard 209D, U.S. GSA 1989.

usually quite difficult to use these systems to reliably detect particles smaller than 0.1 μm. Patterned wafers, when used, have larger minimum detection sizes and produce a high rate of false alarms.

As an alternative to laser scanners, optical inspection systems are used with actual product wafers. These systems compare the optical pattern on the wafer to an expected pattern. The reference image can be either an adjacent die or a database containing the mask information. Any areas with significant differences are shown to the operator for a visual inspection. Not only do these systems detect particles, they can also be used to detect resist defects and for etch line monitoring of high contrast surfaces. Although much slower (6 to 10 wafers per hour) than simple surface scanners, the fact that they can be used on product wafers make them an important tool for some critical levels [17]. A disadvantage of any type of witness wafer particle detection system is that it provides information only about the total amount of particles deposited during a process. It cannot provide real-time information about the particle sources.

Gas and liquid particle detection is done by drawing a particle-laden stream through a laser beam. This type of system can be used to measure particle counts in the clean room, or the devices can be connected to the exhaust of a piece of process equipment to determine the particle creation rate in the process. Figure 20.8 shows a schematic of an optical detector. The amount of light scattered depends on the intensity, wavelength, and polarization of the optical source, the size, shape, orientation, and refractive index of the particle, and the geometry of the optical detection method [18]. As commonly applied to microelectronics, however, simple optical detectors are not reliable for particles much less than 100 nm in diameter. Condensation nucleus counters (CNCs) can be used to detect smaller particles in a gas ambient [19]. In these systems, the particle-laden gas is first drawn through a saturated vapor. Various alcohols are often used to generate the vapor. The vapor condenses on the particles, making them large enough to be detected by optical scattering. Unlike simple optical systems, CNCs are limited to atmospheric pressure gases.

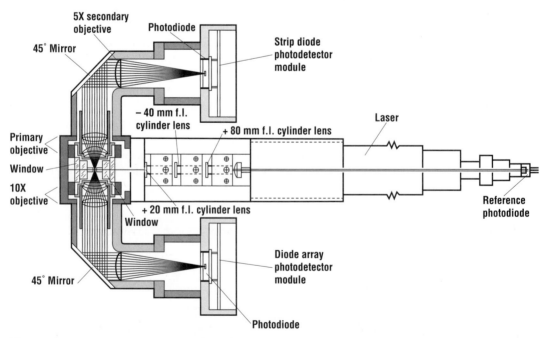

Figure 20.8   A schematic view of the optical path in the Model HVLIS detector made by Particle Measuring Systems, Inc. *(used by permission).*

20.3 Statistical Process Control

Thus far, defects have been thought of as particles or other randomly distributed contamination. If the definition of a defect is a process irregularity that leads to a circuit malfunction, defects may be caused by simple variations in process parameters. As an example, consider the gate level of a MOSFET. Assume that the mean linewidth is 90 nm and that the standard deviation of the linewidth is 5 nm. Finally, assume that the MOSFET will punch through from source to drain if the gate length is less than 65 nm. This is 5 standard deviations from the mean, which would seem to be a rather comfortable distance. Assuming a normal distribution of linewidths, the probability of any transistor having a linewidth of 65 nm or less, is only about 3×10^{-7}. If the circuit contains 10^9 transistors, however, this represents a significant source of "defects." Furthermore, the effective channel length is determined not only by the photolithography itself, but also by the etch process, the photomask that is used, the source/drain implant and diffusion steps, including the possibility of anomalous diffusion into the channel region due to defects, and the channel implant and anneal processes. Each of these processes has characteristic standard deviations. The combination of multiple variances can significantly increase the likelihood of process variation defects.

The effect of these process variations on yield is a very sensitive function of design. Ideally the IC would be designed so that it could operate with the full range of device parameters produced by the fabrication facility. Statistical process simulators exist to help in this exercise. Disturbances in a process can be used to project the range of transistor performance that can be expected. The output can then be fed into a circuit simulator like SPICE to determine the effects of the perturbation on the circuit performance. Since the performance must fall within certain limits in order to be regarded as good, one can then construct a multidimensional box in parameter space within which the IC must be fabricated. The design can be modified to increase the size of the box, but often this trade-off requires a compromise in either performance or yield. Designing the circuit to guarantee operation for any possible set of device parameters would almost certainly require extremely poor performance.

A great deal of emphasis is therefore placed on controlling the individual process steps. A simple model for this control is shown in Figure 20.9. In this diagram, wafers are introduced into the process to accomplish some task. The quality of the result is measured against a series of desired outcomes. As an example, assume that the process step is intended to etch contacts down to silicon through an oxide. The process variables under the direct control of the engineer might include the residual gas pressure in the etch chamber, the process gas flows, the plasma power, the electrode temperature, and the chamber pressure during the etch. Some of the variables that are not controlled during the process include the composition of the material to be etched, the photoresist profile, the gases adsorbed onto the surface of the wafer, the films coating the walls of the chamber and the electrodes deposited during previous etching, and the residual gas composition. The desired outcomes are the completeness of the etch, the amount of erosion of the underlying silicon, and the size and profile of the etched contact.

The ideal solution to obtaining a reliable contact etch process would be real-time adaptive control. This requires the construction of a process tool that measures the process outcomes and continuously adjusts the process variables to obtain the desired

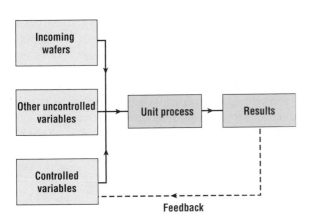

Figure 20.9   A schematic of a semiconductor process. Both controlled and uncontrolled variables must be considered.

outcomes [20]. There is an important difference between this type of control and simply manually controlling the process variables. In the etcher example, as long as the values of the uncontrolled variables are not too far from the norm, the control parameters of power, gas flows, pressure, and electrode temperature can be adjusted automatically to achieve the desired etch outcome. Because the process is adjusted in real time to compensate for the condition of the wafers as they are loaded into the etch chamber, process variations do not accumulate, resulting in a far better process.

For this type of control to be implemented, however, it is necessary to be able to measure all of the process outcomes in real time and in the process ambient. Returning to the etch example one more time, one can construct a system for measuring the remaining film thickness on the wafer. Various end point detectors, which shut off the etch when some signal indicates etch completion (plus some percent overetch), are a simple and very useful form of this sort of control. Imagine, however, using the etch rate information to allow the etch system to adjust itself to maintain the etch rate at some predetermined value. One demonstration of such a system has shown the ability to etch nitride layers down to 503 Å $\pm$ 12 Å over 23 trials when the targeted final thickness was 500 Å. The initial nitride thickness was about 2300 Å [21]. In a similar application *in situ* ellipsometry was to compensate for long-term etch rate drift in a multistep polysilicon gate etching process [22]. Although these systems are currently single-point detectors, it might also be possible to extend them to multipoint sensors, thereby giving the system the ability to automatically adjust the process to improve uniformity. On the other hand, it is very hard to envision any type of system that could continuously measure the etch profiles. As a result, one can expect to see some progress in this area, but for the foreseeable future real-time adaptive control is not a panacea.

Since the adaptive control techniques just describe are still in the demonstration phase, current manufacturing must use another approach. The method commonly used in IC manufacturing to deal with uncontrolled variables has two components. The first, of course, is to control as many variables as possible. Even if the parameter cannot be adjusted to optimize the process result, control of the parameter reduces the process variability. Because of the enormous number of uncontrolled parameters, however, this approach alone cannot solve the problem. The most widely used technique for dealing with poorly controlled process variables is to design a robust process, that is, a process that is insensitive to the uncontrolled parameters.

Since uncontrolled variables are by definition impossible to adjust, a valuable measure of process robustness is its stability or performance over time. To improve reproducibility and process control, statistical process control (SPC) has widely been adopted throughout the IC industry. SPC is a method to provide immediate feedback at each process step. It is often used to try to drive each process step to be as consistent as possible. To institute an SPC program, trend data must first be collected that show the behavior of a process as a function of time. Once enough data have been accumulated, they can be used to control the process. In a typical CVD application, for example, a film thickness is measured on several samples from a batch of wafers immediately after the deposition. The result is stored along with all previous data from the same process. Plots of thickness, wafer-to-wafer and across-the-wafer thickness variation, refractive index and refractive index variation, and other important thin film parameters versus batch number are updated. By presenting trend data, it is immediately obvious if the lot data are significantly different from those of previous runs. It is also apparent if a long-term drift is present in the process.

Of primary interest is the process variance. From elementary statistics, the variance of a set of data that represents a larger population is given by

$$\sigma^2 = \frac{\sum [x - \bar{x}]^2}{N} = \frac{N\sum x^2 - [\sum x]^2}{N^2} \qquad \textbf{(20.8)}$$

The graph of the variance is sometimes called an *s-chart*. If the variance is large, any information regarding the process mean is of very limited value. Increases in the variance call for immediate action to improve the process uniformity and repeatability. If the variation is sufficiently small, the mean (the x-chart) is then examined. Alert conditions exist if the process mean is outside some pre-determined limits, if two consecutive readings are near these limits, or if a linear regression of a number of readings (for example, the last seven) shows a suitably large slope with a high regression coefficient. Generally speaking, any of these conditions requires immediate corrective action, although the statistical nature of random process variations suggests that occasionally such an excursion will occur even when nothing has changed. As more robust processes are developed, the process limits for the x-chart and the acceptable upper limit for the s-chart can be successively tightened. Experience has shown that this approach will generally lead in turn to higher IC yields.

The ramifications of improving process control are sometimes subtle and quite often indirect. In one example, Hewlett-Packard instituted SPC for the polysilicon deposition process in a CMOS technology. The poly thickness variation improved from a standard deviation of about 5% to a variation of about 1%. The next step in the process was gate patterning, which was done at the limit of the lithography tool. The change in the reflected light due to the polysilicon thickness variation was found to be enough to increase the variation of the poly linewidth measurably. This variation resulted in a yield loss. As a result, reducing the variability in the deposition process reduced the excursions of one of the most important uncontrolled variables for the lithography process and in so doing produced more repeatable lithography results. Implementing a full SPC program is a valuable, but time-consuming enterprise. One facility reported requiring 600 charts on line at any time. These charts require not only constant updating, but wholesale modification whenever the technology changes [23].

We are left with the question of how to set the upper and lower control limits. Should one just change both by the same factor to constantly push to the center of the limit space? To be a bit more intelligent, one might want to know the ideal value of a parameter. Take, for example, the poly resistivity just discussed. If one runs a high volume process, the SPC will automatically track the distribution of poly resistivity. One can then try to correlate that to yield. Often the parameter measurements (in this case the values of resistivity) are lumped into a few categories such as low, medium, and high, and the yield is plotted for each of these categories. The wafer count in each category must be high for all other variables to be averaged out. One can fit the data to a simple binomial and predict the ideal resistivity. It is very easy to be misled with this type of analysis. For example, the resistivity may be trending upward with time. At the same time, various process improvements have been implemented that have increased yield. Some of these improvements may have been done casually by the technicians and not documented. The analysis would tell you that the poly resistivity is related to the yield, when the reality is that both the yield and the resistivity are functions of time.

20.4 Full Factorial Experiments and ANOVA

One of the goals of the process engineer is to reduce the variability of the process outcome as evidenced by control charts and to make minor adjustments to a process to compensate for long-term trends. To do this, it is desirable to understand the effect of each of the control variables on the process outcome and possible interactions between variables. Most often these relationships are unknown. To develop a process then, the engineer can experimentally map out the response of a process to a set of variables. Figure 20.10 shows a simple example. An etch rate as a function of RF power and etch gas pressure. All other considerations aside, the etch rate appears to be the least sensitive to the pressure at low power and high pressure. Strictly from a manufacturing control standpoint, therefore, it would be desirable to run the etch in this regime, but clearly trade-offs must be evaluated

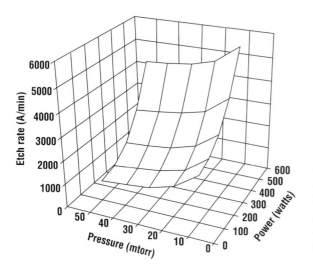

Figure 20.10   Etch rate of a plasma system as a function of power and pressure.

with regard to other important parameters such as etch rate, selectivity, and etch profile.

When there are only two variables it is possible to try a wide range of both variables, including all possible combinations. Generating the data in Figure 20.10 would require measurements at 5 powers times 6 pressures, or 30 experiments. Such an experiment does not allow for the effects of uncontrolled variables. To get some idea of the variability, multiple measurements must be taken for each of the 30 pressure/power combinations. Thus, at least 90 and perhaps as much as 270 measurements must be done. If one takes into account gas flows, bias, substrate temperature, and other variables, several thousand runs are required. This type of experiment is called a *full factorial*.

The full factorial analysis is clearly inappropriate when the number of process variables exceeds three or four. The simplest technique for analyzing data more systematically is the analysis of variance (ANOVA) technique. The method, developed by Sir Ronald Fisher in the 1930s to analyze agricultural data, is a statistical tool to determine the effects of several variables. ANOVA experiments are classified by the number of variables that are controlled. As an example consider a two-way ANOVA. Return to the etching experiment once again, where the experiment varies pressure and power. Each variable is run at a high and a low value, and each combination of parameters is run twice to get some idea about the variability of the data. The results are summarized in Table 20.2.

The total variance for the experiment is given by

$$SS_{\text{total}} = \sum_{i=1}^{N} y_i^2 - N \times \bar{y}^2 \qquad (20.9)$$

where y_i are the individual measurements, $\bar{y}$ is the average of the measured etch rates, and N is the number of observations (eight in this case). Applying the formula SS_{total} is 0.5075 [kÅ/min]^2, and the mean etch rate is 1.10 kÅ/min. Adding all of the results in each column gives a total of the etch rates at power 1 of 4.50 kÅ/min and a total of the etch rates for power 2 of 4.30 kÅ/min. The variance due to the power parameter is given by

$$SS_{\text{power}} = \frac{[\text{total of results at power 1}]^2}{\text{number of observations at power 1}}$$

$$+ \frac{[\text{total of results at power 2}]^2}{\text{number of observations at power 2}} - N \times \bar{y}^2 \qquad (20.10)$$

Table 20.2   The raw data for etch rate (kÅ/min) for a simple two-way experiment

	Power 1	Power 2
Pressure 1	1.15, 1.075	0.675, 0.775
Pressure 2	1.175, 1.10	1.40, 1.45

For these data, SS_{power} is 0.005 [kÅ/min]^2, a very small part of the total variance. Similarly, by summing the data for each row, the $SS_{pressure}$ can be calculated as 0.263 [kÅ/min]^2. Finally, to determine the variance of the etch rate due to the interaction power × pressure, one must calculate

$$SS_{pow \times press} = \sum_{i=1}^{n_{cells}} \frac{(cell)_i^2}{n_{cells}} - SS_{pow} - SS_{press} - N \times \bar{y}^2 \tag{20.11}$$

where $(cell)_i$ is the total of the observed etch rates in each of the cells in Table 20.2. In this case, $SS_{pow \times press}$ is

$$SS_{pow \times press} = \frac{2.225^2}{2} + \frac{1.45^2}{2} + \frac{2.275^2}{2} + \frac{2.85^2}{2} - 0.005 - 0.263 - 8 \times 1.1^2 \tag{20.12}$$

or 0.228 [kÅ/min]^2. The final variability is the error that can be readily determined as the difference between SS_{Total} and the sum of all of the other variances. For the data listed in Table 20.2, SS_e is 0.0115 [kÅ/min]^2.

For this experiment, pressure appears to have the largest effect, followed by the *power × pressure* interaction, but it has not been determined whether the effects are statistically significant. That is, is the variation caused by the change in the control variables significantly larger than the noise level in the experiment? First, one must find the number of degrees of freedom for each component of the variation. In statistics, a degree of freedom (ν) means the number of fair (i.e., independent) comparisons that can be made from the data. This ANOVA is calculating mean etch rates. A mean etch rate can be referenced to only one number, 0. Each variable, therefore, has one degree of freedom. The cross-variable *power × pressure* has the product of the degrees of freedom of the two component variables (*power* and *pressure*). In this case, this is also 1, since $1 \times 1 = 1$. The total number of degrees of freedom is given by the number of observations minus 1 or $N - 1 = 7$. The number of degrees of freedom associated with the error, then, is the difference between the total number of degrees of freedom and the degrees of freedom associated with the variables. In this case, the degrees of freedom associated with the error is $7 - 1 - 1 - 1 = 4$. Dividing the variations by the degrees of freedom determines a V score as summarized in Table 20.3. By dividing the V scores by the error V score, one calculates the F score. The purpose of this score is to determine whether the effect of the variable is statistically significant. In this case, the V score of the error is very small. As a result, some of the F scores are quite large. Statistical significance is always stated at some level of confidence. The F score must be larger than the number tabulated in Appendix VI in order to say that the variable is significant to that confidence level. Each of the upper three parameters has one degree of freedom, while the denominator (the error) has four. According to the appendix, $F_{0.05;1;4}$ and $F_{0.01;1;4}$, which correspond to the 95% and 99% confidence, are 7.71 and 21.2, respectively. Thus both *pressure* and

Table 20.3   ANOVA results for a two-way experiment

	SS	ν	V	F
Power	0.005	1	0.005	1.67
Pressure	0.263	1	0.263	87.67
Power × Press	0.228	1	0.228	76
Error	0.012	4	0.003	
Total	0.508	7		

power $\times$ *pressure* are significant at the 99% level, while *power* alone is not significant at the 95% confidence level.

20.5 Design of Experiments

The primary disadvantage of any full factorial method is the number of measurements required for a realistic system. One alternative to the full factorial experiment is to run two sets of one-at-a-time experiments. For example, a set of experiments could be done by varying pressure with power fixed near the middle of the range to be investigated and then fixing the pressure similarly and varying the power. The primary advantage of one-at-a-time experiments is the relatively small number of data points required. It is common for realistic processes to have 6 to 12 variables not including inter-actions. A full factorial experiment with three levels for each of six parameters would require 3^6, or 728 experiments without repetition. Clearly, running full factorial experiments is not practical. Running a one-at-a-time experiment would require only $6 \times 3 - 1 = 17$ experiments, but would not provide complete information about the parameter space.

Consider performing two one-at-a-time experiments instead of a full factorial experiment for the simple etch experiment described in the last section. The power could first be held constant while the pressure is varied, and then the pressure could be fixed while the power is varied. This method would work fairly well for the data shown in Figure 20.10, but consider the data in Figure 20.11. This experiment tried to minimize the particle counts in an LPCVD polysilicon furnace by trying two silane flows (A1 = 45% of full flow and A2 = 35% of full flow) at three levels of pressure (B1 = nominal, B2 = nominal − 40 mtorr, B3 = nominal − 80 mtorr). At the nominal pressure, increasing the silane flow increases particle production, while the opposite is true during operation at a pressure 80 mtorr below nominal [24]. The two parameters are said to be confounded. The effect of one depends on the value of the other. In the data of Table 20.3, power and pressure are confounded because of the large significance of *power* $\times$ *pressure*. The presence of confounded variables makes the conclusions of any one-at-a-time experiment extremely uncertain.

Clearly, some less haphazard technique must be developed to economically optimize a process with a large number of variables. This can be done using design of experiments (DOE). It must be reemphasized that this is an extremely useful topic that is really a course in itself. The intention of this section is only to describe an example of a simple designed experiment to introduce the reader to the possibilities of DOE. Readers are strongly encouraged to consult more specialized texts on the topic. In addition, a number of software products exist that can lead the engineer through the process of setting up a designed experiment.

The first step in DOE is to identify all of the possible variables as well as all possible combinations of variables. These variables and combinations of variables are called *factors*. At this point, it is important to identify potential variables without regard to whether they are expected to be important. It is not unusual for the first round of experiments to contain 10 or more

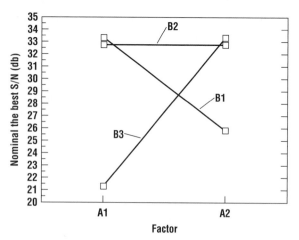

Figure 20.11   Particle counts as a function of silane flow and chamber pressure for the low pressure deposition of polysilicon (*after DePinto*).

factors. A reasonable range for each variable must also be identified. The first round of experiments, called *screening experiments*, is intended to be able to eliminate the factors that have little effect on the desired process outcome. Once this has been done, a second and possibly third round of experiments is done to focus on the important variables.

Table 20.4 shows an example of a set of eight experiments that can be run to investigate the effects of seven factors. The set of experiments is called *orthogonal*. Each factor is set to a low level in four tests and high in the other four tests. Orthogonality means that factors can be evaluated independently, that the total experiments are designed so that one factor does not affect another. The matrix described in Table 20.4 is therefore called an eight-level orthogonal array (L8 OA). One now has to assign the variables and interactions between variables to the factor columns. One method would be to assign seven variables. For an etching experiment, one might choose RF power, pressure, platen temperature, $\%CF_4$ in the gas flow, chamber base pressure, number of wafers in the chamber, and total gas flow. Once the eight measurements have been taken, the results can be analyzed using the ANOVA method. If an estimate of the error is desired, 16 or 24 measurements can be done by repeating each row two or three times. A wide variety of OAs have been tabulated for these applications.

This arrangement would seem to be incredibly efficient. The effects of seven variables can be determined in only eight measurements. The trade-off here is that no information can be obtained about the interaction between variables. More importantly, however, if these interactions exist, they will affect the outcome of the experiment. In this example, the third column of the OA is related to the first two. Using the Boolean laws

$$L \times L = L$$
$$L \times H = H \times L = H$$
$$H \times H = L \tag{20.13}$$

where H represents a high value and L a low value, Column 3 is exactly the product of the first two columns. Similarly, Column 5 is the product of Columns 1 and 4, and Column 6 is the product of Columns 2 and 4. Extending these rules to three terms, one can show that Column 7 is the product of Columns 1, 2, and 4. Thus, if our seven variables are assigned, the experiments run, and Column 3 is found to be significant, it may be that the variable assigned to this column has a significant effect, or it may be that the interaction between the first two variables is important. One can avoid this ambiguity by assigning these columns to the interactions described. Then only three of the factors are variables (Columns 1, 2, and 4). Performing eight trials on two levels of three variables is once again a

Table 20.4 An L8 OA for a two-level experiment

	1	2	3	4	5	6	7
Trial 1	L	L	L	L	L	L	L
Trial 2	L	L	L	H	H	H	H
Trial 3	L	H	H	L	L	H	H
Trial 4	L	H	H	H	H	L	L
Trial 5	H	L	H	L	H	L	H
Trial 6	H	L	H	H	L	H	L
Trial 7	H	H	L	L	H	H	L
Trial 8	H	H	L	H	L	L	H

The H and L indicate high and low values of the factors in the columns.

Example 20.2   **ANOVA Analysis of L8 OA**

Assume that the etching experiment just described was run. The column assignments and levels were as in Table 20.5.

One can now calculate the net effect of the first factor by adding all of the results at low power and comparing them to the sum of all of the results at high power. From Table 20.6 these sums are 2.81 and 6.06 kÅ/min. Since there are eight observations, the sum of the squares for power is

$$SS_{Power} = \frac{[6.06 - 2.81]^2}{8} = 1.32 \tag{20.14}$$

The results of the analysis for each column are given in Table 20.7 and analyzed in Table 20.8.

Notice that the largest effects seem to be caused by the variables assigned to Columns 1, 2, 4, and perhaps 6. Column 3 tests as having a small effect; however, one must also bear in mind that the data in this column can also be affected by the Column 1–Column 2 product. The small response may mean that there is no significant effect of the platen temperature or of the *power–pressure* cross-term, or it may mean that the two effects cancel each other. Since power has by far the largest effect, columns that may involve cross-products of power must be viewed with the most suspicion.

There are several ways available in an OA to get an error estimate. Sometimes a column may be unassigned and so reserved for error estimation. Otherwise, the lowest response columns may be pooled. In doing so, the assumption is made that these parameters have little effect on the etch rate and the results are roughly indicative of the random error in the experiment. In our case, variables 3, 5, and 7 are clearly candidates for error estimation. Adding the sum of the squares for

Table 20.5   Experiment parameters for the L8 OA etch experiment

	Variable	H Level	L Level
1	RF power (W)	500	100
2	Etch pressure (mtorr)	50	10
3	Platen temperature (°C)	80	40
4	%CF$_4$	75	50
5	Base pressure (torr)	1×10^{-4}	1×10^{-5}
6	Number of wafers	4	1
7	Total gas flow (slpm)	2.5	1.0

Table 20.6   Etch results (kÅ/min) for the L8 OA experiment using the factor assignments in Table 20.5

	1	2	3	4	5	6	7	Etch Rate (kÅ/min)
Trial 1	L	L	L	L	L	L	L	0.760
Trial 2	L	L	L	H	H	H	H	0.895
Trial 3	L	H	H	L	L	H	H	0.400
Trial 4	L	H	H	H	H	L	L	0.755
Trial 5	H	L	H	L	H	L	H	1.575
Trial 6	H	L	H	H	L	H	L	1.800
Trial 7	H	H	L	L	H	H	L	1.170
Trial 8	H	H	L	H	L	L	H	1.515

The H and L indicate high and low values of the factors in the columns.

Table 20.7   The raw data of the L8 OA described in Tables 20.5 and 20.6

	Variable	*H* Total	*L* Total	*SS*
1	RF power (W)	6.06	2.81	1.320
2	Etch pressure (mtorr)	3.84	5.03	0.177
3	Platen temperature (°C)	4.53	4.34	0.005
4	%CF$_4$	4.965	3.905	0.140
5	Base pressure (torr)	4.39^5	4.47^5	0.001
6	Number of wafers	4.265	4.605	0.014
7	Total gas flow (slpm)	4.385	4.485	0.001

Table 20.8   Analysis of the data given in Table 20.7

	Variable	*SS*	*df*	*V*	*F*
1	RF power (W)	1.32	1	1.32	603.4
2	Etch pressure (mtorr)	0.177	1	0.177	80.9
3	Platen temperature (°C)	0.005	*	*	*
4	%CF$_4$	0.140	1	0.140	64.0
5	Base pressure (torr)	0.00^1	*	*	*
6	Number of wafers	0.014	1	0.014	6.4
7	Total gas flow (slpm)	0.001	*	*	*

---

these three entries and dividing by 3 (the denominator in Equation 20.12 now is 24 instead of 8) gives an SS_e of 0.0022. Then, dividing the error estimation to get the F scores and comparing these scores to the F scores in Appendix VI assuming 1 degree of freedom in the numerator and 3 in the denominator, the power, pressure, and %CF$_4$ are all significant at the 99% level, while the number of wafers is significant at the 90% level (table not included in Appendix).

full factorial experiment. An advantage of DOE, then, is that it allows the engineer to allocate the experiments to investigate the variables that are considered important.

A final word of caution is in order regarding the use of OAs. The analysis described in this section, often called the Taguchi approach, is a powerful tool that allows one to quickly optimize a process in a complicated parameter space even in the presence of noise. This section has only scratched the surface of the technique. A wide variety of orthogonal arrays exist to fit almost any set of variables that you may wish to investigate. An accurate interpretation of the results, however, requires some skill with the technique. It is easy to be misled, and confirming experiments are extremely important before the results can be relied upon.

20.6 Computer-integrated Manufacturing

Computer-integrated manufacturing (CIM) is an umbrella term that encompasses a number of ways to use computers to improve the manufacturing enterprise. For IC manufacturing, CIM includes the following major components [25]:

Planning and scheduling of factory operations and resources
Modeling of factory operation

Process and product specifications and recipes
Tracking of work in progress (WIP)
Factory performance monitoring
Machine and process monitoring, control, and diagnosis

All of these functions are not present in every CIM system, but the list includes the most common components. Whatever its form, CIM generally involves the installation of a local area network throughout the factory including the clean room. Before each process step, the operator enters the batch ID into the computer. Where possible, the process tools themselves report the process results. Ion implanters, for example, now have *in situ* sensors that determine the implant results and can output

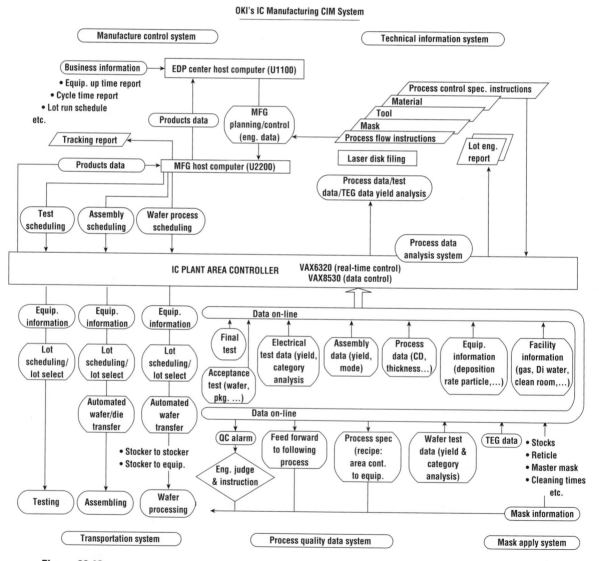

Figure 20.12 A full CIM system implemented by one major semiconductor manufacturer (*from Mizokami*).

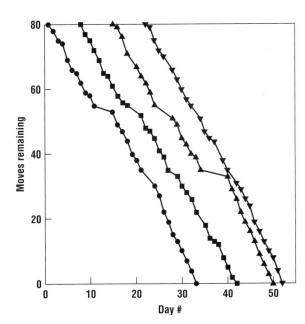

Figure 20.13   Charts showing the remaining lot moves as a function of time are often used to point out bottlenecks in the line. These diagrams are sometimes called chicken scratches.

the results to a LAN interface via the semiconductor equipment communication standard (SECS), which was developed by SEMI [26]. Newer characterization and process tools, such as ellipsometers, linewidth measuring instruments, and etchers, similarly are equipped with SECS or SECS-II interfaces for CIM as well as other developing SEMI standards [27]. Figure 20.12 shows a schematic for a CIM system [28].

A full CIM system represents an enormous amount of information potentially available for use. The issue is not data collection, but data distribution and utilization. The use of relational, distributed databases allows the users easy query of the database without requiring a very large central computer system. In spite of the availability of the hardware and software to operate CIM systems, intelligent applications of the information that they produce are not as forthcoming.

One of the earliest and perhaps most important applications of CIM has been the scheduling of wafers through particular pieces of equipment in the fabrication line. It is generally accepted that the longer it takes to fabricate a batch of wafers, the lower the yield. It is desirable then to schedule the fabrication line to maximize throughput without incurring undo delays. Since the process is sequential, one needs to model the throughput of the various pieces of equipment and then examine the projected effects of a particular wafer move. By projecting ahead to the expected inventory at particular steps, long delays for particular process equipment can be avoided. Perhaps even more importantly, the wafer inventory can be controlled to minimize cycle time and/or maximize tool utilization.

Charts like those shown in Figure 20.13 are often used to point out bottlenecks in the line. The number of steps remaining until the lot is completed is plotted for each batch of wafers as a function of time. Each line is sometimes called a *scratch*. Process steps at which the scratch frequently is horizontal represent fabrication line bottlenecks that limit the overall throughput of the factory. Cycle time is also often measured against the theoretical cycle time, the minimum time that it would take to move a batch of wafers through the line if they were the only wafers in the factory. A well-managed CIM system can produce routine turn times of twice the theoretical as compared to about seven times theoretical for a poorly controlled line [29].

20.7 Summary

Semiconductor IC manufacturing has as one of its primary goals the maximization of the process yield. This involves the reduction of particles both in the clean room ambient and in the process equipment, and the reduction of variability in the various unit processes. Statistical process control is an extremely valuable tool for reducing this variability. ANOVA was presented as a method to analyze experiments that determine the effects of various process parameters on a process. For realistic processes, the number of variables that must be optimized is excessive, and so orthogonal arrays have been developed as a more efficient method for exploring complex parameter spaces.

Problems

1. An IC manufacturer is running a technology at a 40% yield. One of the engineers manages to locate a major contamination source for the technology. The engineer changes the process to eliminate this defect and finds that this has decreased the total defect denstiy of the technology by a factor of 2.
   (a) If the technology has a gross fail area of 5% ($G = 0.05$), find the new yield, based on a simple Poisson model.
   (b) If the manufacturer runs 500 wafers per week on this technology, each wafer has 400 die, and each good die sells for $10, find the net increase in income per year, due to this defect reduction change.

2. The experiment suggested in the text just before Equation 20.4 was done. It is found that for a 3-cm^2 chip area, the single die yield was 55%, while for two adjacent die the composite yield was 43%, and 37% for three adjacent die, and finally 33% for four adjacent die. Find the defect density and the clustering factor.

References

1. J. P. de Gyvez and D. K. Pradhan, eds., *Integrated Circuit Manufacturability*, IEEE, New York, 1999.
2. Y. Nishi and R. Doering, eds., *Handbook of Semiconductor Manufacturing Technology*, Marcel Dekker, New York, 2000.
3. D. Dance and K. Gildersleeve, "Estimating Semiconductor Yield from Equipment Particle Measurements," *Proc. 2092 IEEE/SEMI Int. Semiconductor Manufacturing Sci. Symp.,* 1992, p. 18.
4. F. Lee and S. Smith, "Yield Analysis and Data Management Using Yield Manager™ at Motorola's MOS-12." *Proc. 9th Annual IEEE/SEMI Adv. Semiconductor Manufacturing Conf.*, Boston, 1998.
5. D. Carpenter, "Defect Monitor and Characterization Standard," *Proc. IEEE Int. Workshop Defect, Fault Tolerance,* November 1980, p. 70.
6. S. Swaving, A. Ketting, and A. Trip, "MOS-IC Process and Device Characterization Within Philips," *IEEE Proc. Microelectron. Test Struct.* **1**(1):180 (1988).
7. J. P. de Gyvez, *Integrated Circuit Defect-Sensitivity: Theory and Computational Models,* Kluwer, Boston, 1993.
8. A. V. Ferris-Prabhu, "Role of Defect Size Distribution in Yield Modeling," *IEEE Trans. Electron Dev.* **ED-32**:1727 (1985).
9. R. B. Seeds "Yield and Cost Analysis of Bipolar LSI," IEEE Int. Electron Devices Meeting, 1967.
10. R. Ross and N. Atchison, "Yield Modeling," in *Handbook of Semiconductor Manufacturing Technology*, Y. Nishi and R. Doering, eds., Dekker, New York, 2000.
11. C. H. Stapper, "Fact and Fiction in Yield Modeling," *Microelectron. J.* **20**:129 (1989).
12. C. Kooperberg, "Circuit Layout and Yield," *IEEE J. Solid-State Circuits* **23**:887(1988).
13. G. Cheek and G. O'Donoghue, "Yield Models in a Design for Manufacturability Environment: A Bibliography," *Proc. 1993 IEEE/SEMI Int. Semiconductor Manufacturing Sci. Symp.,* 1993, p. 133.
14. T. Hattori, "Contamination Control: Problems and Prospects," *Solid State Technol.* **33**(7): S1 (1990).

15. P. Lilienfeld, "Optical Detection of Particles on Surfaces: A Review," *Aerosol Sci. Technol.* **5**:145 (1986).

16. B. Tullis, "Particle Contamination by Process Equipment," in *Handbook of Contamination Control in Microelectronics,* D. L. Tollivier, ed., Noyes, Park Ridge, NJ, 1988.

17. P. Burggraaf, "Patterned Wafer Inspection: Now Required!" *Semicond. Int.* December 1994, p. 57.

18. D. W. Cooper, "Contamination Monitoring," in *Microelectronics Manufacturing Diagnostics Handbook,* A. H. Landzberg, ed., Van Nostrand Reinhold, New York, 1993.

19. K. H. Ahn and B. Y. Liu, "Particle Activation and Growth Processes in Condensation Nucleus Counter—I. Theoretical Background, II Experimental Study," *J. Aerosol Sci.* **21**:249, 263 (1990).

20. H. F. Guo, C. J. Spanos, and A. J. Miller, "Real-Time Statistical Process Control for Plasma Etching," *1991 IEEE/SEMI Int. Semiconductor Manufacturing Sci. Symp.,* 1991, p. 113.

21. S. Maung, S. Banerjee, D. Draheim, S. Henck, and S. W. Butler, "Integration of *In-situ* Spectral Ellipsometry with MMST Machine Control," *IEEE Trans. Semiconductor Manufacturing* **7**:184 (1994).

22. S. W. Watts and J. A. Stefani, "Supervisory Run-to-Run Control of Polysilicon Gate Etch Using *In Situ* Ellipsometry," *IEEE Trans. Semiconductor Manufacturing* **7**:193 (1994).

23. R. Abugov, "From Trend Charts to Control Charts: Setup Tests for Making the Leap," *1992 IEEE/SEMI Int. Semiconductor Manufacturing Sci. Technol.,* 1992, p. 3.

24. G. DePinto, "A Methodology for Optimizing a Constant Temperature Polysilicon Deposition Process," *IEEE Trans. Semiconductor Manufacturing* **6**:332 (1993).

25. J. McGehee, J. Hebley, and J. Mahaffey, "The MMST Computer-Integrated Manufacturing Framework," *IEEE Trans. Semiconductor Manufacturing* **7**:107 (1994).

26. SEMI Standards "SEMI Equipment Communications Standards 1 (Message Transfer: SECS-I, E4-917, 1992) and 2 (Message Content: SECS-II, E5-93, 1993), Semiconductor Equipment and Materials Institute.

27. R. Beaver, A. Coleman, D. Draheim, and A. Hoffman, "Architecture and Overview of MMST Machine Control," *IEEE Trans. Semiconductor Manufacturing* **7**:127 (1994).

28. Y. Mizokami, "The Total CIM System for Semiconductor Plants," *Proc. 1990 Int. Semiconductor Manufacturing Sci. Symp.,* IEEE, Piscataway, NJ, 1990, p. 24.

29. M. Adams and B. Smoak, "Managing Manufacturing Improvement Using Computer Integrated Manufacturing Methods," *Proc. 1990 Int. Semiconductor Manufacturing Sci. Symp.,* IEEE, Piscataway, NJ, 1990, p. 9.

Appendix I

Acronyms and Common Symbols

The cgs units listed in parentheses are not necessarily the most commonly used units for a given variable.

a:Si:-H	Amorphous hydrogenated silicon
A	Area (cm^2)
	Coefficient of Deal-Grove equation (cm)
	Absorbance (dimensionless)
ALD	Atomic layer deposition
ANOVA	Analysis of variance
APCVD	Atmospheric pressure chemical vapor deposition
AR	Aspect ratio
B	Magnetic field (gauss)
	Coefficient of Deal-Grove equation (cm^2/sec)
BHF	Buffered hydrofluoric acid
BiCMOS	Bipolar/complementary metal oxide semiconductor
BJT	Bipolar junction transistor
BPSG	Borophosphosilicate glass
C	Concentration (cm^{-3})
	Conductance (cm^3/sec)
C_{ox}	Oxide capacitance per unit area (F/cm^2)
	Solubility of oxygen in silicon (cm^{-3})
CAR	Chemically amplified resist
CEL	Contrast-enhancing layer
CIM	Computer integrated manufacturing
CMOS	Complementary metal oxide semiconductor
CMP	Chemical mechanical polishing
CMTF	Critical modulation transfer function
CNC	Condensation nucleus counter
CV	Capacitance voltage curve
CVD	Chemical vapor deposition
CZ	Czochralski growth
D	Diffusion coefficient (cm^2/sec)
	Defect density (cm^{-2})
D_o	Preexponential for diffusion coefficient (cm^2/sec)
DBR	Distributed Bragg reflector

DCFL	Direct-coupled FET logic
DCS	Dichlorosilane
DI	Dielectric isolation
DOE	Design of experiments
DQN	Diazoquinone novolac resist
DRAM	Dynamic random access memory
DUV	Deep ultraviolet
e	Charge of an electron (coulombs)
E	Energy (ergs)
E_a	Activation energy (erg)
E_G	Energy gap (erg)
EBL	Electron beam lithography
ECL	Emitter-coupled logic
ECR	Electron cyclotron resonance
EDP	Ethylenediamine pyrocatechol
EEPROM	Electrically erasable programmable read-only memory
EOT	Equivalent oxide thickness
EPD	Electroplate deposition
f_T	Unity gain frequency (Hz)
F	View factor (dimensionless)
FET	Field effect transistor
FZ	Float zone growth
g_m	Device transconductance (A/V)
G	Gap between mask and wafer (cm)
	Gibbs free energy (ergs)
	Gross fail area (dimensionless)
G_B	Gummel number of the base (cm^{-2})
GILD	Gas immesion laser doping
GOLD	Gate overlap of drain
GSMBE	Gas source molecular beam epitaxy
h_g	Mass transport coefficient (cm/sec)
HARMST	High aspect ratio microsystems technology
HEMT	High electron mobility transistor
HDP	High density plasma
HF	Hydrofluoric acid
HMDS	Hexamethyldioxysilazane
HOMO	Highest occupied molecular orbital
I_x	Atomic current (sec^{-1})
IC	Integrated circuit
ICP	Inductively coupled plasma
ILD	Interlevel dielectric
IR	Infrared
J	Flux density ($cm^{-2}\text{-}sec^{-1}$)
k	Relative dielectric constant (dimensionless)
	Thermal conductivity (ergs/°C-cm)
	Segregation coefficient (dimensionless)
K_p	Reaction equilibrium constant (units depend on reaction)

k_x	Reaction rate constant (units depend on reaction)
KOH	Potassium hydroxide
LDD	Lightly doped drain
LEC	Liquid-encapsulated Czochralski (growth)
LED	Light-emitting diode
LIGA	*Lithographic galvanoformung abformung*
LOCOS	Local oxidation of silicon
LPCVD	Low pressure chemical vapor deposition
LUMO	Lowest unoccupied molecular orbital
m	Atomic mass (gm)
	Segregation coefficient (dimensionless)
M	Molar mass (gm)
MBE	Molecular beam epitaxy
MEMS	Microelectromechanical systems
MERIE	Magnetically enhanced reactive ion etching
MESFET	Metal semiconductor field effect transistor
MMIC	Millimeter microwave integrated circuit
MOCVD	Metallorganic chemical vapor deposition
MODFET	Modulation-doped field effect transistor
MOS	Metal oxide semiconductor
MTF	Median time to failure
	Modulation-transfer function
$M_\lambda(T)$	Spectral radiant ecxitance
n	Electron concentration (cm^{-3})
n_i	Intrinsic carrier concentration (cm^{-3})
N	Number density (cm^{-3})
N_{it}	Interface state density (cm^{-2})
N_{Re}	Reynolds number (dimensionless)
N_v	Vacancy concentration (cm^{-3})
NIL	Nanoimprint lithography
NGL	Next-generation lithography
NMOS	N-channel metal oxide semiconductor
OA	Orthogonal array
OE	Optoelectronics
OLED	Organic light-emitting diode
OM	Organometallic
OSF	Oxidation-induced stacking faults
p	Hole concentration (cm^{-3})
p_x	Partial pressure of species x (dynes/cm^2)
P	Total pressure (dynes/cm^{-2})
	Probability function (dimensionless)
PAC	Photoactive compound
PAG	Photoacid generator
PECVD	Plasma-enhanced chemical vapor deposition
PII	Plasma immersion implantation
PLED	Polymer light-emitting diode
PMMA	Polymethylmethacrylate

PMOS	P-channel metal oxide semiconductor
PSG	Phosphosilicate glass
PVD	Physical vapor deposition
q	Charge on an electron (coulombs)
Q	Gas throughput (dyne-cm/sec)
Q_f	Fixed charge density (C/cm^2)
R	Growth or deposition rate (cm/sec)
	Resistance (Ω)
	Reflectivity
R_c	Specific contact resistance (Ω-cm^2)
R_{ME}	Mass evaporation rate (gm/cm^2-sec)
R_p	Projected range (cm)
R_s	Sheet resistance ($\Omega/\square$)
RCA	A popular wet chemical clean
RF	Radio frequency
RHEED	Relative high energy electron diffraction
RIE	Reactive ion etch
RR	Removal rate (for CMP)
RTA	Rapid thermal annealing
RTCVD	Rapid thermal chemical vapor deposition
RTP	Rapid thermal processing
S	Sputter yield (dimensionless)
S_P	Pumping speed (cm^3/sec)
SA	Self-aligned bipolar technology
SAGFET	Self-aligned gate field effect transistor
SAINT	Self-aligned implantation N+ technology
SBC	Standard buried collector
SDFL	Schottky diode FET logic
SEM	Secondary electron microscopy
SICOS	Sidewall contacted structure
SILO	Sealed interface local oxidation
SIMOX	Separation by implanted oxygen
SIMS	Secondary ion mass spectroscopy
SMA	Shape memory alloy
SOI	Silicon on insulator
SPC	Statistical process control
SPE	Solid phase epitaxial regrowth
SPICE	Stanford Program for Integrated Circuit Emulation
SRAM	Static random access memory
SS	Sum of squares
SST	Super-self-aligned bipolar transistor
STI	Shallow trench isolation
SUPREM	Stanford University Process Emulation Module
t	Time (sec)
t_{ox}	Oxide thickness (cm)
T	Temperature (centigrade unless otherwise specified)
TCA	Trichloroethane
TCE	Trichloroethylene

TCE	Temperature coefficient of expansion
TDDB	Time dependent dielectric breakdown
TEG	Triethylgallium
TEM	Transmission electron microscopy
TEOS	Tetraethyloxysilane
TFT	Thin film transistor
TMG	Trimethyl gallium
TOC	Total organic content
TTV	Total thickness variation
u	Surface free energy (ergs/cm^2)
U_∞	Gas velocity far from a surface (cm/sec)
UHVCVD	Ultrahigh vacuum chemical vapor deposition
ULSI	Ultra large scale integration
UV	Ultraviolet
v	Velocity (cm/sec)
V_A	Early voltage of bipolar transistor (V)
V_{bi}	Buil-in voltage (V)
V_{po}	Pinchoff voltage of MESFET (V)
V_t	Threshold voltage (V)
VCSEL	Vertical cavity surface-emitting laser
VLSI	Very large scale integration
VOC	Volatile organic compound
VPE	Vapor phase epitaxy
VUV	Vacuum ultraviolet
W	Base width of bipolar transistor (cm)
W_D	Depletion width (cm)
WIP	Work in progress
x_j	Junction depth (cm)
Y	Yield (dimensionless)
Z	Atomic number
α	Inverse absorption length (cm^{-1})
	Ratio of collector and emitter currents in a bipolar transistor (dimensionless)
β	Ratio of collector and base currents in a bipolar transistor (dimensionless)
γ	Resist contrast (dimensionless)
δ	Boundary layer thickness (cm)
ε	Emissivity (dimensionless)
	Relative dielectric constant (dimensionless)
η	Dynamic viscosity (cm^{-1}-sec^{-1})
λ	Wavelength of radiation (cm)
	Mean free path (cm)
μ	Kinematic viscosity (cm^2/sec)
	Mobility (cm^2/volt-sec)
μCP	Microcontact printing
μ_{eff}	Effective mobility (cm^2/V-sec)
ν	Frequency of radiation (sec^{-1})
ρ	Resistivity (Ω-cm)
	Mass density (gm/cm^3)

ρ_D	Sheet resistance ($\Omega/\square$)
σ	Cross section (cm^2)
	Stress in wafer (dynes/cm^2)
	Supersaturation (dimensionless)
σ_o	Saturation (dimensionless)
θ	Fraction of free sites on a surface (dimensionless)
ϕ_{ms}	Metal semiconductor work function (erg)
ϕ_b	Barrier height (erg)
ψ	Critical implant angle (degrees)
$\mathscr{E}$	Electric field (V/cm)
3-D	Triple diffused bipolar technology

Appendix II

Properties of Selected Semiconductor Materials

Table II.1 Properties of various materials I: Semiconductors and insulators

	Si	GaAs	Ge	αSiC	SiO$_2$	Si$_3$N$_4$
Mass density (gm/cm^3)	2.33	5.32	5.32	2.9	2.2	3.1
Breakdown field (MV/cm)	0.3	0.5	0.1	2.3	10	10
Dielectric constant	11.7	12.9	16.2	6.52	3.9	7.5
Energy gap (eV)	1.12	1.42	0.66	2.86	9	5
Electron affinity (eV)	4.05	4.07	4		0.9	
Index of refraction	3.42	3.3	3.98	2.55	1.46	2.05
Melting point (°C)	1412	1240	937	2830	~1700	~1900
Specific heat (J/gm-°C)	0.7	0.35	0.31		1	
Thermal conductivity (W/cm-°C)	1.31	0.46	0.6		0.014	
Thermal diffusivity (cm^2/sec)	0.9	0.44	0.36		0.006	
Thermal expansion (K$^{-1} \times 10^{-6}$)	2.6	6.86	2.2	2.9	0.5	2.7

Table II.2 Properties of various materials II: Metals

	Al	Cu	Au	TiSi$_2$	PtSi
Density (gm/cm^3)	2.7	8.89	19.3	4.043	12.394
Resistivity ($\mu\Omega$-cm)	2.82	1.72	2.44	14	30
Temperature coefficient	0.0039	0.0039	0.0034	4.63	
Barrier to n-type Si (eV)	0.55	0.60	0.75	0.60	0.85
Thermal conductivity (W/cm-°C)	2.37	3.98	3.15		
Melting point (°C)	659	1083	1063	1540	1229
Specific heat (J/gm-°C)	0.90	0.39	0.13		
Thermal expansion (K$^{-1} \times 10^{-6}$)	25	16.6	14.2	12.5	

Appendix III

Physical Constants

Quantity	Symbol	Value
Gravitational constant	G	6.67×10^{-11} N m^2/kg^2 (or m^3/kg s^2)
Avogadro's number	N_0	6.0222×10^{23} particles/mole (or amu/gm)
Boltzmann's constant (microscopic gas constant)	k	1.3806×10^{-23} J/K
	$\dfrac{1}{k}$	8.617×10^{-5} eV/K
		11,605 K/eV
Macroscopic gas constant	$R(= N_0 k)$	8.314 J/mole K
		1.9872 kcal/kmole K
Quantum unit of charge	e	1.60219×10^{-19} C
		4.8033×10^{-10} esu
Faraday constant (1 mole of electricity)	$F(= N_0 e)$	9.6487×10^4 C/mole
		2.8926×10^{14} esu/mole
Permittivity of space		8.85419×10^{-12} C^2/N m^2
	$\varepsilon_0 \left(= \dfrac{1}{\mu_0 c^2} \right)$	
	$4\pi\varepsilon_0 \left(= \dfrac{4\pi}{\mu_0 c^2} \right)$	1.112650×10^{-10} C^2/N m^2
	$\dfrac{1}{4\pi\varepsilon_0} \left(= \dfrac{\mu_0 c^2}{4\pi} \right)$	8.98755×10^9 N m^2/C^2
Permeability constant	μ_0	$4\pi \times 10^{-7}$ N/A^2 *exact, by definition* or 1.256637×10^{-6} N/A^2
	$\dfrac{\mu_0}{4\pi}$	*exactly* 10^{-7} N/A^2
Speed of light	c	2.997925×10^8 m/s
Planck's constant	h	6.6262×10^{-34} J s
		4.1357×10^{-15} eV s
		4.1357×10^{-21} MeV s
	$\hbar \left(= \dfrac{h}{2\pi} \right)$	1.05459×10^{-34} Js
		6.58822×10^{-16} eV s
		6.58822×10^{-22} Me Vs
Charge-to-mass ratio or electron	$\dfrac{e}{m_e}$	1.75880×10^{11} C-kg
		5.2728×10^{17} esu/gm
Mass of electron	m_e	9.1096×10^{-31} kg
		5.4859×10^{-4} amu
Mass of proton	m_p	1.67261×10^{-27} kg
		1.0072766 amu
		$1836.11 m_e$

Quantity	Symbol	Value
Mass of neutron	m_n	1.67492×10^{-27} kg
		1.0086652 amu
		$1838.64 m_e$
Intrinsic energy of electron	$m_e c^2$	0.51100 MeV
Intrinsic energy of proton	$m_p c^2$	938.27 MeV
Intrinsic energy of neutron	$m_n c^2$	939.55 MeV
Rydberg constant for infinitely massive nucleus	$\mathcal{R}_\infty \left[= \left(\dfrac{1}{4\pi\varepsilon_0} \right)^2 \dfrac{m_e e^4}{4\pi\hbar^3 c} \right]$	1.0973731×10^7 m^{-1}
Rydberg constant for hydrogen 1	$\mathcal{R}_H$	1.0967758×10^7 m^{-1}
Fine structure constant	$\alpha \left(= \dfrac{1}{4\pi\varepsilon_0} \dfrac{e^2}{\hbar c} \right)$	7.29735×10^{-3} or $1/137.036$

(This table is adapted from B. N. Taylor, W. H. Parker, and D. N. Langenberg, *The Fundamental Constants and Quantum Electrodynamics* [New York: Academic Press, 1969]. A good popular article on the fundamental constants, by the same authors, is to be found in the October 1970 issue of *Scientific American*. The numbers recorded here have been truncated so that the uncertainty in each is at most ± 1 in the last digit.)

Appendix IV

Conversion Factors

1. Length
10^2 cm/m
10^3 m/km

2.54 cm/in.
12 in./ft
5280 ft/mi

0.3048 m/ft
1.609344×10^3 m/mi
1.609344 km/mi

1.49598×10^{11} m/au
9.461×10^{15} m/light-yr
3.084×10^{16} m/pars

10^{-6} m/μm (or m/micron)
10^{-10} m/Å
10^{-15} m/fm

2. Volume
10^{-3} m³/L
10^3 cm³/L
0.94635 L/qt
3.7854×10^{-3} m³/gal

3. Time
(The day is a mean solar day; the year is a
sidereal year.)

3,600 sec/hr
8.64×10^4 sec/day
365.26 d/yr
3.1558×10^7 sec/yr

4. Speed
0.3048 (m/sec)/(ft/sec)
1.609×10^3 (m/sec)/(mi/sec)

0.4470 (m/sec)(mi/hr)
1.609 (km/hr)/(mi/hr)

5. Acceleration
0.3048 $(m/sec^2)/(ft/sec^2)$

6. Angle
60 sec of arc(")/min of arc(')
60 min of arc(')/deg
$180/\pi (\cong 57.30)$ deg/rad
$2\pi (\cong 6.283)$ rad/revolution

7. Mass
10^3 gm/kg

453.59 gm/lb
0.45359 kg/lb
2.2046 lb/kg

1.66053×10^{-27} kg/amu
6.0222×10^{-26} amu/kg
6.0222×10^{23} amu/gm

8. Density
10^3 $(kg/m^3)/(gm/cm^3)$
16.018 $(kg/m^3)/(lb/ft^3)$
1.6018×10^{-2} $(gm/cm^3)/(lb/ft^3)$

9. Force
10^5 dynes/N
10^{-5} N/dyne
4.4482 N/lb$_f$
(1 lb$_f$ = weight of 1 pound at standard gravity
   [g = 9.80665 m/s^2])

10. Pressure
0.1 $(N/m^2)/(dynes/cm^2)$
10^5 $(N/m^2)/bar$

1.01325×10^5 $(N/m^2)/atm$
1.01325×10^6 $(dynes/cm^2)/atm$
1.01325 bar/atm
133.32 $(N/m^2)/mm$ Hg (0°C)
3.386×10^3 $(N/m^2)/in.Hg$ (0°C)
6.895×10^3 $(N/m^2)/(lbf/in.^2,$ or psi)

11. Energy
10^7 ergs/J
10^{-7} J/erg

4.184 J/Cal
4184 J/kcal
10^3 cal/kcal
(The kilocalorie [kcal] is also known as
   the food calorie, the large calorie, or the Calorie.)

1.60219×10^{-19} J/eV
1.60219×10^{-13} J/MeV
10^6 eV/MeV

3.60×10^6 J/kW hr

4.20×10^{12} J/kton
4.20×10^{15} J/mton

0.04336 (eV/molecule)/(kcal/mole)
23.06 (kcal/mole)/(eV/molecule)

12. Power
746 W/hp

13. Temperature
1.00 °F/°R
1.00 °C/K
1.80 °F/°C
1.80 °F/K
$T(\text{K}) = T(\text{°C}) + 273.15$
$T(\text{°C}) = [T(\text{°F}) - 32]/1.80$
$T(\text{K}) = T(\text{°R})/1.80$

14. Electrical quantities
(Note that 2.9979 is well approximated by 3.00.)
Charge: 2.9979×10^9 esu/C
Current: 2.9979×10^9 (esu/s)/A
Potential: 299.79 V/statvolt
Electric field: 2.9979×10^4
   (V/m)/(statvolt/cm)
Magnetic field: 10^4 G/T
Magnetic flux: 10^8 G cm^2/Wb
Pole strength: 10 cgs unit/mitchell
   (cgs unit = $\sqrt{\text{ergcm}}$; mitchell = Am)

Appendix V

Some Properties
of the Error Function

$$\text{erf } u = \frac{2}{\sqrt{\pi}} \int_0^u e^{-z^2} \, dz = \frac{2}{\sqrt{\pi}} \left(u - \frac{u^3}{3 \times 1!} + \frac{u^5}{5 \times 2!} - \cdots \right)$$

Therefore

$$\text{erf}(-u) = -\text{erf } u$$

$$\text{erfc } u = 1 - \text{erf } u = \frac{2}{\sqrt{\pi}} \int_u^\infty e^{-z^2} \, dz$$

$$\text{erf } u \approx \frac{2u}{\sqrt{\pi}} \quad \text{for} \quad u \ll 1$$

$$\text{erfc } u \approx \frac{1}{\sqrt{\pi}} \frac{e^{-u^2}}{u} \quad \text{for} \quad u \gg 1$$

$$\text{erf}(\infty) = 1, \quad \text{erf}(0) = 0$$

$$\text{erfc}(0) = 1, \quad \text{erfc}(\infty) = 0$$

$$\frac{d \text{ erf } u}{du} = \frac{2}{\sqrt{\pi}} e^{-u^2}$$

$$\int_0^u \text{erfc } z \, dz = u \text{ erfc } u + \frac{1}{\sqrt{\pi}} (1 - e^{-u^2})$$

$$\int_0^\infty \text{erfc } z \, dz = \frac{1}{\sqrt{\pi}}$$

$$\int_0^\infty e^{-u^2} \, du = \frac{\sqrt{\pi}}{2}, \quad \int_0^u e^{-z^2} \, dz = \frac{\sqrt{\pi}}{2} \text{ erf } u$$

Table V.1   Error Function erf(w)

w	erf(w)	w	erf(w)	w	erf(w)	w	erf(w)
0.00	0.000 000	0.44	0.466 225	0.88	0.786 687	1.32	0.938 065
0.01	0.011 283	0.45	0.475 482	0.89	0.719 843	1.33	0.940 015
0.02	0.022 565	0.46	0.484 655	0.90	0.796 908	1.34	0.941 914
0.03	0.033 841	0.47	0.493 745	0.91	0.801 883	1.35	0.943 762
0.04	0.045 111	0.48	0.502 750	0.92	0.806 768	1.36	0.945 561
0.05	0.056 372	0.49	0.511 668	0.93	0.811 564	1.37	0.947 312
0.06	0.067 622	0.50	0.520 500	0.94	0.816 271	1.38	0.949 016
0.07	0.078 858	0.51	0.529 244	0.95	0.820 891	1.39	0.950 673
0.08	0.090 078	0.52	0.537 899	0.96	0.825 424	1.40	0.952 285
0.09	0.101 281	0.53	0.546 464	0.97	0.829 870	1.41	0.953 852
0.10	0.112 463	0.54	0.554 939	0.98	0.834 232	1.42	0.955 376
0.11	0.123 623	0.55	0.563 323	0.99	0.838 508	1.43	0.956 857
0.12	0.134 758	0.56	0.571 616	1.00	0.842 701	1.44	0.958 297
0.13	0.145 867	0.57	0.579 816	1.01	0.846 810	1.45	0.959 695
0.14	0.156 947	0.58	0.587 923	1.02	0.850 838	1.46	0.961 054
0.15	0.167 996	0.59	0.595 936	1.03	0.854 784	1.47	0.962 373
0.16	0.179 012	0.60	0.603 856	1.04	0.858 650	1.48	0.963 654
0.17	0.189 992	0.61	0.611 681	1.05	0.862 436	1.49	0.964 898
0.18	0.200 936	0.62	0.619 411	1.06	0.866 144	1.50	0.966 105
0.19	0.211 840	0.63	0.627 046	1.07	0.869 773	1.51	0.967 277
0.20	0.222 703	0.64	0.634 586	1.08	0.873 326	1.52	0.968 413
0.21	0.233 522	0.65	0.642 029	1.09	0.876 803	1.53	0.969 516
0.22	0.244 296	0.66	0.649 377	1.10	0.880 205	1.54	0.970 586
0.23	0.255 023	0.67	0.656 628	1.11	0.883 533	1.55	0.971 623
0.24	0.265 700	0.68	0.663 782	1.12	0.886 788	1.56	0.972 628
0.25	0.276 326	0.69	0.670 840	1.13	0.889 971	1.57	0.973 603
0.26	0.286 900	0.70	0.677 801	1.14	0.893 082	1.58	0.974 547
0.27	0.297 418	0.71	0.684 666	1.15	0.896 124	1.59	0.975 462
0.28	0.307 880	0.72	0.691 433	1.16	0.899 096	1.60	0.976 348
0.29	0.318 283	0.73	0.698 104	1.17	0.902 000	1.61	0.977 207
0.30	0.328 627	0.74	0.704 678	1.18	0.904 837	1.62	0.978 038
0.31	0.338 908	0.75	0.711 156	1.19	0.907 608	1.63	0.978 843
0.32	0.349 126	0.76	0.717 537	1.20	0.910 314	1.64	0.979 622
0.33	0.359 279	0.77	0.723 822	1.21	0.912 956	1.65	0.980 376
0.34	0.369 365	0.78	0.730 010	1.22	0.915 534	1.66	0.981 105
0.35	0.379 382	0.79	0.736 103	1.23	0.918 050	1.67	0.981 810
0.36	0.389 330	0.80	0.742 101	1.24	0.920 505	1.68	0.982 493
0.37	0.399 206	0.81	0.748 003	1.25	0.922 900	1.69	0.983 153
0.38	0.409 009	0.82	0.753 811	1.26	0.925 236	1.70	0.983 790
0.39	0.418 739	0.83	0.759 524	1.27	0.927 514	1.71	0.984 407
0.40	0.428 392	0.84	0.765 143	1.28	0.929 734	1.72	0.985 003
0.41	0.437 969	0.85	0.770 668	1.29	0.931 899	1.73	0.985 578
0.42	0.447 468	0.86	0.776 110	1.30	0.934 008	1.74	0.986 135
0.43	0.456 887	0.87	0.781 440	1.31	0.936 063	1.75	0.986 672

continued

w	erf(w)	w	erf(w)	w	erf(w)	w	erf(w)
1.76	0.987 190	2.22	0.998 308	2.67	0.999 841	3.13	0.999 990 42
1.77	0.987 691	2.23	0.998 388	2.68	0.999 849	3.14	0.999 991 03
1.79	0.988 641	2.24	0.998 464	2.69	0.999 858	3.15	0.999 991 60
1.80	0.989 091	2.25	0.998 537	2.70	0.999 866	3.16	0.999 992 14
1.81	0.989 525	2.26	0.998 607	2.71	0.999 873	3.17	0.999 992 64
1.82	0.989 943	2.27	0.998 674	2.72	0.999 880	3.18	0.999 993 11
1.83	0.990 347	2.28	0.998 738	2.73	0.999 887	3.19	0.999 993 56
1.84	0.990 736	2.29	0.998 799	2.74	0.999 893	3.20	0.999 993 97
1.85	0.991 111	2.30	0.998 857	2.75	0.999 899	3.21	0.999 994 36
1.86	0.991 472	2.31	0.998 912	2.76	0.999 905	3.22	0.999 994 73
1.87	0.991 821	2.32	0.998 966	2.77	0.999 910	3.23	0.999 995 07
1.88	0.992 156	2.33	0.999 016	2.78	0.999 916	3.24	0.999 995 40
1.89	0.992 479	2.34	0.999 065	2.79	0.999 920	3.25	0.999 995 70
1.90	0.992 790	2.35	0.999 111	2.80	0.999 925	3.26	0.999 995 98
1.91	0.993 090	2.36	0.999 155	2.81	0.999 929	3.27	0.999 996 24
1.92	0.993 378	2.37	0.999 197	2.82	0.999 933	3.28	0.999 996 49
1.93	0.993 656	2.38	0.999 237	2.83	0.999 937	3.29	0.999 996 72
1.94	0.993 923	2.39	0.999 275	2.85	0.999 944	3.30	0.999 996 94
1.95	0.994 179	2.40	0.999 311	2.86	0.999 948	3.31	0.999 997 15
1.96	0.994 426	2.41	0.999 346	2.87	0.999 951	3.32	0.999 997 34
1.97	0.994 664	2.42	0.999 379	2.88	0.999 954	3.33	0.999 997 51
1.98	0.994 892	2.43	0.999 411	2.89	0.999 956	3.34	0.999 997 68
1.99	0.995 111	2.44	0.999 441	2.90	0.999 959	3.35	0.999 997 838
2.00	0.995 322	2.45	0.999 469	2.91	0.999 961	3.36	0.999 997 983
2.01	0.995 525	2.46	0.999 497	2.92	0.999 964	3.37	0.999 998 120
2.02	0.995 719	2.47	0.999 523	2.93	0.999 966	3.38	0.999 998 247
2.03	0.995 906	2.48	0.999 547	2.94	0.999 968	3.39	0.999 998 367
2.04	0.996 086	2.49	0.999 571	2.95	0.999 970	3.40	0.999 998 478
2.05	0.996 258	2.50	0.999 593	2.96	0.999 972	3.41	0.999 998 582
2.06	0.996 423	2.51	0.999 614	2.97	0.999 973	3.42	0.999 998 679
2.07	0.996 582	2.52	0.999 634	2.98	0.999 975	3.43	0.999 998 770
2.08	0.996 734	2.53	0.999 654	2.99	0.999 976	3.44	0.999 998 855
2.09	0.996 880	2.54	0.999 672	3.00	0.999 977 91	3.45	0.999 998 934
2.10	0.997 021	2.55	0.999 689	3.01	0.999 979 26	3.46	0.999 999 008
2.11	0.997 155	2.56	0.999 706	3.02	0.999 980 53	3.47	0.999 999 077
2.12	0.997 284	2.57	0.999 722	3.03	0.999 981 73	3.48	0.999 999 141
2.13	0.997 407	2.58	0.999 736	3.04	0.999 982 86	3.49	0.999 999 201
2.14	0.997 525	2.59	0.999 751	3.05	0.999 983 92	3.50	0.999 999 257
2.15	0.997 639	2.60	0.999 764	3.06	0.999 984 92	3.51	0.999 999 309
2.16	0.997 747	2.61	0.999 777	3.07	0.999 985 86	3.52	0.999 999 358
2.17	0.997 851	2.62	0.999 789	3.08	0.999 986 74	3.53	0.999 999 403
2.18	0.997 951	2.63	0.999 800	3.09	0.999 987 57	3.54	0.999 999 445
2.19	0.998 046	2.64	0.999 811	3.10	0.999 988 35	3.55	0.999 999 485
2.20	0.998 137	2.65	0.999 822	3.11	0.999 989 08	3.56	0.999 999 521
2.21	0.998 224	2.66	0.999 831	3.12	0.999 989 77	3.57	0.999 999 555

w	erf(w)	w	erf(w)	w	erf(w)	w	erf(w)
3.58	0.999 999 587	3.69	0.999 999 820	3.80	0.999 999 923	3.91	0.999 999 968
3.59	0.999 999 617	3.70	0.999 999 833	3.81	0.999 999 929	3.92	0.999 999 970
3.60	0.999 999 644	3.71	0.999 999 845	3.82	0.999 999 934	3.93	0.999 999 973
3.61	0.999 999 670	3.72	0.999 999 857	3.83	0.999 999 939	3.94	0.999 999 975
3.62	0.999 999 694	3.73	0.999 999 867	3.84	0.999 999 944	3.95	0.999 999 977
3.63	0.999 999 716	3.74	0.999 999 877	3.85	0.999 999 948	3.96	0.999 999 979
3.64	0.999 999 736	3.75	0.999 999 886	3.86	0.999 999 952	3.97	0.999 999 980
3.65	0.999 999 756	3.76	0.999 999 895	3.87	0.999 999 956	3.98	0.999 999 982
3.66	0.999 999 773	3.77	0.999 999 903	3.88	0.999 999 959	3.99	0.999 999 983
3.67	0.999 999 790	3.78	0.999 999 910	3.89	0.999 999 962		
3.68	0.999 999 805	3.79	0.999 999 917	3.90	0.999 999 965		

Appendix VI

F Values

Table VI.1 95th percentile values for the *F* distribution
n_1 = degrees of freedom for numerator
n_2 = degrees of freedom for denominator (shaded area = 0.95)

n_2 \ n_1	1	2	3	4	5	6	8	12	16	20	30	40	50	100	∞
1	161.4	199.5	215.7	224.6	230.2	234.0	238.9	243.9	246.3	248.0	250.1	251.1	252.2	253.0	254.3
2	18.51	19.00	19.16	19.25	19.30	19.33	19.37	19.41	19.43	19.45	19.46	19.46	19.47	19.49	19.50
3	10.13	9.55	9.28	9.12	9.01	8.94	8.85	8.74	8.69	8.66	8.62	8.60	8.58	8.56	8.53
4	7.71	6.94	6.59	6.39	6.26	6.16	6.04	5.91	5.84	5.80	5.75	5.71	5.70	5.66	5.63
5	6.61	5.79	5.41	5.19	5.05	4.95	4.82	4.68	4.60	4.56	4.50	4.46	4.44	4.40	4.36
6	5.99	5.14	4.76	4.53	4.39	4.28	4.15	4.00	3.92	3.87	3.81	3.77	3.75	3.71	3.67
7	5.59	4.74	4.35	4.12	3.97	3.87	3.73	3.57	3.49	3.44	3.38	3.34	3.32	3.28	3.23
8	5.32	4.46	4.07	3.84	3.69	3.58	3.44	3.28	3.20	3.15	3.08	3.05	3.03	2.98	2.93
9	5.12	4.26	3.86	3.63	3.48	3.37	3.23	3.07	2.98	2.93	2.86	2.82	2.80	2.76	2.71
10	4.96	4.10	3.71	3.48	3.33	3.22	3.07	2.91	2.82	2.77	2.70	2.67	2.64	2.59	2.54
11	4.84	3.98	3.59	3.36	3.20	3.09	2.95	2.79	2.70	2.65	2.57	2.53	2.50	2.45	2.40
12	4.75	3.89	3.49	3.26	3.11	3.00	2.85	2.69	2.60	2.54	2.46	2.42	2.40	2.35	2.30
13	4.67	3.81	3.41	3.18	3.03	2.92	2.77	2.60	2.51	2.46	2.38	2.34	2.32	2.26	2.21
14	4.60	3.74	3.34	3.11	2.96	2.85	2.70	2.53	2.44	2.39	2.31	2.27	2.24	2.19	2.13
15	4.54	3.68	3.29	3.06	2.90	2.79	2.64	2.48	2.39	2.33	2.25	2.21	2.18	2.12	2.07
16	4.49	3.63	3.24	3.01	2.85	2.74	2.59	2.42	2.33	2.28	2.20	2.16	2.13	2.07	2.01
17	4.45	3.59	3.20	2.96	2.81	2.70	2.55	2.38	2.29	2.23	2.15	2.11	2.08	2.02	1.96
18	4.41	3.55	3.16	2.93	2.77	2.66	2.51	2.34	2.25	2.19	2.11	2.07	2.04	1.98	1.92
19	4.38	3.52	3.13	2.90	2.74	2.63	2.48	2.31	2.21	2.15	2.07	2.02	2.00	1.94	1.88
20	4.35	3.49	3.10	2.87	2.71	2.60	2.45	2.28	2.18	2.12	2.04	1.99	1.96	1.90	1.84
22	4.30	3.44	3.05	2.82	2.66	2.55	2.40	2.23	2.13	2.07	1.98	1.93	1.91	1.84	1.78
24	4.26	3.40	3.01	2.78	2.62	2.51	2.36	2.18	2.09	2.03	1.94	1.89	1.86	1.80	1.73
26	4.23	3.37	2.98	2.74	2.59	2.47	2.32	2.15	2.05	1.99	1.90	1.85	1.82	1.76	1.69
28	4.20	3.34	2.95	2.71	2.56	2.45	2.29	2.12	2.02	1.96	1.87	1.81	1.78	1.72	1.65
30	4.17	3.32	2.92	2.69	2.53	2.42	2.27	2.09	1.99	1.93	1.84	1.79	1.76	1.69	1.62
40	4.08	3.23	2.84	2.61	2.45	2.34	2.18	2.00	1.90	1.84	1.74	1.69	1.66	1.59	1.51
50	4.03	3.18	2.79	2.56	2.40	2.29	2.13	1.95	1.85	1.78	1.69	1.63	1.60	1.52	1.44
60	4.00	3.15	2.76	2.53	2.37	2.25	2.10	1.92	1.81	1.75	1.65	1.59	1.56	1.48	1.39
70	3.98	3.13	2.74	2.50	2.35	2.23	2.07	1.89	1.79	1.72	1.62	1.56	1.53	1.45	1.35
80	3.96	3.11	2.72	2.48	2.33	2.21	2.05	1.88	1.77	1.70	1.60	1.54	1.51	1.42	1.32
100	3.94	3.09	2.70	2.46	2.30	2.19	2.03	1.85	1.75	1.68	1.57	1.51	1.48	1.39	1.28
150	3.91	3.06	2.67	2.43	2.27	2.16	2.00	1.82	1.71	1.64	1.54	1.47	1.44	1.34	1.22

n_1 / n_2	1	2	3	4	5	6	8	12	16	20	30	40	50	100	∞
200	3.89	3.04	2.65	2.41	2.26	2.14	1.98	1.80	1.69	1.62	1.52	1.45	1.42	1.32	1.19
400	3.86	3.02	2.62	2.39	2.23	2.12	1.96	1.78	1.67	1.60	1.49	1.42	1.38	1.28	1.13
∞	3.84	2.99	2.60	2.37	2.21	2.09	1.94	1.75	1.64	1.57	1.46	1.40	1.32	1.24	1.00

Table VI.2 99th percentile values for the F distribution
n_1 = degrees of freedom for numerator
n_2 = degrees of freedom for denominator (shaded area = 0.99)

n_1 / n_2	1	2	3	4	5	6	8	12	16	20	30	40	50	100	∞
1	4052	4999	5403	5625	5764	5859	5981	6106	6169	6208	6258	6286	6302	6334	6366
2	98.49	99.01	99.17	99.25	99.30	99.33	99.36	99.42	99.44	99.45	99.47	99.48	99.48	99.49	99.5
3	34.12	30.81	29.46	28.71	28.24	27.41	27.49	27.05	28.63	26.69	26.50	26.41	26.35	26.23	26.12
4	21.20	18.00	16.69	15.98	15.52	15.21	14.80	14.37	14.15	14.02	13.83	13.74	13.69	13.57	13.46
5	16.26	13.27	12.06	11.39	10.97	10.67	10.27	9.89	9.68	9.55	9.38	9.29	9.24	9.13	9.02
6	13.74	10.92	9.78	9.15	8.75	8.47	8.10	7.72	7.52	7.39	7.23	7.14	7.09	6.99	6.88
7	12.25	9.55	8.45	7.85	7.46	7.19	6.84	6.47	6.27	6.15	5.98	5.90	5.85	5.75	5.65
8	11.26	8.65	7.59	7.01	6.63	6.37	6.03	5.67	5.48	5.36	5.20	5.11	5.06	4.96	4.86
9	10.56	8.02	6.99	6.42	6.06	5.80	5.47	5.11	4.92	4.80	4.64	4.56	4.51	4.41	4.31
10	10.04	7.56	6.55	5.99	5.64	5.39	5.06	4.71	4.52	4.41	4.25	4.17	4.12	4.01	3.91
11	9.05	7.20	6.22	5.67	5.32	5.07	4.74	4.40	4.21	4.10	3.94	3.86	3.80	3.70	3.60
12	9.33	6.93	5.95	5.41	5.06	4.82	4.50	4.16	3.98	3.86	3.70	3.61	3.56	3.46	3.36
13	9.07	6.70	5.74	5.20	4.86	4.62	4.30	3.96	3.78	3.67	3.51	3.42	3.37	3.27	3.16
14	8.86	6.51	5.56	5.03	4.69	4.46	4.14	3.80	3.62	3.51	3.34	3.26	3.21	3.11	3.00
15	8.68	6.36	5.42	4.89	4.56	4.32	4.00	3.67	3.48	3.36	3.20	3.12	3.07	2.97	2.87
16	8.53	6.23	5.29	4.77	4.44	4.20	3.89	3.55	3.37	3.25	3.10	3.01	2.96	2.86	2.75
17	8.40	6.11	5.18	4.67	4.34	4.10	3.79	3.45	3.27	3.16	3.00	2.92	2.86	2.76	2.65
18	8.28	6.01	5.09	4.58	4.25	4.01	3.71	3.37	3.19	3.07	2.91	2.83	2.78	2.68	2.57
19	8.18	5.93	5.01	4.50	4.17	3.94	3.63	3.30	3.12	3.00	2.84	2.76	2.70	2.60	2.49
20	8.10	5.85	4.94	4.43	4.10	3.87	3.56	3.23	3.05	2.94	2.77	2.69	2.63	2.53	2.42
22	7.94	5.72	4.82	4.31	3.99	3.76	3.45	3.12	2.94	2.83	2.67	2.58	2.53	2.42	2.31
24	7.82	5.61	4.72	4.22	3.90	3.67	3.36	3.03	2.85	2.74	2.58	2.49	2.44	2.33	2.21
26	7.72	5.53	4.64	4.14	3.82	3.59	3.29	2.96	2.77	2.66	2.50	2.41	2.36	2.25	2.13
28	7.64	5.45	4.57	4.07	3.76	3.53	3.23	2.90	2.71	2.60	2.44	2.35	2.30	2.18	2.06
30	7.56	5.39	4.51	4.02	3.70	3.47	3.17	2.84	2.66	2.55	2.38	2.29	2.24	2.13	2.01
40	7.31	5.18	4.31	3.83	3.51	3.29	2.99	2.66	2.49	2.37	2.20	2.11	2.05	1.94	1.81
50	7.17	5.06	4.20	3.72	3.41	3.18	2.88	2.56	2.39	2.26	2.10	2.00	1.94	1.82	1.68
60	7.08	4.98	4.13	3.65	3.34	3.12	2.82	2.50	2.32	2.20	2.03	1.93	1.87	1.74	1.60
70	7.01	4.92	4.08	3.60	3.29	3.07	2.77	2.45	2.28	2.15	1.98	1.88	1.82	1.69	1.53
80	6.96	4.88	4.04	3.56	3.25	3.04	2.74	2.41	2.24	2.11	1.94	1.84	1.78	1.65	1.49
100	6.90	4.82	3.98	3.51	3.20	2.99	2.69	2.36	2.19	2.06	1.89	1.79	1.73	1.59	1.43
150	6.81	4.75	3.91	3.44	3.14	2.92	2.62	2.30	2.12	2.00	1.83	1.72	1.66	1.51	1.33
200	6.76	4.71	3.88	3.41	3.11	2.90	2.60	2.28	2.09	1.97	1.79	1.69	1.62	1.48	1.28
400	6.70	4.66	3.83	3.36	3.06	2.85	2.55	2.23	2.04	1.92	1.74	1.64	1.57	1.42	1.19
∞	6.64	4.60	3.78	3.32	3.02	2.80	2.51	2.18	1.99	1.87	1.69	1.59	1.52	1.36	1.00

Index

Periodic table of the elements and element atomic weights

1 IA IA	2 IIA IIA	3 IIIA IIIB	4 IVA IVB	5 VA VB	6 VIA VIB	7 VIIA VIIB	8 VIIIA VIIIB	9 VIIIA VIIIB	10 VIIIA VIIIB	11 IB IB	12 IIB IIB	13 IIIB IIIA	14 IVB IVA	15 VB VA	16 VIB VIA	17 VIIB VIIA	18 VIIIB VIIIA
1 H 1.008																	2 He 4.003
3 Li 6.941	4 Be 9.012											5 B 10.811	6 C 12.011	7 N 14.007	8 O 15.999	1 H 1.008 9 F 18.998	10 Ne 20.180
11 Na 22.990	12 Mg 24.305											13 Al 26.982	14 Si 28.086	15 P 30.974	16 S 32.066	17 Cl 35.453	18 Ar 39.948
19 K 39.098	20 Ca 40.078	21 Sc 44.956	22 Ti 47.88	23 V 50.942	24 Cr 51.996	25 Mn 54.938	26 Fe 55.847	27 Co 58.933	28 Ni 58.693	29 Cu 63.546	30 Zn 65.39	31 Ga 69.723	32 Ge 72.61	33 As 74.922	34 Se 78.96	35 Br 79.904	36 Kr 83.80
37 Rb 85.468	38 Sr 87.62	39 Y 88.906	40 Zr 91.224	41 Nb 92.906	42 Mo 95.94	43 Tc (97.907)	44 Ru 101.07	45 Rh 102.906	46 Pd 106.42	47 Ag 107.868	48 Cd 112.411	49 In 114.818	50 Sn 118.710	51 Sb 121.757	52 Te 127.60	53 I 126.904	54 Xe 131.29
55 Cs 132.905	56 Ba 137.327	57–71	72 Hf 178.49	73 Ta 180.948	74 W 183.84	75 Re 186.207	76 Os 190.23	77 Ir 192.22	78 Pt 195.08	79 Au 196.967	80 Hg 200.59	81 Tl 204.383	82 Pb 207.2	83 Bi 208.980	84 Po (208.982)	85 At (209.987)	86 Rn (222.018)
87 Fr (223.020)	88 Ra 226.025	89–103	104 Unq (261.11)	105 Unp (262.114)	106 Unh (263.118)	107 Uns (262.12)	108 Uno (265)	109 Une (265)									

57 La 138.906	58 Ce 140.115	59 Pr 140.908	60 Nd 144.24	61 Pm (144.913)	62 Sm 150.36	63 Eu 151.965	64 Gd 157.25	65 Tb 158.925	66 Dy 162.50	67 Ho 164.93	68 Er 167.26	69 Tm 168.934	70 Yb 173.04	71 Lu 174.967
89 Ac 227.028	90 Th 232.038	91 Pa 231.036	92 U 238.029	93 Np 237.048	94 Pu (244.064)	95 Am (243.061)	96 Cm (247.070)	97 Bk (247.070)	98 Cf (251.080)	99 Es (252.083)	100 Fm (257.095)	101 Md (258.10)	102 No (259.101)	103 Lr (262.11)